T0321023

Molecular Spectroscopy

Molecular Spectroscopy
Second Edition

Jeanne L. McHale
Washington State University

CRC Press
Taylor & Francis Group
Boca Raton London New York

CRC Press is an imprint of the
Taylor & Francis Group, an **informa** business

CRC Press
Taylor & Francis Group
6000 Broken Sound Parkway NW, Suite 300
Boca Raton, FL 33487-2742

International Standard Book Number-13: 978-1-4665-8658-1 (Hardback)

Library of Congress Cataloging-in-Publication Data

Names: McHale, Jeanne L.
Title: Molecular spectroscopy / Jeanne L. McHale.
Description: Second edition. | Boca Raton : CRC Press, 2017. | Includes bibliographical references.
Identifiers: LCCN 2016038575 | ISBN 9781466586581 (hardback : alk. paper)
Subjects: LCSH: Molecular spectroscopy. | Spectrum analysis.
Classification: LCC QC454.M6 M38 2017 | DDC 539/.60287--dc23
LC record available at https://lccn.loc.gov/2016038575

Visit the Taylor & Francis Web site at
http://www.taylorandfrancis.com

and the CRC Press Web site at
http://www.crcpress.com

Printed and bound by CPI Group (UK) Ltd, Croydon, CR0 4YY

Come forth into the light of things,
Let Nature be your teacher.

William Wordsworth

Contents

Preface

The first edition of *Molecular Spectroscopy* was coaxed from a pile of lecture notes that I had written for my graduate spectroscopy class at the University of Idaho, Moscow, Idaho. At the time it was published in 1999, there were many books about traditional high-resolution spectroscopy of small molecules in the gas phase. Influenced—or, perhaps, biased—by my own research interests, I felt that the topic of condensed phase spectroscopy was not being adequately covered. It became my pleasure and burden to try to provide this coverage, thus spreading my bias as well as my enthusiasm and modest expertise for the study of the interaction of light and matter. Since the publication of the first edition, spectroscopic applications in fields such as materials science, biology, solar energy conversion, and environmental science have intensified the need for a textbook that prepares researchers to handle systems far more complex than isolated gas phase molecules. The second edition of *Molecular Spectroscopy* benefits from many years of vetting of the first edition by students and instructors, whose feedback is gratefully acknowledged. As in the first edition, this book aims to present the theoretical foundations of traditional and modern spectroscopy methods and provide a bridge from theory to experimental applications. Emphasis continues to be placed on the use of time-dependent theory to link the spectral response in the frequency domain to the behavior of molecules in the time domain. This link is made stronger by the addition of two new chapters: Chapter 13 on nonlinear optical spectroscopy and Chapter 14 on time-resolved spectroscopy, both of which go beyond the linear spectroscopy regime that is the subject of the first 12 chapters. Chapter 13 provides the quantum mechanical basis for nonlinear spectroscopy from several different theoretical vantage points, whereas Chapter 14 enters the realm of real time to uncover the dynamics concealed in spectral lineshapes.

In addition to correcting pedagogical speedbumps and other glitches discovered by readers over the years, I have added new material that I hope will smooth the transition from the linear (Chapters 1 through 12) to the nonlinear (Chapters 13 and 14) spectroscopy regime. The density matrix is now featured prominently in Chapter 4 and returned to in Chapters 13 and 14. A discussion of magnetic resonance has been added to Chapters 3 and 4 in order to illustrate how the time-dependent density matrix is manifested experimentally. The concepts of population relaxation and dephasing appear throughout this book. A new section on non-radiative relaxation of polyatomic molecules has been added to Chapter 11. The discussion of electromagnetic radiation in Chapter 2 has been augmented by the consideration of the Fresnel equations, important in spectroscopy of surfaces, and a discussion of widely employed Gaussian beams. A new section on surface-enhanced Raman spectroscopy has been added to Chapter 12. The discussion of the third-order underpinnings of spontaneous Raman spectroscopy has been moved to Chapter 13, where it follows the discussion of the third-order susceptibility.

This book is intended as a resource for researchers and for use in a graduate course in spectroscopy. It would be difficult to cover all 14 chapters in entirety in a single semester. Like the electromagnetic spectrum, the topic of spectroscopy is open ended, and instructors and students can choose where to focus. Depending on their backgrounds, students may be able to skim Chapter 1, which reviews the material taught in a graduate quantum mechanics course. Others may be able to skip Chapter 2 (on electromagnetic radiation) and Chapter 3 (about electric and magnetic properties). I always urge students to read at least the first few sections of Chapter 5 and glean physical intuition from the idea that spectra and dynamics are Fourier-transform pairs. I believe Chapters 4 and 6, which make the link between quantum mechanics and the practice of spectroscopy, are critical to the full appreciation of subsequent chapters. As the order of the field–matter interaction increases, so do the information content, complexity, and number of experiments. I have strived to present the theory that is needed to go beyond what is presented in this book, so that readers can apply spectroscopic tools to their own research interests. Particularly in the last two chapters, the examples I have chosen to illustrate fundamental principles are not intended to constitute an exhaustive review; rather, they reflect my pedagogical goals and personal interests.

There are many people to thank for helping me prepare this second edition. Colleagues who provided encouragement, reviewed chapters, answered questions, or generously gave permission to reprint figures include John Bertie, Gregory Scholes, Robert Boyd, Richard Mathies, Mark Johnson, Martin Moskovits, Eric Vauthey, Mark Maroncelli, Lawrence Ziegler, Nancy Levinger, Thomas Elsaesser, Haruko Hosoi, Andrew Hanst, Gerald Meyer, Robert Walker, Susan Dexheimer, Stephen Doorn, Andrew Shreve, Igor Adamovich, Michael Tauber, Erik Nibbering, Henk Fidder, V. Ara Apkarian, Andrei Tokmakoff, Charles Schmuttenmaer, Mary Jane Shultz, and David W. McCamant. Special thanks go to Jahan Dawlaty for his careful and very helpful review of the two new chapters. On behalf of students and researchers, I thank these colleagues for their commitment to furthering spectroscopy education. I also hold them blameless for any remaining errors, for which I alone am responsible.

In recent years, students in my graduate spectroscopy class and those working in my lab helped to refine draft chapters of this edition. These include McHale group members Christopher Leishman, Riley Rex, Nicholas Treat, Lyra Christianson, Candy Mercado, Greg Zweigle, Deborah Malamen, Christopher Rich, and Stephanie Doan. Extra special thanks are due to Christopher Leishman whose careful review of every chapter was enormously helpful. Graduate students Samuel Battey, Saewha Chong, Sakun Duwal, Elise Held, Adam Huntley, Jason Leicht, Victor Murcia, Junghune Nam, Nathan Turner, and Tiecheng Zhou, who were enrolled in Chem 564 at Washington State University, Pullman, Washington, provided thoughtful comments on draft chapters and helped to find errors.

I am pleased to acknowledge the National Science Foundation for their support of the spectroscopy research that my students, colleagues, and I have had the privilege to engage in. I am grateful to my editor Luna Han for her support and patience. Most of all, I thank my husband and scientific collaborator, Dr. Fritz Knorr. His encouragement and understanding helped to keep me on track. In addition, this book would literally not have been possible without him, because he drew all the figures.

Jeanne L. McHale
Moscow, Idaho
August 2016

Physical constants and conversion factors

PHYSICAL CONSTANTS

Constant	Symbol	Value
Avogadro's number	N_A	6.0221×10^{23} mol^{-1}
Bohr magneton	μ_B	9.2740×10^{-24} J T^{-1}
Bohr radius	a_0	5.2918×10^{-11} m
Boltzmann's constant	k_B	1.3806×10^{-23} J K^{-1}
Electron rest mass	m_e	9.1094×10^{-31} kg
Nuclear magneton	μ_N	5.0508×10^{-27} J T^{-1}
Permeability of free space	μ_0	$4\pi \times 10^{-7}$ N A^{-2}
Permittivity of free space	ε_0	8.8542×10^{-12} C^2 N^{-1} m^{-2}
Planck's constant	h	6.6261×10^{-34} J s
Proton charge	e	1.6022×10^{-19} C
	e'	4.8032×10^{-10} esu (or statC)
Proton rest mass	m_p	1.6726×10^{-27} kg
Rydberg constant	R_H	1.0974×10^7 m^{-1}
Speed of light in a vacuum	c	2.9979×10^8 m s^{-1}

ENERGY CONVERSION FACTORS

	Erg	J	eV	cm^{-1}
erg	1	10^{-7}	6.24150×10^{11}	5.03414×10^{15}
J	10^7	1	6.24150×10^{18}	5.03414×10^{22}
eV	1.60218×10^{-12}	1.60218×10^{-19}	1	8065.6
cm^{-1}	1.98644×10^{-16}	1.98644×10^{-23}	1.23983×10^{-4}	1

MISCELLANEOUS CONVERSION FACTORS

1 amu $= 1.6606 \times 10^{-27}$ kg
1 cal $= 4.184$ J
1 Debye (D) $= 10^{-18}$ esu cm $= 3.336 \times 10^{-30}$ C m
1 Tesla (T) $= 1$ N s C^{-1} m^{-1} $= 1$ Weber m^{-2} $= 10^4$ Gauss

About the author

Jeanne L. McHale obtained a BS in chemistry from Wright State University, Dayton, Ohio, where Prof. Paul Seybold ignited her lifelong interest in optical spectroscopy. She earned a PhD in physical chemistry from the University of Utah, Salt Lake City, Utah, in 1979 under the direction of Prof. Jack Simons and then did postdoctoral research there with Prof. C. H. Wang. She began her academic career as a member of the chemistry faculty at the University of Idaho, Moscow, Idaho, in 1980, where she established a research program emphasizing spectroscopic studies of intermolecular interactions, photoinduced electron transfer, and molecular dynamics of liquids. She is a fellow of the American Association for the Advancement of Science and has been an active member of the American Chemical Society and the Materials Research Society. She organized the first Telluride Science Center workshop on solar solutions to energy and environmental problems. Dr. McHale served for 12 years as a professor in the Chemistry Department and the Materials Science and Engineering Program at Washington State University (WSU), Pullman, Washington. She has been Professor Emerita at WSU since 2016. Her recent research interests highlight applications of spectroscopy to the study of light-harvesting aggregates, semiconductor nanoparticles, and solar energy conversion.

Introduction and review

1.1 HISTORICAL PERSPECTIVE

Spectroscopy is about light and matter and how they interact with one another. The fundamental properties of both light and matter evade our human senses, and so there is a long history to the questions: What is light and what is matter? Studies of the two have often been interwoven, with spectroscopy playing an important role in the emergence and validation of quantum theory in the early twentieth century. Max Planck's analysis of the emission spectrum of a blackbody radiator established the value of his eponymous constant, setting in motion a revolution that drastically altered our picture of the microscopic world. The line spectra of atoms, though they had been employed for chemical analysis since the late 1800s, could not be explained by classical physics. Why should gases in flames and discharge tubes emit only certain spectral wavelengths, while the emission spectrum of a heated body is a continuous distribution? Niels Bohr's theory of the spectrum of the hydrogen atom recognized the results of Rutherford's experiments, which revealed previously unexpected details of the atom: a dense, positively charged nucleus surrounded by the diffuse negative charge of the electrons. In Bohr's atom the electrons revolve around the nucleus in precise paths like the orbits of planets around the sun, a picture that continues to serve as a popular cartoon representation of the atom. Though the picture is conceptually wrong, the theory based on it is in complete agreement with the observed absorption and emission wavelengths of hydrogen! Modern quantum theory smeared the sharp orbits of Bohr's hydrogen atom into probability distributions, and successfully reproduced the observed spectral transition frequencies. Observations of electron emission by irradiated metals led to Einstein's theory of photons as packets of light energy, after many hundreds of years of debate on the wave–particle nature of light. Experiments (such as electron diffraction by crystals) and theory (the Schrödinger equation and the Heisenberg uncertainty principle) gave rise to the idea that matter, like light, has wave-like as well as particle-like properties. Quantum theory and Einstein's concept of photons converge in our modern microscopic view of spectroscopy. Matter emits or absorbs light (photons) by undergoing transitions between quantized energy levels. This relatively recent idea rests on the foundation built by philosophers and scientists who considered the nature of light since ancient times. The technology of recording spectra is far older than the quantum mechanical theories for interpretation of spectra.

Reference [1] gives an excellent historical account of investigations that led to our present understanding of electromagnetic radiation. Lenses and mirrors date back to before the common era, and the ancient Greeks included questions about the nature of light in their philosophical discourses. In 1666, Isaac Newton measured the spectrum of the sun by means of a prism. He speculated that the seven colors he observed (red, orange, yellow, green, blue, indigo, and violet) were somehow analogous to the seven notes of the musical scale. It is interesting that the frequency of violet light is a little less than twice that of red, so we see just less than an octave of this spectrum. We now know that the human eye can discern millions of colors [2] and that the wavelengths spanned by electromagnetic radiation extend indefinitely beyond the boundaries of vision. Newton was a proponent of the corpuscular view of light, a theory that held that light was a stream of particles bombarding the viewer. His contemporary, Christian Huygens, proposed a wave theory and showed how the concept could account for refraction and reflection. Newton had considerable influence, and the corpuscular theory dominated the scene for a long time after his death. (Its proponents may have been more zealous then Newton himself had been.) Just a few years after the discovery of the infrared and ultraviolet ranges of the spectrum, Thomas Young in 1802 made the connection between wavelength and color. Young also investigated the phenomenon that is now known as polarization. A. J. Fresnel made contributions to the emerging wave theory as well, and equations describing the polarization dependence of reflection at a boundary bear his name [1].

The nineteenth century saw many developments in the analysis of spectral lines. Fraunhofer repeated Newton's measurement of the solar spectrum in 1814, using a narrow slit rather than a circular aperture to admit the light onto the prism. The resolution of this experiment was sufficient to reveal a number of dark lines, wavelengths where the solar emission was missing. Numerous "Fraunhofer lines" are now assigned, thanks to the work of Bunsen, Ångström, and others. They originate from the reabsorption of sunlight by cooler atoms and ions in the outer atmosphere of the sun and that of the earth. In 1859, Kirchhoff demonstrated that two of these dark lines occur at the wavelengths of the yellow emission of hot sodium atoms. This helped to establish the notion that absorption and emission wavelengths of atoms coincide. Fraunhofer made further contributions by fabricating the first diffraction gratings, by wrapping fine silver wires around two parallel screws, and later by etching glass with diamond. By 1885 it was firmly established that elements have characteristic spectral wavelengths. The very discovery of the element helium in 1868 was made by analyzing the wavelengths of the solar spectrum.

The wave theory of light attracted many proponents in the nineteenth century, but it posed problems when light waves were compared to waves in matter, such as sound, which require a medium for support. How could light travel in a vacuum? The luminiferous ether was proposed, and efforts to detect it motivated some historic experiments, such as the accurate measurement of the speed of light by Michelson and Morley in 1887. (Michelson had earlier invented the interferometer while still in his twenties.) Their careful work resulted in rapid evaporation of the ether theory and in the recognition that light propagates in free space. The idea of a transverse wave, where the disturbance is perpendicular to the direction of propagation, was an elusive part of the picture, though it had been appreciated by Young. It took James Clerk Maxwell, whose equations we examine in Chapter 2, to refine the picture of light as a transverse electromagnetic wave. Maxwell's equations were fertilized by a large body of existing work on electricity and magnetism, particularly that of Michael Faraday. The crown jewel of Maxwell's theoretical accomplishment was the *derivation* of the speed of light in terms of two known experimental quantities: the permittivity and permeability of free space. These properties and their interrelation are further discussed in Chapters 2 and 3. The point to be made here is that the result was in *agreement with experiment.*

Thus the wave theory of light would seem to have been on pretty firm ground as we entered the twentieth century. So too was the idea that matter was composed of particles, although little was known about atomic structure. The quantum upheaval rattled the complacency of classical physics and built the stage on which the field of spectroscopy continues to perform. Looking at things through quantum mechanical goggles, one eye sees the wave and the other the particle. We have to keep both eyes open.

1.2 DEFINITIONS, DERIVATIONS, AND DISCOVERY

As the study of the interaction of light and matter, spectroscopy embraces a wide range of physical and chemical behavior. There is a certain reciprocity in the light–matter interaction. It is often natural to consider an experimental effect to be the result of matter exerting an influence on light (refraction, scattering, absorption, etc.). In other experiments, we may prefer to consider the effect of light on matter (photochemistry, photobleaching, optical trapping, etc.). The range of experimental situations encompassed by this definition is vast, especially considering that by "light" we mean electromagnetic radiation of *any* frequency, not just the narrow visible region of the spectrum. The emphasis in this book is on spectroscopy as a tool for studying the structure and dynamics of molecules. In Chapters 1 through 12, the experiments to be discussed fall within the range of *linear spectroscopy*, in which the material response is directly proportional to the amplitude of the electric field vector of the radiation. A typical spectrum consists of the *intensity* of a certain response, such as absorption of light, as a function of *frequency* of the light. We shall see that the intensity is a measure of the rate at which molecules make transitions from one energy level to another, while the frequency is directly related to the difference in the initial and final energies of the molecule. In Chapters 13 and 14 we consider spectroscopic effects that come into play with more intense, typically pulsed, sources of electromagnetic radiation. These chapters will delve into nonlinear and time-resolved optical spectroscopy, and the material response will be expressed as a power series expansion in the amplitude of the electric field.

In either the linear or nonlinear regime, we shall take a perturbative approach, in which the zeroth order states are the quantized energy levels of the molecules in the absence of light. These are the stationary states found by solving the time-independent Schrödinger equation. The radiation does not perturb the energy levels themselves; rather, it induces transitions between them. We shall see that the radiation creates a superposition state in which the basis states are those of the system in the dark. In a sense, spectroscopy is applied quantum mechanics: a bridge between experiment and theory. Though in principle the Schrödinger equation permits the energy levels of a molecule to be determined theoretically, exact solutions to chemically interesting problems cannot be attained. Quantum chemists deal with this by developing sophisticated approximation methods to calculate energy levels and wavefunctions. Spectroscopists do their part by using light to discover these energy levels. The two groups keep each other honest, and working together they accomplish more than they could on their own.

Practical applications of spectroscopy routinely deal with large collections of molecules. The interactions among molecules can exert considerable influence on the response of a bulk sample to incident radiation. For example, spectra of isolated (gas phase) molecules reveal numerous spectroscopic transitions in the form of sharp, well-separated lines. Plunk these molecules into a solution, and the lines may broaden or even blur together into a continuous spectrum. How is the Schrödinger equation to help us if we cannot resolve the quantum states? This question begs for a theoretical approach that avoids the need to know molecular eigenstates. The time-dependent theory for interpretation of spectra, introduced in Chapters 5 and 12, will provide such an eigenstate-free approach. Before we can consider this theory, however, we need to understand how isolated molecules respond to light and then see how microscopic physical properties, such as polarizability or dipole moment, sum to give physical properties of matter in bulk. These electric and magnetic properties of molecules and bulk matter are discussed in Chapter 3.

In this chapter, we review some of the basic quantum mechanical principles that enable us to characterize the translational, rotational, vibrational, and electronic energy states of molecules. The topic of statistical mechanics will then be summarized briefly, in order to discuss the behavior of large collections of molecules on the basis of the quantized states of individual molecules.

A note about derivations in the study of spectroscopy is in order. These derivations, which constitute a large part of this book, provide the foundation for getting microscopic information from spectra. Certainly, one can employ a formula or theoretical concept correctly without having derived it, and there are times when we do this. But having gone through a derivation conveys the scientist with additional insight and power. Knowing the theoretical foundations implies knowing the limits and assumptions behind the equations, so one may avoid incorrect application of a model or theory. Skill and familiarity with derivations enable the practicing spectroscopist to make predictions and perhaps extend the current state of knowledge. Whenever possible, spectroscopic formulas are derived from first principles and the reader is urged to follow along with pencil and scratch paper. Occasionally, it will be necessary to simply say "It turns out that...," and the reader will know that the proof of the statement is outside the scope of this book or is just too complex to be worthwhile. When this happens, it is hoped that interested readers will be motivated to consult the references cited at the end of the chapter.

This chapter is an exception in that no derivations will be presented, except in Section 1.3.3, where raising and lowering operators are discussed. It is assumed that the reader has had some previous exposure to quantum mechanics and statistical mechanics. The intent of Section 1.3 is to provide a review and establish some notation and terms that will be employed throughout this book.

1.3 REVIEW OF QUANTUM MECHANICS

Quantum mechanics postulates the existence of a well-behaved wavefunction ψ that describes the state of the system. This function depends on the spatial coordinates of the system, for example, $\psi(x, y, z)$ for a single particle in three dimensions. The wavefunction is not a physical observable; i.e., it cannot be measured experimentally, and it can be a complex function. However, its complex square, $\psi^*\psi \equiv |\psi|^2$, which is real, is interpreted as the probability density and is in principle experimentally observable. In a three-dimensional system, for example, $\psi^*\psi dxdydz$ is the probability of finding the particle in the infinitesimal volume $dxdydz$. It is convenient to use the symbol $d\tau$ as a generic volume element, in order to write down general expressions that do not depend on the dimension or coordinate system. One condition on the wavefunction is that it be normalizable. A normalized wavefunction gives a total probability of unity when integrated over all space:

$$\int \psi^*\psi \, d\tau = 1 \tag{1.1}$$

The wavefunction must also be single-valued and continuous.

The wavefunction is found by solving the time-independent Schrödinger equation $\hat{H}\psi = E\psi$ and applying the appropriate boundary conditions. The Hamiltonian $\hat{H} = \hat{T} + \hat{V}$ is the operator for the energy of the system: the sum of the operators for kinetic energy ($\hat{T}$) and potential energy ($\hat{V}$). It is the latter that makes things interesting, in that it decides whether we are dealing with, say, a harmonic oscillator or a hydrogen atom. For a single particle in three dimensions, the kinetic energy operator is

$$\hat{T} = \frac{-\hbar^2}{2m}\left(\frac{\partial^2}{\partial x^2} + \frac{\partial^2}{\partial y^2} + \frac{\partial^2}{\partial z^2} \right) \tag{1.2}$$

The symbol $\hbar$ represents $h/2\pi$, where h is Planck's constant, and m is the particle mass.

Quantum numbers arise when boundary conditions are applied. The result is that only certain wavefunctions and energies satisfy the time-independent Schrödinger equation.

$$\hat{H}\psi_n = E_n\psi_n \tag{1.3}$$

The index n in Equation 1.3 represents a quantum number or set of quantum numbers. Equation 1.3 is an eigenvalue equation. The allowed states of the system are specified by the solutions ψ_n (the eigenfunctions) and the energy of a system in a specific state is the eigenvalue E_n. If two different states correspond to different energy levels they must be orthogonal:

$$\int \psi_n^*\psi_m \, d\tau = 0 \;\; \text{if } E_n \neq E_m \tag{1.4}$$

Spectroscopy experiments probe differences in these energy levels. The Bohr frequency condition, to be derived in Chapter 4, states that the energy of the photon, $h\nu$, must match the energy level difference of the initial and final states:

$$\nu = \frac{E_2 - E_1}{h} \tag{1.5}$$

Equation 1.5 is just one condition that must be met for the transition between states 1 and 2 to be allowed. Not all transitions are permitted, even when the frequency of the light ν satisfies Equation 1.5. We will derive *selection rules* throughout the book, which enable the allowed transitions to be predicted.

Equation 1.3 is actually a special case that results when the potential energy operator is not a function of time. More generally, we need the time-dependent Schrödinger equation to be discussed in Chapter 4. In later developments, we will account for the time dependence of the applied electromagnetic field. For now, we consider Equation 1.3 to represent the system in the dark.

If the eigenfunctions of Equation 1.3 can be found, then other physical properties can be predicted. It is one of the postulates of quantum mechanics that physical properties are associated with Hermitian operators. For example, the momentum operator in the x direction is

$$\hat{p}_x = -i\hbar \frac{\partial}{\partial x} \tag{1.6}$$

The position operator is just $\hat{x} = x$. Since the potential energy is a function only of position, the quantum mechanical operator $\hat{V}$ looks just like the classical expression for the system's potential energy. A Hermitian operator $\hat{A}$ obeys the turn-over rule, which states that, for any two well-behaved functions f and g,

$$\int f^*(\hat{A}g)d\tau = \int g(\hat{A}f)^* d\tau \qquad (1.7)$$

It can be shown that an operator that obeys Equation 1.7 corresponds to a real physical property.

A physical property may be calculated in one of two ways. If the system happens to be in a state that is an eigenfunction of the operator corresponding to physical property $\hat{A}$, then the only allowed values of the property are the eigenvalues a_n of the operator:

$$\hat{A}\psi_n = a_n\psi_n \qquad (1.8)$$

The a_n's are constants and are real. Now, an eigenvalue equation is a special case. What if the system is in a state for which $\hat{A}\psi_n$ is not equal to a constant times ψ_n? In this case we cannot specify the exact value of the physical property; we must resort to calculating the expectation value:

$$\langle A \rangle = \int \psi_n^* \hat{A}\psi_n \, d\tau \qquad (1.9)$$

$\langle A \rangle$ is the average value of the physical property when the system is in state n. It is often convenient to use Dirac notation in expressions such as 1.9. The wavefunction ψ_n is represented by the ket vector $|\psi_n\rangle$ or $|n\rangle$, and its complex conjugate ψ_n^* is the bra vector $\langle\psi_n|$ or $\langle n|$. Putting the two together to form a bra-ket (bracket) represents integration over the coordinates of the system. For example, normalization and orthogonality can be summarized by the inner product:

$$\delta_{nm} = \langle n | m \rangle \qquad (1.10)$$

where the Kronecker delta function δ_{nm} is equal to 1 if $n = m$ and 0 if $n \neq m$. The expectation value in Dirac notation is

$$\langle A \rangle = \langle n | \hat{A} | n \rangle \equiv A_{nn} \qquad (1.11)$$

In many problems we are interested in matrix elements of the operator $\hat{A}$ connecting two different states: $A_{nm} \equiv \langle n | \hat{A} | m \rangle$. We will also have need for the outer product $|m\rangle \langle n|$. This outer product is just the product of ψ_m and ψ_n^*. Unlike the inner product, no integration is implied. The outer product is "waiting" to meet up with a bra from the left or a ket from the right.

So the good news of quantum mechanics is that we can solve the Schrödinger equation and know all there is to know about a system. The bad news is that this equation can be solved exactly for only a handful of model systems! Fortunately, these exactly solvable models are reasonable approximations to some very relevant chemical problems, and have much to teach us about quantum mechanical systems in general. Let us review the model problems that help us to understand translational, rotational, vibrational and electronic energies of molecules.

1.3.1 THE PARTICLE IN A BOX: A MODEL FOR TRANSLATIONAL ENERGIES

This model assumes that a particle is confined to a region of space by a potential energy that rises to infinity at the walls of the container. The problem may seem very simple and perhaps artificial, but it illustrates some general principles that apply to more realistic quantum mechanical systems. Furthermore, it provides the basis for the analysis of translational energies of ideal gases using statistical thermodynamics. Let us consider the three-dimensional system, in which a particle is considered to occupy a volume $V = abc$, where a, b, and c are the edge lengths of the box along the x, y, and z directions. The picture behind this model is shown in Figure 1.1. The potential energy is zero within the confines of the box and infinite outside the box. This discontinuity in the potential amounts to an infinite force on the particles at the walls of the box, and thus there is no leakage of the probability, that is, no tunneling. The wavefunction must go to zero at the boundaries, as there is no probability of finding the particle outside the box. Since there are three dimensions, there are three quantum numbers: n_x, n_y, and n_z, one for each boundary condition. The wavefunctions are

$$\psi_{n_x n_y n_z} = \sqrt{\frac{8}{abc}} \sin\left(\frac{n_x \pi x}{a}\right) \sin\left(\frac{n_y \pi y}{b}\right) \sin\left(\frac{n_z \pi z}{c}\right) \qquad (1.12)$$

where each of the three quantum numbers ranges independently over the nonzero positive integers. The possibility that any of them could be zero is excluded because this would result in the wavefunction vanishing everywhere. If there *is* a particle in the box, $\psi_{n_x n_y n_z}$ cannot be zero! This has an important consequence, which is revealed by the expression for the quantized energy:

$$E_{n_x n_y n_z} = \frac{h^2}{8m}\left[\left(\frac{n_x^2}{a^2}\right) + \left(\frac{n_y^2}{b^2}\right) + \left(\frac{n_z^2}{c^2}\right)\right] \qquad (1.13)$$

The lowest possible (ground state) energy for this particle is attained when $(n_x n_y n_z)$ is equal to (111), and it is evident that $E_{111} \neq 0$. This existence of zero-point energy (nonzero energy in the ground state) is required by the uncertainty principle:

$$\Delta x \Delta p_x \geq \hbar/2 \qquad (1.14)$$

which states that the product of the uncertainties in the position and momentum of the particle cannot be less than $\hbar/2$. Uncertainty principles such as that of Equation 1.14 result whenever two operators do not commute. In this case, the commutator, defined by $[\hat{x}, \hat{p}_x] = \hat{x}\hat{p}_x - \hat{p}_x\hat{x}$, is equal to $i\hbar$. Inside the box the potential energy is zero, so the energy is entirely kinetic: $T = p^2/2m = (p_x^2 + p_y^2 + p_z^2)/2m$. If the energy could be zero, then the momentum would be precisely known to be zero, in violation of Equation 1.14.

The uncertainty in a quantum mechanical property is defined as

$$\Delta A = \sqrt{\langle A^2 \rangle - \langle A \rangle^2} \qquad (1.15)$$

Equation 1.15 says that the uncertainty in physical property A is the square root of the difference in two expectation values: The first is the average of the square of the operator, and the second is the square of the average. These two expectation values are different unless the state of the system is an eigenfunction of $\hat{A}$. Equation 1.15 is really just the standard deviation of the physical property A, since the wavefunction-squared is a probability distribution. Equation 1.14 springs from the general expression

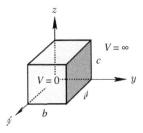

Figure 1.1 The potential energy function for a particle in a three-dimensional box.

$$\Delta A \, \Delta B \geq \frac{1}{2}\left|\left\langle\left[\hat{A},\hat{B}\right]\right\rangle\right| \tag{1.16}$$

In other words, the product of two uncertainties can be no less than one-half the expectation value of the commutator of the operators for the properties. Equation 1.14 follows directly from 1.16.

The three-dimensional particle in a box illustrates degenerate energy levels. When two or more states (represented by wavefunctions, e.g., Equation 1.12) share the same energy level, they are said to be degenerate. Suppose that the box is a cube: $a = b = c$. The ground state, with $n_x = n_y = n_z = 1$, is nondegenerate with energy $3h^2/8ma^2$. The first excited state, however, is triply degenerate, because the energy $6h^2/8ma^2$ can be achieved by three different combinations of the quantum numbers $(n_x n_y n_z)$: (211), (121), and (112). These represent three distinct states since they differ in the position of the nodal plane: ψ_{211} has a nodal plane at $x = a/2$, ψ_{121} has one at $y = a/2$, and ψ_{112} has one at $z = a/2$. Degeneracy is a consequence of symmetry, and if we start with the cubic box and distort it, the degeneracy is lifted and the levels split apart. We can see this effect in molecules; e.g., a degenerate electronic state of a symmetric molecule such as benzene may be split by substitution, which perturbs the sixfold rotation symmetry. An octahedral transition metal complex may tend to distort to a less symmetric form if in doing so it can put electrons in lower energy levels.

Note also the general feature that states of increasing energy have more and more nodes. These nodes, where the wavefunction and thus the probability goes to zero, are one of the many strange aspects of quantum mechanics that would not be expected from the classical analogy to the problem. It is evident that nodes are required in order for two states to be orthogonal to one another. Another lesson is that the energy levels are farther apart for smaller boxes. Conversely, as the box size increases the energy levels get closer together, merging into a continuum in the classical limit. This trend, in which the separation of adjacent energy levels decreases with increasing size, is also seen in atoms, molecules, and nanoparticles.

The particle in a box model is useful for modeling translational energies of molecules in the gas phase. The model is consistent with the ideal gas approximation, in that it neglects intermolecular interactions. As you will show in one of the homework problems, the typical energy level spacings of particles in macroscopic boxes are much smaller than thermal energy at normal temperatures. This means that translational energy levels are essentially continuous and can be treated classically. Translational motion leads to Doppler broadening of gas phase spectra and to light scattering by fluids.

1.3.2 THE RIGID ROTOR: A MODEL FOR ROTATIONAL MOTION OF DIATOMICS

The rigid rotor model is an approximation to the problem of a freely rotating molecule. For a molecule to qualify as a free rotor, there can be no torques, or equivalently, no angular dependence of the potential energy. As such, the model is applicable to the analysis of gas phase samples where the intermolecular interactions are negligible. Angular momentum is a key concept in many quantum mechanical problems, including the topic of electron spin to be discussed in Chapter 3. The simplest example of rotational motion is a diatomic molecule having a rigid bond (Figure 1.2). This model obviously ignores the vibrational motion of the molecule, but it is a good first approximation to the problem if the vibrational energy is not too high. (We shall improve on this picture in Chapters 8 and 9 by accounting for the coupling of vibration and rotation.) In the present model, we picture a dumbbell-like object in which two masses m_1 and m_2 are connected by a rigid (and massless) rod of length R. In this section, we consider angular motion of this body relative to a fixed center of mass. We postpone until

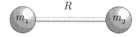

Figure 1.2 The rigid rotor model.

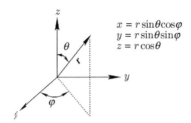

$$x = r\sin\theta\cos\varphi$$
$$y = r\sin\theta\sin\varphi$$
$$z = r\cos\theta$$

Figure 1.3 Spherical polar coordinates.

Chapter 8 the discussion of how this *internal motion* (relative to a coordinate system fixed in the molecule) may be separated from the *external motion* (the translational motion of the center of mass). The reader is referred to the coordinate system shown in Figure 1.3.

In spherical polar coordinates, the distance r is equal to $(x^2 + y^2 + z^2)^{1/2}$, and the vector $\vec{r}$ makes an angle θ with respect to the positive z-axis. The angle φ is the angle of rotation about the z-axis, as defined in Figure 1.3. For the present problem, we consider the origin of the coordinate system to be fixed at the center of mass of the molecule. The orientation of the coordinate system is arbitrary, but it must remain stationary for our analysis. Since the bond distance R is constant, the angles θ and φ are the only coordinates on which the wavefunctions depend. Thus we expect two quantum numbers.

The moment of inertia I of a diatomic is

$$I = \mu R^2 \tag{1.17}$$

where μ is the reduced mass:

$$\mu = \frac{m_1 m_2}{m_1 + m_2} \tag{1.18}$$

The potential energy is independent of angle and may be set equal to zero everywhere. The Hamiltonian then consists only of the operator for kinetic energy:

$$\hat{H} = \frac{\hat{L}^2}{2I} \tag{1.19}$$

where $\hat{L}$ is the operator for angular momentum. The form of the $\hat{L}^2$ operator will be presented in Chapter 8. (The square of an operator means to operate twice with the same operator.) Recall that angular momentum is a vector. The classical expression is $\vec{L} = \vec{r} \times \vec{p}$, the cross-product of the position and momentum vectors, where r is the distance from the axis of rotation (see Appendix A for a review of vector operations). The angular momentum vector of a diatomic spinning in a plane is perpendicular to the plane. It points up or down according to whether the rotation is clockwise or counterclockwise (Figure 1.4), in accordance with the right-hand rule for the cross-product. So the direction of $\vec{L}$ depends on the orientation of the rotational motion.

Let us consider the commutation properties of the operators for the x, y, and z components of angular momentum $\hat{L}_x$, $\hat{L}_y$, and $\hat{L}_z$. It turns out that these operators do not commute with one another:

$$\left[\hat{L}_x, \hat{L}_y\right] = i\hbar L_z$$

$$\left[\hat{L}_y, \hat{L}_z\right] = i\hbar L_x \tag{1.20}$$

$$\left[\hat{L}_z, \hat{L}_x\right] = i\hbar L_y$$

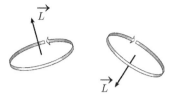

Figure 1.4 Angular momentum vectors: $\vec{L} = \vec{r} \times \vec{p}$.

Thus, we cannot simultaneously specify all three components of $\vec{L}$. The $\hat{L}^2$ operator, however, commutes with each of the three components of angular momentum:

$$\left[\hat{L}_x, \hat{L}^2\right] = \left[\hat{L}_y, \hat{L}^2\right] = \left[\hat{L}_z, \hat{L}^2\right] = 0 \tag{1.21}$$

Equation 1.21 would seem to imply that we can know the magnitude of the angular momentum and each of the three components, but that would violate the uncertainty required by Equation 1.20. The result is that we may simultaneously specify the magnitude and one component of the angular momentum. When two operators commute, they share a set of common eigenfunctions. The form of the $\hat{L}_z$ operator is particularly simple,

$$\hat{L}_z = -i\hbar \frac{\partial}{\partial \varphi} \tag{1.22}$$

It is therefore convenient to find the eigenfunctions shared by $\hat{L}_z$ and $\hat{L}^2$. These are the spherical harmonics $Y_{lm}(\theta, \varphi)$ (see Appendix A). These functions are tabulated in many books, and they will crop up again in many later chapters. (In the hydrogen atom problem, the shapes of these functions determine the angular dependence of atomic orbitals.) Here, we are interested in their eigenvalues:

$$\hat{L}^2 Y_{lm} = l(l+1)\hbar^2 Y_{lm}, \text{ where } l = 0, 1, 2 \ldots$$
$$\hat{L}_z Y_{lm} = m\hbar Y_{lm}, \text{ where } m = 0, \pm 1, \pm 2, \ldots \pm l \tag{1.23}$$

The implications of Equation 1.23 are of considerable importance. The angular momentum vector of a molecule in a particular state has the magnitude

$$|\vec{L}| = \sqrt{l(l+1)}\,\hbar \tag{1.24}$$

For any value of the angular momentum quantum number l, there are $2l + 1$ possible orientations corresponding to the allowed values of $L_z = m\hbar$. Since the x and y components of angular momentum are unspecified, the vector $\vec{L}$ can be considered to lie anywhere on a cone making an angle θ with respect to the z-axis, as shown in Figure 1.5. The permissible angles are given by

$$\cos\theta = \frac{m}{\sqrt{l(l+1)}} \tag{1.25}$$

The angular momentum vector cannot coincide with the z direction, as that would mean $L_x = L_y = 0$, in violation of the uncertainty required by Equation 1.20.

The coordinate frame xyz has an arbitrary orientation in the laboratory. Now, there is nothing special about the z-axis as far as a molecule is concerned, unless there happens to be an electric or magnetic field that makes one direction unique. If so, then it is convenient to call the direction of the field the z direction, since L_z is quantized. Pictures such as Figure 1.5 have an interesting resemblance to a classical mechanical situation:

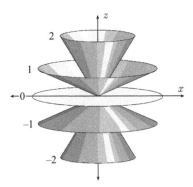

Figure 1.5 Quantized angular momenta for the case $l = 2$.

that of a magnetic moment in a magnetic field. There is a magnetic dipole moment associated with electronic orbital or spin angular momentum, or nuclear spin. (All these examples result in angular momentum equations like Equation 1.23, except that different symbols are used for different types of angular momentum.) The torque on a magnetic dipole in a magnetic field causes the dipole to precess about the field direction, sweeping out a cone like one of those in Figure 1.5.

Returning to the rotating diatomic molecule, the eigenfunctions of Equation 1.19 are readily obtained once we have Equation 1.23, since the inertia I is a constant. It is conventional to use the symbols J and M for the quantum numbers when discussing rotational states of molecules. The wavefunctions are the spherical harmonics Y_{JM} and the energy levels are given by

$$E_J = \frac{\hbar^2}{2I} J(J+1) \tag{1.26}$$

Note that the energy is independent of the quantum number M, since in the absence of a field the energy of a rotating molecule does not depend on its orientation in space. This leads to degeneracy of $2J + 1$, the number of different values of M for a given J. The symbol g is commonly used to denote degeneracy. In this context, we have $g_J = 2J + 1$. The degeneracy of rotational energy levels is taken into account when calculating the equilibrium populations, and it will be seen that these populations contribute to the intensity patterns observed in rotational spectra.

The rigid rotor has no zero-point energy; the ground state has $E = 0$. That might seem like a violation of the uncertainty principle, but it is not. Equation 1.14 does not apply to this problem, but similar uncertainty principles interrelating the components and magnitude of $\vec{L}$ can be derived using Equation 1.16. Does it violate an uncertainty principle if all three components of the angular momentum are zero?

1.3.3 The harmonic oscillator: Vibrational motion

1.3.3.1 CLASSICAL MECHANICS OF HARMONIC MOTION

Imagine the situation pictured in Figure 1.6: a mass m connected to a stationary wall (or something of infinite mass) by a spring that obeys Hooke's law. This law says that the force on the spring is directly proportional to the displacement:

$$F = -kx \tag{1.27}$$

The equilibrium position of the mass is defined to be $x = 0$. The minus sign in Equation 1.27 indicates that F is a restoring force; it acts in a direction opposite to the displacement. k is called the force constant, and it is a measure of the stiffness of the spring. The force is also equal to the negative slope of the potential energy: $F = -dV/dx$. Thus the potential energy is

$$V = \frac{1}{2} kx^2 \tag{1.28}$$

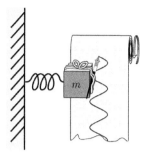

Figure 1.6 Thought experiment for harmonic motion.

This is the equation of a parabola. We refer to V as a harmonic potential. It is clearly an approximation to a real-world spring, as it assumes that compressing the spring a certain distance has the same effect as extending it the same amount. (Think about this. Does it make sense to allow x to go to positive infinity, or to negative infinity?) But if we promise to use this model only when the displacements are small, it will be a fine approximation. We can justify the harmonic approximation by expanding the potential energy in a Taylor series about $x = 0$, as will be discussed in later chapters.

It is worthwhile to consider the classical situation first, and then see how the quantum mechanical problem compares. In classical mechanics, we can find the trajectory of the particle by solving the equations of motion. Alternatively, let us do the following thought experiment. Imagine that the mass rests in the middle of a roll of chart paper, and has a pen stuck in it, as shown in Figure 1.6. We can pull (or push) the mass a certain amount and set it in motion. Meanwhile, let the chart paper roll and the pen will make a graph of position versus time. This graph will be a cosine wave, or a sine wave, since the origin of time is arbitrary. The initial conditions determine the amplitude (maximum displacement) x_0 of the oscillation. Choosing the cosine function (letting $x = x_0$ at $t = 0$), the function drawn on the chart paper is $x = x_0 \cos\omega_0 t$, where ω_0 is the angular frequency. Every time $\omega_0 t$ increases by 2π, the mass returns to the same position. Thus the period of the motion is $2\pi/\omega_0 = 1/\nu_0$, where ν_0 is the frequency in cycles per second (or just plain s^{-1}). You can think of $\omega_0 = 2\pi\nu_0$ as the frequency in radians per second, but the units are still just s^{-1}. The subscript 0 is used here to denote the natural frequency of the oscillator, which you will show in one of the homework problems to be

$$\omega_0 = 2\pi\nu_0 = \sqrt{\frac{k}{m}} \tag{1.29}$$

This relation predicts that stiffer springs and lighter masses result in higher frequency motion.

Once we set the oscillator in motion, it will maintain a constant energy, since there are no frictional losses in the model. It is readily seen that the kinetic and potential energies are both periodic functions:

$$T = \frac{1}{2}m\left(\frac{dx}{dt}\right)^2 = \frac{kx_0^2}{2}\sin^2\omega_0 t \tag{1.30}$$

$$V = \frac{1}{2}kx^2 = \frac{kx_0^2}{2}\cos^2\omega_0 t \tag{1.31}$$

The total energy is constant and decided by the amplitude: $E = T + V = kx_0^2/2$. An important punchline is the following: A classical harmonic oscillator can have any amplitude, and thus any energy. This is in stark contrast to the quantum mechanical harmonic oscillator.

1.3.3.2 THE QUANTUM MECHANICAL HARMONIC OSCILLATOR

Let us base the quantum mechanical problem on a picture that is just slightly different from the mass and spring used previously. Consider two masses (two atoms) m_1 and m_2, connected by a Hooke's law spring as in Figure 1.7. We take $x = R - R_e$ to be the displacement of the bond length R from its equilibrium position R_e. The Hamiltonian for this system is

$$\hat{H} = \frac{-\hbar^2}{2\mu}\frac{d^2}{dx^2} + \frac{1}{2}kx^2 \tag{1.32}$$

Note the appearance of the reduced mass $\mu \equiv m_1 m_2/(m_1 + m_2)$ in the kinetic energy operator. (We will see where this comes from in Chapter 8.) The eigenvalues of this Hamiltonian are the harmonic oscillator energy levels:

$$E_v = \left(v + \frac{1}{2}\right)hv_0 \tag{1.33}$$

where the quantum number v can be 0, 1, 2, …, ∞. Note that there is only one quantum number for this one-dimensional problem, and the energy levels are nondegenerate. The natural frequency is

$$v_0 = \frac{1}{2\pi}\sqrt{\frac{k}{\mu}} \tag{1.34}$$

Once again, the position–momentum uncertainty principle results in zero-point energy: $E_0 = hv_0/2$. The eigenfunctions of Equation 1.32 are sketched in Figure 1.8. The energy levels of Equation 1.33 are represented by horizontal lines in the figure, and each wavefunction is graphed using the energy line as the x-axis. The fact that the wavefunctions extend to infinity is significant, as it leads to the concept of tunneling. Tunneling means that the quantum mechanical system has some probability of being found in a classically forbidden region. In the regions where the tails of the wavefunctions extend outside the parabolic potential well, the potential energy exceeds the total energy. Applying the notion that $E = T + V$, this leads to the disturbing idea that the kinetic energy is negative in the tunneling region! This would never occur for the classical harmonic oscillator, which always turns around at the points $\pm x_0$ (the "turning points") where the total energy and the potential energy are equal. Once again, the quantum mechanical problem gives rise to a weird effect that has no comparison in the classical mechanical world.

Figure 1.7 The harmonic approximation for a diatomic molecule.

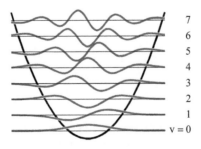

Figure 1.8 Eigenfunctions and energies of a quantum mechanical harmonic oscillator.

1.3.3.3 HARMONIC OSCILLATOR RAISING AND LOWERING OPERATORS

The harmonic oscillator (HO) raising and lowering operators, a^+ and a, provide an elegant way to approach many problems in spectroscopy. The use of these operators and the HO eigenkets can greatly simplify the calculation of expectation values and matrix elements in vibrational problems. In Chapter 2, we will use formally identical operators to describe the Hamiltonian for the quantized radiation field.

The non-Hermitian operators a^+ and a are defined in terms of the operators for position $\hat{q}$ and momentum $\hat{p}$:

$$a^+ \equiv \left(\frac{\mu\omega_0}{2\hbar}\right)^{1/2}\left(\hat{q} - \frac{i\hat{p}}{\mu\omega_0}\right) \tag{1.35}$$

$$a \equiv \left(\frac{\mu\omega_0}{2\hbar}\right)^{1/2}\left(\hat{q} + \frac{i\hat{p}}{\mu\omega_0}\right) \tag{1.36}$$

These equations are definitions. It will now be shown how the effect of the raising and lowering operators on the HO eigenfunctions follows from their commutation properties. We first rewrite the position and momentum operators in terms of a^+ and a:

$$\hat{q} = \left(\frac{\hbar}{2\mu\omega_0}\right)^{1/2}(a + a^+) \tag{1.37}$$

$$\hat{p} = -i\left(\frac{\mu\hbar\omega_0}{2}\right)^{1/2}(a - a^+) \tag{1.38}$$

When these expressions are substituted into the Hamiltonian, $\hat{H}_{vib} = \hat{p}^2/2\mu + \mu\omega_0^2\hat{q}^2/2$, the result is

$$\hat{H}_{vib} = \frac{1}{2}\hbar\omega_0(aa^+ + a^+a) \tag{1.39}$$

A more convenient form of Equation 1.39 is obtained by making use of the commutation relation $[a, a^+] = 1$, which follows from $[\hat{q}, \hat{p}] = i\hbar$:

$$\hat{H}_{vib} = \hbar\omega_0\left(a^+a + \frac{1}{2}\right) \tag{1.40}$$

With our prior knowledge of the eigenfunctions and eigenvalues of the Hamiltonian, this expression leads us to postulate a set of functions $|v\rangle$ that are eigenfunctions of a number operator defined as $\hat{N} \equiv a^+a$; that is, $\hat{N}|v\rangle = v|v\rangle$. From the conventional differential equation solution to the HO problem, we know that the spectrum of eigenvalues of the number operator is the set of all positive integers, including zero. Using operator algebra, we can demonstrate the basis for calling a^+ and a raising and lowering operators. Assume that we can operate on the eigenket $|v\rangle$ with a to generate the new vector $|f\rangle = a|v\rangle$. Let us find the result of applying the number operator to this new function:

$$\hat{N}a|v\rangle = a^+aa|v\rangle = (aa^+ - 1)a|v\rangle$$

$$= (aa^+a - a)|v\rangle \tag{1.41}$$

$$= a(\hat{N} - 1)|v\rangle = (v - 1)a|v\rangle$$

Equation 1.41 shows that $|f\rangle = a|v\rangle$ is an eigenfunction of $\hat{N}$ with eigenvalue $v - 1$; the operator a has lowered the eigenvalue by one. The ket $|f\rangle$ must therefore be proportional to the ket $|v - 1\rangle$; $|f\rangle = a|v\rangle = C|v - 1\rangle$.

The proportionality constant C can be chosen by taking the eigenvectors $|v\rangle$ and $|v-1\rangle$ to be normalized. Since $\langle v|a^+ = (a|v\rangle)^*$,

$$\langle f|f\rangle = \langle v|a^+a|v\rangle = C^*C \tag{1.42}$$

Using $\langle v|a^+a|v\rangle = v\langle v|v\rangle = v$, and choosing the constant C to be real, we get

$$a|v\rangle = \sqrt{v}\,|v-1\rangle \tag{1.43}$$

Notice that Equation 1.43 ensures the lower bound of 0 on the eigenvalues v, since applying the lowering operator to the ket $|0\rangle$ returns the value zero, and no negative values of v can be obtained.

Using the same approach on the vector $|f\rangle = a^+|v\rangle$, one can show that

$$a^+|v\rangle = \sqrt{v+1}\,|v+1\rangle \tag{1.44}$$

Thus a^+ converts an eigenket $|v\rangle$ into a new eigenket having its eigenvalue increased by one. As an example of the utility of this formalism, consider using these operators to derive the following recursion formula for the HO eigenfunctions:

$$q|v\rangle = \left(\frac{\hbar}{2\mu\omega_0}\right)^{1/2}\left[\sqrt{v+1}\,|v+1\rangle + \sqrt{v}\,|v-1\rangle\right] \tag{1.45}$$

Equation 1.45 is obtained by using the position operator q as given in Equation 1.37. It will be useful in future chapters when we consider selection rules for vibrational transitions.

1.3.4 THE HYDROGEN ATOM

As a one-electron atom, the hydrogen atom lacks the pairwise inter-electronic repulsion that prevents an exact solution of the Schrödinger equation in the case of many-electron atoms. By examining the quantum mechanical treatment of one-electron atoms, we obtain some general physical concepts and indeed the basis for approximate treatments of the electronic structure of many-electron atoms and molecules. Hydrogen-like (that is, one-electron) atoms require a set of four quantum numbers to fully specify the wavefunction, and only one to define the energy. The number of quantum numbers is consistent with the electron having three spatial and one spin degree of freedom. The variational principle, on which self-consistent field calculations are based, permits us to build approximate wavefunctions for many-electron atoms which assign individual electrons to hydrogenic orbitals with definite spatial and spin quantum numbers, as will be considered in Chapter 7.

The hydrogen atom wavefunctions (which apply to all one-electron atoms or ions by appropriate choice of the atomic number Z) depend on the spatial variables r, θ, and φ and an abstract spin coordinate often called σ. The spatial coordinates are the spherical polar coordinates which allow the Schrödinger equation to be solved using separation of variables. They define the position of the electron relative to the center of mass of the atom, which is quite close to the position of the more massive nucleus. The polar coordinates r, θ, φ (see Figure 1.3) are related to the Cartesian coordinates x, y, z as follows:

$$x = r\sin\theta\cos\varphi \tag{1.46}$$

$$y = r\sin\theta\sin\varphi \tag{1.47}$$

$$z = r\cos\theta \tag{1.48}$$

The spin coordinate, on the other hand, is a conceptual device to keep track of the two possible states of electron spin, designated α for spin-up and β for spin-down. The wavefunctions for the one-electron atom are represented as follows:

$$\Psi_{nlm_lm_s}\left(r,\theta,\varphi,\sigma\right)=R_{nl}\left(r\right)Y_{lm_l}\left(\theta,\varphi\right)\begin{Bmatrix}\alpha\\\beta\end{Bmatrix} \tag{1.49}$$

where $R_{nl}(r)$ is the radial wavefunction and $Y_{lm_l}\left(\theta,\varphi\right)$ the angular one. The angular wavefunctions are the same spherical harmonics that pertain to the rigid rotator problem. Both are cases of two particles rotating in a coordinate system for which the origin is the center of mass. The two angular variables can be further separated as follows:

$$Y_{lm_l}\left(\theta,\varphi\right)\propto P_{lm_l}\left(\cos\theta\right)e^{im_l\varphi} \tag{1.50}$$

The associated Legendre polynomials $P_{lm_l}(x)$ (Appendix A) are tabulated in various books [3,4] and can be generated by a recursion formula. For one-electron atoms we designate the angular momentum quantum numbers with the letters l, m_l corresponding to the eigenvalues of $\hat{L}^2$ and $\hat{L}_z$ respectively (using Dirac notation):

$$\hat{L}^2\left|Y_{lm_l}\right\rangle=l\left(l+1\right)\hbar^2\left|Y_{lm_l}\right\rangle \tag{1.51}$$

$$\hat{L}_z\left|Y_{lm_l}\right\rangle=m_l\hbar\left|Y_{lm_l}\right\rangle \tag{1.52}$$

Thus the magnitude of the orbital angular momentum vector is $\sqrt{l\left(l+1\right)}\,\hbar$, and its projection onto the z direction is $m_l\hbar$.

The one-electron energy levels are given by

$$E_n=\frac{-Z^2e^2}{\left(4\pi\varepsilon_0\right)2n^2a_0}=\frac{-Z^2}{n^2}\times13.6\text{ eV} \tag{1.53}$$

where $n=1,2,3,\ldots\infty$ is the principal quantum number and $a_0=0.529$ Å is the Bohr radius:

$$a_0=\frac{4\pi\varepsilon_0\hbar^2}{\mu_e e^2} \tag{1.54}$$

The quantity $\mu_e=m_e m_N/(m_e+m_N)$ is the reduced mass computed from the mass of the electron m_e and that of the nucleus m_N. Because m_N is much larger than m_e, μ_e is approximately equal to m_e. The degeneracy of each energy level specified by the principal quantum number n is $g_n=2n^2$, consistent with the number of values of l, m_l, and m_s that are permitted for each value of n, namely:

$$l=0,1,\ldots,n-1 \tag{1.55}$$

$$m_l=-l,-l+1,\ldots,0,\ldots,l-1,l \tag{1.56}$$

$$m_s=\pm\frac{1}{2} \tag{1.57}$$

The states having orbital quantum number $l = 0, 1, 2, 3, 4...$ are referred to by the letters $s, p, d, f, g...$, etc. This quantum number is also called the azimuthal quantum number. The reader may wish to verify that $g_n = 2\sum_{l=0}^{n-1}(2l+1) = 2n^2$ by evaluating the summation. The allowed values for these quantum numbers derive from the boundary conditions on the Schrödinger equation. The spin quantum number s is always equal to 1/2 for any electron, so it does not need to be specified for a one-electron atom. It pertains to the length of the spin angular momentum vector the same way that the quantum number l determines the magnitude of orbital angular momentum, as shown in the eigenvalue relations given below. Just as the quantum number m_l, which decides the z component of orbital angular momentum, ranges from $-l$ to l in steps of one, so too does the quantum number m_s range from $-s$ to s, or from $-1/2$ to $1/2$.

In the case of spin angular momentum, we may use the abstract spin eigenfunctions $|\alpha\rangle$ and $|\beta\rangle$ to write:

$$\hat{S}^2|\alpha\rangle = \frac{1}{2}\left(\frac{1}{2}+1\right)\hbar^2|\alpha\rangle = \frac{3}{4}\hbar^2|\alpha\rangle \tag{1.58}$$

$$\hat{S}^2|\beta\rangle = \frac{1}{2}\left(\frac{1}{2}+1\right)\hbar^2|\beta\rangle = \frac{3}{4}\hbar^2|\beta\rangle \tag{1.59}$$

$$\hat{S}_z|\alpha\rangle = \frac{1}{2}\hbar|\alpha\rangle \tag{1.60}$$

$$\hat{S}_z|\beta\rangle = \frac{-1}{2}\hbar|\beta\rangle \tag{1.61}$$

An essential difference between the physical angular momentum of a rotating two-particle system and spin angular momentum is the occurrence of half-integral quantum numbers for the latter. Since the operators for the x, y, and z components of angular momentum (orbital or spin) do not commute with one another, we cannot specify the x and y components along with the z component. This leads to the sort of pictures shown in Figure 1.9, where the angular momentum vector is considered to lie on the surface of a cone of revolution about the z-axis.

The radial wavefunctions $R_{nl}(r)$ of the hydrogen atom determine the extent of the probabilistic electron cloud that is often visualized as the "size" of the atom. These take the form of a product of a polynomial and an exponential function, with the latter serving to enforce the boundary condition that the probability of finding the electron at a distance r from the nucleus decay asymptotically to zero:

$$R_{nl}(r) = N\left(c_0 + c_1 r + \cdots + c_{n-l-1}r^{n-l-1}\right)r^l e^{-Zr/na_0} \tag{1.62}$$

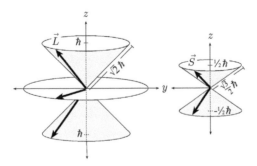

Figure 1.9 Angular momentum vectors for $l = 1$ and $s = \frac{1}{2}$.

The coefficients c_n of each term r^n obey a recursion formula which can be found in most quantum chemistry texts [3,4]. As will be shown in Chapter 7, the allowed transitions between energy levels of the hydrogen atom result in absorption and emission frequencies

$$\tilde{v} = R_H \left(\frac{1}{n_1^2} - \frac{1}{n_2^2} \right) \tag{1.63}$$

The Rydberg constant is $R_\mathrm{H} = 109{,}737$ cm^{-1} and $n_1 < n_2$. R_H can be found using Equations 1.53 and 1.5:

$$R_H = \frac{2\pi^2 \mu_e e^4}{(4\pi\varepsilon_0)^2 h^3 c} \tag{1.64}$$

Equation 1.64 introduces the wavenumber unit $\tilde{v}$, or cm^{-1}, where $\tilde{v}$ is the reciprocal of the wavelength of light in cm. $\tilde{v} = v/c$ is equivalent to the frequency of light v in s^{-1} divided by the speed of light $c = 2.99792 \times 10^{10}$ cm/s. Wavenumbers are convenient frequency units in spectroscopy.

1.3.5 GENERAL ASPECTS OF ANGULAR MOMENTUM IN QUANTUM MECHANICS

Angular momentum of charged particles leads to magnetic properties as discussed further in Chapter 3. In addition to the orbital l and spin s angular momenta of an electron in a hydrogen atom, it will be seen in Chapter 7 that many-electron atoms possess orbital L and spin S angular momenta which can couple further to give a total angular momentum J. The use of upper case letters for these quantum numbers is to distinguish them from their single electron counterparts. Many nuclei also possess angular momentum as designated by the quantum number I, which can take on nonnegative integer or half-integer values depending on the atomic number and isotope. Despite the variety of symbols for the various angular momentum quantum numbers, s, l, S, J, L, I, etc., their quantum mechanical properties, i.e., commutator relations and eigenvalue expressions, are similar. Owing to the former, the magnitude of the angular momentum and its z-component are quantized while the x and y components are uncertain. We state these generalities here in terms of a generic angular moment quantum number j which can be integral (as in the case of orbital angular momentum) or half-integral (as in the case of $s = \frac{1}{2}$ for an electron). The value of j determines the magnitude of the angular momentum vector, referred to here as $|\vec{J}| = \hbar\sqrt{j(j+1)}$. A second quantum number m_j ranges from j to $-j$ in integral steps and specifies the z-component of the angular momentum: $J_z = m_j\hbar$. There are $2j + 1$ values of m_j. We can specify the eigenfunctions by using Dirac notation and indexing the quantum numbers: $|j, m_j\rangle$. For example, when this notation is applied to the case of electron spin, the eigenfunctions are expressed as $|\alpha\rangle = |1/2, 1/2\rangle$ and $|\beta\rangle = |1/2, -1/2\rangle$. We can then write generic eigenvalue relations for any kind of quantized angular momentum as follows:

$$\hat{J}^2|j, m_j\rangle = j(j+1)\hbar^2|j, m_j\rangle$$
$$\hat{J}_z|j, m_j\rangle = m_j\hbar|j, m_j\rangle \tag{1.65}$$

It is useful to define ladder operators for angular momentum as follows:

$$\hat{J}_+ = \hat{J}_x + i\hat{J}_y$$
$$\hat{J}_- = \hat{J}_x - i\hat{J}_y \tag{1.66}$$

The $|j, m_j\rangle$ are not eigenfunctions of these ladder operators. As shown in most quantum mechanics texts, $\hat{J}_+$ and $\hat{J}_-$ act as raising and lowering operators, respectively:

$$\hat{J}_+|j, m_j\rangle = [j(j+1) - m_j(m_j+1)]^{1/2}\hbar|j, m_j+1\rangle$$
$$\hat{J}_-|j, m_j\rangle = [j(j+1) - m_j(m_j-1)]^{1/2}\hbar|j, m_j-1\rangle \tag{1.67}$$

The raising operator has the effect of converting an eigenfunction of $\hat{J}_z$ with eigenvalue $m_j\hbar$ into a new eigenfunction with eigenvalue $(m_j + 1)\hbar$, with no effect on the eigenvalue of $\hat{J}^2$. Similarly, $\hat{J}_-$ lowers the eigenvalue of $\hat{J}_z$ to $(m_j - 1)\hbar$. Notice that if m_j has the maximum value of j, then $\hat{J}_+|j, j\rangle = 0$, e.g. it is impossible to raise the eigenvalue of $\hat{J}_z$ any further. Similarly, $\hat{J}_-|j, -j\rangle = 0$.

Ladder operators are useful for finding matrix elements of the operators for the x and y components of angular momentum, using the following:

$$\hat{J}_x = \frac{1}{2}(\hat{J}_+ + \hat{J}_-)$$

$$\hat{J}_y = \frac{1}{2i}(\hat{J}_+ - \hat{J}_-)$$

(1.68)

To illustrate this using the spin angular momentum operators (using the letter S in place of J for the relevant operators), it is straightforward to demonstrate the following:

$$\hat{S}_x|\alpha\rangle = \frac{\hbar}{2}|\beta\rangle$$

$$\hat{S}_x|\beta\rangle = \frac{\hbar}{2}|\alpha\rangle$$

$$\hat{S}_y|\alpha\rangle = \frac{i\hbar}{2}|\beta\rangle$$

$$\hat{S}_y|\beta\rangle = \frac{-i\hbar}{2}|\alpha\rangle$$

(1.69)

It is seen that the x and y components of the spin operators connect the α and β spin functions, in contrast to $\hat{S}_z$ which is a diagonal operator.

1.4 APPROXIMATE SOLUTIONS TO THE SCHRÖDINGER EQUATION: VARIATION AND PERTURBATION THEORY

We have now reviewed the exactly solvable quantum mechanical problems. (A fifth example, the hydrogen molecule ion H_2^+, can only be solved "exactly" if one makes the Born–Oppenheimer approximation, discussed in Chapter 9.) What of the many problems for which $\hat{H}\psi = E\psi$ cannot be solved exactly? For example, in atoms and molecules with two or more electrons, the interelectronic repulsions make exact solutions to the Schrödinger equation impossible. In other situations, we might need to find the energy for a problem that is close to one for which we already know the solution. Two methods are commonly employed to address these problems. The variation theorem, which states that there is a lower bound on the energy calculated with an approximate wavefunction, is of great utility in electronic structure calculations. Another approach is perturbation theory, which is useful whenever the true Hamiltonian is slightly different from one for which the energies and wavefunctions are known.

1.4.1 VARIATION METHOD

In this approach, one uses an approximate wavefunction φ which has to obey the appropriate boundary conditions. The variation theorem states that the energy expectation value calculated using this trial function cannot be any lower than the true ground-state energy E_0:

$$\langle E \rangle = \frac{\langle \varphi|\hat{H}|\varphi \rangle}{\langle \varphi|\varphi \rangle} \geq E_0$$

(1.70)

Note that $\hat{H}$ has to be the exact Hamiltonian operator and φ has to be a well-behaved function which satisfies the boundary conditions. The denominator in 1.70 is present in case the trial function φ is not normalized. Equation 1.70 says that the lower the energy calculated with some function φ, the closer it is to the exact energy. This suggests that a good way to proceed is to build some flexibility into the trial functions, some adjustable parameters that can be varied to obtain the lowest possible energy with that functional form. One way to do this is to use linear variation theory, in which the trial function is taken to be a linear combination of basis functions, say, $f_1, f_2, \ldots, f_n$. Then we have

$$\varphi = c_1 f_1 + c_2 f_2 + \cdots + c_n f_n \tag{1.71}$$

The c_i's are coefficients (just plain numbers) to be determined by minimizing the energy. The basis functions may be any functions appropriate to the boundary conditions. In the calculation of electronic energy levels of molecules, for example, they may be hydrogen-like wavefunctions localized on the atoms. This is the basis for the LCAO-MO (linear combination of atomic orbitals to get molecular orbitals) method, which will be reviewed in Chapter 11.

Equation 1.71 can be used in Equation 1.70 to get an expression for the energy that depends on the coefficients. Then, to get the lowest possible energy with the given basis set, we set $d\langle E\rangle/dc_i = 0$ for each of the n coefficients. This leads to a set of simultaneous linear equations:

$$\begin{pmatrix} H_{11} - E & H_{12} - ES_{12} & \cdots & H_{1n} - ES_{1n} \\ H_{21} - ES_{21} & H_{22} - E & \cdots & H_{2n} - ES_{2n} \\ \vdots & \vdots & \ddots & \vdots \\ H_{n1} - ES_{n1} & H_{n2} - ES_{n2} & \cdots & H_{nn} - E \end{pmatrix} \begin{pmatrix} c_1 \\ c_2 \\ \vdots \\ c_n \end{pmatrix} = \begin{pmatrix} 0 \\ 0 \\ 0 \\ 0 \end{pmatrix} \tag{1.72}$$

where $H_{ij} = \langle f_i | H | f_j \rangle = H_{ji}^*$ and $S_{ij} = \langle f_i | f_j \rangle = S_{ji}^*$ (Note that, in general, an element of the square matrix on the left-hand side is $H_{ij} - ES_{ij}$, but the overlaps S_{ii} are unity for normalized basis functions. We have dropped the angle brackets around the energy E.) The methods for solving Equation 1.72 are well established (see Appendix A). A solution exists only if the determinant of the matrix of $H_{ij} - ES_{ij}$ is zero:

$$\begin{vmatrix} H_{11} - E & H_{12} - ES_{12} & \cdots & H_{1n} - ES_{1n} \\ H_{21} - ES_{21} & H_{22} - E & \cdots & H_{2n} - ES_{2n} \\ \vdots & \vdots & \ddots & \vdots \\ H_{n1} - ES_{n1} & H_{n2} - ES_{n2} & \cdots & H_{nn} - E \end{vmatrix} = 0 \tag{1.73}$$

Expanding this determinant results in a polynomial of degree n in the energy: thus there are n roots for the energy. We can arrange these in order of increasing energy: $E_0 \leq E_1 \leq E_2 \cdots \leq E_n$. Each root in this sequence is a lower bound to the true energy; that is, E_0 is no lower than the true ground state energy, E_1 is no lower than the energy of the first excited state, etc. For each solution E_k, there is a particular set of coefficients to describe the corresponding wavefunction: $\varphi_k = \sum_i c_{ki} f_i$. (We have to use an additional subscript on the coefficients to keep track of the n different linear combinations.) The c_{ki}'s for a particular root are found by substituting the energy E_k into the matrix Equation 1.72 and solving. When this is done, one finds that the coefficients can only be determined to within a multiplicative constant, so it is customary to impose normalization of each φ_k to fix the values of the c_{ki}'s.

Linear variation methods are frequently used in electronic structure calculations, and the idea of combining wavefunctions (basis states) to get new wavefunctions (trial functions like φ) is a recurring theme. The number of calculated energy levels and states is always equal to the number of basis states combined.

1.4.2 PERTURBATION THEORY

We are often interested in molecules which are subject to weak perturbations, such as an external electric or magnetic field. In the study of spectroscopy, we deal with the perturbation resulting from the application of a time-dependent electromagnetic field (that of light), and the theory required for that problem is presented in Chapter 4. Here, we are concerned with perturbations that do not vary in time, such as static fields. Sometimes, the perturbing field is intrinsic to the molecule, as in spin-orbit coupling where the orbital motion of the electrons presents a magnetic field with which the electron spin interacts. (We will have more to say about this in later chapters.) In general, the Hamiltonian is taken to be

$$\hat{H} = \hat{H}^{(0)} + \lambda \hat{H}' \tag{1.74}$$

where we assume that we know the zero-order eigenfunctions and energies:

$$\hat{H}^{(0)} \psi_i^{(0)} = E_i^{(0)} \psi_i^{(0)} \tag{1.75}$$

The extra term is the perturbation $\hat{H}'$ and λ is just a number that we can use to turn the perturbation on ($\lambda = 1$) and off ($\lambda = 0$). In derivations of the working equations for perturbation theory, λ is a convenient tag for keeping track of the order of the correction. The perturbed wavefunction is written as

$$\psi_i = \psi_i^{(0)} + \lambda \psi_i^{(1)} + \lambda^2 \psi_i^{(2)} + \cdots \tag{1.76}$$

We say that $\psi_i^{(1)}$ is the first-order correction to the wavefunction, and so forth. Similarly, the perturbed energy is a series:

$$E_i = E_i^{(0)} + \lambda E_i^{(1)} + \lambda^2 E_i^{(2)} + \cdots \tag{1.77}$$

In order for the perturbation approach to be useful, the series expansions of the wavefunction and energy must converge. The zero-order energy should be larger than the first-order correction, which should be larger than the second-order correction, and so forth.* So the approach is a good one when the perturbation results in small changes in the energies compared to the original (zero-order) values.

The working equations for perturbation theory are derived by substituting Equations 1.76 and 1.77 into the Schrödinger equation and equating like powers of λ (see Ref. [3] or [4] for a complete discussion). These equations depend on whether the zero-order state of interest is degenerate. In the case that $\psi_i^{(0)}$ is not degenerate, the following first-order corrections are obtained:

$$E_i^{(1)} = \left\langle \psi_i^{(0)} \middle| \hat{H}' \middle| \psi_i^{(0)} \right\rangle \equiv H_{ii}' \tag{1.78}$$

$$\psi_i^{(1)} = \sum_{j \neq i} \frac{\left\langle \psi_j^{(0)} \middle| \hat{H}' \middle| \psi_i^{(0)} \right\rangle}{E_i^{(0)} - E_j^{(0)}} \psi_j^{(0)} = \sum_{j \neq i} \frac{H_{ji}'}{E_i^{(0)} - E_j^{(0)}} \psi_j^{(0)} \tag{1.79}$$

The first-order correction to the energy can be found knowing only the zero-order wavefunctions. The first-order correction to the wavefunction is a linear combination of the zero-order wavefunctions: $\psi_i^{(1)} = \sum_{j \neq i} c_{ji} \psi_j^{(0)}$. The form of the coefficients (Equation 1.79) results in important considerations as to which zero-order states make significant contributions to the perturbed wavefunctions. In order for state j to contribute to perturbed state i, it is required that the matrix element of the perturbation operator connecting them not vanish: $H_{ji}' \neq 0$. If the perturbation is totally symmetric, then according to group theoretical considerations, H_{ji}' vanishes if the two zero-order wavefunctions are of different symmetry (see Appendix C). It is also apparent that the energy difference in the denominator of Equation 1.79 is important. If two zero-order states of the same symmetry are close in energy, then they will mix strongly when the perturbation is turned on. This is a recurring theme in quantum chemistry and spectroscopy: two nearby states of the same

---

* It sometimes happens that a lower order correction vanishes due to symmetry, in which case the higher order correction is larger.

symmetry mix together to give two perturbed states with shifted energies. (Fermi resonance of vibrational states, discussed in Chapter 10, is one such example.)

Equations 1.78 and 1.79 are not valid when the perturbation is applied to a degenerate set of zero-order states. Consider an energy level that is n-fold degenerate. The problem is that the zero-order wavefunctions $\psi_1^{(0)}, \psi_2^{(0)}, \ldots, \psi_n^{(0)}$ are not unique. Whenever two or more eigenfunctions share a set of common eigenvalues, then any linear combination of these functions also has the same eigenvalue. In the zero-order problem, we are free to choose any linear combinations that we like (for example, the real versus complex p orbitals.) In a perturbation calculation, however, we need to find what are called the correct zero-order wavefunctions:

$$\varphi_j = \sum_{k=1}^{n} c_k \psi_k^{(0)} \tag{1.80}$$

The φ_j's are still zero-order wavefunctions; the superscript (0) has just been omitted for clarity. The linear combinations of Equation 1.80 (there are n different combinations) are said to be the correct linear combinations if the perturbed wavefunctions ψ_j go to φ_j in the limit $\lambda \to 0$. Imposing this condition leads to a set of simultaneous linear equations for the coefficients. The first-order energy correction is found by solving a matrix equation that bears much resemblance to Equation 1.72:

$$\begin{pmatrix} H'_{11} - E^{(1)} & H'_{12} & \cdots & H'_{1n} \\ H'_{21} & H'_{22} - E^{(1)} & \cdots & H'_{2n} \\ \vdots & \vdots & \vdots & \vdots \\ H'_{n1} & H'_{n2} & \cdots & H'_{nn} - E^{(1)} \end{pmatrix} \begin{pmatrix} c_1 \\ c_2 \\ \vdots \\ c_n \end{pmatrix} = \begin{pmatrix} 0 \\ 0 \\ \vdots \\ 0 \end{pmatrix} \tag{1.81}$$

Equation 1.81 is based on starting wavefunctions $\psi_k^{(0)}$ that are orthogonal. This is not a problem, because we can always find linear combinations that satisfy orthogonality. The n roots for the first-order energy are found by setting the determinant of the square matrix in Equation 1.81 equal to zero. For each of the resulting values of $E^{(1)}$, there is a correct zero-order combination found from solving the coefficients. If all the energy roots are different, then the perturbation completely lifts the degeneracy. This does not always happen, and much depends on the symmetry of the problem.

1.5 STATISTICAL MECHANICS

Statistical mechanics provides the link between the quantum mechanical properties of individual molecules and the thermodynamic states of collections of molecules. Elementary quantum mechanical problems are usually concerned with isolated molecules, yet most spectroscopy experiments are done on samples containing large numbers of molecules. Quantum mechanics alone does not address the question of how molecules distribute themselves among the quantum states—we need statistical mechanics. Spectral transitions from some initial level i to final level f are only possible if level i is occupied and there is "room" in level f, and the strength of the response depends on the number of molecules in each level. In this section, we use the letter ε to denote single-molecule quantized energies, reserving the uppercase E for the total average energy of the collection. We must be careful to distinguish between *levels* (specified by energies) and *states* (specified by wavefunctions). If a level is degenerate, then we cannot resolve the transitions originating in different states. However, the degeneracy of the level, as we shall see, influences the probability that the level is occupied. The number of molecules n_i in a particular level ε_i is called the occupation number.

The formulas of statistical mechanics are based on the concept of the ensemble average. The average of a physical property could conceivably be obtained by making a large number of repeated measurements on the same molecule at different times. Alternatively, we could measure the property for a large number of identical molecules and average that result. These two types of averages are hypothesized to be the same (the ergodic hypothesis). The ensemble average carries this idea a little further, and imagines a very large number of copies of a system, each consistent with a set of thermodynamic properties. In the commonly employed canonical ensemble, the number of molecules N, the volume V, and the temperature T of all systems are fixed. Each system may be found in any microstate (collection of quantum states of the individual molecules), consistent with the values of N, V, and T. The volume of the system determines the spectrum of translational energies. The principle of equal a priori probabilities is applied, in which it is assumed that all possible microstates are equally probable. As shown, for example, in [5,6], this results in the well-known Boltzmann distribution being the *most probable* one. And, due to the large number of molecules (or systems), this most probable distribution is essentially the only one that is ever observed! The Boltzmann distribution is presented here without derivation.

Consider an ideal gas, where we can neglect the interactions between molecules, and let us suppose we know the quantized energy levels available to each molecule: ε_0, ε_1, ε_2,..., etc. There are numerous ways that a large number N of molecules can be distributed among these levels. The Boltzmann distribution is the set of occupation numbers that is most probable for a system in equilibrium at the temperature T. According to this distribution, the ratio of the populations of a pair of states 1 and 2 is given by

$$\frac{n_1}{n_2} = \frac{g_1 \exp\left(\dfrac{-\varepsilon_1}{k_B T}\right)}{g_2 \exp\left(\dfrac{-\varepsilon_2}{k_B T}\right)} \tag{1.82}$$

where k_B is Boltzmann's constant, $1.3806488... \times 10^{-23}$ J/K, and g_i is the degeneracy of state i. In expressions such as Equation 1.82, it is common to use the following definition to save writing:

$$\beta = \frac{1}{k_B T} \tag{1.83}$$

We see that the population of a level i is proportional to $g_i \exp(-\beta \varepsilon_i)$. With a little effort, we can be more specific and find the number n_i by requiring the populations to sum to the total number of molecules: $\sum_i n_i = N$. This results in a normalized probability distribution:

$$p_i = \frac{n_i}{N} \tag{1.84}$$

where p_i is the probability that level i is occupied. Introducing the concept of a partition function z,

$$z = \sum_{\text{levels } i} g_i \exp(-\beta \varepsilon_i) \tag{1.85}$$

the probability is given by

$$p_i = \frac{g_i \exp(-\beta \varepsilon_i)}{z} \tag{1.86}$$

The partition function is a measure of the number of thermally available states. This can be appreciated by taking the ground state energy to be zero, so that $z = g_0 + g_1 \exp(-\beta \varepsilon_1) + g_2 \exp(-\beta \varepsilon_2) + \cdots$. In the limit of low temperature (large β), the partition function goes to g_0, and only the ground state is available. As the temperature is increased, the exponential terms contribute with greater weight as higher energy states become available, and z increases with temperature. Once the partition function is known, thermodynamic properties, such as energy, entropy, and pressure, may be calculated.

Equation 1.85 is written in terms of a sum over levels. It is sometimes desirable to express the partition function as a sum over states, in which case the degeneracy factor does not appear:

$$z = \sum_{\text{states } n} \exp(-\beta \varepsilon_n) \tag{1.87}$$

z is the molecular partition function; it represents the number of states available to one molecule. Later, we will find the partition function for a collection of identical molecules.

Here is one example of how to calculate a thermodynamic property from the partition function. Suppose that we want to know the average energy per molecule $\bar{\varepsilon}$ and thus the total energy $E = N\bar{\varepsilon}$. The average energy is

$$\bar{\varepsilon} = \sum_i p_i \varepsilon_i \tag{1.88}$$

Using Equation 1.86 in 1.88, it is straightforward to prove the following:

$$\bar{\varepsilon} = k_B T^2 \left(\frac{\partial \ln z}{\partial T} \right)_{N,V} \tag{1.89}$$

The quantum mechanical averaging process, that is, the calculation of expectation values, can be incorporated into the statistical mechanical picture. If we want to know the average value of a property corresponding to operator $\hat{A}$, we take an ensemble average of the expectation values:

$$\bar{A} = \sum_n p_n A_{nn} = \frac{1}{z} \sum_n \exp(-\beta \varepsilon_n) \langle n | \hat{A} | n \rangle \tag{1.90}$$

The bar over A represents an ensemble average, and A_{nn} is the expectation value in quantum state n. Equation 1.90 can be restated more elegantly as follows:

$$\bar{A} = \sum_n \langle n | \frac{\exp(-\beta \hat{H}) \hat{A}}{z} | n \rangle = \sum_n \langle n | \hat{\rho} \hat{A} | n \rangle = Tr(\hat{\rho} \hat{A}) \tag{1.91}$$

The exponential operator in Equation 1.91 may seem odd, but it is not hard to show that, if $\hat{H} | n \rangle = \varepsilon_n | n \rangle$, then

$$\langle n | \exp(-\beta \hat{H}) = \langle n | \exp(-\beta \varepsilon_n) \tag{1.92}$$

Of course, the complex conjugate of Equation 1.92 is also true:

$$\exp(-\beta \hat{H}) | n \rangle = \exp(-\beta \varepsilon_n) | n \rangle \tag{1.93}$$

The density operator in Equation 1.91 is defined as

$$\hat{\rho} = \frac{\exp(-\beta \hat{H})}{z} \tag{1.94}$$

and the trace (Tr) is the sum of the diagonal elements of the matrix representation of the operator, in this case $\hat{\rho} \hat{A}$. In this notation, the partition function can be written as the trace of the *density matrix*:

$$z = Tr(e^{-\beta \hat{H}}) \tag{1.95}$$

The value of using Equation 1.91 instead of 1.90 or Equation 1.95 instead of 1.87, is that there is a particularly important property of the trace: It is independent of representation. In the language of linear algebra, we say that the trace of a matrix is preserved by a unitary transformation. In the present context, a change of representation is a change of basis functions. Any convenient orthonormal set of wavefunctions that obey the appropriate boundary conditions can be used to calculate the trace.

Equation 1.94 gives the equilibrium density operator. In nonequilibrium situations, the Hamiltonian depends on time and so does the density operator. The time evolution of the density operator will be considered briefly in Chapter 4 and in more detail in Chapter 13.

In order to apply the methods of this section, we need to be more specific about the form of the energy levels. A great deal of simplification results when the energy of a molecule can be expressed as the sum of translational, rotational, vibrational, and electronic contributions:

$$\varepsilon = \varepsilon_{tr} + \varepsilon_{rot} + \varepsilon_{vib} + \varepsilon_{el} \tag{1.96}$$

When Equation 1.96 holds, the partition function z can be factored:

$$z = z_{tr} z_{rot} z_{vib} z_{el} \tag{1.97}$$

The energy contributions on the right-hand side of Equation 1.96 are listed in order of increasing separation of the quantum levels, ranging from the essentially continuous translational levels to the widely spaced electronic levels. The electronic partition function is usually very easy to calculate. Since excited electronic states are usually much higher than thermal energy $k_B T$ at ordinary temperatures, z_{el} is equal to the ground state degeneracy. (The exceptions to this occur when molecules have low-lying excited electronic states.) The translational, rotational, and vibrational partition functions can be found using the models discussed in the previous section. For example, the particle in a box model is used to get the translational energies. These are so closely spaced that the sum over n_x, n_y, and n_z can be replaced by an integral. The result is

$$z_{tr} = \left[\frac{2\pi m k_B T}{h^2} \right]^{3/2} V \tag{1.98}$$

Note that the mass m is that of a single molecule. The molecular structure is immaterial since translation just involves motion of the center of mass.

We can use the rigid-rotor model to find z_{rot} for a diatomic molecule. The rotational energy-level spacings for many diatomics (and polyatomics, not considered here) are often, but not always, small enough compared to thermal energy that the sum in z_{rot} can be replaced by an integral. In this limit, we find

$$z_{rot} = \frac{8\pi^2 I k_B T}{\sigma h^2} \tag{1.99}$$

The symmetry number σ is one for heteronuclear and two for homonuclear diatomics. As will be shown in Chapter 8, the latter have only even or odd J states, but not both, resulting in half as many energy levels as for heteronuclear molecules. Equation 1.99 is valid when $h^2/8\pi^2 I k_B$ is much smaller than T, in other words, when the number of thermally available states is large. When this is not the case, then we have to roll up our sleeves and find the partition function by direct summation.

Using the harmonic oscillator model, the vibrational partition function of a diatomic molecule can also be found. The partition function for vibration is

$$z_{vib} = \frac{e^{-\beta h \nu_0/2}}{1 - e^{-\beta h \nu_0}} \tag{1.100}$$

where ν_0 is the vibrational frequency. Equation 1.100 is readily extended to polyatomic molecules, in which case the total vibrational partition function is a product of terms like 1.100, one for each normal mode frequency. Normal modes will be considered in Chapter 10.

The partition functions discussed so far are for individual molecules, and we would like to know the partition function for a collection of molecules. If the intermolecular interactions can be neglected, then it is straightforward to find the partition function Z for N molecules from the molecular partition function z. The simplification results from the fact that the total energy E is the sum of the energies of the individual molecules. The total partition function derives from a sum over all possible N molecule states:

$$Z = \sum_{\text{states } n} \exp(-\beta E_n) \tag{1.101}$$

where E_n is the total energy for a particular set of molecular quantum states. Using $i, j, k, \ldots$ to label the molecular quantum states and $a, b, c, \ldots$ to index the molecules, we have

$$E_n = \varepsilon_{i,a} + \varepsilon_{j,b} + \varepsilon_{k,c} + \cdots \tag{1.102}$$

Inserting Equation 1.102 in 1.101 and separating the exponentials, Z is found to be

$$Z = \sum_{\text{states } i} \exp(-\beta \varepsilon_{i,a}) \sum_{\text{states } j} \exp(-\beta \varepsilon_{j,b}) \sum_{\text{states } k} \exp(-\beta \varepsilon_{k,c}) \ldots$$

$$= \left[\sum_{\text{states } i} \exp(-\beta \varepsilon_{i,a}) \right]^N = z^N \text{ (distinguishable particles)} \tag{1.103}$$

Equation 1.103 derives from the fact that the particles are all the same, so each sum over molecular states gives the same result. However, there is a limitation to this expression. It only applies to distinguishable particles. For example, in a crystal we can distinguish particles by the positions in the lattice, and Equation 1.103 is applicable. But for the ideal gas on which we are focusing in this section, Equation 1.103 overestimates the number of states. The reason is that Equation 1.103 counts as distinct microstates those which differ only in the way that the particles are labeled. For example, if we have the microstate where the total energy is $\varepsilon_{7,a} + \varepsilon_{3,b} + \varepsilon_{4,c} + \cdots$, it is really the same state as, say, $\varepsilon_{4,a} + \varepsilon_{3,b} + \varepsilon_{7,c} + \cdots$ if the particles $a, b, c, \ldots$ are indistinguishable. It might seem like keeping track of all possible permutations presents a horrible bookkeeping problem! Fortunately, there an aspect of the problem that makes it easy to correct for overcounting the states. The simplification results from the fact that the number of states greatly exceeds the number of molecules (thanks to the translational energy contribution), so it is highly unlikely that more than one molecule is assigned to the same quantum state. Once we settle on a particular combination of molecular energies, $\varepsilon_{i,a} + \varepsilon_{j,b} + \varepsilon_{k,c} + \cdots$, there are $N!$ ways to permute the molecule labels. So we know we have overcounted by $N!$ and the partition function should be

$$Z = \frac{z^N}{N!} \text{ (indistinguishable particles)} \tag{1.104}$$

Another condition on the validity of Equation 1.104 is that there be no overlap of the wavefunctions on different molecules. If this condition is met, as we expect for the case of an ideal gas, then the molecules adhere to "Boltzmann statistics." On the other hand, if the wavefunctions overlap, we must account for the indistinguishability of particles quantum mechanically by placing a restriction on the total wavefunction for the N particle system. If we write a total wavefunction for N particles, it is a function of the coordinates of all N particles: $\Psi(1, 2, 3, \ldots, N)$. Since the probability distribution in terms of the square of the wavefunction cannot depend on the particle labels, we have to be able to exchange any two coordinates and obtain the same probability, for example,

$$\left| \Psi(1,2,3,\ldots,N) \right|^2 = \left| \Psi(2,1,3,\ldots,N) \right|^2 \tag{1.105}$$

This leaves two choices for the effect of swapping two coordinates on the wavefunction itself. It may either stay the same or change sign:

$$\Psi(1,2,3,\ldots,N) = \pm\,\Psi(2,1,3,\ldots,N) \tag{1.106}$$

It turns out that there are two kinds of particles in nature. *Bosons*, which have integral spin quantum numbers, have wavefunctions that stay the same when any two particles are exchanged (the plus sign in Equation 1.106). They are said to adhere to Bose–Einstein (BE) statistics. *Fermions*, on the other hand, have half-integral spins and wavefunctions that change sign when the labels for two particles are switched. They follow Fermi–Dirac (FD) statistics. The consequences of this difference are dramatic. Bosons can share quantum states, but fermions cannot. Electrons are fermions, and the familiar Pauli exclusion principle is a result of FD statistics. Photons have spin 1 (as will be discussed in Chapter 2) and are bosons. There is no limit to the number of photons that can occupy a given quantum state.

The molecular partition function derived here does not apply to systems of particles having overlapping wavefunctions. There are, however, expressions for the occupation numbers of states that obey BE and FD statistics. In the limit of high temperature or large volume, the BE and FD occupation numbers go over to the same result as obtained using Boltzmann statistics. For our ideal gas, Boltzmann statistics are just fine. In many spectroscopy experiments, however, the symmetry requirements of Equation 1.106 on the *molecular wavefunction* lead to observable experimental features. We shall examine this in Chapter 8, where rotational spectra are considered.

1.6 SUMMARY

In this chapter, we have reviewed the quantum mechanical and statistical mechanical preliminaries on which much of this book is based. Four simple models provide the basis for approximating the translational, rotational, vibrational, and electronic energies of molecules. The corresponding particle-in-a-box, rigid-rotor, harmonic-oscillator, and hydrogen atom problems provide starting points ("zero-order" approximations) for more sophisticated quantum mechanical treatments of molecular energies. In addition, these models illustrate general quantum mechanical principles, such as degeneracy, symmetry, and the uncertainty principle. The stationary states and their energies will provide the basis for understanding spectroscopic transitions, using the time-dependent perturbation theory to be presented in Chapter 4. The rigid-rotor and harmonic-oscillator models will serve as a point of departure for the analysis of rotational and vibrational spectra. The hydrogenic wavefunctions provide a basis for approximating the electronic wavefunctions of many electron atoms and molecules, and give us the familiar chemical concept of orbitals. The techniques of variation theory and perturbation theory are tools for interpreting molecular properties that are revealed in a myriad of spectroscopy experiments.

The solutions to the Schrödinger equation provide a spectrum of energies that serves as input to the calculation of ensemble average properties, following the prescriptions of statistical mechanics. This averaging procedure is fundamental to the understanding of the spectra of collections of molecules. We have introduced the concept of the density matrix as a component of the ensemble average calculation. In Chapters 4 and 13, this density matrix will be generalized to allow for time dependence, as we will be concerned with the time evolution of the system undergoing a spectroscopic transition.

PROBLEMS

1. Find the difference between the ground- and first excited-state translational energies of a helium atom in a 1 mm³ cubic box. Compare this to room temperature energy $k_B T$ at 300 K.
2. Find all the possible angles with respect to the z-axis for an angular momentum vector of magnitude $\sqrt{6}\hbar$.

3. Show that the uncertainty as defined in Equation 1.15 vanishes when ψ is an eigenfunction of the operator $\hat{A}$.

4. The classical harmonic oscillator obeys the equation of motion: $F = -kx = md^2x/dt^2$. Use this equation to verify Equation 1.29.

5. Can a harmonic oscillator ever be dissociated?

6. The vibrational frequency of $^1H^{35}Cl$ is $\nu_0 = 8.97 \times 10^{13}\,s^{-1}$. Find the force constant for the H–Cl bond.

7. Derive Equation 1.89.

8. Show that if $|n\rangle$ is an eigenfunction of the operator $\hat{A}$ with eigenvalue a_n, then $e^{\hat{A}}|n\rangle = e^{a_n}|n\rangle$. (Hint: Expand the exponential in a series: $e^x = \sum_{k=0}^{\infty} x^k/k!$.)

9. Calculate the fraction of molecules in each of the first ten rotational levels of CO at 300 K, assuming that it is a rigid rotor with a bond length of 1.13 Å.

10. Calculate the translational partition function for a mole of N_2 at 1 atm and 300 K. Verify that the effective number of available states greatly exceeds the number of molecules.

11. Use the partition functions in Equations 1.98, 1.99, and 1.100 to find the translational, rotational, and vibrational contributions to the average energy of a diatomic molecule. Compare each result to the prediction of the equipartition theorem, which states that in the classical limit each degree of freedom contributes $(1/2)k_BT$ to the average energy.

12. Hydrides typically have rotational energies that are too widely spaced to be able to use Equation 1.99 at room temperature. Find z_{rot} for HF at 300 K by summing the energies directly. Try to obtain a value that is accurate to three significant figures. Compare the numerical result to the value obtained using Equation 1.99. (Use 0.92 Å for the bond length.)

REFERENCES

1. E. Hecht, and A. Zajac, *Optics* (Addison-Wesley, Reading, MA, 1974).
2. K. Nassau, *The Physics and Chemistry of Color* (Wiley, New York, 1983).
3. I. N. Levine, *Quantum Chemistry*, 5th ed. (Prentice Hall, Upper Saddle River, NJ, 1999).
4. C. Dykstra, *Quantum Chemistry and Molecular Spectroscopy* (Prentice Hall, Englewood Cliffs, 1992).
5. D. A. McQuarrie, *Statistical Mechanics* (Harper and Row, New York, 1976).
6. P. W. Atkins, *Physical Chemistry*, 6th ed. (W. H. Freeman, New York, 1998).

2

The nature of electromagnetic radiation

2.1 INTRODUCTION

In order to use spectroscopy to study matter through its interaction with electromagnetic radiation, we first have to understand radiation. The electromagnetic spectrum spans a frequency range that varies by many orders of magnitude, as shown in Figure 2.1. We broadly refer to this radiation as light, even though only a very small portion of the spectrum is visible to the human eye. The frequency ν and the wavelength λ of light are inversely proportional to one another: $c = \lambda\nu$ is the speed of light in a vacuum, 2.9979×10^8 m/s. The spectrum ranges from the low-energy radiofrequency region, where the frequencies are in the MHz to kHz range (Hz = Hertz, a unit of frequency equivalent to cycles per second or s^{-1}), to the high frequency gamma-ray region, characterized by frequencies as high as 10^{22} Hz. The corresponding wavelengths go from thousands of kilometers down to less than 10^{-15} m. The wavelengths of visible light, which span only a tiny fraction of an essentially limitless spectrum, range from about 400 to 700 nm, on going from violet to red. These wavelengths correspond to frequencies from 7.5×10^{14} Hz (violet) to 4.3×10^{14} Hz (red), or in wavenumber units, from about 25,000 cm^{-1} to 14,000 cm^{-1}.

In this chapter we consider the properties of electromagnetic radiation in free space and, briefly, the propagation of light in matter. Further discussion of the latter will be postponed until after we have considered the electromagnetic properties of matter itself (Chapter 3). Our point of view will acknowledge both the wave (classical) and particle (quantum mechanical) nature of light. We will rely more on the classical picture of light to describe the radiation itself, yet we will view spectroscopic transitions in terms of absorption and emission of photons of energy $E = h\nu$. The idea of light having dual wave–particle nature, like that of matter, is a mere artifact of our anthropocentric attempt to apply macroscopic concepts to microscopic phenomena. It is the experiment or act of observation that imposes upon light its wave or particle nature. We like to think of waves when considering interference phenomena, but we tend to think of photons when we consider properties such as the photoelectric effect. There is a quantum mechanical theory to describe all the properties of light, whether wave-like or particle-like, called quantum electrodynamics (QED). The interested reader can consult [1] to learn more about QED, while [2] is a good source for the classical description of light. For the spectroscopy experiments considered in this book, we will usually take the semiclassical approach: the light will be treated classically and the matter quantum mechanically.

The classical description of radiation will often suffice for several reasons. The first is due to the correspondence principle, which states that for large quantum numbers the quantum mechanical picture coincides with the predictions of classical mechanics. The intensity of light used in ordinary spectroscopy experiments is high enough to correspond to a very large number of photons (or quanta). In the classical picture, the intensity of light is related to the square of the amplitude of the wave, rather than the number of photons. Another reason that light may be treated classically is that the wavelengths of interest are generally large compared to the size of the spectroscopically active atom or molecule. This statement bears numerical justification, as it is one way to arrive at a number that has long been a source of fascination to physicists: the fine structure constant.* As the name implies, the fine structure constant, $\alpha \approx 1/137$, arises in problems related to the splitting of spectral lines caused by electron spin. Here, we find it when we consider the ratio of the size of the hydrogen atom to the smallest wavelength capable of causing spectral transitions between bound electronic states. This minimum wavelength corresponds to the energy required to ionize a ground state hydrogen atom. Hence,

---

* Physicists' interest in this number has perhaps been a source of fascination to many chemists!

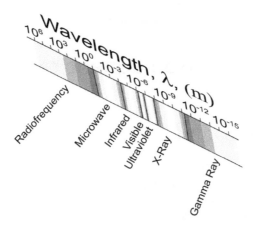

Figure 2.1 The electromagnetic spectrum.

$$\lambda_{min} = \frac{hc}{E_{ion}} \tag{2.1}$$

where the ionization energy is

$$E_{ion} = \frac{e^2}{8\pi\varepsilon_0 a_0} \tag{2.2}$$

and e is the charge on the electron in Coulombs, $a_0 = 0.0529$ nm is the Bohr radius, and ε_0 is the permittivity of free space, equal to 8.854×10^{-12} C^2 m^{-1} J^{-1}.* Equation 2.2 is just the negative of the energy of the ground state hydrogen atom. Using $4\pi a_0$ to represent the size (it is twice the circumference) of the atom, the ratio we obtain is

$$\frac{4\pi a_0}{\lambda_{min}} = \frac{e^2}{2\varepsilon_0 hc} \approx \frac{1}{137} \tag{2.3}$$

Even at the shortest wavelengths of interest, the wavelength of radiation exceeds the size of the particle by a factor of about 137. This allows the particle (quantum) nature of light to be neglected and the wave picture to be used. There are exceptions to this conclusion; for example, the delocalized wavefunctions of electronic bands in metals represent systems that are not small in size compared to optical wavelengths, so the quantum picture of radiation is needed to describe their transitions. In the next section, we look at how to describe light as an electromagnetic wave.

2.2 THE CLASSICAL DESCRIPTION OF ELECTROMAGNETIC RADIATION

2.2.1 MAXWELL'S EQUATIONS

Figure 2.2 displays a snapshot of a linearly polarized ray of light viewed as oscillating electric ($\vec{E}$) and magnetic ($\vec{B}$) fields perpendicular to the propagation direction. While this linearly polarized ray is just one solution to Maxwell's equations, Equations 2.5–2.8 below, we shall see that the transverse nature of electromagnetic radiation is a general property. The frequently invoked concept of an infinite plane wave is a bundle of identical parallel rays such as those in Figure 2.2. The electric and magnetic fields are in phase

---

* In the MKS units that we are using, you can think of ε_0 as a conversion factor. More will be said about the physical meaning of permittivity in Chapter 3.

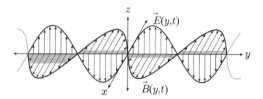

Figure 2.2 Linearly polarized electromagnetic radiation.

and mutually perpendicular as a consequence of Maxwell's equations. Physically, electric fields result from electric charges or from time-dependent magnetic fields, while magnetic fields result from electric currents or time-dependent electric fields. The discussion of fields due to static charges and steady currents is the subject of the Chapter 3, where the electric and magnetic properties of matter are discussed. In this section, we concentrate on the interdependent time-varying electric and magnetic fields that propagate through space in the absence of matter. Maxwell's equations reflect this symmetry in nature through the way that they treat the two related fields.

James Clerk Maxwell (1831–1879) developed the theoretical treatment of light as electromagnetic waves during the time in history when the wave and particle pictures of light were still dueling. The set of fundamental equations presented below, Maxwell's equations, can be manipulated to give classical wave equations for the time and space dependence of the electric and magnetic fields. A significant achievement of this wave picture was the connection derived between the speed of light and the product of the electric permittivity and magnetic permeability of free space:

$$c^2 = \left(\mu_0 \varepsilon_0\right)^{-1} \tag{2.4}$$

The magnetic permeability of free space is defined to be $\mu_0 = 4\pi \times 10^{-7}$ N/amp^2. The calculated value of the speed of light based on the known values of μ_0 and ε_0 was found to be in excellent agreement with the value measured in 1859 by Fizeau, using light aimed through a rotating toothed wheel. Thus the wave picture of light was assured a place in history that would not be invalidated by later developments in the quantum picture of radiation.

Let us take the point of view that the following relationships, known as Maxwell's equations, are postulates, and see what they have to predict about the physical properties of light. In free space, they are*

$$\nabla \cdot \vec{E} = 0 \tag{2.5}$$

$$\nabla \cdot \vec{B} = 0 \tag{2.6}$$

$$\nabla \times \vec{E} = \frac{-\partial \vec{B}}{\partial t} \tag{2.7}$$

$$\nabla \times \vec{B} = \mu_0 \varepsilon_0 \frac{\partial \vec{E}}{\partial t} \tag{2.8}$$

The above expressions apply in vacuum. In the presence of a charge density ρ, Equation 2.5 is rewritten as $\nabla \cdot \vec{E} = \rho/\varepsilon_0$, and to the right-hand side of Equation 2.8 we would add a term proportional to the current density. $\nabla \cdot \vec{B}$ always equals zero because there are no known magnetic monopoles. The "del-dot" operator is sometimes referred to as "divergence," thus Equation 2.5 can also be written div$\vec{E} = 0$ (the divergence of the field is zero), and similarly for Equation 2.6. The vector operation $\nabla \times \vec{E}$ reads curl $\vec{E}$. For a review of these kinds of vector operations, see Appendix A. For example, Equations 2.6 and 2.7 can also be expressed as

---

* The magnetic field $\vec{B}$ is actually the magnetic induction or the magnetic flux density. In free space, it is related to the magnetic field $\vec{H}$ through $\vec{B} = \mu_0 \vec{H}$. The relationship between $\vec{H}$ and $\vec{B}$ is discussed further in Chapter 3.

$$\frac{\partial B_x}{\partial x} + \frac{\partial B_y}{\partial y} + \frac{\partial B_z}{\partial z} = 0 \tag{2.9}$$

$$\frac{\partial E_z}{\partial y} - \frac{\partial E_y}{\partial z} = -\frac{\partial B_x}{\partial t} \tag{2.10}$$

$$\frac{\partial E_x}{\partial z} - \frac{\partial E_z}{\partial x} = -\frac{\partial B_y}{\partial t} \tag{2.11}$$

$$\frac{\partial E_y}{\partial x} - \frac{\partial E_x}{\partial y} = -\frac{\partial B_z}{\partial t} \tag{2.12}$$

Maxwell's equations are mathematical expressions of the statement that a time-dependent magnetic field gives rise to a spatially varying electric field perpendicular to the magnetic field, and vice versa. After some manipulation, which we skip (see, for example, [3]), Maxwell's equations can be recast in the following form:

$$\nabla^2 \vec{E} = \varepsilon_0 \mu_0 \frac{\partial^2 \vec{E}}{\partial t^2} \tag{2.13}$$

$$\nabla^2 \vec{B} = \varepsilon_0 \mu_0 \frac{\partial^2 \vec{B}}{\partial t^2} \tag{2.14}$$

These expressions are significant because they are in the form of the classical wave equation, provided we identify the speed of the light as $c = (\mu_0 \varepsilon_0)^{-1/2}$. The various forms of light that we can envision, such as spherical or plane waves, linear or circular polarization, are just solutions to these equations subject to different boundary conditions.

Equations 2.5–2.8 point out the symmetrical relationship between the magnetic and electric fields. It turns out that there are two quantities, the scalar potential φ and the vector potential $\vec{A}$, from which both $\vec{E}$ and $\vec{B}$ can be derived:

$$\vec{E} = -\nabla \varphi - \frac{\partial \vec{A}}{\partial t} \tag{2.15}$$

$$\vec{B} = \nabla \times \vec{A} \tag{2.16}$$

Note that in free space, where there are no charges to present a change in scalar potential, we can put $\nabla \varphi = 0$. Since for any vector function $\vec{X}$, $\nabla \cdot \nabla \times \vec{X} = 0$, Equation 2.16 also satisfies 2.6. The vector and scalar potentials are not unique. Any function of time could be added to φ, and the gradient of any function of r could be added to $\vec{A}$, without changing $\vec{E}$ or $\vec{B}$. In the frequently used Coulomb gauge, the choice $\nabla \cdot \vec{A} = 0$ is made. The vector and scalar potentials are convenient because they allow the complete specification of the electromagnetic radiation in terms of four quantities, the scalar potential and three components of the vector potential, rather than six quantities, three components each for the magnetic and electric fields. We are also interested in $\vec{A}$ because it appears in the Hamiltonian for a charged particle in a field, to be discussed in Section 2.3.3. It is straightforward to show that the vector potential also obeys the classical wave equation:

$$\nabla^2 \vec{A} = \frac{1}{c^2} \frac{\partial^2 \vec{A}}{\partial t^2} \tag{2.17}$$

Any function of the form $\vec{A} = f(\vec{k} \cdot \vec{r} - \omega t)$ will satisfy Equation 2.17, where $\vec{k}$ is the propagation vector, sometimes called the wave vector, having magnitude $k = 2\pi/\lambda = \omega/c$, where $\omega = 2\pi\nu$ is the angular frequency in radians per second, s^{-1}. One particular solution is

$$\vec{A} = A_0 \hat{e} \cos(\vec{k} \cdot \vec{r} - \omega t) \tag{2.18}$$

where $\hat{e}$ is a unit vector and A_0 is the amplitude of the vector potential. It is often convenient to write expressions like Equation 2.18 in the form

$$\vec{A} = A_0\hat{e}\exp\left[i\left(\vec{k}\cdot\vec{r}\right)-\omega t\right] \tag{2.19}$$

where it is understood that the real part of Equation 2.19 is to be taken. The electric field obtained using Equations 2.15 and 2.18 is then

$$\vec{E} = -\frac{\partial\vec{A}}{\partial t} = -\omega A_0\hat{e}\sin\left(\vec{k}\cdot\vec{r}-\omega t\right) \tag{2.20}$$

So the electric field oscillates in the same direction as the vector potential. Since $\nabla\cdot\vec{A}=0$,

$$\vec{k}\cdot\hat{e}A_0\sin\left(\vec{k}\cdot\vec{r}-\omega t\right)=0 \tag{2.21}$$

Equation 2.21 requires that the propagation vector be perpendicular to the vector potential, and thus also to $\vec{E}$. From $\vec{B}=\nabla\times\vec{A}$, we get the magnetic field:

$$\vec{B} = -A_0(\vec{k}\times\hat{e})\sin\left(\vec{k}\cdot\vec{r}-\omega t\right) \tag{2.22}$$

from which we conclude that the field $\vec{B}$ is perpendicular to both the propagation vector and the field $\vec{E}$. Also, it is apparent that the electric and magnetic fields are in phase and have the same frequency.

From Equations 2.20 and 2.22 we can find the amplitudes of the electric and magnetic fields in terms of the vector potential:

$$E_0 = -\omega A_0 \tag{2.23}$$

$$B_0 = \frac{-\omega A_0}{c} \tag{2.24}$$

More generally, $E_0 = cB_0$ is true for all electromagnetic waves. The energy density of the radiation field is proportional to the square of the amplitude of either the electric or the magnetic field. In units of J/m³, the electric and magnetic field energy densities, which are equal, are given by

$$u_E = \frac{1}{2}\varepsilon_0 E^2 \tag{2.25}$$

$$u_B = \frac{1}{2\mu_0}B^2 \tag{2.26}$$

and the total energy density, also equal to the radiation pressure, is given by $u = u_E + u_B = \varepsilon_0 E^2$.

Since the fields are rapidly varying functions of time, so is the energy density. Devices for measuring this energy typically respond much more slowly than the fields vary, so it is valid to average E^2 and B^2 over a cycle of radiation: $<E^2> = 1/2\,E_0^2$. Thus we write the total average energy density as

$$u = \frac{1}{2}\varepsilon_0 E_0^2 \tag{2.27}$$

We could also have chosen to write the total energy density in terms of the square of the magnetic field, or the product of E and B.

A quantity of great interest in spectroscopy is the intensity, also called the irradiance, of light. It is the energy per unit time per unit area:

$$I = uc = \frac{1}{2}\varepsilon_0 c E_0^2 \tag{2.28}$$

The Poynting vector $\vec{S}$ is defined to point in the direction of propagation and have a magnitude equal to the power per unit area I. Thus

$$\vec{S} = \frac{1}{\mu_0}\vec{E} \times \vec{B} = \varepsilon_0 c^2 \vec{E} \times \vec{B} \tag{2.29}$$

Equation 2.29 can be averaged to obtain

$$\langle S \rangle = I = \frac{1}{2}c\varepsilon_0 E_0^2 \tag{2.30}$$

Let us examine Figure 2.2 with Maxwell's equations in mind. We can write the fields as

$$E_z = E_z^0 \cos\left[\omega\left(t - \frac{y}{c}\right)\right] \tag{2.31}$$

$$B_x = B_x^0 \cos\left[\omega\left(t - \frac{y}{c}\right)\right] \tag{2.32}$$

The minus sign within the parentheses indicates that the wave is moving to the right, in the direction of the positive y-axis. We could also include an arbitrary phase angle within the square brackets, if we wanted to. If we imagine the existence of a time-dependent electric field in the z direction, Maxwell's equations require the presence of the B field in the x direction, since Equation 2.7 reduces to

$$\frac{\partial E_z}{\partial y} = -\frac{\partial B_x}{\partial t} \tag{2.33}$$

Similarly, Equation 2.8 reduces to

$$-\mu_0 \varepsilon_0 \frac{\partial E_z}{\partial t} = \frac{\partial B_x}{\partial y} \tag{2.34}$$

Thus if the only component of the electric field is in the z direction, the only component of the magnetic field is in the x direction. Both E_z and B_x are functions only of time and the direction y, in which the light propagates.

The radiation depicted in Figure 2.2 is said to be polarized in the z direction. (This terminology reflects our tendency to emphasize the electric rather than the magnetic field, since the former interacts with matter more strongly, as we shall see in the next chapter.) It is also monochromatic, having only a single frequency. Neither of these properties is characteristic of ordinary light sources, such as the sun or an incandescent light bulb. Fortunately, light waves obey the principle of superposition, so we can describe more complex types of radiation by adding together the fields due to the component waves.

2.2.2 POLARIZATION PROPERTIES OF LIGHT

We can use the principle of superposition to describe various polarization states of light. For example, suppose that we add a second electric field, of the same frequency but in a perpendicular direction, to the one described by Equation 2.31:

$$E_x = E_x^0 \cos\left[\omega\left(t - \frac{y}{c}\right) + \theta\right] \tag{2.35}$$

If the amplitudes of the two added waves are the same, $E_x^0 = E_z^0$, and if the phase shift θ is 90°, the net electric field vector will spiral about the y-axis as the wave propagates, in what is known as circular polarization. When viewed along the y-axis, looking toward the source, the electric field vector would appear to rotate in a clockwise direction, and the light is called right circularly polarized. Alternatively, a counterclockwise rotation of the field vector is called left circularly polarized. If the amplitudes of the two added waves are not equal, the resulting radiation is elliptically polarized. On the other hand, if there is no phase shift in the two orthogonal fields, the resulting radiation is linearly polarized in a direction that depends on the relative amplitudes of the two added fields. An electric field vector polarized in any direction can be resolved into two orthogonal linearly polarized fields. And a linearly polarized beam can be resolved into two counter-rotating circularly polarized beams. These circularly polarized components correspond to the angular momentum states of light. Quantum mechanically, photons are described as having angular momentum of $\pm\hbar$, meaning the spin quantum number of a photon is one.* In our wave picture, these two states correspond to angular momentum vectors aligned parallel or antiparallel to the propagation vector. This property of light is of interest in the consideration of selection rules for absorption and emission of photons, where conservation of angular momentum imposes restrictions on spectroscopic transitions. A one-photon transition cannot change the angular momentum state of an atom or molecule by more than one unit of $\hbar$. We will see in future chapters that this restriction holds for the most common type of spectroscopic transition, that which is caused by the interaction of the electric dipole with the electric field.

The polarization properties of light are of great interest in a number of spectroscopic techniques, and it would be interesting to discuss here how one could start with a source of natural light and produce polarized light. Although this subject requires some knowledge of how light travels through matter and in this chapter we consider only light traveling in a vacuum, it is worth mentioning briefly how polarized light is obtained. Materials that are dichroic or birefringent are capable of polarizing light. Dichroic materials, as the name suggests, absorb one component of polarization more strongly than the other. The common Polaroid sheets used in the laboratory and in sunglasses contain aligned polyvinyl alcohol molecules onto which iodine molecules, I_2, have been adsorbed. The component of the electric field that is parallel to the alignment axis is strongly absorbed by the I_2 chromophores, but the perpendicular component is transmitted. The effect is naturally wavelength dependent because it depends on the absorption properties of the material. Dichroic crystals such as tourmaline transmit different colors depending on the direction of viewing. Birefringent materials, such as crystalline calcite ($CaCO_3$) and quartz, have two different indices of refraction for light polarized parallel or perpendicular to the optic axis, and are thus capable of splitting an unpolarized beam into two perpendicularly polarized rays. If light impinges on the face of a crystal which is parallel to the optic axis and at normal incidence, the resulting ray is not split but the parallel and perpendicular polarized components travel with different speeds. If the thickness of the crystal and the difference in refractive index are such that the phase shift is 90°, then the crystal acts as a "quarter-wave plate." When linearly polarized light having an electric field vector at 45° to the optic axis impinges on a quarter-wave plate, the emerging light is circularly polarized, and vice versa. For more information on these optical effects and some simple experiments that demonstrate them, the reader is encouraged to consult [2] for a very readable account.

2.2.3 ELECTRIC DIPOLE RADIATION

Unlike material waves such as sound, light needs no medium to support it, not even the "luminiferous ether" sought by early physicists to explain the propagation of light in a vacuum. The laws of physics dictate that accelerating charges broadcast electromagnetic radiation. And our studies of spectroscopy teach us that light interacts with matter through setting charges in motion. The charges responsible for radiation must come from somewhere, so we are faced with a chicken-and-egg sort of question inherent in the symmetrical

---

* It may be surprising that the spin quantum number $s = 1$ leads to two ($m_s = 1, -1$) rather than three ($m_s = 1, 0, -1$) orientations of the angular momentum. The missing component would correspond to angular momentum perpendicular to the propagation direction, which is forbidden.

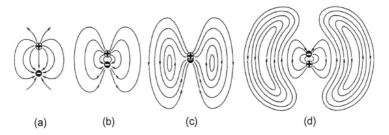

(a) (b) (c) (d)

Figure 2.3 Electric dipole radiation, showing the field lines during successive stages of oscillation, and the formation of electromagnetic waves propagating away from the dipole. (Adapted from E. Hecht, and A. Zajac, *Optics*, 1974.)

properties of nature. Which comes first; the oscillating charges or the radiation associated with them? The electric dipole moment plays a hugely important role in optical spectroscopy, thus we get ahead of ourselves a bit to bring up a topic more carefully treated in the next chapter, the field due to an electric dipole. A dipole $\mu = qd$ is a pair of equal and opposite charges, $\pm q$, separated by a distance d. If such a dipole oscillates with angular frequency ω, it will emit radiation of the same frequency. The mathematics of this system are very messy, but the pictures in Figure 2.3 provide a visualization. According to the principles of electrostatics discussed in Appendix B, the electric field of a static dipole can be represented by lines of force such as those drawn in Figure 2.3a. The curved arrows there represent the direction of the force on a unit positive charge, and their density represents the magnitude of the electric field (the force per unit charge). As you will show in Chapter 3, the field due to a *static* dipole is zero when viewed along the direction of the dipole axis, as is apparent in Figure 2.3. Now consider that the charges that comprise the initially static dipole of Figure 2.3a begin to approach one another, resulting in the changes in the field loops shown in Figure 2.3b, and when the charges are superimposed as in Figure 2.3c, the field lines close on themselves. Upon reversal of the dipole direction, Figure 2.3d, two blobs of field lines are pinched off and a new set of field lines forms, terminating once again on the charges but directed in the opposite sense from the original lines. In the "near-field" region, there is no particular wavelength and no radiation propagates. However, at distances large compared to the size of the dipole, simple expressions for the electric and magnetic fields are obtained. If the dipole oscillates harmonically, $qd = qd_0\cos\omega t$, then the electric field in the radiation zone, where $r \gg d$, is

$$E(r,t) = \frac{qd_0 k^2 \sin\theta}{4\pi\varepsilon_0 r}\cos(kr - \omega t) \tag{2.36}$$

The magnitude of the magnetic field is $B = E/c$. The angle θ is the angle between the vector $\vec{r}$, which points from the center of the dipole to the observation point, and the dipole direction, and qd_0 is the amplitude of the dipole moment. The direction of B is that of $\vec{r} \times \vec{d}$, and it is everywhere perpendicular to E. Notice that the electric field falls off as $1/r$, as it does for a spherical wave. This is a consequence of conservation of energy, since the total energy traversing the surface of a sphere of radius r must remain constant. Solving for the intensity using Equation 2.28, and substituting $k = \omega/c$, we get

$$I = \frac{(qd_0)^2 \omega^4 \sin^2\theta}{32\pi^2 c^3 \varepsilon_0 r^2} \tag{2.37}$$

This is an important result, which we will refer to later in this book. Notice that the intensity is proportional to the fourth power of the frequency (a very strong dependence) and the square of the dipole moment. It is noteworthy that no radiation is emitted in a direction parallel to the dipole. This consideration will be important in later chapters when we consider the polarization of scattered light.

2.2.4 Gaussian beams

The concept of collimated, plane polarized light is an idealization that simplifies the mathematics of electromagnetic radiation and allows for easily drawn pictures such as Figure 2.2. However, the prevalence of laser sources and focusing optics in spectroscopy experiments makes it important to consider Gaussian beams. A perfectly collimated, monochromatic beam would consist of rays that all have the same propagation vector $\vec{k}$, but our Gaussian beam, though generally propagating along the z-direction, will have a spread in the direction of the k-vectors as they emerge from the focus. The Gaussian nature of such a beam indicates that the electric field amplitude decays as e^{-r^2/w^2} in the direction transverse to the propagation direction, where $r^2 = x^2 + y^2$ and w is the distance at which the electric field amplitude has decayed to $1/e$ of its on-axis value. The electric field is written in the form

$$\vec{E}(r,z,t) = \vec{A}(r)e^{-i(kz-\omega t)} + \vec{A}^*(r)e^{i(kz-\omega t)} = \vec{A}(r)e^{-i(kz-\omega t)} + c.c. \tag{2.38}$$

where $c.c.$ stands for complex conjugate. In addition to allowing the amplitude of the field to depend on r, the notation above is a little different from what we have seen so far. We are using the symbol A for electric field amplitude (it is not the vector potential here), and given that $\cos(kz - \omega t) = 1/2 \left(e^{i(kz-\omega t)} + e^{-i(kz-\omega t)}\right)$, we see that A is really one-half of the electric field amplitude. These concerns are not important to the physics of the problem; they just keep our notation in sync with standard treatments in books such as [4]. In the approximation that the wave propagates generally along the z-direction (paraxial ray) and that the amplitude of the electric field varies on a length scale that is large compared to the wavelength (the slowly-varying envelope approximation), the wave equation for a Gaussian beam [5] is found to be

$$A(r,z) = A_0 \frac{w_0}{w(z)} \exp\left(\frac{-r^2}{w^2(z)}\right) \exp\left(\frac{-ikr^2}{2R(z)}\right) \exp\left(i\varphi(z)\right) \tag{2.39}$$

In the above, w_0 designates the beam waist at the focus ($z = 0$), $R(z)$ is the radius of the spherical wave fronts, $w(z)$ is the $1/e$ extent of the beam at z, and $\varphi(z)$ is the phase of the wave front. The z-dependent quantities are function of the Rayleigh range z_R, as follows:

$$R(z) = z\left[1 + \left(\frac{z_R}{z}\right)^2\right] \tag{2.40}$$

$$w(z) = w_0 \sqrt{1 + \left(\frac{z}{z_R}\right)^2} \tag{2.41}$$

$$\varphi(z) = \arctan\left(\frac{z}{z_R}\right) \tag{2.42}$$

$$z_R = \frac{\pi n w_0^2}{\lambda} \tag{2.43}$$

The Rayleigh range represents the distance at which the size of the beam has increased from $2w_0$, the value at the focus, to $2(2)^{1/2}w_0$. n in Equation 2.43 is the refractive index. Figure 2.4 illustrates a Gaussian beam, with the bold lines defining the beam radius and dashed lines representing wave fronts (contours of constant phase) at selected values along the propagation direction.

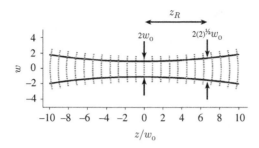

Figure 2.4 A Gaussian beam. The bold lines span $2w(z)$ and the dashed lines are wave fronts.

2.3 PROPAGATION OF LIGHT IN MATTER

2.3.1 REFRACTION AND REFLECTION

Let us briefly describe common properties of light traversing a dielectric (nonconducting) medium within the context of the electromagnetic theory. In the absence of net charge or current density, Maxwell's equations can be modified simply by replacing the electric permittivity ε_0 and magnetic permeability μ_0 of the vacuum by their values in the material, ε and μ. These determine the speed of light in matter, v as[*]

$$v = \frac{1}{\sqrt{\varepsilon\mu}} \tag{2.44}$$

This change in the speed of light from its value of c in the vacuum is associated with the phenomenon of refraction. The refractive index n is the ratio c/v, so it is given by

$$n = \sqrt{\frac{\varepsilon\mu}{\varepsilon_0\mu_0}} \tag{2.45}$$

For most materials, which are not magnetic, the magnetic permeability is little different from the value in free space, so the refractive index can be expressed as $n = \sqrt{\varepsilon/\varepsilon_0} = \sqrt{\varepsilon_r}$, where the ratio of the permittivity in the material to that of free space is defined as the relative permittivity, ε_r. The relative permittivity determines the response of a dielectric material to an applied electric field, and we will consider it further in Chapter 3. For now, we concentrate on the refractive index and note that it is a frequency-dependent quantity; that is, it undergoes dispersion. Furthermore, the frequency-dependent $n(\nu)$ is a complex quantity, the real part of which is responsible for the physical phenomenon of refraction and the imaginary part for absorption and emission. The dispersion of the real part is greatest in the vicinity of an absorption band, as illustrated in Figure 2.5. The connection between refraction and absorption will be examined in Chapter 3.

The refraction of light when it encounters a boundary between materials of two different refractive indices can be viewed in terms of the effect of the medium on the phase of the secondary waves in the medium. The incident (or primary) wave drives the electrons in matter with a frequency given by that of the incident light, which in turn causes them to reradiate light with the same frequency but with a different phase. The phase difference depends on whether the frequency is below or above the resonance frequency (absorption peak) of the material. Below resonance, the phase of the secondary radiation lags behind that of the primary radiation, and refractive index increases with frequency. This is called normal dispersion. The familiar effect of a glass prism on white light corresponds to this situation, where blue light is refracted more than red, since the frequency of visible light is below the absorption band of glass in the ultraviolet region. The effect of a

---

[*] Be careful not to confuse the similar-looking symbols used for frequency ν (a Greek letter nu) and speed of light in a medium v, a script v.

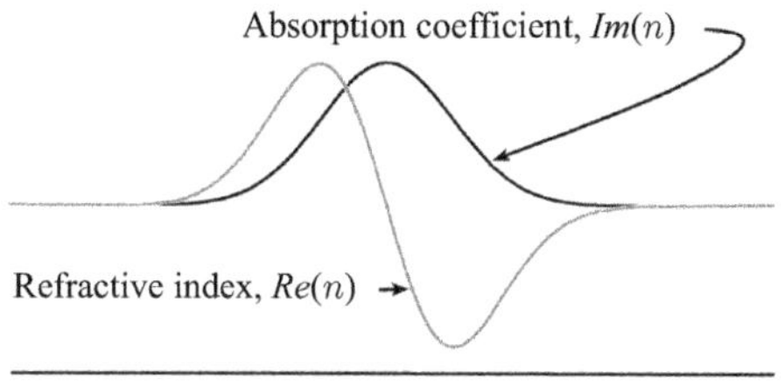

Absorption coefficient, $Im(n)$

Refractive index, $Re(n)$

Frequency, $v \longrightarrow$

Figure 2.5 Real and imaginary parts of the refractive index.

phase lag is to cause the velocity of light, the phase velocity, to be less than the value in free space. Above resonance, the phase of the secondary wave is ahead of the primary wave, refractive index decreases as frequency increases, and the phenomenon is called anomalous dispersion, for historical reasons. In either case, the frequency of light remains unchanged, it is the speed and the wavelength that are altered by the medium, $v = c/n$ and $\lambda = \lambda_0/n$.

We postpone until Chapter 3 a discussion of the microscopic basis for the refractive index, which depends on the response of the constituent charges within matter to the driving field of the incident light. The macroscopic manifestation of refraction is straightforward: the bending of a ray of light when it strikes the boundary between materials, as shown in Figure 2.6. The meaning of a ray of light, while intuitively familiar to the artist or poet, can be made more precise within the wave picture of radiation. A light ray simply points in the direction of energy propagation, so it coincides with $\vec{k}$. Figure 2.6 defines the angles of incidence θ_i and reflectance θ_r as the angles that the propagation vectors of the incident and reflected k-vectors make with respect to the surface normal, and similarly θ_t is the angle of the transmitted (refracted) ray. The idealized linear rays drawn in pictures like this are perfectly collimated, an ideal that can be approached using lenses, and thus the wave fronts are plane surfaces perpendicular to the propagation direction (like the wave fronts far from the focus of a Gaussian beam or a point source of emission.) The phase lag of the light striking the more dense medium is associated with bending of the ray. Boundary conditions on the electric field at the interface lead to specular reflection, $\theta_i = \theta_r$. The angles that the incident and transmitted rays make with respect to the normal to the surface are given by Snell's law, $n_i \sin\theta_i = n_t \sin\theta_t$. When light is incident from a more dense phase, such that $n_i > n_t$, there exists a critical angle θ_c for which θ_t is 90°, when the angle of incidence satisfies $\sin\theta_c = n_t/n_i$. For the air-glass interface, where $n_i/n_t \approx 1.5$, this angle is about 42°.

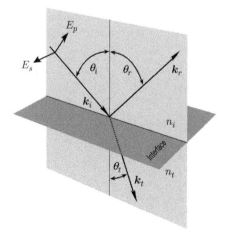

Figure 2.6 Refraction and reflection of light at a boundary.

Above the critical angle, all the incident light undergoes total internal reflection; i.e., there is no light transmitted through the less dense medium. Total internal reflection is a useful technique for the spectroscopy of interfaces, and one can see that incident angles near the critical angle result in probing the interface rather than the bulk.

Any light not transmitted by the material must be reflected, scattered, or absorbed. In transparent materials, we can account for all the light in terms of reflection and transmission. The relationships between the amplitudes of reflected and transmitted light are known as the Fresnel equations. The relative intensity of reflected light compared to incident light is a function of the angle of incidence, the refractive indices, and the polarization of the incident light with respect to the plane of incidence, defined by the incident and reflected rays. We define light as s-polarized if the electric field vector is perpendicular to the plane of incidence, and p-polarized if it is parallel to the plane. (The terms derive from German words for parallel and perpendicular, so it may be helpful to remember that the electric field vector for s-polarized light "sticks" out of the plane.) The ratio of reflected (ref) to incident (inc) light intensity for the two polarizations, in the case of dielectric media,

$$R_s = \frac{I_s, ref}{I_s, inc} = \left[\frac{n_i \cos\theta_i - n_t \cos\theta_t}{n_i \cos\theta_i + n_t \cos\theta_t} \right]^2 \tag{2.46}$$

$$R_p = \frac{I_p, ref}{I_p, inc} = \left[\frac{n_i \cos\theta_t - n_t \cos\theta_i}{n_i \cos\theta_t + n_t \cos\theta_i} \right]^2 \tag{2.47}$$

See [2] for a derivation of the above expressions. For the transparent medium considered here, the incident light intensity must match the sum of the transmitted and reflected intensities. We define the ratio of transmitted to incident light intensity as

$$T_s = \frac{I_s, trans}{I_s, inc}$$

$$T_p = \frac{I_p, trans}{I_p, inc} \tag{2.48}$$

Thus we have $R_s + T_s = R_p + T_p = 1$. Figure 2.7 shows the dependence of R_s and R_p on the angle of incidence for the case $n_t/n_i = 1.5$, i.e., for light incident from air striking glass. There are several aspects of this figure with spectroscopic consequences. One sees that for small angle of incidence $R_s \approx R_p$ (s and p polarizations are indistinguishable for normal incidence, $\theta_i = 0$), but at that larger angles there is a preference for reflecting s-polarized over p-polarized light. Thus in nature and in the spectroscopy lab, reflection can be a source of polarization even when the incident light is not polarized. Another practical consequence of these expressions is the rule of thumb that for light striking glass at normal incidence, about 4% of the intensity is reflected. At Brewster's angle, θ_p, p-polarized light is completely transmitted. Brewster's angle satisfies the equation:

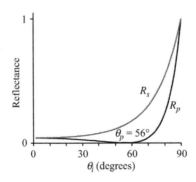

Figure 2.7 Reflection of s- and p-polarized light at a boundary, with $n_t/n_i = 1.5$.

$$\tan\theta_p = \frac{n_t}{n_i} \tag{2.49}$$

This equation is satisfied when the angle of incidence is $\theta_p = 90° - \theta_t$. As will be seen in Chapter 7, optical elements oriented at this special angle with respect to a laser beam are used to produce coherent polarized radiation.

The value of the refractive index in a medium has consequences for the intensity of light. We have seen that in a vacuum, the intensity I is related to the amplitude of the electric field E_0 according to Equation 2.28, $I = c\varepsilon_0 E_0^2/2$. In a medium with refractive index, n, the speed of light is reduced to c/n. In addition, we have to replace ε_0 in Equation 2.27 by the permittivity of the medium, $\varepsilon = \varepsilon_0 n^2$. As a result, the intensity of light in a medium is altered to

$$I = \frac{1}{2}cn\varepsilon_0 E_0^2 \tag{2.50}$$

2.3.2 ABSORPTION AND EMISSION OF LIGHT

As previously mentioned, dispersion in the refractive index is always associated with an absorption band. The details of the absorption of light will be developed at length in this book. The effect reveals itself as a reduction in the intensity of light as it traverses the absorbing medium. This is called attenuation, and it is an exponential function of the path traveled through the absorbing medium:

$$I = I_0 e^{-\gamma x} \tag{2.51}$$

where γ is the absorption coefficient of the medium and x is the distance traversed by the light.

The absorption of light promotes an atom or molecule to an excited state, and there is a tendency to restore the populations to their equilibrium values. An excited-state molecule can return to a lower energy state by means of either a nonradiative or a radiative transition. In the former, the energy is ultimately dissipated into heat by being transferred to degrees of freedom of lower-energy quanta. In the latter, the energy is reemitted as light in the form of luminescence. (Alternatively, the absorbed photon energy could be consumed by a photochemical reaction, but this will not be covered in this book.) Emission of light is either stimulated or spontaneous, the latter being much more common, at least at visible wavelengths, and the former being characteristic of laser emission. Ordinary fluorescence (emission involving states of the same spin multiplicity) and phosphorescence (emission involving states of different spin multiplicity) are forms of spontaneous emission. While stimulated emission requires an incident photon, in addition to the one that may have created the initially excited state, spontaneous emission takes place in the absence of stimulating photons. This presents a problem for the classical picture of light, but it can be dealt with phenomenologically in a way proposed by Einstein and presented in Chapter 6.

The emphasis in Chapters 1 through 12 of this book is on linear spectroscopic techniques, which do not significantly perturb the Boltzmann populations of molecules. It may seem odd that light is ever absorbed or emitted at all if the Boltzmann populations remain unchanged. In fact, the intensities of absorption and emission are related to the *rate* at which upward and downward transitions take place, so it is possible to maintain a steady state of level populations and still observe absorption and emission of light. In nonlinear spectroscopy experiments, the equilibrium populations may be significantly altered by the radiation. For example, in photobleaching experiments the populations of ground and excited states are made equal, and net absorption and emission cease. In nonlinear spectroscopy experiments, it is often the case that the radiation–matter interaction alters the properties of the incident light, as in second harmonic generation, also called frequency doubling. These types of experiments will be considered in Chapters 13 and 14.

2.3.3 EFFECT OF AN ELECTROMAGNETIC FIELD ON CHARGED PARTICLES

The classical expression for the force experienced by a charged particle in an electromagnetic field is known as the Lorentz law:

$$\vec{F} = q\left(\vec{E} + \vec{v} \times \vec{B}\right) \tag{2.52}$$

where q is the charge and $\vec{v}$ its velocity. Classical mechanics provides us with prescriptions for expressing the momenta of particles in terms of the classical Hamiltonian or Lagrangian, both of which are functions of the kinetic energy T and potential energy V of the particle. In the absence of a field, the classical Hamiltonian for a single particle is given by

$$H = T + V = \frac{p^2}{2m} + V \tag{2.53}$$

where p is the momentum, m is the mass and the potential energy V is a function of the position. The Hamiltonian expressed above does not correspond to the Lorentz force. It is possible to show (see for example [6]) that the substitution $\vec{p} \rightarrow \vec{p} - e\vec{A}$ results in a Hamiltonian that is consistent with the force given by Equation 2.52. So, in the presence of an electromagnetic field, the Hamiltonian should be written as

$$H = \frac{1}{2m}(\vec{p} - e\vec{A})^2 + V \tag{2.54}$$

This expression is for a single electron, but it can be generalized by summing over a collection of charges. The classical Hamiltonian is converted to a quantum mechanical operator simply by replacing the momentum $\vec{p}$ by the operator $\hat{p} = -i\hbar\nabla$. The Hamiltonian operator is thus

$$\hat{H} = \frac{1}{2m}(-i\hbar\nabla - e\vec{A})^2 + V \tag{2.55}$$

Performing the square, this is equivalent to

$$\hat{H} = \frac{-\hbar^2}{2m}\nabla^2 + V + \frac{i\hbar e}{2m}(\nabla\cdot\vec{A} + \vec{A}\cdot\nabla) + \frac{e^2}{2m}\vec{A}\cdot\vec{A} \tag{2.56}$$

In the linear spectroscopy regime, we can consider the electromagnetic field to be weak compared to the internal fields due to the charges that comprise the molecule. This allows us to neglect the $\vec{A}\cdot\vec{A}$ term compared to the term in parentheses. The term in parentheses can be cleaned up by noting that $\nabla\cdot(\vec{A}\psi) + \vec{A}\cdot(\nabla\psi) = 2\vec{A}\cdot(\nabla\psi)$. Finally, we recast the ∇ operator in terms of the particle momentum and write the Hamiltonian as the sum of a field-free zero-order term $\hat{H}_0$ and a field-dependent perturbation operator $\hat{H}'$, $\hat{H} = \hat{H}_0 + \hat{H}'$, where

$$\hat{H}_0 = \frac{-\hbar^2}{2m}\nabla^2 + V \tag{2.57}$$

$$\hat{H}' = \frac{-e}{m}\vec{A}\cdot\vec{p} \tag{2.58}$$

These expressions form the basis for the time-dependent perturbation theory to be considered in Chapter 4. The point of Equation 2.58 is that the fundamental basis for spectroscopic transitions is the interaction of the momenta of charged particles with the vector potential of radiation. We shall see in Chapter 4 how, in the long-wavelength limit, this leads to a picture in which the most common spectroscopic transitions are the electric dipole allowed transitions.

2.4 QUANTUM MECHANICAL ASPECTS OF LIGHT

2.4.1 QUANTIZATION OF THE RADIATION FIELD

Although many ordinary techniques in spectroscopy can be understood without treating the light quantum mechanically, we will introduce this topic in order to be able to look at effects for which the classical treatment fails. In the quantum mechanical picture, the energy levels of photons are the same as those of a harmonic oscillator, and photons, like phonons, obey Bose–Einstein statistics. This means that the occupation number of a particular state can take on any integral value: i.e., photons, unlike electrons, do not mind sharing quantum states. The classical limit of the wave picture of light is a consequence of this, since the number of photons in a given state can be large. The superposition principle, which allows us to add waves of the same frequency, polarization, and propagation vector, also follows from the ability of photons to share quantum states.

In Equation 2.18, the vector potential $\vec{A}$ was expressed for a monochromatic, plane polarized wave. Here, we take the vector potential to be the superposition of a number of allowed cavity modes. These cavity modes are subject to boundary conditions that limit the allowed wavelengths, and thus the frequencies, the same way that the length of a violin string decides the pitch of the note. The allowed cavity modes will be indexed by k, the magnitude of the propagation vector, and they are distinguished by their frequencies $\omega_k = 2\pi\nu_k$ and polarizations $\hat{e}_k$. The superposition form of $\vec{A}$ is

$$\vec{A}(r,t) = \sqrt{\frac{1}{\varepsilon_0 V}} \sum_k q_k(t)\hat{u}_k(r) \tag{2.59}$$

The summation in Equation 2.59 also runs over two allowed polarization directions for each allowed propagation vector $\vec{k}$, but the polarization index has been suppressed for notational convenience. Each term in the sum, Equation 2.59, is the product of a time-dependent amplitude $q_k(t) = |q_k|\exp(i\omega_k t)$ and a position-dependent term $\hat{u}_k(r) = \hat{e}_k \exp(i\vec{k}\cdot\vec{r})$. The variable $|q_k|$ is related to the amplitude of the vector potential; the notation is chosen to anticipate the result given below. The factor in front of the summation sign is a convenient normalization term. A classical Hamiltonian $H = T + V$ is obtained by averaging the energy density (see Equations 2.25 and 2.26) over the volume of the cavity. Thus the radiation Hamiltonian is

$$H_{rad} = \frac{\varepsilon_0}{2} \int \left(|E|^2 + c^2|B|^2\right) d\vec{r} \tag{2.60}$$

By writing the electric and magnetic fields in Equation 2.60 in terms of the vector potential given by Equation 2.59, the Hamiltonian for the radiation field is found to be

$$H_{rad} = \frac{1}{2}\sum_k \left(\dot{q}_k^2 + \omega_k^2 q_k^2\right) = \frac{1}{2}\sum_k \left(p_k^2 + \omega_k^2 q_k^2\right) \tag{2.61}$$

Equation 2.61 is derived in one of the problems. (See [6] for more details.) Except for the absence of an explicit mass term, Equation 2.61 is identical to the Hamiltonian for a classical harmonic oscillator, and defines the conjugate "position" (q_k) and "momentum" ($p_k = \dot{q}_k = dq_k/dt$) operators by analogy to the harmonic oscillator problem. We need not worry about the physical meaning of the position or momentum, as it is the mathematical analogy that concerns us here. The form of Equation 2.61 prompts us to express the quantum mechanical Hamiltonian for the radiation field in terms of the raising (b_k^+) and lowering (b_k) operators (also called creation and annihilation operators) for mode k:

$$b_k^+ = \frac{1}{\sqrt{2\hbar\omega_k}}\left(\omega_k q_k - ip_k\right) \tag{2.62}$$

$$b_k = \frac{1}{\sqrt{2\hbar\omega_k}}\left(\omega_k q_k + ip_k\right) \tag{2.63}$$

Except for the absence of a mass term, these operators are the same as those discussed in Section 1.3.3 for the harmonic oscillator. The commutation relations for the raising and lowering operators follow from those for position and momentum, $\left[q_k, p_{k'}\right] = i\hbar\delta_{k,k'}$. They are

$$\left[b_k, b_{k'}^+\right] = \delta_{k,k'} \tag{2.64}$$

$$\left[b_k, b_{k'}\right] = \left[b_k^+, b_{k'}^+\right] = 0 \tag{2.65}$$

These commutator relationships can be used to derive the equations given next, which describe the action of b_k^+ and b_k on a particular eigenstate. Let us say that a state having n_k photons in mode k is designated by $|n_k\rangle$. The effect of applying a creation or annihilation operator is

$$b_k^+|n_k\rangle = \sqrt{n_k+1}\,|n_k+1\rangle \tag{2.66}$$

$$b_k|n_k\rangle = \sqrt{n_k}\,|n_k-1\rangle \tag{2.67}$$

The effect of the raising operator b_k^+ is to increase the number of photons by one, and that of the lowering operator b_k is to decrease the number of photons by one. This means that the raising operator is associated with emission, and the lowering operator with absorption of photons.

The full state of the system must specify the number of quanta in each of the allowed cavity modes. Formally, this is represented as a product wavefunction of the form $|\Psi\rangle = |n_{k_1}, n_{k_2}, \ldots\rangle$. The quantum mechanical Hamiltonian for the radiation field can be expressed in terms of the creation and annihilation operators:

$$\hat{H}_{rad} = \sum_k \hbar\omega_k\left(b_k^+ b_k + \frac{1}{2}\right) = \sum_k \hbar\omega_k\left(n_k + \frac{1}{2}\right) \tag{2.68}$$

where we have defined the occupation number operator $b_k^+ b_k = n_k$. The eigenvalues of this operator are the integers 0, 1, 2,..., ∞. The allowed energy levels of the field are quantized exactly like those of a harmonic oscillator. An important aspect of Equation 2.68 is the zero-point energy associated with the vacuum state, where all $n_k = 0$. The virtual photons associated with the zero-point state can be considered to be responsible for spontaneous emission, as will be discussed in Chapter 4.

The form of the energy-level expression for photons, along with the universal concept of energy conservation, leads naturally to the Bohr frequency condition $\Delta E = nh\nu$. This relation simply states that energy jumps ΔE due to emission or absorption of light by molecules must correspond to the energy gained or lost by the field; that is, the creation or annihilation of an integral number of photons of energy $h\nu$. In the linear regime, only one photon transitions are permitted, so that $\Delta E = h\nu$. Although the time–energy uncertainty principle to be derived in Chapter 4 will qualify the Bohr condition somewhat, it is of central importance to the understanding of spectroscopy.

2.4.2 BLACKBODY RADIATION AND THE PLANCK DISTRIBUTION LAW

The entire discussion of the previous section would be unnecessary if Planck's constant h were equal to zero. The fact that it is indeed quite small, making it possible to neglect it sometimes, may explain why it took so long for it to be noticed! The story of the derivation of Planck's constant is interesting because it explains how the theory of quantum mechanics emerged in order to resolve paradoxes that classical theories were powerless to explain. In addition, the blackbody radiation model teaches us something about matter in equilibrium with radiation, so it is relevant to many interesting spectroscopic situations.

There is of course no such thing as a perfect blackbody, but here is how we imagine constructing one in a gedanken or "thought" experiment. Take a material that absorbs all of the light that impinges upon it and fashion it into a hollow body with a small orifice. Shine light on the orifice and it will be absorbed by the internal walls of the cavity and reemitted. Alternatively, heat the cavity and observe the small portion of the energy emitted through the orifice. When equilibrium is achieved, the rate of energy absorbed must equal that of energy emitted. The distribution of radiant energy will depend on temperature, and we expect that the higher the temperature the brighter will be the glow from the cavity. What is the spectrum of light emitted by this body?

Common experience leads us to expect that the higher the temperature the higher will be the average frequency of emitted light. This is in fact realized experimentally, as is shown in Figure 2.8. The problem is that this is not at all what classical physics predicted. To see why not, let us look at the problem through the eyes of a nineteenth-century physicist.

The wave theory of light restricts the wavelengths of cavity modes within the blackbody. These are modes whose wavelengths correspond to standing waves having nodes at the walls of the cavity. The quantum theory has no problem with the idea of allowed modes, but the classical description goes on to apply the equipartition theorem to each cavity mode. Equipartition theorem assigns the energy $1/2\,k_BT$, where k_B is Boltzmann's constant and T the absolute temperature, to each degree of freedom of the system. Now, each mode of the blackbody resonator is associated with two degrees of freedom, one for the electric and one for the magnetic field. The number of modes (allowed wavelengths) in the interval $d\lambda$ increases rapidly as the frequency of the radiation increases. In the classical theory, each mode is associated with k_BT of energy, so the intensity or energy density of emitted light should increase rapidly with frequency. This effect was referred to as the ultraviolet catastrophe, and it is illustrated in Figure 2.8 along with experimentally observed distributions for various temperatures.

In order to calculate an energy spectrum of the form shown in Figure 2.8, Planck had to make a radical assumption: that the atoms comprising the blackbody could absorb and emit radiation in integral multiples of

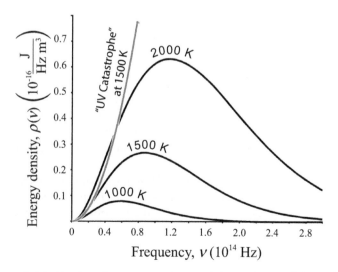

Figure 2.8 Spectrum of a blackbody at three temperatures and the prediction of classical theory at 1500 K.

the energy $h\nu$. Note that this idea predates the quantum mechanical idea of photons, and indeed it makes no mention of the idea of quantized radiation energy. Planck's picture introduced the idea of quantized energy levels of *matter*, and he himself was uncomfortable with it. Nevertheless, by adjusting the value of the constant h to get a calculated energy distribution consistent with experiment, Planck's constant emerged and there was no turning back.

The problem may be examined quantitatively. Since the emission from a macroscopic body does not depend on its shape, we can simplify things by imagining the blackbody to be a cubic box with sides of length L. From the boundary conditions, the three components of the allowed k-vectors are

$$k_x = \frac{n2\pi}{L} \quad k_y = \frac{m2\pi}{L} \quad k_z = \frac{l2\pi}{L} \tag{2.69}$$

where n, m, and l are integers. Equation 2.69 derives from the condition that, in each dimension, a half-integral number of wavelengths should fit into the length L. As $k = 2\pi/\lambda$, this would seem to imply that the allowed k-values are $k_x = \pi/L$ rather than $k_x = n2\pi/L$, and similarly for the y and z components. As explained in [7], the factor of two in Equation 2.69 reduces the mode density by a factor of one-half, since each permitted cavity mode is actually a standing wave, which is a superposition of right- and left-moving waves.

We are interested in an expression for the energy density $\rho = du/d\nu$, that is, the energy per unit volume per unit frequency interval. We can arrive at a mode density by taking the "volume" in k-space, $4/3\pi k^3$, and dividing by the "volume" per mode, $(2\pi/L)^3$, to get the number of modes N_k having propagation vectors less than or equal to k:

$$N_k = \frac{\frac{4}{3}\pi k^3}{(2\pi/L)^3} \times 2 = \frac{k^3 L^3}{3\pi^2} \tag{2.70}$$

The factor of two comes from the two polarization states. The number of allowed frequencies N_ν that correspond to N_k is obtained by making the substitution $k = 2\pi\nu/c$. This gives

$$N_\nu = \frac{8\pi\nu^3 L^3}{3c^3} \tag{2.71}$$

The mode density is defined as the number of modes per unit volume per unit frequency interval. Dividing Equation 2.71 by the volume $V = L^3$ and taking the derivative with respect to ν, we get

$$\frac{d(N_\nu/V)}{d\nu} = \frac{8\pi\nu^2}{c^3} \tag{2.72}$$

It is the energy density $\rho(\nu)$ that is depicted in Figure 2.8. The energy density is the mode density in Equation 2.72 times the average energy per oscillator. The classical approach was to multiply Equation 2.72 by $k_B T$ to get the curve representing the ultraviolet catastrophe in Figure 2.8. All the energy should have gone into the high-frequency modes, instead of falling off at high frequencies as observed. The essential difference between the classical prediction and Planck's approach is in the calculation of the average energy per oscillator.

Planck proposed that the atoms of the blackbody could absorb and emit energy only in integral multiples of $h\nu$. The energy levels available to the blackbody are thus $E_n = nh\nu$. The probability p_n of the nth energy level being occupied is given by Boltzmann's law: $p_n \propto e^{-nh\nu/k_B T}$. The normalized probability is

$$p_n = \frac{e^{-nh\nu/k_B T}}{\sum_n e^{-nh\nu/k_B T}} \tag{2.73}$$

The average energy of a blackbody oscillator is given by $\bar{E} = \sum_n E_n p_n$, thus *

$$\bar{E} = \frac{\displaystyle\sum_n nh\nu e^{-nh\nu/k_B T}}{\displaystyle\sum_n e^{-nh\nu/k_B T}} = \frac{h\nu}{e^{h\nu/k_B T} - 1} \tag{2.74}$$

Taking this average energy per mode and multiplying by the mode density given previously, we get Planck's distribution law for the blackbody energy density:

$$\rho(\nu) = \frac{8\pi h\nu^3}{c^3} \frac{1}{e^{h\nu/k_B T} - 1} \tag{2.75}$$

The exponential function of frequency in this expression pulls the energy density down to zero as the frequency goes to infinity, or as the temperature goes to zero. Physically, this results from the unavailability of higher-energy states when $h\nu \gg k_B T$. The equipartition theorem prediction is obtained from Planck's equation in the limit $h\nu \ll k_B T$, so it coincides with the experimental results at low frequencies.

2.4.3 THE PHOTOELECTRIC EFFECT AND THE DISCOVERY OF PHOTONS

The photoelectric effect was an important experiment that could not be explained using the wave picture of radiation, a situation that led Einstein to postulate the existence of photons. Einstein went one step farther than Planck and hypothesized that the energy of light was quantized. We review the historical development of the theory and experiment in order to point out some essential differences between the wave and particle pictures.

The experiment was to shine light on a photoemissive material, such as potassium, functioning as a cathode, and measure the resulting current. The magnitude of the current was observed to be proportional to the intensity of the light, which the wave picture relates to the square of the amplitude of the field. The maximum kinetic energy of ejected electrons, determined from the magnitude of the potential required to prevent the flow of photocurrent, was found to be proportional to the frequency of the incident light. This fact was at odds with the wave theory of light, since the energy of the electrons should have been related to the intensity of the light. In fact, brighter light increased the number of electrons, but not their energy. Also, light of frequency less than some threshold frequency, which depended on the metal, was incapable of ejecting any photoelectrons at all, no matter how bright the light. It should have been possible for electrons in the metal to eventually acquire enough energy to escape the surface, even if low intensity light of any frequency were used.

The picture proposed by Einstein in 1905 provided a way out of this predicament. He suggested that light energy was quantized in packets of energy, $E = h\nu$, called photons. One photon was capable of ejecting one electron provided the photon energy was at least as great as the binding energy of the metal. The ionization energy of surface metal atoms is called the work function w. Thus the threshold frequency ν_0 is related to the work function: $w = h\nu_0$. The energy of the photon in excess of the binding energy is converted into the kinetic energy of the ejected electron, thus the maximum kinetic energy is

$$T_{max} = h\nu - w \tag{2.76}$$

Equation 2.76 had the exact form of the experimental results for maximum kinetic energy as a function of frequency. It remained only to adjust the slope of the calculated line, T_{max} versus ν, to agree with experiment. The resulting value of h was found to agree with the constant obtained by Planck, and so the idea of quantization of energy of matter and light was firmly established.

---

* Equation 2.74 can be derived with the help of the following summation formulas, which hold for $x < 1$: $\sum_{n=0}^{\infty} x^n = 1/(1-x)$ and $\sum_{n=0}^{\infty} nx^n = x/(1-x)^2$.

2.5 SUMMARY

We have now examined two approaches for understanding the nature of light, classical and quantum mechanical. In the classical picture, electromagnetic radiation is considered to be a transverse wave in which the oscillating electric and magnetic fields and the propagation direction are mutually perpendicular. The intensity of the light is proportional to the square of the electric or magnetic field amplitude. The quantum mechanical treatment of electromagnetic radiation leads to a Hamiltonian that is strikingly similar to that for a harmonic oscillator. The quantum numbers associated with the radiation Hamiltonian are interpreted to be the numbers of photons in the various allowed cavity modes. A cavity mode is a particular wave vector $\vec{k}$, of magnitude $2\pi/\lambda$, that satisfies the boundary conditions, and for each wave vector there are two possible polarizations. The intensity of light is proportional to the number of photons. Throughout this book, we will find it necessary to refer to *both* of these descriptions of light. While it will often be desirable to use the classical (wave) picture, based on Maxwell's equations, we still consider spectroscopic transitions to result from absorption, emission and scattering of photons. In addition, there will be some experimental scenarios, such a spontaneous emission, that cannot be understood without a quantum mechanical treatment of light. We will make frequent reference to the intensity I and the energy density, $\rho(v) = du/dv$ in future chapters, as the rates of spectroscopic transitions are dependent on these quantities. Further discussion of many of the concepts in this chapter can be found in References [8–10].

PROBLEMS

1. The total intensity of sunlight, for all visible wavelengths striking the planet, is about 1000 W/m². In a wavelength interval of about 1 nm in the visible range, the intensity is on the order of 2 W/m². Using a wavelength of 500 nm, find the electric field amplitude and photon flux (number of photons per unit area per unit time) that corresponds to this intensity.

2. Convert the limits of the visible spectrum, 400–700 nm, to kJ/mol, eV, and cm^{-1}.

3. The momentum of a photon is given by the de Broglie relation: $p = h/\lambda = \hbar k$. Suppose that a flat surface 1.00 m² in area completely absorbs 1000 W of 400 nm light. Calculate the radiation pressure.

4. The blackbody energy density given in Equation 2.75 is the energy density per unit frequency interval. In other words, $du = \rho(v)dv$ is the energy per unit volume between v and $v + dv$. Convert the frequency-dependent $\rho(v)$ to a wavelength-dependent function $\rho(\lambda)$ such that $du = \rho(v)dv = \rho(\lambda)d\lambda$.

5. Given that the temperature of the sun is about 5700 K, use your result from Problem 4 to estimate the wavelength of the maximum in the solar emission spectrum.

6. Show that the classical Hamiltonian in Equation 2.54 is consistent with the Lorentz force $F = q(\vec{E} + \vec{v} \times \vec{B})$. To proceed, calculate the force in the x direction, $F_x = m\ddot{x}$, with the help of the following expressions from classical mechanics:

$$\frac{\partial H}{\partial x} = -\dot{p}_x \text{ and } \frac{\partial H}{\partial p_x} = \dot{x}$$

A dot over a quantity indicates a derivative with respect to time, and two dots represents the second derivative. The y and z components of the force can be obtained by cyclic permutation.

(Hint: $\dfrac{dA_x}{dt} = \dfrac{\partial A_x}{\partial t} + \dfrac{\partial A_x}{\partial x}\dot{x} + \dfrac{\partial A_y}{\partial y}\dot{y} + \dfrac{\partial A_z}{\partial z}\dot{z}$.)

7. Use the formula for the vector potential of the quantized radiation field

$$\vec{A}(r,t) = \sqrt{\frac{1}{\varepsilon_0 V}} \sum_k q_k(t)\hat{u}_k(r)$$

to derive the radiation Hamiltonian:

$$H = \frac{1}{2} \sum_k \left(\dot{q}_k^2 + \omega_k^2 q_k^2 \right)$$

See Section 2.4.1 for an outline of the necessary steps.

8. The work function of Cs is 2.14 eV. Find the maximum kinetic energy of electrons ejected by light of wavelength 400 nm.

9. An argon ion laser emits green light at 514.5 nm with a power of 1 W concentrated in a beam of cross section 0.01 cm². Calculate (a) the electric field amplitude and (b) the photon flux.

REFERENCES

1. W. Heitler, *Quantum Theory of Radiation,* 3rd ed. (Oxford University Press, London, 1954).
2. E. Hecht, and A. Zajac, *Optics* (Addison-Wesley, Reading, MA, 1974).
3. W. S. Struve, *Fundamentals of Molecular Spectroscopy* (Wiley, New York, 1989).
4. Boyd, R. W. *Nonlinear Optics,* 3rd ed. (Academic Press, Amsterdam, 2008).
5. Yariv, A. *Quantum Electronics,* 2nd ed. (John Wiley & Sons, New York, 1975).
6. G. C. Schatz, and M. A. Ratner, *Quantum Mechanics in Chemistry* (Prentice Hall, Englewood Cliffs, NJ, 1993).
7. R. P. Feynman, R. B. Leighton, and M. Sands, *The Feynman Lectures on Physics, Vol. III* (Addison-Wesley, Reading, MA, 1965).
8. J. D. Jackson, *Classical Electrodynamics* (Wiley, New York, 1962).
9. A. Yariv, *Quantum Optics* (Wiley, New York, 1967).
10. R. P. Feynman, R. B. Leighton, and M. Sands, *The Feynman Lecture in Physics,* Vol. II, Ch. 21 (Addison-Wesley, Reading MA, 1964).

Electric and magnetic properties
of molecules and bulk matter

3.1 INTRODUCTION

The charge distribution of an atom or molecule provides a handle by which electromagnetic radiation interacts with matter and causes spectroscopic transitions. As discussed in the previous chapter, light consists of oscillating electric and magnetic fields. These fields can exert torques on the electric and magnetic multipoles of molecules, causing energy to be absorbed by matter. Conversely, oscillations in the charge distribution of a molecule result in the emission of electromagnetic radiation. The reciprocity of these two physical phenomena, oscillating charge distributions and electromagnetic radiation, is an elegant example of symmetry in the laws of nature.

Most of the spectroscopic tools presented in this book fall within the realm of linear response. This enables us to treat the incident light as a first-order perturbation, and the reaction of the molecules to this perturbation depends on physical properties of the unperturbed system. This is the essence of the fluctuation-dissipation theorem, to which we will turn in Chapter 5. Of course, many spectroscopy experiments employ intense radiation fields that do more than just tickle the molecules, and the nonlinear response of the system then requires higher order perturbation corrections, as discussed in Chapters 13 and 14.

The overwhelming majority of strong spectroscopic transitions, whether absorption, emission, or scattering, result from the interaction of the electric dipole moment with the time-varying electric field of light. (In magnetic resonance experiments it is the magnetic dipole moment that carries the transition strength. We will not consider these at length in this book.) As will be seen later, it is actually the quantum mechanical transition moment, a matrix element of the electric dipole moment operator connecting two quantum states, that makes spectroscopic jumps possible. In order to discuss physical observables of interest to spectroscopists, such as polarization, permittivity, and refraction, we start with a discussion of the electromagnetic properties of individual molecules. The microscopic and macroscopic electromagnetic properties of matter are strongly linked to the theory and practice, respectively, of spectroscopy. Theoretical tools provided by quantum mechanics provide us with our first expressions for determining the spectroscopic behavior of individual molecules, while practical aspects of interpreting the spectra of matter in bulk depend on the macroscopic electromagnetic response. We consider how molecular properties decide the frequency-dependent dielectric and optical properties of collections of molecules.

The electromagnetic properties of molecules also determine their interactions with one another. When the distance between a pair of molecules is large compared to the size of the electron clouds, they can be considered to interact via classical electrostatic forces. The charge distribution of one molecule presents a field that interacts with the electric moments of another molecule. As a starting point, we consider static charge distributions and the fields that they produce. The concept of polarizability is introduced in order to draw a more realistic picture of molecules as fluctuating collections of charge. If the separation of the two molecules is larger than about 10 nm, the fluctuations in the charge distribution during the propagation time of the field result in interactions that are not instantaneous. These retardation effects are important in systems where the interactions are long ranged, such as colloidal dispersions, but are not be considered here.

In addition to understanding how an external field affects an individual molecule, we want to know how this field is screened by bulk matter. This will enable us to consider how the intensity and frequency of a spectroscopic transition varies with the environment of the active molecule. This is a long-standing problem often referred to as the *local field effect*. We will attempt to approach this nontrivial problem with due regard for its complexity and implications. This chapter concludes with a brief discussion of magnetic properties and electronic spin and orbital angular momentum.

3.2 ELECTRIC PROPERTIES OF MOLECULES

There are two ways to visualize an atom or molecule as a collection of charges. In the classical picture, the nuclei and surrounding electrons are discrete charges with definite positions in space at any given time. In the quantum view, this picture is blurred, and we can only know probability distributions for the positions of particles. In calculations of electric properties, summations involving discrete charges can be replaced by integrals over charge distributions that depend on the appropriate squared wavefunctions. This leads to a more correct approach to familiar quantities such as the dipole moment. Viewed through the eyes of a freshman chemistry student, the dipole moment of hydrochloric acid (HCl) results from a partial positive charge on the hydrogen atom and a partial negative charge of the same magnitude on the chlorine. But quantum mechanics teaches us that there is no unique place where either partial charge resides, the electrons being delocalized according to the electronic wavefunction. The dipole of the HCl molecule merely behaves as if there were two discrete charges separated by a definite distance. (Going even further to be quantum mechanically correct, the dipole moment of a gas-phase HCl molecule vanishes when averaged over its rotational motion! However, we can imagine aligning the dipole in an external field.) The discrete charges of the classical picture are actually the "centers of gravity" of the positive and negative charge distributions. It turns out that the dipole moment is a particular average over the molecular charge distribution, which results when a collection of charges is decomposed into a hierarchy of electric moments (monopole, dipole, quadrupole, etc.). We will show how this decomposition scheme results in a simple picture for the interaction of a molecule with the electric field, the gradient of this field, and higher-order terms.

3.2.1 REVIEW OF ELECTROSTATICS

In order to review basic electrostatic principles, we begin with the idea of discrete charge distributions. Consider the pair of charges q_a and q_b arranged in the coordinate system shown in Figure 3.1. The force on charge b, due to charge a, is given by Coulomb's law:

$$\vec{F}_b = \frac{1}{4\pi\varepsilon_0} \frac{q_a q_b}{r_{ab}^2} \hat{r}_{ab} \tag{3.1}$$

Note that $\hat{r}_{ab}$ is a unit vector pointing in the direction of $\vec{r}_{ab} = \vec{r}_b - \vec{r}_a$. The vector notation ensures that the force will be repulsive for like charges and attractive for unlike charges. Equation 3.1 is written using MKS units, which is apparent from the term $4\pi\varepsilon_0$, where $\varepsilon_0 = 8.854 \times 10^{-12}$ C^2 N^{-1} m^{-2} is the permittivity of free space. The unit of charge in this expression is the Coulomb, and the force F is in Newtons. In the cgs–esu system of units, the charge q would be in electrostatic units (esu or statcoulombs), and the permittivity of free space would be equal to $1/4\pi$, so the factor $4\pi\varepsilon_0$ would not appear. (See Appendix B for a discussion of units in formulas such as Equation 3.1.)

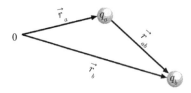

Figure 3.1 A pair of point charges.

The Coulombic force is pairwise additive, obeying a superposition principle:

$$\vec{F}_b = \frac{1}{4\pi\varepsilon_0} \sum_{i\neq b} \frac{q_i q_b}{r_{ib}^2} \hat{r}_{ib}$$

(3.2)

This would present a great computational advantage in the consideration of intermolecular forces, were it possible to treat molecules as static collections of charge. But the charge distribution of a molecule is polarizable: it deforms in the presence of external fields, including those resulting from other molecules. The dispersion forces that result from the interactions of induced moments are not pairwise additive. We will consider polarizability in Section 3.2.5.

The concepts of force and field are important to the discussion of molecular interactions. Fundamentally, the electric field $\vec{E}$ is the force per unit charge. It is more exact to say that it is the force exerted on a test charge in the limit that the test charge is infinitesimally small. The mathematical expression of this is given in Equation 3.3.

$$\vec{E} = \lim_{\delta q \to 0} \frac{\delta\vec{F}}{\delta q}$$

(3.3)

Note that the direction of the field is that of the force on a positive charge: the field lines point in the direction that a positive charge wants to go. Combining Equations 3.2 and 3.3, the electric field experienced by a test charge at point r_a, due to a collection of other charges, is given by

$$\vec{E}(r_a) = \frac{1}{4\pi\varepsilon_0} \sum_{i\neq a} q_i \frac{\hat{r}_{ia}}{r_{ia}^2}$$

(3.4)

The electric field $\vec{E}$ is the negative gradient of the scalar potential φ, introduced in Chapter 2:

$$\vec{E} = -\nabla\varphi$$

(3.5)

The potential at a distance r from charge q is thus

$$\varphi(r) = \frac{q}{4\pi\varepsilon_0 r}$$

(3.6)

While the electric field is the force per unit charge, the potential represents the work per unit charge, such that the work done on a charge to move it between two points depends on the difference in potential: $W_{1\to 2} = q(\varphi_2 - \varphi_1)$. In MKS units, the potential is expressed in volts, where 1 V is equivalent to 1 J/C.

Let us use the idea of a simple dipole, composed of two point charges, to illustrate these concepts. For two equal and opposite charges q separated by a distance d, the dipole moment is $\vec{\mu} = q\vec{d}$, where the dipole direction points from the negative to the positive charge.[*] It can be shown (see Problem 1 and Appendix B) that at a distance r which is large compared to dipole length d, the potential due to the dipole is given by

$$\varphi(r) = \frac{\vec{\mu} \cdot \vec{r}}{4\pi\varepsilon_0 r^3}$$

(3.7)

The dependence on the dot product of $\vec{\mu}$ and $\vec{r}$ means that the potential in a direction normal to the dipole is zero, as it must be because the contributions of the positive and negative charges would cancel. It is

---

[*] Although many chemistry texts use the opposite convention, the one used here is consistent with the expression given later for the energy of a dipole in an electric field.

important to note that the potential falls off as the square of the distance r, which is a more rapid decrease than the potential due to a point charge.

Taking the gradient of the potential in Equation 3.7, the electric field due to a dipole is found as follows:

$$\vec{E} = \frac{1}{4\pi\varepsilon_0}\left[\frac{3(\vec{\mu}\cdot\vec{r})\vec{r}}{r^5} - \frac{\vec{\mu}}{r^3}\right] \tag{3.8}$$

The energy of a dipole moment in a field depends on its orientation, through the expression $W = -\vec{\mu}\cdot\vec{E}$. You can use Equation 3.8 to find the interaction energy* of two dipoles:

$$W = \frac{1}{4\pi\varepsilon_0}\left[\frac{\vec{\mu}_1\cdot\vec{\mu}_2}{r^3} - \frac{3(\vec{\mu}_1\cdot\vec{r})(\vec{\mu}_2\cdot\vec{r})}{r^5}\right] \tag{3.9}$$

Try using this equation to determine which configuration of a pair of neighboring dipoles is more stable, $\rightarrow\rightarrow$ or $\uparrow\downarrow$. The answer may surprise you!

In the next sections we will consider higher order multipole moments and their interaction with an electric field.

3.2.2 ELECTRIC MOMENTS

It would be a difficult problem to calculate the electrostatic potential φ due to a molecule if one had to know all the details of its charge distribution. Fortunately, there is a way to circumvent this problem. Consider for the moment that there is a collection of discrete charges q_i located at distances r_i from an arbitrary origin. The potential at point $\vec{R}$ due to this collection is

$$\varphi(\vec{R}) = \frac{1}{4\pi\varepsilon_0}\sum_i\frac{q_i}{|\vec{R}-\vec{r}_i|} \tag{3.10}$$

Equation 3.10 is exact, but if we are willing to limit consideration to distances that are large compared to the extent of the charge distribution, that is, $R \gg r_i$, then it is acceptable to approximate Equation 3.10 by expanding the inverse of the distance $|\vec{R}-\vec{r}_i|$ in a Taylor series:

$$\frac{1}{|\vec{R}-\vec{r}_i|} = \frac{1}{R} - \vec{r}_i\cdot\nabla\left(\frac{1}{R}\right) + \frac{1}{2}\vec{r}_i\vec{r}_i : \nabla\nabla\left(\frac{1}{R}\right) + \cdots \tag{3.11}$$

Then the potential φ is given by

$$4\pi\varepsilon_0\varphi(\vec{R}) = \frac{q}{R} - \vec{\mu}\cdot\nabla\left(\frac{1}{R}\right) + \frac{1}{2}\Theta : \nabla\nabla\left(\frac{1}{R}\right) + \cdots \tag{3.12}$$

(The tensor contractions $A\cdot B$ and $A:B$ are discussed in Appendix A, and the gradient operations involving $1/R$ are considered in Appendix B.) The above equation introduces a hierarchy of multipole moments, the charge (or monopole) q, the dipole moment $\vec{\mu}$, and the quadrupole moment Θ. Not shown are the higher-order terms containing the octupole, hexadecapole, etc., moments. The first three multipole moments are enough to consider for now. They are defined as follows:

$$q = \sum_i q_i = \int \rho(\vec{r})d\vec{r} \tag{3.13}$$

---

* To avoid confusing energy and electric field, the symbol W is used for energy in this chapter.

$$\vec{\mu} = \sum_i q_i \vec{r}_i = \int \vec{r} \rho(\vec{r}) d\vec{r} \tag{3.14}$$

$$\Theta = \sum_i q_i \vec{r}_i \vec{r}_i = \int \vec{r}\vec{r} \rho(\vec{r}) d\vec{r} \tag{3.15}$$

Note that the first of these is a scalar, the dipole moment is a vector (or a first-rank tensor), and the quadrupole moment is a matrix (or a second-rank tensor). Equations 3.13 through 3.15 show how these quantities may be expressed either in terms of sums over discrete charges or integrals over continuous charge density, where $\rho(\vec{r})$ is the charge per unit volume. The MKS units for dipole moment are coulomb-meters, C m, but the Debye unit of the cgs system, equal to 10^{-18} esu cm, is commonly used. The conversion from Debye to C m is 1 D = 3.336 × 10^{-30} C m. It is helpful to remember that an electron and a positron separated by a distance of 1 Å lead to a dipole moment of 4.8 D. Equation 3.12 points out that increasingly higher order moments lead to potentials of increasingly shorter range.

As a vector, the dipole moment requires three components to define it in a reference frame fixed in the laboratory: either three Cartesian components or the magnitude and two angles in a spherical polar coordinate system. In a molecule-fixed coordinate system, the symmetry of a molecule dictates the form of $\vec{\mu}$. For example, a molecule belonging to one of the C_{nv} point groups has a dipole moment coincident with the n-fold rotation axis, and a spherical molecule, such as one belonging to the T_d or O_h point group, has no dipole moment at all.

Just as a dipole is the union of two monopoles, a quadrupole is formed by joining two dipoles. Figure 3.2 shows two ways to do this, along with pictorial representations of some of the other moments. Note that the octupole moment is pictured as the union of two quadrupole moments, and so on for higher order moments.

As a second rank Cartesian tensor, Θ could have nine components: Θ_{xx}, Θ_{xy}, etc. However, the tensor is symmetric: $\Theta_{xy} = \Theta_{yx}$. And as shown in Appendix B, if an arbitrary term is added to each diagonal element of Θ, there is no change in the field due to the quadrupole. We can thus choose that the quadrupole tensor have zero trace; that is, the sum of the diagonal elements vanishes. There are thus only five components needed to specify the quadrupole moment. With the convention that Θ be traceless, we have

$$\Theta = \frac{1}{2} \sum_i q_i (3\vec{r}_i \vec{r}_i - r_i^2 \mathbf{I}) \tag{3.16}$$

where $\mathbf{I}$ is the unit tensor. Hereafter, the definition given here will be used when referring to the quadrupole moment. It is always possible to find a molecule-fixed coordinate system, that is, one that rotates and translates with the molecule, known as the principal axes, which diagonalizes Θ. But since the tensor is chosen to be traceless, only two diagonal elements need to be specified. Since three angles are needed to orient the principal axes of the molecule in the laboratory, there are still five components needed to specify the quadrupole moment in a laboratory-based coordinate system.

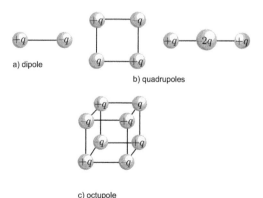

a) dipole

b) quadrupoles

c) octupole

Figure 3.2 Electric moments: (a) dipole, (b) quadrupole, and (c) octupole.

When the dipole moment vanishes, the existence of a quadrupole moment is especially important. Consider the intermolecular interactions in liquid benzene or supercritical CO_2, in which the considerable quadrupole moment operates to favor T-shaped relative orientations of neighboring pairs of molecules.

Problem 2 of this chapter explores the possibility that the dipole and quadrupole moments depend on the origin of the coordinate system. It can be shown that the dipole moment of a molecule lacking a net charge is independent of the choice of origin. Similarly, if the dipole moment also vanishes, then the quadrupole moment does not depend on the origin.

Another way to express a general multipole moment is as a spherical, rather than Cartesian, tensor. In this approach, the generalized multipole moment Q_{lm} is expressed as

$$Q_{lm} = \sum_i q_i r_i^l Y_{lm}(\theta_i, \varphi_i) = \int r^l \rho(\vec{r}) Y_{lm}(\theta, \varphi) d\vec{r} \tag{3.17}$$

Y_{lm} is a spherical harmonic, and Q_{lm} is the mth component of the moment of order l. Recall that for every value of l there are $2l + 1$ values of m, ranging from $-l$ to l. The lowest-order moment, $l = m = 0$, is the monopole moment or total charge as defined previously. The dipole moment is represented by $l = 1$, for which there are three components ($m = -1, 0, 1$). The $l = 2$ moment is the quadrupole, for which there are five components, and so on. Equations 3.18 and 3.19 give some examples of how spherical and Cartesian tensors may be interconverted.

$$l = 0, \ Q_{00} = \frac{1}{\sqrt{4\pi}} q \tag{3.18}$$

$$l = 1, \ Q_{1,\pm 1} = -\sqrt{\frac{3}{8\pi}}(\mu_x \mp i\mu_y), \ Q_{10} = \sqrt{\frac{3}{4\pi}}\mu_z \tag{3.19}$$

Using spherical tensors, the potential is written as follows:

$$\varphi(r) = \sum_{l=0}^{\infty} \sum_{m=-l}^{l} \frac{4\pi}{2l+1} \frac{Q_{lm} Y_{lm}^*(\theta, \varphi)}{4\pi\varepsilon_0 r^{l+1}} \tag{3.20}$$

Spherical tensors are advantageously employed when making transformations from one frame of reference to another, usually from a molecule-fixed to a space-fixed coordinate system. (See the discussion of Wigner rotation matrices in Appendix A.) Physically, the multipole moments beyond the monopole, in any representation, can be interpreted as indicators of the nonspherical nature of the charge distribution. The equipotential surfaces surrounding a multipole of order $l = 0, 1, 2, \ldots$ have the same angular symmetry as the hydrogen atom $s, p, d, \ldots$ atomic orbitals.

3.2.3 QUANTUM MECHANICAL CALCULATION OF MULTIPOLE MOMENTS

Given the wavefunction for a molecule in a particular quantum state, the expectation value of any multipole moment can be calculated. The operator for the dipole moment, for example, is

$$\hat{\mu} = e \sum_\alpha Z_\alpha \vec{R}_\alpha - e \sum_i \vec{r}_i \tag{3.21}$$

The first summation runs over the positions of the nuclei and the second over those of the electrons. The charge on each nucleus is eZ_α, where Z_α is the atomic number of nucleus α.

Following the usual quantum mechanical recipe for calculating average physical properties, the corresponding operator is sandwiched between the wavefunction and its complex conjugate, and the result is integrated over all space:

$$\langle \mu_k \rangle = \int \Psi_k^* \hat{\mu} \Psi_k d\tau \tag{3.22}$$

The wavefunction Ψ_k represents the total (electronic plus nuclear) state of the system, and the volume element $d\tau$ indicates integration over the positions of all the electrons and nuclei. We refer to $\langle \mu_k \rangle$ as the state dipole moment of the molecule in state k.

Within the Born–Oppenheimer approximation, to be considered in Chapter 9, the nuclei can be considered fixed at the positions R_α, so we can treat them as discrete charges. The electrons, on the other hand, present a charge distribution according to the electronic wavefunction ψ_k for state k. This leads to an expression containing two terms, the nuclear and electronic contributions to the total dipole moment:

$$\langle \mu_k \rangle = e \sum_\alpha Z_\alpha \vec{R}_\alpha - e \int \psi_k^* \left(\sum_i \vec{r}_i \right) \psi_k d\tau_{el} \tag{3.23}$$

Equation 3.23 represents the dipole moment for a fixed configuration of the nuclei, such as the equilibrium geometry of the molecule. For the ground electronic state, this is the permanent dipole moment that would be found in a handbook. In later chapters, we will go beyond the simple expression for this average moment and look at how the dipole moment varies as the molecule vibrates and rotates. Higher order multipole moments can be calculated using expressions similar to Equation 3.22, on substitution of the appropriate operator.

3.2.4 INTERACTION OF ELECTRIC MOMENTS WITH THE ELECTRIC FIELD

The multipole moments have been introduced as a convenient way to represent the electrostatic potential due to a collection of charges. They also present a physically appealing way to visualize the interaction of light with matter. For example, the energy of a collection of charges subjected to a field is given by

$$W = q\varphi - \vec{\mu} \cdot \vec{E} - \frac{1}{3} \Theta : \nabla \vec{E} + \cdots \tag{3.24}$$

Equation 3.24 is found by taking the energy as the sum $W = \Sigma_i q_i \varphi(r_i)$ and using a Taylor series expansion for the potential. The resulting picture is that of the total charge interacting with the potential, the dipole interacting with the field, the quadrupole interacting with the field gradient, etc.

An electric field, whether due to radiation or another collection of charges, can exert a force $\vec{F}$ or torque $\vec{T}$ on a charge assembly. The expressions for force and torque are

$$\vec{F} = \sum_i q_i \vec{E}(\vec{r}_i)$$

$$\vec{T} = \sum_i \vec{r}_i \times q_i \vec{E}(\vec{r}_i) \tag{3.25}$$

The electric field can be expanded in a Taylor series about the point $\vec{r} = 0$:

$$\vec{E}(\vec{r}) = \vec{E}_0 + \vec{r} \cdot (\nabla \vec{E})_0 + \frac{1}{2} \vec{r}\vec{r} : (\nabla\nabla\vec{E})_0 + \cdots \tag{3.26}$$

Using Equation 3.26 in Equation 3.25 leads to

$$\vec{F} = q\vec{E}_0 + \vec{\mu} \cdot (\nabla\vec{E})_0 + \frac{1}{3} \Theta : (\nabla\nabla\vec{E})_0 + \cdots \tag{3.27}$$

$$\vec{T} = \vec{\mu} \times \vec{E}_0 + \frac{2}{3} \Theta \times (\nabla\vec{E})_0 + \cdots \tag{3.28}$$

The consequences of Equation 3.28 are pictured in Figure 3.3. A homogeneous field $\vec{E}_0$, for which the gradient is zero, exists well within the space confined by two oppositely charged plates (Figure 3.3a). Clearly, such a field exerts a force on a point charge, but not on a dipole, quadrupole, or higher order multipole. The field $\vec{E}_0$

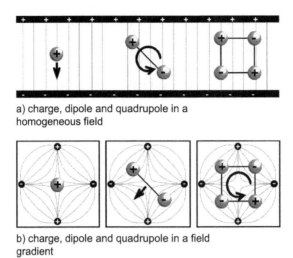

a) charge, dipole and quadrupole in a
homogeneous field

b) charge, dipole and quadrupole in a field
gradient

Figure 3.3 Electric moments interacting with (a) a field and (b) a field gradient.

does exert a torque on the dipole; it will tend to align with the field as shown. The quadrupole moment, on the other hand, is not affected by a homogeneous field. The field gradient $(\vec{\nabla}\vec{E})_0$ can be pictured as resulting from four charged wires arranged as shown in Figure 3.3b. A point charge symmetrically situated at the center of this arrangement would experience no net force, but the dipole would be accelerated as shown in the figure and indicated in Equation 3.27. Figure 3.3b reveals how the quadrupole, shown in this example as a set of alternating charges arranged in a square, would experience a torque; that is, it would tend to be reoriented by the field gradient.

What is the significance of this for the study of spectroscopy? We can view molecules as little antennae that can receive or transmit electromagnetic radiation, in absorption and emission of light, respectively. Equation 3.24 provides a way to view the interaction between the charge distribution of a molecule and the electromagnetic field. For a neutral molecule, the lead term is the interaction of the dipole moment with the field at the molecule. The second term accounts for the interaction of the quadrupole moment with the field gradient. According to Equation 3.28, a static electric field exerts a torque on the dipole moment, while the field gradient rotates the quadrupole moment. The wavelength of light used in a typical spectroscopy experiment is much larger than the size of an atom or a molecule. This means that the molecule sees a time-varying electric field, but is not very sensitive to the spatial variations (gradients) of the field. The net result, as we shall see, is that the spectroscopic transitions permitted by the dipole operator tend to be stronger than those permitted by the quadrupole moment operator. In the quantum mechanical treatment of the next chapter, we will account for the rate of a "dipole-allowed" spectroscopic transition by taking matrix elements of the operator equivalent of $W = -\vec{\mu} \cdot \vec{E}$. When this matrix element happens to vanish, the transition is said to be dipole forbidden, and we consider the possibility of higher order terms coming into play. Quadrupole-allowed transitions are weaker than dipole transitions due to the small variation of the electric field over the extent of the molecule.

3.2.5 POLARIZABILITY AND INDUCED MOMENTS

In the presence of an external field, the flexibility of the molecular charge distribution leads to revised multipole moments. In yet another use of the ubiquitous Taylor series expansion, the dipole moment can be expanded in a power series in the applied field:

$$\vec{\mu} = \vec{\mu}_0 + \alpha \cdot \vec{E} + \frac{1}{2}\beta : \vec{E}\vec{E} + \cdots \tag{3.29}$$

The lead term $\vec{\mu}_0$ is the previously discussed permanent dipole moment, and subsequent terms represent induced moments. The linear term introduces the polarizability α, a second-rank tensor whose components, say the xy term, can be defined through

$$\alpha_{xy} = \left(\frac{\partial \mu_x}{\partial E_y} \right)_0 \qquad (3.30)$$

The zero subscript means that the derivative is evaluated at zero field. The term quadratic in the field depends on the hyperpolarizabilty β, a third-rank tensor with components β_{xxx}, β_{xxy}, etc. Lest one is tempted to think that Taylor series only ever have three terms, rest assured that higher-order terms involving increasingly higher rank tensors could be expressed, and may come into play at higher field strengths.

The polarizability of an atom or a molecule represents the tendency for an electric moment to be induced by an external field. We can view this polarizability in terms of the softness of the electron cloud; larger atoms are more polarizable because the outer shell electrons are farther from the nucleus. In an atom, or in a molecule with spherical symmetry such as CCl_4, the induced dipole moment is necessarily in the same direction as the field that induces it, and the polarizability α is a scalar: $\vec{\mu}_{ind} = \alpha \vec{E}$. But nonspherical molecules need not have induced moments which are parallel to the electric field, as expressed by Equation 3.30. The tendency of electrons to follow the field is influenced by the paths provided by chemical bonds. This necessitates expressing the polarizability as a second-rank tensor, for example, as a Cartesian tensor with nine components:

$$\alpha = \begin{pmatrix} \alpha_{xx} & \alpha_{xy} & \alpha_{xz} \\ \alpha_{yx} & \alpha_{yy} & \alpha_{yz} \\ \alpha_{zx} & \alpha_{zy} & \alpha_{zz} \end{pmatrix} \qquad (3.31)$$

The three components of the induced moment are given by

$$\begin{pmatrix} \mu_x \\ \mu_y \\ \mu_z \end{pmatrix} = \begin{pmatrix} \alpha_{xx} & \alpha_{xy} & \alpha_{xz} \\ \alpha_{yx} & \alpha_{yy} & \alpha_{yz} \\ \alpha_{zx} & \alpha_{zy} & \alpha_{zz} \end{pmatrix} \begin{pmatrix} E_x \\ E_y \\ E_z \end{pmatrix} \qquad (3.32)$$

Now each component of the induced moment can depend on all three components of the field; for example, $\mu_x = \alpha_{xx} E_x + \alpha_{xy} E_y + \alpha_{xz} E_z$.

Like the quadrupole tensor, α can be made diagonal in the principal axes of the molecule. When the polarizability tensor is expressed within a coordinate system embedded in the molecule, it must reflect the molecular symmetry. In a laboratory frame of reference, the induced moments depend on the orientation of the molecule. This is a topic of great importance to the subject of light scattering, and we will return to it in Chapters 8 and 12. For the time being, we note that the mean polarizability $\bar{\alpha}$ is independent of the orientation, just as the length of a dipole moment vector is invariant to orientation. The mean polarizability is one-third the trace of the polarizability tensor: $\bar{\alpha} = 1/3 \left(\alpha_{xx} + \alpha_{yy} + \alpha_{zz} \right) = 1/3 \, \mathrm{Tr}\,\alpha$.

The total energy of the molecule can also be expanded in terms of powers of the field. It can be shown that

$$W = W_0 - \vec{\mu}_0 \cdot \vec{E} - \frac{1}{2} \vec{E} \cdot \alpha \cdot \vec{E} - \cdots \qquad (3.33)$$

This leads to yet another fundamental expression for a particular component of the polarizability, for example:

$$\alpha_{yz} = -\left(\frac{\partial^2 W}{\partial E_y \partial E_z} \right)_0 \qquad (3.34)$$

The polarizability increases with the number of electrons in the molecule, or the volume of the charge distribution. In a classical picture, it can be shown that an atom of radius a has $\alpha \approx 4\pi\varepsilon_0 a^3$. (See Appendix B.) This expression emphasizes the unusual units that result when the polarizability is expressed in the MKS system: C^2 m N^{-1}. It is sometimes desirable to refer to the volume polarizability $\alpha/4\pi\varepsilon_0$, which has units of m^3. In cgs units, where $4\pi\varepsilon_0 = 1$, the units of polarizability are cm^3. For example, the polarizability of CH_4 is 2.6×10^{-24} cm^3, while that of CCl_4 is 11.2×10^{-24} cm^3.

The induced dipoles caused by fluctuating molecular charge distributions set up fields that induce dipoles in neighboring molecules. The interaction is automatically attractive, and the resulting forces are called dispersion forces. The polarizability is actually a frequency-dependent property, as discussed in the next section.

3.2.6 FREQUENCY DEPENDENCE OF POLARIZABILITY

The perturbing field that induces a dipole moment may be time-dependent, as it is in a spectroscopy experiment. The ability of the field to induce a dipole moment depends on the frequency because the molecular motions that respond to the field have their own natural frequencies. This is an important point to which we will frequently return. For the moment, we envision the induced moment to result from motion of electrons, ignoring but anticipating the contributions of nuclear motions. For visible and ultraviolet wavelengths, for example, the electrons are capable of keeping up with the driving field, but the more sluggish nuclei are not.

A classical view of the molecule and its electron cloud results in a phenomenologically correct expression for $\alpha(\omega)$, which we will show is comparable to the quantum mechanical expression. The Lorentz model of matter was the first attempt to explain atomic spectra, and it predates the emergence of quantum mechanics. This classical picture rests on the idea that an electron is bound to the nucleus by a Hooke's law force; that is, the restoring force is proportional to the displacement. This may seem to contradict the expression for Coulomb's law given at the start of the chapter, but if the entire charge due to the electrons is considered to be smeared out into a continuous sphere, Gauss' law (Appendix B) can be used to show that the force resulting from a displacement of this charge distribution, relative to the positive center, is indeed proportional to the displacement.

Picture a single electron in one dimension subject to a field E in the x direction, resulting in a displacement x from the center of positive charge. The induced moment is given by

$$\mu_{ind} = -ex = \alpha E \tag{3.35}$$

By analogy to the harmonic oscillator, Hooke's law for the restoring force is

$$F = -kx = -m\omega_0^2 x \tag{3.36}$$

The force constant k is related to the mass m and the harmonic frequency ω_0. In the static case the Hooke's law force is balanced by the electric force $-eE$. The net force is zero, so $-m\omega_0^2 x = eE$. Solving for x and substituting in Equation 3.35 gives

$$\mu_{ind} = \frac{e^2}{m\omega_0^2} E \tag{3.37}$$

and the polarizability is

$$\alpha = \frac{e^2}{m\omega_0^2} \tag{3.38}$$

This expression is generalized by allowing for a total of N electrons, divided into groups of $f_j N$ having harmonic frequencies ω_j:

$$\alpha = \frac{e^2}{m} \sum_j \frac{f_j}{\omega_j^2} \tag{3.39}$$

The quantity f_j is known as the oscillator strength; a quantum mechanical definition will be presented in Chapter 4. Equation 3.39 actually gives the static polarizability, valid when the external field is constant in time. In the case where the field $E(t)$ is a function of time, we have to solve the equation of motion for a driven harmonic oscillator. Summing the forces in this case leads to

$$F = m\frac{d^2x}{dt^2} = -eE(t) - m\omega_0^2 x(t) - \Gamma\frac{dx}{dt}$$ (3.40)

The three contributions to the right-hand side of Equation 3.40 are the electrical force, the Hooke's law force, and a frictional force, respectively. The constant Γ is a damping factor that allows for a slowing down of the oscillation. The solution to this differential equation can be found with the help of physical intuition. First, assume a time-dependent electric field of the form $E(t) = E_{0x}\exp[i(ky - \omega t)]$. (Here, k is the magnitude of the wave vector $2\pi/\lambda$ and not the force constant.) This represents radiation polarized in the x direction and traveling in the y direction. It seems logical to assume that the displacement will follow this field, so we try a solution of the form

$$x(t) = x_0 \exp[i(ky - \omega t)]$$ (3.41)

When the time derivatives of $x(t)$ are evaluated and plugged into Equation 3.40, the result is

$$x(t) = \frac{-(e/m)E(t)}{\omega_0^2 - \omega^2 - i\omega/\tau}$$ (3.42)

The time $\tau = m/\Gamma$ represents a relaxation time for damping of the oscillating dipole. The definition of the polarizability $\alpha = -ex/E$ can now be introduced to produce

$$\alpha(\omega) = \frac{e^2}{m}\sum_j \frac{f_j}{\omega_j^2 - \omega^2 - i\omega/\tau}$$ (3.43)

This equation has been generalized to allow for a collection of electrons, as in the previous expression.

The polarizability becomes large when the frequency of the driving field ω is resonant with one of the natural frequencies ω_j. Note that clearly the polarizability has both real and imaginary parts:

$$\alpha(\omega) = \alpha'(\omega) + i\alpha''(\omega)$$ (3.44)

The real part $\alpha'(\omega)$ represents the part of the induced polarization that is in phase with the driving field, and the imaginary part $\alpha''(\omega)$ pertains to that which is 90° out of phase. To justify this statement, consider the function $\exp(i\theta)$. Addition of $\pm\pi/2$ to the angle θ amounts to multiplication of the original function by $\pm i$. The real and imaginary parts of $\alpha(\omega)$ correspond, respectively, to the dispersion and absorption or emission of light. Note that the out-of-phase part would vanish in the absence of damping.

The idea of a physical property having an imaginary part can be disconcerting when it is first introduced. But keep in mind that it is $\alpha''(\omega)$, a *real* quantity, that is experimentally accessible. Equation 3.44 has the form common to linear response functions. The real and imaginary parts of a linear response function such as $\alpha(\omega)$ are related to one another through the Kramers–Kronig relations, which are

$$\alpha'(\omega) = \frac{2}{\pi}P\int_0^\infty \frac{s\alpha''(s)ds}{s^2 - \omega^2}$$

$$\alpha''(\omega) = \frac{-2\omega}{\pi}P\int_0^\infty \frac{\alpha'(s)ds}{s^2 - \omega^2}$$ (3.45)

where P indicates the principal part of the integral. Thus, the real part depends on knowledge of the imaginary part at all frequencies, and vice versa. The Kramers–Kronig relations are not limited to the polarizability, but apply to the real and imaginary parts of other response functions as well.

3.2.7 QUANTUM MECHANICAL EXPRESSION FOR THE POLARIZABILITY

The quantum mechanical derivation of the frequency-dependent polarizability is presented in Chapter 4. The result is merely quoted here in order to compare it to the classical expressions given previously:

$$(\alpha_k)_{\rho\sigma} = \frac{1}{\hbar} \sum_r \frac{2\omega_{rk} \langle \psi_k | \mu_\rho | \psi_r \rangle \langle \psi_r | \mu_\sigma | \psi_k \rangle}{\omega_{rk}^2 - \omega^2} \tag{3.46}$$

Equation 3.46 gives the frequency-dependent polarizability of a molecule in quantum state k, expressed as a particular element of the Cartesian polarizability tensor. The subscripts ρ and σ designate two of the directions x, y, and z. The sum is over all excited electronic (more precisely, rovibronic or rotational plus vibrational plus electronic) states r, and the matrix elements in the numerator are transition dipole moments. The transition dipole moment figures heavily in the discussion of spectroscopy and will be considered further in the next chapter. For now, we note that it signifies a spectroscopic connection between two states.

Equation 3.46 lacks the damping term responsible for the imaginary part of the polarizability. In Chapter 4, we will revise the quantum mechanical expression for the polarizability to include such a damping term. It will be shown that this damping derives from the finite lifetime of excited states, which contribute to the sum-over-states form of the polarizability. Equation 3.46 shows the same resonance behavior as the classical expression: the polarizability becomes large when the frequency of light is tuned to match the frequency of an allowed electronic transition, $\omega \approx \omega_{rk}$, where

$$\omega_{rk} \equiv \frac{E_r - E_k}{\hbar} \tag{3.47}$$

is the frequency of the $k \rightarrow r$ transition. In the quantum mechanical point of view, it is the tendency of electrons to be excited to higher-energy states that gives rise to the polarizability. The larger the atom or molecule the closer together these energy levels are, so increasingly accessible energy levels make for more polarizable molecules, in agreement with reasoning based on the size of the electron cloud.

A final point of interest is that the quantity referred to as oscillator strength in the classical expression for $\alpha(\omega)$ can be compared to the product of the transition dipole moments appearing in the numerator of the quantum mechanical expression. The concept of oscillator strength will be further explored in Chapter 4.

3.3 ELECTRIC PROPERTIES OF BULK MATTER

3.3.1 DIELECTRIC PERMITTIVITY

In this section, we examine the collective response of matter to an applied field, which may be time-dependent. We begin by dividing all materials into two groups: conductors, which have free charges, and insulators, which do not. We are more concerned with the latter in this section, as we consider the effect of an electric field on the bound charges that make up the insulator.

When an external field is applied to a conductor, the free charges respond in such a way as to set up a field inside the conductor that exactly cancels the externally applied field. In an insulator subjected to an external field, on the other hand, the matter responds with an induced polarization that *partially* cancels the field due to outside charges. Although the charges in an insulator are not mobile, as they are in a conductor, an external field has the ability to displace the charges through distortion of the electron cloud, orientation of permanent dipoles, and changes in the lengths of polar bonds. The resulting net dipole moment per unit volume is called the polarization, and in linear dielectric materials it is proportional to the electric field.

The historical approach to the concept of dielectric properties begins with the idea of a parallel plate capacitor, like the arrangement envisioned in Figure 3.3a depicting a homogeneous electric field. The surface charge density on each plate is $\sigma = q/A$, where A is the area and q the absolute value of the charge. As shown in Appendix B, the electric field is given by

$$E_0 = \frac{\sigma}{\varepsilon_0}\hat{z} = \frac{\varphi}{d}\hat{z} \tag{3.48}$$

where $\hat{z}$ is a unit vector in the direction normal to the plates. The zero subscript on E signifies the field that would result if a vacuum existed between the plates. This field is the voltage φ across the plates divided by the distance d between them. The capacitance is defined as the ratio of the charge stored by the capacitor, q, to the voltage φ, $C = q/\varphi$. (The units of C are farads; one F is equal to one C/V.) Thus the capacitance depends on the geometry:

$$C_0 = \frac{A\varepsilon_0}{d} \tag{3.49}$$

This expression is for an empty capacitor, and ε_0 is the usual permittivity of free space. Now imagine that the space between the plates is filled with a dielectric material. The external field will cause the bound charges to be distorted such that a layer of induced negative charge resides next to the positively charged plate and vice versa. Microscopically, these induced charges result from induced dipoles (via the polarizability) and from the alignment of permanent dipoles. We refer to these as the electronic and the orientational polarization, respectively. While the former exists in all dielectrics, the latter requires the component molecules to possess permanent dipole moments. The induced polarization leads to a decrease in the field and thus an increase in capacitance in the presence of the dielectric. In the presence of a dielectric, the permittivity of free space ε_0 is replaced by the permittivity of the material ε:

$$C = \frac{A\varepsilon}{d} \tag{3.50}$$

The relative permittivity ε_r is the ratio of the permittivity of the material to that of the vacuum:

$$\varepsilon_r = \frac{\varepsilon}{\varepsilon_0} \tag{3.51}$$

The relative permittivity is often referred to as the dielectric constant. It is more precise to say that the dielectric constant is the value of ε_r for a static field, since ε_r is a function of frequency. The relative permittivity is related to appropriate ratios of the field and capacitance, with and without dielectric:

$$\varepsilon_r = \frac{E_0}{E} = \frac{C}{C_0} \tag{3.52}$$

It is also the factor by which the Coulombic force between charges is reduced when they are immersed in a dielectric medium; hence, the ease with which many salts are dissolved in polar solvents such as water.

The relative permittivity of a dielectric material is always greater than unity. Some representative values of ε_r are given in Table 3.1, along with the permanent dipole moments μ_0. It is clear that the more polar the molecules in a liquid, the larger the dielectric constant, but Table 3.1 indicates that ε_r does not consistently increase with μ_0. The reason is that ε_r depends on the relative alignment of dipoles in the liquid. Hydrogen bonding in liquids like water and methanol leads to an enhancement of the permittivity over what would be

Table 3.1 Dielectric constants and dipole moments for some liquids at room temperature

	ε_r	μ_0, **D**		ε_r	μ_0, **D**
Benzene	2.28	0	Acetone	20.7	2.88
Ethyl ether	4.34	1.15	Methanol	32.7	1.70
Chloroform	4.81	1.01	Dimethylsulfoxide	46.7	3.96
Ethyl acetate	6.02	1.78	Water	80.2	1.86
Dichloromethane	8.93	1.60	Formamide	111.0	3.73

expected in the case of more random relative alignment of neighboring dipoles. Note that ε_r is greater than unity even if the dipole moment vanishes, due to the electronic contribution to the induced polarization.

The partial cancellation of the applied field is determined by the polarization $\vec{P}$, defined as the net dipole moment per unit volume. Polarization, not to be confused with polarizability, is a vector quantity and comprises both induced dipole moments and aligned permanent moments. Gauss' law (Appendix B) enables us to visualize a field due to the polarization as the result of bound charges induced in the layers adjacent to the conducting plates. This field opposes the applied field such that the net field is

$$\vec{E} = \vec{E}_0 - \frac{\vec{P}}{\varepsilon_0} \tag{3.53}$$

$\vec{E}$ is called the macroscopic or average field. The polarization is proportional to this average field:

$$\vec{P} = \varepsilon_0(\varepsilon_r - 1)\vec{E} = \chi_e \varepsilon_0 \vec{E} \tag{3.54}$$

Equation 3.54 introduces the electric susceptibility $\chi_e = \varepsilon_r - 1$, where the subscript e distinguishes the electric susceptibility from its magnetic counterpart. In expressing the susceptibility as a scalar, we have assumed that the dielectric is isotropic. More generally, a crystalline material would require that χ_e be expressed as a second-rank tensor.

It is conventional to introduce a quantity referred to as the electric displacement, $\vec{D}$, defined by $\vec{D} = \varepsilon_0\vec{E} + \vec{P} = \varepsilon\vec{E}$. The displacement is related to the free charges, through Gauss' law, the same way that $\vec{E}$ is related to the net charge.

At this point, the reader might want to ponder the following dilemma. The polarization of the dielectric depends on the field between the plates of the capacitor, which depends in turn on the polarization of the medium. How are we to break out of this circular relationship and determine the molecular basis for the relative permittivity?

3.3.2 FREQUENCY DEPENDENCE OF PERMITTIVITY

The external field applied to a parallel plate capacitor could be an alternating field rather than a DC field. With a microscopic point of view, we can see why the resulting polarization depends on frequency. At low frequency, less than about 10^{12} s^{-1}, both the electronic and the nuclear polarization contribute. At microwave frequencies and lower, permanent dipoles respond by reorienting. At higher frequencies, into the infrared, the field switches so rapidly that the rotational motion of the molecules cannot keep up, but the induced dipoles that accompany vibrational motion still contribute. At frequencies characteristic of visible and ultraviolet light, 10^{15} to 10^{16} s^{-1}, the electronic contribution to the response remains while that due to nuclear motion is frozen out. Thus we have spectral windows in the microwave, infrared, and visible-ultraviolet regions, through which we view rotational, vibrational, and electronic motions, respectively.

The relative permittivity, like the molecular properties on which it is based, is a function of frequency and has both real and imaginary parts:

$$\varepsilon_r(\omega) = \varepsilon_r'(\omega) + i\varepsilon_r''(\omega) \tag{3.55}$$

Like the real and imaginary parts of the polarizability (Equation 3.45), $\varepsilon_r'(\omega)$ and $\varepsilon_r''(\omega)$ are related to one another via the Kramers–Kronig relations:

$$\varepsilon_r'(\omega) = \varepsilon_\infty + \frac{2}{\pi} \int_0^\infty \frac{\varepsilon_r''(s)s\,ds}{s^2 - \omega^2}$$

$$\varepsilon_r''(\omega) = \frac{-2\omega}{\pi} \int_0^\infty \frac{\varepsilon_r'(s) - \varepsilon_\infty}{s^2 - \omega^2} ds$$

(3.56)

where ε_∞ is the relative permittivity at infinite frequency.

The refractive index is also complex:

$$n(\omega) = n_r(\omega) + i\kappa(\omega)$$

(3.57)

As discussed in Chapter 2, the relative permittivity of a nonmagnetic material is the square of the complex refractive index, $n^2(\omega) = \varepsilon_r(\omega)$. Equating real and imaginary parts of both sides of this expression leads to

$$\varepsilon_r'(\omega) = n_r^2(\omega) - \kappa^2(\omega)$$

(3.58)

$$\varepsilon_r''(\omega) = 2n_r(\omega)\kappa(\omega)$$

(3.59)

Let us explore the physical significance of these expressions. The familiar refractive index, that is, the real part n_r, is the factor by which the speed of light in a medium is reduced from its value in free space. Consider a plane electromagnetic wave traveling in the x direction. The electric field is given by

$$\vec{E} = \mathrm{Re}\,\vec{E}_0 \exp\left[i(kx \pm \omega t)\right]$$

(3.60)

The wave vector $k = 2\pi/\lambda$ should be corrected for the change in wavelength. If λ is the vacuum wavelength, then we should write

$$k = \frac{2\pi n}{\lambda} \Rightarrow \frac{2\pi}{\lambda}(n_r + i\kappa)$$

(3.61)

Thus the electric field in the medium is given by

$$\vec{E} = \mathrm{Re}\,\vec{E}_0 \exp\left[i\left(\frac{2\pi n_r x}{\lambda} \pm \omega t\right)\right] \exp\left(\frac{-2\pi\kappa x}{\lambda}\right)$$

(3.62)

This expression shows that the real part of the refractive index serves to modify the wavelength as expected, while the imaginary part leads to an exponential decay in the electric field as it travels through the medium, when κ is positive. Since the intensity is proportional to the square of the amplitude of the electric field, $I \propto |E|^2$, the intensity of light depends exponentially on the distance traveled:

$$I = I_0 \exp(-\gamma x) = I_0 \exp\left(\frac{-4\pi\kappa x}{\lambda}\right)$$

(3.63)

The absorption coefficient γ has been introduced. Using Equation 3.59, we find that the absorption coefficient is

$$\gamma(\omega) = \frac{\omega\varepsilon_r''}{n(\omega)c}$$

(3.64)

It is related to the more familiar molar absorptivity ε_M (units of L mol^{-1} cm^{-1}) by comparing Equation 3.63 to

$$I = I_0 \exp(-2.303\varepsilon_M Cx) = I_0 10^{-\varepsilon_M Cx} \tag{3.65}$$

(Unfortunately, the symbol epsilon, ε, is widely used for both the dielectric function and molar absorptivity, so the subscript M has been added to the latter.) The absorbance $A = \varepsilon_M Cx$ is given by Beer's law, that is, it is proportional to the concentration C in moles per liter, and the path length x, usually expressed in cm. Combining the three previous equations, we find that the imaginary part of the relative permittivity is related to the molar absorptivity by the expression

$$\varepsilon_r''(\omega) = \frac{2303\varepsilon_M N c n_r}{\omega N_A} \tag{3.66}$$

where N is the number of absorbing molecules per cm^3, N_A is Avogadro's number, c is the speed of light, and the conversion factor between liters and cubic centimeters has been absorbed into the number 2303.

As the frequency ω approaches zero, the relative permittivity becomes purely real. The quantity commonly referred to as the dielectric constant is $\varepsilon_r'(0)$. In transparent materials, the square of the refractive index measured at the wavelength 586 nm (the yellow sodium D line), n_D, is sometimes called the optical dielectric constant. More precisely, the optical dielectric constant is $\varepsilon_r'(\infty)$, but it is often assumed that this is close to the value n_D^2. The Kramers–Kronig equations connect the real and imaginary parts of the refractive index; thus n_r undergoes dispersion (is frequency dependent) in the vicinity of an absorption transition ($\kappa \neq 0$). (See Figure 2.5.)

3.3.3 RELATIONSHIPS BETWEEN MACROSCOPIC AND MICROSCOPIC PROPERTIES

The collective response of bulk matter to an applied electric field depends upon the permittivity ε or the susceptibility χ_e, while that of an individual molecule is dictated by the polarizability α and the permanent dipole moment μ_0. In a bulk sample, the interactions of molecules with one another result in collective electromagnetic properties that are not merely the sum of single molecule quantities. Thus the exact treatment of the molecular basis for the screened electric field in the condensed phase is quite complicated. Nevertheless, we can use simple models to appreciate how the bulk permittivity derives from molecular properties such as dipole moment and polarizability. The simplest case will be considered first: a low-pressure gas consisting of nonpolar molecules. We will then look at liquids composed of nonpolar and then polar molecules, for which the discussion leads naturally to a consideration of the local field problem to be discussed in Section 3.3.4.

In the following discussion, we must be careful to distinguish between the previously mentioned macroscopic field and the quantity known as the local field. The local field at the site of an individual molecule contains the contributions of the fields due to the molecules that surround it. The induced moment of a single molecule is proportional to this local field. The polarization $\vec{P}$, however, is proportional to the macroscopic field $\vec{E}$, which depends on the fields due to *all* the molecules in the sample.

3.3.3.1 NONPOLAR MOLECULES IN THE GAS PHASE

At sufficiently low number density N/V, the macroscopic field within the sample is approximately the same as the local field. This amounts to neglecting the interactions between molecules. The polarization, which is purely electronic for nonpolar systems, is just the number density times the induced moment. Thus

$$\vec{P} = \frac{N}{V}\alpha\vec{E} = \chi_e \varepsilon_0 \vec{E} \tag{3.67}$$

and using $\varepsilon_r = 1 + \chi_e$, the relative permittivity is given by

$$\varepsilon_r = 1 + \frac{N\alpha}{\varepsilon_0 V} \tag{3.68}$$

Since $\varepsilon_r = n^2$ is close to unity for a low-pressure gas, the refractive index is approximated by

$$n \approx 1 + \frac{N\alpha}{2\varepsilon_0 V} \tag{3.69}$$

Equation 3.69 predicts the refractive index of a gas to be a linear function of density, provided the density is not too high.

3.3.3.2 NONPOLAR MOLECULES IN THE CONDENSED PHASE

In a condensed phase, the interactions between molecules cannot be neglected, and we must consider how the local field $\vec{F}$ is different from the macroscopic field $\vec{E}$. Imagine a spherical region within the sample which is large enough to be representative of the whole sample. There are several ways to arrive at the Lorentz local field for the region within this sphere. The resulting expressions apply to an isotropic medium, such as a cubic crystalline lattice. Imagine that the spherical region has a radius a, large compared to the size of a molecule. The field due to the matter inside a uniformly polarized sphere behaves as if it were due to a dipole given by

$$\vec{\mu} = \frac{4\pi a^3}{3}\vec{P} \tag{3.70}$$

This is just the volume of the sphere times the dipole moment per unit volume. The field due to the matter inside the sphere, let us call it $\vec{E}_{int}$, is given by $-\vec{P}/3\varepsilon_0$ (see [1]). The local field $\vec{F}$ inside the sphere is the macroscopic field $\vec{E}$ less the contribution of the field due to the matter within the sphere. That is,

$$\vec{F} = \vec{E} - \vec{E}_{int} = \vec{E} + \frac{\vec{P}}{3\varepsilon_0} \tag{3.71}$$

Using Equation 3.54 for $\vec{P}$, the Lorentz local field is found:

$$\vec{F} = \frac{1}{3}\left(\varepsilon_r + 2\right)\vec{E} \tag{3.72}$$

The polarization is the number density times the polarizability times the local field. Thus

$$\vec{P} = \frac{N\alpha}{V}\vec{F} = \frac{N\alpha}{3V}(\varepsilon_r + 2)\vec{E} = \varepsilon_0(\varepsilon_r - 1)\vec{E} \tag{3.73}$$

Rearranging and eliminating $\vec{E}$ leads to the Clausius–Mossotti equation:

$$\frac{\varepsilon_r - 1}{\varepsilon_r + 2} = \frac{N\alpha}{3\varepsilon_0 V} \tag{3.74}$$

When the relative permittivity ε_r in Equation 3.74 is replaced by n^2, the expression is referred to as the Lorenz–Lorentz equation. The number density can be related to the molecular weight M and the mass density ρ through $N/V = N_A\rho/M$ and the Lorenz–Lorentz expression cast in the following form:

$$R_M = \frac{N_A\alpha}{3\varepsilon_0} = \frac{M}{\rho}\frac{(n^2 - 1)}{(n^2 + 2)} \tag{3.75}$$

where the molar refractivity R_M has been introduced. Equation 3.75 connects the bulk property, refractive index, to the molecular property, polarizability. Both are functions of frequency.

3.3.3.3 POLAR MOLECULES IN CONDENSED PHASES

The polarization discussed previously is properly referred to as the electronic polarization $\vec{P}_e$. In the case of collections of polar molecules, the total polarization is the sum of the electronic and orientational contributions. The latter takes into account the average component of the permanent dipole moment in the direction of the applied field: $<\vec{\mu}>$. The alignment of the permanent moments with the field is opposed by the thermal motion of the molecules. The orientational polarization is the number density times the thermally averaged component of the dipole moment in the field direction:

$$\vec{P}_o = \frac{N<\vec{\mu}>}{V} \tag{3.76}$$

The angle brackets indicate an equilibrium average. Let the angle between the permanent moment and the field direction be θ. The average needed to compute Equation 3.76 is

$$\langle\vec{\mu}\rangle = \mu_0\langle\cos\theta\rangle \tag{3.77}$$

The energy of the dipole in the local field is $W = -\vec{\mu}_0\cdot\vec{F} = -\mu_0 F\cos\theta$. A Boltzmann average of $\cos\theta$ is performed as follows:

$$\langle\cos\theta\rangle = \frac{\displaystyle\int_0^\pi \cos\theta \exp\left(\frac{\mu_0 F\cos\theta}{k_B T}\right)\sin\theta d\theta}{\displaystyle\int_0^\pi \exp\left(\frac{\mu_0 F\cos\theta}{k_B T}\right)\sin\theta d\theta} \tag{3.78}$$

With the substitutions $u = \mu_0 F/k_B T$ and $x = \cos\theta$, the integral can be expressed as the Langevin function, $L(u)$:

$$\langle\cos\theta\rangle = \frac{\displaystyle\int_{-1}^1 xe^{ux}dx}{\displaystyle\int_{-1}^1 e^{ux}dx} \equiv L(u) = \coth u - \frac{1}{u} \tag{3.79}$$

The Langevin function approaches the value $u/3$ for small values of the argument u, and the limit of 1 when u is large. For moderate temperatures and typical values of the field strength and dipole moment, $\mu_0 F/k_B T \ll 1$, so $L(u)$ can be replaced by $u/3$. The orientational polarization in this limit is

$$\vec{P}_o = \frac{N\mu_0^2}{3Vk_B T}\vec{F} \tag{3.80}$$

The total polarization is given by

$$\vec{P}_{tot} = \vec{P}_e + \vec{P}_o = \frac{N}{V}\left[\alpha + \frac{\mu_0^2}{3k_B T}\right]\vec{F} \tag{3.81}$$

Using $\vec{P}_{tot} = \varepsilon_0(\varepsilon_r - 1)\vec{E}$ and the expression for the Lorentz local field, the Debye equation for the molar polarization P_M is obtained:

$$P_M = \frac{M}{\rho}\frac{\varepsilon_r - 1}{\varepsilon_r + 2} = \frac{N_A}{3\varepsilon_0}\left[\alpha + \frac{\mu_0^2}{3k_B T}\right] \tag{3.82}$$

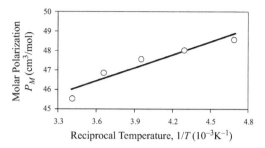

Figure 3.4 Molar polarization of $CHCl_3$ versus reciprocal temperature.

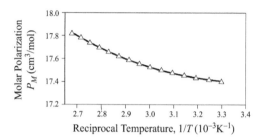

Figure 3.5 Molar polarization of H_2O versus reciprocal temperature.

Figure 3.4 illustrates the temperature-dependent molar polarization of liquid $CHCl_3$ plotted versus $1/T$. Equation 3.82 suggests that the dipole moment and polarizability can be obtained from the slope and intercept of such a plot, respectively. Application of the Debye equation to the data shown in Figure 3.4 yields a volume polarizability $\alpha/4\pi\varepsilon_0$ of 1.5×10^{-23} cm^3 and a dipole moment of 0.5 D, compared to literature values of 0.9×10^{-23} cm^3 and 1.0 D, respectively. The agreement is rather poor, although there is considerable scatter in the data. For more insight into the limitations of the Debye equation, consider the plot of P_M versus $1/T$ for liquid water, shown in Figure 3.5. The graph is clearly nonlinear, and the slope is in fact negative! Why does the Debye model fail for liquid water?

Equation 3.82 is based on a simple model and can be expected to yield dipole moments and polarizabilities in rough agreement with gas-phase quantities. There is a serious flaw in the Debye equation, however, in that it incorrectly predicts a temperature below which the dielectric constant becomes infinite. Were this a real phenomenon, it would represent permanent electrical polarization, or "ferroelectricity," of a liquid. In the next section we look at why the Debye model fails.

3.3.4 THE LOCAL FIELD PROBLEM: THE ONSAGER AND KIRKWOOD MODELS

The Debye equation works better for dilute solutions of polar molecules in nonpolar solvents than for neat polar liquids. The reason lies in the simple approach to the local field that was used to obtain Equation 3.82. The Lorentz local field is the field within a volume that is large enough to represent the macroscopic properties of the sample. What is really needed is the field experienced by a single molecule, which takes into account the microscopic nature of the surroundings. The exact calculation of such a field would require knowledge of the structure of the liquid and a sufficiently accurate approach to calculating the intermolecular interactions. This is a daunting task; an exact solution to the local field problem would not be practical. One of the difficulties is the pesky long-range nature of the dipole-dipole interaction, which leads to a troublesome dependence of the energy of a finite collection of dipoles on the shape of the sample.

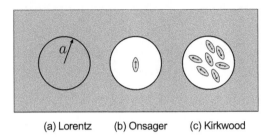

(a) Lorentz　　　(b) Onsager　　　(c) Kirkwood

Figure 3.6 Spherical cavities embedded in a dielectric continuum for three models of the local field: (a) Lorentz, (b) Onsager, and (c) Kirkwood.

There are, however, some approximate approaches that improve upon the Debye equation. The first of these, due to Onsager, replaces the macroscopically representative spherical region of the Lorentz picture with a spherical cavity containing a single molecule. The Kirkwood model is also based on a spherical cavity, but one that contains a number of molecules. In either picture, the liquid outside the sphere is treated as a dielectric continuum; that is, the molecular nature of the medium is ignored. In what follows, the basic ideas of the two models will be presented, in order to clarify the physical picture, without going into all the details of the original derivations. Figure 3.6 depicts the simple concepts on which the Lorentz, Onsager, and Kirkwood models are based. More details on the local field problem can be found in references [2], [3], [4] and [5].

3.3.4.1 ONSAGER MODEL

Imagine an isotropically polarizable molecule in a spherical cavity of radius a surrounded by a dielectric continuum having a bulk dielectric constant ε_r. The short-range interactions of the molecule with the surroundings are neglected. The problem is to find the total moment of the molecule, $\vec{m}$, which is the sum of the permanent and induced parts:

$$\vec{m} = \vec{\mu}_0 + \alpha \vec{F} \tag{3.83}$$

The first term is the dipole moment of the isolated molecule, and the second is the induced dipole, which depends on the local field. The local field can be thought of as the sum of two contributions, the cavity field $\vec{G}$ that would be present in the empty sphere, and the reaction field $\vec{R}$ that results from the polarization of the surroundings due to the field of the molecule within the cavity:

$$\vec{F} = \vec{G} + \vec{R} \tag{3.84}$$

The cavity field is a function of the dielectric constant of the surroundings and the macroscopic field:

$$\vec{G} = \frac{3\varepsilon_r}{2\varepsilon_r + 1} \vec{E} \tag{3.85}$$

The reaction field is proportional to the dipole moment of the molecule in the cavity:

$$\vec{R} = \left(\frac{\varepsilon_r - 1}{2\varepsilon_r + 1} \right) \left(\frac{\vec{m}}{2\pi a^3 \varepsilon_0} \right) \equiv g\vec{m} \tag{3.86}$$

If the dipole moment in Equation 3.86 is replaced by that due to a macroscopic region of radius a and polarization $\vec{P}$; that is, if the substitution $\vec{m} = (4\pi a^3/3)\vec{P}$ is made, the Lorentz local field is obtained for $\vec{F}$. This points to the error in using the Lorentz formula to obtain the Debye equation: Only the cavity field acts to exert a torque on the dipole, since the reaction field is always parallel to the dipole direction. Certainly, both fields affect the value of the induced moment. Using Equations 3.85 and 3.86 in Equation 3.84 and eliminating $\vec{m}$ leads to the Onsager local field:

$$\vec{F} = \left(\frac{3\varepsilon_r}{2\varepsilon_r + 1}\right)\left(\frac{\vec{E}}{1 - \alpha g}\right) + \frac{g\vec{\mu}_0}{1 - \alpha g} \tag{3.87}$$

The total moment is given by

$$\vec{m} = \left(\frac{3\varepsilon_r}{2\varepsilon_r + 1}\right)\left(\frac{\alpha\vec{E}}{1 - \alpha g}\right) + \frac{\vec{\mu}_0}{1 - \alpha g} \tag{3.88}$$

In order to get the polarization, Equation 3.88 needs to be averaged over the orientation of the molecule. As in the derivation of the Debye equation, we require $<\vec{\mu}_0> = \mu_0 <\cos\theta>$, where θ is the angle between $\vec{\mu}_0$ and $\vec{F}$. When the local field $\vec{F} = \vec{G} + \vec{R}$ is used in the Boltzmann weighting factor required for this average (see Equation 3.78), the contribution due to the reaction field vanishes, as reasoned earlier. The result is

$$<\cos\theta> = \left(\frac{3\varepsilon_r}{2\varepsilon_r + 1}\right)\left(\frac{1}{1 - \alpha g}\right)\frac{\mu_0 E}{3k_B T} \tag{3.89}$$

The limit $\mu_0 E/3k_B T \ll 1$ has been taken. The average total moment of the molecule in the cavity is then

$$<\vec{m}> = \frac{3\varepsilon_r}{2\varepsilon_r + 1}\left[\frac{\mu_0^2}{3k_B T(1 - \alpha g)^2} + \frac{\alpha}{1 - \alpha g}\right]\vec{E} \tag{3.90}$$

This equation still depends on the as yet undefined Onsager radius a, through the proportionality factor g. It can be eliminated by expressing the polarizability according to the Clausius–Mossotti formula, Equation 3.74, taking the volume per molecule to be $(4/3)\pi a^3$. Then, using the constitutive relationship between dielectric constant and polarization,

$$\varepsilon_r - 1 = \frac{\vec{P}}{\varepsilon_0\vec{E}} = \frac{N<\vec{m}>}{V\varepsilon_0\vec{E}} \tag{3.91}$$

the Onsager formula is obtained:

$$\varepsilon_r - n^2 = \frac{N}{V\varepsilon_0}\frac{3\varepsilon_r}{\left(2\varepsilon_r + n^2\right)}\left(\frac{n^2 + 2}{3}\right)^2\frac{\mu_0^2}{3k_B T} \tag{3.92}$$

This is an improvement over the Debye equation, but it still neglects the microscopic intermolecular interactions. Experimentally, the Onsager equation has been shown to work fairly well for dilute solutions of polar solutes in nonpolar solvents or for not too polar liquids.

3.3.4.2 KIRKWOOD MODEL

The Kirkwood model goes beyond that of Onsager by allowing for the dipolar interactions of neighboring molecules. The analogous expression for the dielectric constant is only slightly different from the previous one:

$$\varepsilon_r - n^2 = \frac{N}{V\varepsilon_0}\frac{3\varepsilon_r}{\left(2\varepsilon_r + n^2\right)}\left(\frac{n^2 + 2}{3}\right)^2\frac{\mu_0^2}{3k_B T}(1 + z<\cos\gamma>) \tag{3.93}$$

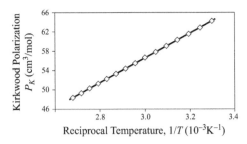

Figure 3.7 Kirkwood polarization of H_2O versus reciprocal temperature.

Here, $<\cos \gamma>$ is the average cosine of the angle between a pair of neighboring dipoles, and z is the effective coordination number, e.g., the number of molecules in the first coordination sphere. If there is no preferred dipolar alignment, then $<\cos \gamma>$ vanishes and the Onsager expression results. The final quantity in parentheses in Equation 3.93 can be viewed as a correction factor to the squared dipole moment. It is known as the Kirkwood g-factor:

$$g_K = 1 + z < \cos \gamma > \qquad (3.94)$$

The physical meaning of this is straightforward: If neighboring molecules tend to align their dipoles in antiparallel fashion, then $g_K < 1$ and the average square dipole moment is less than μ_0^2. Conversely, parallel alignment leads to $g_K > 1$ and enhancement of the effective squared dipole moment.

Equation 3.93 can be arranged in a way that suggests how temperature dependent dielectric data can be exploited. Let us define the Kirkwood polarization P_K as follows:

$$P_K = \frac{\left(\varepsilon_r - n^2\right)\left(2\varepsilon_r + n^2\right)}{3\varepsilon_r \left(n^2 + 2\right)^2} \frac{M}{\rho} = \frac{N_A}{9\varepsilon_0} \frac{\mu_0^2 g_K}{3k_B T} \qquad (3.95)$$

The polarization of water as defined in Equation 3.95 is plotted versus $1/T$ in Figure 3.7. The slope obtained from linear regression can be combined with the value of $\mu_0 = 1.86$ D to obtain a Kirkwood g-factor g_K of about 3.7. Water is rather unusual; the Kirkwood g-factor for nonassociated polar liquids is typically very close to one.

3.4 MAGNETIC PROPERTIES OF MATTER

While electric fields arise from static charges, magnetic fields are the result of moving charges. Electric fields tend to displace charges along the field lines, whereas magnetic fields cause charges to spin or curve around the field. In spite of these distinctions, the beautiful symmetry displayed by the fields $\vec{E}$ and $\vec{B}$ in Maxwell's equations points out that there are certain analogies between electric and magnetic phenomena. In general, magnetic forces are weaker then electrostatic ones by a factor of v/c, the ratio of the speed of the particle to the speed of light. We want to understand the consequences of spin and orbital angular momenta, which are magnetic phenomena leading to fine structure in high-resolution spectra of atoms and molecules. In addition, magnetic dipole moments of electrons and nuclei are central to electron paramagnetic resonance (EPR) and nuclear magnetic resonance (NMR) spectroscopy, respectively. Though neither of these techniques is discussed at length in this book, we wish to mention them briefly here to provide a physical picture for later discussions of analogous phenomena in optical spectroscopy. Before exploring the microscopic basis for the magnetic behavior of matter, we briefly summarize the fundamentals of magnetic fields and discuss macroscopic magnetic properties.

3.4.1 BASIC PRINCIPLES OF MAGNETISM

In Section 3.2.1, the electric field was described as the negative gradient of the scalar potential φ. The magnetic field $\vec{B}$, on the other hand, is the curl of the vector potential $\vec{A}$, $\vec{B} = \nabla \times \vec{A}$. The MKS units for magnetic field are Tesla, which are related to the commonly used cgs units of Gauss through $1\ T = 10^4\ G$. One Tesla is the same as one N/amp m or one Weber/m². To provide some physical insight into the source of static magnetic fields, consider the expression for the magnetic field due to a charge q moving with velocity $\vec{v}$:

$$\vec{B}(r) = \frac{\mu_0}{4\pi} \frac{q(\vec{v} \times \vec{r})}{r^3} \qquad (3.96)$$

The field is normal to the plane defined by the vectors $\vec{v}$ and $\vec{r}$. The familiar right hand rules encountered in the study of magnetism are the consequence of cross-products such as the one in Equation 3.96. The constant μ_0 (not to be confused with the electric dipole moment) is the permeability of free space, related to the permittivity ε_0 and the speed of light: $(\varepsilon_0 \mu_0)^{-1} = c^2$. Equation 3.96 comes with the caveat that the charge q is part of a steady current. (Changing currents are associated with electromagnetic radiation, which we are not considering at present.) An experimentally more relevant expression would express the magnetic field in terms of electric current I rather than a single charge q:

$$d\vec{B} = \frac{\mu_0}{4\pi} \frac{I(d\vec{s} \times \vec{r})}{r^3} \qquad (3.97)$$

where $d\vec{s}$ is an infinitesimal section of the length of the wire carrying the current, pointing in the direction of the flow of positive charge. Equation 3.97 is known as the law of Biot and Savart, and it is as essential to problems of magnetostatics as Coulomb's law is to electrostatics.

The force on a charged particle is given by the previously introduced Lorentz law: $\vec{F} = q(\vec{E} + \vec{v} \times \vec{B})$. The first term is responsible for the fact that a charged particle follows the electric field, and the second gives rise to the tendency of the charge to spiral around the magnetic field lines. As an illustration of magnetic interactions, Equations 3.96 and 3.97 can be used to derive an expression for the force per unit length between two parallel wires, separated by a distance r and each carrying current I:

$$F = \frac{\mu_0}{4\pi} \frac{2I^2}{r} \qquad (3.98)$$

This force is attractive when the currents in the two wires are parallel and repulsive when they are antiparallel.

Experimental evidence, or perhaps lack of it, suggests that magnetic monopoles do not exist. (One consequence of this is that, unlike the lines of force due to electric fields, those due to magnetic fields have no beginning or end! The field lines due to the current in a straight line, for example, form circles around the wire.) The most significant magnetic moment is thus the magnetic dipole moment, $\vec{\mu}_{mag}$. To introduce it, consider the application of the Biot–Savart law to the case of a current loop of area A. The field far from the loop and along the axis is given by

$$\vec{B}(r) = \frac{\mu_0}{4\pi} \frac{2IA}{r^3} \hat{n} \qquad (3.99)$$

where $\hat{n}$ is a unit vector normal to the plane of the current loop. The similarity of this expression to that for the electric field due to an electric dipole, Equation 3.8, suggests that the magnetic moment be defined as the product of the current times the area: $\mu_{mag} = IA$. (The more general expression for the field $\vec{B}$ due to a magnetic dipole looks just like Equation 3.8, but with $1/4\pi\varepsilon_0$ replaced by $\mu_0/4\pi$ and the magnetic dipole moment replacing the electric dipole moment. To see the comparison of Equations 3.99 and 3.8, note that the dipole moment in the former is parallel to the direction of observation.) The magnetic moment has units of amp m² or J/T. The classical result expressed above has spectroscopic implications: we conclude that orbiting charge is associated with a magnetic dipole moment perpendicular to the plane of the orbit. Thus, as will be discussed in Section 3.4.3, net electronic orbital and spin angular momentum are always accompanied by magnetic dipole moments, and the intrinsic angular momentum of some nuclei gives rise to nuclear spin effects. Imagine an orbiting charge q having angular frequency ω. The current due to this motion is $q\omega/2\pi$, so the magnetic moment is

$$\vec{\mu}_{mag} = IA\hat{n} = \frac{q\omega}{2\pi}(\pi r^2)\hat{n} = \frac{q}{2m}\vec{J} \tag{3.100}$$

where m is the mass of the spinning charge and $|J| = m\omega r^2$ is its angular momentum. Like the torque associated with the interaction of an electric dipole with an electric field, that due to a magnetic dipole in a magnetic field is given by $\vec{T} = \vec{\mu}_{mag} \times \vec{B}$. Similarly, the energy is given by $W = -\vec{\mu}_{mag} \cdot \vec{B}$. The expression for the torque leads to the phenomenon of precession, in which the magnetic dipole moment rotates about the magnetic field sweeping out a cone. Such motion leads to a torque which is always at right angles to both the dipole moment and the field as required by the cross-product. This classical physical picture is strikingly similar to those drawn to depict the quantum mechanical angular momentum vector, as in Figure 1.9.

3.4.2 MAGNETIC PROPERTIES OF BULK MATTER

We have seen that when an external electric field is applied to a dielectric the result is an induced polarization that tends to cancel the applied field. We can distinguish the field due to the free charges from that due to the bound charges, at least conceptually, even though the net field depends on both. The bound charges are only displaced by the field or, in the case of permanent moments, aligned by it. Similarly, when ordinary matter is placed in an external magnetic field, an induced magnetization $\vec{M}$, the magnetic dipole moment per unit volume, results. Unlike the polarization $\vec{P}$, the magnetization can either reinforce or oppose the applied field. This distinction will lead to the categorization of matter as diamagnetic, paramagnetic, or ferromagnetic, as will be discussed shortly. It turns out that there are magnetic parallels to the induced and permanent electric dipole moments. The bound and free charges in electric polarization are analogous to the bound and free currents in magnetic polarization.

First, we must introduce another magnetic field $\vec{H}$ to which the magnetization is directly proportional. The distinction between $\vec{H}$ and $\vec{B}$ can be a source of some confusion, and indeed the relationship between the two even depends on the system of units! The following discussion will be given in the MKS system, where the defining relationship for the H field* is

$$\vec{H} = \frac{\vec{B}}{\mu_0} - \vec{M} \tag{3.101}$$

Naturally, in free space the relationship would be $\vec{B} = \mu_0\vec{H}$. Note that the units of $\vec{H}$ are amp/m. (In the cgs system of units, $\vec{H}$ and $\vec{B}$ are both in Gauss.) The physical significance, and reason for encumbering ourselves with two different magnetic fields, is related to the previously mentioned bound and free currents. It turns out that the field $\vec{H}$ is a function of the latter, the currents in ordinary circuits which can be easily measured. In this way, the magnetic field $\vec{H}$ is related to $\vec{B}$ much like the electric displacement $\vec{D}$ is related to the field $\vec{E}$. The bound currents are those due to the induced magnetic moments and the alignment of permanent moments, possessed by the atoms and molecules that comprise the sample. These determine the magnetization $\vec{M}$, which is proportional to the $\vec{H}$ field:

$$\vec{M} = \chi_M \vec{H} \tag{3.102}$$

---

* In some books, $\vec{B}$ is called the magnetic induction and $\vec{H}$ is called the magnetic field. We will refer to them both as the magnetic field and rely on the symbol or the units to distinguish between them.

Equation 3.102 introduces the dimensionless magnetic susceptibility χ_M. We are ready now to classify all matter on the basis of three types of magnetic behavior: diamagnetic, paramagnetic, and ferromagnetic. Most matter falls into the first category. Diamagnetic substances have negative susceptibilities and are very weakly repelled by a magnetic field. Paramagnetic substances have positive χ_M and are weakly attracted by a magnetic field. In either case, the magnitude of the susceptibility is quite small compared to one. Ferromagnetic substances, for example iron and magnetite (Fe_3O_4), have positive χ_M, larger than that of a paramagnetic material, and are strongly attracted into a magnetic field.

It is the field B that exists inside the material. (Fortunately, the typically small values of magnetic interactions enable us to consider this the same as the local magnetic field.) Using Equations 3.101 and 3.102, this is

$$\vec{B} = \mu_0(\vec{H} + \vec{M}) = \mu_0(1 + \chi_M)\vec{H} \tag{3.103}$$

It still remains to examine the microscopic basis for the behavior of bulk matter. Just as all atoms and molecules are polarizable (α is never zero), all matter has a diamagnetic response to an applied magnetic field. The basis for this response is the tendency of the field to induce microscopic currents; that is, a circulation of electronic charge. These induced currents result in a magnetic field that opposes the direction of the applied field. The magnetic susceptibility due to this effect is proportional to the expectation value $<r^2>$ for the electronic ground state. If the constituent atoms and molecules possess net angular momentum, due to orbital or spin motion, the intrinsic magnetic moments will tend to align with the field and reinforce it. This is somewhat like the alignment of permanent electric dipoles in an electric field. This paramagnetic response is temperature dependent (the Curie law is approximately obeyed: $\chi_M \propto 1/T$) because thermal motion opposes this alignment. If, in addition to net spin angular momentum due to unpaired electrons, there is a tendency for all the spins in a microscopic neighborhood to line up in the same direction (such regions are called domains), then the material is ferromagnetic. The behavior of ferromagnets is beyond the scope of this book, but we are very interested in the concept of spin and angular momentum in general. We turn to this topic in the next section.

3.4.3 MAGNETIC MOMENTS AND INTRINSIC ANGULAR MOMENTA

Just as a current loop has a magnetic dipole moment, so too does an atom or molecule having net electronic angular momentum. There are two types of intrinsic angular momentum due to electrons: orbital and spin. In atoms, where the orbital angular momentum is quantized (i.e., L is a good quantum number), atoms having $L \neq 0$ possess magnetic moments due to orbital angular momentum. Diatomic molecules can also have quantized orbital angular momentum, along the bond axis. Atoms and molecules having unpaired electrons have magnetic moments due to spin. Nuclei can also have intrinsic angular momentum, but the resulting magnetic moments are smaller than those due to electronic angular momentum, due to the greater mass of the nucleus. We will begin our discussion by considering electron spin.

Electron spin is a relativistic effect that permits us to imagine that the electron behaves as if it has intrinsic angular momentum and therefore a magnetic moment. The relationship between the spin magnetic dipole moment and the spin angular momentum vector $\vec{S}$ is

$$\vec{\mu}_{mag} = \frac{-g_e e}{2m_e}\vec{S} = \gamma_e \vec{S} \tag{3.104}$$

The electronic g-factor is $g_e = 2.0023$. The negative sign in Equation 3.104 comes from the charge $-e$ on the electron. Equation 3.104 introduces the gyromagnetic ratio for the electron γ_e, sometimes called the magnetogyric ratio. In general, the gyromagnetic ratio of a particle is the ratio of the magnetic dipole moment to the angular momentum, with units of $T^{-1} s^{-1}$. Alternatively, the units are stated as rad $T^{-1} s^{-1}$ to highlight the relationship of the gyromagnetic ratio to the (angular) Larmor precession frequency ω_0 discussed below. Equation 3.104 should be compared to the classical expression Equation 3.100. Is anything strange? Of course, it is the extra factor of approximately two that enters into the quantum mechanical picture.

A single electron has spin quantum number $s = 1/2$, characterizing the length of the spin angular momentum vector: $|s| = \sqrt{s(s+1)}\hbar = \sqrt{3/2}\hbar$. The quantum number $m_s = \pm 1/2$ determines the component of s along an arbitrary direction in the laboratory frame, taken to be the z direction. The z component of spin angular momentum is $s_z = \pm 1/2\,\hbar$. We can generalize to the case of a many electron atom in a state with total spin S, the vector sum of the individual spin angular momenta, for which the preceding expressions hold by replacing s by S and m_s by M_S. The quantum number M_S ranges from S to $-S$ in integral steps. This leads to the following expression for the magnitude of the spin magnetic moment:

$$\left|\vec{\mu}_{mag}\right| = \frac{g_e e}{2m_e}[S(S+1)]^{1/2}\hbar = g_e \mu_B [S(S+1)]^{1/2} \tag{3.105}$$

We have introduced the convenient constant known as the Bohr magneton:

$$\mu_B = \frac{e\hbar}{2m_e} = 9.274\times10^{-24}\ \text{J/T} \tag{3.106}$$

(The reader is cautioned that the Greek letter μ appears here with several different meanings, distinguished by the subscript.) The magnetic susceptibility due to spin is given by

$$\chi_M(\text{spin}) = \frac{4N}{3Vk_B T}\mu_0\mu_B^2 S(S+1) \tag{3.107}$$

The susceptibility of Equation 3.107 exhibits the previously mentioned Curie law temperature dependence. The total susceptibility should also include the diamagnetic contribution, which we have not treated quantitatively (see [6] for a discussion of this part).

In the presence of atomic orbital angular momentum, the total magnetic moment is given by

$$\vec{\mu}_{mag} = \frac{-e}{2m_e}(\vec{L}+2\vec{S}) \tag{3.108}$$

The most important spectroscopic consequence of this is that magnetic fields affect the energy levels of atoms or molecules having net angular momentum. This comes about through a splitting due to the energy of interaction $W = -\vec{\mu}_{mag}\cdot\vec{B}$ of a magnetic dipole and a magnetic field. We will return to this topic in Chapter 7.

For comparison, we next present analogous relations connecting magnetic dipole moments of nuclei to their intrinsic nuclear angular momenta $\vec{I}$.

$$\vec{\mu}_{mag} = \frac{g_N e}{2m_p}\vec{I} = \gamma_N\vec{I} = \frac{g_N\mu_N}{\hbar}\vec{I} \tag{3.109}$$

This equation introduces the nuclear magneton:

$$\mu_N = \frac{e\hbar}{2m_p} = 5.0505\times10^{-27}\ \text{J/T} \tag{3.110}$$

The nuclear magneton μ_N is smaller than the Bohr magneton μ_B by three orders of magnitude as a result of the larger proton mass m_p compared to electron mass m_e. The gyromagnetic ratio γ_N is defined for nuclei possessing nonzero spin quantum number I. The values of γ_N and I depend on the atomic number as well as the mass number, and both positive and negative values of the former are possible, meaning that the magnetic dipole may be either parallel or antiparallel to the direction of the angular momentum.

For a general type of angular momentum such as nuclear or electron spin, for example, $j = I$ or S, the energy $W = -\vec{\mu}_{mag}\cdot\vec{B}$ translates into the quantum mechanical expression $W = -\gamma\hbar m_j B$. We arrive at this equation easily by aligning the laboratory z direction with that of the magnetic field. This results in removal of the $(2j + 1)$-fold degeneracy of the angular momentum in the presence of a static magnetic field, as in magnetic resonance. An oscillating applied magnetic field can then induce transitions between adjacent states when the frequency of the field matches the energy level separation. Though the quantum mechanical basis for these spectroscopic transitions is not revealed until the next chapter, we wish to consider in the next section a semiclassical picture for magnetic resonance. This physical picture has implications in a variety of contexts within the subject of optical spectroscopy as well as magnetic resonance spectroscopy.

3.4.4 MAGNETIC RESONANCE PHENOMENA

Consider for illustration the case of proton magnetic resonance. The ^{1}H nucleus has $I = 1/2$ and a positive value of the gyromagnetic ratio. The $m_I = \pm 1/2$ spin states are degenerate in the absence of a magnetic field and thus there is no net magnetization M for a collection of these spins. When a static field B_0 is turned on and specified to be in the z direction, the two spin states are split and have energies

$$\varepsilon_1 = -\frac{1}{2}\hbar\gamma B_0 = -\frac{1}{2}\hbar\omega_0$$

$$\varepsilon_2 = \frac{1}{2}\hbar\gamma B_0 = \frac{1}{2}\hbar\omega_0$$

(3.111)

We have introduced the Larmor precession frequency $\omega_0 \equiv \gamma B_0$, a function of the field strength. (The subscript on γ is omitted here, as the formal expressions could just as easily be applied to electron spin.) The spin 1/2 state represents a magnetic dipole aligned with the field and thus lower in energy, ε_1, while the higher energy state ε_2 pertains to the spin $-1/2$ state for which the dipole is antiparallel to the field. Now there is a net magnetization in a sample of N spins that is proportional to the population difference between the two states. The calculation of the equilibrium magnetization M_0 using statistical mechanics is particularly simple in this two-state case, and we obtain

$$M_0 = N\langle \mu_{mag} \rangle = \frac{1}{2}N\gamma\hbar \tanh\left(\frac{\hbar\omega_0}{2k_BT}\right)$$

(3.112)

This magnetization is aligned with the laboratory z direction, while the precessional motion of the spins averages the x and y components to zero. For purposes of later comparisons to optical phenomena involving two states, it is stressed that the magnetization is a function of the population difference.

Next, we imagine an oscillating magnetic field in the x direction:

$$\vec{B}_x = \hat{i}2B_1\cos\omega t$$

(3.113)

The amplitude $2B_1$ is weak compared to B_0, so that the oscillating field is a weak perturbation. The time-varying magnetic field induces spectroscopic transitions between the spin states when its frequency is resonant with the Larmor frequency. (This is the Bohr frequency condition, to be derived in the next Chapter.) Since the spins precess in a clockwise fashion when viewed from the positive z direction, it is helpful to resolve this linear field into its clockwise ($\hat{i}B_1\cos\omega t - \hat{j}B_1\sin\omega t$) and counterclockwise ($\hat{i}B_1\cos\omega t + \hat{j}B_1\sin\omega t$) components. Only the former is able to interact with the spins. Adding the clockwise component of the field in the xy plane to the static field in the z direction, the total field that interacts with the spins is

$$\vec{B} = \hat{i}B_1\cos\omega t - \hat{j}B_1\sin\omega t + \hat{k}B_0$$

(3.114)

The time rate of change of the magnetization follows from the expression for the torque $\vec{T}$, which is the rate of change of the angular momentum:

$$\vec{T} = \vec{\mu}_{mag} \times B = \frac{d\vec{I}}{dt} = \frac{1}{\gamma}\frac{d\vec{\mu}_{mag}}{dt} \tag{3.115}$$

Multiplying the average magnetic moment by the number of spins per unit volume N to get the magnetization, we obtain

$$\frac{d\vec{M}}{dt} = -\gamma\vec{B} \times \vec{M} \tag{3.116}$$

We can use this equation to write the equations of motion for the x, y, and z components of $\vec{M}$, however, this would ignore relaxation of the spins; i.e., interactions that damp out the precession. The Bloch equations consider these effects by introducing two phenomenological relaxation times T_1 and T_2. T_1 is the so-called longitudinal relaxation time of the magnetization M_z in the direction of the field B_0. In future chapters, we will see that in general T_1 is the population relaxation time, consistent with the connection between M_z and the population difference. Hence, an alternative name for T_1 in the context of magnetic resonance is the spin-lattice relaxation time as it represents the decay of the nonequilibrium spin population through energy dissipation to the surrounding "lattice." T_2 is called the transverse or spin-spin relaxation time and pertains to the decay of magnetization M_x and M_y. It represents the tendency of spins to get out of sync with one another as they precess about the field. Later in this book we will encounter a more general view of T_2 as the dephasing time, pertinent to the relaxation of coherent superpositions of basis states that comprise the time-dependent wavefunction. For the present discussion, the inverse times $1/T_1$ and $1/T_2$ are the first order rate constants for relaxation of the magnetization in the longitudinal and transverse directions, respectively. We will add these relaxation terms to the equations of motion, but first we make a transformation to a coordinate system x', y', and z that rotates about the z-axis in a clockwise direction with frequency ω, the frequency of the oscillating field, as depicted in Figure 3.8a. The components of the magnetization in this reference frame are

$$u = M_x \cos\omega t - M_y \sin\omega t$$
$$v = M_x \sin\omega t + M_x \cos\omega t \tag{3.117}$$

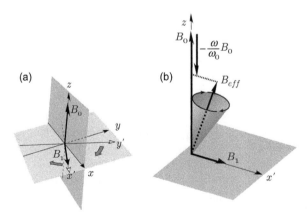

Figure 3.8 (a) Rotating coordinate system for magnetic resonance, showing the static field B_0 and the clockwise-rotating component of the oscillating field B_1. (b) Precession of magnetization about the effective magnetic field B_{eff}.

The component u is the magnetization in the xy plane that is in phase with the clockwise component of B_1, i.e., coincident with the x'-axis, while v is the component which is $90°$ out of phase, aligned with y'. We now arrive at the Bloch equations:

$$\frac{dM_z}{dt} = \gamma B_1 v - \frac{M_z - M_0}{T_1}$$

$$\frac{du}{dt} = (\omega_0 - \omega)v - \frac{u}{T_2} \tag{3.118}$$

$$\frac{dv}{dt} = -(\omega_0 - \omega)v - \gamma B_1 M_z - \frac{v}{T_2}$$

If we consider the cross-product in Equation 3.116 in this new coordinate system, we find the non-relaxing part of the equation of motion for the magnetization in the x' directions is

$$\frac{du}{dt} = \gamma B_{z,eff} v = (\omega_0 - \omega)v \text{ (nonrelaxing part)}$$

$$B_{z,eff} = B_0 \left(1 - \frac{\omega}{\omega_0}\right) \tag{3.119}$$

In the rotating frame, there is now an effective magnetic field in the z direction which depends on the relationship of the oscillating field frequency to the Larmor frequency. As shown in Figure 3.8b, the magnetization now precesses about an effective field direction in the $x'z$ plane; i.e., in the rotating frame the field is tipped away from the z-axis toward the x'-axis. Interesting things happen on resonance, when $\omega = \omega_0$. Now $B_{z,eff} = 0$ and the spins precess about the x'-axis with frequency $\omega_1 = \gamma B_1$. This motion describes a circle in the $y'z$ plane. In the absence of relaxation, this motion represents a continuous flow of energy into and out of the spin system as the magnetization is alternately aligned parallel and antiparallel to B_0. This effect is called transient nutation in analogy to gyroscopic motion in which a slow precession about a perpendicular axis (in this case about B_1, which is weak compared to B_0) is superimposed on a faster precession. This motion is eventually damped out by T_1 and T_2 relaxation.

In the pulsed magnetic resonance experiment known as spin echo, a spin system equilibrated to the static field B_0 is irradiated with a pulse of electromagnetic radiation (typically in the radiofrequency range for nuclear magnetic resonance), polarized perpendicular to the z direction and resonant with ω_0. By adjusting the duration τ and amplitude B_1 of this pulse, the magnetization can be tipped by $90°$. This requires that $\gamma B_1 \tau$ satisfy $\pi/2$ (or $2\pi n + \pi/2$, where n is an integer). This is known as a $\pi/2$ pulse. Immediately following this pulse, the spin vectors are all aligned with the y'-axis as seen in Figure 3.9a. Due to heterogeneities in the

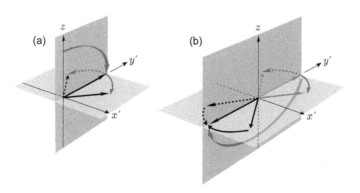

Figure 3.9 (a) A $\pi/2$ pulse tips the magnetization from the z to the y' direction, and spins begin to dephase in the rotating frame. (b) A π pulse tips the magnetization to the $-y'$ direction, and the spins begin to rephase.

sample and the field, the spins begin to fan out in the rotating frame as some precess slower and some faster than γB_1. This is dephasing or T_2 relaxation. After some waiting time t', a second pulse in the x' direction is applied that rotates the spins by 180°, equivalent to reflecting the spin vectors through the $x'z$ plane. Now, the spin vectors that were previously moving away from one another reverse course and begin to merge instead, as depicted in Figure 3.9b. After the same period of time t' over which they were allowed to dephase, the spins have coalesced and a spin echo signal is detected a time $2t'$. The reduced amplitude of this echo as a function of the waiting time t' is determined by the dephasing time T_2.

To understand this process better, consider an analogy where runners take off on a linear track with a distribution of speeds. At the start of the race, the runners are lined up as are the spins immediately following the $\pi/2$ pulse. During the race, i.e., during the waiting time, the runners spread out because they run at different speeds. The π pulse is simulated by a referee who fires a gun at time t' signaling that the runners must reverse course and head back to the starting line. Faster runners get the signal to turn around at a greater distance from the starting line, while slower runners have the advantage of needing to cover a shorter distance to return. If each individual runner had a constant speed throughout the race, they would all return to the starting line at the same time, and the time required to return would be the same as that for the forward course, i.e., there will be a runner echo. However, the runners get "dephased" because they do not all run at constant speeds. If we assume that fluctuations in their speeds result from interactions with their environment (a friend cheers from the stands, a squirrel on the course causes a distraction, etc.), then we have a good analogy to what is called "pure dephasing": the fluctuations in a spectroscopic frequency that result from interactions with the surroundings. As a result of these fluctuations, though most runners return to the start at the same time, some will get there sooner and some later, resulting in a spread of this human pulse such that the amplitude (number of runners who return at t') decreases as waiting time increases. Worse yet, some of the runners have dropped out (T_1, population relaxation), further reducing the amplitude of the pulse. The total dephasing rate $1/T_2$ depends on both population relaxation and the rate of pure dephasing:

$$\frac{1}{T_2} = \frac{1}{2T_1} + \frac{1}{T_2^*} \tag{3.120}$$

Here, $1/T_2^*$ is the pure dephasing rate.

The above ideas will be revisited in later chapters within the context of optical spectroscopy. In the next chapter, we will consider a quantum mechanical density matrix treatment for a two-state system that is analogous to the Bloch equations found here. In later chapters we consider how the rates of population relaxation and dephasing can be determined experimentally with time-resolved spectroscopy. Further discussion of magnetic resonance phenomena can be found in [7].

3.5 SUMMARY

In this chapter, we have given much attention to the electric properties of individual molecules, such as the dipole and quadrupole moments. In the next chapter, we will see how the quantum mechanical matrix elements of the operators for these properties, along with that for the angular momentum, give rise to spectroscopic transitions. In some spectroscopy experiments, such as absorption and emission in the microwave, the activity of a molecule depends on the existence of a permanent dipole moment, that is, a diagonal matrix element of the dipole operator. Spectroscopic transitions in condensed phases depend on bulk properties such as the permittivity, and the frequency-dependent real and imaginary parts of this response function determine the intensity and peak frequency of a transition. With the exception of the discussion of spin, much of the material in this chapter could be treated classically. In the next chapter, we present the quantum mechanical theory that will enable us to interpret spectroscopy experiments.

PROBLEMS

1. Derive Equation 3.7 for the potential φ far from a dipole. Then use $\vec{E} = -\nabla\varphi$ to derive Equation 3.8 for the electric field due to a dipole.

2. Show that the dipole moment of a collection of charges is independent of origin, provided that the net charge is zero. Also, show that the quadrupole moment is independent of origin if the dipole moment is zero.

3. Consider four charges, equal in magnitude. Two positive charges are located at $(x, y) = (1,1)$ and $(-1,-1)$ and two negative charges are at $(1,-1)$ and $(-1,1)$. Calculate the nonzero components of the Cartesian quadrupole tensor, using arbitrary units. Then find a new coordinate system for these charges that diagonalizes the quadrupole tensor.

4. Estimate the polarizability of a ground state hydrogen atom from its size, in MKS and cgs units. Compare to the literature value, $\alpha/4\pi\varepsilon_0 = 0.667$ Å^3.

5. The refractive index of CCl_4 is $n_D = 1.4601$ and the density is 1.594 g/ml. Use the Lorenz–Lorentz equation to estimate the polarizability of CCl_4. Compare to the literature value $\alpha/4\pi\varepsilon_0 = 1.25 \times 10^{-29}$ m^3.

6. The molar polarization P_M of methanol decreases from about 36 to 33 cm^3/mol on increasing the temperature from 298 to 333 K. Use the Debye equation to estimate the polarizability and dipole moment of methanol and compare to literature values.

7. Below what temperature does the Debye equation predict ferroelectric behavior for water? Use $\mu_0 = 1.86$ D and $\alpha/4\pi\varepsilon_0 = 1.5 \times 10^{-24}$ cm^3.

8. The molar absorptivity of the dye rhodamine 6G is about 10^5 L/mol cm at 540 nm. Estimate ε'' for a 10^{-3} M solution of this dye in ethanol.

9. Consider an electron moving parallel to an electric field of 100 V/m and perpendicular to a magnetic field of 1 T. How fast would the electron have to travel for the electric and magnetic forces to be equal?

10. Derive Equation 3.112 for the equilibrium magnetization M_0 and find the limiting expression for $\hbar\omega_0 \ll k_B T$. Use your result to find the Curie law for the magnetic susceptibility χ_M, Equation 3.107. Given that NMR transitions are resonant in the radiofrequency range of the electromagnetic spectrum, justify the use of the high temperature limit.

REFERENCES

1. J. D. Jackson, *Classical Electrodynamics* (Wiley, New York, 1962).
2. H. Fröhlich, *Theory of Dielectrics*, 2nd ed. (Oxford University Press, Oxford, 1958).
3. A. Hinchliffe and R. W. Munn, *Molecular Electromagnetism* (Wiley, New York, 1985).
4. J. G. Kirkwood, The dielectric polarization of polar liquids, *J. Chem. Phys.* 7, 911 (1939).
5. L. Onsager, Electric moments of molecules in liquids, *J. Am. Chem. Soc.* 58, 1486 (1936).
6. P. W. Atkins, *Molecular Quantum Mechanics*, 2nd ed. (Oxford University Press, New York, 1983).
7. J. D. Macomber, *The Dynamics of Spectroscopic Transitions* (Wiley, New York, 1976).

Time-dependent perturbation theory
of spectroscopy

4.1 INTRODUCTION: TIME DEPENDENCE IN QUANTUM MECHANICS

In this chapter we derive the key equation of linear spectroscopy, Fermi's Golden Rule, using time-dependent perturbation theory. We will use as a basis set the stationary states that are the eigenfunctions of the time-independent Schrödinger equation. This leads naturally to the idea that the field–matter interaction induces transitions between states that are solutions to the time-independent Schrödinger equation. The Golden Rule relates the transition rate (and thus the intensity of absorption or emission) to the transition dipole moment and the frequency of radiation. Various formulas that spring from the Golden Rule apply to common experimental situations. We shall use the Golden Rule as a starting point in Chapter 5 to derive the fundamental expressions for the time-dependent view of spectroscopy, which will prove to be of great utility in the description of condensed phase spectra.

We begin by reviewing time-dependence in quantum mechanics in general. There are two ways to view the time evolution of the expectation value of an operator. In the Schrödinger picture, the wavefunctions are considered to evolve in time, and in the Heisenberg picture, the operators evolve in time. We now show how these two pictures are related.

The time-independent Schrödinger equation, $\hat{H}\psi = E\psi$, which results when the Hamiltonian does not depend on time, is a special case of the more general time-dependent equation $\hat{H}\psi = i\hbar(\partial\psi/\partial t)$. The formal solution to the time-dependent Schrödinger equation is

$$\psi_S(t) = e^{-i\hat{H}t/\hbar}\psi_S(0) = e^{-iEt/\hbar}\psi_S(0) \tag{4.1}$$

where the second equality follows if ψ_S is an eigenfunction of $\hat{H}$. The subscript S designates the Schrödinger picture. In the study of time-independent problems, the factor $\exp(-iEt/\hbar)$ is but a phase factor that has no effect on the expectation value of an operator, so it can be dropped. But if we wish to consider how the expectation value of an operator $\hat{A}$ evolves in time, we need to use the more general form of the wavefunction in Equation 4.1. Using bra-ket notation for convenience:

$$\left|\psi_S(t)\right\rangle = e^{-i\hat{H}t/\hbar}\left|\psi_S(0)\right\rangle \tag{4.2}$$

$$\left\langle\psi_S(t)\right| = \left\langle\psi_S(0)\right|e^{i\hat{H}t/\hbar} \tag{4.3}$$

The expectation value at time t is expressed as $\left\langle A_t \right\rangle = \left\langle\psi_S(t)\right|\hat{A}_S\left|\psi_S(t)\right\rangle$. This gives

$$\left\langle A_t \right\rangle = \left\langle\psi_S(0)\right|e^{i\hat{H}t/\hbar}\hat{A}_S e^{-i\hat{H}t/\hbar}\left|\psi_S(0)\right\rangle \tag{4.4}$$

The form of Equation 4.4 suggests that we could just as easily associate the exponential functions of time with the operator as with the wavefunctions. Considering the operator to be time-dependent leads to the definition of the Heisenberg representation of $\hat{A}$:

$$\hat{A}_H \equiv e^{i\hat{H}t/\hbar} \hat{A}_S e^{-i\hat{H}t/\hbar} \tag{4.5}$$

Going back to the previous expression, Equation 4.4, and taking the time derivative, we have

$$\frac{d}{dt}\langle A_t \rangle = \left\langle \psi_S(0) \left| \left(\frac{d}{dt} e^{i\hat{H}t/\hbar} \right) \hat{A}_S e^{-i\hat{H}t/\hbar} \right| \psi_S(0) \right\rangle + \left\langle \psi_S(0) \left| e^{i\hat{H}t/\hbar} \frac{d\hat{A}_S}{dt} e^{-i\hat{H}t/\hbar} \right| \psi_S(0) \right\rangle$$
$$+ \left\langle \psi_S(0) \left| e^{i\hat{H}t/\hbar} \hat{A}_S \left(\frac{d}{dt} e^{-i\hat{H}t/\hbar} \right) \right| \psi_S(0) \right\rangle \tag{4.6}$$

The Hamiltonian $\hat{H}$ commutes with $\exp(\pm i\hat{H}t/\hbar)$, so Equation 4.6 can be rearranged to give

$$\frac{d}{dt}\langle A_t \rangle = \left\langle \psi_S(0) \left| e^{i\hat{H}t/\hbar} \frac{d\hat{A}_S}{dt} e^{-i\hat{H}t/\hbar} \right| \psi_S(0) \right\rangle$$
$$+ \left\langle \psi_S(0) \left| e^{i\hat{H}t/\hbar} [\hat{H}, \hat{A}_S] e^{-i\hat{H}t/\hbar} \right| \psi_S(0) \right\rangle \tag{4.7}$$

Evaluating the above expression at $t = 0$ and using angle brackets to represent expectation values taken with respect to $\psi_S(0)$, we get

$$\frac{d}{dt}\langle A \rangle_{t=0} = \left\langle \frac{d\hat{A}_S}{dt} \right\rangle + \frac{i}{\hbar}\left\langle \left[\hat{H}, \hat{A}_s \right] \right\rangle \tag{4.8}$$

Equation 4.8 is called the Heisenberg equation of motion. In the frequently encountered case that the operator does not depend explicitly on time, it reduces to

$$\frac{d}{dt}\langle A \rangle_{t=0} = \frac{i}{\hbar}\left\langle \left[\hat{H}, \hat{A}_s \right] \right\rangle \tag{4.9}$$

The conclusion drawn from the above is that the time evolution of the expectation value is determined by the commutator of the Hamiltonian with the operator for that physical property. From Equation 4.9, we conclude that operators which commute with the Hamiltonian correspond to constants of motion; that is, physical properties which are constant in time, such as the energy of a stationary state. Operators such as those for position and momentum, which do not commute with the Hamiltonian, correspond to physical properties that evolve in time even though the operators themselves do not depend on time. Equation 4.9 will prove useful in the next chapter, where we will examine the time-domain view of spectroscopy. By taking the time derivative of Equation 4.9 $n - 1$ times, the following more general form is obtained:

$$\frac{d^n}{dt^n}\langle A \rangle_{t=0} = \left(\frac{i}{\hbar} \right)^n \left\langle \left[\hat{H}, \left[\hat{H}, \left[\hat{H}, \dots \left[\hat{H}, \hat{A}_s \right] \dots \right] \right] \right] \right\rangle \tag{4.10}$$

where the commutator is applied n times.

In the next section, we will see how a time-dependent perturbation can cause transitions from one stationary state to another. How can these stationary states persist in the presence of the perturbation? How does one describe the wavefunction for a molecule during a transition; that is, when it is "between states?" Let us see what answers to these questions can be provided by the theory.

4.2 TIME-DEPENDENT PERTURBATION THEORY

4.2.1 FIRST-ORDER SOLUTION TO THE TIME-DEPENDENT SCHRÖDINGER EQUATION

Our goal here is to find an approximate solution to the time-dependent Schrödinger equation, in the case that the Hamiltonian can be expressed as the sum of a zero-order part $\hat{H}_0$ that does not depend on time, and a perturbation $\hat{H}'$ that does. The Schrödinger equation then takes the form

$$\left[\hat{H}_0 + \hat{H}'(t)\right]|\Psi\rangle = i\hbar\frac{\partial}{\partial t}|\Psi\rangle \tag{4.11}$$

We assume that the zero-order eigenfunctions and eigenvalues are known: $\hat{H}_0|n\rangle = E_n|n\rangle$, and we use the zero-order wavefunctions $\psi_n(0) \equiv |n\rangle$ as a basis for expanding the perturbed wavefunction:

$$|\Psi\rangle = \sum_n c_n(t)e^{-iE_n t/\hbar}|n\rangle \tag{4.12}$$

Equation 4.12 presents the perturbed state as a superposition of the stationary states expressed in the Schrödinger representation. The phase factors $\exp(-iE_n t/\hbar)$ and the coefficients $c_n(t)$ convey the time dependence of $|\Psi\rangle$. The coefficient $c_n(t)$ is given by the projection of the total wavefunction onto the nth basis state:

$$c_n(t) = \langle\Psi(t)|n\rangle e^{iE_n t/\hbar} \tag{4.13}$$

Equation 4.13 is presented because of its conceptual value. It is not useful to us in our quest to find the coefficients, because we do not know $|\Psi\rangle$. (If we did, we would not need perturbation theory!) Our next step is to substitute the superposition wavefunction into Equation 4.11:

$$\sum_n c_n(t)e^{-iE_n t/\hbar}(\hat{H}_0 + \hat{H}')|n\rangle = i\hbar\sum_n \frac{\partial}{\partial t}\left\{c_n(t)e^{-iE_n t/\hbar}|n\rangle\right\} \tag{4.14}$$

The sum runs over an infinite number of eigenstates indexed by the letter n. Let us take one of these states, call it state m, multiply both sides of Equation 4.14 by the complex conjugate of the wavefunction for state m, and then integrate over all space. The result is

$$\sum_n c_n(t)e^{-iE_n t/\hbar}\left\{\langle m|\hat{H}_0|n\rangle + \langle m|\hat{H}'|n\rangle\right\} = i\hbar\sum_n \frac{\partial}{\partial t}\left\{c_n(t)e^{-iE_n t/\hbar}\langle m|n\rangle\right\} \tag{4.15}$$

The eigenfunctions are orthonormal: $\langle n|m\rangle = \delta_{nm}$; so in the infinite sum on the right-hand side, only the term $n = m$ survives. And since $\langle n|\hat{H}_0|m\rangle = E_m\delta_{nm}$ the first sum on the left-hand side is similarly reduced to one term. Thus

$$c_m(t)e^{-iE_m t/\hbar}E_m + \sum_n c_n(t)e^{-iE_n t/\hbar}\langle m|\hat{H}'|n\rangle$$

$$= i\hbar\frac{\partial c_m}{\partial t}e^{-iE_m t/\hbar} + i\hbar\left(\frac{-iE_m}{\hbar}\right)e^{-iE_m t/\hbar}c_m(t) \tag{4.16}$$

This can be rearranged to give

$$\frac{dc_m(t)}{dt} = \frac{-i}{\hbar} \sum_n c_n(t) e^{-i\omega_{nm}t} V_{mn}(t) \tag{4.17}$$

where $V_{nm}(t) \equiv \langle m | \hat{H}'(t) | n \rangle$ and $\omega_{nm} \equiv (E_n - E_m)/\hbar = -\omega_{mn}$. Equation 4.17 is exact, but disappointing. It says that the time dependence of any one coefficient is a function of all the other time-dependent coefficients. A way around this obstacle is provided by first-order perturbation theory, in which we replace all the coefficients on the right-hand side of Equation 4.17 by their values at time zero. This is obviously a weak perturbation limit, as it assumes that $\hat{H}'$ is small enough that the coefficients never depart greatly from their unperturbed values. We imagine the perturbation to be turned on at $t = 0$, and we probe the state of the system at time t later. We assume that before the start of the perturbation the system was in a definite initial state, indexed by i. So we can replace the coefficients on the right-hand side of Equation 4.17 by $c_n(t) \approx c_n(0) = \delta_{ni}$, eliminating all but one term in the summation. We will use the index f (for final state, where $f \neq i$) for the state whose coefficient we wish to calculate. The resulting differential equation is then integrated to give

$$c_f^{(1)}(t) = \frac{-i}{\hbar} \int_0^t dt' e^{i\omega_{fi}t'} V_{fi}(t') \tag{4.18}$$

The probability that the system is in state f at time t, given that it was in state i at time zero, is given by the square of the amplitude of the coefficient:

$$P_f(t) = \frac{1}{\hbar^2} \left| \int_0^t dt' e^{i\omega_{fi}t'} V_{fi}(t') \right|^2 \tag{4.19}$$

4.2.2 Perturbation due to electromagnetic radiation: Momentum versus dipole operator

Equation 4.18 applies to any problem involving a perturbation that varies in time. We are now ready to consider the case where $V_{fi}(t)$ is due to a time-varying electromagnetic field. As shown in Chapter 2, the time-dependent operator that we need is

$$\hat{H}'(t) = \frac{ie\hbar}{m} \vec{A} \cdot \nabla = \frac{-e}{m} \vec{A} \cdot \vec{p} \tag{4.20}$$

We write the vector potential as follows:

$$\vec{A} = \frac{\vec{A}_0}{2} \left[e^{i(\vec{k} \cdot \vec{r} - \omega t)} + e^{-i(\vec{k} \cdot \vec{r} - \omega t)} \right] = \text{Re} \vec{A}_0 e^{i(\vec{k} \cdot \vec{r} - \omega t)} \tag{4.21}$$

As discussed in Chapter 3, the size of a molecule is typically much smaller than the wavelength of light used in spectroscopy, so we can often neglect the gradient of the field and in fact its spatial dependence altogether. This results in the primary mechanism for the interaction of light and matter being that due to the operator $\hat{H}'(t) = -\vec{\mu} \cdot \vec{E}(t)$, corresponding to energy of a dipole in a time-varying, but spatially constant, electric field. We now show how this dipole operator follows from the momentum-dependent perturbation operator given in Equation 4.20, in the long wavelength limit. As $\vec{k} \cdot \vec{r} \approx 2\pi r/\lambda \ll 1$ over the typical dimensions of a molecule, the exponential function in Equation 4.21 can be expanded about $\vec{k} \cdot \vec{r} = 0$,

$$e^{i\vec{k} \cdot \vec{r}} \approx 1 + (i\vec{k} \cdot \vec{r}) + \frac{1}{2}(i\vec{k} \cdot \vec{r})^2 + \cdots \tag{4.22}$$

We shall see that the first term in this series leads to electric dipole-allowed transitions, and the second to magnetic dipole and electric quadrupole-allowed transitions. The electric dipole term is sufficient for most of the spectroscopy experiments discussed in this book. However, when the extent of the system undergoing a spectroscopic transition is not small compared to the wavelength of light, as is the case for delocalized excitations in metals, semiconductors, and conjugated polymers, the k-dependence of the field cannot be neglected.

Taking the first term of Equation 4.22 gives the following form for the matrix element of Equation 4.20:

$$V_{fi}(t) = \frac{ie\hbar}{m} \text{Re}\left(e^{-i\omega t}\right)\langle f|\vec{A}_0 \cdot \nabla|i\rangle \ (k \to 0 \text{ limit}) \tag{4.23}$$

Suppose that the vector potential points in the z direction. The relevant matrix element is then

$$V_{fi}(t) = \frac{ie\hbar A_0}{m} \text{Re}\left(e^{-i\omega t}\right)\left\langle f\left|\frac{\partial}{\partial z}\right|i\right\rangle \tag{4.24}$$

where we could also write

$$i\hbar\left\langle f\left|\frac{\partial}{\partial z}\right|i\right\rangle = -\langle f|p_z|i\rangle \tag{4.25}$$

With the help of the commutator relation $[z, p_z] = i\hbar$, one can prove the following (Problem 2):

$$p_z = \frac{im}{\hbar}\left(\hat{H}_0 z - z\hat{H}_0\right) = \frac{im}{\hbar}\left[\hat{H}_0, z\right] \tag{4.26}$$

The matrix element we need is thus

$$\begin{aligned}
\langle f|p_z|i\rangle &= \frac{im}{\hbar}\left\{\langle f|\hat{H}_0 z|i\rangle - \langle f|z\hat{H}_0|i\rangle\right\} \\
&= \frac{im(E_f - E_i)}{\hbar}\langle f|z|i\rangle + im\omega_{fi}\langle f|z|i\rangle
\end{aligned} \tag{4.27}$$

Since the z component of the dipole moment operator is $\mu_z = -ez$, the matrix element can be written

$$V_{fi}(t) = \frac{-e}{m}\text{Re}\left(e^{-i\omega t}\right)\langle f|\vec{A}_0 \cdot \vec{p}|i\rangle = i\omega_{fi}\text{Re}\left(e^{-i\omega t}\right)\langle f|\vec{A}_0 \cdot \vec{\mu}|i\rangle \tag{4.28}$$

Switching from the vector potential to the electric field, with the relationship $E_0 = -\omega A_0$, and putting $\text{Re}[\exp(-i\omega t)] = 1/2[\exp(-i\omega t) + \exp(i\omega t)]$, we get

$$V_{fi}(t) = \frac{-i\omega_{fi}}{2\omega}(\vec{\mu}_{fi} \cdot \vec{E}_0)(e^{i\omega t} + e^{-i\omega t}) \text{ (E1 term)} \tag{4.29}$$

where we have abbreviated the matrix element: $\langle f|\vec{\mu} \cdot \vec{E}_0|i\rangle = \vec{\mu}_{fi} \cdot \vec{E}_0$. Anticipating that the frequency ω of the light will have to match that of the transition ω_{fi}, Equation 4.29 clearly gives the same result that we would have obtained had we started with a perturbation operator of the form $-\vec{\mu} \cdot \vec{E}(t)$. Transitions which are made possible by a nonzero value of the *transition dipole*, $\mu_{fi} \equiv \langle f|\mu|i\rangle$, are referred to as *electric dipole-allowed transitions*, or E1 transitions. Though we have approached this derivation with a single charged particle in mind, we can readily generalize the result by allowing the dipole moment operator to consist of a sum over all the charges, electrons and nuclei, as in Equation 3.21.

Next, consider the second term in the expansion of $\exp(i\vec{k}\cdot\vec{r})$. This term gives rise to a contribution to the perturbation operator of the form $(e\hbar/2m)(e^{i\omega t}-e^{-i\omega t})(\vec{k}\cdot\vec{r})(\vec{A}_0\cdot\nabla)$. Again, taking the light to propagate in the y direction with electric field (and vector potential) oscillating in the z direction, we have to find the matrix element $\langle f|y\partial/\partial z|i\rangle$. This term can be manipulated as follows:

$$\left\langle f\left|y\frac{\partial}{\partial z}\right|i\right\rangle = \frac{1}{2}\left\{\left\langle f\left|y\frac{\partial}{\partial z}-z\frac{\partial}{\partial y}\right|i\right\rangle + \left\langle f\left|y\frac{\partial}{\partial z}+z\frac{\partial}{\partial y}\right|i\right\rangle\right\}$$

$$= \frac{i}{2\hbar}\langle f|\hat{L}_x|i\rangle - \frac{m\omega_{fi}}{\hbar}\langle f|zy|i\rangle \tag{4.30}$$

The first term on the righthand side of Equation 4.30 follows from the definition of the angular momentum operator $\hat{L}_x$. Since the magnetic dipole moment is proportional to the orbital angular momentum, this term gives magnetic dipole (M1) allowed transitions. The second term, obtained with the help of the same commutator relationship used above to get the E1 term, gives electric quadrupole (E2) allowed transitions. (As you will show in one of the homework problems, the resolution of the identity can be used to prove that $\langle f|y\partial/\partial z|i\rangle = -m\omega_{fi}/\hbar\langle f|yz|i\rangle$.) The relevant matrix elements are thus:

$$V_{fi}(t) = \frac{ieE_0}{2mc}(e^{-i\omega t}-e^{i\omega t})\langle f|\hat{L}_z|i\rangle \qquad \text{M1}$$

$$V_{fi}(t) = \frac{eE_0\omega_{fi}}{2c}(e^{-i\omega t}-e^{i\omega t})\langle f|yz|i\rangle \qquad \text{E2} \tag{4.31}$$

Recalling that the ratio of the size of the atom to the smallest wavelength capable of causing transitions between bound states is on the order of $1/137$, the intensities of E2- and M1-allowed transitions are only about 10^{-4} as strong as those of E1 allowed transitions. Henceforth, we will emphasize E1-allowed transitions.

Using the matrix element given in Equation 4.29 in Equation 4.18 leads to two integrals of the type

$$\int_0^t e^{i(\omega_{fi}\pm\omega)t'}dt' = \frac{e^{i(\omega_{fi}\pm\omega)t}-1}{i(\omega_{fi}\pm\omega)} \tag{4.32}$$

The coefficient of the state f at time t is thus

$$c_f^{(1)}(t) = \frac{i\omega_{fi}}{2\hbar\omega}\left(\vec{\mu}_{fi}\cdot\vec{E}_0\right)\left[\frac{e^{i(\omega+\omega_{fi})t}-1}{\omega+\omega_{fi}}-\frac{e^{-i(\omega-\omega_{fi})t}-1}{\omega-\omega_{fi}}\right] \tag{4.33}$$

Equation 4.33 reveals that the amplitude of the final state is expected to be large whenever $\omega \approx -\omega_{fi}$ or $\omega \approx \omega_{fi}$. The frequency of the light ω is considered positive, so whether the first or second resonance condition can be met depends on whether the final state is higher or lower in energy than the initial state. If the final state is below the initial state, then ω_{fi} is negative, and when the first condition is met, emission of light is possible. Conversely, a positive value of ω_{fi} corresponds to a transition to a higher-energy state, and when the second condition is met, absorption of light can take place.

It is important to recognize that the emission we are talking about here is the exact opposite of absorption; thus it is stimulated emission and not spontaneous emission, which falls out of the treatment here. Stimulated emission is exemplified by the output of lasers. As the name implies, this emission is stimulated by incident photons, and the properties of the incident photons (wavelength, phase, polarization, and propagation direction) are imparted to the emitted photons. Spontaneous emission, exemplified by ordinary luminescence, cannot be accounted for using a classical view of the radiation. Spontaneous emission proceeds on its own, once the excited state is prepared, and it is therefore independent of the incident radiation. We will see in

Section 4.5 that a quantized radiation Hamiltonian gives rise to spontaneous as well as stimulated emission. Stimulated and spontaneous emission are discussed further in Chapter 6.

Suppose that $\omega \approx \omega_{fi}$. We can neglect the first term and square the second to get the probability of the $i \to f$ transition.

$$P_f(t) = \frac{\omega_{fi}^2 \left| \left(\vec{\mu}_{if} \cdot \vec{E}_0 \right) \right|^2}{4\hbar^2 \omega^2} \frac{\sin^2\left(\frac{\Delta\omega t}{2} \right)}{\left(\frac{\Delta\omega}{2} \right)^2} \tag{4.34}$$

where $\Delta\omega \equiv \omega - \omega_{fi}$ is the difference between the frequency of light and that of the transition. The properties of the function

$$f(t,\Delta\omega) \equiv \frac{\sin^2\left(\frac{\Delta\omega t}{2} \right)}{\left(\frac{\Delta\omega}{2} \right)^2} \tag{4.35}$$

are of great interest here. As shown in Figure 4.1, where $f(t,\Delta\omega)/t^2$ is represented at an instant in time t, the function reaches its maximum value of one at $\Delta\omega = 0$. The full width at half-maximum is $2\pi/t$. Using the integral

$$\int_{-\infty}^{\infty} \frac{\sin^2 x}{x^2} \, dx = \pi \tag{4.36}$$

The area under $f(t,\Delta\omega)/t^2$ is found to be $2\pi/t$. Thus $f(t,\Delta\omega)$ is an increasingly sharply peaked function as time increases. In the limit that $t \to \infty$, $f(t,\Delta\omega) = 2\pi t \delta(\Delta\omega)$. We can use the following result to convert the delta function when the argument is changed:

$$\delta(kx) = \frac{1}{|k|} \delta(x) \tag{4.37}$$

We now consider some physically meaningful limits of Equation 4.34.

4.2.3 FERMI'S GOLDEN RULE AND THE TIME–ENERGY UNCERTAINTY PRINCIPLE

4.2.3.1 CASE 1. THE LIMIT $t \to \infty$

In the limit that the perturbation persists for a time that is long compared to the inverse of the frequency mismatch $\Delta\omega$, we can take $f(t,\Delta\omega) \Rightarrow 2\pi t \delta(\omega - \omega_{fi})$. This gives a transition probability that is linear in time:

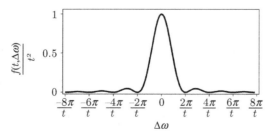

Figure 4.1 Graph of $f(t,\Delta\omega)/t^2$ at time t, versus $\Delta\omega$.

$$P_f(t) = \frac{2\pi \left| \left(\vec{\mu}_{if} \cdot \vec{E}_0 \right) \right|^2 t}{4\hbar^2} \delta \left(\omega - \omega_{fi} \right) \tag{4.38}$$

The significance of this result is that the transition rate w_{if}, which is the transition probability per unit time, is independent of time:

$$w_{if} = \frac{P_f(t)}{t} = \frac{\left| \left(\vec{\mu}_{if} \cdot \vec{E}_0 \right) \right|^2}{4\hbar^2} \left[\delta \left(v - v_{fi} \right) + \delta \left(v + v_{fi} \right) \right] \tag{4.39}$$

The second delta function in the above expression results from adding in the contribution of stimulated emission. The substitution $\delta(v - v_{fi}) = 2\pi\delta(\omega - \omega_{fi})$ has been made. (See Equation 4.37.) Equation 4.39 is sometimes called Fermi's Golden Rule of spectroscopy. It is of central importance to relating experimentally observable intensities to quantum mechanical quantities such as the transition dipole. In Chapter 6 we will use the Golden Rule as a starting point for showing how the integrated molar absorptivity depends on $|\mu_{if}|^2$. Equation 4.39 says that the longtime limit results in the familiar Bohr frequency condition: $\hbar\omega = E_f - E_i$; that is, the energy of the photon must match that of the energy difference of the two states. Since w_{if} is independent of time, we conclude that the transition rate in this limit is constant. This is the behavior that we expect with light sources of ordinary intensity. For example, the amount of visible light absorbed by a dye, which is proportional to the transition rate, remains constant in time upon illumination with a source of given intensity.

4.2.3.2 CASE 2. EXACT RESONANCE, MONOCHROMATIC RADIATION

In the limit that $\Delta\omega \to 0$, then $\omega = \omega_{fi}$ and we can take $f(t, \Delta\omega) \Rightarrow t^2$. This gives a transition probability that increases as t^2, and a transition rate that is a linear function of time:

$$w_{if}(t) = \frac{P_f(t)}{t} = \frac{\left| \vec{\mu}_{if} \cdot \vec{E}_0 \right|^2 t}{4\hbar^2} \tag{4.40}$$

Thus monochromatic radiation is capable of pumping the state f, since the transition rate evidently increases with time. We must be careful, however, not to overstep the limits of first-order perturbation theory, which requires that $|\vec{\mu}_{if} \cdot \vec{E}_0| t \ll \hbar$. Equation 4.40 is valid for short times and weak fields. We next ask whether sufficiently monochromatic sources exist for which the short-time condition, and thus Equation 4.40 can be fulfilled. The short-time condition requires the use of a pulsed rather than a steady state source of radiation. The uncertainty in the frequency of the light, Δv, cannot be any smaller than about $1/t$, where t is the duration of the light pulse. But monochromatic radiation, in view of the behavior of $f(t, \Delta\omega)$, implies that $\Delta\omega = 2\pi\Delta v$ is small compared to the spread of $f(t, \Delta\omega)$, which is about $2\pi/t$. Thus we cannot simultaneously fulfill the conditions of monochromaticity and short time. The correct treatment of optical pumping in the presence of intense resonant radiation, for example in the case of a two-level system, leads to state populations that oscillate in time with a frequency given by $|\vec{\mu}_{12} \cdot \vec{E}_0|/\hbar$, known as the Rabi frequency, Ω. The short-time limit of the transition rate obtained using higher-order perturbation theory is the same as that given in Equation 4.40.

4.2.3.3 CASE 3. INTERMEDIATE TIMES AND THE TIME–ENERGY UNCERTAINTY PRINCIPLE

For finite durations of the light pulse that excites absorption or emission, we must concede that the finite width of $f(t, \Delta\omega)$ permits transitions even when the photon frequency does not exactly match the transition frequency. Since frequencies within about $\pm\pi/t$ of ω_{fi} account for a considerable fraction of the probability, the associated spread of frequencies $\delta\omega$ is on the order of $2\pi/\delta t$, where δt is the duration of the radiation–matter interaction. The time duration of the interaction may be limited by the duration of the light pulse or by the lifetime of the states involved in the transition. Hence, the uncertainty in the energy obeys

$$\delta E = \hbar\delta\omega = \hbar \left(2\pi / \delta t \right) \tag{4.41}$$

which gives $\delta E \delta t \approx h$ or $\delta \nu \delta t \approx 1$. Since the high-frequency wings of $f(t, \Delta \omega)$ allow for transitions to take place at even higher frequency shifts than $\pm \pi/t$, albeit with small probability, Equation 4.41 represents the minimum energy uncertainty. Thus the time-energy uncertainty principle is $\delta E \delta t \geq h$ or $\delta \nu \delta t \geq 1$. The second statement agrees with common sense expectations: we cannot specify the frequency of light if the pulse duration is less than one period. As will be discussed in Chapter 14, when we take up the subject of time-resolved spectroscopy, the Fourier transform relationship between the frequency distribution of a light source and its time duration translates into an inverse correlation between the widths of the pulse in the time and frequency domains. Shorter pulses result in broader frequency distributions and longer-lived pulses have a more narrow spread of frequencies. But we can also look at the problem from the point of view of the molecule, where the finite lifetimes δt of the states themselves limit the duration of the interaction, resulting in a spread in frequencies inversely proportional to the lifetime. This lifetime broadening represents the minimum linewidth for a spectroscopic transition. Other contributions that cause the observed linewidth to exceed the inverse lifetime will be discussed throughout this book.

4.2.3.4 CASE 4. ACCOUNTING FOR THE INTENSITY DISTRIBUTION OF THE SOURCE

Finally, we consider the situation of interest in many practical applications of the Golden Rule. We have a source that spans some frequency range that encompasses the transition frequency, with some energy density $\rho(\nu)$, which is particular to the experimental arrangement. By analogy to the expression for the energy density of monochromatic radiation (see Equation 2.27), we can replace the square of the amplitude of the electric field in Equation 4.39 by making the correspondence, $u = 1/2\,\varepsilon_0 E_0^2 \Leftrightarrow \int \rho(\nu) d\nu$ and integrating over all frequencies. Now, in general the energy density of the source $\rho(\nu)$ is a slowly varying function of frequency compared to $f(t, \Delta \nu)$, so the energy density can be evaluated at the transition frequency and left out of the integral over frequency. The result is

$$w_{if}(t) = \frac{P_f(t)}{t} = \frac{\left| \vec{\mu}_{if} \cdot \hat{e} \right|^2 \rho(\nu_{fi})}{2\varepsilon_0 \hbar^2} \tag{4.42}$$

where $\hat{e}$ is a unit vector in the direction of the electric field. The transition dipole moment that the experiment sees is actually projected onto one of the laboratory-fixed directions determined by the polarization of the light, say the x direction. In a randomly oriented sample, such as a gas or liquid, the x, y, and z components of the transition moment squared must be equal, since there is no preference for any one lab direction. We can choose, say, the x component and put $|\mu_{if}|^2 = 3(\mu_{if})_x^2$ to get

$$w_{if}(t) = \frac{\left| \mu_{if} \right|^2 \rho(\nu_{fi})}{6\varepsilon_0 \hbar^2} \tag{4.43}$$

The transition rate is proportional to the energy density $\rho(\nu)$ of the incident field evaluated at the transition frequency.

4.3 RATE EXPRESSION FOR EMISSION

4.3.1 PHOTON DENSITY OF STATES

In this section we consider how the transition rate depends on the photon density of states. The expression to be obtained is particularly useful when considering the experimental parameters on which the rate of emission depends. (We will reconsider the absorption rate in Chapter 6 when we relate the Golden Rule rate expression to the Beer's law molar absorptivity.) We need to know the density of states for photons of a particular energy or frequency. Recall from Equation 2.69 how the components of the propagation vector are quantized. This is a consequence of the boundary conditions for a cubic box of length L, which require that

$\exp[ik_x(x + L)] = \exp[ik_x x]$, and similarly for the y and z components. As a result, the number of photon states dN in an infinitesimal volume in k-space is

$$dN = \left(\frac{L}{2\pi}\right)^3 dk_x\, dk_y\, dk_z = \frac{V}{(2\pi)^3} k^2 dk \sin\theta\, d\theta\, d\varphi \tag{4.44}$$

Here, we prefer to write the angular part in terms of the solid angle $d\Omega$, where $d\Omega = \sin\theta\, d\theta\, d\varphi$, and substitute $\omega = kc$ to get

$$dN = \frac{V}{(2\pi c)^3} \omega^2 d\omega\, d\Omega \tag{4.45}$$

The number density of states per unit energy, dN/dE, is

$$dN = \frac{dN}{dE} dE = \frac{V}{(2\pi c)^3} \frac{\omega^2}{\hbar} d\Omega\, dE \tag{4.46}$$

So dN/dE is given by

$$\frac{dN}{dE} = \frac{V}{(2\pi c)^3} \frac{\omega^2}{\hbar} d\Omega = \frac{V}{c^3} \frac{v^2}{h} d\Omega \tag{4.47}$$

4.3.2 FERMI'S GOLDEN RULE FOR STIMULATED AND SPONTANEOUS EMISSION

Now let us reexamine the Golden Rule rate expression given in Equation 4.39. Taking the case where the final energy is lower than the initial (emission), we can neglect the first delta function and write

$$w_{if} = \frac{2\left|\left(\vec{\mu}_{if} \cdot \vec{E}_0\right)\right| \pi^2}{4\hbar} \delta\left(E + E_{fi}\right) \tag{4.48}$$

We have recast the delta function in terms of energy rather than frequency by using the relationship $\delta(\omega + \omega_{fi}) = \hbar\delta(E + E_{fi})$. We now make two changes to Equation 4.48: We replace the square of the amplitude of the electric field by a term proportional to the number density of photons, and we replace the delta function by the number of photon states per unit energy considered above. To accomplish the first change, recall the classical expression for the energy density u discussed in Chapter 2 (Equation 2.27), proportional to the square of the electric field. Quantum mechanically, the energy density is the number of photons per unit volume times the energy per photon. Hence

$$u = \frac{1}{2} \varepsilon_0 E_0^2 \Leftrightarrow \frac{N}{V} \hbar\omega \tag{4.49}$$

So we can make the substitution

$$E_0^2 = \frac{2N\hbar\omega}{\varepsilon_0 V} \tag{4.50}$$

in Equation 4.48, keeping track of the direction of the electric field by projecting the transition dipole onto the direction of polarization:

$$w_{if} = \frac{2\pi\left[\vec{\mu}_{if}\cdot\hat{e}\right]^2}{4\hbar} \frac{2N\hbar\omega}{\varepsilon_0 V}\delta\left(E+E_{fi}\right) \tag{4.51}$$

Next, we consider the delta function to be better expressed as the number of photon states having $E = E_{if}$ per unit energy:

$$\delta(E+E_{fi}) = \delta(E-E_{if}) \Rightarrow \left(\frac{dN}{dE}\right)\bigg|_{\omega=\omega_{if}} \tag{4.52}$$

Thus, putting Equation 4.46 in Equation 4.51,

$$dw_{if} = \frac{2\pi\left[\vec{\mu}_{if}\cdot\hat{e}\right]^2}{4\hbar} \frac{2N\hbar\omega}{\varepsilon_0 V} \frac{V}{(2\pi c)^3} \frac{\omega^2}{\hbar}d\Omega \tag{4.53}$$

where we have admitted that the rate is really a differential rate per unit solid angle. Simplifying and rearranging gives

$$\frac{dw_{if}}{d\Omega} = \frac{N\omega_{if}^3}{2\hbar\varepsilon_0} \frac{1}{(2\pi)^2 c^3}\left|\vec{\mu}_{if}\cdot\hat{e}\right|^2 = \frac{N\nu_{if}^3}{\hbar\varepsilon_0} \frac{\pi}{c^3}\left|\vec{\mu}_{if}\cdot\hat{e}\right|^2 \tag{4.54}$$

We can also recast this equation, as we did previously, for randomly oriented systems:

$$\frac{dw_{if}}{d\Omega} = \frac{N\nu_{if}^3}{3\hbar\varepsilon_0} \frac{\pi}{c^3}\left|\mu_{if}\right|^2 \tag{4.55}$$

The importance of Equation 4.55 is that the emission rate is proportional to the cube of the frequency, and to the number of photons N that stimulate the emission event. This expression does not allow for spontaneous emission of light. We shall see in Section 4.5 that a quantum mechanical treatment of the radiation–matter interaction leads to a slight modification of Equation 4.55: the number of photons N is replaced by $N+1$. The consequence of the modification is that emission can take place in the absence of photons, as it does in the case of spontaneous emission. Anticipating the quantum mechanical result, we can rewrite Equation 4.55 as

$$\frac{dw_{if}}{d\Omega} = \frac{(N+1)\nu_{if}^3}{3\hbar\varepsilon_0} \frac{\pi}{c^3}\left|\mu_{if}\right|^2 \tag{4.56}$$

The rate of absorption, on the other hand, can only depend on the number of photons N, not $N+1$, because there is no such thing as "spontaneous absorption."

4.4 PERTURBATION THEORY CALCULATION OF POLARIZABILITY

4.4.1 DERIVATION OF THE KRAMERS–HEISENBERG–DIRAC EQUATION

The transition rates calculated in the previous sections apply to one-photon transitions: absorption and stimulated emission between two states coupled by a transition dipole moment. (Equation 4.56 covers spontaneous emission as well.) Light scattering, on the other hand, requires two photons: an incident and a scattered

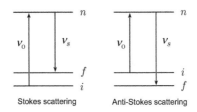

Figure 4.2 State diagram for light scattering.

photon. We view the transition from the initial to final state as taking place by way of one or more intermediate states. Figure 4.2 represents the conceptual picture associated with Raman scattering involving two vibrational or rotational states i and f. The upward and downward arrows represent the incident and scattered photon energies, respectively, but are not to be interpreted as distinct absorption and emission processes. In the case of Stokes scattering, the initial state is lower in energy than the final state, and the scattered frequency v_s is lower than the incident frequency v_0. Anti-Stokes scattering, on the other hand, results when the initial state is higher in energy than the final, and consequently v_s is greater than v_0.

The picture shown in Figure 4.2 correlates well with the quantum mechanical representation of the transition polarizability, given by the Kramers–Heisenberg–Dirac (KHD) equation:

$$(\alpha_{\rho\sigma})_{if} = \frac{1}{\hbar} \sum_n \left[\frac{\langle i | \mu_\rho | n \rangle \langle n | \mu_\sigma | f \rangle}{\omega_{ni} - \omega} + \frac{\langle i | \mu_\sigma | n \rangle \langle n | \mu_\rho | f \rangle}{\omega_{nf} + \omega} \right] \tag{4.57}$$

In Equation 4.57, $(\alpha_{\rho\sigma})_{if}$ is a component of the transition polarizability tensor connecting states i and f, where ρ and σ are Cartesian directions. The intensity of the transition $i \to f$ is proportional to the square of the amplitude of $(\alpha_{\rho\sigma})_{if}$. In this section, we derive Equation 4.57 using the first-order perturbation theory approach presented previously. We calculate the transition moment $\langle \psi_i(t) | \hat{\mu} | \psi_f(t) \rangle$ to first order using time-dependent wavefunctions that are correct to first order in the radiation–matter perturbation. We expect to get a result of the form

$$\langle \psi_i(t) | \hat{\mu} | \psi_f(t) \rangle = e^{i\omega_{if}t} \left[\mu_{if}(\text{perm}) + \mu_{if}(\text{ind}) \right] \tag{4.58}$$

where $\mu_{if}(\text{perm})$ is the previously discussed transition dipole moment evaluated with the zero-order states: $\mu_{if}(\text{perm}) \equiv \langle i | \mu | f \rangle$. We call it permanent because it persists in the absence of an applied field, while the induced term, which depends on the transition polarizability, is directly proportional to the amplitude of the field. The factor $\exp(i\omega_{if}t)$ comes from the phase factors attached to the Schrödinger representation wavefunctions and is not particularly interesting except to anticipate its appearance as we go through the derivation. The polarizability operator corresponds to a physical property and must be Hermitian. Therefore

$$\alpha_{if} = \alpha_{fi}^* \tag{4.59}$$

Similarly, the dipole operator is Hermitian, so we expect

$$\mu_{if} = \mu_{fi}^* \tag{4.60}$$

The induced transition moment will satisfy Equation 4.60 if it is of the form

$$\mu_{if}(\text{ind}) = \frac{1}{2} \left[\alpha_{if} e^{i\omega t} + \alpha_{fi}^* e^{-i\omega t} \right] \cdot E_0 \tag{4.61}$$

We are neglecting vector notation here for simplicity. Let us now write the first-order wavefunctions that we need to evaluate Equation 4.58.

$$\left|\psi_f(t)\right\rangle = \sum_n c_n^f(t)e^{-iE_nt/\hbar}\left|n\right\rangle \tag{4.62}$$

The superscript on the coefficient reminds us that state f is the unperturbed state (the state at $t = 0$). A similar expression can be written for the perturbed state i. The first-order coefficients are, for $j = i$ or f,

$$c_n^j(t) = \delta_{nj} - \frac{i}{\hbar}\int_0^t dt' e^{i\omega_{nj}t'}V_{nj}(t') \tag{4.63}$$

The delta function in this expression allows for the unperturbed wavefunction to be the lead term in the case that $n = j$. (Compare to Equation 4.18.) The matrix element of the perturbation will be taken to be

$$V_{nj}(t) = \frac{-1}{2}\langle n|\mu\cdot E_0|j\rangle\left(e^{i\omega t}+e^{-i\omega t}\right) \equiv -V_{nj}^0\left(e^{i\omega t}+e^{-i\omega t}\right) \tag{4.64}$$

We simply factored out the amplitude of the perturbation, given by

$$V_{nj}^0 = \frac{1}{2}(\mu_{nj}\cdot E_0) \tag{4.65}$$

As was done previously, the coefficients are found by direct integration:

$$\begin{aligned}
c_n^f(t) &= \delta_{nf} + \frac{iV_{nf}^0}{\hbar}\int_0^t dt' e^{i\omega_{nf}t'}\left(e^{i\omega t'}+e^{-i\omega t'}\right)\\
&= \delta_{nf} + \frac{V_{nf}^0}{\hbar}\left[\frac{e^{i(\omega+\omega_{nf})t}-1}{\omega+\omega_{nf}}-\frac{e^{-i(\omega-\omega_{nf})t}-1}{\omega-\omega_{nf}}\right]
\end{aligned} \tag{4.66}$$

and similarly for the coefficient $c_n^i(t)$. The perturbed wavefunction is thus

$$\left|\psi_f(t)\right\rangle = e^{-iE_ft/\hbar}\left|f\right\rangle + \sum_{n\neq f}\frac{V_{nf}^0}{\hbar}\left[\frac{e^{i(\omega+\omega_{nf})t}-1}{(\omega+\omega_{nf})}-\frac{e^{-i(\omega-\omega_{nf})t}-1}{(\omega-\omega_{nf})}\right]e^{-iE_nt/\hbar}\left|n\right\rangle \tag{4.67}$$

We want to get the phase factor $\exp(-iE_ft/\hbar)$ for the wavefunction taken care of, so we factor it out of the expression to write

$$\left|\psi_f(t)\right\rangle = e^{-iE_ft/\hbar}\left\{\left|f\right\rangle + \sum_{n\neq f}\frac{V_{nf}^0}{\hbar}\left[\frac{e^{i\omega t}-e^{-i\omega_{nf}t}}{(\omega+\omega_{nf})}-\frac{e^{-i\omega t}-e^{-i\omega_{nf}t}}{(\omega-\omega_{nf})}\right]\left|n\right\rangle\right\} \tag{4.68}$$

The term $\exp(-iE_ft/\hbar)$ is going to combine with the term $\exp(iE_it/\hbar)$ in $\langle\psi_i(t)|$ to give the phase factor $\exp(i\omega_{if}t)$ that we anticipated in Equation 4.58. Now, we are only interested in that part of the induced moment that follows the field, so it is the part of the perturbed wavefunction that depends on $\exp(\pm i\omega t)$ that matters. The part that depends on the transition frequency $\exp(i\omega_{nf}t)$ does not contribute to the induced transition moment that follows the field, so it is dropped. This is called the rotating wave approximation. The result is

$$\left|\psi_f(t)\right\rangle = e^{-iE_ft/\hbar}\left\{\left|f\right\rangle + \sum_{n\neq f}\frac{V_{nf}^0}{\hbar}\left[\frac{e^{i\omega t}}{(\omega+\omega_{nf})}-\frac{e^{-i\omega t}}{(\omega-\omega_{nf})}\right]\left|n\right\rangle\right\} \tag{4.69}$$

Taking the complex conjugate of Equation 4.69 and changing the index f to i results in

$$\langle \psi_i(t) | = e^{iE_it/\hbar} \left\{ \langle i | + \sum_{n \neq i} \langle n | \frac{V_{in}^0}{\hbar} \left[\frac{e^{-i\omega t}}{(\omega + \omega_{ni})} - \frac{e^{i\omega t}}{(\omega - \omega_{ni})} \right] \right\}$$

(4.70)

Next we use Equations 4.69 and 4.70 in Equation 4.58. The zero-order wavefunctions combine to give the ordinary transition moment $\mu_{if}(\text{perm}) = \langle i | \mu | f \rangle$. The cross-terms connecting zero-order states with the first-order correction terms are linear in the perturbation and therefore give us the induced transition dipole moment. The second-order term proportional to the square of the field is dropped. Thus we get

$$\mu_{if}(\text{ind}) = \frac{1}{\hbar} \sum_n \mu_{in} V_{nf}^0 \left[\frac{e^{i\omega t}}{(\omega + \omega_{nf})} - \frac{e^{-i\omega t}}{(\omega - \omega_{nf})} \right]$$
$$+ \frac{1}{\hbar} \sum_n \mu_{nf} V_{in}^0 \left[\frac{e^{-i\omega t}}{(\omega + \omega_{ni})} - \frac{e^{i\omega t}}{(\omega - \omega_{ni})} \right]$$

(4.71)

Note that the induced moment is the sum of two terms, either of which can be obtained from the other by swapping the indices i and f and taking the complex conjugate. This is the form expressed in Equation 4.61. Substituting for the matrix elements of V^0, using Equation 4.65, we get:

$$\mu_{if}(\text{ind}) = \frac{1}{2\hbar} \sum_n \mu_{in} \mu_{nf} \left[\frac{e^{i\omega t}}{(\omega + \omega_{nf})} - \frac{e^{-i\omega t}}{(\omega - \omega_{nf})} \right] \cdot E_0$$
$$+ \frac{1}{2\hbar} \sum_n \mu_{nf} \mu_{in} \left[\frac{e^{-i\omega t}}{(\omega + \omega_{ni})} - \frac{e^{i\omega t}}{(\omega - \omega_{ni})} \right] \cdot E_0$$

(4.72)

Comparing to Equation 4.61 gives the following expression for the transition polarizability:

$$\alpha_{if} = \frac{1}{\hbar} \sum_n \left[\frac{\mu_{in} \mu_{nf}}{\omega + \omega_{nf}} - \frac{\mu_{nf} \mu_{in}}{\omega - \omega_{ni}} \right]$$

(4.73)

Recall that each of the transition dipole moments in this expression is actually a vector. Thus the $\rho\sigma$-component of α_{if} is

$$\left(\alpha_{\rho\sigma} \right)_{if} = \frac{1}{\hbar} \sum_n \left[\frac{\langle i | \mu_\rho | n \rangle \langle n | \mu_\sigma | f \rangle}{\omega + \omega_{nf}} - \frac{\langle i | \mu_\sigma | n \rangle \langle n | \mu_\rho | f \rangle}{\omega - \omega_{ni}} \right]$$

(4.74)

Equation 4.74 is the Kramers–Heisenberg–Dirac (KHD) expression that we sought to derive. It has a physically appealing interpretation in that the polarizability is related to excited states connected to the initial and final states by transition dipoles. When the frequency of the incident light is close to that of a dipole-allowed transition $i \rightarrow n$, then the transition polarizability becomes large (through the second term in Equation 4.74) and we speak of resonance enhancement. In that case the first term, called the anti-resonance term, is unimportant and can be dropped.

When the initial and final states are rotation–vibration levels within the same electronic state, the KHD expression refers to Raman scattering. Alternatively, if i and f designate the same quantum state, the resulting expression refers to Rayleigh scattering, and the polarizability is connected to the refractive index as discussed in Chapter 3. The ground-state polarizability is

$$\left(\alpha_{\rho\sigma} \right)_{00} = \frac{2}{\hbar} \sum_{n \neq 0} \frac{\omega_{n0} \langle 0 | \mu_\rho | n \rangle \langle n | \mu_\sigma | 0 \rangle}{\omega_{n0}^2 - \omega^2}$$

(4.75)

4.4.2 Finite State Lifetimes and Imaginary Component of Polarizability

Equations 4.74 and 4.75 are not quite complete, as is evident when comparing them to the expression obtained in Chapter 3 using the Lorentz model of the atom, Equation 3.43. We need to amend these expressions, especially in the case of resonance, to allow for the finite lifetime and thus energy width of the states. As discussed previously, the imaginary component of polarizability is associated with absorption and emission of light. In the Lorentz model, the damping factor for electron oscillations results in an imaginary component of the polarizability.

Let us explore a quantum mechanical approach to obtaining this damping factor by introducing the idea of a quasi-stationary state. A quasi-stationary state is one that decays in time due to radiation of energy associated with transitions to lower energy states. Consider the probability of stationary state $|n\rangle$, which decays exponentially in time:

$$P_n(t) = |c_n(t)|^2 \propto e^{-\Gamma t} \tag{4.76}$$

The coefficient $c_n(t)$ therefore has a time dependence such that

$$|c_n(t)| \propto e^{\frac{-1}{2}\Gamma t} \tag{4.77}$$

The rate Γ at which the state probability decays is just the inverse of the lifetime of the state. The decay is due to transitions to all lower energy levels. From Fermi's Golden Rule, we have

$$\Gamma_n = \frac{1}{\hbar^2} \sum_{m<n} \left| \langle n | V^0 | m \rangle \right|^2 \rho(\nu_{nm}) \tag{4.78}$$

where the sum is over all states m having energies less than that of state n. The perturbation operator V^0 is often the electric dipole moment operator, but other mechanisms that permit transitions, such as electric quadrupole- or magnetic dipole-allowed transitions can also contribute. Note that the more strongly allowed the transition, the faster is the rate of decay of the state. This is why excited states that are connected to the ground state by weakly allowed transitions are longer lived than excited states that make strongly allowed transitions to the ground state. Compare the typical lifetime of phosphorescence, which corresponds to a spin-forbidden transition, to that of fluorescence which is spin-allowed. Phosphorescence lifetimes can be microseconds or longer, while fluorescence lifetimes are typically only on the order of nanoseconds or less.

The form of Equation 4.78 suggests a simple way to revise the expression for the superposition wavefunction discussed previously in order to account for the finite lifetime of the basis states. In the phase factor of the Schrödinger representation wavefunction, Equation 4.1, we make the following change:

$$E_n \Rightarrow E_n - \frac{i}{2}\hbar\Gamma_n \tag{4.79}$$

The result will be probabilities (squared coefficients) that have the appropriate time decay. This suggests a modification of the KHD equation as follows:

$$\left(\alpha_{\rho\sigma}\right)_{if} = \frac{1}{\hbar} \sum_n \frac{\langle i | \mu_\sigma | n \rangle \langle n | \mu_\rho | f \rangle}{\omega_{ni} - \omega - \frac{1}{2}i\Gamma_n} \tag{4.80}$$

We have written only the resonance term and assumed that the lifetime of the initial state is infinite. The inverse lifetimes are often treated phenomenologically, that is, they are based on the agreement of Equation 4.80 with experimental data. In future chapters, we will consider other factors that affect the lifetime and the spectral linewidth, in addition to the radiative contribution Γ.

4.4.3 Oscillator strength

The polarizability expression obtained previously in Chapter 3 was

$$\alpha(\omega) = \frac{e^2}{m} \sum_n \frac{f_{0n}}{\omega_{0n}^2 - \omega^2 - i\omega/\tau} \tag{4.81}$$

where m is the electron mass. Comparing this to Equation 4.75 leads to the following quantum mechanical interpretation of the oscillator strength for the $0 \rightarrow n$ transition:

$$f_{0n} = \frac{2m|\mu_{0n}|^2 \omega_{n0}}{3e^2 \hbar} \tag{4.82}$$

Thus the oscillator strength is proportional to the transition dipole squared. We can generalize this expression by writing it for any pair of states, say m and n. Oscillator strengths are convenient for specifying the intensity of a transition. Problem 3 offers the chance to prove the Thomas–Reiche–Kuhn sum rule, which says that if all the oscillator strengths connecting a given state to other states are summed the result is

$$\sum_m f_{nm} = 1 \tag{4.83}$$

Summing *all* the oscillator strengths for allowed electronic transitions results in the number of electrons in the molecule. This is consistent with the classical interpretation of the meaning of oscillator strength.

4.5 QUANTUM MECHANICAL EXPRESSION FOR EMISSION RATE

In this section we revisit Fermi's Golden Rule using a quantum mechanical version of the radiation–matter perturbation. We wish to write the long-wavelength form of the perturbation operator, $\hat{H}'(t) = -\vec{\mu} \cdot \vec{E}(t)$, in terms of a quantized radiation field. Recall from Chapter 2 how the vector potential was expressed as a superposition of cavity modes:

$$A(r,t) = \sqrt{\frac{1}{\varepsilon_0 V}} \sum_k q_k(t)\hat{u}_k(r) \tag{4.84}$$

We can find the electric field as follows:

$$\vec{E}(r,t) = \frac{-\partial \vec{A}(r,t)}{\partial t} = -\sqrt{\frac{1}{\varepsilon_0 V}} \sum_k \dot{q}_k(t)\hat{u}_k(r) \tag{4.85}$$

In the long wavelength limit, we can replace $\hat{u}_k(r) = \hat{e}_k \exp(i\vec{k} \cdot \vec{r})$ by $\hat{e}_k$, the unit vector in the direction of polarization. (Recall that there are two allowed polarizations for every k-vector.) The variable $\dot{q}_k = p_k$ can be expressed in terms of the raising and lowering operators with the help of Equations 2.62 and 2.63:

$$\dot{q}_k = i\sqrt{\frac{\hbar\omega_k}{2}} \left(b_k^+ - b_k \right) \tag{4.86}$$

What is needed in Equation 4.85 is

$$\dot{q}_k(t) = i\sqrt{\frac{\hbar\omega_k}{2}}\left[b_k^+(t) - b_k(t)\right] \qquad (4.87)$$

The time-dependent raising and lowering operators in Equation 4.87 are found using the Heisenberg equation of motion, Equation 4.9:

$$\frac{db_k}{dt} = \frac{i}{\hbar}\left[\hat{H}, b_k\right] = i\omega_k\left[b_k^+ b_k, b_k\right] = -i\omega_k b_k \qquad (4.88)$$

The Hamiltonian employed here is the radiation Hamiltonian presented in Equation 2.68. The commutator is reduced using Equations 2.64 and 2.65. Remember that b_k is the operator at time zero. We can integrate the preceding equation to get the value of $b_k(t)$. The solution to Equation 4.88 is

$$b_k(t) = e^{-i\omega_k t} b_k \qquad (4.89)$$

And taking the complex conjugate we also have

$$b_k^+(t) = e^{i\omega_k t} b_k^+ \qquad (4.90)$$

Thus, in the long wavelength limit, the quantum mechanical radiation-matter Hamiltonian is

$$\hat{H}'(t) = -\vec{\mu}\cdot\vec{E}(t) = -i\sqrt{\frac{\hbar}{2\varepsilon_0 V}}\sum_k \omega_k^{1/2}\left[e^{i\omega_k t}b_k^+ - e^{-i\omega_k t}b_k\right](\vec{\mu}\cdot\hat{e}_k) \qquad (4.91)$$

Compare this to our previous expression using the classical view of radiation:

$$\hat{H}'(t) = -\vec{\mu}\cdot\vec{E}(t) = \frac{-1}{2}\vec{\mu}\cdot\vec{E}_0(e^{i\omega t} + e^{-i\omega t}) \qquad (4.92)$$

Next we need to evaluate matrix elements of this operator with respect to initial and final radiation–matter states. The quantum mechanical matrix element $\langle i|\hat{H}'(0)|f\rangle$ is comparable to the matrix element $1/2\,\vec{\mu}_{if}\cdot\vec{E}_0$ obtained previously. We write each total wavefunction as the product of wavefunctions for radiation and matter. That is

$$\langle\Psi_i| = \langle n_{k1}n_{k2}n_{k3}\ldots|\langle i| \qquad (4.93)$$

$$|\Psi_f\rangle = |n_{k1}'n_{k2}'n_{k3}'\ldots\rangle|f\rangle \qquad (4.94)$$

where $|i\rangle$ and $|f\rangle$ are the wavefunctions for the matter states, and the radiation states are designated by the number of quanta (photons) in each mode. The unprimed n's are the initial and the primed n's are the final photon numbers. When the operator of Equation 4.91 is sandwiched between the bra and ket vectors given above, it is apparent that one part of the resulting (product) expression will just be the previously obtained transition dipole moment, projected onto the field direction: $\vec{\mu}_{if}\cdot\hat{e}_k$. The photon operators lead to matrix elements of the type

$$\langle n_{k1}n_{k2}n_{k3}\ldots|b_{k2}|n_{k1}'n_{k2}'n_{k3}'\ldots\rangle = \sqrt{n_{k2}+1}\;\delta_{n_{k2},n_{k2}+1}\delta_{n_{k1},n_{k1}}\delta_{n_{k3},n_{k3}}\ldots \qquad (4.95)$$

where we are considering a particular term in the sum over modes where $k = k_2$. Equation 4.95 says that the operator b_{k2} permits transitions in which the number of photons in mode 2 is increased by one, and the number of photons in all the other modes is unchanged. Thus this operator corresponds to one-photon emission.

The matrix element of the creation operator accounts for one-photon absorption transitions. It is

$$\langle n_{k1} n_{k2} n_{k3} \ldots | b_{k_2}^+ | n'_{k1} n'_{k2} n'_{k3} \ldots \rangle = \sqrt{n_{k2}}\, \delta_{n'_{k2}, n_{k2}-1} \delta_{n'_{k1}, n_{k1}} \delta_{n'_{k3}, n_{k3}} \ldots \tag{4.96}$$

The probability of a transition is proportional to the square of the amplitude of $\langle i | \hat{H}'(0) | f \rangle$. The part that corresponds to emission is given by

$$\left| \langle i | \hat{H}'(0) | f \rangle \right|^2 = \frac{\hbar}{2\varepsilon_0 V} \sum_k \omega_k (n_k + 1) \left| \vec{\mu}_{if} \cdot \hat{e}_k \right|^2 \tag{4.97}$$

This is analogous to the expression obtained in Section 4.3, where we made the substitution given in Equation 4.50 to replace the square of the electric field amplitude with a function of the number density of photons. The summation over modes in Equation 4.97 takes care of the photon density of states. The importance of Equation 4.97 is that the emission rate is proportional to the number of photons plus one; the transition that takes place when the number of photons is zero is spontaneous emission. The square of the matrix element corresponding to absorption, as expected, is proportional to the number of photons. Note that just as the creation and annihilation operators for the harmonic oscillator can only connect states that differ by one vibrational quantum, those for the quantized radiation field can only connect states of the field that differ by one photon in one of the allowed modes. This is a consequence of our first-order perturbation approach. If instead of using the electric dipole approximation in the above discussion we had used the perturbation operator $-e/m(\vec{A} \cdot \vec{p})$ of Equation 4.20 in order to account for E2- and M1-allowed transitions, keeping the $i\vec{k} \cdot \vec{r}$ term in the expansion of Equation 4.22, that would have changed the nature of the matrix element connecting the matter states $|i\rangle$ and $|f\rangle$. We would still be limited, however, to one-photon transitions as this restriction results from the matrix element involving the radiation field. In Chapter 13, we will see how two-photon transitions require a third-order perturbation approach.

4.6 TIME DEPENDENCE OF THE DENSITY MATRIX

In this section, we introduce the density matrix formalism, which provides a convenient way to describe the relaxation dynamics of a molecule when it is coupled to the surroundings. Ideally, we could describe the time evolution of the system by solving the Schrödinger equation for the molecule *plus* the surroundings. Since this is not possible, we take a phenomenological approach to account for the effect of the surroundings on the dynamics of the molecule of interest. Time-dependent perturbation theory enables us to use a superposition wavefunction to describe the molecule when it is in the process of making a transition.

To get a physical handle on the density matrix, we start by considering a two-state picture. In addition to spin-1/2 magnetic resonance phenomena, some optical spectroscopy experiments can be adequately described in a two-state basis. The wavefunction can be written

$$|\Psi(t)\rangle = a_1(t)|1\rangle + a_2(t)|2\rangle \tag{4.98}$$

where the phase factors are absorbed into the time-dependent coefficients: $a_n(t) = c_n(t)\exp(-iE_n t/\hbar)$. (Compare to Equation 4.12.) We have used this approach to treat a system that starts out in one state before the perturbation is turned on, $a_1(0) = 1$, $a_2(0) = 0$, and then makes a transition to state 2 under the influence of light. We can also use the same picture to discuss the relaxation of a system as it returns to equilibrium, after an absorption transition has taken place. During the transition, both coefficients are in the range $0 < |a_n| < 1$, and normalization requires that $a_1 a_1^* + a_2 a_2^* = 1$. Thus $a_n^*(t)a_n(t)$ is the quantum mechanical probability of finding the system in state n at time t. If this hypothetical two-level molecule is isolated, then once it is promoted to level 2 it can return to level 1 by making a radiative transition. This was the point of view taken in Section 4.4.2, when we accounted for the damping term in the polarizability. A molecule interacting with its environment can also lose energy by making nonradiative transitions. These transitions take place via collisions and interactions with other molecules, resulting in energy transfer to other degrees of freedom. The time decay of the probability $a_2^*(t)a_2(t)$ thus represents population relaxation by both radiative and nonradiative means. The terms $a_1^*(t)a_2(t)$ and $a_2^*(t)a_1(t)$, on the other hand, represent the *coherence* of the system. If a perturbation affects states 1 and 2 the same way, we refer to the effect as coherent, and states 1 and 2 would bear a definite phase relationship with respect to one another. Relaxation of the terms $a_1^*(t)a_2(t)$ and $a_2^*(t)a_1(t)$ is called phase relaxation or *dephasing*. At equilibrium, these off-diagonal terms vanish.

The density operator may be defined as $\hat{\rho}(t) = |\Psi(t)\rangle\langle\Psi(t)|$. In the two-dimensional basis used here, the matrix representation of this operator is

$$\rho(t) = \begin{pmatrix} a_1(t)a_1^*(t) & a_1(t)a_2^*(t) \\ a_2(t)a_1^*(t) & a_2(t)a_2^*(t) \end{pmatrix} \tag{4.99}$$

The diagonal elements, e.g., $\rho_{11}(t) = \langle 1|\Psi(t)\rangle\langle\Psi(t)|1\rangle = a_1(t)a_1^*(t)$, represent populations and the off-diagonal elements ρ_{12} and ρ_{21} represent coherences. Borrowing the language used in the previous chapter in our discussion of magnetic resonance, population (or energy) relaxation is referred to as a T_1 process. The populations are assumed to decay exponentially with time constant T_1. The dephasing time T_2 is the time constant for phase relaxation. It is said that "all T_1 processes are also T_2 processes." This is so because any time dependence of the diagonal elements of the density matrix is always associated with time dependence of the off-diagonal parts. There is, however, something called pure dephasing, with relaxation time T_2^*, which derives from medium-induced phase fluctuations. (These will be considered further in Chapters 6 and 12.) The physical picture here is that the surroundings cause the transition frequency to fluctuate without changing populations, leading to pure dephasing. As previously stated (Equation 3.120), these relaxation times are related as follows:

$$\frac{1}{T_2} = \frac{1}{2T_1} + \frac{1}{T_2^*} \tag{4.100}$$

Note that in the absence of pure dephasing one expects $T_2 = 2T_1$. More generally, we have $T_1 \geq T_2/2$.

The density matrix can be generalized to any dimension. Consider the superposition state

$$|\Psi(t)\rangle = \sum_n a_n(t)|n\rangle \tag{4.101}$$

The density matrix elements are then $\rho_{nm}(t) = \langle n|\hat{\rho}(t)|m\rangle = a_n(t)a_m^*(t)$. The expectation value of any operator $\hat{O}$ at time t is

$$\langle O(t)\rangle = \langle\Psi(t)|\hat{O}|\Psi(t)\rangle = \sum_{n,m} a_n(t)a_m^*(t)\langle m|\hat{O}|n\rangle$$

$$= \sum_{n,m} \rho_{nm}(t)\langle m|\hat{O}|n\rangle = Tr(\hat{\rho}(t)\hat{O}) \tag{4.102}$$

This is the time-dependent equivalent of Equation 1.91.

Density matrix formalism is advantageously applied when we want to focus on a system that interacts weakly with its surroundings, often referred to as the bath. For example, we might have a spectroscopically active molecule in solution, and we want to know how the states of the molecule evolve under the influence of bath perturbations. The density matrix is then an ensemble average:

$$\rho_{nm}(t) = \overline{a_n(t)a_m^*(t)} \tag{4.103}$$

where the bar represents the ensemble average. From here on we will consider the density operator $\hat{\rho}$ to be that for which the matrix elements are given by Equation 4.103.

The equilibrium density operator, discussed in Chapter 1, is assumed to apply in the absence of the time-dependent perturbation: $\hat{\rho}_0 = z^{-1} \exp(-\beta \hat{H}_0)$. Perturbations to the zero-order Hamiltonian include that due to the radiation field as well as the coupling between the system and the bath.

Let us return to the two-state picture and use the above expressions to find the time-dependence of the density matrix, using the following equation of motion:

$$\frac{\partial \hat{\rho}}{\partial t} = \frac{-i}{\hbar} \left[\hat{H}, \hat{\rho} \right] \tag{4.104}$$

The Hamiltonian is the sum of three contributions: $\hat{H} = \hat{H}_0 + \hat{H}'(t) + \hat{H}_R$. The stationary states $|1\rangle$ and $|2\rangle$ are eigenfunctions of the zero-order Hamiltonian with eigenvalues ε_1 and ε_2, hence the matrix representation of $\hat{H}_0$ is diagonal. Specializing to electric dipole transitions, the perturbation operator is taken as $\hat{H}'(t) = -\vec{\mu} \cdot \vec{E}(t)$ and its matrix representation has only nondiagonal elements. Though this may seem like we are ignoring the possibility of a permanent dipole moment of the ground or excited state, on average there is no component of the permanent moment in the direction of the time-varying electric field. Finally, the "relaxation Hamiltonian" is just phenomenological; we cannot evaluate its commutator with the density matrix. Instead, we add terms to the equation of motion that allow for T_1 and T_2 relaxation of the diagonal and nondiagonal elements, respectively, of the density matrix. Let us call **H** the matrix representation of $\hat{H} = \hat{H}_0 + \hat{H}'(t)$ and evaluate the matrix representation of the commutator in Equation 4.104:

$$i\hbar \frac{d\rho}{dt} = [\mathbf{H}, \rho] = \begin{pmatrix} \varepsilon_1 & -\mu_{12}E(t) \\ -\mu_{12}E(t) & \varepsilon_2 \end{pmatrix} \begin{pmatrix} \rho_{11} & \rho_{12} \\ \rho_{21} & \rho_{22} \end{pmatrix} - \begin{pmatrix} \rho_{11} & \rho_{12} \\ \rho_{21} & \rho_{22} \end{pmatrix} \begin{pmatrix} \varepsilon_1 & -\mu_{12}E(t) \\ -\mu_{12}E(t) & \varepsilon_2 \end{pmatrix} \tag{4.105}$$

For simplicity, we omit vector notation on the transition dipole and field, such that μ_{12} represents the projection of the transition dipole onto the field direction. We also assume that $\mu_{12} = \mu_{21}$. After adding in the relaxation, we obtain the following:

$$\frac{d\rho_{11}}{dt} = \frac{-2\mu_{12}E(t)}{\hbar} \text{Im}(\rho_{21}) - \frac{\rho_{11} - \rho_{11}^{eq}}{T_1}$$

$$\frac{d\rho_{22}}{dt} = \frac{2\mu_{12}E(t)}{\hbar} \text{Im}(\rho_{21}) - \frac{\rho_{22} - \rho_{22}^{eq}}{T_1}$$

$$\frac{d\rho_{12}}{dt} = i\omega_{21}\rho_{12} - \frac{i}{\hbar}\mu_{12}E(t)(\rho_{11} - \rho_{22}) - \frac{\rho_{12}}{T_2} \tag{4.106}$$

$$\frac{d\rho_{21}}{dt} = -i\omega_{21}\rho_{21} + \frac{i}{\hbar}\mu_{12}E(t)(\rho_{11} - \rho_{22}) - \frac{\rho_{21}}{T_2}$$

As usual the Bohr frequency is defined as $\omega_{21} = (\varepsilon_2 - \varepsilon_1)/\hbar$. We have used the fact that the ρ_{12} and ρ_{21} are complex conjugates of one another. The equilibrium values of ρ_{12} and ρ_{21} are zero, while those of ρ_{11} and ρ_{22} are given by Boltzmann's law. We take the time-dependent electric field to be $E(t) = E_0(e^{i\omega t} + e^{-i\omega t})$. As E_0 is then one-half the amplitude of the electric field, the Rabi frequency is $\Omega = 2\mu_{12}E_0/\hbar$. In the absence of this field or any other perturbation, the solution to the equation of motion for the off-diagonal elements of the density matrix would be

$$\rho_{12}(t) = e^{i\omega_{21}t}\rho_{12}(0)$$

$$\rho_{21}(t) = e^{-i\omega_{21}t}\rho_{21}(0) \tag{4.107}$$

On resonance, $\omega_{21} \approx \omega$, thus there is a rapidly varying part of ρ_{12} and ρ_{21} that we are not interested in. This suggests a modification analogous to the rotating reference frame used in our treatment of magnetic resonance in Chapter 3. We define the off-diagonal elements in this frame as follows:

$$\tilde{\rho}_{12} \equiv \rho_{12} e^{-i\omega t}$$

$$\tilde{\rho}_{21} \equiv \rho_{21} e^{i\omega t}$$

(4.108)

The equations of motion for these transformed quantities are

$$\frac{d\tilde{\rho}_{12}}{dt} = -i(\omega - \omega_{21})\tilde{\rho}_{12} - i\frac{\Omega}{2}(\rho_{11} - \rho_{22}) - \frac{\tilde{\rho}_{12}}{T_2}$$

$$\frac{d\tilde{\rho}_{21}}{dt} = i(\omega - \omega_{21})\tilde{\rho}_{21} + i\frac{\Omega}{2}(\rho_{11} - \rho_{22}) - \frac{\tilde{\rho}_{21}}{T_2}$$

(4.109)

In arriving at the above, we discarded the terms that vary as $\exp(\pm 2i\omega t)$ in keeping with the rotating wave approximation (RWA). This is entirely justified because ultimately the density matrix will be found by integrating over time. The time-dependence of the rest of the integrand is much slower than that of $\exp(\pm 2i\omega t)$, so that integration would give a zero result. The transformation of Equation 4.108 and the RWA also affect our calculation of the time-dependence of the diagonal elements of the density matrix. Consider the product of the time-varying field and the imaginary part of ρ_{12} that appears in the first two equations of 4.106. The RWA leads to

$$E(t)\operatorname{Im}\rho_{21} = \frac{-iE_0}{2}(\tilde{\rho}_{21} - \tilde{\rho}_{12})$$

(4.110)

This results in

$$\frac{d\rho_{11}}{dt} = \frac{i\Omega}{2}(\tilde{\rho}_{21} - \tilde{\rho}_{12}) - \frac{\rho_{11} - \rho_{11}^{eq}}{T_1}$$

$$\frac{d\rho_{22}}{dt} = \frac{-i\Omega}{2}(\tilde{\rho}_{21} - \tilde{\rho}_{12}) - \frac{\rho_{22} - \rho_{22}^{eq}}{T_1}$$

(4.111)

By subtracting the second equation from the first, we obtain an equation of motion for the population difference $\rho_{11} - \rho_{22}$. Further, we add and subtract Equation 4.109 to obtain terms that are respectively proportional to the real and imaginary parts of ρ_{12}. The result is the three optical Bloch equations:

$$\frac{d}{dt}(\rho_{11} - \rho_{22}) = i\Omega(\tilde{\rho}_{21} - \tilde{\rho}_{12}) - \frac{(\rho_{11} - \rho_{11}^{eq}) - (\rho_{22} - \rho_{22}^{eq})}{T_1}$$

$$\frac{d}{dt}(\tilde{\rho}_{12} + \tilde{\rho}_{21}) = -i(\omega - \omega_{21})(\tilde{\rho}_{12} - \tilde{\rho}_{21}) - \frac{(\tilde{\rho}_{12} + \tilde{\rho}_{21})}{T_2}$$

$$\frac{d}{dt}(\tilde{\rho}_{12} - \tilde{\rho}_{21}) = -i(\omega - \omega_{21})(\tilde{\rho}_{12} + \tilde{\rho}_{21}) - i\Omega(\rho_{11} - \rho_{22}) - \frac{(\tilde{\rho}_{12} - \tilde{\rho}_{21})}{T_2}$$

(4.112)

To see the correspondence between the above expressions and those in Chapter 3 for magnetic resonance, Equation 3.118, the following analogies are made:

$$M_z \Rightarrow (\rho_{11} - \rho_{22})$$

$$u \Rightarrow \operatorname{Re}\tilde{\rho}_{12} \propto \tilde{\rho}_{12} + \tilde{\rho}_{21}$$

$$v \Rightarrow \operatorname{Im}\tilde{\rho}_{12} \propto \tilde{\rho}_{12} - \tilde{\rho}_{21}$$

(4.113)

As previously pointed out (and shown in homework problem 10 of Chapter 3), the magnetization M_z is proportional to the population difference, so the first correspondence is no surprise. In addition, we now have the following handle on the real and imaginary parts of ρ_{12}. The real part is analogous to the part of the response (magnetization u) that is in phase with the driving field, and the imaginary part corresponds to the component in quadrature with the field (magnetization v). It is the latter which gives rise to the signal in NMR. Since the time-dependent polarization is found from $Tr\langle \rho(t)\mu \rangle$, the real and imaginary parts of the density matrix give rise to the parts of the response that are respectively in phase and 90° out of phase with the driving field.

Finally, before generalizing the density matrix formalism, we wish to use the two-state model to get further insight into the concept of Rabi oscillations in the resonant case: $\omega = \omega_{21}$. We assume the boundary conditions $\rho_{11}(0) = 1, \rho_{22}(0)$; i.e., the system starts out in the lower energy state before the perturbing radiation is turned on. We further neglect the relaxation terms and obtain the simple equations of motion $\rho_{11}(t) = \cos^2(\Omega t)$, $\rho_{22}(t) = \sin^2(\Omega t)$. We see that the population oscillates back and forth between "normal" ($\rho_{11} > \rho_{22}$) and "inverted" ($\rho_{11} < \rho_{22}$), as in the transient nutation experiment of magnetic resonance. Figure 4.3a illustrates these Rabi oscillations in the case of exact resonance and ignoring all relaxation terms. If dephasing is considered, but population relaxation is ignored, the oscillations damp with time until the populations of the two states eventually become equal, as shown in Figure 4.3b.

Rabi oscillations are much easier to observe in magnetic resonance than in optical spectroscopy owing to the much slower dephasing rate of spin states compared to electronic and vibrational states. The reason is that the spin states interact weakly with their surroundings. Both dephasing and population relaxation rates contribute to the linewidth of a spectroscopic transition, as will be explored throughout this book. Hence the slow relaxation in an NMR experiment is responsible for the sharp lines that make the technique a powerful structural tool.

Next, we outline a general perturbation formalism for finding the density operator without regard to the nature of the perturbation. For details, the reader should consult [1–4]. Suppose that we have the full Hamiltonian $\hat{H} = \hat{H}_0 + \hat{V}$, where $\hat{V}$ is a perturbation that may depend on time. It is convenient to introduce what is called the interaction representation of the density operator $\hat{\rho}^I(t)$, defined by

$$\hat{\rho}^I(t) \equiv \exp\left(\frac{i\hat{H}_0 t}{\hbar} \right) \hat{\rho}(t) \exp\left(\frac{-i\hat{H}_0 t}{\hbar} \right) \tag{4.114}$$

Similarly, the interaction representation of $\hat{V}$ is

$$\hat{V}^I(t) \equiv \exp\left(\frac{i\hat{H}_0 t}{\hbar} \right) \hat{V}(t) \exp\left(\frac{-i\hat{H}_0 t}{\hbar} \right) \tag{4.115}$$

Differentiating both sides of Equation 4.114 with respect to time gives

$$\frac{\partial \hat{\rho}^I}{\partial t} = \frac{-i}{\hbar} \left[V^I, \hat{\rho}^I \right] \tag{4.116}$$

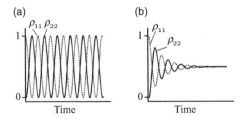

Figure 4.3 Rabi oscillations in a two level system without (a) and with (b) dephasing.

The above equation can be compared to Equation 4.104. The advantage of transforming to the interaction representation is that the density operator then evolves only according to the perturbation operator rather than the full Hamiltonian. The solution to Equation 4.116 is

$$\hat{\rho}^I(t) = \hat{\rho}^I(0) - \frac{i}{\hbar} \int_0^t dt' \left[\hat{V}^I(t'), \hat{\rho}^I(t') \right] \tag{4.117}$$

Equation 4.117 is the result of direct integration of 4.116, but it has ρ^I on both sides. This suggests an iterative solution:

$$\hat{\rho}^I(t) = \hat{\rho}^I(0) - \frac{i}{\hbar} \int_0^t dt' \left[\hat{V}^I(t'), \hat{\rho}^I(0) \right]$$
$$+ \left(\frac{-i}{\hbar} \right)^2 \int_0^t dt' \int_0^{t'} dt'' \left[\hat{V}^I(t'), \left[\hat{V}^I(t''), \hat{\rho}^I(0) \right] \right] + \cdots \tag{4.118}$$

The perturbation series of Equation 4.118 is comparable to that which could be obtained from iteration of Equation 4.17. The difference is that here we are calculating products of the state coefficients, rather than the coefficients themselves. The former are more directly related to experiment. In the linear response regime, the perturbation $\hat{V}$ is the dipole operator and Equation 4.118 is truncated after the second term. The treatment of nonlinear spectroscopy requires higher-order terms in this series. Solutions to this equation depend on the nature of the interaction and the level of approximation. The interaction representation is also useful in the analysis of time-correlation functions, to be discussed in the next chapter. We return to the discussion of the time-dependent density matrix in Chapters 13 and 14 where it is indispensable to the discussion of nonlinear optical spectroscopy.

4.7 SUMMARY

The key result of this chapter is the derivation, using time-dependent perturbation theory, of Fermi's Golden Rule for the rate of a spectroscopic transition. One significant consequence of this equation is the time-energy uncertainty principle, which reduces to the Bohr frequency condition, $v = (E_f - E_i)/h$, in limits appropriate to many experimental situations. We have seen that the transition dipole operator derives from the lead term in an expansion of the function $\exp(i\vec{k} \cdot \vec{r})$ about $\vec{k} = 0$. This limit is valid when the wavelength is large compared to the size of the spectroscopically active system, and results in what are called electric dipole (E1) selection rules. The electric quadrupole (E2) and magnetic dipole (M1) transitions spring from the second term in the expansion of $\exp(i\vec{k} \cdot \vec{r})$. These types of transitions are generally less intense than those permitted by the electric dipole operator. In future chapters, we will apply symmetry considerations to deduce whether matrix elements of the electric dipole, electric quadrupole, and magnetic dipole moment operators are zero or not, thus determining selection rules for various spectroscopy experiments.

Two forms of the Golden Rule were obtained that are convenient to the discussion of absorption and emission. The transition rate in absorption is proportional to the energy density of light $\rho(v)$ at the frequency of the transition. In Chapter 6, we will compare this expression to a phenomenological kinetic expression, due to Einstein, and uncover the relationship between experimentally determined intensities and the theoretical transition rate. In emission, the transition rate depends on the photon density of states, which must be treated quantum mechanically in order to recover spontaneous as well as stimulated emission. The resulting transition rate is proportional to the cube of the frequency.

As a preface to future discussions of light scattering, the quantum mechanical expression for the transition polarizability was derived: the KHD equation of Equation 4.57. The theoretical picture behind this two-photon experiment invokes transitions to numerous intermediate states. This sum-over-states expression for the transition polarizability will serve as the basis for selection rules in rotational and vibrational Raman scattering, to be considered in future chapters. The KHD expression also provides the necessary link to the Lorentz theory expression for $\alpha(\omega)$, obtained in Chapter 3. The result is a quantum mechanical expression for the oscillator strength f, which is proportional to the square of the transition moment.

The time-dependent density matrix was introduced in order to discuss transitions of a molecule coupled to the surroundings. The energy relaxation and phase relaxation are characterized by times T_1 and T_2, respectively. In Chapter 6, we will consider a model for pure dephasing, with time scale T_2^*, due to environment-induced fluctuations in the spectral frequency. The next chapter will also make use of some of the quantum mechanical tools presented in this chapter, in order to derive the time-dependent theory of spectral intensities.

PROBLEMS

1. Solve the Heisenberg equations of motion for the momentum and position of a harmonic oscillator:

$$\frac{d}{dt}\langle p\rangle_{t=0} = \frac{i}{\hbar}\left\langle\left[\hat{H},p\right]\right\rangle$$

$$\frac{d}{dt}\langle q\rangle_{t=0} = \frac{i}{\hbar}\left\langle\left[\hat{H},q\right]\right\rangle$$

Verify that the expectation values evolve in time like those for a classical harmonic oscillator; that is, they oscillate sinusoidally in time with angular frequency ω.

2. Show that $\langle f|y\partial/\partial z|i\rangle = -m\omega_{fi}/\hbar\langle f|yz|i\rangle$. Hint: Insert the resolution of the identity, $\sum_j|j\rangle\langle j|=1$, between the operators y and $\partial/\partial z$.

3. Derive the Thomas–Reiche–Kuhn sum rule for the oscillator strength:

$$\sum_{i\neq j}f_{ij}=1$$

Hint: This is a job for commutator algebra. Derive an expression for the commutator $[\hat{H},\hat{\mu}]$, where $\hat{\mu}=-e\vec{r}$ and $\vec{r}=\hat{i}x+\hat{j}y+\hat{k}z$. Then use it to convert

$$\nu_{ji}\left|\mu_{ij}\right|^2 = \frac{1}{h}\left\{\langle i|\hat{\mu}\hat{H}|j\rangle\langle j|\hat{\mu}|i\rangle - \langle i|\hat{H}\hat{\mu}|j\rangle\langle j|\hat{\mu}|i\rangle\right\}$$

to a more useful quantity. The resolution of the identity is also helpful on this problem.

4. Treat a one-electron atom according to the Lorentz model and calculate the oscillator strength for the $v=0 \rightarrow v=1$ transition. The relevant wavefunctions are

$$\psi_0 = \left(\frac{\alpha}{\pi}\right)^{1/4}\exp\left(\frac{-\alpha x^2}{2}\right)$$

$$\psi_1 = \left(\frac{4\alpha^3}{\pi}\right)^{1/4}x\exp\left(\frac{-\alpha x^2}{2}\right)$$

where $\alpha \equiv (m\omega/\hbar)$.

5. Derive the E1 selection rules for an electron in a one-dimensional box of length L. In other words, for what changes in the quantum number n will the transition dipole $\mu_{nn'}$ be nonzero? The wavefunctions are $\psi_n = \sqrt{\frac{2}{L}}\sin\frac{n\pi x}{L}$.

6. Calculate the oscillator strength for the $1s \rightarrow 2p_z$ transition of the hydrogen atom. The wavefunctions are

$$\psi_{1s} = \frac{1}{\sqrt{\pi a_0^3}} \exp\left(\frac{-r}{a_0}\right)$$

$$\psi_{2p_z} = \frac{1}{4}\sqrt{\frac{1}{2\pi a_0^3}} \frac{r}{a_0} \exp\left(\frac{-r}{2a_0}\right) \cos\theta$$

where $a_0 = 0.0529$ nm is the Bohr radius. Hint: By symmetry, only the z component of the dipole moment operator, $\mu_z = -er\cos\theta$, contributes to the transition moment.

7. Consider a single molecule with an electronic transition at 530 nm that is exposed to a source having an energy density of 10^{-19} J/m^3 Hz at that wavelength. The transition dipole moment of the molecule is 3.0 Debye. Assume the light source is polarized along the direction of the transition dipole moment. Find the transition rate w_{if}.

REFERENCES

1. J. D. Macomber, *The Dynamics of Spectroscopic Transitions* (John-Wiley and Sons, New York, 1976).
2. C. H. Wang, *Spectroscopy of Condensed Media* (Academic Press, Orlando, FL, 1985).
3. G. C. Schatz and M. A. Ratner, *Quantum Mechanics in Chemistry* (Prentice-Hall, Englewood Cliffs, NJ, 1993).
4. W. G. Rothschild, *Dynamics of Molecular Liquids* (Wiley, New York, 1984).

5

The time-dependent approach to spectroscopy

5.1 INTRODUCTION

Conventional spectroscopic measurements are concerned with the intensity of absorption, emission, or scattering as a function of the frequency of radiation. In such steady-state, frequency-domain approaches, the emphasis is on the eigenstates and their energies, and the spectral response conveys structural information. Complementary to the behavior of the system as a function of frequency is the time-dependent response of matter. The focus of this chapter will be on the intimate relationship between the frequency and time domains. The formalism presented here will give rise to the point of view that sufficiently weak electromagnetic fields can be exploited to probe the molecular dynamics that take place in the unperturbed system. The fluctuation–dissipation theorem, to be discussed in Section 5.4, states that the energy imparted to the system by the probing field is dissipated by fluctuations characteristic of the system at equilibrium. The mathematical essence of this theory is that equilibrium time-correlation functions (TCFs) can be obtained by Fourier transformation of spectral intensity data.

One advantage of the time-dependent view is insight into how the same molecular dynamics can affect various types of spectra. For example, the lineshapes of infrared absorption, depolarized Rayleigh scattering, and dielectric relaxation are all influenced by molecular reorientation, and vibrational relaxation contributes to the lineshapes of both infrared and Raman spectra. The combination of different spectral tools to study dynamics can provide more information than would be afforded by a single spectrum. The use of TCFs to interpret spectra is especially useful in the liquid phase, where discrete rotational structure is not observed. The time-dependent approach avoids the need to consider the eigenstates of the system and is thus well-suited to the interpretation of diffuse spectra. The tools of classical statistical mechanics can be brought to bear on the treatment of rotational and translational motion in liquids.

Though the dynamical response of the system can be measured directly using pulsed excitation sources and time-resolved detection, in this chapter we are concerned with uncovering the time-dependent behavior from the frequency spectrum. The time and frequency responses are connected by a Fourier transform relationship. By understanding the theoretical basis for this connection, practicing spectroscopists can extract far more information from experimental data than is realized in common "phrenological"* approaches to interpreting spectra. The application of the time-dependent theory has both advantages and disadvantages. The advantage is in the practice of modeling or interpreting the data without having to know the eigenfunctions of the system, and in direct determination of the timescales for relaxation of the perturbed system. Unfortunately, there are impediments to accurate determination of either experimental or calculated correlation functions. The truncation of what ought to be an integral over all frequencies, and baseline errors in the wings of the spectrum, lead to errors in the experimental TCF. Overlapping bands, such as those due to naturally occurring isotopes, obscure the meaning of the TCF obtained by direct Fourier transformation. Modeling of spectra and correlation functions in condensed phases is fraught with the problem of accounting for the influence of intermolecular interactions. On the other hand, there is the potential for the approach to reveal these interactions. The numerous physical mechanisms that can simultaneously perturb the lineshape; reorientation, vibrational energy transfer, fluctuations in the transition energy, collisions, etc. can be difficult to disentangle in practice. Nevertheless, the approaches described here provide a powerful spectroscopic

---

* Phrenology was a nineteenth century fad of deducing character traits and intellectual abilities by measuring the bumps on a person's skull.

vantage point, and have given rise to a modern view of spectroscopy that has been applied to both time- and frequency-domain experiments.

In this chapter, we confine our discussion to spectroscopic transitions taking place on a single potential energy surface, that of the ground electronic state. The time-dependent view can also be fruitfully applied to electronic spectroscopy (absorption, fluorescence, and resonance Raman). This time-dependent view of spectroscopy will be discussed in Chapter 12.

In Section 5.2 it will be shown how the Golden Rule derived in Chapter 4 can be transformed to an expression in which the spectral intensity is the Fourier transform of an equilibrium-averaged time-correlation function. The resulting formalism is well suited to the study of condensed phase dynamics through the measurement of rotational and vibrational spectra. The more general approach outlined in Section 5.4 considers the full wave vector and frequency resolved response to depend respectively on correlations in space and in time. In Chapter 12, we will extend the approach to account for dynamics on more than one potential energy surface. The result will be a theory for calculating absorption, emission, and resonance Raman scattering spectra from the time-dependent overlaps of vibrational wavefunctions.

5.2 TIME-CORRELATION FUNCTIONS AND SPECTRA AS FOURIER TRANSFORM PAIRS

In this section, we review the derivation presented by Gordon in [1], in order to find the relationship between the intensity spectrum $I(\omega)$ and the time-correlation function $C(t) \equiv \langle A^*(0) \cdot A(t) \rangle$, where A is the dynamical variable associated with the spectrum of interest (in absorption spectra it will be the dipole moment or transition dipole moment) and the angle brackets indicate an equilibrium average. The name correlation function indicates that $C(t)$ is a measure of the extent to which the dynamic variable $A(t)$ is correlated with the value it had at time zero. A correlation function of the type $\langle A^*(0) \cdot A(t) \rangle$ is called an autocorrelation function, in contrast to a cross-correlation function, $\langle B^*(0) \cdot A(t) \rangle$.

We shall find that the physical property A of interest to a particular TCF may depend on molecular orientation or the coordinate defining a molecular vibration. It is often the case that the operator corresponding to this property is real, and we can dispense with the complex conjugate notation. If we singled out one molecule in an ensemble and tried to track, say, its orientation, then $A(t)$ would appear to fluctuate in an unpredictable manner, as illustrated in Figure 5.1. It may seem that little information is obtainable from such a single trajectory. However, if we compare the value of $A(t)$ at some time t with its value $t + \tau$ later, for short enough time intervals τ the two measurements of A are correlated.

The equilibrium average $\langle A(t)A(t + \tau) \rangle$ does not depend on the choice of starting time t, which is arbitrary, but does depend on the interval τ. To find the average $\langle A(t)A(t + \tau) \rangle = \langle A(0)A(\tau) \rangle$, imagine making pairs of measurements of A, separated in time by τ, over a large time interval and then averaging them:

$$\langle A(0)A(\tau) \rangle = \frac{1}{T} \int_0^T dt A(t)A(t + \tau) \tag{5.1}$$

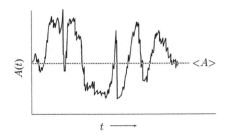

Figure 5.1 Trajectory of a dynamic variable $A(t)$.

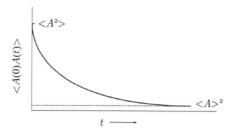

Figure 5.2 Time-correlation function $\langle A(0)A(t) \rangle$.

Equation 5.1 corresponds to averaging a number of trajectories such as that of Figure 5.1. The result is a smooth function $\langle A(0)A(\tau) \rangle = C(\tau)$, as shown in Figure 5.2, whose time decay is a measure of the loss of correlation between repeated measurements of A. According to the ergodic hypothesis, the time average is equivalent to an ensemble average. It is this ensemble average which will be obtained in the derivation which follows. The initial value of $C(\tau)$ at $\tau = 0$ is the equilibrium average $\langle A^2 \rangle$. As time proceeds, $C(\tau)$ must decrease overall due to cancellation effects: contributions to the average of 5.1 where $A(t)$ and $A(t + \tau)$ have opposite sign. Therefore $\langle A(0)A(\tau) \rangle \leq \langle A^2 \rangle$. At times much longer than the correlation time τ_c, there is no relation between $A(t)$ and $A(t + \tau)$, and the TCF tends toward $\langle A(0) \rangle \langle A(\tau) \rangle = \langle A \rangle^2$. In other words, at long times the two measurements can be averaged separately. The net result is the relation:

$$\langle A^2 \rangle \geq C(t) \geq \langle A \rangle^2 \tag{5.2}$$

We shall see that the rate at which $C(t)$ decays determines the width of the corresponding spectrum.

In the case where A is the dipole moment, we expect $C(t)$ to go from an initial value of $\langle A^2 \rangle = \langle \mu^2 \rangle$ to the value of zero at times long enough that $\vec{\mu}(t)$ has "forgotten" its original value. The long-time limit of $\langle \vec{\mu}(0) \cdot \vec{\mu}(t) \rangle$ vanishes because $\vec{\mu}(t)$ is just as likely to be antiparallel to $\vec{\mu}(0)$ as it is to be parallel, so the dot product averages to zero.

Let us recall the Golden Rule derived in the previous chapter, Equation 4.39, in which the rate of transitions is given by

$$w_{if} = \frac{\pi}{2\hbar^2} \left| \left\langle f \left| \vec{\mu} \cdot \vec{E}_0 \right| i \right\rangle \right|^2 \left\{ \delta\left(\omega_{fi} - \omega \right) + \delta\left(\omega_{fi} + \omega \right) \right\} \tag{5.3}$$

This is the probability per unit time of the transition $i \rightarrow f$. Notice that Equation 5.3 is unchanged if the indices i and f are interchanged. The inherent transition rate per molecule is the same for upward (absorption) and downward (stimulated emission) transitions. To obtain the net energy change in the radiation field we take the sum, over all possible initial and final states, of the transition rate times the energy of the transition, weighted by the probability p_i that the initial state is occupied:

$$\frac{-du}{dt} = -\dot{u} = \frac{N}{V} \sum_{i,f} p_i w_{if} \hbar \omega_{fi} \tag{5.4}$$

The energy density u is the energy per unit volume, and there are N molecules in the irradiated volume V. Note that in Equation 5.4 the sign on the left-hand side and the sign of the frequency ω_{fi} work together to account for absorption, which subtracts energy from the radiation field, and stimulated emission, which adds to it. The probability is given by the Boltzmann distribution, which is negligibly perturbed for the types of experiments for which the Golden Rule applies. Therefore, $p_i = z^{-1} \exp[-(\hbar \omega_{i0}/k_B T)]$, where z is the partition function, $\hbar \omega_{i0}$ is the energy of state i less that of the ground state, and k_B is Boltzmann's constant. We now substitute Equation 5.3 in Equation 5.4 and separate the two delta functions to get

$$-\dot{u} = \frac{\pi}{2\hbar} \frac{N}{V} \sum_{i,f} \left\{ p_i \omega_{fi} \left| \left\langle f \left| \vec{\mu} \cdot \vec{E}_0 \right| i \right\rangle \right|^2 \delta\left(\omega_{fi} - \omega\right) \right.$$

$$\left. + p_i \omega_{fi} \left| \left\langle f \left| \vec{\mu} \cdot \vec{E}_0 \right| i \right\rangle \right|^2 \delta\left(\omega_{fi} + \omega\right) \right\}$$

(5.5)

Since i and f are only dummy indices, we can interchange them in the second sum, and use the fact that $\omega_{if} = -\omega_{fi}$ to obtain

$$-\dot{u} = \frac{\pi}{2\hbar} \frac{N}{V} \sum_{i,f} \left(p_i - p_f\right) \omega_{fi} \left| \left\langle f \left| \vec{\mu} \cdot \vec{E}_0 \right| i \right\rangle \right|^2 \delta\left(\omega_{fi} - \omega\right)$$

(5.6)

Using Boltzmann's law, we can replace p_f by $p_i \exp(-\hbar\omega_{fi}/k_B T)$ and drop the subscripts on ω, since the delta function requires that ω equal ω_{fi}. The imaginary part of the relative permittivity* is given by $\varepsilon_r''(\omega) = -\dot{u}/\omega u$, where $u = (1/2)\varepsilon_0 E_0^2$ is the time averaged energy density. Dividing Equation 5.6 by ωu gives

$$\varepsilon_r''(\omega) = \frac{\pi}{\varepsilon_0 \hbar} (1 - e^{-\hbar\omega/k_B T}) \frac{N}{V} \sum_{i,f} p_i \left| \left\langle f \left| \vec{\mu} \cdot \hat{e} \right| i \right\rangle \right|^2 \delta\left(\omega_{fi} - \omega\right)$$

(5.7)

The unit vector $\hat{e}$ points in the direction in which the radiation is polarized. In the absence of a static field, this direction is arbitrary, so suppose that the radiation is x-polarized and it is the quantity $\left(\mu_{if}\right)_x^2$ that is needed in Equation 5.7. For an isotropic sample such as a liquid or gas, all three components of the transition moment are equal, so we can replace $\left(\mu_{if}\right)_x^2$ by $(1/3)\, \vec{\mu}_{if} \cdot \vec{\mu}_{fi}$.

The intensity $I(\omega)$ is *defined* to be

$$I(\omega) = \frac{3\hbar\varepsilon_0 \varepsilon_r''(\omega)}{\pi(1 - e^{-\hbar\omega/k_B T}) \frac{N}{V}} = \sum_{i,f} p_i \left\langle f \left| \vec{\mu} \right| i \right\rangle \cdot \left\langle i \left| \vec{\mu} \right| f \right\rangle \delta\left(\omega_{fi} - \omega\right)$$

(5.8)

For frequencies large compared to $k_B T/h$, the exponential function can be neglected compared to one.[†] In the opposite limit, $\hbar\omega \ll k_B T$, we can make the approximation: $\exp(-\hbar\omega/k_B T) \approx 1 - \hbar\omega/k_B T$. The quantity $\omega^2 I(\omega)$ is then proportional to the molar absorptivity.

Next, we introduce the integral representation of the delta function:

$$\delta(\omega) = \frac{1}{2\pi} \int_{-\infty}^{\infty} e^{i\omega t}\, dt$$

(5.9)

Here, it is convenient to write the above as follows:

$$\delta(\omega_{fi} - \omega) = \frac{1}{2\pi} \int_{-\infty}^{\infty} e^{i\left[(E_f - E_i/\hbar) - \omega\right]t}\, dt$$

(5.10)

After inserting Equation 5.10 into Equation 5.8, we can use the exponential functions of energy to express one of the dipole moment operators in the Heisenberg representation:

$$e^{i(E_f - E_i/\hbar)t} \left\langle f \left| \vec{\mu} \right| i \right\rangle \left\langle i \left| \vec{\mu} \right| f \right\rangle = \left\langle f \left| e^{i\hat{H}t/\hbar} \vec{\mu} e^{-i\hat{H}t/\hbar} \right| i \right\rangle \left\langle i \left| \vec{\mu} \right| f \right\rangle$$

$$= \left\langle f \left| \vec{\mu}(t) \right| i \right\rangle \left\langle i \left| \vec{\mu} \right| f \right\rangle$$

(5.11)

---

* Recall that the molar absorptivity ε_M is proportional to $\omega \varepsilon_r''$.

[†] Thermal energy $k_B T/hc$ is around 200 cm^{-1} at room temperature.

This step is valid when the states i and f are eigenfunctions of the *same* Hamiltonian. This will not be the case for electronic spectroscopy, which we do not consider at present. Note that $\vec{\mu} \equiv \vec{\mu}(0)$. The intensity can now be expressed as

$$I(\omega) = \frac{1}{2\pi} \sum_{i,f} \int_{-\infty}^{\infty} p_i \langle i|\vec{\mu}|f\rangle \cdot \langle f|\vec{\mu}(t)|i\rangle e^{-i\omega t} dt \tag{5.12}$$

The two matrix elements in Equation 5.12 are just numbers, so it is okay to interchange the order of multiplication. Next, we take advantage of the resolution of the identity, $\sum_f |f\rangle\langle f| = 1$, to remove the sum over f in Equation 5.12, resulting in

$$I(\omega) = \frac{1}{2\pi} \sum_{i} \int_{-\infty}^{\infty} p_i \langle i|\vec{\mu} \cdot \vec{\mu}(t)|i\rangle e^{-i\omega t} dt \tag{5.13}$$

This is the Fourier transform of an equilibrium average. Recall that:

$$\langle A\rangle_{eq} = \sum_{i} p_i \langle i|\hat{A}|i\rangle = Tr(\hat{\rho}\hat{A}) \tag{5.14}$$

Thus, Equation 5.13 is equivalent to

$$I(\omega) = \frac{1}{2\pi} \int_{-\infty}^{\infty} \langle \vec{\mu} \cdot \vec{\mu}(t)\rangle_{eq} e^{-i\omega t} dt \tag{5.15}$$

Equation 5.15 is the key result of this chapter. The spectral intensity $I(\omega)$ is the Fourier transform of the time-correlation function $C(t) \equiv \langle \vec{\mu}(0) \cdot \vec{\mu}(t)\rangle_{eq}$. We will henceforth drop the subscript eq as is done in the conventional notation. A very useful property of Fourier transforms is that they can be turned inside-out. The correlation function is the inverse Fourier transform of $I(\omega)$:

$$C(t) = \int_{-\infty}^{\infty} e^{i\omega t} I(\omega) d\omega \tag{5.16}$$

Note the limits are from minus to plus infinity, which is impossible to achieve experimentally. Nevertheless, the dynamics which take place on a certain timescale will influence the intensity at frequencies which are reciprocally related to the characteristic times. Applications of Equations 5.15 and 5.16 frequently involve vibrational bands in Raman and infrared spectra, in which case the band center is considered to be the origin of frequency. By using $\omega - \omega_0$ as the transform variable, high frequency oscillations in the correlation function are eliminated and one can concentrate on the dynamics which contribute to the width (rather than the position) of the spectrum.

By a suitable expression of the dipole moment operator in Equation 5.15, various spectroscopic experiments are accounted for as follows:

1. Microwave or far-infrared (far-IR) spectroscopy: $\vec{\mu} = \vec{\mu}_0$, the permanent dipole moment.
2. Infrared spectroscopy: $\vec{\mu} = (\partial\vec{\mu}/\partial Q)_0 Q$, where Q is a normal coordinate for the vibrational mode, to be discussed in Chapter 10.
3. Rayleigh scattering: $\vec{\mu} = \vec{\mu}_{ind} \propto \hat{e}_i \cdot \alpha \cdot \hat{e}_s$, where α is the polarizability, a second rank tensor, and $\hat{e}_i$ and $\hat{e}_s$ are the polarization directions of the incident and scattered radiation.
4. Raman scattering: $\vec{\mu} = \vec{\mu}_{ind} \propto (\hat{e}_i \cdot \alpha' \cdot \hat{e}_s) Q$, where α', also a second rank tensor, is the polarizability derivative with respect to the normal coordinate, $(\partial\alpha/\partial Q)_0$.

More generally, for an operator $\hat{A}$ which is responsible for the spectroscopic intensity, we should write $C(t) = \langle A^*(0) \cdot A(t)\rangle$, allowing for the possibility that $\hat{A}$ is complex. Since the dipole moment operator is real, we did not need to worry about the complex conjugate in the above derivation.

5.3 PROPERTIES OF TIME-CORRELATION FUNCTIONS AND SPECTRAL LINESHAPES

In this section we explore the mathematical features of correlation functions and intensity spectra. As correlation functions cannot necessarily be measured directly, they do not have to be real: $C(t) = C_{Re}(t) + iC_{Im}(t)$. As will be shown, classical systems have purely real correlation functions while quantum mechanical TCFs are complex. The real and imaginary parts are related to the even and odd parts of the intensity, respectively. To show this, we need the help of the detailed balance theorem, which states that $I(-\omega) = \exp(-\hbar\omega/k_BT)I(\omega)$.

To obtain the detailed balance theorem, we begin be deriving a relationship between a correlation function and its complex conjugate. Without assuming that $C(t)$ is real, we write

$$C(t) = \left\langle A^*A(t) \right\rangle = \sum_{i,f} p_i \left\langle i \middle| A^* \middle| f \right\rangle \left\langle f \middle| A(t) \middle| i \right\rangle \tag{5.17}$$

Then the complex conjugate of $C(t)$ is

$$C^*(t) = \left\langle AA^*(t) \right\rangle = \sum_{i,f} p_i \left\langle i \middle| A \middle| f \right\rangle \left\langle f \middle| A^*(t) \middle| i \right\rangle \tag{5.18}$$

Because the operator A is Hermitian, we can use the turn-over rule: $\left\langle i|A|f \right\rangle = \left\langle f|A^*|i \right\rangle$:

$$C^*(t) = \sum_{i,f} p_i \left\langle f \middle| A^* \middle| i \right\rangle \left\langle i \middle| A(t) \middle| f \right\rangle \tag{5.19}$$

The probabilities that state i and f are occupied satisfy $p_f = \exp(-\hbar\omega_{fi}/k_BT)p_i$, therefore:

$$C^*(t) = \sum_{i,f} p_f \exp(\hbar\omega_{fi}/k_BT) \left\langle f \middle| A^* \middle| i \right\rangle \left\langle i \middle| A(t) \middle| f \right\rangle \tag{5.20}$$

We want to relate $I(\omega)$, as given by

$$I(\omega) = \frac{1}{2\pi} \int_{-\infty}^{\infty} \left\langle A^*A(t) \right\rangle e^{-i\omega t} dt \tag{5.21}$$

to $I(-\omega)$:

$$I(-\omega) = \frac{1}{2\pi} \int_{-\infty}^{\infty} \left\langle A^*A(t) \right\rangle e^{i\omega t} dt \tag{5.22}$$

We know that the intensity is real; thus, $I(\omega) = I^*(\omega)$. Taking the complex conjugate of Equation 5.21, and using Equation 5.20, we obtain

$$I(\omega) = \frac{1}{2\pi} \int_{-\infty}^{\infty} \left\langle AA^*(t) \right\rangle e^{i\omega t} dt$$

$$= \frac{1}{2\pi} \int_{-\infty}^{\infty} \sum_{i,f} p_f \exp(\hbar\omega_{fi}/k_BT) \left\langle f \middle| A^* \middle| i \right\rangle \left\langle i \middle| A(t) \middle| f \right\rangle e^{i\omega t} dt \tag{5.23}$$

$$= \exp(\hbar\omega/k_BT)\left(\frac{1}{2\pi}\right)\int_{-\infty}^{\infty} \sum_{f} p_f \left\langle f \middle| A^*A(t) \middle| f \right\rangle e^{i\omega t} dt$$

$$= \exp(\hbar\omega/k_BT)I(-\omega)$$

In the third line of Equation 5.23, it has been recognized that ω_{fi} has to equal ω. This can be shown by writing $\left\langle i|A(t)|f \right\rangle$ in the Heisenberg representation, and expressing $\delta(\omega-\omega_{fi})$ in the integral representation as in Equation 5.10. This allows closure (the resolution of the identity) to be used to remove the sum over i. Since summing over all states f, $\sum_f p_f \langle f | A^* A(t) | f \rangle$ gives $\langle A^* A(t) \rangle$ of Equation 5.22, we recover the desired relationship:

$$I(-\omega) = \exp(-\hbar\omega/k_B T)I(\omega) \tag{5.24}$$

Detailed balance is a consequence of equilibrium; the rate of transitions into the state is balanced by transitions out of the state. The factor $\exp(-\hbar\omega/k_B T)$ is the ratio of the Boltzmann population of the upper state to that of the lower state. In the classical limit, $\hbar$ can be put equal to zero, and the intensity is an even function of frequency.

As with any function whose argument can range over positive and negative values, the intensity can be broken up into odd and even parts as follows:

$$I_{odd}(\omega) = \frac{1}{2}\left[I(\omega) - I(-\omega)\right] \tag{5.25}$$

$$I_{even}(\omega) = \frac{1}{2}\left[I(\omega) + I(-\omega)\right] \tag{5.26}$$

such that $I(\omega) = I_{odd}(\omega) + I_{even}(\omega)$. The intensity can also be written as

$$I(\omega) = \frac{1}{2\pi}\int_{-\infty}^{\infty}\left[C_{Re}(t) + iC_{Im}(t)\right]\left[\cos\omega t - i\sin\omega t\right]dt \tag{5.27}$$

Since $I(\omega)$ is real, the imaginary part of the right-hand side of Equation 5.27 vanishes. Therefore

$$0 = \frac{1}{2\pi}\int_{-\infty}^{\infty}\left[\cos\omega t\, C_{Im}(t) - \sin\omega t\, C_{Re}(t)\right]dt \tag{5.28}$$

Equation 5.28 will hold if the imaginary part of $C(t)$ is an odd function of time and the real part an even function: that is Re $C(t) = $ Re $C(-t)$ and Im $C(t) = -$Im $C(-t)$. Equivalently, we can write $C^*(t) = C(-t)$. The classical part of the correlation function displays the time reversal symmetry dictated by Newton's laws. Equations 5.27 and 5.28 lead to the conclusion that the even part of the intensity depends on the real part of the correlation function and the odd part of the intensity on the imaginary part. With the help of the detailed balance theorem, it can be shown that $I_{odd}(\omega) = \tanh(\hbar\omega/2k_B T)I_{even}(\omega)$. Since $\lim_{x\to 0}\tanh x = 0$, the odd part of the intensity vanishes as $\hbar \to 0$, meaning that $I_{odd}(\omega)$ and $C_{Im}(t)$ are purely quantum mechanical quantities. Let us concentrate on classical correlation functions and explore the form of the intensity for various functional forms for $C(t)$. For example,

1. If $C(t) = C(0)\cos\omega_0 t$, then $I = I_0\delta(\omega-\omega_0)$.
2. If $C(t) = C(0)\exp(-t^2/\tau^2)$, then $I = I_0\exp(-\omega^2\tau^2/2)$.
3. If $C(t) = C(0)\exp(-|t|/\tau)$, then $I = I_0/(1 + \omega^2\tau^2)$.

The first example would apply to a hypothetical case of a molecule rotating with a constant frequency of ω_0 without interruption. This is physically unrealistic, but if such a situation were possible, the resulting spectrum would be an infinitely sharp spike at the frequency of rotation. The second example applies to a Gaussian correlation function, a limit observed at short times in the case of a classical linear free rotor (to be discussed later). In this case, the intensity is also a Gaussian function, having a frequency distribution which is inversely related to the decay time for the correlation function. In other words, the Fourier transform of a fat Gaussian is a skinny Gaussian, and vice versa. It is a quite general and important result that the more rapidly the correlations decay, the broader will be the frequency distribution, and vice versa. This is another manifestation of the time-energy uncertainty principle. In the last example, an exponential correlation function gives rise to a Lorentzian function of frequency. The Lorentzian lineshape applies to the case of a rotational Brownian diffuser, a long-time limit that is valid after a rotating molecule has suffered a large number of reorienting collisions. Since this picture holds at relatively long times, it can only account for the low-frequency part of the intensity.

5.4 THE FLUCTUATION–DISSIPATION THEOREM

The use of correlation functions in chemical physics is not limited to the spectroscopic examples discussed above. Whenever one is interested in the effect of a weakly perturbing field or a mechanical or thermal probe, the use of linear response theory is appropriate. In this section we summarize the key equations of linear response theory without deriving them. The interested reader is referred to the review article by Berne and Harp [2] for more details on this approach. The goal here is to present a general form for the wave vector- and frequency-dependent spectral response, which can be employed when collective motions influence the spectrum. The spatial correlations in the dynamics give rise to the k-dependence of the spectrum, just as time correlations affect the frequency dependence.

We consider a generalized susceptibility function χ_{AB} having real χ'_{AB} and imaginary χ''_{AB} parts related by the Kramers–Kronig transforms. The notation is based on the idea that there is a perturbing field $\mathbf{F}$, the perturbation Hamiltonian is of the form $\hat{H}' \propto B \cdot F$, and A is some physical property whose response to the perturbation will be probed. The average value of the physical property $A(\mathbf{R}, t)$ depends on the perturbation at all positions $\mathbf{R}'$ and at all times t' prior to t. This dependence is expressed with the help of the after-effect function Φ_{AB}, where

$$\langle A(\mathbf{R},t) \rangle = \int_{-\infty}^{t} dt' \int d\mathbf{R}' \Phi_{AB}(\mathbf{R},\mathbf{R}';t,t')F(\mathbf{R}',t') \tag{5.29}$$

Vector quantities are in bold typeface in this section. Equation 5.29 is a mathematical statement of *causality*. The after-effect function, a real quantity, conveys all the information about the response of the system to the perturbing field. It takes into account any time lag between the perturbation and the measured response. For frequency-dependent fields of interest to spectroscopy, we know that the imaginary part of the susceptibility determines the out-of-phase response of the system, associated with energy loss due to absorption. As written, the after-effect function depends on two space and two time coordinates, but a physically reasonable function for Φ_{AB} should depend only on displacements in space and time: $\Phi_{AB}(\mathbf{R}, \mathbf{R}';t, t') = \Phi_{AB}(\mathbf{R} - \mathbf{R}';t - t')$. We want to show how the after-effect function relates to the susceptibility by considering a monochromatic perturbing field $\mathbf{F}_{k\omega}$. For linear response we expect that

$$\langle A(\mathbf{R},t) \rangle = \chi_{AB}(\mathbf{k},\omega)\mathbf{F}_{k\omega} \exp[i(\mathbf{k} \cdot \mathbf{R} - \omega t)] \tag{5.30}$$

The susceptibility depends on both wave vector and frequency. To be consistent with the previous expression, it must be the Fourier transform of the after-effect function

$$\chi_{AB}(\mathbf{k},\omega) = \int_{0}^{\infty} dt \int d\mathbf{R} \Phi_{AB}(\mathbf{R},t) \exp[i(\mathbf{k} \cdot \mathbf{R} - \omega t)] \tag{5.31}$$

Since $\Phi_{AB}(\mathbf{R}, t)$ is real, the real and imaginary parts of $\chi_{AB}(\mathbf{k},\omega)$ are cosine and sine transforms, respectively, of the after-effect function. Equation 5.31 reveals why the real and imaginary parts of χ must be related, since they are cosine and sine transforms, respectively, of the same function. As shown in [2], the spatial Fourier transform of Φ_{AB} is related to a commutator of the Fourier components of the operators A and B:

$$\Phi_{AB}(\mathbf{k},t) = \int \exp(i\mathbf{k} \cdot \mathbf{R}) \Phi_{AB}(\mathbf{R},t) d\mathbf{R}$$

$$= \left\langle \frac{i}{\hbar}[A_k(t), B_{-k}] \right\rangle \tag{5.32}$$

In Equation 5.32, the Fourier components of the operators are defined by

$$A_k = \sum_{j} \frac{1}{2}[A_j, e^{i\mathbf{k} \cdot \mathbf{R}_j}]_+ \tag{5.33}$$

where the sum is over all the molecules, and $\mathbf{R}_j$ is the position of molecule j. The subscript $+$ indicates an anticommutator: $[\hat{A},\hat{B}]_+ = \hat{A}\hat{B} + \hat{B}\hat{A}$. If the operator $\hat{A}$ commutes with the position variable, then the Fourier component is just $A_k = \sum_j \hat{A}_j e^{i\mathbf{k}\cdot\mathbf{R}_j}$.

We can express χ_{AB} as

$$\chi_{AB}(\mathbf{k},\omega) = \int_0^\infty \Phi_{AB}(\mathbf{k},t)e^{-i\omega t}\,dt \tag{5.34}$$

In addition to the k-dependence of the operators in Equation 5.32, the correlation function expressed here differs from the previously discussed TCFs in that a commutator of two operators is involved, and the potential for considering cross-correlations $(A \neq B)$ exists.

Now consider the susceptibility χ_{AA} of interest in a spectroscopy experiment. The perturbation Hamiltonian is of the form $\hat{H}' = -\vec{\mu}\cdot\vec{E}$. The measured response is the polarization $\vec{P} = \chi_e\varepsilon_0\vec{E}$, which is the dipole moment per unit volume. So $A = B = \sum_i \mu_i$ in the calculation of the susceptibility. The imaginary part of the susceptibility determines the rate of change of the radiation energy density.

$$\frac{-du}{dt} = \frac{1}{2}\omega\chi_{AA}''(\mathbf{k},\omega)|E_{k,\omega}|^2 \tag{5.35}$$

As shown in [2], χ_{AA}'' is given by

$$\chi_{AA}''(\mathbf{k},\omega) = \frac{1}{\hbar}\tanh\left(\frac{\hbar\omega}{2k_BT}\right)\int_{-\infty}^{\infty} dt\, e^{i\omega t}\left\langle \frac{1}{2}[A_k(t),A_{-k}]_+\right\rangle \tag{5.36}$$

Equation 5.36 is the fluctuation–dissipation theorem. In the limit that $k \to 0$, it is equivalent to Equation 5.15, as will be shown in Problem 5.

Linear response theory has been applied to many spectroscopic techniques, such as Mossbauer spectroscopy, neutron scattering, and magnetic resonance. In addition, it is a useful way to approach transport problems, such as translational and rotational diffusion and thermal and electrical conductivity. In the case where the correlation function can be considered classical and independent of wave vector, the fluctuation–dissipation theorem is equivalent to the TCF derived in Section 5.2.

In the next two sections, the low-frequency reorientational spectroscopy of liquids is considered, where the k-dependence of light plays a potentially important role in the analysis, due to the collective nature of the molecular dynamics.

5.5 ROTATIONAL CORRELATION FUNCTIONS AND PURE ROTATIONAL SPECTRA

The rotational transitions of molecules in the gas phase are observed using microwave absorption or emission, or Raman scattering, as will be discussed in Chapter 8. Here, we would like to consider the spectra of molecules which do not rotate freely and therefore do not have resolvable rotational eigenstates. The pure rotational spectra of molecules in the liquid phase can be observed by means of dielectric relaxation, far-IR absorption, or depolarized Rayleigh scattering. In recent years, new terahertz sources [14–16] operating in the frequency range 0.1 – 10 THz (about 3 – 300 cm^{-1}) have expanded opportunities for measuring reorientational relaxation and low-frequency intermolecular vibrations of liquids in the far-IR region of the spectrum. The dielectric relaxation experiment, to be discussed in Section 5.6.1, involves a measurement of the real and imaginary parts of the relative permittivity at frequencies in the microwave and radio-frequency range. The far-infrared absorption spectrum is essentially an extension of the dielectric relaxation experiment to higher frequencies, though the real and imaginary parts of the response are not always measured separately. Both these techniques observe the dynamics of rotating dipoles and, to a first approximation, the observed signal depends on the existence of a permanent dipole moment. The observation of depolarized Rayleigh scattering requires that the molecular polarizability tensor be anisotropic. Other techniques for measuring rotational relaxation times, not considered here, include polarized fluorescence and nuclear magnetic resonance.

5.5.1 CORRELATION FUNCTIONS FOR ABSORPTION AND LIGHT SCATTERING

The correlation function appropriate to far-IR absorption is $\langle \vec{\mu}_0 \cdot \vec{\mu}_0(t) \rangle$, where $\vec{\mu}_0$ is the permanent dipole moment. This form of the TCF neglects the possibility of collective motion, which will be considered later. The magnitude of the dipole moment influences the total intensity but not the lineshape. Thus it is convenient to work with the normalized correlation function $C_1(t) = \langle \hat{u} \cdot \hat{u}(t) \rangle$, where $\hat{u}$ is a unit vector in the dipole direction. This form of the TCF has the value of one at time zero, and results in a spectral lineshape $I(\omega)$, which is normalized to unit area. The subscript 1 reminds us that this correlation function is the same as $\langle P_1(\cos\theta_t) \rangle$, where θ_t is the angle through which the dipole has rotated after time t, and $P_1(x) = x$ is the first Legendre polynomial. The reason for this notation is that the corresponding correlation function for depolarized light scattering of axially symmetric molecules is $\langle P_2(\cos\theta_t) \rangle$, where the second Legendre polynomial is $P_2(x) = (1/2)(3x^2 - 1)$.

In depolarized Rayleigh scattering, the scattered light is detected having a polarization perpendicular to that of the incident radiation. The polarized Rayleigh spectrum detects light scattered with the same polarization direction as the incident light. Rayleigh scattering detects rotational and translational dynamics, while Raman scattering is due to vibrational transitions. The experimental configuration for the polarized and depolarized light scattering experiment (Rayleigh or Raman) will be discussed in Chapter 6. The intensity of depolarized light scattering depends on the anisotropy of the polarizability tensor in the molecular frame of reference. The polarizability tensor of a nonvibrating molecule, i.e., one that is frozen in its equilibrium geometry, reflects its symmetry, and one can always find a reference frame that diagonalizes this tensor:

$$\alpha = \begin{pmatrix} \alpha_{xx} & 0 & 0 \\ 0 & \alpha_{yy} & 0 \\ 0 & 0 & \alpha_{zz} \end{pmatrix} \tag{5.37}$$

Equation 5.37 represents the polarizability in a molecule frame of reference, whereas the light scattering spectrum depends directly on the polarizability tensor in the laboratory frame of reference. In Chapter 8, we will examine the mathematical details of converting the polarizability tensor from one reference frame to another.

Consider spherical molecules, such as atoms and molecules of tetrahedral or octahedral symmetry. These have isotropic polarizability tensors: $\alpha_{xx} = \alpha_{yy} = \alpha_{zz}$. In this case, the induced dipole moment, which acts as a source for scattered radiation (see Equation 2.37), is always in the same direction as the electric field vector of the incident radiation. Thus there is no depolarized Rayleigh scattering for molecules with isotropic polarizability tensors. Isotropic molecules, for example CCl_4 or SF_6, do give rise to polarized Rayleigh scattering, but in these systems the light scattering is insensitive to the orientation of the molecule. The observation of a pure rotational spectrum in the form of depolarized Rayleigh scattering requires that the molecular polarizability be anisotropic. This is a called a *gross selection rule*, and it is analogous to the requirement that a molecule have a permanent dipole moment in order to be active in pure rotational absorption and emission.

Now, the truth is that weak depolarized Rayleigh scattering is observed even for spherical molecules, and nonpolar liquids absorb far-IR radiation weakly. The more liberal selection rules are a consequence of collisions and intermolecular interactions in the liquid, which result in changes to the polarizability and dipole moment of the isolated molecule. The correlation functions of this section account for the "allowed" part of the pure rotation spectrum.

We focus here on molecules having axial symmetry, in which case $\alpha_{xx} = \alpha_{yy} \neq \alpha_{zz}$. The polarizability anisotropy is then defined as $\beta = (\alpha_{zz} - \alpha_{xx})$, the square of which scales the intensity of depolarized light scattering. Similarly, the far-IR intensity is proportional to the square of the permanent dipole moment, μ_0, in the absence of interaction-induced effects. The normalized TCF for the pure rotational spectrum, in the absence of collective dynamics, is given by

$$C_l(t) = \langle P_l(\cos\theta_t) \rangle \tag{5.38}$$

where $l = 1$ for absorption and $l = 2$ for depolarized Rayleigh scattering.

5.5.2 CLASSICAL FREE–ROTOR CORRELATION FUNCTION AND SPECTRUM

To get some insight into the nature of correlation functions, let us consider an ensemble of classical free rotors having axial symmetry. The dipole moment, if it exists, must coincide with the symmetry axis, taken to be the z direction in the molecule frame.

We write the correlation function for pure rotational spectroscopy as $\langle P_l(\cos\theta_t)\rangle$, where the angle brackets denote an equilibrium average. To compute the classical equilibrium average, we need the normalized probability distribution $p(\Omega)$ for the rotational velocity $\Omega = \dot{\theta}$ [3]:

$$p(\Omega) = \frac{I_z\Omega}{k_BT}\exp\left(\frac{-I_z\Omega^2}{2k_BT}\right) \tag{5.39}$$

I_z (not to be confused with the intensity) is the moment of inertia for rotation *of* the symmetry axis (the molecule z-axis). (Rotation *about* the symmetry axis does not change the dipole moment or polarizability of an axially symmetric molecule.) The angle θ_t though which the molecule has rotated after time t is Ωt. The desired correlation functions are found by evaluating the following integral:

$$C_l(t) = \langle P_l(\cos\theta_t)\rangle = \frac{I_z}{k_BT}\int_0^\infty P_l[\cos(\Omega t)]\exp\left(\frac{-I_z\Omega^2}{2k_BT}\right)d\Omega \tag{5.40}$$

The resulting correlation functions, obtained by numerical integration of Equation 5.40, are displayed in Figure 5.3 as a function of dimensionless reduced time: $t^* = (k_BT/I_z)^{1/2}t$. There are several significant features of $C_l(t)$. One is that $C_2(t)$ decays twice as rapidly as $C_1(t)$. The basis for this is that, for axially symmetric molecules, the polarizability rotates twice as fast as the dipole moment. Rotation by 180° returns the polarizability to its original value, while the dipole moment must rotate 360° to be restored to its original value. This feature is related to the selection rules to be discussed in Chapter 8, where the frequency spacing of pure rotational spectra will be found to be twice as large in Raman scattering as in microwave absorption. The present discussion, however, is concerned with classical rotors, for which the spectrum does not consist of a series of discrete lines, in contrast to the situation for quantum mechanical rotors.

Another interesting feature is that each correlation function starts out looking like a Gaussian function at short times. As a result, the wings of the spectrum should also approach Gaussian functions of frequency. Furthermore, the $l = 2$ correlation function, unlike that for $l = 1$, does not decay to zero at long time. This is a consequence of the difference in the functional forms of the Legendre polynomials. In any Fourier transform pair, the area under one function is proportional to the value of the other at the origin. The persistence of $C_2(t)$ at long time results in a delta function spike in the intensity at zero frequency. If translational motion were accounted for, $C_2(t)$ would eventually decay and this zero frequency peak would have a finite width.

Taking the Fourier transform of Equation 5.40, we can evaluate the classical spectra of the free rotors as follows (see Problem 1):

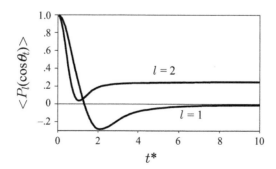

Figure 5.3 Free-rotor correlation functions.

$$I_1(\omega) = \frac{I_z|\omega|}{2k_BT} e^{-I_z\omega^2/2k_BT} \tag{5.41}$$

$$I_2(\omega) = \frac{3I_z|\omega|}{16k_BT} e^{-I_z\omega^2/8k_BT} + \delta(\omega) \tag{5.42}$$

Notice that the absolute value of the frequency appears in each expression. In the case of Equation 5.41, negative frequencies correspond to stimulated emission. In 5.42, they correlate with anti-Stokes scattering, for which the frequency of the scattered radiation is higher than that of the incident radiation. However, since the above expressions are for classical rotors, the intensities are even functions of frequency: $I(\omega) = I(-\omega)$. This is expected from the detailed balance relation if we take the limit $\hbar \to 0$.

5.6 REORIENTATIONAL SPECTROSCOPY OF LIQUIDS: SINGLE-MOLECULE AND COLLECTIVE DYNAMICS

In this section we examine in more detail the nature of pure rotational spectra of liquids. The hindered rotational motions of molecules in liquids and solutions are responsible for several varieties of low-frequency spectroscopic techniques. In none of these methods is rotational structure observed; information about reorientational dynamics must be obtained by analysis of the spectral lineshape. We consider three experimental approaches and the corresponding theory for interpreting the data. The first of these, dielectric relaxation, is a very old and very low-frequency (less than about 1 cm^{-1}) experiment. Far-IR absorption and depolarized Rayleigh scattering, in the vicinity of 1 to 100 cm^{-1}, also reflect reorientational dynamics, and the time-domain views of these two types of spectra will be presented.

5.6.1 DIELECTRIC RELAXATION

In Chapter 3, we looked at the relative permittivity $\varepsilon_r(\omega)$ as a means of understanding the polarization induced in a sample by an applied electric field: $\vec{P} = (\varepsilon_r - 1)\varepsilon_0\vec{E}$. Naturally, the frequency dependence of the permittivity (or dielectric function) depends on the time-dependent response of the system. Imagine that a static electric field is suddenly applied to a sample. The polarization will experience an initial instantaneous response, dependent on the high-frequency permittivity ε_∞ and then increase more slowly to its final value, a function of the static permittivity ε_s. Alternatively, the field could be turned off instantaneously and the system allowed to relax back to its unpolarized state. The response function (or relaxation function) which governs this process is called $\Phi(t)$, and the connection between the time and frequency responses is given by

$$\varepsilon_r(\omega) = \varepsilon_\infty + \int_0^\infty e^{i\omega t}\Phi(t)dt \tag{5.43}$$

Equation 5.43 can be compared to Equation 5.34. Recall that susceptibility and relative permittivity are related through $\varepsilon_r = 1 + \chi_e$. Equation 5.43 comprises the expressions for the real and imaginary parts of the permittivity, which are related via the Kramers–Kronig expressions given in Chapter 3 (Equation 3.56). They are

$$\varepsilon_r'(\omega) = \varepsilon_\infty + \int_0^\infty \cos\omega t\, \Phi(t)dt \tag{5.44}$$

$$\varepsilon_r''(\omega) = \int_0^\infty \sin\omega t\Phi(t)dt \tag{5.45}$$

The real and imaginary parts of the permittivity determine the part of the polarization that is respectively in phase or in quadrature with the time varying field. A typical dielectric relaxation experiment measures both

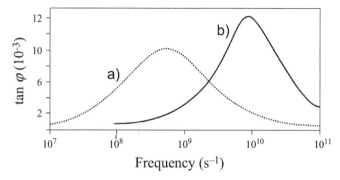

Figure 5.4 Loss angle of benzophenone (a) in paraffin and (b) in benzene. (From H. Fröhlich, *Theory of Dielectrics* 1986, by permission of Oxford University Press.)

ε_r' and ε_r'' at frequencies from about 10^7 to 10^{11} Hz. One way to represent the data is to plot $\varepsilon_r''(\omega)$ versus $\varepsilon_r'(\omega)$. This is known as a Cole–Cole plot. Another representation of the data invokes the definition of the loss angle φ:

$$\tan\varphi = \frac{\varepsilon''(\omega)}{\varepsilon'(\omega)} \tag{5.46}$$

The loss angle is the phase lag between the induced polarization and the applied field. Figure 5.4 shows some dielectric relaxation data for benzophenone in terms of this loss angle, taken from [4].

The simplest treatment of rotational relaxation in liquids is known as the Debye model. This picture is based on the idea that rotation can be treated as diffusional. The connection between rotational and translational diffusion can be made by describing the former in terms of the orientation vector of the molecule tracing a path on the surface of a sphere. If the path consists of many small steps, with random angles between successive steps, then we speak of rotational diffusion. In reality, this picture is valid only at long times, where the net displacement of the orientation vector can be viewed as the sum of many small steps, and the direction of each small step is uncorrelated with that of the previous step. This long-time caveat implies that the Debye model is valid only at low frequencies, and we shall see that this is the case. In the diffusional limit, the relaxation function is exponential:

$$\Phi(t) \propto e^{-t/\tau_D} \tag{5.47}$$

The Debye theory yields the following expressions for the relative permittivity:

$$\varepsilon_r'(\omega) = \varepsilon_\infty + \frac{\varepsilon_s - \varepsilon_\infty}{1 + \omega^2 \tau_D^2} \tag{5.48}$$

$$\varepsilon_r''(\omega) = \frac{(\varepsilon_s - \varepsilon_\infty)\omega\tau_D}{1 + \omega^2 \tau_D^2} \tag{5.49}$$

Hydrodynamic theory can be applied to estimate the Debye relaxation time τ_D. For example, in a dilute solution of polar molecules in a nonpolar solvent, the Debye time is given by

$$\tau_D = \frac{4\pi\eta a^3}{k_B T} \tag{5.50}$$

where η is the viscosity and a is the radius of the solute molecule. Typical Debye times are in the range of about 0.1 to 100 ps. For example, τ_D is 4 ps for acetonitrile and 20.6 ps for dimethylsulfoxide. The Cole–Cole plot for a Debye liquid with a single relaxation time is a semicircle. Deviations from a semicircle indicate more complex behavior. In alcohols, for example, torsional motion of the hydroxyl group and the breaking of hydrogen bonds contribute to the polarization relaxation, in addition to overall rotation of the molecule. The dielectric relaxation of alcohols is typically characterized by three, rather than one, relaxation times.

Recalling that the absorption coefficient is proportional to $\omega\varepsilon_r''(\omega)$, it can be shown that the Debye model predicts that the absorption tends to a constant value at high frequency ($\omega^2\tau_D^2 \gg 1$). This "Debye plateau" is completely unphysical; it is an artifact of the assumption of rotational diffusion. The short-time behavior of rotational motion must be accounted for properly in order to get the high-frequency response right. This leads naturally to a discussion of far-infrared absorption spectroscopy, which may be considered to be an extrapolation of $\omega \varepsilon_r''(\omega)$ measured in a dielectric relaxation experiment.

In truth, dielectric relaxation is a collective phenomenon, which means that we cannot neglect the wave vector (k) dependence of the relevant correlation functions. The variable of interest is thus the Fourier component of the total dipole moment:

$$\vec{M} = \sum_i \vec{\mu}_i \exp\left(i\vec{k} \cdot \vec{R}_i\right) \tag{5.51}$$

where $\vec{R}_i$ is the position of molecule i and k is the wave vector. The correlation function for the total dipole moment is

$$\left\langle \vec{M}^* \cdot \vec{M}(t) \right\rangle = \sum_{ij} \left\langle \vec{\mu}_i \cdot \vec{\mu}_j(t) \exp\left[i\vec{k} \cdot \left(\vec{R}_j(t) - \vec{R}_i \right) \right] \right\rangle \tag{5.52}$$

Equation 5.52 applies to both dielectric relaxation and far-IR experiments.

We consider here the longitudinal and transverse relaxation behavior. The former refers to the component of the polarization which is parallel to the propagation vector $\vec{k}$ and the latter represents that part which is perpendicular to $\vec{k}$. The long-range order in a dipolar liquid results in different polarizations in the longitudinal and transverse directions. The Debye time actually corresponds to transverse relaxation. Consider the qualitative picture of dipolar alignment shown in Figure 5.5. (The orientational and translational order in this Figure is exaggerated for illustration purposes.) Relative to a central dipole, dipoles in the direction parallel to the central one tend to be aligned in parallel fashion, while those in a perpendicular direction tend to be parallel or antiparallel with equal probability. Let us examine the effect of this long-range order on the longitudinal (parallel to $\vec{k}$) and transverse (perpendicular to $\vec{k}$) polarizations.

The phase factor $\exp(i\vec{k} \cdot \vec{R})$ plays an important role in the relaxation times for the longitudinal and transverse polarization. Consider the longitudinal component of $\langle |M|^2 \rangle$, where M is the collective dipole moment:

$$\left\langle |M_Z|^2 \right\rangle = \sum_{i,j} \left\langle \mu_Z^i \mu_Z^j \exp(i\vec{k} \cdot \vec{R}_{ij}) \right\rangle \tag{5.53}$$

Imagine that $\vec{k}$ propagates in the Z direction and consider molecule i to be aligned with Z. The phase factor $\exp(i\vec{k} \cdot \vec{R}_{ij})$ oscillates rapidly and tends to cancel out contributions to the sum over j unless $\vec{R}_{ij}$ is orthogonal to $\vec{k}$. In directions perpendicular to the alignment of μ_i, about half of the neighboring dipoles, on the average, are aligned parallel and the other half antiparallel to the dipole moment of i. The result is that $\langle |M_Z|^2 \rangle$ is reduced from the value expected in the absence of order, by an amount dependent on the static dielectric constant. Since

Figure 5.5 Relative alignment of dipoles in a polar liquid.

individual molecules do not need to rotate a full 180° in order to change the direction of the longitudinal polarization, the longitudinal relaxation time τ_L is actually less than the single-molecule rotational relaxation time.

To consider the transverse components $\langle |M_X|^2 \rangle$ and $\langle |M_Y|^2 \rangle$, imagine that the wave vector is perpendicular to the central dipole. Now the main contributions to the sum over j will be from molecules in directions parallel to μ_i. These dipoles tend to align parallel to the central dipole, enhancing the values of $\langle |M_X|^2 \rangle$ and $\langle |M_Y|^2 \rangle$ compared to $\langle |M_Z|^2 \rangle$. This reinforcement of neighboring dipoles results in a transverse relaxation time which is greater than that of a single molecule. The ratio of the transverse to longitudinal relaxation times is given by

$$\frac{\tau_T}{\tau_L} = \frac{\varepsilon_0}{\varepsilon_\infty} \tag{5.54}$$

It is the long-range nature of the dipolar interactions that causes these two relaxation times to differ. In polar liquids, $\varepsilon_0 > \varepsilon_\infty$, so the transverse relaxation time exceeds the longitudinal relaxation time. The magnitudes of τ_L and τ_D are important in theories of solvent effects on electron transfer and other reactions.

5.6.2 FAR-INFRARED ABSORPTION

In contrast to the plateau predicted by the Debye theory, the far-IR spectrum of a polar liquid consists of a broad and strong absorption with a peak frequency in the vicinity of about 50 cm⁻¹. Early literature reports of this type of spectrum, once referred to as the Poley absorption, attempted to assign the peak absorption frequency as if it were a discrete vibration, which it is not. The word libration is often used to refer to a hindered harmonic motion of the permanent dipole moment. We will use the time-dependent view of spectroscopy to take a more complete view of the basis for far-IR spectra of liquids. While it is mostly the reorientational motion of permanent dipoles which contributes to the response, the dipoles induced by collisions and by the fields of neighboring molecules are also represented. This greatly complicates the analysis of far-IR lineshapes. For a review that encompasses both dielectric phenomena and far-IR spectra, see [5].

The molar absorptivity ε_M is related to the intensity $I(\omega)$ as follows:

$$\varepsilon_M(\omega)n(\omega) \propto \omega(1 - e^{-\hbar\omega/k_BT})I(\omega) \tag{5.55}$$

The factor $[1 - \exp(-\hbar\omega/k_BT)]$ has the effect of subtracting the stimulated emission rate from the absorption rate to get the net intensity of light absorbed by the sample. In far-IR spectroscopy at room temperature, where the frequency $\tilde{\nu}$ of the source is less than 100 cm⁻¹ and $k_BT/hc \approx 200$ cm⁻¹, this term plays a role in deciding the frequency at which the measured absorption is a maximum. Also note that the expected dispersion in the refractive index, $n(\omega)$, becomes important whenever the absorption is quite strong. In this case, the frequency dependence of n contributes to the observed band shape and should be measured independently in order to extract the intensity. It is this intensity that is directly related to the time-correlation function:

$$I(\omega) = \frac{1}{2\pi} \int_{-\infty}^{\infty} \langle \vec{M} \cdot \vec{M}(t) \rangle e^{-i\omega t} \, dt \tag{5.56}$$

The variable $\vec{M}$ is the collective dipole moment of the sample, defined in Equation 5.51. Neglecting the k-dependence of this variable, the correlation function appropriate to the far-IR spectrum is

$$C_{FIR}(t) = \sum_i \langle \vec{\mu}_i \cdot \vec{\mu}_i(t) \rangle + \sum_{i \neq j} \langle \vec{\mu}_i \cdot \vec{\mu}_j(t) \rangle$$

$$= N \langle \vec{\mu}_1 \cdot \vec{\mu}_1(t) \rangle + N(N-1) \langle \vec{\mu}_1 \cdot \vec{\mu}_2(t) \rangle \tag{5.57}$$

The summations in Equation 5.57 run over all molecules in the irradiated volume, and the second line follows from the equivalence of all molecules in a fluid sample. The $\langle \vec{\mu}_1 \cdot \vec{\mu}_2(t) \rangle$ term involves pairs of molecules, and while it complicates the analysis of far-IR spectra, it indicates the importance of intermolecular interactions in the interpretation of such spectra.

Absorption is frequently expressed in terms of the absorption coefficient $\gamma(\omega)$, introduced in Chapter 3,

$$\gamma(\omega) = \frac{\omega \varepsilon_r''(\omega)}{n(\omega)c} \tag{5.58}$$

It can be checked that the units on γ are cm^{-1}. To avoid confusion with wavenumber units, however, the power absorption coefficient is usually reported in neper/cm, where the word "neper" connotes a logarithmic scale analogous to the decibel scale but referred to the natural rather than the common logarithm. (Many references represent the absorption coefficient of Equation 5.58 by the symbol α, but we have reserved that Greek letter for the polarizability.) Figure 5.6 shows the far-IR absorption spectrum of CH_3CN in the neat liquid and in two solutions with CCl_4. The corresponding TCFs are also shown. The bulk of the spectral response in these data is due to the reorientation of permanent dipole moments. Note that, compared to the relaxation in neat acetonitrile, in more concentrated solution $C(t)$ decays more slowly. In contrast, the relaxation in dilute solution is more rapid than in the neat liquid. It is interesting to speculate on the microscopic properties that could be responsible for this behavior. The importance of the pair terms in Equation 5.57 should increase with concentration of the polar molecule.

Equation 5.57 assumes that the intensity of far-IR absorption is due only to permanent dipole moments. The fact that weak far-IR absorption is observed in nonpolar liquids indicates that induced moments lead to rotational spectra as well. These moments may result from the internal field due to the surrounding molecules and from collisional impacts which distort the molecular framework. Similarly, the weak depolarized Rayleigh scattering of spherical molecules such as CCl_4 is a consequence of interactions that distort the polarizability. The observation of nominally forbidden spectra suggests that allowed rotational spectra depend on these interaction-induced effects as well. If the total correlation function could be expressed as a sum of the allowed and forbidden parts, then the total intensity would also be a sum of these two contributions.

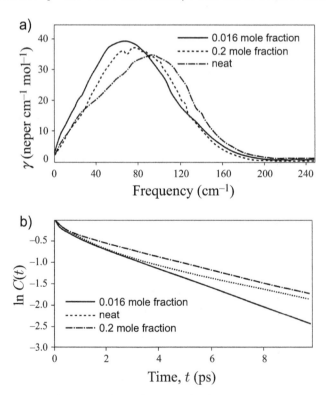

Figure 5.6 Far-IR spectra (a) and time-correlation functions (b) of acetonitrile, neat and in solution with carbon tetrachloride. (Reprinted from *Spectroscopy and Relaxation of Molecular Liquids*, D. Steele and J. Yarwood, eds, Elsevier Science Publishers, Amsterdam, 1991, with permission.)

Unfortunately, cross correlations of permanent and induced quantities prevent the spectrum from being considered to be the sum of pure rotational and interaction-induced components. For more insight into this problem, see [5–7].

Consider the effect of the intermolecular potential that hinders reorientational motion. The torque is defined as the angular derivative of the potential, $\hat{O}V$, and we would like to compare the mean square torque $<(\hat{O}V)^2>$ to the rotational kinetic energy, on the order of k_BT. In the low-torque or high-temperature limit, $<(\hat{O}V)^2>^{1/2}<<k_BT$, free rotation is expected. Even in the presence of torques, molecules rotate freely for short times, before collisions are suffered. This leads to a Gaussian correlation function as $t \to 0$, $C_{FIR}(t) \propto \exp(-k_BTt^2/2I_z)$, and a lineshape which ought to be Gaussian in the wings. In the high-torque limit, $<(\hat{O}V)^2>^{1/2}>>k_BT$, then perhaps one would observe a librational transition whose frequency depends on the mean square torque and the inertia about an axis perpendicular to the dipole direction. For such a librational frequency to show up as a discrete feature, the relaxation rate of the torque would have to be sufficiently slow compared to the librational frequency, a condition which does not generally hold. Thus the overall far-IR lineshape is a function of the torque, the torque relaxation rate, and the inertia for dipolar reorientation. Add to this the complication of pair dynamics, the consequences of collisions and induced moments, and the analysis of far-IR absorption can be a formidable problem. Computer simulation has been advantageously applied to the interpretation of far-IR spectra of liquids.

Recent developments in pulsed sources of terahertz radiation (1 THz = 10^{12} s^{-1}) have made the far-IR region of the spectrum more accessible than previously. The typical frequency range of these sources, from about 3–300 cm^{-1}, is indeed that of the far-IR, but the term THz connotes radiation associated with coherent sub-picosecond pulses generated by accelerating electrons or nonlinear mixing crystals. With pulse durations of only one cycle of radiation or less, the frequency range of the pulse spectrum is large. The technique is capable of probing the low-frequency motions of liquids as well as those of molecular solids and biological molecules, and the excitations of charge carriers in semiconductors. In contrast to ordinary incoherent sources of light, both the amplitude and the phase of the transmitted signal can be detected in a THz experiment. On Fourier transformation of this signal, both the real and imaginary parts of the dielectric function are obtained. Figure 5.7 presents an example of the determination of the absorption coefficient γ and real part of the refractive index n for some neat dipolar liquids [16]. Using molecular dynamics simulations, the authors determined

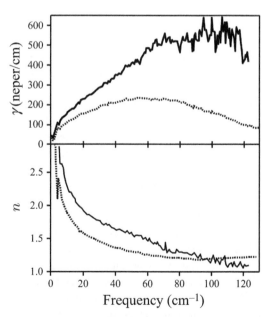

Figure 5.7 THz measurement of the absorption coefficient and refractive index of neat acetonitrile and neat acetone. (Reproduced from D. S. Venables, A. Chiu, and C. A. Schmuttenmaer, Structure and Dynamics of Nonaqueous Mixtures of Dipolar Liquids: I. Infrared and Far-IR Spectroscopy, *J. Chem. Phys. 113*, 3243 (2000), with the permission of AIP Publishing.) The full lines are the data for acetonitrile, and dashed lines are for acetone.

that the absorption spectrum, with a peak at about 60–90 cm^{-1}, arises mostly from librational motion of the molecules, with some contributions from translational motions which influence induced dipole moments.

5.6.3 Depolarized Rayleigh scattering

Depolarized light scattering and far-infrared absorption of liquids are related to one another just as rotational Raman and microwave absorption of gases, to be discussed in Chapter 8, are connected. The combined frequency range of dielectric relaxation and far-IR absorption ($\tilde{v}$ from less than 1 cm^{-1} to about 100 cm^{-1}) is the same as the range of frequency shifts observed in depolarized Rayleigh scattering. Thus these experiments explore similar dynamics. In the case of low-frequency Rayleigh scattering by liquids, as in dielectric relaxation, we must account for the fact that the response of the system is collective. The collective polarizability should be expressed as

$$A(t) = \sum_j \alpha_{\rho\sigma}^j(t) \exp[i\vec{k} \cdot \vec{R}_j(t)] \tag{5.59}$$

where the sum is over all the molecules in the scattering volume, at positions $\vec{R}_j$, and $\alpha_{\rho\sigma}^j$ is the polarizability of molecule j projected on to the laboratory directions, ρ and σ, of the polarization of the incident and scattered light. The scattering vector $\vec{k}$ is the difference between the wave vectors for the incident and scattered light. Its magnitude is given by

$$k = \left| \vec{k}_s - \vec{k}_0 \right| \approx \frac{4\pi n}{\lambda_0} \sin\frac{\varphi}{2} \tag{5.60}$$

The scattering angle is φ, n is the refractive index, and the assumption $\lambda_0 \approx \lambda_s$ has been made. The wavelength of the incident light is λ_0 and that of the scattered light is λ_s, but for Rayleigh scattering, more so than for Raman scattering, the wavelength shift is very small.

The correlation function whose time and space Fourier transform determines the depolarized Rayleigh scattering (DRS) intensity is

$$C_{DRS}(t) = \sum_{i,j} \left\langle \alpha_{XZ}^i(0)\alpha_{XZ}^j(t)e^{-i\vec{k} \cdot \left[\vec{R}_j(t) - \vec{R}_i(0) \right]} \right\rangle \tag{5.61}$$

where the incident radiation is assumed to be polarized in the Z direction, and the scattered light is detected with X polarization. The sum over i and j includes both one- and two-molecule terms, and the vector $\vec{R}_j(t)$ locates the position of molecule j at time t. Equation 5.61 reveals the role of translational motion and positional correlations in Rayleigh scattering. The influence of reorientational motion may not be obvious from Equation 5.61, but it is implicit in the projection of the molecule frame polarizability onto the lab frame directions Z and X, as will be shown in Chapter 8. The translational motion, which contributes through the exponential factor in Equation 5.61, is much slower than rotational motion. The scattered light in a Rayleigh scattering experiment is shifted in frequency from the incident light by only fractions of a wavenumber. This requires interferometric techniques and theoretical approaches which we will not cover (see [3]).

The far wings of depolarized Rayleigh scattering have been observed to decay exponentially with frequency, an effect which has been attributed to the influence of collisions at short timescales. Attempts to subtract this collisional contribution have had some success, but are dependent on the assumption that the allowed and collision-induced spectra are additive. Figure 5.8 shows an experimentally determined correlation function $C_2(t)$ obtained from the depolarized Rayleigh spectrum of benzene, taken from [8]. At short times, less than the mean time between collisions, the correlation function resembles that of a free rotor. The exponential behavior at long times is characteristic of a diffusional process.

Neglecting pair correlations for the moment, and other complications such as collision-induced effects, we can compare the correlation functions relevant to far-IR (FIR) and depolarized Rayleigh scattering (DRS) by axially symmetric molecules:

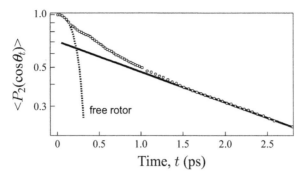

Figure 5.8 Reorientational correlation function from the depolarized Rayleigh spectrum of benzene at room temperature. (Reprinted with permission from H. D. Dardy, V. Volterra, and T. A. Litovitz, *J. Chem. Phys. 59*, 4491 (1972), copyright 1973, American Institute of Physics.)

$$l=1 \quad C_{FIR}(t) \propto (\mu_0)^2 \langle P_1(\cos\theta_t) \rangle \tag{5.62}$$

$$l=2 \quad C_{DRS}(t) \propto (\alpha_{zz} - \alpha_{xx})^2 \langle P_2(\cos\theta_t) \rangle \tag{5.63}$$

The Legendre polynomial P_l depends on the angle through which the molecule has rotated at time t. Thus both types of spectra reveal rotational dynamics and it should be possible to compare them. Unfortunately, collisional contributions and interaction-induced effects perturb DRS and FIR spectra in different ways, complicating the comparison of data from the two experiments. In the absence of these complications, the intensity can be expressed as

$$I_l(\omega) = \frac{1}{2\pi} \int_{-\infty}^{\infty} dt e^{-i\omega t} C_l(t) \tag{5.64}$$

where $l = 1, 2$ corresponds to the intensity and correlation function for far-IR absorption and depolarized Rayleigh scattering, respectively. In Rayleigh scattering experiments, this intensity is measured directly; that is, the scattering cross-section is directly proportional to $I_2(\omega)$. The far-IR absorptivity, $\varepsilon_{FIR}(\omega)$, on the other hand, is roughly proportional to $\omega[1 - \exp(-\hbar\omega/k_B T)]I_1(\omega)$, neglecting the dispersion in the refractive index. The frequency dependent prefactor can cause a maximum in $\varepsilon_{FIR}(\omega)$ when none appears in the intensity, which again points out the error of assigning the peak frequency in a far-IR spectrum to a discrete vibration. For the general purpose of comparing far-IR and DRS data, the representation $R(\omega) \equiv \omega[1 - \exp(-\hbar\omega/k_B T)]I_2(\omega)$ is sometimes used. This amounts to subtracting the anti-Stokes from the Stokes scattering and multiplying by the frequency to get something comparable to the net energy absorbed. The result is a maximum in $R(\omega)$, which has been observed for a number of liquids to be equal to or somewhat greater than the maximum observed in the far-IR spectrum. A simple theory [9] has been presented showing that the peak in $R(\omega)$ equals that in $\varepsilon_{FIR}(\omega)$ for high-torque liquids. As the torque decreases, the ratio of the maximum in $R(\omega)$ to that in the far-IR spectrum increases, approaching a value of two for freely rotating molecules.

The discussion so far has assumed axially symmetric molecules, where the dipole direction coincides with the parallel component of the polarizability and the two perpendicular components are equal. Anisotropic molecules are more difficult to treat. The central part of the DRS lineshape may consist of up to five superimposed Lorentzian functions [3]. Since even spherical molecules, such as CCl_4, exhibit some degree of depolarized light scattering, interaction-induced effects are also important. Such effects contribute strongly to the spectral wings of "allowed" depolarized Rayleigh scattering. In spite of these drawbacks, light scattering spectroscopy provides a useful way to determine the rotational dynamics of small molecules in liquids.

5.7 VIBRATION-ROTATION SPECTRA

In the case of infrared or Raman scattering, the lineshape is influenced by both rotational and vibrational dynamics. The latter will be discussed in more detail in Chapter 12, after the basic principles of vibrational spectra have been considered. Our goal here is to consider how to extract rotational information from liquid phase vibrational spectra. Vibrational dynamics include energy relaxation (decay of the population of the excited vibrational state) as well as fluctuations in the phase of the vibrational motion (dephasing). Solvent perturbations to the vibrational energy also appear in the lineshape, and in the limit that these perturbations are long lived compared to the overall vibrational relaxation, the resulting effect on the lineshape is referred to as inhomogeneous broadening. This language invokes a picture in which the spectroscopic transition is considered to occur instantaneously, capturing the active molecules in an ensemble of solution environments all having slightly different perturbations to the vibrational frequency. In truth, the limit of inhomogeneous broadening is achieved when the rate of solvent-induced frequency fluctuations is slow compared to their amplitude. In the opposite limit of fast fluctuations, the effect is referred to as motional narrowing. These limits and the middle ground between them will be considered in Chapter 6.

Just as the isotropic Rayleigh scattering depends on translational but not rotational dynamics, the isotropic Raman spectrum reflects vibrational dynamics and is independent of rotation. The lineshapes of the infrared and depolarized Raman spectra depend on both vibrational and rotational motion. The combination of infrared and Raman spectroscopy provides a powerful approach for exploring these dynamics. In the case of an axially symmetric molecule with a permanent dipole moment, the totally symmetric vibration is allowed in both Raman and infrared, as will be shown in Chapter 9.

The operator responsible for IR spectroscopy is $\vec{\mu} = (\partial \vec{\mu} / \partial Q)_0 \cdot Q$, the dipole moment derivative times the normal coordinate. The form of the normal coordinates will be discussed in Chapter 10. In the case of a simple bond stretch, we can view Q as the displacement of the bond from its equilibrium position. We will write $\vec{\mu}$ as $\mu' \hat{u} Q$, where μ' is the magnitude and $\hat{u}$ the direction of $(\partial \vec{\mu} / \partial Q)_0$. The square of μ' determines the integrated intensity of the band, and we can drop it for now as we consider the normalized intensity. The standard assumption is to consider the vibrational and reorientational dynamics to be independent, so that the total infrared TCF factors into rotational and vibrational parts:

$$C_{IR}(t) = \langle \hat{u} \cdot \hat{u}(t) \rangle \langle QQ(t) \rangle \tag{5.65}$$

and similarly for the anisotropic Raman scattering:

$$C_{anis}(t) = \langle P_2[\hat{u} \cdot \hat{u}(t)] \rangle \langle QQ(t) \rangle \tag{5.66}$$

The angle brackets in Equations 5.65 and 5.66 represent separate averages over the rotational and vibrational degrees of freedom, which are considered to be independent. The expression given here for $C_{anis}(t)$ applies to the special case of a totally symmetric Raman band of a molecule with axially symmetric polarizability. In both Equations 5.65 and 5.66, $\hat{u}$ is a unit vector along the symmetry axis of the molecule. More complicated correlation functions result for lower symmetry molecules and nontotally symmetric vibrations. Part of the time-dependence of the normal coordinate is that of the periodic vibrational motion. Let the peak vibrational frequency be called ω_0; then the phase factor $\exp(i\omega_0 t)$ in $Q(t)$ can be combined with $\exp(-i\omega t)$ in the Fourier transform of $C(t)$. The net result is that the vibration–rotation correlation function can be obtained by transforming the measured spectrum $I(\omega - \omega_0)$, where the peak frequency is considered to be the origin. Then $I(\omega - \omega_0) = \int \exp[-i(\omega - \omega_0)t] C(t) dt$ and the vibrational part of $C(t)$ depends on fluctuations in the mean frequency and population relaxation.

The isotropic Raman spectrum depends on the mean polarizability, which is independent of orientation. Thus

$$C_{iso}(t) = \langle QQ(t) \rangle \tag{5.67}$$

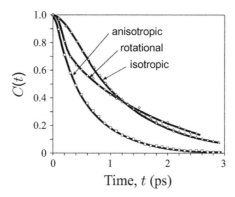

Figure 5.9 Time-correlation functions obtained from the polarized and depolarized Raman spectra of the C–D stretch of CDCl₃, in the neat liquid. (Courtesy of Dr. Douglas C. Daniel, PhD thesis, University of Idaho, 1996.)

As will be shown in Chapter 8, the anisotropic and isotropic intensities are obtained from the polarized and depolarized spectra as follows:

$$I_{pol}(\omega) = I_{iso}(\omega) + \frac{4}{3} I_{anis}(\omega) \tag{5.68}$$

$$I_{dep}(\omega) = I_{anis}(\omega) \tag{5.69}$$

Thus the measured depolarized spectrum depends only on the anisotropy of the polarizability, while the polarized spectrum depends on both the isotropic and anisotropic parts.

Equations 5.66 and 5.67 suggest that the vibrational and rotational correlation functions may be separated from one another by measuring the depolarized and polarized Raman spectra. This separation is possible for molecules possessing at least a threefold rotational axis. However, when the orientational and vibrational coordinates are strongly correlated, as they are in the case of stretching of very polar bonds, the factorization of the TCF into rotational and vibrational parts may not be valid.

Figure 5.9 shows the experimentally determined correlation functions obtained from the isotropic and anisotropic Raman spectra of the C–D stretch of CDCl₃. The rotational correlation function $\left\langle P_2 \left[\hat{u} \cdot \hat{u}(t) \right] \right\rangle$ was found by dividing the anisotropic by the isotropic TCF. The faster decay of $C_{anis}(t)$ compared to $C_{iso}(t)$ reflects the additional contribution of rotational motion to the former. This translates into a broader Raman peak in the depolarized spectrum compared to the polarized spectrum.

One easily obtained measure of the relaxation rate is the area under the normalized TCF. The relaxation time τ_c is defined by $\int C(t)dt$. For the data given in Figure 5.9, the rotational relaxation time is about 1.2 ps.

The correlation functions of Equations 5.65, 5.66, and 5.67 assume that only single-particle correlations contribute to the spectrum. This is a valid assumption for vibrational spectra in liquids. For lower frequency spectra, however, such as depolarized Rayleigh scattering and far-IR absorption, the collective nature of the response requires that the correlation function include two-particle terms as in Equation 5.57. The result is that the reorientational relaxation time depends on the experiment used to determine it. In neat liquids, longer correlation times τ_{Ray} are found using depolarized Rayleigh scattering, compared to those from depolarized Raman, τ_{Ram}. The reason is that the collective nature of the response in the Rayleigh scattering experiment means that longer times are required for the memory of the initial orientation of the particles to be lost. It has been experimentally verified that the values of τ_{Ray} approach the reorientational times τ_{Ram} upon dilution. Nuclear magnetic resonance and fluorescence depolarization can also be used to determine reorientational relaxation times. In these experiments, the single-particle relaxation time is determined.

5.8 SPECTRAL MOMENTS

If the complete lineshape $I(\omega)$ could be determined at all frequencies, the dynamics at all times could be determined, and vice versa. Alternatively, we could take the lineshape and decompose it into moments M_n as follows:

$$M_n = \frac{\int\limits_{-\infty}^{\infty} \omega^n I(\omega)d\omega}{\int\limits_{-\infty}^{\infty} I(\omega)d\omega} = \int\limits_{-\infty}^{\infty} \omega^n \hat{I}(\omega)d\omega \tag{5.70}$$

where $\hat{I}(\omega)$ is the intensity normalized to unit area. M_0 is just unity, and M_1 vanishes for symmetric bands. Thus the lowest interesting moment is the second moment, which conveys information about the width of the band. Notice that the computation of spectral moments according to Equation 5.70 is insensitive to scaling factors such as the square of the dipole moment.

As you will show in Problem 4, the nth spectral moment is related to the nth time derivative of the TCF. This can be derived by beginning with the equation $C(t) = \int_{-\infty}^{\infty} e^{i\omega t} I(\omega)d\omega$ and expanding $e^{i\omega t}$ in a Taylor series. Similarly, the correlation function can also be expressed as a Taylor series: $C(t) = C(0) + (dC/dt)_0 t + \cdots$. Equating like powers of time in the two series leads to the following expression for the nth spectral moment:

$$M_n = (-i)^n \frac{d^n}{dt^n} \langle A \cdot A(t) \rangle \Big|_{t=0} \tag{5.71}$$

Using the tools presented in the previous chapter (see Equation 4.10), the quantum mechanical equivalent of this expression is

$$M_n = \frac{1}{\hbar^n} \left\langle \hat{A}\left[\hat{H},\left[\hat{H},...\left[\hat{H},\hat{A}\right]...\right]\right]\right\rangle \tag{5.72}$$

where the nested commutator is repeated n times. For classical systems, this commutator is replaced by Poisson brackets (see [10]). One consequence of this result is that odd moments vanish for even (classical) correlation functions, resulting in spectra which are symmetric about the peak frequency. Thus, as shown earlier, the asymmetry of $I(\omega)$ is a quantum mechanical effect. For frequencies such that $\hbar\omega$ is small compared to k_BT, it can be shown that $M_{2n-1} \approx (\hbar/2k_BT)M_{2n}$.

The low-order moments can be evaluated for simple systems with the help of commutator algebra and statistical mechanics [11]. The rotational second and fourth moments for absorption (microwave or infrared) have been found to be

$$M_2 = \frac{2k_BT}{I_z} \tag{5.73}$$

$$M_4 = 8\left(\frac{k_BT}{I_z}\right)^2 + \frac{\left\langle \left(\hat{O}V\right)^2\right\rangle}{I_z^2} \tag{5.74}$$

For depolarized Raman or Rayleigh scattering, they are

$$M_2 = \frac{6k_BT}{I_z} \tag{5.75}$$

$$M_4 = 4!\left[4\left(\frac{k_BT}{I_z}\right)^2 + \frac{\left\langle \left(\hat{O}V\right)^2\right\rangle}{8I_z^2}\right] \tag{5.76}$$

In either absorption or scattering, the second moment is calculated to be independent of intermolecular interactions (since the intermolecular potential V commutes with any orientation variable). I_z is the moment of inertia for rotation about an axis perpendicular to the symmetry axis, M_2 is thus a function of the mean rotational velocity.* The fourth moments depend on interactions through the mean-square torque term $<(\hat{O}V)^2>$.

The above equations for M_2 and M_4 were obtained by considering the rotational motion of permanent dipole moments (in the case of absorption) or fixed polarizability (in the case of scattering). In numerous experimental investigations [6], the second moments have been found to exceed those calculated using these expressions. This additional line width is often attributed to the effects of collisions or internal fields, which perturb the dipole moment and polarizability on a short time scale and thus affect the spectral wings, increasing M_2. In this case the second and fourth moments cannot be used to accurately determine the mean-square torque. From the Raman data used to calculate the TCFs in Figure 5.9, the rotational second moment of $CDCl_3$ was found to be 264 cm^{-2}, compared to a theoretical value of 268 cm^{-2}.

Equations 5.73 to 5.76 assume that the intensity is normalized to one, such that $M_0 = 1$. In the case where the correlations of different molecules become important, we should write

$$C_l(t) = \sum_i \left\langle P_l\left(\hat{u}_i \cdot \hat{u}_i(t)\right)\right\rangle + \sum_{i \neq j}\left\langle P_l\left(\hat{u}_i \cdot \hat{u}_j(t)\right)\right\rangle \tag{5.77}$$

The zeroth moment is equal to the correlation function at time zero: $M_0 = C_l(0)$. Thus static orientational order from the second term in Equation 5.77 can cause M_0 to differ from unity:

$$M_0 = 1 + \sum_{i \neq j}\left\langle P_l\left(\hat{u}_i \cdot \hat{u}_j\right)\right\rangle \tag{5.78}$$

When $l = 1$, Equation 5.78 is equivalent to the Kirkwood–g factor presented in Chapter 3 (Equation 3.94). In principle, integrated far-IR intensities should permit the determination of g_K. Two extenuating circumstances, not necessarily independent, make this determination difficult: the uncertain correction for the local field and the contribution of interaction-induced effects to the spectrum.

5.9 SUMMARY

In this chapter, we have shown that the intensity as a function of frequency is the Fourier transform of an appropriate time-correlation function. The faster the decay of the relevant correlation function, the broader is the frequency spectrum. Using the fluctuation–dissipation theorem, it was shown that spatial correlations in the molecular dynamics result in response functions which depend on wave vector. These correlations are important in experiments such as dielectric relaxation and Rayleigh scattering, which depend on collective molecular dynamics. Higher-frequency spectra, such as infrared absorption and Raman scattering, are not sensitive to collective motion and thus do not depend on wave vector. The importance of rotational dynamics in the lineshapes of both pure rotation and vibration–rotation spectra has been stressed. For totally symmetric vibrations of axially symmetric molecules, measurement of both the polarized and depolarized Raman spectra allows vibrational and rotational relaxation to be separated.

It is also possible to measure the time-dependent response directly, and the inverse Fourier transform of the time-dependent signal gives the frequency spectrum. Theory and experiment have explored rotational and translational dynamics in liquids and solutions on femtosecond timescales [12,13]. Time-resolved experimental techniques such as transient birefringence and time-resolved fluorescence reveal solvent relaxation dynamics comparable to that obtained from frequency-domain experiments. The combined use of complementary time-domain and frequency-domain experiments will no doubt continue to shed light on molecular dynamics.

---

* According to the equipartition theorem, the mean kinetic energy $(1/2)I_z\Omega^2$ for each rotational degree of freedom is $(1/2)k_BT$.

PROBLEMS

1. Derive Equations 5.41 and 5.42 for the spectra of a classical free rotor, and plot them using reduced units: $\omega^* = (I/k_BT)^{1/2}\omega$. Do not worry about finding a closed-form expression for the TCF of Equation 5.40; you can take advantage of the integral representation of the delta function.

2. Prove that the free-rotor correlation function is Gaussian at short times.

3. Show that the inverse Fourier transform of $\omega^2 I(\omega)$ gives the correlation function $\langle \dot{A}(0) \cdot \dot{A}(t) \rangle$. You can use the fact that an equilibrium average is independent of the origin of time: $\langle A(0) \cdot A(t) \rangle = \langle A(-t) \cdot A(0) \rangle$.

4. Derive the expression given in Equation 5.71 for the nth spectral moment.

5. Verify that the imaginary part of the susceptibility χ'' given by Equation 5.36 is equivalent to the expression of Equation 5.15 in the long-wavelength limit.

6. Graph a Gaussian and Lorentzian function with the same width at half the maximum intensity and comment on the essential difference between the two. What is the second moment in each case?

7. Prove that $M_{2n-1} \approx (\hbar/2k_BT)M_{2n}$ for $\hbar\omega$ small compared to k_BT.

8. The inertia of a diatomic molecule is $I_z = \mu R^2$. Calculate the rotational second moment of CO in cm^{-2}, taking the bond distance to be 1.13 Å.

9. Show that the Debye model for the imaginary part of the permittivity, Equation 5.49, leads to a Lorentzian lineshape for the intensity $I(\omega)$.

REFERENCES

1. R. G. Gordon, Correlation functions for molecular motion, *Adv. Mag. Res. 3*, 1 (1968).

2. B. J. Berne and G. D. Harp, On the calculation of time correlation functions, *Adv. Chem. Phys. 17*, 109 (1970).

3. B. J. Berne and R. Pecora, *Dynamic Light Scattering* (Wiley-Interscience, New York, 1976).

4. H. Fröhlich, *Theory of Dielectrics, 2nd ed.* (Oxford University Press, New York, 1986).

5. P. A. Madden and D. Kivelson, A consistent molecular treatment of dielectric phenomena, *Adv. Chem. Phys. 56*, 467 (1984).

6. W. G. Rothschild, *Dynamics of Molecular Liquids* (Wiley, New York 1984).

7. D. Steele and J. Yarwood, eds., *Spectroscopy and Relaxation of Molecular Liquids*, (Elsevier Science Publishers, Amsterdam, 1991).

8. H. D. Dardy, V. Volterra, and T. A. Litovitz, Rayleigh scattering: Orientational motion in highly anisotropic liquids, *J. Chem. Phys. 59*, 4491 (1973).

9. R. Rodriguez and J. L. McHale, First moment of the liquid phase far-infrared absorption cross-section and the $R(\omega)$ representation of depolarized Rayleigh scattering, *J. Chem. Phys. 88*, 2264 (1988).

10. D. A. McQuarrie, *Statistical Mechanics* (Harper & Row, New York, 1976).

11. R. G. Gordon, Molecular motion in infrared and Raman spectra, *J. Chem. Phys. 43*, 1307 (1965).

12. G. R. Fleming and M. Cho, Chromophore-solvent dynamics, *Ann. Rev. Phys. Chem. 47*, 109 (1996).

13. R. M. Stratt and M. Maroncelli, Nonreactive dynamics in solution: The emerging view of solvation dynamics and vibrational relaxation, *J. Phys. Chem. 100*, 12981 (1996).

14. M. C. Beard, G.M. Turner, and C. A. Schmuttenmaer, Terahertz spectroscopy, *J. Phys. Chem. B 106*, 7146 (2002).

15. J. T. Kindt and C. A. Schmuttenmaer, Far-infrared dielectric properties of polar liquids probed by femtosecond THz pulse spectroscopy, *J. Phys. Chem. 100*, 10373 (1996).

16. D. S. Venables, A. Chiu, and C. A. Schmuttenmaer, Structure and dynamics of nonaqueous mixtures of dipolar liquids: I. Infrared and far-IR spectroscopy, *J. Chem. Phys. 113*, 3243 (2000).

Experimental considerations: Absorption, emission, and scattering

6.1 INTRODUCTION

The previous chapters have explored the molecular and bulk properties that are responsible for the dielectric and optical behavior of matter. For example, the real and imaginary parts of the complex refractive index were found to correspond to refraction and absorption/emission, respectively. We have also examined the quantum mechanical treatment of the radiation–matter interaction and found theoretical expressions for the transition dipole moment (for absorption and emission) and transition polarizability (for light scattering). In this chapter, we want to understand the connection between measured spectra and quantum mechanical properties. It will be shown how to extract properties such as transition dipole moments and excited-state radiative lifetimes from experiment. Of central importance to the discussions of this chapter is the idea that the *rates* of spectroscopic transitions decide the *intensity* of the spectral response.

We begin with a phenomenological treatment of absorption and emission that was developed by Einstein, in which the relative rates of stimulated and spontaneous emission are derived. The resulting expressions are fundamental to the understanding of laser emission. We then examine the measurement of absorption and fluorescence spectra to obtain quantum mechanical transition moments and radiative lifetimes. The experimental considerations relevant to light scattering are discussed, where the polarizations of the incident and scattered light are of interest. Finally, we briefly consider some approaches to interpreting the lineshapes of spectra.

6.2 EINSTEIN *A* AND *B* COEFFICIENTS FOR ABSORPTION AND EMISSION

As we have seen, the classical treatment of the radiation field accounts only for the effects of absorption and stimulated emission, which are essentially inverses of one another. A quantum mechanical treatment introduces the additional concept of spontaneous emission, the emission of light in the absence of a radiation field. Since absorption is always stimulated by the incident radiation, there is no absorption counterpart to spontaneous emission. In this section, we follow Einstein's treatment and consider these three spectroscopic events to be experimentally observable transitions for which we can write rate equations. It is convenient to consider the radiation in which the sample is bathed to be that of a blackbody absorber–emitter. Although the blackbody is an ideal (sometimes referred to as an ideal gas of photons), the use of a convenient expression for the energy density $\rho(\nu)$ will not restrict the applicability of the general rate expressions obtained. The rate constants for absorption, spontaneous and stimulated emission are molecular properties, so the relationships among them do not depend on the use of the blackbody model. The rates of both absorption and stimulated emission are proportional to the energy per unit volume per unit frequency interval: $\rho(\nu) = du/d\nu$.

Figure 6.1 shows a simple two-state diagram which represents the three possible processes: absorption, stimulated emission, and spontaneous emission. The phenomenological rate constants for these events are called B_{12}, B_{21}, and A_{21}, respectively. The rate of transitions out of a state is proportional to the number of molecules in that state. Designating the populations of the lower and upper levels by N_1 and N_2, and letting $\rho(\nu)$ represent the energy density of the radiation, we have the following expressions for the rates of upward (W_{12}) and downward (W_{21}) transitions:

$$W_{12} = N_1 B_{12} \rho(\nu) \tag{6.1}$$

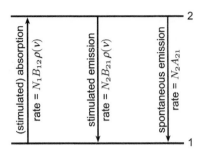

Figure 6.1 Einstein coefficients for absorption and emission.

$$W_{21} = N_2 B_{21} \rho(v) + N_2 A_{21} \tag{6.2}$$

The energy density is that at the frequency for which $hv = E_2 - E_1$. Note that in the absence of radiation, Equation 6.2 predicts a first-order decay of the excited state populations. That is, $-dN_2/dt = A_{21}N_2$ and $N_2(t) = N_2(0)\exp(-A_{21}t)$. This leads to the important conclusion that the radiative lifetime is the inverse of the A coefficient:

$$\tau_{rad} = \frac{1}{A_{21}} \tag{6.3}$$

Equation 6.3 assumes that the upper state can only decay to one lower state. If there is more than one downward transition, the radiative rate is the sum of the rates for all downward transitions (see Equation 4.78). We show here that A_{21} is proportional to the square of the transition dipole moment.

In linear spectroscopy experiments, the Boltzmann populations of the two states are unperturbed:

$$\frac{N_2}{N_1} = \frac{g_2}{g_1} \exp\left[\frac{-(E_2 - E_1)}{k_B T}\right] \tag{6.4}$$

$$\frac{N_2}{N_1} = \frac{g_2}{g_1} \exp\left[\frac{-hv}{k_B T}\right] \tag{6.5}$$

We have allowed for the possibility of degeneracy; g_i is the number of states at the energy level E_i. To maintain equilibrium, the rates of upward and downward transitions must be balanced:

$$N_1 B_{12} \rho(v) = N_2 B_{21} \rho(v) + N_2 A_{21} \tag{6.6}$$

We now assume that $\rho(v)$ is that of a blackbody. As shown in Chapter 2,

$$\rho(v) = \frac{8\pi hv^3 n^3}{c^3} \frac{1}{e^{hv/k_B T} - 1} \tag{6.7}$$

Equation 6.7 was obtained from Equation 2.75 by replacing c by the speed of light in the medium, c/n, where n is the real part of the refractive index. Solving Equation 6.6 for the energy density and using Boltzmann's law gives

$$\rho(v) = \frac{A_{21}N_2}{N_1 B_{12} - N_2 B_{21}} = \frac{A_{21}}{B_{12}\dfrac{g_1}{g_2}\exp\left(\dfrac{hv}{k_B T}\right) - B_{21}} \tag{6.8}$$

Equations 6.8 and 6.7 are in agreement if

$$g_1 B_{12} = g_2 B_{21} \tag{6.9}$$

and

$$\frac{A_{21}}{B_{21}} = \frac{8\pi h \nu^3 n^3}{c^3} \tag{6.10}$$

Equation 6.10 has the important consequence that the ratio of the rate constants for spontaneous to stimulated emission is a strong function of frequency. As the frequency increases, spontaneous emission competes more and more effectively with stimulated emission. It is not really fair to compare A_{21} and B_{21}, however, as they are not dimensionally equivalent. A better comparison is of the rates A_{21} and $B_{21}\rho(\nu)$, the ratio of which is $A_{21}/B_{21}\rho(\nu) = e^{h\nu/k_B T} - 1$, in the case where the radiation comes from a blackbody. It is clear from this ratio that when $h\nu \gg k_B T$ the rate of spontaneous emission greatly exceeds that of stimulated emission. Thus in systems at equilibrium at room temperature, the spontaneous emission of light at optical frequencies is greatly favored over stimulated emission. But as the frequency decreases into the far-infrared (far-IR) and microwave regions of the spectrum, the process of stimulated emission becomes more favorable. The properties of the two types of radiation are quite different. In the case of stimulated emission, the stimulated photon has the same properties as the incident radiation, resulting in emission which is collimated and coherent. Spontaneous emission, such as ordinary fluorescence, is emitted in all directions with random phase.

In attempting to achieve laser emission at frequencies for which $h\nu \gg k_B T$, one needs to get around the constraints of equilibrium that were invoked in the preceding discussion. Sustainable laser emission requires an inverted excited state population, a nonequilibrium situation for which $N_2 > N_1$. Even in this case, the inherent rate coefficients A_{21} and B_{21} still obey Equation 6.10, and one can appreciate why the first sustained stimulated emission was achieved in the microwave region of the spectrum. This first "laser" was actually a "maser," an acronym which stands for "microwave amplification by stimulated emission of radiation." In the next section we will see how the B_{21} coefficient can lead to amplification of the incident radiation, and how the Einstein coefficients may be found from experiment.

6.3 ABSORPTION AND STIMULATED EMISSION

In this section, we consider the attenuation or amplification of radiation as it traverses a sample. These two processes are represented by the diagrams in Figure 6.2, where the amplitude of the electromagnetic field either increases or decreases on passing through the sample. Since spontaneous emission is emitted in all

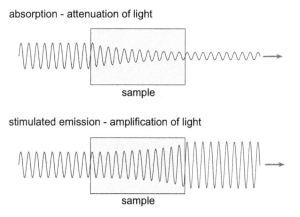

Figure 6.2 Attenuation and amplification of light.

directions, we can neglect the small contribution of spontaneous emission in the same direction as the beam. The processes of absorption and stimulated emission, respectively, subtract from and add to the energy of the radiation field. In Chapter 3, it was shown that the imaginary component of the relative permittivity is associated with absorption and emission. The susceptibility $\chi_e(\omega)$ is equal to $\varepsilon_r(\omega) - 1$. It follows that $\chi_e''(\omega)$ is equal to $\varepsilon_r''(\omega)$. The transmitted intensity is $I = I_0 e^{-\gamma x}$, where x is the path length, I_0 is the incident intensity, and γ is the power absorption coefficient: $\gamma = \omega \chi_e''/cn$. Recall that n is the real part of the refractive index. As will now be shown, this "absorption coefficient" takes on a negative value when the populations are inverted. The differential form of Beer's law is

$$dI = -\gamma I \, dx \qquad (6.11)$$

Positive values of γ lead to attenuation and negative values lead to amplification of the light intensity. $dI/dx = -\gamma I$ can be equated to the net power created or dissipated per unit volume, $\langle P \rangle / V$. If N_2 and N_1 are the number of molecules per unit volume in the upper and lower states, then the Einstein rate picture yields

$$\frac{\langle P \rangle}{V} = \frac{dI}{dx} = (N_2 - N_1) w_{12}(v) hv \qquad (6.12)$$

The intensity I is the power per unit area, and dI/dx is the change in power per unit volume. To see this, imagine carving out a section of the sample having a unit cross-sectional area $A = 1$ cm^2 perpendicular to the direction of propagation of the light, and account for the change in energy within the volume $A dx = dx$. If N is the number of molecules per cm^3, then $N dx$ is the number of molecules in that volume. Equation 6.12 expresses the net power dissipation per unit volume in terms of the difference in the number of molecules N_2 in the upper state, which add to the power via stimulated emission, minus the number of molecules N_1 in the lower state, which subtract power through absorption. Equation 6.12 contains the intrinsic rate $w_{12}(v)$ of transitions between states 1 and 2, which is equal to $B_{12}\rho(v)$. The lowercase symbol w_{12} is the transition rate per molecule, and $W_{12} = N_1 w_{12}$ is the total transition rate. The two states are assumed here to be nondegenerate, so that $B_{12} = B_{21}$. In the previous chapter, we found the transition rate w_{12} in a form convenient for absorption spectroscopy, shown in Equation 4.43. We reproduce that below after replacing the permittivity of free space, ε_0, by that in the medium, $\varepsilon_r \varepsilon_0 = n^2 \varepsilon_0$:

$$w_{12} = \frac{|\mu_{12}|^2 \rho(v_{21})}{6n^2 \varepsilon_0 \hbar^2} = B_{12} \rho(v_{21}) \qquad (6.13)$$

In this correction for the permittivity of the medium, we are neglecting the dispersion in the refractive index, which is justified when the absorbing molecule is part of a dilute solution or other matrix which is transparent in the region where the molecule absorbs. Thus in typical applications, n is the refractive index of the solvent. By virtue of Equation 6.10, the transition rate is also given by

$$w_{12}(v) = \frac{A_{21} c^3 \rho(v)}{8\pi n^3 hv^3} = \frac{c^3 \rho(v)}{8\pi n^3 hv^3 \tau_{rad}} \qquad (6.14)$$

The coefficient of spontaneous emission has been replaced by the reciprocal of the radiative lifetime. The lack of subscripts on v foreshadows the need to integrate over a range of transition frequencies. The energy density $\rho(v)$ is related to the intensity I and the energy per unit volume u by $u = \int \rho(v) dv = nI/c$. There is in general some characteristic lineshape function $g(v)$ that describes a transition, taking into account all the broadening mechanisms including the minimum lifetime broadening, such that $\int g(v) dv = 1$. Thus we can replace $\rho(v)$ by $nI(v)g(v)/c$. Then the transition rate is

$$w_{12} = \frac{c^2 I(v) g(v)}{8\pi n^2 hv^3 \tau_{rad}} \qquad (6.15)$$

and the rate of change in power is

$$\frac{dI(v)}{dx} = \frac{\langle P \rangle}{V} = \frac{(N_2 - N_1)c^2 g(v)}{8\pi n^2 v^2 \tau_{rad}} I(v) \tag{6.16}$$

Equation 6.16 emphasizes that the relative populations decide whether light is attenuated (if $N_2 < N_1$) or amplified (if $N_2 > N_1$). It is also clear that the rate of either process is slowed by increasing the radiative lifetime. In attempting to achieve amplification, a fast rate of radiative decay is desired in that it increases $\langle P \rangle / V$, but such an increase in rate makes achieving a population inversion more difficult. We will examine some examples of laser systems in the next chapter, where it will be shown how this population inversion may be created.

Equation 6.16 can be compared to Equation 6.11, and using $\gamma = \omega \chi_e'' / cn$ the imaginary part of the susceptibility is obtained:

$$\chi_e''(v) = \frac{c^3 g(v)(N_1 - N_2)}{16\pi^2 n v^3 \tau_{rad}} \tag{6.17}$$

Equation 6.17 was derived taking the degeneracies of the two states to be equal. If this is not the case, one should replace N_1 by $N_1 g_2 / g_1$.

6.4 ELECTRONIC ABSORPTION AND EMISSION SPECTROSCOPY

Although many principles presented in this section are quite general, we now focus on applications in electronic spectroscopy. Thus it is necessary to summarize a few important features of this topic that will be presented in more detail in later chapters, particularly Chapter 11. Suppose that the ground electronic state of a molecule is a singlet state; that is, all electrons are paired as shown in Figure 6.3. This is a common situation, and the ground state of any closed-shell molecule (in which all electrons are paired) is called the S_0 ("singlet-zero") state. There are two ways to generate an excited electronic state by promoting an electron from one of the occupied to one of the virtual orbitals. The first way preserves the pairing of spins and generates a singlet excited state, called S_1 if it is the lowest excited state. Alternatively, if the electrons in the two half-occupied orbitals have the same spin, a triplet state (T_1) is generated. The triplet state is generally lower in energy than the singlet state arising from the same occupied orbitals, due to lower interelectronic repulsion in the triplet state. Transitions between the T_1 and S_0 states are forbidden by spin but become allowed by virtue of spin–orbit coupling, to be discussed in the next chapter. Emission involving two states of the same spin is called fluorescence, while emission between states of different spin is phosphorescence. Emission from a triplet excited electronic state to a singlet ground state is thus phosphorescence. It is generally of longer wavelength and longer lifetime than the fluorescence emission from the singlet excited state. In this section, we concentrate on the absorption and fluorescence spectra that are typical of transitions between two singlet electronic states.

Here, we compare the transition rate in the Einstein picture to the quantum mechanical transition rate given by the Golden Rule. The latter, $w_{if} = |\mu_{if}|^2 \rho(v_{fi}) / 6n^2 \varepsilon_0 \hbar^2$, Equation 6.13, is a transition rate per molecule,

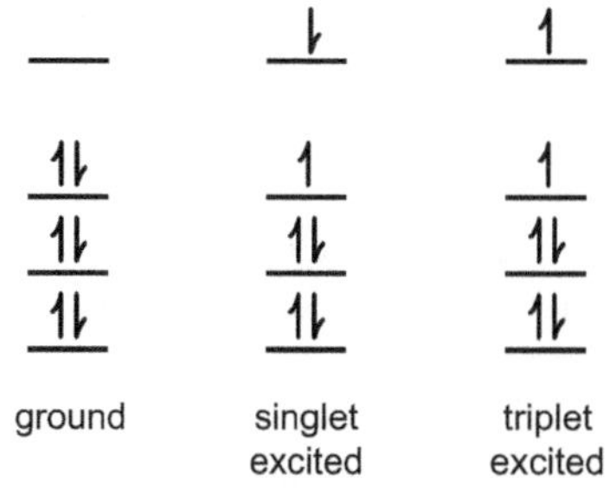

Figure 6.3 Electronic configurations for a molecule having a closed-shell ground state.

so on multiplying by the number of molecules in the initial state, we get the total transition rate $N_1 w_{12} = N_1 B_{12} \rho(v_{12})$. Thus the Einstein rate is the same as that given by the Golden Rule if

$$B_{12} = \frac{|\mu_{12}|^2}{6n^2 \varepsilon_0 \hbar^2} \tag{6.18}$$

Using the relationship between A_{21} and $g_2 B_{21} = g_1 B_{12}$, we have

$$A_{21} = \frac{g_1}{g_2} \frac{16\pi^3 |\mu_{12}|^2 n v^3}{3\varepsilon_0 h c^3} = \frac{1}{\tau_{rad}} \tag{6.19}$$

Equations 6.18 and 6.19 tell us that the rates of absorption and emission are proportional to the square of the transition dipole moment. It is also seen that for spontaneous emission, the more allowed the transition, the shorter the radiative lifetime. In this section we consider how these quantities may be determined from the integrated absorption and fluorescence spectra. We assume equilibrium populations of the initial and final states and that $hv \gg k_B T$, so that the effect of stimulated emission can be neglected.

The differential form of Beer's law can be written in terms of the molar absorptivity $\varepsilon_M(v)$ as follows:

$$-dI = 2.303 \varepsilon_M(v) C I dx \tag{6.20}$$

The molar absorptivity is generally expressed in units of L mol⁻¹ cm⁻¹, so the concentration C is in mol L⁻¹ and the path length x is in cm. The factor of 2.303 appears in Equation 6.20 because it is derived from the form of Beer's law preferred by spectroscopists: $I = I_0 10^{-\varepsilon_M C x}$. If N is the number of molecules per cm³ in the initial state i (essentially equal to the total number of molecules for electronic spectroscopy), then another way to write Equation 6.20 is

$$-dI = N h v w_{if} dx \tag{6.21}$$

Equating Equations 6.20 and 6.21 gives the transition rate per molecule as

$$w_{if} = \frac{2.303 \varepsilon_M(v) C I}{N h v} \tag{6.22}$$

It is convenient to replace the intensity I by cu/n and the molar concentration C by $1000 N/N_A$, where N is the number of molecules per cm³ and N_A is Avogadro's number:

$$w_{if} = \frac{2303 c \varepsilon_M(v) u}{N_A h v n} \tag{6.23}$$

The total transition rate ought to be integrated over the band, replacing u by $\int \rho(v) dv$ and including $\varepsilon_M(v)/v$ in the integrand:

$$w_{if} = \frac{2303 c}{N_A h n} \int_{band} \frac{\rho(v) \varepsilon_M(v)}{v} dv \tag{6.24}$$

If the bandwidth of the source $\rho(v)$ is broad compared to the range of frequencies for which $\varepsilon_M(v)$ is significant, we can take $\rho(v) \approx \rho(v_{fi})$ as constant and move it outside the integral. This gives a transition rate which can be directly compared to $B_{if} \rho(v_{fi})$, and we obtain

$$B_{if} = \frac{2303 c}{N_A h n} \int_{band} \frac{\varepsilon_M(v)}{v} dv = \frac{|\mu_{if}|^2}{6n^2 \varepsilon_0 \hbar^2} \tag{6.25}$$

This is now rearranged to get an equation which relates the square of the transition moment to the integrated molar absorptivity:

$$|\mu_{fi}|^2 = \frac{3(2303)hnc\varepsilon_0}{2\pi^2 N_A} \int\limits_{band} \frac{\varepsilon_M(v)}{v} dv \cong 9.19 \times 10^{-3} n \int\limits_{band} \frac{\varepsilon_M(v)}{v} dv \tag{6.26}$$

Equation 6.26 shows how the integrated molar absorptivity can be used to find the absolute value of the transition moment. The units on this expression require some care. The number 2303 is actually 2.303 times 1000 cm^3 L^{-1}. The reader should verify that the conversion factor in front of the last integral is appropriate for calculating the transition dipole moment μ_{fi} in Debye, where 1 Debye = 10^{-18} esu cm, when the molar absorptivity is in L mol^{-1} cm^{-1}. In typical applications of this expression, i and f are the ground and first excited electronic states, and the width of the absorption band results from a progression of vibrational transitions within a single electronic transition. As will be shown in Chapter 11, the vibrational progression serves to distribute the total intensity over a range of frequencies while preserving the total area of the band. The integral over frequency in Equation 6.26 is sometimes expressed as $\int \varepsilon_M(v) d\ln v$. If cgs rather than MKS units are desired, then ε_0 should be replaced by $1/4\pi$.

Equation 6.26 can be combined with Equation 4.82 to relate the oscillator strength f to the integrated intensity. In doing so, the frequency $\omega = 2\pi v$ in the numerator of Equation 4.36 is moved inside the integral over molar absorptivity. The result is

$$f = \frac{4(2303)m_e cn\varepsilon_0}{e^2 N_A} \int\limits_{band} \varepsilon(v) dv \tag{6.27}$$

The oscillator strength is a convenient way to express the strength of a spectroscopic transition. Very small values of the oscillator strength often result when a transition is forbidden to a first approximation, or when it is permitted by electric quadrupole or magnetic dipole selection rules. Oscillator strengths larger than one can be a sign of overlapping electronic transitions.

The intensity of emission $f \to i$ is also proportional to $|\mu_{fi}|^2$ and to A_{fi}. A measurement of fluorescence intensity ought to reveal the radiative lifetime τ_{rad}. There are several reasons why this is not necessarily a practical way to determine τ_{rad} or the transition dipole moment. One reason is that fluorescence intensities, unlike absorption intensities, are difficult to quantify. As discussed in Chapter 4, it is necessary in the case of emission to consider the intensity of light collected into a solid angle. The measured intensity depends on the solid angle subtended by the detector, so it is sensitive to experimental conditions. One does not usually have the advantage of comparing a sample and reference beam as in absorption spectroscopy.

One can measure fluorescence lifetimes using time-resolved detection of fluorescence intensity. This gives the total rate of decay τ_{fluor} (the fluorescence lifetime) of the excited electronic state. In general, a state may decay by both radiative and nonradiative pathways, thus the total decay rate is

$$\frac{1}{\tau_{fluor}} = \frac{1}{\tau_{rad}} + \frac{1}{\tau_{nonrad}} \tag{6.28}$$

The nonradiative decay pathway accounts for the conversion of the energy of the excited electronic state into other degrees of freedom, essentially degrading the energy as heat. If nonradiative decay is negligible, then for every photon that is absorbed in order to create the excited state, one is emitted. The fluorescence quantum yield φ_{fluor} is the ratio of the number of photons emitted to that absorbed. It is also equal to the ratio of the radiative decay rate to the total decay rate:

$$\varphi_{fluor} = \frac{1/\tau_{rad}}{1/\tau_{fluor}} = \frac{\tau_{fluor}}{\tau_{rad}} \tag{6.29}$$

The fluorescence yield can be determined experimentally by comparing the observed fluorescence intensity to that of a dye for which φ_{fluor} is known. Only if the fluorescence yield approaches one can a measurement of the fluorescence lifetime alone reveal the radiative lifetime. However, if one determines the transition moment from absorption spectroscopy, then it is possible to calculate the A-coefficient and thus τ_{rad}. The procedure for this is straightforward for the case of atomic spectra, but requires additional care in the case of molecular electronic spectra. These two cases are considered next.

6.4.1 ATOMIC SPECTRA

Unlike molecular electronic spectra, which are broadened by vibrational and rotational transitions, atomic emission and absorption lines are sharp. The frequency $\nu_{fi} = -\nu_{if}$ of absorption is equal to that of emission when the same two states are involved. Thus one can evaluate $|\mu_{if}|^2$ from the absorption spectrum using Equation 6.26, then substitute it in Equation 6.19 to obtain τ_{rad} without making any assumptions. Putting all this together, we have

$$\frac{1}{\tau_{rad}} = \frac{8\pi(2303)cn^2\tilde{\nu}^2}{N_A}\frac{g_1}{g_2}\int \varepsilon_m(\tilde{\nu})d\tilde{\nu} \tag{6.30}$$

In obtaining Equation 6.30, the frequency has been converted to wavenumbers, $\tilde{\nu} = \nu/c$, and it has been assumed that, since the atomic spectral lines are quite sharp, the frequency originally part of the integrand can be considered constant and factored out. Thus atomic absorption spectra lead directly to the determination of radiative lifetimes.

6.4.2 MOLECULAR ELECTRONIC SPECTRA

Equation 6.30 fails for molecular electronic spectra because of the contribution of vibrational transitions to the distribution of intensity. As will be discussed in detail in Chapter 11, molecular electronic spectra typically comprise a number of vibrational subbands due to transitions such as those indicated schematically in Figure 6.4. In solution-phase electronic spectra at room temperature, the absorption transitions often originate from the ground vibrational level of the ground electronic state ($v'' = 0$ in Figure 6.4), because higher vibrational states tend not to be occupied at room temperature. This will not be true for low-frequency vibrations, but we will neglect these for now.

After being excited to vibrational levels within the excited electronic states, nonradiative vibrational relaxation takes place within the excited electronic state. For large molecules in solution, such as the dyes often used in fluorescence spectroscopy, this radiationless relaxation process is fast compared to the rate of emission. As a result, the fluorescence originates from the $v' = 0$ level within the excited electronic state. This is known as

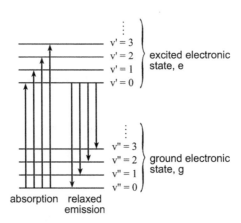

Figure 6.4 Vibrational transitions within electronic absorption and emission bands.

relaxed fluorescence. The total intensity of absorption or emission, proportional to $|\mu_{if}|^2$, is spread out over the vibrational progression such that the intensity of each vibronic transition is proportional to the Franck–Condon factor *FC*, which is the square of the overlap of vibrational wavefunctions within the ground and excited electronic states:

$$FC = \langle 0|v\rangle\langle v|0\rangle = \left|\langle 0|v\rangle\right|^2 \tag{6.31}$$

where v = v″ or v′ for emission or absorption, and in each scalar product one vibrational wavefunction corresponds to the ground and the other to the excited electronic state, which have different potential energy functions. We shall see in Chapter 11 that the distribution of electronic spectral intensity across this progression of vibrational transitions is a strong function of the change in geometry of the molecule. For now, we are interested in summing the coefficient for spontaneous emission over the bands within the Franck–Condon progression. Thus

$$A_{eg} = \sum_{v''} A_{e0,gv''}$$

$$\propto \sum_{v''} v_{e0,gv''}^3 \left|\langle 0|v''\rangle\right|^2 \tag{6.32}$$

The Einstein *A* coefficient is proportional to the cube of the frequency, which varies according to initial and final vibrational quantum numbers. By applying the completeness theorem to the set of vibrational states within any electronic state, we have $\sum_{v''} \left|\langle 0|v''\rangle\right|^2 = 1$. Thus the *A* coefficient is proportional to

$$\frac{\sum_{v''} v_{e0,gv''}^3 \left|\langle 0|v''\rangle\right|^2}{\sum_{v''} \left|\langle 0|v''\rangle\right|^2} = \frac{\int I_{fluor}(v)dv}{\int v^{-3} I_{fluor}(v)dv} = \left\langle v_{fluor}^{-3}\right\rangle^{-1} \tag{6.33}$$

Recall that the fluorescence intensity I_{fluor} is proportional to the cube of the frequency. The angle brackets in Equation 6.33 represent an average over the fluorescence spectrum. We therefore rewrite Equation 6.10 as follows:

$$A_{ge} = \frac{8\pi h n^3}{c^3}\left\langle v_{fluor}^{-3}\right\rangle^{-1} B_{ge} \tag{6.34}$$

and using Equations 6.25 and 6.26, we have

$$A_{ge} = \frac{1}{\tau_{rad}} = \frac{8\pi(2303)n^2}{c^2 N_A}\left\langle v_{fluor}^{-3}\right\rangle^{-1}\int \varepsilon_M(v)d\ln v \tag{6.35}$$

Equation 6.35 shows how fluorescence and absorption data may be combined to determine the radiative lifetime. Notice that the averaging process used to compute $\left\langle v_{fluor}^{-3}\right\rangle^{-1}$ avoids having to know the absolute fluorescence intensity. Strickler and Berg [1] demonstrated the validity of the above expression by comparing the radiative lifetimes so determined to those measured using phase-sensitive fluorescence spectroscopy, which involves exciting the fluorescence spectrum with a modulated intensity source and detecting the phase lag between the exciting light and fluorescence radiation. Good agreement was obtained in the case of molecules having fluorescence yields near unity. The Strickler–Berg equation, Equation 6.35, can be combined with a direct measurement of the fluorescence lifetime in order to determine the fluorescence yield.

6.5 MEASUREMENT OF LIGHT SCATTERING: THE RAMAN AND RAYLEIGH EFFECTS

Light scattering spectra depend strongly on the polarization of the incident and scattered beams, and as such we must take into account the projection of the polarizability tensor onto the lab frame in order to analyze the experimental data. In addition, the nature of the experimental setup introduces special concerns about the quantitative measurement of intensities. The discussion about relating the molecule and lab frame polarizability tensors will be postponed until Chapters 8 and 12. For now, we concentrate on the experimental considerations related to the scattering geometry.

Figure 6.5 shows the coordinate system that we envision for a typical $90°$ scattering arrangement. The incident light is considered to propagate along the X-direction, and the scattered light is detected along the Y-direction. Uppercase letters XYZ will always indicate the laboratory frame of reference (the "space-fixed" coordinate system) and lowercase letters will be used when referring to a coordinate system fixed in the molecule (the "body-fixed" coordinate system). Conventional Raman or Rayleigh spectra are often measured using incident light polarized in the Z-direction ("vertically polarized"), which is perpendicular to the scattering plane. A polarization analyzer between the sample and detector enables the scattered light to be resolved into its two polarization components. The polarized spectrum is measured by observing the scattered light having the same polarization direction as the incident light. This intensity is referred to as $I_\parallel$, I_{ZZ}, or I_{pol}. If, instead of detecting the Z-polarized light, one measures the scattered light polarized in the X-direction (but still excited with Z-polarized incident radiation), then the depolarized light scattering spectrum is obtained. This is referred to as $I_\perp$, I_{XZ}, or I_{dep}. The depolarization ratio ρ is defined by

$$\rho = \frac{I_{dep}}{I_{pol}} \tag{6.36}$$

(The Greek letter rho is a popular one! The ρ defined in Equation 6.36 should not be confused with the energy density discussed in the preceding sections.) We shall see in future chapters that ρ is a revealing quantity when it comes to the symmetry of the transition responsible for light scattering. For now, note that the depolarization ratio is almost always less than one. And if the polarizability were perfectly isotropic, then no induced moment could result in a direction other than that of the incident polarization, so ρ would vanish for spherically symmetric molecules, frozen at the equilibrium geometry. We have some work to do (in Chapter 8) before this statement can be proven.

First, let us consider how the measured intensity of scattered light depends on molecular properties. As in fluorescence, we must account for the radiation being scattered into a sphere of 4π steradians. This net intensity will be determined by the cross-section for scattering, σ, a molecular property which depends on the square of the transition polarizability. The part of the cross-section $d\sigma$ which contributes to the detected scattered light is the ratio of the power measured at the detector to the incident intensity:

$$d\sigma = \frac{dP}{I_0} \tag{6.37}$$

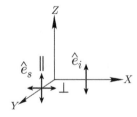

Figure 6.5 $90°$ scattering geometry.

The differential power dP is proportional to the solid angle $d\Omega$ subtended by the detector. Introducing the irradiance I_s in W/sr, the differential power dP is $I_s d\Omega$. The differential cross-section is then

$$\frac{d\sigma}{d\Omega} = \frac{I_s}{I_0} \tag{6.38}$$

The typical units of cm²/sr for the differential cross-section reflect the use of different ways of quantifying the scattered and incident intensities, the former in W/sr and the latter in W/cm². The depolarization ratio in the 90° scattering experiment is the ratio:

$$\rho = \frac{(d\sigma/d\Omega)_\perp}{(d\sigma/d\Omega)_\parallel} \tag{6.39}$$

In general, the differential cross-section depends on the polarization directions $\hat{e}_s$ and $\hat{e}_i$ of the scattered and incident radiation. It is necessary to project the polarizability tensor onto these directions:

$$\left(\frac{d\sigma}{d\Omega}\right)_{s,i} = \frac{\pi^2 v_s^3 v_i}{\varepsilon_0^2 c^4} \left| \hat{e}_s \cdot \alpha \cdot \hat{e}_i \right|^2 \tag{6.40}$$

In the case of Rayleigh scattering, the polarizability tensor is a state property that varies in the lab frame due to the translational and rotational motion of the molecule. In Raman scattering, the elements of this tensor are matrix elements of the polarizability operator connecting rotational and/or vibrational states. Thus, just as E1-allowed absorption and emission depend on the square of the transition dipole moment, the Raman scattering intensity is proportional to the square of the transition polarizability. This will be explored further in Chapter 12. Notice that the cross-section for scattering depends on the incident light frequency times the cube of the scattered light frequency. For small frequency shifts, the scattering intensity is proportional to the fourth power of frequency, as predicted by the form of the expression for dipole radiation given in Equation 2.37. The polarizability tensor α must be expressed in the lab frame XYZ, whereas it is natural to start from a consideration of the polarizability in the molecular frame.

To figure out the total cross-section σ_R, the differential cross-section must be integrated over all possible scattering angles:

$$\sigma_R = \int_0^{2\pi} d\varphi \int_0^\pi \left(\frac{d\sigma}{d\Omega}\right)_{\theta,\varphi} \sin\theta \, d\theta \tag{6.41}$$

To do this, we need to consider how the intensity of light varies with the scattering angles (θ,φ) and polarization. Figure 6.6 shows a generalized scattering geometry in which the incident radiation is considered

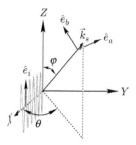

Figure 6.6 Generalized scattering geometry.

to propagate along the X-direction and have Z-polarization, as in Figure 6.5, but the wave vector $\vec{k}_s$ of the scattered radiation may have any orientation, specified by polar and azimuthal angles θ and φ. The scattered light polarization may be resolved into two components $\hat{e}_a$ and $\hat{e}_b$ as shown in Figure 6.6. The first of these is chosen to be perpendicular to $\hat{e}_i$ and $\vec{k}_s$, while the second is fixed by the requirement that $\vec{k}_s$, $\hat{e}_a$ and $\hat{e}_b$ be mutually perpendicular. The cross-section at a given scattering angle is the sum of that for the two polarization components:

$$\left(\frac{d\sigma}{d\Omega}\right)_{\theta,\varphi} = \left(\frac{d\sigma}{d\Omega}\right)_a + \left(\frac{d\sigma}{d\Omega}\right)_b \tag{6.42}$$

It is possible to show [2] using trigonometry that this is equivalent to

$$\left(\frac{d\sigma}{d\Omega}\right)_{\theta,\varphi} = \left[1 - \left(\frac{1-\rho}{1+\rho}\right)\cos^2\theta\right]\left(\frac{d\sigma}{d\Omega}\right)_{90°} \tag{6.43}$$

The depolarization ratio in Equation 6.43 is that evaluated at the 90° scattering geometry, as is the differential cross-section that appears on the right-hand side. Using Equation 6.43 in Equation 6.41 leads to the desired result:

$$\sigma_R = \frac{8\pi}{3}\left(\frac{1+2\rho}{1+\rho}\right)\left(\frac{d\sigma}{d\Omega}\right)_{90°} \tag{6.44}$$

The total cross-section can be found by measuring the differential cross-section and the depolarization ratio using a 90° scattering geometry. Note that the differential cross-section $(d\sigma/d\Omega)_{90°}$ is the sum of the polarized and depolarized components.

6.6 SPECTRAL LINESHAPES

We have seen in Chapter 5 that molecular dynamics are reflected in the frequency distribution of the spectral response $I(\omega)$. Furthermore, we have seen that the time–energy uncertainty principle dictates that the minimum spectral width be inversely related to the lifetime of the state: $\delta\omega \geq \tau^{-1}$. Linewidth information is often quoted by citing the width of the band which corresponds to one-half the maximum intensity. This is called the "full width at half maximum," or FWHM for short. A myriad of dynamical effects, in addition to the lifetime broadening, contribute to the total observed width. Such dynamics are quite specific to the type of spectrum and the sample under consideration. Nevertheless, there are some frequently encountered lineshape motifs that are often referred to in the literature, such as the characterization of *homogeneous* versus *inhomogeneous* broadening. In truth, these are limits that are not always achieved, but it is convenient nonetheless to classify lineshapes in this way.

Homogeneous broadening refers to a mechanism that affects the lineshape of every molecule in the sample in the same way. The radiative lifetime is one example of a homogeneous broadening effect. As we have seen, radiative decay is a first-order rate process. The Fourier transform of an exponential is a Lorentzian function. In general, for a correlation function that decays in time according to $\exp(-t/\tau)$, the resulting Lorentzian lineshape form is

$$I_L(\omega) = \frac{I_{max}\left(\frac{1}{4\tau^2}\right)}{(\omega - \omega_0)^2 + \left(\frac{1}{4\tau^2}\right)} \tag{6.45}$$

where ω_0 is the peak frequency and the FWHM is $\delta\omega_{1/2} = 2\pi\delta\nu_{1/2} = \tau^{-1}$.

The Franck–Condon progression observed in molecular electronic spectra is also an example of homogeneous broadening, in that every molecule experiences the same type of broadening, but in this case we don't expect to observe an overall Lorentzian profile because no exponential decay is involved. Instead, the intensity distribution depends on the Franck–Condon factors, which in turn depend on the change in geometry of the molecule in going from the ground to the excited electronic state, as will be discussed in Chapter 11.

Inhomogeneous broadening is the result of having a collection of molecules with a distribution of transition frequencies. It is a statistical effect and therefore leads to a Gaussian (bell-shaped) distribution of spectral intensities:

$$I_G(\omega) = I_{max} \exp\left[\frac{-(\omega - \omega_0)^2}{2\sigma^2}\right] \tag{6.46}$$

The symbol σ has been used (reluctantly, since it has just been used to refer to the scattering cross-section) because it is the conventional way to represent the standard deviation of a normal probability distribution. The FWHM for the lineshape of Equation 6.46 is $\delta\omega_{1/2} = 2\sqrt{2\ln 2}\,\sigma$.

The concept of inhomogeneous broadening occurs frequently in the discussion of solvent effects on spectral lineshapes. There is a gas-phase phenomenon, however, that also leads to Gaussian lineshapes. The velocity distribution of molecules in the gas phase obeys the temperature-dependent Boltzmann distribution. The relative motion of molecules with respect to the source (in absorption) or detector (in emission) leads to an apparent frequency shift due to the Doppler effect. For molecular velocity v along a line connecting the observer (or source) to the molecule, the apparent transition frequency is $v = v_0(1 \pm (v/c))$, where v_0 is the unshifted frequency. Molecules moving toward the detector have blue-shifted transitions, while those moving away have their transition frequencies shifted to the red. The resulting FWHM is

$$\delta v_{1/2} = 2v\left(\frac{2k_B T \ln 2}{mc^2}\right)^{1/2} \tag{6.47}$$

where the molecular mass is m. Suitable experimental conditions can minimize the Doppler broadening, for example, by observing the spectrum of a molecular beam in a direction orthogonal to the beam. The translational motion of molecules in dense phases also leads to Doppler broadening, but typical Doppler shifts are quite small compared to other mechanisms contributing to $\delta v_{1/2}$, such as collisions and intermolecular interactions, so one can neglect this effect in low resolution spectra.

It is often desired to fit spectral lineshapes to functional forms, for example to characterize the area or spectral moments. A convenient lineshape expression which accounts for both homogeneous and inhomogeneous effects is the Voigt profile, which is a convolution of Gaussian and Lorentzian functions:

$$I_{voigt}(\omega) = \int I_G(\omega') I_L(\omega - \omega') d\omega' \tag{6.48}$$

The Voigt profile is a useful way to resolve the overall linewidth into homogeneous and inhomogeneous components. Although there is no closed-form expression for $I_{voigt}(\omega)$, the integral can be evaluated numerically.

The successful application of a function such as the Voigt profile suggests that the homogeneous or inhomogeneous limits of lineshapes need not be achieved. It is often desirable to account for $I(\omega)$ with a theory that is capable of interpolating between the limits which correspond to fast and slow modulation of the spectral frequency. The Kubo lineshape formulation is one such theory [3]. In this approach, the transition frequency is considered to be a random function of time:

$$\omega = \omega_0 + \delta\omega(t) \tag{6.49}$$

The frequency shift $\delta\omega(t)$ results from perturbations caused by interactions of the active molecule with the environment (solvent). It is considered to be a stochastic variable; i.e., it fluctuates in a random fashion, like the variable $A(t)$ in Figure 5.1. The average frequency is ω_0. The Kubo model assumes that the correlation function for $\delta\omega$ decays exponentially in time:

$$\left\langle \delta\omega(0)\delta\omega(t) \right\rangle = \Delta^2 \exp\left(-t/\tau_c\right) \tag{6.50}$$

The variable $\Delta = <|\delta\omega|^2>^{1/2}$ is the amplitude of the solvent-induced perturbations to the frequency, and $1/\tau_c$ is the rate at which these perturbations relax. τ_c is the pure dephasing time. The lineshape function $\varphi(t)$ whose Fourier transform determines $I(\omega)$ is

$$\varphi(t) = \left\langle \exp \int_0^t i\delta\omega(t')dt' \right\rangle \tag{6.51}$$

The normalized intensity spectrum is

$$I(\omega) = \frac{1}{2\pi} \int_{-\infty}^{\infty} \varphi(t)e^{-i\omega t}\, dt \tag{6.52}$$

If Equation 6.50 is assumed to govern the frequency fluctuations, the lineshape function is found to be

$$\varphi(t) = \exp\left\{ -\Delta^2\tau_c \left(t - \tau_c\left[1 - e^{-t/\tau_c}\right] \right) \right\} \tag{6.53}$$

The lineshape $I(\omega)$ obtained from the Fourier transform of Equation 6.53 is neither Gaussian nor Lorentzian. But there are two physically meaningful limits in which these functions are obtained. If $\Delta \gg \tau_c^{-1}$, the amplitude of frequency fluctuations is large compared to the rate of their decay. In this limit, $\varphi(t)$ goes over to a Gaussian, as does the lineshape, and we may refer to the spectrum as inhomogeneously broadened. In the opposite limit, $\Delta \ll \tau_c^{-1}$, the frequency perturbations relax on a timescale that is fast compared to the amplitude, and the resulting $\varphi(t)$ decays exponentially in time. This results in a Lorentzian intensity spectrum $I(\omega)$ typical of homogeneous broadening. This limit, referred to as motional narrowing, is frequently observed in liquid-phase NMR spectra. In the limit of motional narrowing, fluctuations in the frequency are too short-lived to contribute to the spectral width.

Figure 6.7 shows the lineshape function $\varphi(t)$ and resulting spectra $I(\omega)$ for the case where $\tau_c = 100$ fs and Δ takes on three different values. Note that with increasing magnitude of Δ, compared to $1/\tau_c$, the function $\varphi(t)$ narrows and the lineshape broadens. The Kubo formalism has been advantageously applied to vibrational and electronic spectra to determine the extent of homogeneous and inhomogeneous broadening. We will consider it further in Chapter 12.

6.7 SUMMARY

In this chapter, we have seen how to relate quantum mechanical molecular properties, such as the transition moment and transition polarizability, to measured spectra. The spectral intensity that we have called $I(\omega)$ is an inherent transition rate, whereas the measured spectral intensities, such as molar absorptivity or scattering cross-section, comprise frequency-dependent weighting factors which take experimental considerations into account. It was shown that the radiative and nonradiative lifetimes can be extracted from measurements of the absorption and emission spectra. We have examined the processes of stimulated and

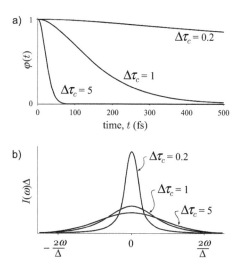

Figure 6.7 (a) Kubo lineshape function $\varphi(t)$ and (b) intensity spectra $I(\omega)$ for $\tau_c = 100$ fs and $\Delta = 5/\tau_c$, $1/\tau_c$, and $1/5\tau_c$.

spontaneous emission, and though the relative rate constants for these processes are molecular properties, the experimental rates are dependent on state populations, and these are subject to experimental control. For stimulated emission to dominate the more common spontaneous emission process, a population inversion is required. (Some approaches for achieving population inversions, in atomic lasers, will be considered in Chapter 7.) Several mechanisms contributing to spectral linewidths were considered, and it was shown how the spectrum can be analyzed to reveal the homogeneous and inhomogeneous contributions to the broadening. The equations presented in this and previous chapters present a set of spectroscopic tools, which we would now like to apply to rotational, vibrational, and electronic spectra. The entire electromagnetic spectrum lies ahead of us.

PROBLEMS

1. Consider an electronic absorption band having an oscillator strength of 0.5 and an absorption maximum at 500 nm. Find (a) the transition dipole moment, (b) the Einstein B coefficient, and (c) the radiative lifetime. What has to hold for your answers to be valid?
2. Consider a molecule having a vibrational frequency of 2000 cm^{-1} and a transition dipole of 0.1 D. Calculate the radiative lifetime of the v = 1 state. What does this lifetime tell you about the radiative decay of states in the infrared?
3. Compare the radiative linewidth obtained in the previous problem to the Doppler width that would be observed at room temperature. Assume a molecular mass equal to that of a CO molecule.
4. For rhodamine B in ethanol, the integrated absorbance $\int \varepsilon_M(\tilde{v})d\ln\tilde{v}$ was found to be 5937 L mol^{-1} cm^{-1} and $\langle \tilde{v}_f^{-3} \rangle^{-1}$ was found to be 5.1×10^{12} cm^{-3} [1]. Find the transition moment and radiative lifetime of rhodamine B. (The refractive index of ethanol is 1.36 at visible wavelengths.)
5. Find expressions for the Kubo lineshape formula $\varphi(t)$ in the limits $\Delta \ll \tau_c^{-1}$ and $\Delta \gg \tau_c^{-1}$.
6. Calculate the lifetime of a $2p_z$ state of hydrogen atom. Start with the result of Problem 6 from Chapter 4.
7. Estimate the ratio of the rate of stimulated to spontaneous emission at three wavelengths: (a) 10 cm, in the microwave (b) 10^{-2} cm, in the far-IR, and (c) 10^{-5} cm, in the near infrared. Assume a blackbody source at 298 K.
8. Verify the conversion factor in front of the integral in Equation 6.26. Derive the conversion factor for finding the oscillator strength f from the integrated molar absorptivity according to Equation 6.27 using (a) cm^{-1} and (b) s^{-1} as frequency units. Does the conversion factor for finding the transition dipole moment depend on the choice of frequency units?

REFERENCES

1. S. J. Strickler and R. A. Berg, Relationship between absorption intensity and fluorescence lifetimes of molecules, *J. Chem. Phys. 37*, 818 (1962).
2. O. Sonnich-Mortensen and S. Hassing, *Advances in Infrared and Raman Spectroscopy 6*, Ch. 1 (Heyden, New York, 1980).
3. R. Kubo, M. Toda, and N. Hashitsume, *Statistical Physics II*, (Springer-Verlag, New York, 1978).

Atomic spectroscopy

7.1 INTRODUCTION

Why should a book about molecular spectroscopy include a chapter on atomic spectroscopy? There are a number of reasons why we want to consider atomic spectra before moving on to the chemically more interesting subject of molecular spectra. Atoms present the simplest possible examples for applying some of the principles laid out in the previous chapters. In this chapter, we consider aspects of selection rules which also apply to more complicated systems. We are concerned with the *electronic spectra* of atoms, and we can concentrate on this topic without worrying about the rotational and vibrational structure that accompanies an electronic transition in a molecule. The absorption wavelengths of an atom coincide with those for emission, which is not necessarily the case in molecular electronic spectra, as we shall see. We also discuss topics such as the effect of internal fields (spin–orbit coupling and hyperfine effects due to nuclear spin) and external fields (the Stark and Zeeman effects) on atomic spectra. These concepts are relevant to molecular spectra as well, but are easier to discuss quantitatively in the atomic case. This chapter also provides an opening into the very interesting topic of laser emission, and we consider the operating principles at work in some common atomic lasers. Lasers are important not only because they exemplify stimulated emission, but also because they are frequently employed as sources in spectroscopic experiments. The field of atomic spectroscopy is an old one, but still of vital interest in fields ranging from analytical chemistry to quantum optics.

7.2 GOOD QUANTUM NUMBERS AND NOT SO GOOD QUANTUM NUMBERS

Our starting point for considering atomic spectra is a review of the quantum numbers that define electronic states of atoms. Hydrogen-like (that is, one-electron) atoms require a set of four quantum numbers to fully specify the wavefunction, and only one to define the energy. The number of quantum numbers is consistent with the electron having three spatial and one spin degree of freedom. The variational principle, on which self-consistent field calculations are based, permits us to build approximate wavefunctions for many-electron atoms which assign individual electrons to hydrogenic orbitals with definite spatial and spin quantum numbers. How far can we carry this approximation and account for experimental spectra of many-electron atoms? And if the hydrogen atom orbitals are *not* exact eigenfunctions in the many-electron case, what *are* the good quantum numbers and how are they determined? To address these questions, we delve into the quantum mechanical bookkeeping which enables us to label the electronic states of many-electron atoms with term symbols. Term symbols represent atomic states and designate the good quantum numbers for the atom. "Good" quantum numbers are eigenvalues of operators which commute with the Hamiltonian and thus correspond to physical properties which can be specified exactly. In the presence of a perturbation that mixes states with different eigenvalues, the corresponding quantum number is no longer strictly good, but may be considered to be "approximately good" in the case of a sufficiently weak perturbation. We shall examine this sort of mixing for the case of angular momentum coupling in many-electron atoms, but first we review the one-electron problem. The one-electron wavefunctions are of interest because they form the basis for expanding approximate wavefunctions of many-electron atoms.

7.2.1 THE HYDROGEN ATOM: ENERGY LEVELS AND SELECTION RULES

The hydrogen atom wavefunctions were reviewed in Chapter 1:

$$\Psi_{nlm_lm_s}(r,\theta,\varphi,\sigma)=R_{nl}(r)Y_{lm_l}(\theta,\varphi)\begin{Bmatrix}\alpha\\\beta\end{Bmatrix} \tag{7.1}$$

where $R_{nl}(r)$ is the radial part and the spherical harmonic function $Y_{lm_l}(\theta,\varphi)\propto P_{lm_l}(\cos\theta)e^{im_l\varphi}$ is the angular part of the wavefunction. The symbol σ specifies the "spin coordinate" which is designated α or β for $m_s=1/2$ or $-1/2$, respectively.

The one-electron energy levels are decided by the principal quantum number n:

$$E_n=\frac{-Z^2e^2}{(4\pi\varepsilon_0)2n^2a_0}=\frac{-Z^2}{n^2}\times13.6\text{ eV} \tag{7.2}$$

where $a_0=0.529$ Å is the Bohr radius. Recall that the degeneracy of a level is $g_n=2n^2$ and that states having orbital quantum number $l=0,1,2,3,4\dots$ are referred to by the letters $s, p, d, f, g\dots$, etc. As reviewed in Chapter 1, the spherical harmonics are eigenfunctions of the operators $\hat{L}^2$ and $\hat{L}_z$ (Equations 1.51 and 1.52), and the spin functions are eigenfunctions of $\hat{S}^2$ and $\hat{S}_z$ (Equations 1.58–1.61).

We now wish to derive selection rules for the spectrum of the hydrogen atom. We characterize a spectroscopic transition by the difference in the initial and final values of the quantum numbers n, l, m_l, and m_s. Our goal is to determine what changes in the quantum numbers, Δn, Δl, Δm_l, and Δm_s, lead to a nonzero electric dipole (E1) or electric quadrupole (E2) transition moment, or a magnetic dipole (M1) transition moment. Since the energy depends only on one quantum number, n, we could neglect selection rules concerning changes in l, m_l, and m_s. However, we *will* consider the selection rules for these quantum numbers, as they become important in the presence of external fields, and similar considerations arise in the discussion of many-electron atoms, where the hydrogen atom wavefunctions are used as a basis for approximate wavefunctions. For the E1 selection rules, our task is to find the matrix elements of the operator for the electric dipole moment, which is given by $\hat{\mu}=-e\vec{r}=-er(\hat{i}\sin\theta\cos\varphi+\hat{j}\sin\theta\sin\varphi+\hat{k}\cos\theta)$. Similarly, the M1 (magnetic dipole) selection rules arise from considering matrix elements of the operator $\vec{L}$ for orbital angular momentum. The E2 selection rules result from matrix elements of the quadrupole moment operator, which has the form $e\vec{r}\vec{r}$. Thus E2 transitions are allowed when operators such as x^2, xy, etc. connect two states.

Let us start with the E1 selection rules, and suppose that the initial and final states are specified by the sets of quantum numbers n, l, m_l, m_s and n', l', m_l', m_s'. Consider the transition dipole moment $\left\langle\Psi_{nlm_lm_s}\left|\hat{\mu}\right|\Psi_{n'l'm_l'm_s'}\right\rangle\equiv\mu_{if}$. The derivation of selection rules consists of figuring out which possible values of initial and final quantum numbers will result in $\mu_{if}\neq0$. Separation of variables allows us to write the transition moment as

$$\left\langle\Psi_{nlm_lm_s}\left|\hat{\mu}\right|\Psi_{n'l'm_l'm_s'}\right\rangle=-e\int_0^\infty rR_{nl}(r)R_{n'l'}(r)r^2dr$$

$$\times\int_0^\pi\sin\theta\,d\theta\int_0^{2\pi}d\varphi\,Y_{lm_l}^*(\theta,\varphi)\begin{Bmatrix}\sin\theta\cos\varphi\\\sin\theta\sin\varphi\\\cos\theta\end{Bmatrix}Y_{l'm_l'}(\theta,\varphi)\times\delta_{m_sm_s'} \tag{7.3}$$

The delta function $\delta_{m_sm_s'}$ comes from the orthonormality of the spin functions:

$$\langle\alpha|\alpha\rangle=\langle\beta|\beta\rangle=1$$
$$\langle\alpha|\beta\rangle=\langle\beta|\alpha\rangle=0 \tag{7.4}$$

The orthonormality relations for the spin functions can also be denoted by $\int\alpha^*\alpha d\sigma=1$, $\int\alpha^*\beta d\sigma=0$, etc. These lead directly to the conclusion that m_s must equal m_s' for an allowed transition. Thus $\Delta m_s=0$ is our first selection rule. It holds for M1 and E2 transitions as well, because the operators for these transitions do not depend on electron spin. The exceptions to this selection rule for spin are notable, for when spin–orbit coupling mixes up the quantum numbers for spin and orbital angular momentum, then transitions between states of different spin become allowed. This will be considered in due time.

Due to separation of variables, a transition becomes forbidden if the integral over any one of the three spatial variables r, θ, or φ vanishes. Consider the angular part of Equation 7.3, which permits transitions whenever the integral involving one of the three functions in the curly brackets is nonzero. The three components of the dipole operator apply when the radiation field is polarized in the x, y, or z direction in the laboratory frame. In the absence of a static field, these three directions are equivalent, so any of these three components may cause a transition, regardless of the polarization of the incident light. Using Equation 1.50 for Y_{lm_l} and extracting the φ-dependent integral, we obtain

$$\int_0^{2\pi} d\varphi\, e^{-im_l\varphi} \begin{Bmatrix} \cos\varphi \\ \sin\varphi \\ 1 \end{Bmatrix} e^{im_l'\varphi} \tag{7.5}$$

For the z component, arising from the "1" term in Equation 7.5, we have

$$\int_0^{2\pi} d\varphi\, e^{-im_l\varphi} e^{im_l'\varphi} = 2\pi\delta_{m_l m_l'} \tag{7.6}$$

This is another integral representation of the delta function. Thus for z-polarized radiation, the selection rule on m_l is $\Delta m_l = 0$. For k equal to any integer, the integral $\int_0^{2\pi} e^{ik\varphi} d\varphi$ is equal to $2\pi\delta_{k0}$. To take advantage of this, the cosine and sine functions of Equation 7.5 are expressed as follows:

$$\cos\varphi = \frac{1}{2}\left(e^{i\varphi} + e^{-i\varphi}\right) \tag{7.7}$$

$$\sin\varphi = \frac{1}{2i}\left(e^{i\varphi} - e^{-i\varphi}\right) \tag{7.8}$$

We find that the x ($\cos\varphi$) and y ($\sin\varphi$) components of the dipole operator lead to the selection rule $\Delta m_l = \pm 1$ for x- or y-polarized radiation. The combined selection rule is $\Delta m_l = 0, \pm 1$.

The selection rule for changes in the orbital quantum number l can be found with the help of the orthogonality relation for the spherical harmonics,

$$\int_0^{2\pi} d\varphi \int_0^{\pi} Y_{l'm_l'}^*(\theta, \varphi) Y_{lm_l}(\theta, \varphi) \sin\theta\, d\theta = \delta_{l'l}\delta_{m_l'm_l} \tag{7.9}$$

combined with the relation:

$$\sin\theta Y_{lm_l} = c_+ Y_{l+1,m_l} + c_- Y_{l-1,m_l} \tag{7.10}$$

A similar expression holds for $\cos\theta Y_{lm_l}$, as shown in [1]. The coefficients c_+ and c_- depend on l and m_l, but their functional forms are not of interest to the result we are pursuing, because here we just want to know if an integral is zero or not. Substituting Equation 7.10 into Equation 7.3 and then using Equation 7.9 leads to the conclusion that l must equal $l' \pm 1$; i.e., $\Delta l = \pm 1$, for E1 transitions.

Finally, what about the changes in the principal quantum number n? These depend on the radial wavefunctions $R_{nl}(r)$, which have the form:

$$R_{nl}(r) = N(c_0 + c_1 r + \cdots + c_{n-l-1} r^{n-l-1}) r^l e^{-Zr/na_0} \tag{7.11}$$

The coefficients c_n of each term r^n obey a recursion formula which can be found in most quantum chemistry texts [2,3]. The point here is that there are no symmetry restrictions that cause the radial integral in Equation 7.3

to vanish for certain combinations of n and n'. Thus Δn can be any positive (for absorption) or negative (for emission) integer. This results in the very rich form of the hydrogen atom spectrum, where the frequencies (in wavenumbers) of the absorption and emission lines are represented by the Rydberg formula:

$$\tilde{v} = R_H \left(\frac{1}{n_1^2} - \frac{1}{n_2^2} \right) \tag{7.12}$$

Here, we require $n_1 < n_2$. The expression for the Rydberg constant R_H, 109,700 cm^{-1}, is given in Equation 1.64.

Emission lines of atomic hydrogen terminating on various lower quantum states (indexed by n_1) fall in characteristic spectral regions as follows:

$n_1 = 1$, $n_2 = 2,3,4...\infty$ The Lyman series in the ultraviolet

$n_1 = 2$, $n_2 = 3,4,5...\infty$ The Balmer series in the visible

$n_1 = 3$, $n_2 = 4,5,6...\infty$ The Paschen series in the near IR, and so on.

Absorption frequencies coincide with those for emission, and transitions having initial quantum number $n \neq 1$ are observed only after preparation of the excited state by some means, such as electric discharge. The ionization limit in absorption corresponds to a final quantum number $n_2 = \infty$. From the ground state, the ionization energy is 109,700 cm^{-1} or 13.6 eV.

Suppose that an excited hydrogen atom finds itself in a $2s$ state. Can it return to the $1s$ ground state via a radiative transition? Such a transition could satisfy the $\Delta m_s = 0$ and $\Delta m_l = 0, \pm 1$ selection rules, but the $\Delta l = \pm 1$ selection rule would be violated. The net result is that the $2s$ excited state of hydrogen is metastable. It can relax nonradiatively, of course, but as far as emission is concerned, the return path to the ground state must take place by means of a more weakly allowed transition than E1. Let us look at the selection rules for the weaker M1 and E2 transitions.

The orbital angular momentum has x, y, and z components. Hence an M1 transition is allowed when the matrix element of the operator $\hat{L}_x$, $\hat{L}_y$, or $\hat{L}_z$ exists. These operators depend only on θ and φ, so we need to consider the angular part of the wavefunction. However, since there is no r dependence of the operator for magnetic dipole-allowed transitions, the orthogonality of the radial wavefunctions requires that $\Delta n = 0$. Since the energy levels depend only on n, this means that magnetic dipole selection rules are rather uninteresting for the hydrogen atom! However, in the case where perturbations lift the degeneracy of states arising from the same principal quantum number (as in many-electron atoms), the M1 selection rules are of interest. So we continue this analysis. The spherical harmonics Y_{lm_l} are eigenfunctions of $\hat{L}_z$ with eigenvalues $m_l \hbar$, but they are not eigenfunctions of $\hat{L}_x$ or $\hat{L}_y$. The z component of the angular momentum operator has only diagonal matrix elements:

$$\left\langle Y_{lm_l} \left| \hat{L}_z \right| Y_{l'm_l'} \right\rangle = m_l \hbar \delta_{ll'} \delta_{m_l m_l'} \tag{7.13}$$

Thus from the z component of the *magnetic* field we get the selection rule $\Delta m_l = 0$, $\Delta l = 0$. For the x- and y-polarized transitions, we can take advantage of the raising and lowering operators defined as $\hat{L}_\pm = \hat{L}_x \pm i\hat{L}_y$. The desired matrix elements are written as

$$\left\langle Y_{lm_l} \left| \hat{L}_x \right| Y_{l'm_l'} \right\rangle = \frac{1}{2} \left\langle Y_{lm_l} \left| \hat{L}_+ + \hat{L}_- \right| Y_{l'm_l'} \right\rangle \tag{7.14}$$

$$\left\langle Y_{lm_l} \left| \hat{L}_y \right| Y_{l'm_l'} \right\rangle = \frac{1}{2i} \left\langle Y_{lm_l} \left| \hat{L}_+ - \hat{L}_- \right| Y_{l'm_l'} \right\rangle \tag{7.15}$$

The action of the raising operator $\hat{L}_+$ on Y_{lm_l} is to create a function proportional to Y_{l,m_l+1}. The exception to this is that when m_l takes the maximum value of l, then $\hat{L}_+ Y_{l,l} = 0$. Similarly, application of $\hat{L}_-$ results in a function proportional to Y_{l,m_l-1}, except when $m_l = -l$, in which case $\hat{L}_- Y_{l,-l} = 0$. Consequently, both the x and y components of the magnetic field result in transitions having $\Delta l = 0$ and $\Delta m_l = \pm 1$.

Table 7.1 Selection rules for transitions of one-electron atoms

		Allowed transitions		
E1	$\Delta l = \pm 1$	$\Delta m_l = 0, \pm 1$	$\Delta m_s = 0$	Δn = any integer
M1	$\Delta l = 0$	$\Delta m_l = 0, \pm 1$	$\Delta m_s = 0$	$\Delta n = 0$
E2	$\Delta l = 0, \pm 2$	$\Delta m_l = 0, \pm 1, \pm 2$	$\Delta m_s = 0$	Δn = any integer

The E2 transitions are straightforward to evaluate on the basis of knowing the E1 selection rules. Consider a general matrix element of the quadrupole operator:

$$\left\langle \Psi_{nlm_lm_s} \left| \hat{\Theta} \right| \Psi_{n'l'm_l'm_s'} \right\rangle = -e\left\langle \Psi_{nlm_lm_s} \left| \vec{r}\,\vec{r} \right| \Psi_{n'l'm_l'm_s'} \right\rangle$$

$$= -e \sum_{n''l''m_l''m_s''} \left\langle \Psi_{nlm_lm_s} \left| \vec{r} \right| \Psi_{n''l''m_l''m_s''} \right\rangle \left\langle \Psi_{n''l''m_l''m_s''} \left| \vec{r} \right| \Psi_{n'l'm_l'm_s'} \right\rangle \tag{7.16}$$

In the second line of Equation 7.16, we have inserted the resolution of the identity, $\sum_n |n\rangle\langle n| = 1$, where "$n$" specifies the set of all four quantum numbers. We already know from considering the E1 selection rules that matrix elements of the type $\left\langle \Psi_{nlm_lm_s} \left| \vec{r} \right| \Psi_{n''l''m_l''m_s''} \right\rangle$ are nonzero for $\Delta l = \pm 1$, $\Delta m_l = 0, \pm 1$ and $\Delta m_s = 0$. Since this is the case for both of the matrix elements in Equation 7.16, the net selection rule for electric quadrupole transitions is $\Delta l = 0$, ± 2, $\Delta m_l = 0, \pm 1, \pm 2$ and $\Delta m_s = 0$. The set of selection rules that we have obtained is summarized in Table 7.1.

7.2.2 MANY-ELECTRON ATOMS

The conventional discussion of electronic states of many-electron atoms is fraught with apparent inconsistencies. We know that the Schrödinger equation cannot be solved exactly, even for helium, and that the hydrogen atom wavefunctions are only approximations which we use to describe electronic configurations qualitatively. The many-electron atom is intractable because the interelectronic repulsion term in the Hamiltonian prevents separation of variables from being used to solve the Schrödinger equation. The central-field approximation, on which the Hartree–Fock calculation is based, deals with this problem by calculating the potential energy for each electron as a spherical average over the distribution of all other electrons. As a result, the potential energy of an electron in a many-electron atom is approximately spherically symmetric, and the functions Y_{lm_l} are appropriate angular functions for the one-electron basis states. The self-consistent field approach is based on the variation theorem, and trial wavefunctions are frequently taken to be linear combinations of functions (spin–orbitals) which resemble those for a one-electron atom (though Gaussian-type orbitals are also employed). A spin–orbital is a product of a spatial function ϕ and a spin function α or β. The angular dependence of each spatial orbital is described by the spherical harmonic Y_{lm_l}. Thus, the quantum numbers l and m_l for each individual orbital can be characterized. If it can be established that an electronic state of the atom is well characterized by one set of occupied orbitals, we can sum the values of m_s and m_l for all occupied orbitals to obtain the quantum numbers for the z component of the total orbital L and spin S angular momentum: M_L and M_S. The capital letters L and S will be reserved for many-electron quantum numbers, analogous to the one-electron quantum numbers l and s. To a first approximation, an atomic energy level can be designated by a *term symbol*, ^{2S+1}L. In this section, we consider the rules for adding the angular momentum quantum numbers of individual electrons to get the quantum numbers L, M_L, S, and M_S for states of the atom.

Because electrons are fermions, the total wavefunction must change sign if the spatial and spin coordinates of any two electrons are exchanged. For a closed-shell atom, all spatial orbitals are doubly occupied, and the total wavefunction can be represented by a single Slater determinant:

$$\Psi = \frac{1}{\sqrt{(2n)!}} \begin{vmatrix} \phi_1(1)\alpha(1) & \phi_1(2)\alpha(2) & \phi_1(3)\alpha(3) & \dots & \phi_1(2n)\alpha(2n) \\ \phi_1(1)\beta(1) & \phi_1(2)\beta(2) & \phi_1(3)\beta(3) & \dots & \phi_1(2n)\beta(2n) \\ \phi_2(1)\alpha(1) & \phi_2(2)\alpha(2) & \phi_2(3)\alpha(3) & \dots & \phi_2(2n)\alpha(2n) \\ \vdots & \vdots & \vdots & \ddots & \vdots \\ \phi_n(1)\beta(1) & \phi_n(2)\beta(2) & \phi_n(3)\beta(3) & \dots & \phi_n(2n)\beta(2n) \end{vmatrix} \tag{7.17}$$

In writing this Slater determinant, it has been assumed that there are $2n$ electrons in the atom; thus n spatial orbitals, ϕ_1, ϕ_2, ..., ϕ_n, are required to accommodate them. Each spatial orbital appears in each column of the determinant twice: once with spin function α and once with spin function β. Along the diagonal of the determinant in Equation 7.17, the electrons 1, 2, 3, ..., $2n$ are placed in consecutively numbered spin–orbitals. In each column of the determinant, the same electron is placed in each of the $2n$ different spin–orbitals, while across a row, all the electrons are placed in the same spin–orbital.

Open-shell atoms have one or more singly occupied orbitals and cannot necessarily be described by a single Slater determinant; rather, one must write a total wavefunction Ψ which is a linear combination of Slater determinants. The appropriate linear combination is that which results in specific values for the angular momentum quantum numbers. The rules for finding these appropriate linear combinations are introduced in Section 7.2.3.

When we write, say, the electronic configuration of carbon as $1s^2 2s^2 2p^2$, we are using a qualitative picture that summarizes a vast array of experimental observations about the element carbon and is consistent with its placement in the periodic table and tendency to form four bonds with other atoms. Yet the ground electronic state of carbon is *not* specified by $1s^2 2s^2 2p^2$, as "states" and "configurations" are precise terms which, unlike electrons, are not interchangeable! A configuration generally implies a way of loading electrons into spin–orbitals in a manner consistent with the Pauli exclusion principle. This principle states that if two electrons in an atom share the same quantum numbers n, l, and m_l, then they must have opposite signs for the spin quantum number m_s. This requirement is enforced by the form of Equation 7.17, which vanishes if two rows (or two columns) are equal. If two electrons could share the same set of all four hydrogen-atom quantum numbers, this would correspond to the same spin–orbital being used twice, and Ψ would be zero.

As we shall see, there are 15 ways to place two electrons in the same p-shell, and there are thus 15 states associated with $1s^2 2s^2 2p^2$. The ground-state configuration of carbon atom can be specified more precisely as follows:

$$\text{C atom ground state } 1s(\downarrow\uparrow)2s(\downarrow\uparrow)2p_{-1}(\uparrow)2p_0(\uparrow)2p_1 \tag{7.18}$$

In accordance with Hund's rule of maximum spin multiplicity, we have placed the two $2p$ electrons in two different p orbitals, with the same spin, leaving the third p orbital unoccupied. The subscript on each p orbital designates the value of m_l. The real p orbitals p_x, p_y, and p_z, which are linear combinations of p_1, p_0, and p_{-1}, are not all eigenfunctions of $\hat{L}_z$ and are thus not convenient for the discussion of this section.* The configuration shown in 7.18 is still somewhat arbitrary: the two unpaired electrons could be spin down rather than spin up, and any one of the three p orbitals could have been left unoccupied. Thus Equation 7.18 only specifies one possible configuration of the ground state of carbon. We will refer to $1s^2 2s^2 2p^2$ as the *ground configuration*. This ground configuration gives rise to 15 different pictures like 7.18, and thus encompasses 15 Slater determinants which contribute to 15 electronic states. The ground state in the case of carbon atom is degenerate and must be expressed as a linear combination of Slater determinants.

The quantum numbers that specify the orbitals to which we assign electrons in a many-electron atom apply to the hydrogen atom; thus they are *not* good quantum numbers for carbon. To a good approximation, we can couple the angular momentum quantum numbers of occupied hydrogen-like atomic orbitals to generate quantum numbers which *are* good quantum numbers for the many-electron atom. The rules for adding angular momenta acknowledge the properties of quantum mechanical vectors. That is, the z component of the orbital or spin angular momentum is quantized, and when we add vectors, we add the z components. The x and y components of the vectors remain unspecified in accordance with the uncertainty depicted in Figure 1.9. The rule which takes all this into consideration is often called the triangle rule. It states that, for any two angular momentum quantum numbers j_1 and j_2, the permitted values of *net* angular momentum J which result from adding them are

$$J = j_1 + j_2, \quad j_1 + j_2 - 1, ..., \quad |j_1 - j_2| \tag{7.19}$$

---

* The p_0 orbital is identical to p_z, while p_x and p_y are each a linear combination of p_{-1} and p_{+1}.

Here, the lowercase symbols $j_{1,2}$ represent either the l or s quantum numbers for individual electrons 1 and 2, while J is the net angular momentum quantum number of the two-electron state. The symbols j_1, j_2 and J are used here as generic quantum numbers for any type of angular momentum. For example, Equation 7.19 applies to the case of adding the net orbital L and spin S angular momenta to get the total electronic angular momentum J. If a three-electron state were under consideration, we could first apply the triangle rule to two of the electrons, and for each of the resulting J's we would then add in the value j_3 to get $J + j_3, J + j_3 - 1, ..., |J - j_3|$. It is worth emphasizing that the type of quantum number under discussion here is that which specifies the length of the angular momentum vector; $|\vec{j}_i| = \hbar\sqrt{j_i(j_i+1)}$ and $|\vec{J}| = \hbar\sqrt{J(J+1)}$. The permitted values of the resultant angular momentum in Equation 7.19 are those that result from vector addition of $\vec{j}_1$ and $\vec{j}_2$ with all possible quantized relative orientations.

We now apply this rule to the determination of the net angular momentum state of a many-electron atom. The orbital and spin angular momenta of individual electrons can first be coupled to get $L = \sum l_i$ and $S = \sum s_i$. When spin–orbit coupling is considered, the net orbital and spin angular momenta are then further coupled to give J, the total angular momentum quantum number. This approach, called LS or Russell–Saunders coupling, is actually only appropriate for lighter atoms, for which the spin–orbit perturbation is weak compared to the Coulombic interactions of the electrons. For heavier atoms, it is more correct to couple the l_i and s_i of individual electrons to get j_i for each electron, and then add these j_i to get J. For more on this so-called jj-coupling scheme, see [4]. We illustrate the case of LS coupling in this discussion.

Our goal is to generate the term symbol for an electronic energy level, ^{2S+1}L, which designates an electronic state of an atom. The set of states belonging to the ^{2S+1}L energy level is called a term. The superscript $2S + 1$ to the left is the spin multiplicity of the term; it results from the fact that for a particular value of S the quantum number for the z component of the spin, M_S, ranges from S to $-S$ by integral steps. Similarly, there are $2L + 1$ degenerate orbital angular momentum states, corresponding to the values of $M_L = -L, -L + 1, ...L$, where M_L is the quantum number for the z component of net orbital angular momentum. The degeneracy of an energy level designated by the symbol ^{2S+1}L is given by

$$g_{LS} = (2S+1)(2L+1) \tag{7.20}$$

The degeneracy g_{LS} accounts for all the allowed values of M_L and M_S.

It may seem a daunting task to apply the triangle rule $N - 1$ times for an N electron atom with large N. Fortunately, we need only consider the electrons in open shells, as the contribution to L and S from the closed shells is zero. The m_l values of occupied orbitals add to give the net M_L, and m_s values also add to give M_S. Closed shells always lead to $M_L = 0$ and $M_S = 0$ because for every plus value of m_l or m_s there is a negative value to cancel it. Another simplification is that "holes" in a shell are equivalent to electrons; a p^5 configuration gives the same term symbols as a p^1. The triangle rule does not take into account the fact that some of the possible values of L and S may violate the Pauli exclusion principle. Such states have to be discarded.

To illustrate this, we consider the case of coupling two p electrons. If these two electrons reside in different p orbitals (e.g., $2p^1 3p^1$), then we do not need to worry that any of the resulting L and S values would violate the Pauli exclusion principle. Using Equation 7.19 with $l_1 = l_2 = 1$, we find that $L = 2, 1$ and 0. Analogous to the designation $s, p, d, f,...$ for $l = 0, 1, 2, 3,...$, states with $L = 0, 1, 2, 3...$ are specified by the capital letters $S, P, D, F...$ etc. The configuration $2p^1 3p^1$ thus gives rise to S, P and D states. For each of these states the values of S are $s_1 + s_2 = 1$ or $s_1 - s_2 = 0$. (For any electron, the quantum number $s = 1/2$ specifies the magnitude of the spin. Do not confuse the letter s being used as the spin quantum number with the same letter used to specify an orbital having $l = 0$. Similarly, the quantum number S for the many-electron spin state is not the same as the symbol S used to designate an electronic state in which $L = 0$.) The values $S = 0$ and 1 belong to singlet ($2S + 1 = 1$) and triplet ($2S + 1 = 3$) states. So there are six states, ^1S, ^3S, ^1P, ^3P, ^1D and ^3D which derive from the $2p^1 3p^1$ configuration. How is it that so many states can result from one choice of occupied orbitals? It is actually a rather straightforward result of quantum mechanical bookkeeping. There are six ways to load an electron into a $2p$ subshell, because there are three equivalent $2p$ orbitals and in each one the electron may be spin up or spin down. For each of these six ways to place the first electron in $2p$, there are six similar ways to add the second electron to $3p$. There are thus $6 \times 6 = 36$ states, but only six energy levels (one for each term), in the absence of spin–orbit coupling. The degeneracies of each of the

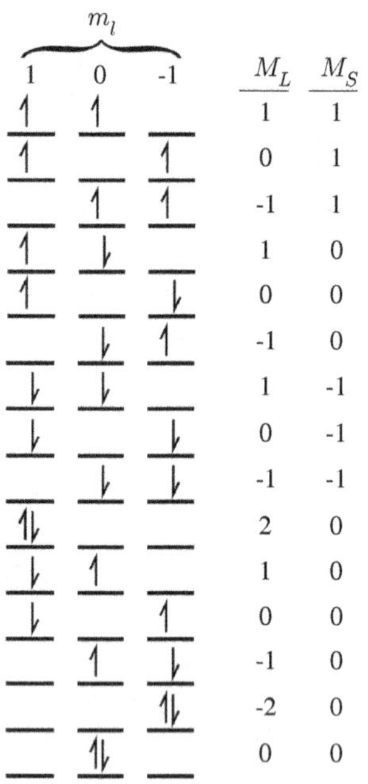

m_l			M_L	M_S
1	0	-1		
↑	↑		1	1
↑		↑	0	1
	↑	↑	-1	1
↑	↓		1	0
↑		↓	0	0
	↓	↑	-1	0
↓	↓		1	-1
↓		↓	0	-1
	↓	↓	-1	-1
↑↓			2	0
↓	↑		1	0
↓		↑	0	0
	↑	↓	-1	0
		↑↓	-2	0
	↑↓		0	0

Figure 7.1 Configurations associated with p^2.

six terms are determined from Equation 7.20, and it can be readily confirmed that they add up to 36. So there is a conservation law at work, which says that the total number of states is the same before and after coupling.

We would like to know the energies of the resulting terms, but that information must be found from a quantum mechanical calculation or a spectroscopy experiment. We can, however, take a chance on predicting the order of the energy levels. States of higher spin multiplicity usually fall at lower energy than singlets with the same value of L. This is the basis for Hund's rule of maximum spin multiplicity, which derives from the difference in the exchange interaction between, for example, singlet and triplet states. Electrons in states having higher S experience less Coulombic repulsion, in general, and are of lower energy. The ordering of the L values is usually such that for the same S, states with higher L are lower in energy due to less Coulombic repulsion.

Next consider the coupling of two equivalent electrons, such as the two $2p$ electrons, in the outer shell of carbon. There are six ways to place the first electron in the $2p$ subshell, and five ways to place the second electron in the same subshell. Since these two electrons are indistinguishable, there are $(6 \times 5) \div 2 = 15$ states associated with this configuration. These are depicted in Figure 7.1. As shown in the figure, the total quantum numbers $M_L = \sum_i m_{l,i}$ and $M_S = \sum_i m_{s,i}$ are found by summing the values for the individual electrons. Each term symbol that we derive for the $2p^2$ configuration comprises a set of configurations from that figure, and the number in the set is given by the degeneracy of the term: g_{LS}. Some of the above states derived for two inequivalent p electrons must be thrown away, because they violate the Pauli exclusion principle. The remaining terms must account for all the diagrams in Figure 7.1. It is a good practice to consider the possible terms in order of decreasing S and then decreasing L. For example, the ^3D term is considered first, but it is impossible for a $2p^2$ configuration, because it would require two electrons to be placed in an orbital with $m_l = \pm 1$ and with the same spin. A ^3P term can be achieved, however, by placing the parallel spin electrons in p orbitals having $m_l = \pm 1$ and $m_l = 0$. For this state the net M_L would be ± 1, as expected for a P state. This ^3P term accounts for $3 \times 3 = 9$ of the 15 states. We should therefore cross out the first nine configurations in Figure 7.1, having M_L values of 1, 0, −1, and M_S values of 1, 0, −1. The possible ^3S term now has to be discarded because all of the remaining terms have $M_S = 0$ and thus correspond to singlet states. So we move on to the ^1D term which we expect to contribute because there are configurations in Figure 7.1 for which $M_L = \pm 2$ and $M_S = 0$. We now have $9 + 5 = 14$ states accounted for by ^3P and ^1D, so there is one more singly degenerate term, which has to be the ^1S state. The $2p^2$ configuration thus gives rise to the energy levels ^3P, ^1D, and ^1S. Using the rules of thumb outlined above, we predict the energy level ordering to be $^3P < {}^1D < {}^1S$.

It is not possible to associate each configuration of Figure 7.1 with a particular state, except in special cases. In order to write a total wavefunction Ψ which is an eigenfunction of the orbital and spin angular momentum operators, we must take particular linear combinations of Slater determinants. The rules for finding these combinations are introduced in the next section.

7.2.3 The Clebsch–Gordan series

When two or more angular momenta are added to get a resulting value, there is a corresponding prescription to describe the resulting states as linear combinations of the basis states. Consider the coupling of two angular momentum states represented by the ket vectors $|j_1 m_1\rangle$ and $|j_2 m_2\rangle$, where j_1 and j_2 can be any kind of angular momentum quantum number pertaining to single- or many-electron states. The eigenvalue relations are

$$\hat{J}^2 |j_1 m_1\rangle = j_1(j_1 + 1)\hbar^2 |j_1 m_1\rangle \tag{7.21}$$

$$\hat{J}_z |j_1 m_1\rangle = m_1 \hbar |j_1 m_1\rangle \tag{7.22}$$

$$\hat{J}^2 |j_2 m_2\rangle = j_2(j_2 + 1)\hbar^2 |j_2 m_2\rangle \tag{7.23}$$

$$\hat{J}_z |j_2 m_2\rangle = m_2 \hbar |j_2 m_2\rangle \tag{7.24}$$

The basis states that span the total angular momentum states are products of $|j_1 m_1\rangle$ and $|j_2 m_2\rangle$, which we write as $|j_1 m_1 j_2 m_2\rangle$. For each resultant J out of the set spanning $j_1 + j_2, \ldots, |j_1 - j_2|$, we can write the eigenfunction as

$$|JM; j_1 j_2\rangle = \sum_{m_1} \sum_{m_2} C(j_1 j_2 J; m_1 m_2 M) |j_1 m_1 j_2 m_2\rangle \tag{7.25}$$

Equation 7.25 expresses the eigenfunction of the combined angular momentum state, having eigenvalues pertaining to J and M and formed from the combination of states with eigenvalues j_1 and j_2, as a linear combination of the product states $|j_1 m_1 j_2 m_2\rangle$. The Clebsch–Gordan (CG) coefficients $C(j_1 j_2 J; m_1 m_2 M)$ are mere numbers. We will not concern ourselves here with how to generate them; they can be found in tables (see [1] or [5]). We are, however, interested in the following properties. The coefficient $C(j_1 j_2 J; m_1 m_2 M)$ vanishes unless J can be formed from adding j_1 and j_2 according to the triangle rule. Also, the CG coefficient is zero unless $m_1 + m_2 = M$, consistent with the fact that the z component of the resultant vector $\vec{J}$ is the sum of those for the vectors $\vec{j}_1$ and $\vec{j}_2$. Formally, the CG coefficient is the projection of the resultant state onto the basis state:

$$C(j_1 j_2 J; m_1 m_2 M) = \langle j_1 m_1 j_2 m_2 | JM; j_1 j_2 \rangle \tag{7.26}$$

The CG coefficients can be related to the $3j$ symbols (Appendix A), defined as follows:

$$C(j_1 j_2 J; m_1 m_2 M) = (-1)^{j_1 - j_2 - M}(2J + 1)^{1/2} \begin{pmatrix} j_1 & j_2 & J \\ m_1 & m_2 & \bar{M} \end{pmatrix} \tag{7.27}$$

A bar over a quantity signifies a negative sign: $\bar{m} = -m$. The $3j$ symbol is a number represented by the array in parentheses in Equation 7.27. The CG coefficients and $3j$ symbols have some particularly convenient symmetry properties, which we will exploit in Chapter 8 when rotational selection rules are considered. (Rotational states have the same eigenfunctions as orbital angular momentum states.)

Consider, for example, the two-electron configurations listed in Figure 7.1. The ^1D state comprises five of these configurations. Each of the five ^1D wavefunctions is some linear combination of states $|1 m_1 1 m_2\rangle$, where m_1 and m_2 range independently over $-1, 0, 1$ and add to give $M_L = 2, 1, 0, -1, -2$. Let us use Equation 7.25 to write a wavefunction which represents the ^1D state having $M_L = 0$. This state is a linear combination of three basis states, corresponding to the three combinations of m_1 and m_2 which sum to zero:

$$|20;11\rangle = a|1-111\rangle + b|111-1\rangle + c|1010\rangle \tag{7.28}$$

The coefficients a, b, c stand for the appropriate CG coefficients. Including spin in the picture, the basis states of Equation 7.28 have to be written as Slater determinants representing singlet states:

$$|1-111\rangle = \frac{1}{\sqrt{2}} \begin{vmatrix} p_{-1}(1)\alpha(1) & p_{-1}(2)\alpha(2) \\ p_1(1)\beta(1) & p_1(2)\beta(2) \end{vmatrix}$$

$$|111-1\rangle = \frac{1}{\sqrt{2}} \begin{vmatrix} p_1(1)\alpha(1) & p_1(2)\alpha(2) \\ p_{-1}(1)\beta(1) & p_{-1}(2)\beta(2) \end{vmatrix} \tag{7.29}$$

$$|1010\rangle = \frac{1}{\sqrt{2}} \begin{vmatrix} p_0(1)\alpha(1) & p_0(2)\alpha(2) \\ p_0(1)\beta(1) & p_0(2)\beta(2) \end{vmatrix}$$

Each of the determinants in Equation 7.29 can be represented by a diagram from Figure 7.1. The linear combination of Equation 7.28 is the wavefunction for which $L = 2$ and $M_L = 0$. More generally, the states $|JM; j_1 j_2\rangle$ have the eigenvalue properties

$$\hat{J}^2|JM; j_1 j_2\rangle = J(J+1)\hbar^2|JM; j_1 j_2\rangle \tag{7.30}$$

$$\hat{J}_z|JM; j_1 j_2\rangle = M\hbar|JM; j_1 j_2\rangle \tag{7.31}$$

The point here is that the quantum mechanical rules for vector addition of angular momentum are formalized by the Clebsch–Gordan series. When figuring out term symbols and the number of states that each one comprises, we are accounting for the number of required basis states, which is equal to the number of resulting combined states. The total number of states $|JM; j_1 j_2\rangle$ for all possible J's satisfies

$$\sum_{J=j_1-j_2}^{j_1+j_2} (2J+1) = (2j_1+1)(2j_2+1) \tag{7.32}$$

In the next section, we consider how the spin–orbit interaction results in coupling of the L and S quantum numbers, an effect which can lead to splitting of states corresponding to a given term.

7.2.4 Spin–orbit coupling

A classic experimental demonstration of spin–orbit coupling is provided by the spectrum of sodium atom, illustrated in Figure 7.2. The transition responsible for the yellow emission can be described as a jump of a single electron from a $3p$ to a $3s$ level. In term symbol language, this would be a $^2P \rightarrow {}^2S$ transition. Under higher resolution, however, it becomes clear that this emission is a doublet of lines separated by about 6 Å. The reason is that the higher energy state is actually split into two energy levels: $^2P_{3/2}$ and $^2P_{1/2}$. The ground state $^2S_{1/2}$ is not split by spin–orbit coupling because it has $L = 0$, $S = 1/2$, and only one possible value of J, namely $1/2$ (see Equation 7.33 below). In this section, we examine the physical basis for this effect without going into much computational detail.

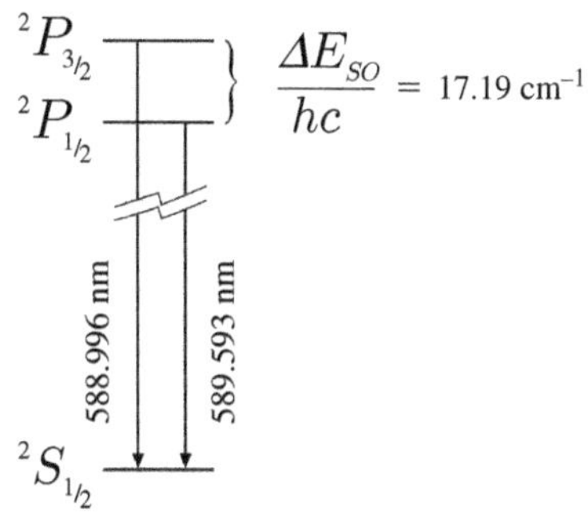

Figure 7.2 The yellow emission doublet of the sodium atom.

If you were an electron riding a quantum mechanical orbit about the nucleus, then from your Ptolemaic view of the atomic universe, the nucleus would appear to orbit about *you*. As such, it would present a magnetic field $\vec{B}$ as suggested by the classical analogy of the field due to a current in a loop (see Equations 3.99 and 3.100). Now, as an electron, you have a magnetic moment $\vec{\mu}_B$ due to your own intrinsic spin, so you would experience a "spin–orbit" interaction of energy $E_{SO} = -\vec{\mu}_B \cdot \vec{B}$. When this energy is cast in the form of a quantum mechanical operator, it is found to depend on the operator $\vec{L} \cdot \vec{S}$, since the field $\vec{B}$ is proportional to the orbital angular momentum $\vec{L}$ and the magnetic moment $\vec{\mu}_B$ is proportional to the spin angular momentum $\vec{S}$. The vectors $\vec{L}$ and $\vec{S}$ add to give the total angular momentum $\vec{J}$, and the energy of the perturbed states depends on a new quantum number J found by applying the triangle rule to the addition of L and S. In other words, each term ^{2S+1}L is split by spin–orbit coupling into a number of new states indexed by $^{2S+1}L_J$, where J is found from

$$J = L+S, L+S-1, \ldots, |L-S| \tag{7.33}$$

The spin–orbit interaction depends strongly on the atomic number of the atom, becoming more important for heavier atoms. For a one-electron atom, the perturbation operator for spin–orbit coupling is

$$\hat{H}'_{SO} = \frac{1}{2m_e^2 c^2 r}\frac{dV}{dr}\left(\vec{l}\cdot\vec{s}\right) \tag{7.34}$$

where $V = -Ze^2/4\pi\varepsilon_0 r$ is the electron–nucleus potential. For the case of a many-electron atom, the operator $\hat{H}'_{SO}$ is the sum over all the electrons of terms like Equation 7.34. It can be shown that in the many-electron case the perturbation operator is proportional to $\vec{L}\cdot\vec{S}$, and that the magnitude of the spin–orbit perturbation is approximately proportional to Z^4. This is a strong dependence on atomic number, and leads to the heavy atom effect, whereby the mixing of the spin and orbital angular momentum quantum numbers in molecules containing heavy atoms is so strong that the usual spin selection rule, $\Delta S = 0$, is called off. (Selection rules for many-electron atoms are considered in the next section.)

Let us designate a state corresponding to the term symbol $^{2S+1}L_J$ by the ket vector $|JM_J; LS\rangle$. The notation for this ket vector is consistent with that used on the left-hand side of the Clebsch–Gordan series in Equation 7.25. The notation $|JM_J; LS\rangle$ keeps track of the good quantum numbers. The quantum numbers M_L and M_S are no longer good because the state $|JM_J; LS\rangle$ is a linear combination of states with different values of M_L and M_S. The operators $\hat{L}^2, \hat{S}^2$, and $\hat{J}^2$ commute with $\hat{H} = \hat{H}^0 + \hat{H}'_{SO}$, so L, S and J are all good quantum numbers, provided the spin–orbit perturbation is weak compared to the energy separation of different terms. Thus

$$\hat{L}^2\left|JM_J; LS\right\rangle = \hbar^2 L(L+1)\left|JM_J; LS\right\rangle \tag{7.35}$$

$$\hat{S}^2\left|JM_J; LS\right\rangle = \hbar^2 S(S+1)\left|JM_J; LS\right\rangle \tag{7.36}$$

$$\hat{J}^2\left|JM_J; LS\right\rangle = \hbar^2 J(J+1)\left|JM_J; LS\right\rangle \tag{7.37}$$

This gives us a way to evaluate the spin–orbit splitting, using the following trick:

$$\hat{J}^2 = \vec{J}\cdot\vec{J} = (\vec{L}+\vec{S})\cdot(\vec{L}+\vec{S})$$
$$= \hat{L}^2 + \hat{S}^2 + 2(\vec{L}\cdot\vec{S}) \tag{7.38}$$

The above equation derives from the fact that the total angular momentum $\vec{J}$ is a vector sum of $\vec{L}$ and $\vec{S}$. It is useful because

$$\hat{H}'_{SO} \propto (\vec{L}\cdot\vec{S}) = \frac{1}{2}(\hat{J}^2 - \hat{L}^2 - \hat{S}^2)$$
$$E^{(1)}_{SO} \propto \frac{1}{2}\Big[J(J+1) - L(L+1) - S(S+1)\Big] \tag{7.39}$$

We can now use perturbation theory to find the change in energy due to the spin–orbit interaction. The zero-order states are the wavefunctions specified by $|JM_J; LS\rangle$, and we evaluate the expectation value of $\hat{H}'_{SO}$ with respect to this state to get the first-order correction to the energy:

$$E^{(1)}_{SO} = \langle JM_J; LS | \hat{H}'_{SO} | JM_J; LS \rangle$$
$$= \frac{1}{2}A\hbar^2\Big[J(J+1) - L(L+1) - S(S+1)\Big] \tag{7.40}$$

The spin–orbit coupling constant A has been introduced. It depends on L and S and the electronic configuration, but not J. Thus A determines the spin–orbit splitting within a multiplet ^{2S+1}L. It is generally positive for terms arising from less than half-filled shells, and negative for shells which are more than half-filled. The former are referred to as regular multiplets and the latter are called inverted multiplets. The effect of spin–orbit coupling on the energy levels associated with the p^2 configuration is shown in Figure 7.3.

We have avoided much quantum mechanical detail in writing this constant A, for it depends on the electron–nuclear potential and the electronic structure of the atom. For spectroscopic purposes, we consider it to be a number which can be determined from experiment. In order to see how, we need to look at selection rules for many-electron atoms.

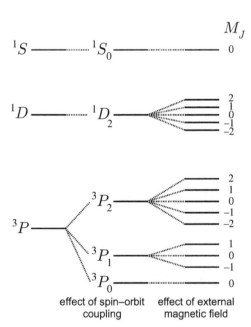

effect of spin–orbit
coupling

effect of external
magnetic field

Figure 7.3 The effect of spin–orbit coupling and an external magnetic field on energy levels derived from a p^2 configuration.

7.3 SELECTION RULES FOR ATOMIC ABSORPTION AND EMISSION

7.3.1 E1, M1, AND E2 ALLOWED TRANSITIONS

Now that we have a procedure for labeling a state with its good quantum numbers, we need to consider what changes are permitted in those quantum numbers for E1, E2, and M1 selection rules. In other words, we consider a transition $^{2S+1}L_J \leftrightarrow {}^{2S'+1}L'_{J'}$ and find which matrix elements $\langle JM_J; LS | \hat{O} | J'M_J'; L'S' \rangle$ are nonzero, where $\hat{O}$ is the appropriate operator, such as $\hat{\mu}$ for an E1 transition. Having worked out the rules for the case of the one-electron atom, and knowing how to combine one-electron states to get wavefunctions for the many-electron case, it is straightforward, though tedious, to derive the selection rules. A key idea behind the selection rules for many-electron atoms is that the dipole moment operator $\vec{\mu} = -e \sum_i \vec{r}_i$ is the sum of one-electron contributions. Here, we summarize the selection rules (see Table 7.2) and try to make them plausible.

For E1 selection rules, we have $\Delta L = 0, \pm 1$, except that $L = 0$ cannot go to $L' = 0$ and Δl cannot equal zero for the electron that jumps. These selection rules are similar to the one-electron case, with the additional permission for L to change by 0. In the case of a one-electron atom, $\Delta l = 0$ would violate conservation of momentum for an E1 transition and is therefore forbidden. Recall that the photon has angular momentum, so an absorbed or emitted photon must impart or destroy one unit of angular momentum in the atom. So how does a transition having $\Delta L = 0$ become allowed in the many-electron atom? Consider a transition $^3P \rightarrow {}^3P$, where one state derives from a configuration $s^1 p^1$ and the other from p^2. The transition dipole moment operator is the sum of $-e\vec{r}_i$ for all electrons i. The transition can be thought of as promoting the s electron to a p orbital, which does conserve angular momentum and for which the one-electron selection rule $\Delta l = \pm 1$ is satisfied. The transition from one S state to another S state remains forbidden; there is simply no way to accomplish this if only one electron undergoes a change of $\Delta l = \pm 1$.

The selection rules for J are similar to those for L. For E1 allowed transitions, we have $\Delta J = 0, \pm 1$, except that $J = 0$ cannot go to $J' = 0$. The reasoning derives from angular momentum considerations as in the preceding paragraph. As for spin, the selection rule $\Delta S = 0$ can be considered to be less and less firm as the atomic number increases. The reason is that as Z gets larger spin–orbit coupling gets stronger, and the quantum numbers L and S become less good. The breakdown in the spin selection rule can be considered to result from mixing of the pure spin states. For a light atom such as helium, we expect $\Delta S = 0$ to hold for E1, E2 and M1 transitions. One of the strongest lines in the emission spectrum of mercury is that due to the transition $^3P_1 \rightarrow {}^1S_0$ at 253.7 nm. However, the spin-allowed $^1P_1 \rightarrow {}^1S_0$ transition at 184.9 nm is stronger still. Some of the mercury atom transitions connecting the ground and valence electron excited states are displayed in the *Grotrian* diagram of Figure 7.4. For comparison, the emission spectrum of a high-pressure mercury vapor lamp is displayed in Figure 7.5. The atomic transitions seen there show the effects of pressure broadening, resulting from collisions that shorten the excited state lifetime and cause instantaneous shifts in the transition frequency.

The M1 selection rules are also similar to the one-electron case and are based on matrix elements of the operators $\hat{O} = \hat{L}_x, \hat{L}_y,$ and $\hat{L}_z$. The result is that $\Delta J = 0, \pm 1$ and $\Delta L = 0, \pm 1$, except that it is forbidden for $J = 0$ to go to $J' = 0$. The E2 selection rules are obtained by consideration of products of E1 matrix elements. This gives $\Delta L = 0, \pm 1, \pm 2$, and $\Delta J = 0, \pm 1, \pm 2$. As for the E1 case, whether or not $\Delta S = 0$ is a strict selection rule depends on the strength of the spin–orbit coupling.

Electric dipole selection rules forbid transitions between states arising from the same configuration, due to the condition that Δl must equal ± 1 for the electron that makes the transition. Take the previous example

Table 7.2 Selection rules for transitions of many-electron atoms (See text for exceptions)

	Allowed transitions		
E1	$\Delta L = 0, \pm 1$	$\Delta J = 0, \pm 1$	$\Delta S = 0$
M1	$\Delta L = 0, \pm 1$	$\Delta J = 0, \pm 1$	$\Delta S = 0$
E2	$\Delta L = 0, \pm 1, \pm 2$	$\Delta J = 0, \pm 1, \pm 2$	$\Delta S = 0$

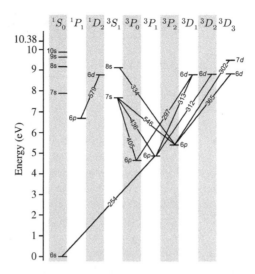

Figure 7.4 Partial Grotrian diagram displaying Hg transitions between the ground state and excited states deriving from $6s^1nl^1$ configurations, where n and l are the principal and orbital quantum numbers of the valence-excited electron. Transition wavelengths observed in the spectrum of Figure 7.5 are indicated in nm.

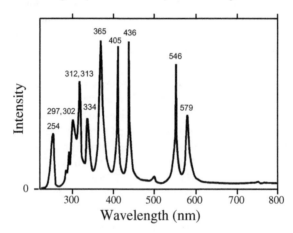

Figure 7.5 Emission spectrum of a high-pressure mercury vapor lamp.

of the carbon atom, where the ground configuration gives rise to the terms $^1S_0$, $^3P_0$, $^3P_1$, $^3P_2$, and $^1D_2$. The spin selection rule prevents all but the $^1S \leftrightarrow {}^1D$ transition, but this is forbidden because $\Delta L = 2$ is E1 forbidden. Note that the transitions within the ^3P multiplet are permitted by M1 selection rules, which allow $\Delta l = 0$, but these should be quite low in frequency for a light atom such as carbon.

Figure 7.6 shows some of the energy levels and transitions for Ar$^+$, which will be discussed in Section 7.5. This example illustrates the selection rules as well as violations thereof. The splittings within each multiplet can be used to find the spin–orbit coupling constant A of Equation 7.40, as will be shown in Problem 6.

7.3.2 HYPERFINE STRUCTURE

At still higher resolution than that required to observe spin–orbit coupling, splittings are observed that may be attributed to nuclear effects. In atomic samples which contain naturally abundant isotopes of an element, fine splittings can be observed which are the result of the effect of nuclear mass. In the case of the hydrogen atom, for example, weak companion lines less than 0.2 nm from the main transitions are due to deuterium and, in fact, were of historical importance in demonstrating the existence of this isotope of hydrogen. For the one-electron atom, the small effect of nuclear mass enters into the evaluation of the reduced mass that appears in the

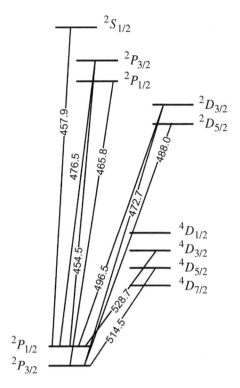

Figure 7.6 Energy levels of Ar+ and some laser transitions. The transition wavelengths are given in nm. (Adapted with corrections from A. Yariv, *Introduction to Optical Electronics*, Holt, Rinehart and Winston, New York, 1976.)

expression for the Rydberg constant. For many-electron atoms the effect of nuclear mass is not so easily predicted, but experiment bears evidence of its importance. Isotope effects in atomic spectra can also result not just from the mass but also from the finite size and nonspherical shape of the nucleus. As an example, the emission spectrum of Zn shows a strong line at 621.5 nm due to an electronic transition of ^{64}Zn. Weaker emission lines 0.095 and 0.189 nm to the red of this line are due to ^{66}Zn and ^{68}Zn, respectively. It is perhaps misleading to refer to these isotope effects as "splittings," since the various lines arise from distinct chemical species.

A nuclear effect which *does* give rise to splittings is that due to nuclear spin. Many nuclei possess intrinsic angular momentum characterized by a nuclear spin quantum number I. Nuclei having even atomic numbers Z and even numbers of neutrons have zero spin. As was discussed in Section 3.4.3, intrinsic angular momentum of a charged particle is associated with a magnetic moment, $\vec{\mu}_{mag}$. For nuclear spin, this is given by

$$\vec{\mu}_{mag} = \frac{g_N e}{2m_p} I = \gamma_N I \tag{7.41}$$

Equation 7.41 contains the nuclear g-factor g_N and gyromagnetic ratio γ_N, introduced in Chapter 3, which depend on the atomic number Z and the mass of the proton. The inverse dependence of the magnetic moment on mass means that nuclear spin splittings are smaller than electronic spin splittings. The nuclear (I) and electronic (J) angular momentum can couple according to the triangle rule, resulting in a total angular momentum quantum number $F = I + J, I + J - 1, \ldots, |I - J|$. Transitions between these states must satisfy the selection rules $\Delta F = 0, \pm 1$ except that $F = 0$ cannot go to $F' = 0$. The hyperfine energy correction due to the coupling of electronic and nuclear spin is given by

$$E_{hfs}^{(1)} = \frac{a}{2}\left[F(F+1) - J(J+1) - I(I-1)\right] \tag{7.42}$$

Note the similarity between Equations 7.42 and 7.40.

As an example, consider ^{23}Na for which $I = 3/2$. A high-resolution spectrum of the previously mentioned yellow doublet reveals that each of the two lines is actually split into two by 0.022 Å. The hyperfine splitting of the lower $^2S_{1/2}$ state is considerably larger than that of the upper $^2P_{3/2}$ and $^2P_{1/2}$ states. (Can you think of a reason why this might be so? Hint: Consider the form of the radial wavefunctions R_{nl}.)

7.4 THE EFFECT OF EXTERNAL FIELDS

The splitting of orbitals having different l and the same n, in many-electron atoms, results from the effect of interelectronic repulsion. For lighter atoms, this is a large effect compared to that of the internal magnetic field responsible for spin–orbit coupling. Now we consider what happens when an atom is placed in an external static field. The result, as we will see, is that some of the degeneracy of the term energies is lifted.

7.4.1 THE ZEEMAN EFFECT

The Zeeman effect pertains to the spectroscopy of an atom in an external magnetic field. The field is assumed to be small enough to constitute a weak perturbation, so we can use first-order perturbation theory to evaluate the effect on the energy levels. The perturbation operator is $\hat{H}_B' = -\vec{\mu}_{mag} \cdot \vec{B}$ where $\vec{\mu}_{mag}$ is the magnetic dipole moment operator and $\vec{B}$ is the magnetic field. We have used a subscript mag to distinguish the magnetic dipole moment from the electric dipole moment, but from here on we will drop it in favor of the subscripts L, S, and J to label the orbital, spin, and total magnetic moments, respectively.

It is most convenient to decide that whatever the direction of the field in the laboratory, we should call that the z direction. The reason is that the z component of total angular momentum is quantized. The projection of the net angular momentum J onto the field direction is quantized as in Figure 7.7. The quantum mechanical uncertainty in the x and y components of angular momentum is in complete accord with the classical expression for the torque $\vec{T}$ which is equal to $\vec{\mu}_B \times \vec{B}$. The torque is always perpendicular to the plane defined by the magnetic moment and the magnetic field, so the magnetic moment must precess about the field on the surface of a cone. As Figure 7.7 suggests, we can view the L and S vectors as precessing about the direction of J, while the net angular momentum J precesses about the direction of the field. The projection of the vector $\vec{J}$ onto the z direction is $M_J \hbar$, and we need to know the component of the magnetic moment in this direction, μ_z. Here, μ_z denotes the projection of $\vec{\mu}_J$ onto the field.

By choosing the z direction to be that of the field, the perturbation operator reduces to $\hat{H}_B' = -\mu_z B$. Now, from Chapter 3, recall that the orbital and spin components of the magnetic moment are given by

$$|\vec{\mu}_L| = \frac{e\hbar}{2m_e}\sqrt{L(L+1)} \tag{7.43}$$

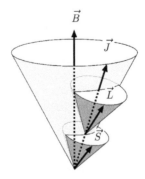

Figure 7.7 Precession of $\vec{J}$ about an external magnetic field $\vec{B}$.

$$|\vec{\mu}_S| = \frac{e\hbar}{m_e}\sqrt{S(S+1)} \tag{7.44}$$

Because the proportionality connecting the magnetic moment to the angular momentum is different for the orbital and spin contributions, and since the net angular momentum J is the vector sum of L and S, the net magnetic moment does not have the same direction as J. We need to find the connection between the net angular momentum and the total magnetic moment $\vec{\mu}_J$. This requires what is known as the Landé g-factor. It will allow Equations like 7.43 and 7.44 to be written for the total magnetic moment:

$$|\vec{\mu}_J| = \frac{ge\hbar}{2m_e}\sqrt{J(J+1)} = g\mu_B\sqrt{J(J+1)} \tag{7.45}$$

where the Bohr magneton μ_B, introduced in Chapter 3, has been substituted for $e\hbar/2m_e$. The total magnetic moment has a magnitude $|\mu_J|$, which can be found by projecting the components due to orbit and spin onto the direction of $\vec{J}$:

$$|\mu_J| = |\mu_L|\cos(L, J) + |\mu_S|\cos(S, J) \tag{7.46}$$

where (L, J) represents the angle between the $\vec{L}$ and $\vec{J}$ vectors, and similarly for (S, J). Putting the previous equations together yields

$$g\sqrt{J(J+1)} = \sqrt{L(L+1)}\cos(L, J) + 2\sqrt{S(S+1)}\cos(S, J) \tag{7.47}$$

The cosines can be found as follows:

$$\vec{L}\cdot\vec{J} = \sqrt{L(L+1)}\sqrt{J(J+1)}\cos(L, J)$$
$$= \vec{L}\cdot(\vec{L} + \vec{S}) = \frac{1}{2}\left[L(L+1) + J(J+1) - S(S+1)\right] \tag{7.48}$$

$$\vec{S}\cdot\vec{J} = \sqrt{S(S+1)}\sqrt{J(J+1)}\cos(S, J)$$
$$= \vec{S}\cdot(\vec{L} + \vec{S}) = \frac{1}{2}\left[J(J+1) + S(S+1) - L(L+1)\right] \tag{7.49}$$

Equations 7.48 and 7.49 are obtained with the help of the expression for $\vec{L}\cdot\vec{S}$ given in 7.39. Solving for the cosines and inserting the resulting expressions into Equation 7.47 gives the Landé g-factor:

$$g = 1 + \frac{J(J+1) + S(S+1) - L(L+1)}{2J(J+1)} \tag{7.50}$$

The operator for the total electronic magnetic moment is

$$\vec{\mu}_J = g\frac{-e}{2m_e}\hat{J} \tag{7.51}$$

and the z component is

$$\mu_z = g\frac{-e}{2m_e}\hat{J}_z \tag{7.52}$$

We can now find the first-order correction to the energy by taking the expectation value of the perturbation operator with respect to the zero-order states. The perturbation operator is $\hat{H}'_B = -\mu_z B = g\mu_B B\hat{J}_z/\hbar$.

$$E_B^{(1)} = \langle JM_J; LS|\hat{H}'_B|JM_J; LS\rangle = \mu_B gBM_J \tag{7.53}$$

Equation 7.53 shows that the degeneracy of the $2J + 1$ states having the same J is lifted by the application of a magnetic field. It also predicts that the splitting of energy levels increases with field strength. The effect of a

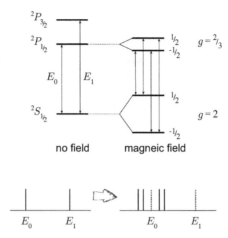

Figure 7.8 The "anomalous" Zeeman effect for the transition between $^2P_{1/2}$ and $^2S_{1/2}$. The effect of the field on the $^2P_{3/2} \rightarrow {}^2S_{1/2}$ transition is left as an exercise.

magnetic field on states belonging to the ^1D and ^3P terms is illustrated in Figure 7.3. Figure 7.8 illustrates the Zeeman effect for the transition between the $^2P_{1/2}$ and $^2S_{1/2}$ states of an alkali metal atom. Transitions between the split states adhere to the selection rule $\Delta M_J = 0, \pm 1$. The splitting of the transition between the $^2P_{3/2}$ and $^2S_{1/2}$ states is left as an exercise. Note that the value of g is different for the upper and lower states, leading to a complex pattern in the spectrum. The type of splitting shown in Figure 7.8 was at one time referred to as the "anomalous Zeeman effect." It was not really anomalous, just more complicated than the splitting pattern that results when the net spin is zero. When $S = 0$, J and L are equal and $g = 1$. We then refer to the "normal" Zeeman effect. For the example of Figure 7.9, the $^1D_2 \rightarrow {}^1P_1$ transition is split into only three lines in the presence of the magnetic field.

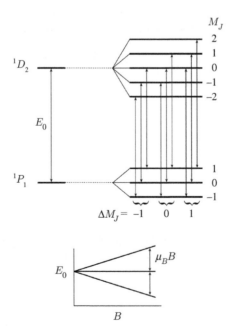

Figure 7.9 The "normal" Zeeman effect ($S = 0$, $J = L$, $g = 1$).

7.4.2 THE STARK EFFECT

The Stark effect refers to the application of an external electric field in spectroscopy. In the case of atomic spectra, there is no permanent dipole moment with which an electric field can interact. The field thus interacts with the induced dipole moment and, as such, the perturbed energy levels are accounted for within second-order perturbation theory. Choosing the z direction to be that of the external field, the operator responsible for the Stark effect is $\hat{H}'_{Stark} = -eE_z\Sigma_i z_i$, where the sum is over the z-coordinates of all the electrons in the atom. This is similar to the operator which causes E1 transitions, except that the Stark field E_z is static*. Since an atom can have no permanent dipole moment, the first-order correction to the energy in the presence of the field E_z is zero, except for one-electron atoms. In the hydrogen atom, degenerate states such as $2s$ and $2p$ are mixed by the perturbation and result in a first-order splitting.

Without going into the theoretical details, it is worth noting that the perturbed energy levels are proportional to the square of M_J. States with the same magnitude but opposite signs for M_J are perturbed in the same way. The effect of an electric field on the transitions characterizing the sodium-D emission is illustrated in Figure 7.10.

7.5 ATOMIC LASERS AND THE PRINCIPLES OF LASER EMISSION

As was shown in the previous chapter, amplification of light intensity by stimulated emission is possible when the excited-level population N_2 exceeds the population N_1 of the lower energy level. In this section we discuss how this population inversion is achieved in some common atomic lasers, and how the design of a laser cavity enables light amplification (or "gain") to be sustained. The end result of laser emission is to create light with the following special properties: it is coherent, meaning that the emitted photons are in phase with one another; it is collimated, meaning that the emitted photons all share the same propagation direction (wave vector); it is intense, much brighter than ordinary sources of spontaneous emission; and it is, at least in the case of atomic lasers, very nearly monochromatic, meaning all photons have the same color. This tendency for emitted photons to share properties such as wavelength and wave vector is a natural

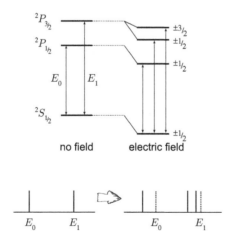

Figure 7.10 The Stark effect.

---

* Practical considerations make it preferable that E_z vary in time in order to employ phase-sensitive detection. However, the oscillation frequency is quite slow compared to the time scale for electronic motion, so the external field can be considered constant in time as far as the atom is concerned.

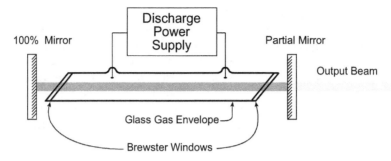

Figure 7.11 A typical gas laser.

consequence of the stimulated nature of the emission: the stimulated photons cause photons to be emitted with the same properties, and these photons in turn can stimulate other photons with the same properties to be emitted, and so on.

Sustained laser emission requires energy input to maintain the nonequilibrium state of a population inversion. This is referred to as pumping. The energy to pump a laser can be provided by an electric discharge, an intense source of light, such as a flashlamp, or sometimes by another laser. The optical cavity of a typical gas laser is sketched in Figure 7.11. The gas is confined to a plasma tube, so named because the electric discharge used to provide the energy input results in a gas of ions. Two mirrors, one of which reflects slightly less than 100% of the incident light, permit most of the photons to bounce back and forth within the confines of the cavity, while a small portion of them are transmitted by the partially reflecting mirror (called the output coupler). The windows of the plasma tube are oriented at Brewster's angle with respect to the long axis of the tube. As discussed in Chapter 2 and shown in Figure 2.7, the vertically polarized (i.e., p-polarized) light is completely transmitted by the Brewster window, while the horizontal (s-polarized) component is strongly reflected. The result is that horizontally polarized light is rejected by reflections at the tube windows, while the vertically polarized light is transmitted and can undergo repeated transits between the mirrors. The light emerging from the laser is thus vertically polarized. For continuous laser emission, the gain that results from amplification by stimulated emission must exceed the loss of light energy such as that due to imperfect reflectivity of the mirrors.

How does one thwart the natural tendency of the state populations to obey Boltzmann's law? It would not be a good idea for only two levels to be involved in the pumping and lasing process. If one tries to achieve a population inversion by direct promotion of atoms from the ground to an excited state, eventually the populations will be equalized, and further net absorption and emission of light will cease. This is why lasers are usually based on at least three or four energy levels that participate in the upward (pumping) and downward (lasing) transitions. The lifetimes of these levels are of interest, because it is desirable for the lifetime of the upper state of the laser transition to exceed that of the lower state, in order to permit the population inversion to build up. The following examples of atomic lasers illustrate these ideas.

The helium–neon laser is a very common one; it is frequently used as an alignment laser in optical systems. The light emission is in fact due to neon, while helium is present to help to create the requisite population inversion in Ne. Figure 7.12 shows the energy levels of both atoms and outlines some of the important events in the operation of the He–Ne laser. The initial excitation (the pump) is provided by an electric discharge. Electron impact excitation of electronic states does not adhere to the same selection rules as those imposed on optical transitions, so this is somewhat of a brute force approach to creating high-energy excited electronic states. When He atoms find themselves in the excited ^1S and ^3S levels, they are forbidden to return to the ground state by E1 selection rules and are thus metastable. The lifetime of the excited He ^1S state is on the order of a microsecond, while that of the ^3S state is about 10^{-4} s. (Note that the spin-forbidden character of the emission from the triplet state leads to a longer lifetime than the singlet state, and that the triplet is lower in energy than the singlet.) Coincidentally, the ^1S and ^3S excited states of He are very close in energy to excited electronic states of Ne. Thus the excitation energy can be readily transferred from He to Ne. The terminal state of the Ne laser transitions is

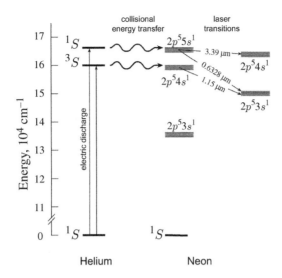

Figure 7.12 Helium and neon energy levels and the He/Ne laser transitions. (Adapted from A. Yariv, *Introduction to Optical Electronics*, Holt, Rinehart and Winston, New York, 1976.)

not the ground state. This makes achieving a population inversion easier than it would be if one had to overcome the tendency of the ground state to have the largest population. The lifetime of the initial state of the laser transition exceeds that of the terminal state, another feature which helps lead to inverted populations.

The energy levels for excited neon atoms exhibit a type of angular momentum coupling which is peculiar to rare gas atoms. The configurations giving rise to excited states of Ne are of the type $2p^5ns^1$ and $2p^5np^1$. The problem reduces to one of two coupled electrons, since the hole in the $2p$ subshell contributes the same angular momentum as a single electron would. This core hole couples to the ns^1 or np^1 valence electron according to a scheme which reflects the fact that the electron in the excited orbital is only weakly coupled to the core. See [6] for a discussion of the pair-coupling scheme adhered to by rare gas atoms. Each $2p^5ns^1$ configuration gives rise to four energy levels (and 12 states) while a $2p^5np^1$ configuration comprises ten energy levels (and 36 states).

The lifetimes of the states pertaining to $2p^5ns^1$ configurations are somewhat longer than those which derive from $2p^5np^1$. This facilitates achieving population inversions of the $2p^5ns^1$ states with respect to $2p^5np^1$. The He–Ne laser is commonly employed to provide red light at 632.8 nm, but several lines in the infrared are also available.

Another laser that is widely employed for spectroscopy is the argon ion laser. The energy levels and laser transitions of the argon ion are shown in Figure 7.6. Note that the spin–orbit states are inverted (higher J states are at lower energies) as expected, since the p-shell is more than half-filled. The ground state $^2P_{3/2}$ and $^2P_{1/2}$ levels derive from the $3p^5$ valence configuration. Promotion of one of the $3p^5$ electrons to higher energy orbitals results in doublet and quartet states as shown in Figure 7.6. An electric discharge of 4 to 5 eV provides the energy to ionize Ar atoms and populate highly excited states of Ar$^+$. These high-energy states of Ar$^+$ then tumble down the energy ladder to levels within the spin–orbit split ^2S, ^2P, ^2D, and ^4D multiplets. Two of the strongest transitions of the Ar$^+$ laser, the green line at 514.5 nm and the blue one at 488.0 nm, result from spin-forbidden transitions from ^4D to the ^2P ground state! It may be surprising that strong laser lines derive from E1-forbidden transitions, but recall the discussion of lifetimes in the previous example. In addition to the nine visible transitions shown in Figure 7.6, the argon ion laser can provide several lines in the ultraviolet as well. The krypton ion laser operates on the same principles as the argon ion laser, providing wavelengths in the red, the most intense being 647.1 nm.

Another widely used laser based on atomic transitions is the neodymium–YAG laser. This laser oscillates on a transition of the Nd^{3+} ion in a matrix of yttrium aluminum garnet (YAG) or $Y_3Al_5O_{12}$. Since Nd is a very heavy atom, the spin–orbit interaction produces quite large splittings. And since the ground electronic configuration includes the partially filled f orbitals (f^3), quite high values of L result. Figure 7.13 shows the energy

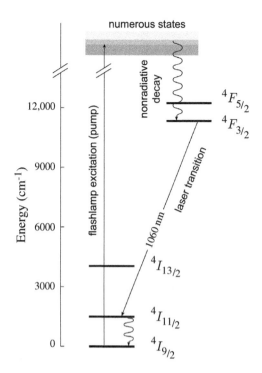

Figure 7.13 Nd^{+3} states relevant to the Nd–YAG laser.

levels relevant to the Nd–YAG laser. The ground state term is ^4I, where the letter I indicates that $L = 6$ for this term. This term is split into states with $J = 15/2$, $13/2$, $11/2$, and $9/2$. The $^4I_{11/2}$ state is the terminal state in the 1.06 μm laser transition, and it lies about 2000 cm^{-1} above the ground state $^4I_{9/2}$. The laser transition from $^4F_{3/2}$ to $^4I_{11/2}$ is forbidden by E1, M1, and E2 selection rules! Figure 7.13 only shows the energy levels of the free ion, and in the crystalline host there are splittings that arise from the local field of the Nd^{3+} ion.

The Ti:sapphire laser has come to be indispensable in time-resolved and nonlinear optical spectroscopy experiments. This laser provides tunable emission from the red to the infrared, ranging from about 650 to 1100 nm. The laser medium consists of the mineral corundum (Al$_2$O$_3$) doped by a small amount of Ti^{3+}, on the order of 0.1% or less. Ti^{3+} has a d^1 outer-shell configuration, hence the free ion has a ^2D ground state. While spin–orbit coupling does result in splitting of this level, more significant is the so-called crystal field splitting of the ion resulting from interactions with surrounding ions. Group theoretical considerations of the splitting of transition metal ions in metal-ligand complexes are discussed in more detail in Chapter 11. Here, we note that Ti^{3+} substitutes for Al^{3+} at sites with three-fold symmetry. Interactions with the surrounding atoms split the ^2D level of Ti^{3+} into a triply degenerate lower level, with group theoretical designation $^2T_2$, and a doubly degenerate ^2E level, separated by about 19,000 cm^{-1}. (We are referring to the spatial component of the degeneracy here, which is multiplied by two to get the total degeneracy.) However, each of these levels is further split by Jahn–Teller interactions which lift the degeneracy of orbital-degenerate electronic states via a nontotally symmetric vibration, also discussed in Chapter 11. The absorption spectrum of Ti:Al$_2$O$_3$ is further broadened by vibrational transitions as well as Jahn–Teller splitting and consists of two maxima at about 490 and 550 nm. Laser emission is initiated by a green pump laser, such as an Ar ion laser operating at 514 nm or the frequency-doubled output of a Nd–YAG laser. The broad fluorescence spectrum of Ti:Al$_2$O$_3$ peaks at about 760 nm and has a lifetime of about 3.2 μs. This fluorescence is the basis for the laser emission. However, the terminal state of the laser transition is not the ground state, but a higher lying component of the split $^2T_2$ level. This excited state has a much shorter lifetime than the 3.2 μs lifetime of the initial state, facilitating population inversion. The breadth of the laser emission is a key to the tunability of the laser and the ability to generate pulses as short as about 6 fs. The latter feature arises from the Fourier-transform relationship between the temporal profile of the pulse and its frequency spectrum, to be discussed in Chapter 14.

The boundary conditions presented by a laser cavity, such as the one pictured in Figure 7.11, lead to restrictions on the allowed wavelengths and also dictate the intensity distribution of the beam profile. The standing waves supported by a cavity of length L are those having nodes at the ends of the cavity. This leads to longitudinal cavity modes, which are the permitted wavelengths $\lambda_n = 2L/n$, where n is an integer. The frequency spacing of the longitudinal modes in wavenumbers is therefore $\Delta\tilde{v} = 1/2L$. For a typical cavity length of one meter, the spacing of the longitudinal modes is 0.005 cm^{-1} or 150 MHz. A typical Doppler width in a gas laser might be 1500 MHz, and a typical radiative lifetime of about 10^{-7} s contributes only 10 MHz to the linewidth. Using Equation 6.47 for the Doppler width, this translates into quite a high temperature, as you will show in Problem 11. The Doppler profile spans a number of longitudinal modes, and the laser can emit all of these simultaneously. For high-resolution work, however, one can select one of these very sharp lines by inserting an etalon in the cavity. An etalon is much like a smaller version of the optical cavity of the laser, in that it consists of two partially transmitting mirrors separated by a distance l that is small compared to L. Constructive interference is possible only for those wavelengths which satisfy the boundary conditions of the etalon, $\lambda_n = 2l/n$. These etalon modes are much more widely spaced than the longitudinal modes of the laser, thus the etalon can be tuned to select one of the allowed modes within the Doppler profile.

The transverse modes are intensity profiles that result from the boundary conditions in the two directions perpendicular to the long axis of the laser. It is usually preferred for a laser to operate in what is called TEM$_{00}$, where TEM stands for transverse electromagnetic mode. TEM$_{00}$ is also called a Gaussian mode because the intensity decreases as $\exp(-r^2)$ where r is the distance from the center of the beam in the transverse direction, as discussed in Chapter 2. The image of the TEM$_{00}$ mode is a round spot, while the intensity profiles of higher mode TEM$_{lm}$ have nodes. The subscripts l and m indicate the number of these nodes in each of the two transverse directions, such that TEM$_{10}$ and TEM$_{01}$ images show two spots, TEM$_{11}$ shows four spots, and so on.

Lasers which provide continuous output are referred to as cw (continuous wave) lasers. Of great utility to time-resolved spectroscopy are pulsed lasers, which provide laser emission in short pulses of nanosecond, picosecond, and femtosecond duration. Pulses which are shorter in time are inherently broader in frequency. Pulsed lasers are of interest when high instantaneous powers are desired, since the emitted energy is of very short duration. They are therefore used in nonlinear spectroscopy experiments. The technology for producing short laser pulses is discussed in [7].

7.6 SUMMARY

In this chapter, we have considered the selection rules for electronic transitions in one-electron and many-electron atoms. The spectra of common gas-phase atoms and ions have been available for some time, as noted in the introduction, but continue to hold our interest. The study of highly excited Rydberg states, for which the outermost electron is so far removed from the core that hydrogen-like energy levels result, is one example. These Rydberg states may be found in molecules and in nanoparticles such as semiconductor quantum dots, providing further motivation for the study of atomic spectra. The splittings induced by internal and external fields are also of interest in molecular spectra, though most of the good quantum numbers of this chapter do not apply to molecular electronic states. An exception is the spin quantum number, which is a good quantum number in molecules having sufficiently weak spin–orbit coupling. We now proceed to the study of molecular spectra, for which the rotational and vibrational states, in addition to electronic, provide additional interest and complexity compared to atomic spectra. For more details on atomic spectroscopy, the reader is encouraged to consult [8] and [9].

PROBLEMS

1. Compute the transition dipole and oscillator strength for the following hydrogen atom transitions $1s \leftrightarrow 3p_0$ and $1s \leftrightarrow 4p_0$.

2. Explain why hydrogen atom E1 transitions having $\Delta l = 0$ are forbidden by symmetry.

3. Compute the three lowest energy transitions in the Lyman series for deuterium and compare to the same lines in hydrogen.

4. Figure out the term symbols associated with the following configurations: (a) $s^1 p^5$ (b) p^3 (c) $p^1 d^1$ and (d) $2p^1 3p^2$. For each case, decide the order of the term energies, and sketch a diagram showing how each term is split by spin–orbit coupling.

5. Are two-electron transitions permitted by electric dipole selection rules? To arrive at your answer, first consider the wavefunction to be a simple product of spin–orbitals. (This is called a Hartree product.) Is your conclusion different if the state is represented by one or more Slater determinants?

6. Refer to the energy level diagram for Ar^+ given in Figure 7.6 and find the spin–orbit coupling constant A for the ground state ^2P and excited state ^4D terms. Predict the wavelength of the transition from the $^4D_{3/2}$ level to the $^2P_{3/2}$ level of the ground state term.

7. The $^2P \rightarrow {}^2S$ transition of Li is split by 0.34 cm^{-1}. Find the spin–orbit coupling constant for the ^2P multiplet and compare to the values found in the previous problem for Ar^+.

8. For each transition in Figure 7.6 state whether or not it is permitted by E1, M1, and E2 selection rules.

9. Compute the Landé g-factor for the $^2P_{3/2}$ state of Figure 7.8 and complete the sketch of the field-induced splittings. Calculate the splittings, in wavenumbers, for a magnetic field of 1.50 Tesla.

10. Sketch the effect of a 10 kG magnetic field on the transition responsible for the 514.5 nm line of the argon ion laser. Compute the wavelengths of the allowed transitions.

11. Compute the effective temperature of Ne atoms emitting at 632.8 nm if the Doppler broadening is 1500 MHz.

REFERENCES

1. A. R. Edmonds, *Angular Momentum in Quantum Mechanics* (Princeton University Press, Princeton, NJ, 1974).

2. C. E. Dykstra, *Quantum Mechanics and Molecular Spectroscopy* (Prentice Hall, Englewood Cliffs, NJ, 1992).

3. I. N. Levine, *Quantum Chemistry*, 5th ed., (Prentice Hall, Upper Saddle River, NJ, 1999).

4. J. I. Steinfeld, *Molecules and Radiation* (MIT Press, Cambridge, MA, 1985).

5. R. N. Zare, *Angular Momentum, Understanding Spatial Aspects in Chemistry and Physics* (Wiley, New York, 1988).

6. W. S. Struve, *Fundamentals of Molecular Spectroscopy* (Wiley, New York, 1989).

7. A. Yariv, *Introduction to Optical Electronics* (Holt, Rinehart and Winston, New York, 1976).

8. E. U. Condon, and G. H. Shortly, *Atomic Spectroscopy* (Macmillan Company, New York, 1935).

9. G. Herzberg, *Atomic Spectra and Atomic Structure* (Dover Publications, New York, 1944).

Rotational spectroscopy

8.1 INTRODUCTION

The rotational energy levels of molecules are active in microwave and far-infrared (IR) spectroscopy and in light scattering experiments. The absorption or emission of light due to rotational motion is allowed only for polar molecules, although nonpolar ones can absorb weakly through induced moments. Light scattering due to rotations is permitted whenever the polarizability is anisotropic. In this chapter, we discuss two types of pure rotational spectra, microwave/far-IR and rotational Raman scattering. With these techniques, it is possible to resolve quantized rotational energy levels for freely rotating (gas phase) molecules. Rotational spectroscopy is often applied to the determination of structure and dipole moments of small molecules, the latter through the use of the Stark effect. Being more closely spaced than vibrational and electronic energy levels, rotational quantum levels can contribute to fine structure in vibrational and electronic spectroscopy of gases. In addition to rotational motion of the entire molecule, some molecules undergo internal rotation about single bonds, which ranges from being quite hindered to relatively free. Molecules in liquids do not rotate freely, as a rule, and their reorientational motion is more difficult to treat quantum mechanically and does not lead to discrete spectral lines. An exception to this rule is discussed in Section 8.4, where the rotational Raman spectrum of H_2 dissolved in water is analyzed. Still, there are spectroscopic manifestations of reorientational motion in liquids, which will be considered in the discussion of depolarized Rayleigh scattering. (See also Chapter 5.)

8.2 ENERGY LEVELS OF FREE RIGID ROTORS

The quantum mechanical solution to the rigid rotor problem is an example of a model for which the Schrödinger equation is exactly solvable. In this section, we review the treatment for a diatomic molecule and extend it to the case of polyatomic molecules having different degrees of symmetry. There are two important caveats to the treatment of this section. The first is that it applies to rigid molecules, while real molecules undergo vibrational motion, even at absolute zero. What is derived for rigid rotors is a good approximation provided the amplitude of vibrational motion is small. The second point to keep in mind when comparing theory to reality is that a free rotor experiences no angular dependent forces (torques), so what is predicted by theory does not apply for liquids, solids, or even dense gases where intermolecular interactions come into play.

The first mathematical chore at hand is to separate the external motion of the molecule (translation) from the internal motion (vibration and rotation). This can be accomplished without approximation and will be described for the simplest case, that of a diatomic rotor. The further separation of vibration (to be treated later) from rotation is exact only for the rigid rotor.

8.2.1 DIATOMICS

Imagine a diatomic molecule as two masses, M_A and M_B, connected by a rigid rod and moving in a laboratory frame of reference (XYZ) as shown in Figure 8.1.

The kinetic energy operator is given by

$$\hat{T}_N = -\frac{\hbar^2}{2M_A}\nabla_A^2 - \frac{\hbar^2}{2M_B}\nabla_B^2 \tag{8.1}$$

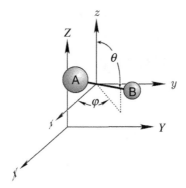

Figure 8.1 Internal and external coordinate systems.

where $\nabla_A^2 = \partial^2/\partial X_A^2 + \partial^2/\partial Y_A^2 + \partial^2/\partial Z_A^2$ operates on the lab-frame coordinates of atom A and similarly ∇_B^2 operates on those for atom B. The subscript N on the operator stands for nuclear motion (we ignore the electrons for now), which comprises translation, rotation, and vibration. Equation 8.1 is correct but not convenient, because the two nuclei do not move independently of one another. So imagine a second coordinate system (xyz) which is parallel to the lab-frame Cartesian directions, but has as its origin the center of mass of the molecule. The position of the center of mass, which translates with the molecule, is given by

$$\vec{R}_{CM} = \frac{M_A \vec{R}_A + M_B \vec{R}_B}{M_A + M_B} \tag{8.2}$$

The internal coordinate system embedded in the molecule is more easily treated in polar coordinates, $(x, y, z) \Rightarrow (R, \theta, \varphi)$, with the radial coordinate designated as the internuclear distance:

$$\vec{R} = \vec{R}_A - \vec{R}_B \tag{8.3}$$

As will be shown in Problem 1, we can transform the kinetic energy operator from the coordinate system $(X_A, Y_A, Z_A, X_B, Y_B, Z_B)$ to one that depends on the internal coordinates and the position of the center of mass. The result is

$$\hat{T}_N = -\frac{\hbar^2}{2M} \nabla_{CM}^2 - \frac{\hbar^2}{2\mu} \nabla_{int}^2 \tag{8.4}$$

where the reduced mass is

$$\mu = \frac{M_A M_B}{M_A + M_B} \tag{8.5}$$

and the total mass is $M = M_A + M_B$. The operators ∇_{CM}^2 and ∇_{int}^2 operate on the center of mass and internal coordinates, respectively.

The importance of Equation 8.4 is that the external and internal coordinates have been separated. The first term applies to the translational motion of the molecule as a whole, which behaves as if the total mass M were concentrated at the center of mass. We are not interested in this at the moment, although translational motion has spectroscopic consequences (such as Doppler broadening). We concentrate instead on the second term. If we were inclined to treat vibrational motion at this point, we could add to the operator for internal kinetic energy one that represents the potential energy as a function of internuclear distance, $V(R)$. But since our rotor is rigid, this is a constant which we take as zero for convenience. The problem of converting the operator ∇_{int}^2 from Cartesian to polar coordinates is an exercise that may be familiar because it is encountered in the solution to the Schrödinger equation for the hydrogen atom or any other two-body problem. We will just refer to the result of that exercise:

$$\nabla^2_{int} = \frac{\partial^2}{\partial x^2} + \frac{\partial^2}{\partial y^2} + \frac{\partial^2}{\partial z^2} = \frac{\partial^2}{\partial R^2} + \frac{2}{R}\frac{\partial}{\partial R} - \frac{\hat{L}^2}{R^2\hbar^2} \tag{8.6}$$

For the rigid rotor, we take $R = R_e$, the equilibrium internuclear distance. Since this is constant, the derivatives with respect to R can be dropped, leaving us with the last term, which contains the square of the angular momentum operator:

$$\hat{L}^2 = -\hbar^2\left[\frac{1}{\sin\theta}\frac{\partial}{\partial\theta}\left(\sin\theta\frac{\partial}{\partial\theta}\right) + \frac{1}{\sin^2\theta}\frac{\partial^2}{\partial\varphi^2}\right] \tag{8.7}$$

The eigenfunctions of the $\hat{L}^2$ operator are the ubiquitous spherical harmonic functions, $Y_{JM}(\theta\varphi)$, which arise in problems having spherically symmetric potential energy. The important thing to recall is that $Y_{JM}(\theta\varphi)$ can be separated into a product of two functions: a function of θ (one of the associated Legendre polynomials, a power series in $\cos\theta$), and a simple function of φ, $e^{iM\varphi}$. The quantum number J (compare to the quantum number l for the hydrogen atom problem) pertains to the magnitude of the angular momentum, while M designates the Z component of angular momentum in the laboratory frame. The eigenvalue relationships are

$$\hat{L}^2 Y_{JM}(\theta\varphi) = \hbar^2 J(J+1)Y_{JM}(\theta\varphi) \tag{8.8}$$

$$\hat{L}_Z Y_{JM}(\theta\varphi) = M\hbar Y_{JM}(\theta\varphi) \tag{8.9}$$

The quantum number J can equal 0, 1, 2, ..., ∞, and for each value of J, M ranges from $-J$ to J in integral steps. It is now straightforward to express the Hamiltonian for the rigid rotor:

$$\hat{H}_{rot} = \frac{\hat{L}^2}{2\mu R_e^2} = \frac{\hat{L}^2}{2I_e} \tag{8.10}$$

where the moment of inertia $I_e = \mu R_e^2$ has been introduced. Thus the rotational energy levels are given by

$$E_{rot} = \frac{\hbar^2}{2I_e}J(J+1) \tag{8.11}$$

and the degeneracy of each level, $g_J = 2J + 1$, is equal to the number of allowed values of M. This degeneracy reflects the fact that the energy does not depend on the orientation of the angular momentum in the lab frame. The rotational constant for a diatomic molecule is defined as $B_e = h/(8\pi^2 I_e)$ in s^{-1}. Alternatively, $\tilde{B}_e = h/(8\pi^2 I_e c)$ in cm^{-1}, so the energy can be expressed as $E_{rot} = hB_e J(J+1)$ or $E_{rot} = hc\tilde{B}_e J(J+1)$. The subscript e reminds us that the expressions hold for $R = R_e$. Sometimes, we will drop the subscript to be more general. The spectroscopic determination of the rotational constant enables the bond length of a diatomic to be calculated. We shall see that a vibrating molecule in its lowest vibrational state has an average value of R which exceeds R_e, due to anharmonicity, causing the observed rotational constant for a nonrigid rotor to be less than B_e.

8.2.2 POLYATOMIC ROTATIONS

The specification of the orientation of a nonlinear molecule requires three angles, and thus three rotational quantum numbers are associated with the wavefunctions. (Compare to two angles and two quantum numbers for a linear molecule.) The orientation is most conveniently expressed in terms of the Euler angles ($\varphi\theta\chi$) as shown in Figure 8.2.

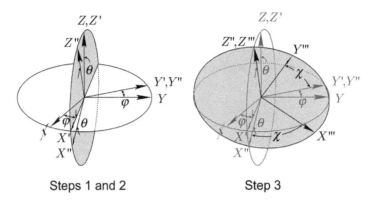

Figure 8.2 The Euler angles.

Again, the coordinate system designated by uppercase letters (XYZ) represents the space-fixed or laboratory frame, while the lowercase letters (xyz) refer to a coordinate system fixed in the molecule. If the molecule has some symmetry, we would be smart to take the z direction as one of the symmetry axes, for example the n-fold rotation axis of a molecule belonging to a C_{nv} point group. The orientation of the molecule with respect to the lab is envisioned in terms of the three-step process of rotating (XYZ) into (xyz) as follows:

Step 1. Imagine that both coordinate systems coincide at the start, and rotate the molecule-fixed system by the angle φ about the Z axis. This takes (XYZ) into $(X'Y'Z')$.

Step 2. The $(X'Y'Z')$ frame is rotated about the Y' axis by the angle θ, which results in the frame $(X''Y''Z'')$. If we were dealing with a linear molecule, we would be done at this point, but a nonlinear one requires one more step.

Step 3. Rotation by the angle χ about Z'' takes $(X''Y''Z'')$ into $(X'''Y'''Z''') = (xyz)$.

In order to appreciate the Euler angles, it may be necessary to stare at Figure 8.2 for a while. The following conclusions have physical significance in the discussion of angular momentum: the angles θ and φ are the polar angles which specify the direction of the molecule z axis, in the spaced-fixed frame, and the angle χ is the angle of rotation of the molecule frame about its own z axis. Angular momentum operators for rotations about the specific axes are

$$\hat{L}_Z = -i\hbar \frac{\partial}{\partial \varphi} \quad \hat{L}_z = -i\hbar \frac{\partial}{\partial \chi} \quad \hat{L}_{Y'} = -i\hbar \frac{\partial}{\partial \theta} \tag{8.12}$$

Components of the angular momentum along directions other than Z, z and Y' are slightly more complicated. For example, the two remaining components in the molecule frame are

$$\hat{L}_x = -i\hbar \left[-\cos \chi \csc \theta \frac{\partial}{\partial \varphi} + \cos \chi \cot \theta \frac{\partial}{\partial \chi} + \sin \chi \frac{\partial}{\partial \theta} \right] \tag{8.13}$$

$$\hat{L}_y = -i\hbar \left[\sin \chi \csc \theta \frac{\partial}{\partial \varphi} - \sin \chi \cot \theta \frac{\partial}{\partial \chi} + \cos \chi \frac{\partial}{\partial \theta} \right] \tag{8.14}$$

The magnitude of the total angular momentum does not depend on the coordinate system in which it is expressed. The sum of the squares of the three components is the same in either frame of reference:

$$\hat{L}^2 = \hat{L}_x^2 + \hat{L}_y^2 + \hat{L}_z^2 = \hat{L}_X^2 + \hat{L}_Y^2 + \hat{L}_Z^2 \tag{8.15}$$

The operators $\hat{L}^2$, $\hat{L}_Z^2$, and $\hat{L}_z^2$ commute with one another, so they share a set of common eigenfunctions, known as the Wigner rotation functions (Appendix A):

$$D_{MN}^J(\Omega) = D_{MN}^J(\theta\varphi\chi) \tag{8.16}$$

We employ the shorthand notation $(\Omega) = (\theta\varphi\chi)$ to designate the Euler angles. The Wigner functions are generalizations of the spherical harmonics. In fact, when either quantum number M or N is zero, the D's reduce to Y's. The form of the Wigner functions is

$$D_{MN}^J(\theta\varphi\chi) = e^{-iM\varphi} d_{MN}^J(\theta) e^{-iN\chi} \tag{8.17}$$

where $d_{MN}^J(\theta)$ is a real function whose form will not concern us here. (See [1], where many useful properties of the Wigner functions are discussed.) The eigenvalue equations are

$$\hat{L}^2 D_{MN}^J(\Omega) = J(J+1)\hbar^2 D_{MN}^J(\Omega) \tag{8.18}$$

$$\hat{L}_Z D_{MN}^J(\Omega)^* = M\hbar D_{MN}^J(\Omega)^* \tag{8.19}$$

$$\hat{L}_z D_{MN}^J(\Omega)^* = N\hbar D_{MN}^J(\Omega)^* \tag{8.20}$$

The quantum numbers N and M each range independently from $-J$ to J in integral steps. As eigenfunctions of Hermitian operators, the Wigner functions form a complete orthonormal set, although they are not normalized to unity, as shown by the relationship:

$$\int d\Omega D_{MN}^J(\Omega)^* D_{M'N'}^{J'}(\Omega) = \left(\frac{8\pi^2}{2J+1}\right)\delta_{NN'}\delta_{MM'}\delta_{JJ'} \tag{8.21}$$

Equation 8.21 employs the following shorthand notation for the triple integral:

$$\int d\Omega = \int_0^{2\pi} d\varphi \int_0^\pi \sin\theta\, d\theta \int_0^{2\pi} d\chi = 8\pi^2 \tag{8.22}$$

Equations 8.18–8.20 clarify the physical meaning of the quantum numbers associated with $D_{MN}^J(\Omega)$. J is associated with the total angular momentum, and M gives the Z component in the lab frame. The quantum number N determines the z component of the angular momentum in the frame of the molecule, so it represents the molecule spinning about its own symmetry axis (if it has one, otherwise the z direction is arbitrary).

The inertia of a three-dimensional figure is actually a second-rank tensor, with components I_{xx}, I_{xy}, etc., but a proper choice of axes diagonalizes the tensor:

$$\begin{bmatrix} I_{xx} & I_{xy} & I_{xz} \\ I_{yx} & I_{yy} & I_{yz} \\ I_{zx} & I_{zy} & I_{zz} \end{bmatrix} \Rightarrow \begin{bmatrix} I_a & 0 & 0 \\ 0 & I_b & 0 \\ 0 & 0 & I_c \end{bmatrix} \tag{8.23}$$

The moments of inertia, I_a, I_b, and I_c, one for rotation about each of three mutually perpendicular axes that intersect at the center of mass, are given by

$$I_a = \sum_i M_i R_{ai}^2 \tag{8.24}$$

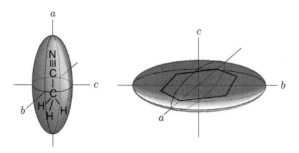

Figure 8.3 Inertial ellipsoid of (a) a prolate top, CH_3CN and (b) an oblate top, C_6H_6.

where R_{ai} is the perpendicular distance of atom i from the a axis of rotation and similarly for the b and c components. By convention, $I_a \leq I_b \leq I_c$. The inertia of a generally shaped molecule can be visualized with the help of its ellipsoid of inertia, as shown in Figure 8.3. The lengths of the axes of this ellipsoid are proportional to the inverse square root of the inertia for rotation about that axis. (How would you picture the inertial ellipsoid for a linear molecule?) We will classify the rotational behavior of nonlinear molecules based on the symmetry of the inertial ellipsoid.

The classical expression for angular momentum about the i-th axis is $L_i = I_i \Omega_i$, where Ω_i is the angular velocity in radians per second. The kinetic energy of rotation is

$$T_{rot} = \frac{1}{2} \sum_i I_i \Omega_i^2 = \sum_i \frac{L_i^2}{2I_i} \tag{8.25}$$

Now we have what we need to write the rotational Hamiltonian for a generally shaped molecule; we just replace the classical functions for the squared angular momentum with the corresponding operators:

$$\hat{H}_{rot} = \frac{\hat{L}_a^2}{2I_a} + \frac{\hat{L}_b^2}{2I_b} + \frac{\hat{L}_c^2}{2I_c} \tag{8.26}$$

For molecules having two or three equal moments of inertia, the eigenfunctions of $\hat{H}_{rot}$ are readily found. Unfortunately, many molecules of interest are asymmetric rotors, and only approximate solutions to the eigenvalue problem can be obtained. Let us look at symmetric molecules first. Spherical tops have three equal moments of inertia, so naturally the inertial ellipsoid is a sphere. Symmetric tops have two equal moments of inertia, a property shared by all molecules having at least threefold rotation symmetry or two twofold rotation axes. There are two types of symmetric tops, prolate symmetric tops, for which the unique moment of inertia is the smallest, and oblate symmetric tops, for which the unique moment of inertia is the largest. Examples of symmetric tops are illustrated in Figure 8.3. A symmetric top may or may not have a dipole moment, but when one exists its direction must coincide with the symmetry axis. It is convenient to define rotational constants corresponding to each inertial axis:

$$A = \frac{h}{8\pi^2 I_a} \tag{8.27}$$

$$B = \frac{h}{8\pi^2 I_b} \tag{8.28}$$

$$C = \frac{h}{8\pi^2 I_c} \tag{8.29}$$

Sometimes, these constants are expressed in units of cm^{-1}, $\tilde{A} = A/c$, etc.

8.2.2.1 SPHERICAL TOPS

Spherically symmetric molecules, such as those belonging to the T_d or O_h point group, have $I_a = I_b = I_c \equiv I$. CCl_4 and SF_6 are spherical tops. The rotational Hamiltonian is

$$\hat{H}_{rot} = \frac{\hat{L}_a^2 + \hat{L}_b^2 + \hat{L}_c^2}{2I} = \frac{\hat{L}^2}{2I} \tag{8.30}$$

The rotational eigenfunctions are the previously introduced Wigner functions: $\psi_{JMK}(\theta\varphi\chi) = D_{MK}^J(\Omega)$. The rotational energy expression is identical to that for a diatomic molecule, $E_{rot} = J(J+1)\hbar^2/2I$, but the degeneracy is greater because both M and K take on values from $-J$ to $+J$. Thus the degeneracy is $g_J = (2J+1)^2$. Spherical tops never have permanent dipole moments, so they are microwave inactive. In addition, the polarizability of a spherical top is isotropic, so pure rotational Raman scattering is forbidden.

8.2.2.2 PROLATE SYMMETRIC TOPS

For these molecules, the inertial components satisfy $I_a < I_b = I_c$. A cigar-shaped object serves as a caricature of the shape of a prolate top molecule. Axially symmetric molecules such as CH_3CN and NH_3 are prolate tops; it is easier to spin the molecule about the symmetry axis than about the two equivalent perpendicular axes. The Hamiltonian simplifies as follows:

$$\hat{H}_{rot} = \frac{\hat{L}_a^2}{2I_a} + \frac{\hat{L}_b^2 + \hat{L}_c^2}{2I_b} = \frac{\hat{L}_a^2}{2I_a} + \frac{\hat{L}^2 - \hat{L}_a^2}{2I_b} \tag{8.31}$$

We can find the eigenfunctions of this Hamiltonian with the help of Equations 8.18–8.20. The symmetry axis of the prolate top is the molecule frame a-axis, which plays the role of the z axis in Equation 8.20. By convention, the quantum number pertaining to rotation about the symmetry axis is called K, so the eigenfunctions of 8.31 are written D_{MK}^J, and the eigenvalues are found to be

$$E_{JK} = J(J+1)\frac{\hbar^2}{2I_b} + K^2\hbar^2\left(\frac{1}{2I_a} - \frac{1}{2I_b}\right) \tag{8.32}$$

Note that compared to the previous example the lowered symmetry leads, as expected, to some lifting of the degeneracy. The energy levels of the symmetric top depend on two, rather than one, quantum numbers. Because the energy depends on K^2, there is a twofold degeneracy associated with each level having $K \neq 0$, in addition to the $2J+1$ factor that arises from the range of values for M. The degeneracy of states having the same absolute value of K reflects the fact that the rotational energy is independent of the sense of rotation about the molecular axis. The total degeneracy is $g_J = 2J+1$ for $K = 0$ and $g_J = 2(2J+1)$ for $K \neq 0$. Notice that, since $I_a < I_b$, the energy increases with K^2, as shown in Figure 8.4.

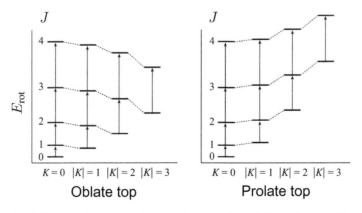

Figure 8.4 Energy level diagrams for symmetric tops, showing allowed absorption transitions.

8.2.2.3 OBLATE SYMMETRIC TOPS

For these molecules, $I_a = I_b < I_c$. An oblate top molecule is disk shaped; benzene is an excellent example. The mathematics of finding the energy levels is identical to the prolate top problem, so we will skip the details and present the result:

$$E_{JK} = J(J+1)\frac{\hbar^2}{2I_b} + K^2\hbar^2\left(\frac{1}{2I_c} - \frac{1}{2I_b}\right) \tag{8.33}$$

The degeneracies of the energy levels are of course the same as for the prolate top. However, since $I_c > I_b$, the energy levels decrease as K^2 increases. Equation 8.33 could also have been written $E_{JK} = hBJ(J+1) + K^2h(C-B)$, and Equation 8.32 as $E_{JK} = hBJ(J+1) + K^2h(A-B)$. Figure 8.4 shows the energy level scheme for prolate and oblate tops. The arrangement of the energy levels anticipates the selection rule $\Delta K = 0$, which will be derived in Sections 8.5 for microwave and 8.6 for Raman scattering spectra. In the meantime, it should be logical that rotation about the symmetry axis is both microwave and Raman inactive, since such motion has no effect on the lab-frame component of dipole moment or polarizability. This leads to the question: How can the moment of inertia about the symmetry axis be determined?

8.2.2.4 ASYMMETRIC TOPS

Even a molecule as simple as H_2O is classified as an asymmetric rotor, for which $I_a \neq I_b \neq I_c$. There is nothing we can do to simplify the rotational Hamiltonian expressed in Equation 8.26. Since the operators $\hat{L}_a^2$, $\hat{L}_b^2$, and $\hat{L}_c^2$ do not commute with one another, they do not share a set of common eigenfunctions. Approximate solutions to the problem take the Wigner functions as a basis (they form a complete set) and express a wavefunction for the asymmetric rotor as a linear combination of wavefunctions with different values of K:

$$\psi_{JM}(\Omega) = \sum_{K'=-J}^{J} c_{K'} D_{MK'}^J(\Omega) \tag{8.34}$$

Since the operators $\hat{L}^2$ and $\hat{L}_z$ both commute with $\hat{L}_a$, $\hat{L}_b$, and $\hat{L}_c$, J and M are still good quantum numbers, but K is no longer a good quantum number. Each approximate wavefunction is a linear combination of $2J+1$ rotational wavefunctions with the same values of J and M, but different values of K. Some molecules with less than threefold rotational symmetry can have two inertial components which are close, making them "accidental" symmetric tops. The deviation of a molecule from being a symmetric top is quantified by the asymmetry parameter κ:

$$\kappa = \frac{2B - A - C}{A - C} \tag{8.35}$$

The value of κ ranges from -1 for a prolate top to $+1$ for an oblate top. The asymmetry parameter can be used to qualitatively estimate the energy levels for a molecule with the help of the correlation diagram of Figure 8.5. This diagram displays the energy of the prolate rotor on the left side and that of the oblate rotor on the right, connected by lines that sketch how the energy varies with κ.

8.2.2.5 LINEAR MOLECULES

Linear polyatomic molecules, regardless of the number of atoms, have rotational energy levels just like those of diatomics. The moment of inertia is found using Equation 8.24, which for a linear molecule gives $I_a = 0$ and $I_b = I_c$. Linear triatomics such as CO_2 and HCN have moments of inertia that depend on two bond lengths. In the case of a symmetric triatomic like CO_2, a single measurement of B would allow the bond distance to be determined. For HCN, however, the inertia depends on two different bond lengths, so one measurement alone cannot determine the structure. The measurement of B for different isotopic derivatives, however, permits the bond lengths to be determined, as you will show in Problem 6. The invariance of bond length to isotopic substitution is a consequence of the Born–Oppenheimer approximation, to be discussed in Chapter 9.

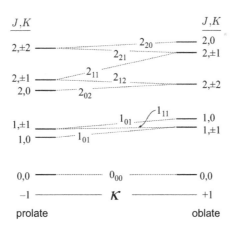

Figure 8.5 Correlation diagram for energy levels of asymmetric rotors.

8.3 ANGULAR MOMENTUM COUPLING IN NON-$^1\Sigma$ ELECTRONIC STATES

In Chapter 11, we will consider the electronic angular momentum of diatomic molecules. Term symbols of the type $^{2S+1}\Sigma$, $^{2S+1}\Pi$, etc. will be introduced to designate the orbital and spin angular momentum of an electronic state. The uppercase Greek letter designates the orbital angular momentum, as described below. The term symbol also corresponds to the symmetry species of the electronic wavefunction in either the $D_{\infty h}$ or $C_{\infty v}$ character table, for homonuclear and heteronuclear diatomics, respectively. Closed-shell states of diatomics have no net electron spin or orbital angular momentum and are designated by the symbol $^1\Sigma$. In these states, the only component of the angular momentum is that due to motion of the nuclei.

Let us now look at diatomic molecules having term symbols other than $^1\Sigma$. Molecules such as O_2 in its triplet ground state, all molecules having odd numbers of electrons, and many excited states of closed-shell molecules provide examples of non-$^1\Sigma$ species. The angular momentum due to electronic motion can couple (add vectorially) to that due to rotation, and the resulting effects show up in electronic or rotational spectra. In very high resolution spectroscopy, the coupling of the nuclear spin to other angular momenta (the hyperfine effect) must also be considered. The discussion here will be limited to the coupling schemes for electronic and rotational angular momenta that are commonly referred to as Hund's cases (a) through (d). The calculation of angular momentum coupling in diatomic molecules is a job for perturbation theory, and it is possible to have strong enough coupling that "good" quantum numbers are lost. The range of various interactions among angular momenta leads to different mental pictures on which the process of coupling vectors is based. The four Hund's cases (and a fifth one not discussed here because of its physical improbability) are idealizations which result when certain matrix elements of the coupling Hamiltonian can be neglected. In what follows, we will call **J** the total angular momentum, excluding nuclear spin, and **O** will represent the angular momentum due to rotation of the nuclear framework. Since the nuclei are point masses and we are treating linear molecules, the vector **O** is always perpendicular to the bond. Bold typeface will designate vector quantities and regular typeface the corresponding quantum numbers. For example, the absolute value of **J** is $\sqrt{J(J+1)}\,\hbar$. The cones shown in Figure 8.6 represent precession, consistent with our uncertainty of the components of the vector perpendicular to the rotation axis, as discussed in Chapter 7.

To review electronic angular momenta in diatomics, recall that although the electronic wavefunctions are not eigenfunctions of $\hat{L}^2$, the angular momentum about the bond is quantized. (**L** precesses about the bond axis.) We say that M_L is a good quantum number, and we designate it by term symbols Σ, Π, Δ, Φ, ..., corresponding to $M_L = 0$, ± 1, ± 2, ± 3,.... The symbol Λ is used to designate the absolute value $|M_L|$. Energy levels having $M_L \neq 0$ have twofold orbital degeneracy, because the energy of the electronic state is

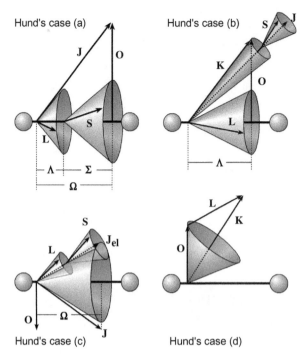

Figure 8.6 Hund's cases (a) through (d).

independent of the sense of rotation of the electrons about the internuclear axis. If the molecule also has spin angular momentum, $S \neq 0$, this can couple to the orbital motion, through spin–orbit coupling, as discussed in Chapter 7 for atoms. This causes the electronic spin $\mathbf{S}$ to couple to the internuclear axis, as shown in Figure 8.6a, where the components along the bond are designated by $\Sigma = S, S - 1, \ldots, -S$. (It is an unfortunate convention to use the symbol Σ to represent the z component of total spin angular momentum as well as the term symbol for states having $|M_L| = 0$.) The quantum number Ω, the z component of the total angular momentum, is equal to $\Lambda + \Sigma$. It represents the result of spin–orbit coupling and takes on integral and half-integral values for even and odd numbers of electrons, respectively.

Hund's case (a) occurs when the coupling of electronic and nuclear angular momentum is weak. In this case, as shown in Figure 8.6a, spin–orbit coupling causes the spin vector to precess about the bond along with $\mathbf{L}$. The total angular momentum about the bond is $\Omega = \Lambda + \Sigma$. The total angular momentum $\mathbf{J}$ is the vector sum of $\mathbf{\Omega}$ and $\mathbf{O}$. The rotational energies are given by

$$E_{rot} = hBJ(J+1) + h(A - B)\Omega^2 \tag{8.36}$$

where B is the usual rotational constant, and A is related to the spin–orbit coupling constant. Equation 8.36 resembles the expression for a prolate symmetric top, with Ω replacing K. Indeed, the problem is just like that, except that the inertia about the bond is entirely due to electrons. If A is visualized to be of the same form as the rotational constant, then it must be inversely proportional to something akin to an electronic moment of inertia; hence A is much larger than B. The allowed values of J are $\Omega, \Omega + 1, \Omega + 2, \ldots$, since the magnitude of the vector $\mathbf{J}$ cannot be any less than that of $\mathbf{\Omega}$. The value of J can be half-integral when there is an odd number of electrons. For example, a $^2\Pi_{1/2}$ state, where the subscript designates that $\Omega = 1/2$, has a series of doubly degenerate rotational states corresponding to $J = 1/2, 3/2, 5/2, \ldots$, etc. The ground state of NO is $^2\Pi_{1/2}$. If the electronic motion couples strongly to the rotation of the nuclear frame, this degeneracy will be lifted. The splitting, referred to as Lambda-type doubling, is typically quite small, much less than one cm^{-1}.

Hund's case (b) applies either when $\Lambda = 0$ or when spin–orbit coupling is very weak, as in light atom diatomics such as the OH radical, which has a $^2\Pi$ state. The spin vector $\mathbf{S}$ is no longer coupled to the bond axis, as it is for cases (a) and (c). We imagine adding the vectors $\mathbf{\Lambda}$ and $\mathbf{O}$ to get $\mathbf{K}$. The quantum number associated with this vector is $K = \Lambda, \Lambda + 1, \Lambda + 2,\ldots$, and so on. (Some authors refer to this quantum number as N rather than K.) The total angular momentum $\mathbf{J}$ results from adding $\mathbf{S}$ and $\mathbf{K}$, and the associated values of J range from $K + S$ to $|K - S|$. Rotational levels having $K \geq S$ are split into $2S + 1$ components.

Hund's case (c) is similar to case (a) except that the spin–orbit coupling is much stronger than the coupling of the orbital motion to the bond axis. This means that the quantum numbers Λ and Σ are no longer good. We imagine a total electronic angular momentum $\mathbf{J}_{el}$ that would be obtained by coupling $\mathbf{L}$ and $\mathbf{S}$. The projection of $\mathbf{J}_{el}$ onto the bond axis is $\mathbf{\Omega}$, which adds to $\mathbf{O}$ to give the total angular momentum $\mathbf{J}$. The energy levels are the same as for Hund's case (a), but the constant A is much larger. The rotational energies can be expressed as $E_{rot} = hB[J(J + 1) - \Omega^2]$, since the A part can be incorporated into the electronic energy. Hund's case (c) behavior is observed for heavier diatomics, for example excited states of I_2.

Hund's case (d) is found in highly excited electronic states known as Rydberg states. In these states, the excited electron is so far removed from the nuclei that the orbital angular momentum $\mathbf{L}$ is no longer quantized along the internuclear axis, and the excited electronic state resembles that of a one-electron atom. The vectors $\mathbf{L}$ and $\mathbf{O}$ are added, and if spin is present, the vector $\mathbf{S}$ is added to that result to give the total angular momentum $\mathbf{K}$. The coupling between $\mathbf{S}$ and $\mathbf{K}$ is usually weak. The rotational levels are split into $2L + 1$ closely spaced levels.

8.4 NUCLEAR STATISTICS AND *J* STATES OF HOMONUCLEAR DIATOMICS

The rotational spectra of symmetric molecules, and notably homonuclear diatomics, display the quantum mechanical constraints imposed on wavefunctions for indistinguishable particles. In this section we discuss the nuclear statistics of homonuclear diatomics and their effect on the statistical weights of rotational levels. The basic ideas also apply to symmetric polyatomics, for example, the equivalent hydrogens in CH_4 or CH_3Cl, but the analysis is more complicated, so only diatomics will be considered. A mixed isotope species such as HD would not be subject to the constraints discussed here, because the two nuclei are distinguishable.

Recall the Pauli exclusion principle, which requires electronic wavefunctions to be antisymmetric with respect to exchange of any two electrons. As discussed in Chapter 1, there are two kinds of particles in nature, fermions and bosons. Fermions, such as electrons and nuclei having odd mass numbers, have half-integral spin, and bosons, such as nuclei with even mass numbers, have integral spin. The wavefunctions for fermions and bosons differ in their symmetry with respect to exchange of equivalent particles. Using the symbol $\hat{P}_{12}$ to denote the result of exchanging two particles, the symmetry constraint on the total wavefunction is

$$\hat{P}_{12}\Psi = \pm\Psi \tag{8.37}$$

where the plus sign holds for bosons and the minus sign for fermions. (There is no reason why the wavefunction should not change sign or phase as a result of a symmetry operation, since it is $\Psi^*\Psi$ which has a physical meaning.) The wavefunctions of homonuclear diatomics are required be eigenfunctions of the exchange operator $\hat{P}_{12}$. If the nuclear spin quantum number I is integral, then the plus sign of Equation 8.37 applies. Conversely, nuclei having half-integral values of I require that the wavefunction be antisymmetric with respect to $\hat{P}_{12}$. The wavefunction of Equation 8.37 is the total wavefunction: $\Psi = \psi_{el}\psi_{vib}\psi_{rot}\psi_{nuc}$. The vibrational wavefunction is always symmetric with respect to $\hat{P}_{12}$ because ψ_{vib} depends on the displacement of the nuclei from equilibrium (to be discussed in Chapter 9), and not on the coordinates of the two nuclei. We therefore concern ourselves with the other contributions to the total wavefunction. ψ_{nuc} here represents the nuclear spin function, which may be either symmetric or antisymmetric with respect to relabeling of the nuclei.

For a given quantum number I, as for any other angular momentum quantum number, there are $2I + 1$ values of the projection of the spin onto the Z direction. For spin-1/2 particles, this gives just two orientations commonly called α (spin up) and β (spin down). More generally, there are a total of $(2I + 1)^2$ ways to combine the spins on the two nuclei, to generate different nuclear spin wavefunctions ψ_{nuc}. Some of these wavefunctions are symmetric and the rest are antisymmetric with respect to nuclear exchange. It turns out that the numbers are given by

$$(2I+1)^2 = (2I+1)(I+1)+(2I+1)I$$

(8.38)

$$\text{total} = \text{symmetric} + \text{antisymmetric}$$

For example, the ^{1}H nucleus has spin $I = 1/2$. Using Equation 8.38 for diatomic hydrogen, we find three symmetric spin functions and one antisymmetric spin function. The three symmetric ones, analogous to the electronic spin functions for a triplet state, are $\alpha(1)\alpha(2)$, $\beta(1)\beta(2)$, and $\alpha(1)\beta(2) + \beta(1)\alpha(2)$. These correspond to a form of hydrogen called ortho-hydrogen. The antisymmetric spin function, $\alpha(1)\beta(2) - \beta(1)\alpha(2)$, which is analogous to a singlet electronic state, corresponds to para-hydrogen. In nature, these two forms of H_2, which differ in their magnetic properties, exist in a 3:1 equilibrium. We shall use these two forms of hydrogen to illustrate the consequences of nuclear spin on rotational energy levels.

To perform the interchange of the two nuclei, it might seem that a mere $\hat{C}_2$ operation would do the job, but that would rotate the electrons as well. We just want to exchange the nuclei and swap their spin functions. The $\hat{P}_{12}$ operation can in fact be achieved by the four-step process depicted in Figure 8.7. The first step is a $\hat{C}_2$ rotation of the molecule, which rotates the electrons and the nuclei. This operation affects the rotational wavefunction as follows:

$$\hat{C}_2\psi_{rot} = (-1)^J \psi_{rot}$$

(8.39)

To convince yourself of the validity of Equation 8.39, recall that the wavefunctions for $J = 0, 1, 2, 3, \ldots$ have the symmetries of the $s, p, d, f, \ldots$ orbitals. The next two steps operate only on the electronic coordinates: $\hat{i}_{el}$ inverts the electronic coordinates through the center of symmetry, and $\hat{\sigma}_{el}$ reflects them through a plane perpendicular to the twofold rotation axis. The effects of these two operations are deduced from the electronic term symbol. The molecule H_2 has a $^1\Sigma_g^+$ ground electronic state, as does any closed-shell homonuclear diatomic, where the subscript g indicates that the wavefunction is symmetric with respect to inversion, and

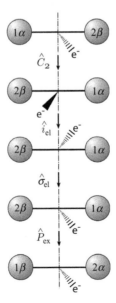

Figure 8.7 Four-step process equivalent to $\hat{P}_{12}$.

the + superscript reveals the symmetry on reflection. (Symmetry species for electronic states of diatomics are discussed in more detail in Chapter 11.) Consequently, the $^1\Sigma_g^+$ electronic wavefunction is preserved by the second and third steps, and thus ψ_{el} is symmetric with respect to the $\hat{P}_{12}$ operation. A ground electronic state having the term symbol $^1\Sigma_u^-$ would also be symmetric with respect to this operation, while $^1\Sigma_u^+$ and $^1\Sigma_g^-$ wavefunctions would change sign. The final step in Figure 8.7 is the application of the $\hat{P}_{ex}$ operator, which swaps the nuclear spin functions. The effect of this operator is to change the sign of ψ_{nuc} if the nuclear spin function is antisymmetric, and preserve it if ψ_{nuc} is symmetric.

The net effect of these four operations on the ortho and para forms of H_2 is summarized in Table 8.1. For the ortho form, the symmetric nuclear spin functions must be combined with odd J states in order for the overall wavefunction to be antisymmetric with respect to exchange. Conversely, the para form of H_2 has only even J states. The rotational Raman spectrum of an equilibrium mixture of ortho- and para-H_2 reveals the effects of quantum statistics in the approximately 3:1 ratio of alternate lines. The selection rules for rotational Raman, to be discussed in Section 8.6, happen to allow $\Delta J = \pm 2$, so each form of H_2 undergoes transitions within its own manifold of even or odd J states.

Figure 8.8 shows the pure rotational Raman spectrum of H_2 dissolved in water, taken from [2]. Molecules in the liquid phase do not generally rotate freely, owing to the torques exerted by the solvent. However, in this example the cavities in the water structure are large enough compared to the size of H_2 to permit rotational Raman transitions comparable to, but much broader than, those observed for H_2 in the gas phase. The nuclear spin statistics discussed above, combined with spectroscopic constants presented in Table 9.2, enable the frequency shifts and relative intensities of the transitions shown in Figure 8.8 to be accounted for, as you will do in Problem 10 of this chapter.

As a final example, consider the diatomic formed by two ^{12}C atoms. Like the hydrogen molecule, the ground electronic state is $^1\Sigma_g^+$. The nuclear spin of ^{12}C is zero, so there is a single nuclear spin function and it is symmetric. Since the overall wavefunction must be symmetric, we conclude that odd J states do not exist. Any homonuclear diatomic composed of spin-zero nuclei will have either even or odd rotational states, but not both.

Table 8.1 Symmetry of ortho- and para-H_2 wavefunctions under nuclear exchange

Symmetry under $\hat{P}_{12}$	Ψ_{tot}	ψ_{el}	ψ_{vib}	ψ_{rot}	ψ_{nuc}	Allowed J
Ortho-H_2 (weight 3)	−	+	+	−	+	1,3,5,7,...
Para-H_2 (weight 1)	−	+	+	+	−	0,2,4,6,...

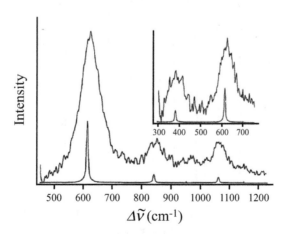

Figure 8.8 Pure rotational Raman spectrum of H_2 dissolved in water at 200 psi (broad bands), compared to that of gas phase H_2 at 2000 psi (sharp bands). (Reprinted from D. G. Taylor III and H. L. Strauss, The rotational Raman spectrum of H_2 in water, *J. Chem. Phys.* 1989, *90*, 768, with the permission of AIP Publishing.)

8.5 ROTATIONAL ABSORPTION AND EMISSION SPECTROSCOPY

Microwave spectroscopy provides a powerful approach for determining the structures and dipole moments of gas-phase molecules. The typical frequency range of interest is about 3 to 300 GHz (0.1 to 10 cm⁻¹). A great experimental advantage is that the source used in a microwave spectrometer, an electronic tube called a klystron, provides tunable, monochromatic, coherent radiation at frequencies accurate to fractions of a MHz. In Stark-modulated spectroscopy, a square-wave voltage is applied to the cell to shift the absorption lines in and out of resonance with the source, and the signal is detected with a lock-in amplifier. The amplitude of the Stark field is varied to permit highly accurate measurement of the dipole moment. Even the small dipole moments (on the order of 0.01 to 0.1 D) of saturated hydrocarbons have been measured. Conventional chemical applications employ microwave absorption, though astronomers use microwave emission to observe molecules in interstellar space. Microwave spectra can also be used to determine barriers to internal rotation (Section 8.8). Practical applications of microwave spectroscopy are limited to the study of rather small molecules. This drawback results from the requirement that the sample be volatile and free from the complications posed by numerous low frequency vibrational states. Microwave spectroscopy has been applied to determine the structure of weakly bound complexes of small molecules known as Van der Waals complexes.

Pure rotational transitions may also be observed in the far-IR. The rotational constants for diatomic hydrides, for example, are larger than those for heavier molecules. The first 14 transitions in the rotational absorption of HCl, for example, span the frequency range from about 20 to 300 cm⁻¹. The rotational constant of HCl is about 10.4 cm⁻¹, which is rather large due to the low inertia. For comparison, that of CO is about 1.9 cm⁻¹. Rotational transitions involving larger values of J also appear in the far-IR, even for heavier molecules. The far-IR absorption spectrum of CO is shown in Figure 8.9. In this section, we derive the selection rules for pure rotational absorption and emission, and we see how transitions such as those in Figure 8.9 may be assigned.

Before moving on to consider selection rules and spectral analysis, it is worthwhile to mention an important caveat. The structural parameters that are extracted from microwave data are always averages over the appropriate vibrational states; ordinarily the ground vibrational state. So it is not B_e which is obtained directly, but rather B_0, where the zero subscript indicates the ground vibrational state. Still, B_0 does not directly reveal the internuclear distance averaged over the ground vibrational state, R_0. That is because B_0 is proportional to

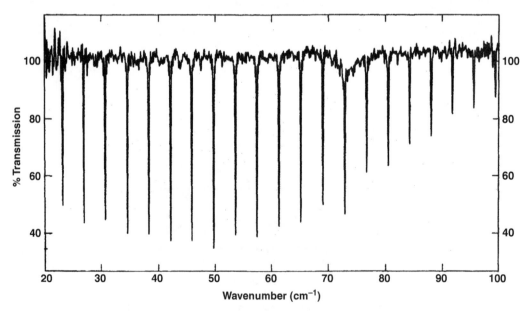

Figure 8.9 Pure rotational spectrum of CO. (Reproduced with permission from G. W. Chantry ed., *Modern Aspects of Microwave Spectroscopy*, Academic Press, New York, 1979.)

$<1/R^2>$, which is not the same as $1/<R>^2$, where the brackets indicate an average over the zero-point vibration. In the next chapter, we will consider how to find R_e from vibration–rotation spectroscopy.

Consider a diatomic molecule going from initial state i to final state f. The wavefunction for each state is written in the form of a product of electronic and nuclear (rovibrational) parts: $\Psi = \psi_{el}\psi_{vJM}$. (The nuclear spin wavefunction is dropped because we are considering spectra for which the nuclear spin state remains constant.) In this Born–Oppenheimer form,* it is implied that the electronic wavefunction depends parametrically on the geometry of the nuclear framework. The rovibrational part is further separated into vibrational and rotational functions: $\psi_{vJM} = \chi_v(q)Y_{JM}(\theta,\varphi)$, where $q \equiv R - R_e$ is the displacement of the bond length from its equilibrium value. Let us examine E1-allowed rovibrational transitions, $vJM \to v'J'M'$, within a given electronic state, usually the ground state. The transition moment as usual is given by $\mu_{if} = \int \psi_i^* \hat{\mu}\psi_f d\tau$. This takes the form

$$\mu_{if} = \int \psi_{vJM}^* \left(\int \psi_{el}^* \hat{\mu}\psi_{el}d\tau_{el} \right) \psi_{v'J'M'}d\tau_{nuc}d\tau_{el} \tag{8.40}$$

The integrals in Equation 8.40 are nested to represent first finding the expectation value of the dipole moment in a particular electronic state, then taking the matrix element of this with respect to two rovibrational states. The dipole moment in a particular electronic state is some function of the internuclear distance $\mu(q)$ and also depends on the orientation in the lab frame. If we knew this function, we could substitute for the inner integral and then perform the outer integration over nuclear coordinates. Unfortunately, the function $\mu(q)$ is not generally available, but if we are justified in starting out from the rigid rotor picture, then the amplitude of vibration is small and maybe we can get away with another truncated Taylor series approximation. The magnitude of the dipole moment is expanded about the value at the equilibrium internuclear distance, $q = 0$:

$$\mu(q) = \mu_0 + \frac{\partial \mu}{\partial q}\bigg|_0 q + \cdots \tag{8.41}$$

The first term is the permanent dipole moment. Its magnitude is constant, but its orientation in the lab frame is not. Clearly, the operator μ_0 cannot connect two different vibrational states, due to the orthogonality of vibrational wavefunctions. However, due to its dependence on orientation, the dipole operator does connect different rotational states within a vibrational level. The permanent dipole moment is thus responsible for pure rotational transitions. The second term, a function of the dipole moment derivative with respect to internuclear distance, permits vibrational transitions. We shall examine these in later chapters and concern ourselves here with pure rotational spectra. Keeping only the first term in Equation 8.41, then, the dipole moment is specified in a space-fixed coordinate system:

$$\vec{\mu}_0 = \mu_X \hat{i} + \mu_Y \hat{j} + \mu_Z \hat{k} \tag{8.42}$$

$$\vec{\mu}_0 = \mu_0(\sin\theta\cos\varphi\,\hat{i} + \sin\theta\sin\varphi\,\hat{j} + \cos\theta\,\hat{k}) \tag{8.43}$$

In the absence of an external electric field (such as that employed in Stark effect experiments), the X, Y, and Z directions in the lab are equivalent. The three components of the dipole operator, which allow absorption by radiation polarized in the corresponding directions, are given by

$$\mu_{if} = \mu_0 \int_0^{2\pi} d\varphi \int_0^\pi \sin\theta\,d\theta\,Y_{JM}^*(\theta\varphi) \begin{Bmatrix} \sin\theta\cos\varphi \\ \sin\theta\sin\varphi \\ \cos\theta \end{Bmatrix} Y_{J'M'}(\theta\varphi) \tag{8.44}$$

---

* The Born–Oppenheimer approximation is discussed in more detail in Chapter 9, where it is shown that it enables the electronic and nuclear wavefunctions to be factored.

The first obvious requirement for an allowed transition is a nonvanishing dipole moment: microwave activity requires that a molecule possess a permanent dipole moment. To derive the selection rules, we consider what values of $JM \rightarrow J'M'$ result in a nonzero value for any of the three integrals in Equation 8.44. These selection rules were discussed in the previous chapter, where the spherical harmonic functions constituted the angular part of the hydrogen atom wavefunction. Proceeding in the same way, the selection rules for microwave transitions of diatomics are found:

$$\mu_0 \neq 0 \quad \Delta J = \pm 1 \quad \Delta M = 0, \pm 1 \tag{8.45}$$

These selection rules apply to linear polyatomics as well. For linear molecules having net electronic angular momentum, the selection rules are amended to include $\Delta J = 0$, which is meaningful in vibration–rotation, but not pure rotation spectra. The origin of this exception is revealed below. The selection rule for M is unimportant unless an external field is applied to lift the degeneracy of states with different M, as in Stark spectroscopy. The rotational absorption spectrum consists of a series of lines corresponding to $J \rightarrow J + 1$, one for each thermally populated initial J state, at frequencies $2\tilde{B}(J + 1)$. Thus the pure rotational spectrum exhibits a series of lines at frequencies $2\tilde{B}, 4\tilde{B}, 6\tilde{B}, \ldots$. In the rigid rotor approximation, the spacing of adjacent lines in the microwave absorption or emission spectrum is predicted to be constant, $\Delta \tilde{v} = 2\tilde{B}$. On closer inspection, the spacing $\Delta \tilde{v}$ is not constant; it decreases slightly with increasing initial J. We return to this issue in Section 8.7.

It is sometimes erroneously stated that the intensities of the lines in a pure rotational spectrum are proportional to the populations of the initial states. There are several reasons why this is incorrect. First, as shown in Chapter 6, the absorption coefficient depends on the population difference of the initial and final states. Since rotational levels are closely spaced compared to thermal energy, $k_B T$, it is not permissible to neglect the population of the $J + 1$ level compared to that of the J level. In addition, the absorption strength is proportional to the frequency of the transition and the square of the transition moment, both of which vary with J. As shown in [3], the transition dipole should be summed over all the M components as well as the three directions:

$$\mu_{J \rightarrow J'}^2 = \sum_{M'=-J'}^{J'} \left| \langle JM | \mu_X | J'M' \rangle \right|^2 + \left| \langle JM | \mu_Y | J'M' \rangle \right|^2 + \left| \langle JM | \mu_Z | J'M' \rangle \right|^2 \tag{8.46}$$

The transition dipole for downward and upward transitions works out to be

$$\mu_{J+1 \rightarrow J}^2 = \frac{\mu_0^2 (J + 1)}{2J + 3} \tag{8.47}$$

$$\mu_{J \rightarrow J+1}^2 = \frac{\mu_0^2 (J + 1)}{2J + 1} \tag{8.48}$$

These two transition moments are not equal because the degeneracies of the two levels are not the same. When the degeneracies are factored in, it is apparent that the inherent rates of upward and downward transitions are equal:

$$g_J \mu_{J \rightarrow J+1}^2 = g_{J+1} \mu_{J+1 \rightarrow J}^2 = \mu_0^2 (J + 1) \tag{8.49}$$

When all the variables are considered, the intensity of an absorption line is found to be approximately proportional to $(J + 1)^2 \exp(-E_J/k_B T)$, compared to the Boltzmann population of the initial energy level,* which is

---

* It is important to use the word "level" rather than "state" in this discussion; e.g., the $J = 2$ level comprises five different states.

proportional to $(2J + 1)\exp(-E_J/k_BT)$. To illustrate the difference, consider the rotational spectrum of HCl at room temperature, where the value of J that maximizes the absorption strength is 3.7, making the transition $J = 4 \to J = 5$ the strongest. The value of J which maximizes the population, however, is about 2.7, so the $J = 3$ level is most populated. It turns out that the line intensities roughly follow the populations of initial states, but not in exact proportion.

The selection rules of Equation 8.45 hold for symmetric tops as well, along with an additional selection rule on ΔK. The dipole moment operator μ_0 is expressed as in Equation 8.43, where the angles θ and φ orient the symmetry axis of the molecule, along which the dipole moment must lie, if it exists. The selection rules are straightforward to derive if we exploit the properties of the Wigner rotation functions, which are the symmetric top eigenfunctions. The equation that we need here is

$$\int d\Omega D_{m_3n_3}^{l_3}(\Omega)^* D_{m_2n_2}^{l_2}(\Omega)D_{m_1n_1}^{l_1}(\Omega) = \frac{8\pi^2}{2l_3+1}C(l_1l_2l_3;m_1m_2m_3)C(l_1l_2l_3;n_1n_2n_3) \tag{8.50}$$

The right-hand side of Equation 8.50 is a product of two Clebsch–Gordan (CG) coefficients (Appendix A), $C(l_1l_2l_3; m_1m_2m_3)$. As discussed in Chapter 7, the properties of the CG coefficients form the basis for rules for adding two quantum mechanical vectors, such as spin and orbital angular momentum. The triangle rule results from the fact that $C(l_1l_2l_3; m_1m_2m_3)$ vanishes unless l_3 takes on one of the values $l_1 + l_2, l_1 + l_2 - 1,...,$ $|l_1 - l_2|$. The coefficient is also zero unless $m_1 + m_2 = m_3$, which just means that the z component of a vector sum is the sum of the z components of the vectors added. To use Equation 8.50 to derive selection rules, we will use the trick of expressing the components of the dipole moment vector as Wigner functions. To get the matrix element of μ_z connecting two states, we use the fact that $\cos\theta = D_{00}^1$. The matrix elements of μ_X and μ_Y can be set up by recognizing that both $\sin\theta\cos\varphi$ and $\sin\theta\sin\varphi$ are linear combinations of D_{10}^1 and D_{-10}^1. The result is that the selection rules for the transition $J, K, M \leftrightarrow J'K'M'$ are based on the condition:

$$\int d\Omega D_{M'K'}^{J'}(\Omega)^* D_{N0}^1(\Omega)D_{MK}^{J}(\Omega) \neq 0 \tag{8.51}$$

where $N = 0$ for the Z component and ± 1 for X and Y. Comparing this to Equation 8.50 and using what is known about the CG coefficients, the selection rules are found to be

$$\mu_0 \neq 0 \quad \Delta J = 0, \pm 1 \quad \Delta M = 0, \pm 1 \quad \Delta K = 0 \tag{8.52}$$

The rule for the change in J results from the triangle condition, $J' = J + 1, J, J - 1$, where all three values of J' can result only when $J \geq 1$. The coefficient $C(j_1j_2j_3;000)$ vanishes whenever $j_1 + j_2 + j_3$ is odd. This leads to the caveat that $\Delta J = 0$ is forbidden when $J = 0$. This is of no consequence in pure rotational spectroscopy but it matters when vibration–rotation spectra are considered. As in the case of linear molecules, the Z component of the dipole operator gives $\Delta M = 0$ and the X and Y components permit $\Delta M = \pm 1$. Note that for linear molecules we can put $K = 0$ in the above analysis and recover the previously described selection rules. However, when there is electronic angular momentum (as in open-shell diatomics), then the angular momentum quantum number K is not zero, and transitions having $\Delta J = 0$ are allowed. This will be illustrated by vibration–rotation spectra to be presented in Chapter 9.

The spectra of asymmetric rotors are more complicated than those of symmetric ones. The selection rules on J and M are still valid, but K is no longer a good quantum number. The notation $J_{K_{-1}K_1}$ is used to label the energy levels of asymmetric rotors, where K_{-1} and K_1 are the limiting values of K as the asymmetry parameter κ goes to -1 and 1, respectively. (See Figure 8.5.) When the direction of the dipole moment coincides with one of the inertial axes (as in H_2O for example), selection rules on ΔK_{-1} and ΔK_1 result. When the dipole moment has projections onto more than one inertial axis, the selection rules are more liberal and the spectra are more complex. See [4] for a discussion.

8.6 ROTATIONAL RAMAN SPECTROSCOPY

Raman spectroscopy is another approach to the measurement of pure rotational spectra. The Raman intensity depends on the induced electric dipole moment, and since all molecules are polarizable, it might be reasonable to conclude that all molecules are active in rotational Raman. As we shall see, however, what is required is anisotropic polarizability. The recalcitrant spherical tops do not show pure rotational scattering or absorption spectra, although rotational structure can be observed, in some cases, as fine structure in vibrational Raman and infrared spectra, to be discussed in the next chapter. For now, we concentrate on the rotational Raman spectra of linear molecules and symmetric tops, both of which can be classified as axially symmetric molecules as far as polarizability is concerned. We want to know the selection rules for rotational transitions, the appearance of the spectra, and so forth. To a first approximation, rotational Raman spectra, like rotational absorption and emission spectra, consist of a series of equally spaced lines, but is the frequency spacing $2B$ or not $2B$?

The polarizability, like the dipole moment, is a function of molecular geometry and, as such, can be expanded in a Taylor series about the equilibrium geometry. For rotational transitions, we only need the first term, the polarizability of the molecule in its equilibrium geometry, which is a diagonal tensor in the frame of the molecule. Except in cases of accidental symmetry, the form of α reflects the same symmetry as the inertial tensor. For example, spherical tops have isotropic polarizability; $\alpha_{xx} = \alpha_{yy} = \alpha_{zz}$. Both symmetric tops and linear molecules have axially symmetric polarizability: $\alpha_{xx} = \alpha_{yy} \neq \alpha_{zz}$. As usual, asymmetric tops, for which $\alpha_{xx} \neq \alpha_{yy} \neq \alpha_{zz}$, are more difficult to treat. The observed light scattering depends indirectly on the molecule frame polarizability, and directly on the lab-frame induced moments. Just as the molecular dipole moment was projected onto the X, Y, and Z directions in the laboratory, the molecular polarizability must also be projected onto the lab frame.

The selection rules for rotational Raman can be derived by considering the Kramers–Heisenberg–Dirac expression for an element of the polarizability tensor, $(\alpha_{\rho\sigma})_{if}$. The notation at hand specifies a particular element of the tensor, where ρ and σ are Cartesian components, in the lab frame, and i and f as usual specify the initial and final states. The quantum mechanical expression (the KHD formula) is

$$(\alpha_{\rho\sigma})_{if} = \frac{1}{\hbar} \sum_n \left\{ \frac{\langle i | \mu_\rho | n \rangle \langle n | \mu_\sigma | f \rangle}{\omega + \omega_{nf}} - \frac{\langle i | \mu_\sigma | n \rangle \langle n | \mu_\rho | f \rangle}{\omega - \omega_{ni}} \right\} \tag{8.53}$$

The initial, intermediate, and final states are specified as follows:

$$|i\rangle \equiv |evJKM\rangle, \quad |n\rangle \equiv |e'v'J'K'M'\rangle, \quad |f\rangle \equiv |evJ''K''M''\rangle \tag{8.54}$$

We let $K = 0$ when linear molecules are considered. The sum over intermediate states includes all possible electronic (e'), vibrational (v'), and rotational ($J'K'M'$) levels. Looking more closely at one of the transition dipoles, let us consider first taking the matrix element with respect to the electronic states, for example,

$$\langle i | \mu_\rho | n \rangle = \langle vJKM | (\mu_\rho)_{ee'} | v'J'K'M' \rangle \tag{8.55}$$

The electronic transition moment $(\mu_\rho)_{ee'}$, like the permanent dipole moment responsible for microwave absorption, is considered to be evaluated at the equilibrium geometry of the ground electronic state. (This is called the Condon approximation, and we will elaborate on it in Chapters 11 and 12.) So now we see how a homonuclear diatomic can be Raman active, through the transition dipoles for allowed electronic transitions. A linear or symmetric top molecule, whether polar or nonpolar, has two kinds of allowed excited electronic states: nondegenerate ones polarized along the bond (or symmetry axis) and doubly degenerate ones polarized perpendicular to the bond. The electronic transition moment is projected onto the lab frame to get the components $(\mu_X)_{ee'}$, $(\mu_Y)_{ee'}$, and $(\mu_Z)_{ee'}$. Therefore, the matrix elements of the electronic transition dipole, connecting initial rotational states with intermediate rotational states, or intermediate with final states, are the same ones previously considered in the analysis of microwave selection rules. To get an allowed rotational Raman transition, $JMK \rightarrow J''M''K''$, the following must hold:

$$\left\langle D_{MK}^{J*} D_{N0}^1 D_{M'K'}^{J'} \right\rangle \left\langle D_{M'K'}^{J'*} D_{N'0}^1 D_{M''K''}^{J''} \right\rangle \neq 0 \tag{8.56}$$

where the angle brackets represent integration over the Euler angles, and N and N' can be 0 or ± 1 depending on which components of the lab frame polarizability are under consideration.

Since $\Delta J = 0, \pm 1$ to get nonzero values of each of these matrix elements, the net selection rule for Raman scattering is $\Delta J = 0, \pm 1, \pm 2$, except that $\Delta J = \pm 1$ is forbidden when $K = 0$ or when $J = 0$. Recall that $C(j_1 j_2 j_3; 000)$ is zero whenever $j_1 + j_2 + j_3$ is odd. Thus $\Delta J = \pm 1$ is forbidden for linear molecules, which have $K = 0$. For symmetric tops, we also obtain the rule $\Delta K = 0$, as we did for microwave absorption. The Stokes side of the spectrum is associated with scattered light shifted to lower frequency than that of the incident radiation, due to transitions having positive ΔJ, while the anti-Stokes transitions have negative ΔJ and lead to positive frequency shifts. The result is a series of lines, on either side of the excitation frequency, referred to as the R and S branches, for which the frequency shift of the scattered light is

$$\text{for } \Delta J = \pm 1, \ \Delta \nu_R = \mp 2B(J+1), \ J = 1,2,3,\ldots \tag{8.57}$$

$$\text{for } \Delta J = \pm 2, \ \Delta \nu_S = \mp 2B(2J+3), \ J = 0,1,2,\ldots \tag{8.58}$$

The quantum number J in Equations 8.57 and 8.58 is the lower value. Linear molecules lack transitions having $\Delta J = \pm 1$. For rigid symmetric tops, the S series consists of equally spaced lines at frequency shifts of $6B$, $10B$, $14B$,... and the R series consists of lines at $4B$, $6B$, $8B$, $10B$,.... The coincidence of every other R branch line with an S branch line leads to an intensity alternation that is not a consequence of nuclear statistics. (For a discussion of nuclear statistics in symmetric tops, see [3] or [4].)

The Raman spectra of homonuclear diatomics, on the other hand, are subject to the constraints of nuclear exchange symmetry. The pure rotational Raman spectrum of N_2 ($\tilde{B}_e = 1.998\,\text{cm}^{-1}$) is shown in Figure 8.10. The nuclear spin of ^{14}N is $I = 1$, so Bose–Einstein statistics apply to nuclear exchange. Following the procedures of Section 8.4, we find that the even J states carry twice the statistical weight of odd J states, resulting in the observed alternating intensities of the Raman spectrum. The frequency separation of the first Stokes and anti-Stokes transitions, on either side of the excitation line, can be compared to that of adjacent lines in the Stokes or anti-Stokes branch. This ratio takes on the value $12B/4B = 3$ if both even and odd J states exist, the value $20B/8B = 5/2$ in the case where only odd J states exist, and the value $12B/8B = 3/2$ when only even J states exist. Thus the rotational Raman spectrum is diagnostic of nuclear exchange symmetry.

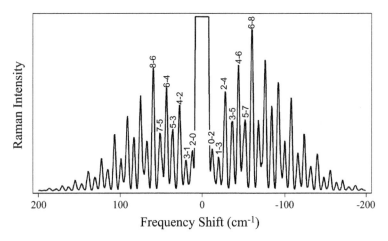

Figure 8.10 Rotational Raman spectrum of N_2.

What does the preceding analysis have to say about spherical tops? They have allowed electronic transitions, which could permit Raman scattering. The problem is that the allowed electronic states of octahedral and tetrahedral molecules belong to triply degenerate representations. This means that the x, y, and z (molecule frame) components of $\mu_{ee'}$ are all equal, so there is no dependence of the transition dipole on the orientation in the lab frame. So pure rotational scattering is forbidden. In other words, their polarizability ellipsoids are spherical, so the induced dipole moment is independent of molecular rotation. In vibrational Raman scattering of nontotally symmetric modes, rotational fine structure is possible due to the symmetry lowering induced by the vibrational motion.

There is an alternative way to look at rotational Raman spectra, and it is one that is more convenient for discussing scattering by liquids, where the quantum numbers J, K, and M no longer apply. Let us assume that all we know is the form of the molecular polarizability, and recall the two experimental arrangements that permit the measurement of polarized and depolarized scattering (see Figure 6.5). The incident light is imagined to propagate along the X direction with polarization vector in the Z direction. Scattered light is viewed in the Y direction, with the electric field vector pointing either in the Z direction (the polarized or VV spectrum) or in the X direction (the depolarized or VH spectrum). In the time-domain picture (Chapter 5), these two spectra are Fourier transforms of time-correlation functions involving the appropriate lab-frame components of the polarizability:

$$I_{VV} \propto \int_{-\infty}^{\infty} dt\, e^{-i\omega t} \left\langle \alpha_{ZZ}^{*} \alpha_{ZZ}(t) \right\rangle \tag{8.59}$$

$$I_{VH} \propto \int_{-\infty}^{\infty} dt\, e^{-i\omega t} \left\langle \alpha_{XZ}^{*} \alpha_{XZ}(t) \right\rangle \tag{8.60}$$

The problem is to relate the lab-frame polarizabilities, α_{ZZ} and α_{XZ}, to the molecule-frame components, α_{xx}, α_{yy}, and α_{zz}. The orientation dependence of this relationship contributes to the time dependence of the correlation functions. One could use direction cosines to project the polarizability onto the lab frame, but it is much more elegant to use spherical rather than Cartesian polarizability tensors and then take advantage of their transformation properties. (See Appendix A for a general discussion of spherical tensor transformations.) Spherical tensor elements of the polarizability are expressed as α_M^J, where M ranges from J to $-J$ in integral steps. The equations for converting Cartesian tensor components to spherical components [6] are given in Table 8.2. Although we have used uppercase subscripts for the Cartesian tensor components, the same expressions connect molecule frame Cartesian and spherical tensors. However, the off-diagonal

Table 8.2 Formulas for spherical tensor components of the polarizability

$$\alpha_0^0 = \frac{1}{\sqrt{3}}\left[\alpha_{XX} + \alpha_{YY} + \alpha_{ZZ}\right]$$

$$\alpha_0^1 = \frac{1}{2}\left[\alpha_{XY} - \alpha_{YX}\right]$$

$$\alpha_{\pm 1}^1 = \frac{\pm 1}{2\sqrt{2}}\left[\left(\alpha_{YZ} - \alpha_{ZY}\right) \pm i\left(\alpha_{ZX} - \alpha_{XZ}\right)\right]$$

$$\alpha_0^2 = \frac{1}{\sqrt{6}}\left[3\alpha_{ZZ} - \left(\alpha_{XX} + \alpha_{YY} + \alpha_{ZZ}\right)\right]$$

$$\alpha_{\pm 1}^2 = \pm\frac{1}{2}\left[\left(\alpha_{ZX} + \alpha_{XZ}\right) \pm i\left(\alpha_{ZY} + \alpha_{YZ}\right)\right]$$

$$\alpha_{\pm 2}^2 = \frac{1}{2}\left[\left(\alpha_{XX} - \alpha_{YY}\right) \pm i\left(\alpha_{XY} + \alpha_{YX}\right)\right]$$

molecule-frame elements such as α_{xy}, α_{yz}, etc., are all zero, in the case of a pure rotational transition. The reason for this is that the polarizability is that of the nonvibrating molecule, and the tensor thus reflects the molecular symmetry.

Consider first the polarized spectrum, for which the required tensor component is

$$\alpha_{ZZ} = \frac{1}{\sqrt{3}}\alpha_0^0 + \sqrt{\frac{2}{3}}\alpha_0^2 \tag{8.61}$$

The lab-frame values of α_M^J, designated by $\alpha_M^J(L)$, are functions of the orientation, as follows:

$$\alpha_M^J(L) = \sum_{M'} \alpha_{M'}^J(B) D_{M'M}^J(\Omega) \tag{8.62}$$

where $\alpha_M^J(B)$ is the body (molecule) frame polarizability and the Wigner rotation functions depend on the orientation Ω of the body with respect to the lab. The value of $\alpha_M^J(B)$ is constant; it is the time dependence of the orientation which is of interest. Combining Equations 8.61 and 8.62 and introducing the time dependence of Ω, we have

$$\alpha_{ZZ}(L,t) = \frac{1}{\sqrt{3}}\alpha_0^0(B) + \sqrt{\frac{2}{3}}\sum_M \alpha_M^2(B) D_{M0}^2(\Omega_t) \tag{8.63}$$

The correlation function that we need is thus

$$\left\langle \alpha_{ZZ}^* \alpha_{ZZ}(t) \right\rangle = \frac{1}{3}\left|\alpha_0^0\right|^2 + \frac{2}{3}\sum_{MM'} \alpha_M^2(B)^* \alpha_{M'}^2(B)\left\langle D_{M0}^2(\Omega_0)^* D_{M'0}^2(\Omega_t)\right\rangle \tag{8.64}$$

Note that $D_{00}^0 = 1$. The absence of cross-terms proportional to $\alpha_0^0 \alpha_M^2$ is due to the fact that $\left\langle D_{MN}^J(\Omega_t)\right\rangle = 0$, except when $J = M = N = 0$. Equation 8.64 states that the polarized light scattering depends in part on the isotropic polarizability (through α_0^0) and in part on the anisotropy of the polarizability (through α_M^2). Our next task is to recast the above correlation function in terms of the change in orientation as a function of time, $\delta\Omega_t$. This is done with the help of the addition theorem for Wigner functions of successive rotations. If the Euler angle Ω_t is viewed as the result of starting at the initial orientation Ω_0 and rotating through $\delta\Omega_t$, then, using the addition theorem of Appendix A, we have

$$D_{M'0}^2(\Omega_t) = \sum_N D_{M'N}^2(\delta\Omega_t) D_{N0}^2(\Omega_0) \tag{8.65}$$

This expression would seem to complicate things when it is inserted into the previous version of the correlation function. But there is a trick, of course. In a collection of randomly oriented molecules, the initial orientation Ω_0 can be averaged over all possible values. The average is performed by integrating the function of Ω_0 over all Euler angle space and dividing by $\int d\Omega = 8\pi^2$. Then the orthogonality of the rotation functions,

$$\frac{1}{8\pi^2}\int D_{M0}^2(\Omega_0)^* D_{N0}^2(\Omega_0) d\Omega_0 = \frac{1}{5}\delta_{MN} \tag{8.66}$$

results in our final version of the correlation function for VV scattering:

$$\begin{aligned} C_{VV}(t) &\equiv \left\langle \alpha_{ZZ}^* \alpha_{ZZ}(t)\right\rangle \\ &= \frac{1}{3}\left|\alpha_0^0\right|^2 + \frac{2}{15}\sum_{MM'} \alpha_M^2(B)^* \alpha_{M'}^2(B)\left\langle D_{M'M}^2(\delta\Omega_t)\right\rangle \end{aligned} \tag{8.67}$$

Equation 8.67 is quite general. If we limit consideration to axially symmetric molecules, it is further simplified:

$$C_{VV}(t) = \frac{1}{3}\left|\alpha_0^0\right|^2 + \frac{2}{15}\left|\alpha_0^2\right|^2 \left\langle D_{00}^2(\delta\Omega_t)\right\rangle \qquad (8.68)$$

The isotropic part, the first term in Equation 8.67 or 8.68, is independent of orientation, so it does not contribute to rotational Raman scattering. Turning next to the depolarized scattering, we require the lab-frame component $\alpha_{XZ} = 1/2(\alpha_{+1}^2 - \alpha_{-1}^2)$. Proceeding as we did above:

$$\alpha_{XZ}(t) = \frac{1}{2}\sum_M \alpha_M^2(B)\left[D_{M1}^2(\Omega_t) - D_{M\bar{1}}^2(\Omega_t)\right] \qquad (8.69)$$

(Note that the bar over the subscript integer specifies a negative value.) The correlation function is then obtained:

$$\begin{aligned}
\left\langle \alpha_{XZ}^* \alpha_{XZ}(t)\right\rangle = \frac{1}{4}\sum_{MM'} \alpha_M^2(B)^* \alpha_{M'}^2(B) &\times \left\{\left\langle D_{M1}^2(\Omega_0)^* D_{M'1}^2(\Omega_t)\right\rangle\right. \\
&- \left\langle D_{M1}^2(\Omega_0)^* D_{M'\bar{1}}^2(\Omega_t)\right\rangle + \left\langle D_{M\bar{1}}^2(\Omega_0)^* D_{M'\bar{1}}^2(\Omega_t)\right\rangle \\
&\left.- \left\langle D_{M\bar{1}}^2(\Omega_0)^* D_{M'1}^2(\Omega_t)\right\rangle\right\}
\end{aligned} \qquad (8.70)$$

This is considerably simplified using our previous tricks. The result is

$$\begin{aligned}
C_{VH}(t) &\equiv \left\langle \alpha_{XZ}^* \alpha_{XZ}(t)\right\rangle \\
&= \frac{1}{10}\sum_{MM'} \alpha_M^{2*}(B)\alpha_{M'}^2(B)\left\langle D_{M'M}^2(\delta\Omega_t)\right\rangle
\end{aligned} \qquad (8.71)$$

And, for axially symmetric molecules,

$$C_{VH}(t) = \frac{1}{10}\left|\alpha_0^2(B)\right|^2 \left\langle D_{00}^2(\delta\Omega_t)\right\rangle \qquad (8.72)$$

The depolarized spectrum is a function of the anisotropy of the polarizability, which vanishes in the case that α is spherically symmetric. We again conclude that pure rotational transitions of spherical tops are inactive in light scattering. We have further shown that there is a rotationally invariant contribution to the scattering, which survives even for spherical tops, called the isotropic scattering. This orientation independent part of the polarizability contributes to quasi-elastic light scattering. Referring to the depolarized scattering as the anisotropic part, we have derived two of the key equations for light scattering:

$$I_{VV} = I_{iso} + \frac{4}{3}I_{anis} \qquad (8.73)$$

$$I_{VH} = I_{anis} \qquad (8.74)$$

These expressions follow from the fact that the integrated intensity is equal to the correlation function at zero time; they are obtained by putting $t = 0$ in Equations 8.67 and 8.71. Equations 8.73 and 8.74 also hold true in subsequent discussions of vibrational Raman scattering. The practical consequences of these results are of great value in the study of condensed phases. The measurement of both VV and VH spectra enables the reorientational contribution to be extracted, as discussed in Chapter 5. Fourier transformation of the aniso-tropic spectrum for a symmetric top molecule results in

$$\left\langle D_{00}^2(\delta\Omega_t) \right\rangle = \left\langle P_2(\cos\theta_t) \right\rangle = \frac{3}{2}\left\langle \cos^2\theta_t \right\rangle - \frac{1}{2} \tag{8.75}$$

8.7 CORRECTIONS TO THE RIGID-ROTOR APPROXIMATION

There are two physical reasons why real molecules cannot be true rigid rotors. The first is that vibrational motion alters the geometry and thus the inertia. Second, the rotational motion itself tends to fling the nuclei apart and change the average geometry. In this section, some of the steps of making corrections to the harmonic oscillator-rigid rotor approximation are discussed for diatomic molecules, and briefly generalized to the case of symmetric tops. In order to treat the problem at hand, we need to consider vibrational and rotational energy at the same time. The vibrational problem will be considered in more detail in the next chapter.

The harmonic oscillator approach to vibrational motion results from approximating the potential function as a parabola, $V(q) = 1/2\,kq^2$, where $q = R - R_e$ and k is the force constant. Adding the vibrational and rotational energies together gives the zero-order approximation to the vibration–rotation energy of a diatomic:

$$E_{vJ}^{(0)} = V(R_e) + h\nu_e\left(v + \frac{1}{2}\right) + hB_e J(J+1) \tag{8.76}$$

$V(R_e)$ is the energy at the bottom of the well, which can be taken to be zero for convenience. The rotational constant $B_e = h/(8\pi^2\mu R_e^2)$ applies to a molecule having a fixed internuclear distance, R_e. The quadratic form of the potential energy results from truncating a Taylor series expansion of $V(R)$ about R_e:

$$V(R - R_e) = V(R_e) + \frac{1}{2}V''(R_e)(R - R_e)^2 + \frac{1}{3!}V'''(R_e)(R - R_e)^3 + \cdots \tag{8.77}$$

Note the first derivative term is absent because the slope is zero at the minimum in $V(R)$. The second derivative term,

$$V''(R_e) = \left(\frac{d^2V}{dR^2}\right)_{R_e} = k \tag{8.78}$$

is the force constant k. Perturbation theory calculations that include the third, fourth, etc. derivatives make corrections for anharmonicity. These will be considered further in Chapter 9. To go beyond the rigid-rotor approximation, we also need to correct the rotational kinetic energy operator (Equation 8.10) by allowing the internuclear distance R to be flexible:

$$\hat{H}_{rot} = \frac{\hat{L}^2}{2R^2} = \frac{\hat{L}^2}{2\mu(R_e + q)^2} \tag{8.79}$$

For small amplitude vibrational motion, $q \ll R_e$, Equation 8.79 can be expanded about R_e:

$$\frac{1}{(R_e + q)^2} = \frac{1}{R_e^2}\left[1 - \frac{2q}{R_e} + \frac{3q^2}{R_e^2} - \cdots\right] \tag{8.80}$$

The details of calculating the perturbed vibration–rotation energy are tedious, so we cut to the result. (See [7] or [8] for details.) The energy, corrected to second order, is

$$E_{vJ} \approx E_{vJ}^{(0)} + E_{vJ}^{(1)} + E_{vJ}^{(2)}$$

$$= V(R_e) + h\nu_e\left(v + \frac{1}{2}\right) + hB_e J(J+1) \tag{8.81}$$

$$- h\nu_e x_e\left(v + \frac{1}{2}\right)^2 - h\alpha_e\left(v + \frac{1}{2}\right)J(J+1) - h\bar{D}_e J^2(J+1)^2$$

Let us explore the physical significance of the correction terms. The first correction term, $h\nu_e x_e(v + 1/2)^2$, allows for anharmonicity. The value of x_e is a function of the third and fourth derivatives of the potential. It is generally positive, and, if perturbation theory is justified, small compared to one. It has the effect of causing the separation of vibrational energy levels to decrease as vibrational energy increases, as they should since the vibrational energy levels of a real molecule must converge as the dissociation limit is approached. The next correction allows for vibration–rotation coupling. It is easiest to see the effect of this coupling by defining a new rotational constant which depends on the vibrational state:

$$B_v = B_e - \alpha_e\left(v + \frac{1}{2}\right) \tag{8.82}$$

The rotational energy (neglecting centrifugal distortion) is $E_{rot} = B_v J(J+1)$, where B_v is a function of the average value of $1/R^2$ in vibrational state v:

$$B_v = \frac{h}{8\pi^2\mu}\langle v|R^{-2}|v\rangle \tag{8.83}$$

Since the vibration–rotation coupling constant α_e is generally positive, the rotational constant B_v decreases with increasing vibrational energy due to the increase in average internuclear distance and thus increase in inertia.

The final term in Equation 8.81 allows for the effect of centrifugal distortion, the tendency of the internuclear distance to increase as the rotational velocity increases. The centrifugal distortion constant is given by

$$\bar{D}_e = \frac{4B_e^3}{\nu_e^2} \tag{8.84}$$

Since rotational energies are typically small compared to vibrational energies, this is a small correction. Nevertheless, it has the observable effect of causing the separation of adjacent lines in microwave spectra to decrease at higher values of J. For pure rotational absorption, where $J \rightarrow J+1$, the corrected transition frequencies are found from $\nu_{J\rightarrow J+1} = (E_{v,J+1} - E_{v,J})/h$:

$$\nu_{J\rightarrow J+1} = 2B_v(J+1) - 4\bar{D}_e(J+1)^3 \tag{8.85}$$

Typical centrifugal distortion constants are about 10^{-4} the size of the rotational constant. The effect on the spacings of pure rotational transitions is thus rather small. The influence of vibration–rotation coupling is more significant, but it is not observed in pure rotational transitions, except to modify the rotational constant. We will explore the effect of vibration–rotation coupling in Chapter 9.

Corrections to the rigid rotor-harmonic oscillator model are more complex for larger molecules. For example, the vibrationally averaged effective rotational constant depends on all $3N - 6$ normal modes:

$$B_{[v]} = B_0 - \sum_{i=1}^{3N-6} \alpha_i^B \left(v_i + \frac{1}{2} \right) \tag{8.86}$$

where [v] represents a set of vibrational quantum numbers $[v_1, v_2, \ldots, v_{3N-6}]$ defining the vibrational state. There are also vibration–rotation coupling constants α_i^A and α_i^C for making corrections to the A and C rotational constants. Similarly, there is an anharmonicity correction factor x_i for each normal mode, as well as coupling terms which allow for the interaction of normal modes due to anharmonicity. For a symmetric top, centrifugal distortion depends on the quantum numbers J and K, leading to this expression for the rotational energy of an oblate top [8]:

$$\frac{E_{vJK}}{h} = B_{[v]}J(J+1) - \left(C_{[v]} - B_{[v]} \right)K^2 - D_J J^2 (J+1)^2$$

$$- D_{JK}J(J+1)K^2 - D_K K^4 \tag{8.87}$$

8.8 INTERNAL ROTATION

In addition to the effects described in the previous section, many molecules deviate from the rigid rotor model by virtue of their fluxional behavior. These "floppy molecules" have more than one stable conformation, separated by energy barriers which are not insurmountable. The possibility of multiple conformations increases with molecular size, but even small molecules can display these effects.

Consider a molecule such as ethane, which can undergo hindered rotation about the C–C single bond. This torsional motion is actually one of the $3N - 6 = 18$ normal modes of the molecule and, as such, we could postpone discussion of internal rotation until Chapter 10. We prefer to discuss hindered rotation in this chapter because the frequency of this type of motion can be quite low compared to the frequency of other normal modes, and indeed it can sometimes be observed in the microwave spectrum.

Figure 8.11 pictures two conformations of a molecule such as ethane, staggered and eclipsed, for which the potential energy as a function of torsional angle is respectively a minimum or maximum. Let us define the torsional angle as φ, where $\varphi = 0$ corresponds to the staggered and $\varphi = \pi$ to the eclipsed conformation. One mathematical form of the potential energy which has the correct threefold symmetry is the following:

$$V(\varphi) = \frac{1}{2} V_0 (1 - \cos n\varphi) \tag{8.88}$$

where $n = 3$ in this case but more generally is equal to the number of equivalent configurations achieved during one full rotation about the bond. The form of this function is sketched in Figure 8.12. V_0 is the barrier height, that is, the energy of the eclipsed configuration relative to the staggered. Equation 8.88 is really only the lead term in a Fourier series expansion of the n-fold potential, and more exact treatments of hindered rotation

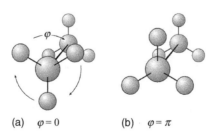

(a) $\varphi = 0$ (b) $\varphi = \pi$

Figure 8.11 (a) Staggered and (b) eclipsed geometries.

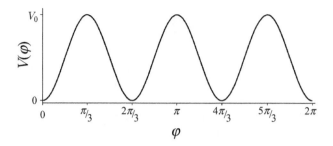

Figure 8.12 Potential energy for internal rotation.

would include more terms in the series. We will use the simple form of $V(\varphi)$ in order to examine the physical aspects of the problem. Depending on the height of the barrier compared to the thermal energy k_BT, internal rotational motion ranges from freely spinning in the low-barrier limit to harmonic torsional oscillation, in the high-barrier limit. We solve the Schrödinger equation in these two limiting cases. The Hamiltonian for torsion about a bond is

$$\hat{H}_{tor} = \frac{-\hbar^2}{2I_r}\frac{\partial^2}{\partial\varphi^2} + V(\varphi) \tag{8.89}$$

where the reduced moment of inertia about the bond is a function of the inertia of the two attached groups:

$$I_r = \frac{I_1 I_2}{I_1 + I_2} \tag{8.90}$$

8.8.1 FREE ROTATION LIMIT, $K_BT \gg V_0$

When the thermal energy greatly exceeds the barrier to rotation, the molecule barely sees the ripples in the bottom of the potential well. We might as well replace the true potential function by a constant, say the average value, $V(\varphi) \approx V_0/2$. The Schrödinger equation then takes the form:

$$\frac{\partial^2\psi}{\partial\varphi^2} = \frac{-8\pi^2 I_r}{h^2}\left(E - \frac{1}{2}V_0\right)\psi \equiv -k^2\psi \tag{8.91}$$

The solution to Equation 8.91 is

$$\psi(\varphi) = \frac{1}{\sqrt{2\pi}}e^{ik\varphi} \tag{8.92}$$

The boundary condition is $\psi(\varphi+2\pi) = \psi(\varphi)$, which restricts k to integer values. The energy levels are therefore,

$$E_k = \frac{k^2 h^2}{8\pi^2 I_r} + \frac{1}{2}V_0, \quad k = 0, \pm1, \pm2,\ldots \tag{8.93}$$

Free internal rotation results in a series of doubly degenerate (except for $k = 0$) energy levels. Equation 8.93 is not valid unless the first term is much larger than the second. Notice how the energy levels resemble those of a free rotor, since for large J, $J(J+1) \approx J^2$, but the degeneracy is not the same since the internal rotor has only one degree of freedom. As you might expect, this rotational motion is microwave active only if the internal rotation changes the direction of the dipole moment.

8.8.2 HARMONIC OSCILLATOR LIMIT, $K_B T \ll V_0$

When the thermal energy is small compared to the barrier height, the system finds itself close to the bottom of one of the wells pictured in Figure 8.12 and undergoes only small excursions from the angle for which the energy is a minimum. This permits the potential function to be expanded about $\varphi = 0$. Taking cos $n\varphi \approx 1 - n^2\varphi^2/2$, the Schrödinger equation takes the form:

$$\frac{-\hbar^2}{2I_r} \frac{\partial^2 \psi}{\partial^2 \varphi} + \frac{n^2 V_0}{4} \varphi^2 \psi = E\psi \tag{8.94}$$

Equation 8.94 is identical to the Schrödinger equation for a harmonic oscillator; the usual variables have merely been renamed. The angle φ is analogous to the coordinate q, the reduced inertia replaces the usual reduced mass, and the force constant, ordinarily given by $k = 4\pi^2 \mu \nu^2$, is in this case equal to $n^2 V_0/2$. Thus the harmonic frequency of torsional motion is

$$\nu = \frac{n}{2\pi} \left(\frac{V_0}{2I_r} \right)^{1/2} \tag{8.95}$$

and the energy levels are $E_v = h\nu(v + 1/2)$ with v = 0, 1, 2, ..., ∞. In this infinite barrier limit, the groups attached to the bond undergo harmonic torsional oscillations of small amplitude, and one would expect to observe transitions in the far-IR or Raman scattering spectrum. Measurement of the frequency of torsional oscillation permits the barrier height to be calculated. For example, the torsional frequency of $Cl_3C–CCl_3$ is 68 cm^{-1}, implying a 45 kJ/mol rotational barrier. This barrier should be compared to room temperature thermal energy, about 2.5 kJ/mol.

There is of course no such thing as an infinite barrier, and we must admit the possibility that a given torsional angle of, say, a threefold symmetric molecule could find the system in any of the three equivalent potential wells. Let us refer to the wavefunctions centered on each of these wells as ψ_1, ψ_2, and ψ_3. If these wavefunctions did not interact, the v = 0 state would be triply degenerate, and we could picture three Gaussian wavefunctions, one in each well. But harmonic oscillator wavefunctions tunnel into the classically forbidden area outside the potential well, and this enables the three functions to interact with one another. The correct wavefunctions are linear combinations of the three localized ones:

$$\Psi_I = \frac{1}{\sqrt{3}} \left(\psi_1 + \psi_2 + \psi_3 \right)$$

$$\Psi_{II} = \frac{1}{\sqrt{3}} \left(\psi_1 + e^{2\pi i/3} \psi_2 + e^{4\pi i/3} \psi_3 \right) \tag{8.96}$$

$$\Psi_{III} = \frac{1}{\sqrt{3}} \left(\psi_1 + e^{4\pi i/3} \psi_2 + e^{2\pi i/3} \psi_3 \right)$$

The tunneling interaction partially lifts the degeneracy and causes the torsional oscillator levels to split into one nondegenerate and one doubly degenerate level.

A classic example of tunneling splitting is provided by the inversion of ammonia. Although this is not a case of rotation about a bond, the quantum mechanical considerations are similar. Inversion in this case refers to the ammonia molecule turning inside out, going through a planar geometry at the transition state. The potential function for this motion is pictured in Figure 8.13. The barrier to inversion is 24 kJ/mol, or about 2000 cm^{-1}. This represents the energy of trigonal planar ammonia (0 in Figure 8.13) relative to either of the two equivalent pyramidal forms (I and II). This barrier is not huge compared to $k_B T/hc \approx 200$ cm^{-1} at room temperature, so the tunneling splitting is considerable. For each zero-order state with quantum number v,

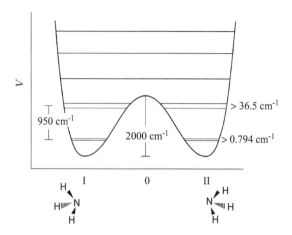

Figure 8.13 Inversion potential of ammonia.

the tunneling interaction lifts the twofold degeneracy. The new wavefunctions for these split states are symmetric and antisymmetric combinations of the two localized wavefunctions, $\Psi_s = \psi_I + \psi_{II}$ and $\Psi_a = \psi_I - \psi_{II}$. The tunneling splitting increases with increasing energy, until, as you might expect, a series of equally spaced levels results at high energy. Well above the barrier, the ammonia molecule interconverts between the two forms and is in fact planar on the average (Figure 8.13).

The tunneling splitting of the v = 0 level is 0.794 cm^{-1}, or 2.38 × 10^{10} Hz, in the microwave region of the spectrum. It represents the frequency of interconversion of the two forms. The first maser (microwave laser) was based on transitions between these tunneling states.

8.9 SUMMARY

We have examined the rotational transitions of molecules in absorption and scattering. For linear and symmetric top molecules, the rigid-rotor approximation and the selection rules result in pure rotational spectra consisting of a series of equally spaced lines. The relative intensities of these lines are dominated by the Boltzmann populations of the initial states, but also reflect the J-dependence of the transition moment and transition frequency. The rotational spectra of asymmetric rotors are more complex, but can be interpreted using perturbation theory. Numerous perturbations may influence rotational spectra. Electronic spin and/ or orbital angular momentum can couple to the rotational angular momentum, leading to splittings not observed in closed-shell molecules. Exchange symmetry in molecules containing equivalent nuclei influences the statistical weights of even and odd J states. When the rigid rotor and harmonic oscillator approximations are relaxed, we find perturbations to the spacing of adjacent rotational transitions, resulting from the rotational- and vibrational-state dependence of the inertia. We have also seen that nonrigid molecules present the interesting possibility of internal rotation, and other potentially microwave active transitions such as the inversion tunneling in ammonia.

The considerations of this chapter carry over to the study of vibrational and electronic spectra, since rotational transitions contribute to the fine structure. To the extent that the rotational wavefunction can be separated from the electronic and vibrational wavefunctions, the selection rules derived in this chapter also apply to the analysis of this rotational substructure. The difference is in the gross selection rule. For example, in infrared absorption, the permanent dipole moment of microwave spectroscopy will be replaced by the dipole moment derivative with respect to the vibrational coordinate (the "normal coordinate" to be presented in Chapter 10). In electronic spectroscopy, the nonvanishing electronic transition dipole moment provides the basis for the gross selection rule. Fortunately, vibrational and electronic spectroscopy often permit the observation of rotational transitions of molecules for which pure rotational absorption or scattering is forbidden. We will keep this in mind as we proceed.

PROBLEMS

1. This problem illustrates the separation of internal and external motions for a one-dimensional rigid rotor. Consider a diatomic molecule consisting of masses m_1 and m_2 at positions x_1 and x_2. Using the notation $\dot{x} \equiv dx/dt$, show that the kinetic energy $T = 1/2\, m_1 \dot{x}_1^2 + 1/2\, m_2 \dot{x}_2^2$ can be expressed as $T = 1/2\, M\dot{x}_{cm}^2 + 1/2\, \mu \dot{x}^2$, where x_{cm} is the position of the center of mass, $x = x_2 - x_1$, M is the total mass, and μ is the reduced mass.

2. The rotational constant $\tilde{B}_e$ for $^{12}C^{16}O$ is 1.93127 cm^{-1} in the ground electronic state and 1.3099 cm^{-1} in the excited triplet electronic state. Calculate the bond length of CO in both the ground and excited electronic states.

3. Calculate the rotational constant for $^{13}C^{16}O$ in its ground electronic state.

4. The barrier to rotation about the C–C bond in CH_3CH_2Cl is about 15 to 20 kJ/mol. Estimate the torsional frequency and predict how it could be observed experimentally. (Consult a table of bond distances in order to estimate the moment of inertia.)

5. In this problem you will calculate the Stark effect on the rotational spectrum of a symmetric top molecule. The perturbation operator for a dipole μ_0 in an electric field E is $\hat{H}' = -\vec{\mu}_0 \cdot \vec{E}$. Take the direction of the field as Z, and note that the dipole moment direction coincides with that of the angular momentum vector K for spinning about the symmetry axis. Show that the first-order correction to the energy W of a rotational state is

$$W_{JKM}^{(1)} = \frac{-\mu_0 KM}{J(J+1)} E$$

   Make a sketch showing how the transition $J = 1$, $K = 1 \rightarrow J = 2$, $K = 1$ would be split by a 100 V/cm field, for a molecule having a dipole moment of 1 D.

6. The rotational constants B_0 have been determined for three isotopic derivatives of chloroacetylene: H–C≡C–^{37}Cl, 5572.3 MHz; D–C≡C–^{37}Cl, 5084.2 MHz; and H–C≡C–^{35}Cl, 5684.2 MHz. Assume that the carbon isotope is ^{12}C in all cases. Calculate the three bond distances in chloroacetylene. What approximations or assumptions do you have to make to work this problem with the information given?

7. The ground electronic state of O_2 is $^3\Sigma_g^-$, and the nuclear spin of ^{16}O is $I = 0$. The rotational constant $\tilde{B}_e$ is 1.4456 cm^{-1} and the vibration–rotation coupling constant α_e is 0.0158 cm^{-1}. Sketch the rotational Raman spectrum of O_2 in the ground vibrational state. Label the transitions with the initial and final rotational quantum numbers and indicate the separation of adjacent lines in cm^{-1}. Neglect centrifugal distortion and coupling of the rotational and spin angular momenta.

8. The rotational constants for HF are $\tilde{B}_e = 20.956$ cm^{-1}, $\tilde{\alpha}_e = 0.796$ cm^{-1}, and $\bar{D}_e/hc = 0.022$ cm^{-1}. (a) Find the initial J value and the frequency for the most intense rotational transition taking place in the ground vibrational state at room temperature. (b) What wavenumber accuracy would be required in order to discern the effect of centrifugal distortion in the vicinity of this transition? Repeat the problem for CO, for which $\tilde{B}_e = 1.9313$ cm^{-1}, $\tilde{\alpha}_e = 0.0175$ cm^{-1}, and $\bar{D}_e/hc = 6.2 \times 10^{-6}$ cm^{-1}.

9. Prove that the moments of inertia I_a and I_b (see Figure 8.3) are equal for the benzene molecule. You do not need to know the bond distances, just invoke the hexagonal symmetry, and for simplicity, just consider the carbon atoms.

10. Assign the H_2 pure rotational transitions in Figure 8.8, using data from Table 9.2 to calculate the predicted frequencies. Assuming the spectra were taken at room temperature, account for the relative intensities of each transition.

REFERENCES

1. A. R. Edmonds, *Angular Momentum in Quantum Mechanics* (Princeton University Press, Princeton, NJ, 1974).
2. D. G. Taylor III, and H. L. Strauss, The rotational Raman spectrum of H_2 in Water *J. Chem. Phys. 90*, 768 (1989).
3. G. Herzberg, *Molecular Spectra and Molecular Structure, Vol. I, Spectra of Diatomic Molecules* (Van Nostrand Rienhold, New York 1950).
4. C. H. Townes, and A. L. Schalow, *Microwave Spectroscopy* (McGraw-Hill, New York, 1955).
5. G. W. Chantry ed., *Modern Aspects of Microwave Spectroscopy*, (Academic Press, New York, 1979).
6. B. J. Berne, and R. Pecora, *Dynamic Light Scattering* (Wiley, New York, 1976).
7. C. E. Dykstra, *Quantum Chemistry and Molecular Spectroscopy* (Prentice Hall, Englewood Cliffs, NJ, 1992).
8. I. N. Levine, *Molecular Spectroscopy* (Wiley-Interscience, New York, 1975).

Vibrational spectroscopy of diatomics

9.1 INTRODUCTION

The vibrational motion of molecules as probed by infrared (IR) absorption and Raman scattering can be analyzed to determine molecular structure. In gas-phase studies of fairly small molecules, the rotational substructure reveals bond lengths and angles. In the solid and liquid phases, this rotational structure is generally lost, but molecular symmetry and the presence of various functional groups can still be obtained from vibrational spectra. In this chapter, we consider diatomic molecules in order to examine the basic principles of rovibrational spectra, many of which apply to polyatomic molecules as well. The study of polyatomic vibrational spectra will be undertaken in Chapter 10, where it will be shown how the $3N - 5$ or $3N - 6$ vibrational degrees of freedom* can be represented by a set of collective atomic displacements known as normal modes. By using this representation, the otherwise intractable overall vibrational motion of a large molecule can be decomposed into a set of one-dimensional coordinates, each of which is analogous to that for a diatomic vibrator. Thus our motivation for studying diatomics is quite strong; in doing so we prepare to understand more complex molecules.

The Born–Oppenheimer approximation is a powerful concept on which much of the material in the next few chapters is based. It provides a physical picture in which nuclear motion takes place on potential energy curves (or surfaces, for polyatomics) belonging to distinct electronic states. The interesting spectroscopic consequences of the breakdown in this approximation will be considered in Chapters 11 and 12. A further approximation to be made is that of the harmonic oscillator (HO) model, which provides potential energy surfaces which are quadratic functions of the vibrational coordinate. The HO model provides a point of departure for considering more realistic potential functions. These more appropriate potentials are said to account for anharmonicity. Using diatomics as the simplest possible examples, some of the interesting consequences of anharmonicity are illustrated. The study of polyatomics reveals additional signatures of these effects.

9.2 THE BORN–OPPENHEIMER APPROXIMATION AND ITS CONSEQUENCES

The vibrational problem of a diatomic is simplified by the fact that there exists only a single normal mode, the displacement of the internuclear distance† R from its equilibrium value R_e. The first question of interest is the form of the potential function that governs the vibrational motion, $V(R)$, and how it can be obtained in a quantum mechanical calculation. The full Hamiltonian for the molecule is represented by $\hat{H} = \hat{T}_N + \hat{T}_e + \hat{V}_{ee} + \hat{V}_{eN} + \hat{V}_{NN}$, where the first two terms are the kinetic energy operators for the nuclei and the electrons, respectively, and the last three terms are the potential energy operators for the Coulombic interactions: electron–electron, electron–nuclear, and nuclear–nuclear. The problem in finding the eigenfunctions of this Hamiltonian is that the coordinates of the electrons are not separable from those of the nuclei: the operator $\hat{V}_{eN}$ prevents these variables from being separated exactly. The Born–Oppenheimer approximation, however, allows the total wavefunction to be written as a product of the electronic and nuclear wavefunctions. It is based on a very reasonable physical picture, in which the nuclear motion is several orders of magnitude slower than that of the electrons. Thus, for the purpose of finding the electronic energy, we consider the

---

* An N-atom molecule has $3N - 5$ vibrational degrees of freedom if linear and $3N - 6$ if nonlinear.
† In this chapter, we use an uppercase letter R to represent the internuclear distance and a lowercase r to signify the positions of the electrons.

nuclear positions fixed. The operator $\hat{T}_N$ (Equation 8.1) is then neglected (as if the nuclear masses were infinite) and the potential V_{NN} is considered a constant. The electronic Schrödinger equation then involves the Hamiltonian $\hat{H}_e = \hat{T}_e + \hat{V}_{ee} + \hat{V}_{eN}$:

$$\hat{H}_e\psi_i(r;R) = E_i(R)\psi_i(r;R) \tag{9.1}$$

Since the electron–nuclear potential energy depends on the fixed positions of the nuclei, the wavefunction for the electronic state i depends explicitly on the electronic coordinates r and parametrically on the internuclear distance R. Imagine finding the solution to Equation 9.1 for a large number of different values of R. This generates an electronic energy $E_i(R)$ which can be added to the nuclear repulsion energy $V_{NN}(R)$ to generate a potential function $V_i(R)$ for the nuclear motion in electronic state i:

$$V_i(R) = E_i(R) + V_{NN}(R) = E_i(R) + \frac{Z_A Z_B e^2}{4\pi\varepsilon_0 R} \tag{9.2}$$

where Z_A and Z_B are the atomic numbers of the two nuclei. It is physically reasonable to imagine $E_i(R)$ to decrease as two atoms approach one another, since the electrons begin to experience simultaneous attraction to both nuclei. Clearly, the internuclear repulsion V_{NN} opposes this attraction. If a bond forms, there is a distance at which the energy is minimum and beyond which further decrease in internuclear distance causes $V_i(R)$ to rise sharply. The electronic state is then said to be bound, and the distance at which $V_i(R)$ is a minimum is the bond length, R_e. A dissociative electronic state, on the other hand, is characterized by a potential energy curve which does not have a minimum at finite internuclear distance, but continues to rise as the nuclei are brought together.

The idea that to each electronic state corresponds a unique potential energy curve (or surface) is an important consequence of the Born–Oppenheimer approximation. Figure 9.1 shows a number of potential energy curves for C_2, for which many spectroscopic measurements and theoretical calculations have been made. The changes in the potential function for different electronic states, as exemplified by the data for C_2, are quite important in electronic spectroscopy. In this chapter, we emphasize vibrational transitions within one electronic state, usually the ground state.

Several physical properties are readily obtained from $V_i(R)$. The equilibrium separation (i.e., bond length) R_e is the distance for which the potential energy is a minimum. The curvature at the bottom of the potential

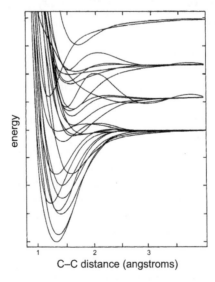

Figure 9.1 Potential energy curves for C_2. (Reprinted from P. F. Fougere, and R. K. Nesbet, Electronic Structure of C_2, J. Chem. Phys. 44, 285 (1966), with the permission of AIP Publishing.)

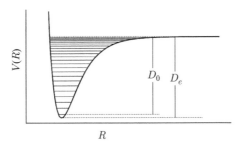

Figure 9.2 Potential energy function and vibrational energy levels for a diatomic molecule.

well is called the force constant k, and it is one measure of the strength of the bond. Another such measure is the bond dissociation energy D_e, shown in Figure 9.2 as the energy required to take the system from its equilibrium position to infinitely separated atoms. The experimental dissociation energy is always less than D_e because the uncertainty principle requires the molecule to have a finite zero-point energy, the lowest possible vibrational energy. The energy required to dissociate the molecule from the ground vibrational level is called D_0. It is the maximum energy required for dissociation; at higher temperatures dissociation also takes place from excited vibrational levels. In Figure 9.2, a typical ladder of quantum vibrational levels is superimposed on the sketch of $V(R)$. Notice that the spacing between adjacent levels decreases with increasing vibrational energy, converging to zero at the dissociation limit, where the energy levels become continuous. This behavior is in accordance with the Bohr correspondence principle, which states that quantum mechanics yields to classical mechanics for large quantum numbers.

In solving Equation 9.1, the nuclear kinetic energy is omitted from the Hamiltonian, so the resulting potential energy function does not depend on the mass of the nuclei. This means that once $V_i(R)$ has been found for a given molecule, for example H_2, it applies to all isotopic variations, such as HD and D_2. Isotopomers therefore have the same bond length within the Born–Oppenheimer approximation. Similarly, they share the same value of D_e, but have different values of D_0, since D_0 depends on the vibrational frequency and thus the reduced mass. Generalizing to polyatomics, we conclude that isotopic substitution preserves the overall geometry. This is very useful to remember, as it allows the isotopic dependence of spectroscopic constants to be readily predicted. It is often possible to exploit isotopic substitution to determine bond lengths and force constants from spectroscopic data.

We next consider how vibrational energy levels and wavefunctions are obtained. We seek to solve the Schrödinger equation for nuclear motion within a particular electronic state. The nuclear Hamiltonian is the sum of the potential energy $V_i(R)$ and the nuclear kinetic energy operator:

$$\hat{H}_N = \hat{T}_N + \hat{V}_i(R) \tag{9.3}$$

The eigenvalues of this Hamiltonian are the vibration–rotation energies. Within the rigid rotor approximation, the vibrational and rotational wavefunctions factor: $\psi_{vib/rot} = \psi_{vib}\psi_{rot}$. The vibrational wavefunction is referred to as χ_v^i, where the superscript i labels the electronic state and the subscript v is a vibrational quantum number. The total Born–Oppenheimer wavefunction for a vibronic (i.e., vibrational plus electronic) state is

$$\Psi_{iv}(r;R) = \psi_i(r;R)\chi_v^i(R) \tag{9.4}$$

Equation 9.4 represents what is called the adiabatic approximation. The word adiabatic is used to describe nuclear motion that takes place on a single potential energy surface pertaining to a particular electronic state. In the more drastic *crude* Born–Oppenheimer approximation, the electronic part of the wavefunction is evaluated at the equilibrium internuclear distance: $\Psi_{iv}(r;R) \cong \psi_i(r;R_e)\chi_v^i(R)$. In this case all the R dependence arises from the vibrational wavefunction. For the remainder of this chapter, we concentrate on the vibrational wavefunctions $\chi_v^i(R)$, and we omit the superscript, and the subscript on $V_i(R)$, when vibrational levels within a single electronic state are under consideration.

9.3 THE HARMONIC OSCILLATOR MODEL

The wavefunctions $\chi_v(R)$ are eigenfunctions of the Hamiltonian of Equation 9.3. To find them requires a functional form for the potential energy $V(R)$. Near the equilibrium position, it is reasonable to expand the potential in a Taylor series about R_e. Introducing the displacement $q = R - R_e$, this series takes the form:

$$V(q) = V(0) + \left(\frac{dV}{dq}\right)_0 q + \frac{1}{2}\left(\frac{d^2V}{dq^2}\right)_0 q^2 + \frac{1}{3!}\left(\frac{d^3V}{dq^3}\right)_0 q^3 + \cdots \tag{9.5}$$

Equation 9.5 is equivalent to Equation 8.77, with a change in notation. The energy at the minimum $V(0)$ can be taken as zero, and the slope of the potential at the minimum is automatically zero, so the first nonzero term in the above series is the quadratic term. If we drop the cubic, quartic, and all higher terms, the potential is said to be harmonic and can be expressed as

$$V(q) = \frac{1}{2}\left(\frac{d^2V}{dq^2}\right)_0 q^2 = \frac{1}{2}kq^2 \tag{9.6}$$

Equation 9.6 defines the force constant k as the second derivative of the potential energy evaluated at the minimum energy position. The harmonic potential of Equation 9.6 is a good approximation to the true potential for energies near the bottom of the well, corresponding to small displacements q, and becomes poorer as the energy increases. One big failing of the harmonic approximation is that it does not permit dissociation because the walls of the confining potential rise to infinity. Figure 9.3 compares the harmonic oscillator potential function to a more realistic anharmonic potential energy curve.

The kinetic energy operator that is needed was derived in the previous chapter, where the internal (vibration–rotation) and lab-frame (translation) degrees of freedom were separated (see Equation 8.4). The kinetic energy operator for internal motion was found to be

$$\hat{T}_N = \frac{-\hbar^2}{2\mu}\nabla_{int}^2 = \frac{-\hbar^2}{2\mu}\left(\frac{\partial^2}{\partial R^2} + \frac{2}{R}\frac{\partial}{\partial R} - \frac{\hat{L}^2}{R^2\hbar^2}\right) \tag{9.7}$$

where μ is the reduced mass as defined in Equation 8.5. In the rigid rotor approximation used in Chapter 8, R was a constant and the two derivatives in Equation 9.7 were discarded. Now we want to retain them, along with the orientation dependent term. The angular momentum operator $\hat{L}$ depends on the angles θ and φ, so $\hat{T}_N$ includes the kinetic energy of both vibration and rotation. When the potential energy $V(R) = V(q + R_e)$ is added to the kinetic energy operator, the result is the Hamiltonian whose eigenvalues are the vibration–rotation energies E_{vr}:

$$\hat{H}_N\Psi = \frac{-\hbar^2}{2\mu}\nabla_{int}^2\Psi + V(R)\Psi = E_{vr}\Psi \tag{9.8}$$

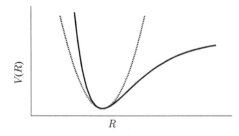

Figure 9.3 Comparison of harmonic (dashed) and anharmonic (solid) potential curves.

We have already solved the rigid rotor problem. Since the commutator $[\hat{H}_N, \hat{L}^2]$ is zero, the two operators share a set of common eigenfunctions. The trick of separation of variables enables the total wavefunction to be factored into a product of two terms which depend separately on the vibrational and rotational coordinates:

$$\Psi(R,\theta,\varphi) = \psi(R) Y_{JM}(\theta,\varphi) \qquad (9.9)$$

By substituting Equation 9.9 into Equation 9.8 and using the fact that the spherical harmonics $Y_{JM}(\theta, \varphi)$ are eigenfunctions of $\hat{L}^2$, we can cancel the Y_{JM}'s and obtain an equation that depends only on the vibrational coordinate:

$$\left[\frac{-\hbar^2}{2\mu} \left(\frac{\partial^2}{\partial R^2} + \frac{2}{R}\frac{\partial}{\partial R} - \frac{J(J+1)}{R^2} \right) + V(R) \right] \psi(R) = E_{vr}\psi(R) \qquad (9.10)$$

Next, we make the substitution $\chi(q) \equiv R\psi(R)$, and use

$$\frac{\partial^2\psi}{\partial R^2} + \frac{2}{R}\frac{\partial\psi}{\partial R} = \frac{1}{R}\frac{\partial^2\chi}{\partial q^2} \qquad (9.11)$$

to reduce Equation 9.10 to

$$\frac{-\hbar^2}{2\mu} \left(\frac{d^2\chi(q)}{dq^2} - \frac{J(J+1)}{(q+R_e)^2}\chi(q) \right) + V(q)\chi(q) = E_{vr}\chi(q) \qquad (9.12)$$

We have made the switch from partial derivatives to ordinary derivatives, since the vibrational wavefunction $\chi(q)$ depends only on the variable q. As it stands, Equation 9.12 is the Schrödinger equation for the vibration–rotation energies of a nonrigid rotor. For small enough displacements, $q \ll R_e$, the rigid-rotor model permits us to make the approximation:

$$\frac{\hbar^2}{2\mu}\left[\frac{J(J+1)}{(q+R_e)^2} \right] \cong \frac{\hbar^2}{2\mu}\left[\frac{J(J+1)}{R_e^2} \right] = E_{rot} \qquad (9.13)$$

The rigid-rotor approximation results in the total energy being the sum of vibrational and rotational parts, $E_{vr} = E_{vib} + E_{rot}$. The term $E_{rot}\chi(q)$ can be subtracted from both sides of 9.12, leaving

$$\frac{-\hbar^2}{2\mu}\frac{\partial^2\chi(q)}{\partial q^2} + V(q)\chi(q) = E_{vib}\chi(q) \qquad (9.14)$$

This is the vibrational Schrödinger equation in the rigid-rotor approximation. It remains to replace $V(q)$ by the harmonic expression $(1/2)kq^2$ to obtain the harmonic oscillator (HO) Hamiltonian:

$$\hat{H}_{vib} = \frac{-\hbar^2}{2\mu}\frac{d^2}{dq^2} + \frac{1}{2}kq^2 \qquad (9.15)$$

The solutions to the quantum mechanical HO problem are well known. (See Chapter 1.) The method for finding the eigenfunctions and eigenvalues of the HO Hamiltonian is discussed in most introductory books on quantum mechanics, so we will just summarize them here. The eigenvalues are

$$E_v = (v + \frac{1}{2})h\nu_e \qquad (9.16)$$

where the vibrational frequency $\nu_e = \omega_e/2\pi$ is related to the force constant k through $k = 4\pi^2\mu\nu_e^2 = \mu\omega_e^2$. The energy in Equation 9.16 is E_{vib}, henceforth we will identify it with the subscript v, the vibrational quantum

Table 9.1 Some Hermite polynomials

$H_0(y) = 1$	$H_3(y) = 8y^3 - 12y$
$H_1(y) = 2y$	$H_4(y) = 16y^4 - 48y^2 + 12$
$H_2(y) = 4y^2 - 2$	$H_5(y) = 32y^5 - 160y^3 + 120y$

number. Boundary conditions permit integral values of the quantum number v ranging from 0 to ∞. The harmonic oscillator eigenfunctions are given by

$$\chi_v(q) = N_v \exp\left(\frac{-\alpha q^2}{2}\right) H_v(\alpha^{1/2} q) \tag{9.17}$$

where $\alpha \equiv \mu\omega/\hbar$, and the Hermite polynomial $H_v(y)$ contains only even or odd powers of $y \equiv \alpha^{1/2} q$, according to whether v is even or odd, up to y^v. See Table 9.1 for a list of the first few Hermite polynomials. The energy levels and wavefunctions for the harmonic oscillator are illustrated in Chapter 1 (Figure 1.8). The alternating symmetries of the wavefunctions, which are even functions of y for v = 0, 2, 4,... and odd functions of y for v = 1, 3, 5,..., are important in the consideration of selection rules. For example, an operator which is an even function of y can only connect two HO states if they are both even or both odd, while an odd function of y can connect even and odd states. As will be shown, further restrictions may apply. The Hermite polynomials can be generated by means of the recursion formula:

$$H_{v+1}(y) = 2yH_v(y) - 2vH_{v-1}(y) \tag{9.18}$$

Note that the order of the Hermite polynomial is the quantum number v, and thus there are v nodes in the wavefunction. The normalization constant for the harmonic oscillator wavefunctions is given by

$$N_v = \left(\frac{\alpha^{1/2}}{2^v v! \pi^{1/2}}\right)^{1/2} \tag{9.19}$$

In many problems, it is possible to find matrix elements of harmonic oscillator states without using the wavefunctions: one can employ Dirac notation for the eigenstates and harmonic oscillator raising and lowering operators, discussed in Chapter 1. We shall see an example of this in the next section.

9.4 SELECTION RULES FOR VIBRATIONAL TRANSITIONS

9.4.1 INFRARED SPECTROSCOPY

The analysis of gas-phase vibration–rotation spectra requires matrix elements of the dipole moment operator with respect to vibration–rotation wavefunctions. Using the harmonic oscillator-rigid rotor approximation, we can represent the rovibrational wavefunctions as $|vJM\rangle = \chi_v(q)Y_{JM}(\theta, \varphi)$. To deduce the selection rules for a transition $vJM \to v'J'M'$, consider the transition moment $\langle vJM|\mu|v'J'M'\rangle \equiv \mu_{if}$, where the usual subscripts denote initial and final states. The dipole moment operator μ is a vector operator; it depends on the orientation of the molecule in the lab frame. It also depends on the coordinate q, which for the diatomic molecule presently under consideration is just the bond length displacement. The IR activity of a molecule derives from the q dependence of the permanent dipole moment. Think about a polar molecule such as CO in its ground electronic state. As the distance between the atoms increases, *perhaps* the dipole moment will also increase, since it is proportional to the charge separation. But this trend cannot continue at large displacements, because the molecule dissociates to neutral atoms and the dipole moment must go to zero. Similarly, we expect the dipole moment to decrease to zero as the internuclear distance shrinks to zero. On going from $R = 0$ to $R = \infty$, we expect the dipole moment to increase from zero and then decrease back to zero, but, in the absence of a calculation, that is really all we know about the function $\mu(q)$.

As in our previous treatment of pure rotational selection rules (Section 8.5), we expand the operator for the magnitude of the dipole moment about its value at the equilibrium position: $\mu(q) = \mu_0 + (\partial\mu/\partial q)_0 q + \cdots$.

As shown in Chapter 8, the first term is responsible for microwave activity. Since μ_0 does not depend on q, it cannot connect different vibrational states, because they are orthogonal, and thus μ_0 does not contribute to IR activity. As shown below, it is the second term which permits vibrational transitions in which $\Delta v = \pm 1$. The transition from $v = 0$ to $v = 1$ is referred to as the *fundamental*. When the initial populations of states having $v \neq 0$ are significant, *hot bands* such as $1 \rightarrow 2$, $2 \rightarrow 3$, etc., are observed. The higher order terms in the expansion of $\mu(q)$, proportional to q^2, q^3, etc., permit *overtones*, for which $\Delta v = \pm 2, \pm 3, \ldots$, etc. The dipole moment derivative, defined as $(\partial \mu / \partial q)_0 \equiv \mu'$, is the slope of $\mu(q)$ at the equilibrium position. The intensity of an IR transition scales as the square of the dipole moment derivative.

With the wavefunction expressed as a product of vibrational and rotational contributions, we can now evaluate the matrix elements of the operator $\vec{\mu}' = \mu' q (\hat{i} \sin\theta \cos\varphi + \hat{j} \sin\theta \sin\varphi + \hat{k} \cos\theta)$. The transition moment factors into a product of two terms:

$$\mu_{if} = \langle \chi_v | \mu' q | \chi_{v'} \rangle \langle Y_{JM} | \begin{Bmatrix} \sin\theta\cos\varphi \\ \sin\theta\sin\varphi \\ \cos\theta \end{Bmatrix} | Y_{J'M'} \rangle \tag{9.20}$$

Vibrational selection rules come from the first matrix element, $\langle \chi_v | \mu' q | \chi_{v'} \rangle$, and the rotational structure within a vibrational transition derives from the matrix element of the spherical harmonics. Using the raising and lowering operators of Section 1.3.3 and the notation $|\chi_v\rangle = |v\rangle$, we have

$$\langle v' | q | v \rangle = \left(\frac{\hbar}{2\mu\omega} \right)^{1/2} \left[\sqrt{v+1} \, \delta_{v',v+1} + \sqrt{v} \, \delta_{v',v-1} \right] \tag{9.21}$$

(See Equation 1.45.) We have factored out the dipole moment derivative μ', which is just a number, albeit an important one through the gross selection rule $\mu' \neq 0$. The dipole moment must change during the vibration, thus, only heteronuclear diatomic molecules can be IR active. (Note that when this analysis is extended to polyatomics in Chapter 10, the existence of a permanent dipole moment is no longer required for IR activity, since a vibrating molecule may experience a change in dipole moment even if there is no dipole moment in the equilibrium geometry.) Equation 9.21 leads to the selection rule $\Delta v = \pm 1$. Note that the same selection rule could be derived using the recursion formula of Equation 9.18 and the fact that the HO wavefunctions are orthonormal. The plus and minus signs of Δv of course correspond to absorption and emission. Ignoring the rotational energy for the moment, the frequency of this transition is the harmonic vibrational frequency ν_e. Next let us see how the second matrix element in Equation 9.20 determines the rotational substructure within a vibrational transition.

The rotational part was evaluated in Chapter 8, where it was shown to lead to the microwave selection rules $\Delta J = \pm 1$ and $\Delta M = 0, \pm 1$. In addition, transitions having $\Delta J = 0$ are permitted for diatomic molecules with nonzero electronic angular momentum. Although such transitions are of no interest in the discussion of pure rotational spectra, they are possible in vibration–rotation spectra and are referred to as the Q branch of the spectrum.[*] In the absence of an external field the selection rules on M can be neglected, so the selection rules on J will decide the rotational structure. For the fundamental transition in IR absorption, the selection rules $\Delta J = -1, 0, +1$ lead, respectively, to transitions at frequencies less than, equal to, and greater than the fundamental frequency ν_e. These are referred to as the P, Q, and R branches, respectively. With the energy levels expressed as $E_{vJ} = (v + 1/2)h\nu_e + hB_e J(J+1)$, the frequencies of the lines in the vibration–rotation spectrum, in the rigid rotor-harmonic oscillator approximation, are determined to be

$$
\begin{aligned}
&\text{P branch} && \nu_{J \rightarrow J-1} = \nu_e - 2B_e J \\
&\text{Q branch} && \nu_{J \rightarrow J} = \nu_e \\
&\text{R branch} && \nu_{J \rightarrow J+1} = \nu_e + 2B_e (J+1)
\end{aligned}
\tag{9.22}
$$

---

[*] In the absence of vibration–rotation coupling, the Q branch is not really a branch, but rather a line corresponding to all transitions having $\Delta J = 0$.

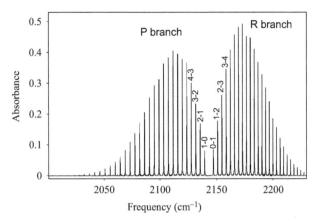

Figure 9.4 Vibration–rotation spectrum of CO. The initial and final quantum numbers are labeled for a few of the transitions.

Note that J is the initial rotational quantum number, and the vibrational state dependence of the rotational constant $B \approx B_e$ has been neglected. Figure 9.4 shows a typical rotationally resolved vibrational spectrum for a diatomic molecule, CO. As in pure rotation spectra, the initial state populations of a number of the lowest J states within the initial vibrational level are significant, so the P and R branches consist of a series of lines separated by twice the rotational constant. The intensity of each line is proportional to the Boltzmann population of the initial state. (Though the absorption intensity is in general proportional to the initial state population minus that of the final state, the latter is negligible for frequencies v_e that are large compared to $k_B T/h$.) The intensities also depend on the frequency and on the square of the J-dependent rotational transition moment. The transition moment is expressed as in Equations 8.47 and 8.48, except that the dipole derivative μ' replaces the permanent dipole moment μ_0. Note that for the same initial value of J the R branch line is stronger than the P branch line.

As in microwave spectra, the bond lengths of gas phase diatomics can be determined from the spacing of the rotational lines in the IR. The prediction that this spacing be constant is a consequence of the rigid rotor and harmonic oscillator approximations. Close inspection of Figure 9.4 reveals that the spacing of adjacent lines in the P branch increases at higher J values, while the R branch lines move closer together at higher J. In Section 9.5, we show how a more realistic approach to the vibration–rotation problem explains this picture.

Carbon monoxide is a closed-shell molecule and therefore the Q branch is forbidden. In contrast, the IR absorption spectrum of NO, shown in Figure 9.5, displays some interesting features that are not observed in closed-shell molecules. The electronic angular momentum about the bond in NO introduces a rotational quantum number analogous to K for a prolate symmetric top. The selection rules of Equation 8.52 apply, and the $\Delta J = 0$ transitions are allowed except for $J = 0$. The ground state of NO is a doublet, having the term symbol $^2\Pi$. This term is split by spin–orbit coupling into two states, $^2\Pi_{1/2}$ and $^2\Pi_{3/2}$, separated by about 120 cm^{-1}. Each of these states displays angular momentum coupling intermediate between Hund's cases (a) and (b), resulting in half-integral rotational quantum numbers. Interaction of the rotational angular momentum with the electronic angular momentum lifts the orbital degeneracy of either Π state through a process called Lambda-type doubling. This leads to a very small splitting, on the order of 0.01 cm^{-1} for NO, which increases with rotational quantum number. This splitting is much less than the resolution of Figure 9.5 and is not responsible for the structure observed there. However, at room temperature the population of the higher-energy $^2\Pi_{3/2}$ state is a little more than half that of the ground $^2\Pi_{1/2}$ state. Each electronic state has its own vibrational frequency and rotational constant, and therefore the spectrum of Figure 9.5 is a superposition of the rovibrational spectra of the $^2\Pi_{1/2}$ and $^2\Pi_{3/2}$ states. Note the poorly resolved Q branch lines which are permitted in this molecule. Vibration–rotation coupling causes the Q branch frequencies to depend on J, as shown in Section 9.5.

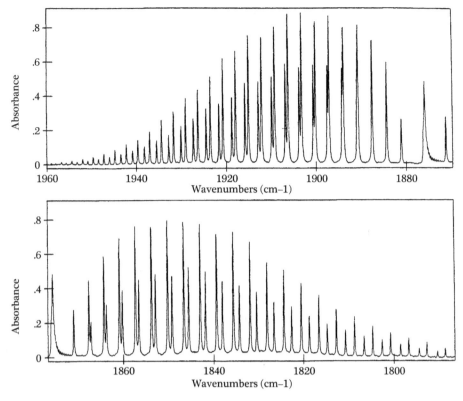

Figure 9.5 Vibration–rotation spectrum of NO. (Reprinted with permission from *Infrared Spectra for Quantitative Analysis of Gases*, P. L. Hanst, and S. T. Hanst. Copyright 1992, Infrared Analysis, Inc.)

9.4.2 Raman scattering

The selection rules for vibration–rotation Raman scattering can be derived from the Kramers–Heisenberg–Dirac expression of Equation 8.53. This is a useful starting point for describing resonance Raman spectra, where the incident laser is resonant with an allowed electronic transition, and we will employ it in Chapter 12. Here, we take a view of Raman scattering that is appropriate for nonresonance Raman scattering. We find that the gross selection rule in Raman scattering is like that for IR absorption with the polarizability replacing the dipole moment. Since vibrational activity in Raman scattering depends on the change in polarizability during vibration, both homonuclear and heteronuclear diatomics are active.

Once again, we expand the quantity of interest about its value at the equilibrium position: $\alpha = \alpha_0 + \left(\partial \alpha / \partial q\right)_0 q + \cdots$. The first term, which cannot connect different vibrational states, leads to Rayleigh scattering or pure rotational Raman, while the second is the one of interest here. Recognizing that the polarizability derivative $\alpha' \equiv (\partial \alpha / \partial q)_0$ is a constant, the vibrational selection rules for Raman scattering depend on the ability of the operator q to connect vibrational states that differ by one quantum. The gross selection rule in Raman scattering is more liberal than that for IR spectra: both homonuclear and heteronuclear diatomics are active in the Raman because $(\partial \alpha / \partial q)_0 \neq 0$. As in IR spectra, the selection rule for ordinary Raman scattering is $\Delta v = \pm 1$, corresponding to Stokes ($\Delta v = +1$) and anti-Stokes ($\Delta v = -1$) transitions. For incident light of frequency ν_0, the scattered light due to the fundamental transition is centered at $\nu_0 - \nu_e$ for the Stokes transition and $\nu_0 + \nu_e$ for the anti-Stokes transition. The latter are less intense than the former due to the lower Boltzmann population of the initial state, typically $v = 1$ in the anti-Stokes case and $v = 0$ for Stokes scattering.

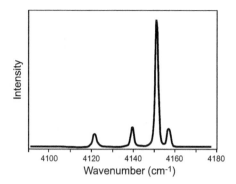

Figure 9.6 Q branch lines of H_2 Raman spectrum at 460 bars. (Courtesy of Dr. Frank Baglin, Department of Chemistry, University of Nevada, Las Vegas.)

To derive the rotational substructure, we need only combine the preceding selection rules with those derived in Chapter 8 for pure rotational transitions. Transitions having $\Delta J = -2, -1, 0, +1, +2$ belong, respectively, to the O, P, Q, R and S branches. Diatomics and linear polyatomics lack P and R branches, because $\Delta J = \pm 1$ is forbidden. Note that the Q branch is permitted for closed-shell molecules in Raman scattering (see the discussion of Section 8.6 to verify this statement), as is apparent in the example spectrum of H_2 shown in Figure 9.6. The bond length can be determined from the rovibrational Raman spectrum, a feature that is especially useful in the case of homonuclear diatomics as they are silent in the IR and microwave.

The Raman spectrum of O_2, shown in Figure 9.7, displays the effects of nuclear spin statistics discussed in Chapter 8. The nuclear spin of ^{16}O is $I = 0$, and the ground state term symbol is $^3\Sigma_g^-$. Using the analysis discussed in Section 8.4, we conclude that only odd J states of $^{16}O_2$ exist. The spacing of adjacent lines in the O or S branch is thus $8\tilde{B}$, where $\tilde{B}$ is 1.446 cm^{-1} for O_2. As in the case of pure rotational Raman scattering, the spacing of the nearest Stokes and anti-Stokes transitions, relative to the separation of adjacent lines in either branch, reveals whether even or odd J states, or both, are present.

The analysis so far has been based on the rigid rotor-harmonic oscillator (RR-HO) model, but the spectra of Figures 9.4–9.7 show evidence of more complex behavior. To better understand experimental spectra, we need a more realistic approach. How indeed can a rigid rotor have any vibrational motion at all? And since real molecules always vibrate (even at absolute zero), what is the effect of this vibration on the rotational energy levels? In the next section, we examine how the selection rules and transition frequencies are altered for nonrigid rotors and anharmonic oscillators.

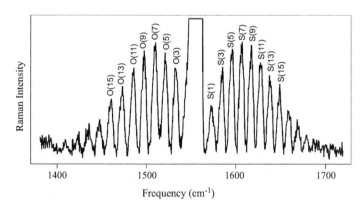

Figure 9.7 Vibration–rotation Raman spectrum of O_2 in air at atmospheric pressure. The initial quantum numbers of the O and S branch lines are indicated, and the Q branch is off-scale.

9.5 BEYOND THE RIGID ROTOR-HARMONIC OSCILLATOR APPROXIMATION

In this section we discuss some effects of anharmonicity from the point of view of diatomic molecules. The qualitative features of this discussion also apply to the case of anharmonicity in polyatomics, but we postpone the discussion of anharmonic effects unique to polyatomics until Chapter 10.

The vibrational absorption spectrum of a molecule is generally dominated by the fundamental transition when it is allowed, but it often happens that weak transitions having $\Delta v = 2, 3, 4, \ldots$ are also observed. These are called overtones, and usually the first overtone, $\Delta v = 2$, is stronger than the second overtone, $\Delta v = 3$, which is in turn more intense than the third overtone, and so on. There are two ways such overtones can appear, and they may be called mechanical and electrical anharmonicity.

Mechanical anharmonicity, or just plain anharmonicity as it is usually referred to, means that the potential function $V(q)$ is not a quadratic function, but rather a more realistic potential function! To appreciate how anharmonicity allows overtones to become active, we can treat some of the higher terms in the Taylor series of Equation 9.5, for example the cubic and quartic terms, as perturbations. The effect of the perturbation on a particular harmonic state $|v\rangle$ is to mix in some of the nearby states. Just as the operator q connects states which differ by one vibrational quantum, q^3 and q^4 can connect states which differ by up to three or four quanta. (This is the kind of statement that is readily proven using raising and lowering operators, as shown in Problem 5.) The perturbation treatment of the anharmonic oscillator is considered in more detail in the next section. Here, we note that this mixing of the zero-order states allows the operator q (from the transition moment operator $\mu' q$) to connect states whose zero-order descriptions differ by more than one vibrational quantum.

The Taylor series expansion of the dipole operator may be carried out beyond the linear term: $\mu(q) = \mu_0 + (\partial \mu / \partial q)_0 q + (\partial^2 \mu / \partial q^2)_0 q^2 \cdots$. The second derivative of the dipole moment, for example, permits transitions between states differing by two quanta, and higher order derivatives allow still higher overtones. This effect is called electrical anharmonicity. When an overtone is observed, the relative contributions made by electrical and mechanical anharmonicity are not readily apparent. However, mechanical anharmonicity leads to decreasing spacing between adjacent vibrational levels as the quantum number increases. For example, the frequency of the first overtone is less than twice the fundamental frequency. We examine two quantitative approaches to interpreting this anharmonicity: perturbation theory and the phenomenological Morse oscillator.

9.5.1 PERTURBATION THEORY OF VIBRATION–ROTATION ENERGY

In this approach, terms are added to the Hamiltonian to account for both nonrigid rotor and anharmonic effects. The first correction improves upon the approximation made in Equation 9.13. Rather then neglecting q compared to R_e, we expand the term in a power series in the ratio q/R_e and keep the first two terms beyond the rigid-rotor approximation:

$$\frac{-\hbar^2}{2\mu}\left[\frac{J(J+1)}{(q+R_e)^2}\right] = \frac{-\hbar^2}{2\mu}\left[\frac{J(J+1)}{R_e^2}\right]\left\{1 - \frac{2q}{R_e} + \frac{3q^2}{R_e^2} - \cdots\right\} \tag{9.23}$$

This gives a perturbation Hamiltonian of the form $\hat{H}'_{rot} = aq + bq^2$, where

$$a \equiv \frac{-2J(J+1)hB_e}{R_e} \tag{9.24}$$

$$b \equiv \frac{3J(J+1)hB_e}{R_e^2} \tag{9.25}$$

At the same time, we correct for anharmonicity by considering a vibrational perturbation of the form $\hat{H}'_{vib} = cq^3 + dq^4$, where the cubic and quartic force constants are defined as follows:

$$c \equiv \frac{1}{3!}\left(\frac{\partial^3 V}{\partial q^3}\right)_0 \equiv \frac{1}{6}V_e^{(3)} \tag{9.26}$$

$$d \equiv \frac{1}{4!}\left(\frac{\partial^4 V}{\partial q^4}\right)_0 \equiv \frac{1}{24}V_e^{(4)} \tag{9.27}$$

The first-order correction to the v-th vibrational level is found by evaluating $E_{vJ}^{(1)} = \langle v | aq + bq^2 + cq^3 + dq^4 | v \rangle$. Notice that, due to the fact that $\chi_v = R\psi(R)$, we can replace the integral $\int \psi^*(R)\hat{O}\psi(R)R^2 dR$ by $\int \chi_v^*(q)\hat{O}\chi_v(q)dq$. The limits of the integral over q really ought to be from $-R_e$ to ∞, but there is a little error in replacing the lower limit with $-\infty$, since the wavefunction is negligibly small for q less than $-R_e$. The linear and cubic terms do not contribute to $E_{vJ}^{(1)}$ because of the symmetry of the wavefunctions. Alternatively, one can use raising and lowering operators to write q and q^3 and prove that these operators are nondiagonal (meaning that they only connect states with different v). The quadratic and quartic terms do contribute to the first-order correction, and the second-order correction depends on all four perturbation terms. The details of obtaining the vibration–rotation energy levels through second order are shown in [1] and are explored further in Problem 9. Here, we state the result and discuss its physical implications. When the perturbed energy is expressed as $E_{vJ} \cong E_{vJ}^{(0)} + E_{vJ}^{(1)} + E_{vJ}^{(2)}$, the final expression is

$$E_{vJ} = (v + \frac{1}{2})hv_e + hB_e J(J+1) - (v+\frac{1}{2})^2 x_e hv_e$$
$$- h\alpha_e (v+\frac{1}{2})J(J+1) - h\bar{D}_e J^2(J+1)^2 \tag{9.28}$$

Equation 9.28 introduces three corrections to the RR-HO model. The first of these (third term on the right-hand side) is the correction for anharmonicity. The anharmonicity constant x_e depends on the cubic and quartic force constants:

$$x_e v_e = \frac{B_e^2 R_e^4}{4hv_e^2}\left\{\frac{10B_e R_e^2 \left[V_e^{(3)}\right]^2}{3hv_e^2} - V_e^{(4)}\right\} \tag{9.29}$$

The sign of x_e is usually positive, so the result of the anharmonic correction is to bring the vibrational energy levels closer and closer together as the energy increases. This is as it should be, but since it is a perturbation calculation, we should not be surprised to reach a maximum value of v where the energy levels turn around and start to decrease with increasing quantum number. This is an artifact and indicates that we are outside the regime where the cubic and quartic force constants suffice to correct for anharmonic effects.

The fourth term in Equation 9.28 represents vibration–rotation coupling, which was mentioned in Chapter 8. The result of this correction term is to cause the rotational constant to depend on the vibrational quantum number: $B_v = B_e - \alpha_e(v + 1/2)$. The physical significance of this term was considered in Chapter 8, where it was noted that the effective rotational constant decreases as vibrational energy increases. Here, we present the expression for the vibration–rotation coupling constant α_e (not to be confused with the polarizability):

$$\alpha_e = \frac{-2B_e^2}{v_e}\left\{\frac{2B_e R_e^3 V_e^{(3)}}{hv_e^2} + 3\right\} \tag{9.30}$$

The sign of $V_e^{(3)}$ is typically negative, and since the first term in parentheses generally outweighs the second, the net result is a vibration–rotation coupling constant which is positive, causing the rotational constant B_v to decrease as v increases. Notice that even a harmonic oscillator, for which $V_e^{(3)}$ vanishes, exhibits vibration–rotation coupling, but the coupling constant turns out to be negative. (See Problem 4.)

The last term in Equation 9.28 accounts for centrifugal distortion, the tendency of the average internuclear distance to increase as J increases. The centrifugal distortion constant, also discussed in Chapter 8, is found to be

$$\bar{D}_e = \frac{4B_e^3}{v_e^2} \tag{9.31}$$

Following Levine [2], a bar has been placed over this symbol to distinguish it from the dissociation energy defined previously. The centrifugal distortion term is generally small compared to the perturbation from vibration–rotation coupling.

Assuming that the selection rules derived for the RR-HO model still apply,* we now look at the influence of vibration–rotation coupling on the spectrum, neglecting the smaller contribution from centrifugal distortion. Taking $\bar{D}_e = 0$, the frequencies of the P, Q, and R branches of the fundamental ($0 \rightarrow 1$) transition are found to be

$$v_P = v_{01} - (2B_e - 2\alpha_e)J - \alpha_e J^2 \tag{9.32}$$

$$v_Q = v_{01} - \alpha_e J(J+1) \tag{9.33}$$

$$v_R = v_{01} + (2B_e - 2\alpha_e)(J+1) - \alpha_e(J+1)^2 \tag{9.34}$$

where J is the initial quantum number. The frequency $v_{01} = v_e - 2x_e v_e$ is that of the pure vibrational transition ($v = 0, J = 0 \rightarrow v = 1, J = 0$). By taking the difference in the frequencies of adjacent lines, we obtain the line spacing as a function of rotational quantum number:

$$\Delta v_P = v_P(J \rightarrow J-1) - v_P(J+1 \rightarrow J)$$
$$= 2(B_e - \alpha_e) + 2\alpha_e(J+1) \tag{9.35}$$

$$\Delta v_R = v_R(J \rightarrow J+1) - v_R(J-1 \rightarrow J)$$
$$= 2(B_e - \alpha_e) - \alpha_e(2J+1) \tag{9.36}$$

These equations indicate that the separation of adjacent lines increases with increasing J for the P branch and decreases with increasing J for the R branch. This effect can be observed in the spectra shown in the previous section. Equations 9.35 and 9.36 suggest that the rotational constants α_e and B_e can be obtained from a plot of line spacing versus J. Spectroscopic constants for some diatomic molecules are given in Table 9.2.

Table 9.2 Spectroscopic constants for diatomics in the ground electronic state

	$\tilde{v}_e$, cm^{-1}	$x_e\tilde{v}_e$, cm^{-1}	$\tilde{B}_e$, cm^{-1}	α_e, cm^{-1}	R_e, Å	D_0, eV
$^{12}C_2$	1641.35	11.67	1.6326	0.01683	1.3117	3.6
$^{12}C^{16}O$	2170.21	13.46	1.9313	0.01896	1.1281	10.96
$^1H_2$	4395.20	117.9	60.80	2.993	0.7416	4.476
$^1H^{35}Cl$	2989.74	52.05	10.59	0.3019	1.2746	4.430
$^{127}I_2$	214.57	0.6127	0.03735	0.000117	2.6666	1.5417
$^{14}N_2$	2359.61	14.456	1.998	0.01731	1.094	7.373
$^{16}O_2$	1580.36	12.073	1.4456	0.01579	1.20739	5.080
$^{16}O^1H$	3735.21	82.81	18.871	0.714	0.9706	4.35

* The reader should try to justify the statement that the selection rules on J still apply to the perturbed system.

9.5.2 The Morse oscillator and other anharmonic potentials

The Morse function is an empirical potential that has the general appearance of the anharmonic potential for a real molecule. It is given by

$$V(R) = D_e \left[1 - e^{-a(R-R_e)} \right]^2 \tag{9.37}$$

This function has the following desirable features: $V(R)$ goes to zero at R_e, approaches the dissociation energy D_e at large R and takes on a very large (though finite) value as R goes to zero. There are three adjustable parameters, D_e, R_e, and a, which are chosen to get agreement with experimental data. The parameter a can be fixed by equating the force constant with the second derivative of the potential evaluated at R_e, $k = 2a^2 D_e$. Since the force constant is related to the frequency, $k = 4\pi^2 \mu v_e^2$, the result is

$$v_e = \frac{a}{2\pi} \sqrt{\frac{2D_e}{\mu}} \tag{9.38}$$

Having only three adjustable parameters does not make the Morse function flexible enough to accurately model the potential function for a real molecule over the complete range of $V(R)$. However, this potential function has the attractive feature that the vibrational eigenfunctions can be found exactly. These turn out to be

$$E_v = (v + \frac{1}{2})hv_e - (v + \frac{1}{2})^2 x_e hv_e \tag{9.39}$$

The anharmonicity constant is given by

$$x_e = \frac{\hbar \omega_e}{4D_e} \tag{9.40}$$

Note that x_e is inversely proportional to the number of harmonic oscillator states that would fit into the potential well. The Morse function provides a convenient way to estimate the anharmonicity from the knowledge of the vibrational frequency and dissociation energy.

A very general expression for the energy levels of an anharmonic oscillator is the series $E_v = (v + 1/2)hv_e - (v + 1/2)^2 x_e hv_e + (v + 1/2)^3 y_e hv_e + \cdots$. The anharmonicity constants x_e, y_e,... can be determined from experimentally observed overtone frequencies.

A numerical approach for determining the potential energy function $V(R)$ from spectroscopic data is the Rydberg–Klein–Rees (RKR) procedure. The method is based on the Bohr–Sommerfield quantization condition. This early version of quantum theory, also used to derive the Bohr model of the atom, was based on quantization of the "action integral," $\oint p\,dq = nh$, where p is the momentum conjugate to the position q, n is an integer, and the integration is over a period of motion with fixed energy E. Applying this to the case of vibration of a diatomic, this condition gives

$$(v + \frac{1}{2})h = 2(2\mu)^{1/2} \int_{R_1}^{R_2} [E_v - V(R)]^{1/2} dR \tag{9.41}$$

where E_v is the (spectroscopically determined) energy of a particular level, and R_1 and R_2 are the inner and outer turning points, where the potential energy becomes equal to the total energy E_v. The RKR method permits these turning points to be determined numerically from Equation 9.41. The potential energy $V(R)$ is then generated by drawing a smooth curve through the turning points. For more details on the RKR method, see [3].

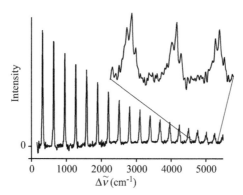

Figure 9.8 Resonance Raman spectrum of liquid Br_2, excited at 405 nm, showing the fundamental and the first 18 overtones. Three of the bands are enlarged in the inset to reveal the isotopic structure. (Reprinted from E. T. Branigan, et al., Solidlike Coherent Vibronic Dynamics in a Room Temperature Liquid: Resonance Raman and Absorption Spectroscopy of Liquid Bromine, *J. Chem. Phys.* 132, 144503 (2010), with the permission of AIP Publishing.)

An example of using vibrational overtone data to derive the potential energy curve is provided in [4] and depicted in Figures 9.8 and 9.9. Branigan et al. measured the resonance Raman spectrum of liquid Br_2 and observed as many as 30 overtones! As will be discussed in Chapter 12, the selection rules for resonance Raman spectroscopy, which uses laser excitation within the electronic absorption band, permit overtones that are not necessarily the result of mechanical and electrical anharmonicity. Figure 9.8 shows a portion of the overtone progression observed on exciting the Raman spectrum of liquid Br_2 at a wavelength within the optical absorption spectrum. The authors used Equation 9.41 to find the turning points for each vibrational level. The potential function $V(R - R_e)$ was then fit to a Morse function for $R < R_e$ and to a fifth-order polynomial for $R > R_e$. As seen in Figure 9.9, the potential function in the liquid phase is softer than that of the gas phase for vibrational quantum numbers larger than about $v = 10$, which the authors attribute to an attractive solvent cage. The dissociation energy of Br_2 was found to be considerably lower in the liquid than in the gas phase.

Branigan et al. also extracted vibrational dephasing times from the widths of the Raman bands, after accounting for isotope shifts and the instrumental linewidth. They found T_2 to range from about 25 ps for the $v = 0 \rightarrow v = 1$ transition to about 2.4 ps for the $v = 0 \rightarrow v = 25$ transition. Recall that the dephasing rate includes population relaxation (with time scale T_1) as well as pure dephasing (T_2^*) from frequency fluctuations. We have seen (Chapter 6) that radiative decay rates at IR frequencies are slow. Thus, excited vibrational levels (within the ground electronic state in the example here) decay radiationlessly via resonant energy transfer to degrees of freedom of the surroundings (the "bath"). In the liquid phase, this transfer is

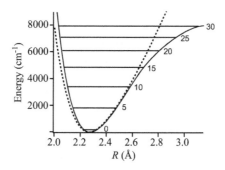

Figure 9.9 Potential energy function for Br_2 in the liquid phase, showing the vibrational levels for $v = 0,5,...30$, derived from analysis of the vibrational overtones observed in the resonance Raman spectrum (Figure 9.8). The dashed curve shows the potential energy curve for Br_2 in the gas phase. (Reprinted from E. T. Branigan et al., Solidlike Coherent Vibronic Dynamics in a Room Temperature Liquid: Resonance Raman and Absorption Spectroscopy of Liquid Bromine, *J. Chem. Phys.* 132, 144503 (2010), with the permission of AIP Publishing.)

envisioned to take place by converting one quantum of the vibrational energy of the solute to an overtone of something akin to an intermolecular vibrational mode of the liquid. As we have seen in Chapter 5, the intermolecular vibrational and reorientational motion of liquids takes place over a range of frequencies in the far-IR. For higher frequency solute vibrations, increasingly higher order overtones of solvent vibrations are required to soak up the energy given off by population relaxation, and the efficiency of energy transfer decreases. Theory and experiment suggest that the rate of energy transfer falls off exponentially with increasing vibrational frequency. This exponential trend is reflected in the level-dependent relaxation times reported in [4]. The vibrational frequency of Br_2, 317 cm^{-1}, is comparable to k_BT at room temperature, permitting facile vibrational population relaxation via one-quantum jumps that transfer energy to the bath degrees of freedom. As the vibrational quantum number increases, the energy difference of these one-quantum jumps decreases as a result of anharmonicity, and the population relaxation rate increases. The large quadrupole moment of diatomic bromine was concluded to favor a herringbone local alignment of molecules similar to what is seen in the solid phase.

9.6 SUMMARY

This chapter has examined the basic features of vibration–rotation spectra of diatomic molecules. Along the way, we have introduced some topics which apply to polyatomic molecules as well. The Born–Oppenheimer approximation is a key concept on which the discussion of electronic states of molecules is based, and we return to it in Chapters 11 and 12. The rotational structure of vibrational absorption or Raman scattering spectra is readily observed in gas-phase samples, and diatomic molecules present opportunities for the straightforward analysis of rotational constants from vibration–rotation spectra. Vibration–rotation coupling causes the spacing of adjacent lines in the P branch ($\Delta J = -1$) of the IR spectrum to increase as J increases, while those in the R branch ($\Delta J = +1$) move closer together. The effects of anharmonicity (mechanical or electrical) show up in the appearance of overtone transitions. Mechanical anharmonicity results in decreasing spacing of adjacent vibrational energies with increasing quantum number v. The Morse function is an empirical function which is capable of accounting for this anharmonicity. In Chapter 10, we will begin our discussion of vibrations in polyatomics with the harmonic approximation, and then we shall see that anharmonicity introduces additional interesting effects compared to those observed in diatomics.

PROBLEMS

1. The vibrational frequency of $^1H^{79}Br$ is $\tilde{v}_e = 2649.7$ cm^{-1} and the anharmonicity is $x_e\tilde{v}_e = 45.2$ cm^{-1}. Find the frequencies of the fundamental, first overtone and second overtone for $^1H^{79}Br$, $^2H^{79}Br$, and $^1H^{80}Br$.
2. Use the data given in Table 9.2 for $^1H^{35}Cl$ to estimate the cubic and quartic force constants: $V^{(3)}$ and $V^{(4)}$.
3. Use the data given in Table 9.2 to find the dissociation energies $\tilde{D}_e$ and $\tilde{D}_0$ (in units of cm^{-1}) for $^2H^{35}Cl$.
4. According to Equation 9.30, the vibration–rotation constant for a harmonic oscillator is not zero and is in fact negative. Consider the rotational constant for the v = 0 state of a harmonic oscillator, which is proportional to the expectation value $\langle 1/R \rangle_{v=0}$. Show that this expectation value is greater than $1/R_e^2$, thus B_0 for a harmonic oscillator is greater than B_e, resulting in a negative vibration–rotation coupling constant.
5. Use the harmonic oscillator raising and lowering operators to derive the selection rules for vibrational transitions that result from electrical anharmonicity in the form $(\partial^2 \mu / \partial q^2)_0 q^2$.
6. The vibrational frequency of NO is 1890 cm^{-1}, the bond length is 1.1508 Å, and the dissociation energy is $D_0 = 6.5$ eV. Use this information to find the Morse function for NO. Make a sketch of the Morse potential and the vibrational energy levels.
7. Estimate the radiative lifetime of a diatomic molecule in the v = 1 vibrational state, given a dipole derivative $(\partial \mu / \partial q)_0$ of about 10 D/Å and a vibrational frequency of 2000 cm^{-1}.
8. Using data from Table 9.2, find the force constants k for O_2, I_2, and N_2. Is there a physical explanation for the relative magnitudes of these values?

9. Calculate the first-order correction to the energy of a harmonic oscillator, $E_v^{(1)}$, using the perturbation operator $\hat{H}' = bq^2 + dq^4$ as defined in Section 9.5. Compare your result to that obtained using perturbation theory to second order, Equation 9.28.

10. Use the data in Table 9.2 to assign the Q branch lines in Figure 9.6. Account for the relative intensities.

11. The two naturally occurring isotopes of Br are ^{79}Br (50.7%, 78.928 amu) and ^{81}Br (49.3%, 80.916 amu). In [4], the vibrational frequency $\tilde{\nu}_e$ was found to be 317.52 cm^{-1} for the ^{79}Br^{81}Br isotopomer. For each of the isotopomers ^{79}Br^{81}Br, ^{79}Br$_2$ and ^{81}Br$_2$, calculate the percentage in a natural sample. Find the vibrational frequency $\tilde{\nu}_e$ of ^{79}Br$_2$ and of ^{81}Br$_2$. Can you account for the fine structure in the inset of Figure 9.8? See [4] for an explanation.

REFERENCES

1. C. E. Dykstra, *Quantum Chemistry and Molecular Spectroscopy* (Prentice Hall, Englewood Cliffs, New Jersey, 1992).

2. I. N. Levine, *Molecular Spectroscopy* (Wiley, New York, 1975).

3. W. S. Struve, *Fundamentals of Molecular Spectroscopy* (Wiley, New York, 1989).

4. E. T. Branigan, M. N. van Staveren, V. A. Apkarian, Solidlike coherent vibronic dynamics in a room temperature liquid: Resonance Raman and absorption spectroscopy of liquid bromine, *J. Chem. Phys.* *132*, 144503 (2010).

5. P. F. Fougere, and R. K. Nesbet, Electronic structure of C$_2$ *J. Chem. Phys.* *44*, 285 (1966).

6. P. L. Hanst, and S. T. Hanst, *Infrared Spectra for Quantitative Analysis of Gases* (Infrared Analysis, Inc., Anaheim, CA, 1992).

Vibrational spectroscopy of polyatomic molecules

10.1 INTRODUCTION

In this chapter, we explore how infrared (IR) absorption and Raman scattering spectra may be applied to elucidate molecular structure. Group theoretical analysis of selection rules makes the combination of Raman and IR data especially useful for deducing symmetry. The method of normal coordinate analysis enables the characterization of vibrational modes and their frequencies, based on a set of force constants and structural information.

Anyone who has survived an undergraduate class in organic chemistry can probably identify the characteristic frequencies of, say, a carbonyl group (about 1700 cm^{-1}) or a C–H bond (about 3000 cm^{-1}). But what makes these stretches appear where they do in the spectrum, and what determines their intensity? And what is really meant by the terms "group frequency" and "normal mode?" These concepts are examined in detail in this chapter. The unsophisticated use of the group frequency concept can lead to erroneous conclusions about molecular structure, and the breakdown of the normal mode approximation can result in unexpected spectral features. To make the best use of spectral information, we must examine the models on which the working principles of vibrational spectroscopy are based.

Suppose that you could actually view a vibrating polyatomic molecule. What would you see? Could the normal modes be identified? As an aid to this thought experiment, consider an ORTEP* representation of a molecule as determined from X-ray diffraction, for example the structure shown in Figure 10.1. The ellipsoids centered at each atom represent the average root-mean-square deviation of the atom from its equilibrium position. The sizes of these ellipsoids increase with temperature as the amplitude of vibrational motion increases. If you visualize each atom jiggling about within the volume of each little ORTEP ellipsoid, the concept of normal modes is hardly apparent. The normal modes are a set of $3N - 6$ collective atomic displacements which may be superimposed to describe the overall vibrational motion of the molecule.† They represent a way of resolving the total vibrational motion (the ORTEP picture) into *independent* modes of motion. If the motion in one of these normal modes is mostly concentrated in a particular bond or part of the molecule, then the idea of a characteristic group frequency is valid. More generally, a normal mode is a combination of motions involving *all* the atoms to varying degrees. The set of atomic displacements that characterizes the normal mode is a normal coordinate, which we call Q_i for the i-th normal mode. This decomposition of the overall motion into $3N - 6$ normal modes is made possible by the harmonic oscillator approximation, which allows the total potential energy for a given electronic state to be expressed as a sum of $3N - 6$ quadratic functions of the normal coordinates. A harmonic potential written in terms of a set of coordinates other than the normal coordinates would contain mixed quadratic terms. The normal coordinates allow separation of variables to be employed, greatly simplifying both the classical and quantum mechanical treatments. The assumption is made that the forces between atoms are harmonic; that is, they obey Hooke's law. Hooke's law states that the restoring force of the spring is proportional to its displacement from equilibrium, the proportionality constant being the force constant k. The stronger the bond is, the stiffer the "spring" and the larger the force constant. So the cartoon picture of a molecule with bonds represented by little springs, such as the one in Figure 10.2, is at the heart of the matrix algebra which gives us the normal mode picture!

---

* ORTEP stands for Oak Ridge thermal ellipsoid plot.
† For the sake of brevity, we will use $3N - 6$ to refer to the number of vibrational degrees of freedom of a general molecule. In the case of a linear molecule, this would be replaced by $3N - 5$.

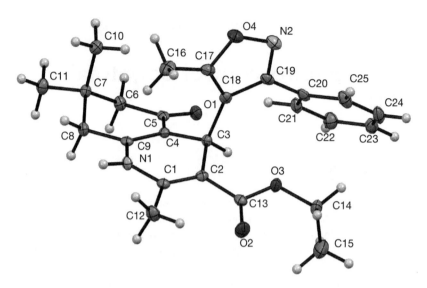

Figure 10.1 ORTEP plot of a molecular structure determined using X-ray diffraction. (Reproduced with permission from Lanthanide and asymmetric catalyzed syntheses of sterically hindered 4-isoxazolyl-1,4-dihydropyridines and 4-isoxazolyl-quinolines, S. A. Steiger et al., *Tet. Lett.* 57, 423 (2016).)

Surely this idea of little springs connecting atoms is only an approximation, and we should not be too surprised when experiments reveal effects which would be unexpected for a harmonic oscillator. We refer to such effects as resulting from anharmonicity, and we use perturbation theory to account for them. There are some common motifs associated with anharmonic vibrational motion, such as the appearance of overtones and combination bands, Fermi resonance, and the Coriolis effect, to be discussed in Section 10.7.

10.2 NORMAL MODES OF VIBRATION

In this section we set up the problem of finding the frequencies and coordinates of the normal modes of a polyatomic molecule. The first task at hand is to derive the classical equations of motion for the normal coordinates. The normal coordinates are expressed as linear combinations of the Cartesian atomic displacements. Then the quantum mechanical operators can be substituted for the corresponding classical expressions for positions and momenta, and a set of $3N - 6$ independent, one-dimensional equations is obtained, each of which is a familiar harmonic oscillator Schrödinger equation. The outcome of performing the normal mode analysis is a set of normal mode frequencies that depend on the masses of the atoms and the force constants for the bonds which connect them. We presume that these force constants can be known and used as input into the calculation. In practice, they are not available from experiment except to the extent that they can be determined by fitting the calculated to the observed vibrational frequencies. The validity of this fitting procedure depends on the existence of a sufficient number of isotopic derivatives, for which the force constants are the same within the Born–Oppenheimer approximation. Force constants may also be determined from quantum mechanical calculations of the energy of a given electronic state (most often the ground state) as a function of molecular geometry or from gradient methods, which avoid the need to make calculations at a

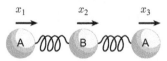

Figure 10.2 The harmonic oscillator model considers atoms to be connected by springs that obey Hooke's law.

large number of geometries. The practicality of performing such calculations, or obtaining a sufficient number of isotopomers, becomes more limited as the size of the molecule increases. It is often necessary to call on chemical intuition to estimate normal mode frequencies of large molecules, for example by assuming the transferability of the stretching force constants of certain bonds in a series of chemically similar compounds.

10.2.1 CLASSICAL EQUATIONS OF MOTION FOR NORMAL MODES

We begin by setting up a classical expression for the total energy, kinetic plus potential, of a collection of N atoms. The equilibrium position of each atom is at the origin of a Cartesian coordinate system embedded in the atom. Then, at any instant in time, the geometry of the system is specified by a set of $3N$ atomic displacement coordinates indexed by a number which specifies the atom: $\{x_1, y_1, z_1, \ldots, x_N, y_N, z_N\}$. An example of such a scheme is shown in Figure 10.3 for the ammonia molecule. The kinetic energy is

$$T = \frac{1}{2}\sum_{i=1}^{N} m_i\left(\dot{x}_i^2 + \dot{y}_i^2 + \dot{z}_i^2\right) = T_{trans} + T_{rot} + T_{vib} \tag{10.1}$$

Note that this kinetic energy comprises the translational, rotational, and vibrational degrees of freedom. The potential energy, as you might expect, is expanded in a series which is truncated after the quadratic term:

$$\begin{aligned}
V = V_0 &+ \sum_{i=1}^{N}\left[\left(\frac{\partial V}{\partial x_i}\right)_0 x_i + \left(\frac{\partial V}{\partial y_i}\right)_0 y_i + \left(\frac{\partial V}{\partial z_i}\right)_0 z_i\right] \\
&+ \frac{1}{2}\sum_{i=1}^{N}\left[\left(\frac{\partial^2 V}{\partial x_i^2}\right)_0 x_i^2 + \left(\frac{\partial^2 V}{\partial y_i^2}\right)_0 y_i^2 + \left(\frac{\partial^2 V}{\partial z_i^2}\right)_0 z_i^2\right] \\
&+ \frac{1}{2}\sum_{i<j}^{N}\left[\left(\frac{\partial^2 V}{\partial x_i \partial x_j}\right)_0 x_i x_j + \left(\frac{\partial^2 V}{\partial y_i \partial y_j}\right)_0 y_i y_j + \left(\frac{\partial^2 V}{\partial z_i \partial z_j}\right)_0 z_i z_j\right] \\
&+ \frac{1}{2}\sum_{i,j=1}^{N}\left[\left(\frac{\partial^2 V}{\partial x_i \partial y_j}\right)_0 x_i y_j + \left(\frac{\partial^2 V}{\partial y_i \partial z_j}\right)_0 y_i z_j + \left(\frac{\partial^2 V}{\partial x_i \partial z_j}\right)_0 x_i z_j\right]
\end{aligned} \tag{10.2}$$

Without making any approximation, we can discard the terms on the right hand side of the first line, since V_0 can be taken as zero and the first derivatives vanish at the equilibrium geometry. The second line contains diagonal quadratic terms and third and fourth lines contain the nondiagonal quadratic terms. Our goal is to eliminate the nondiagonal terms by a suitable coordinate transformation.

It is convenient to express the above energies in mass-weighted Cartesian coordinates. These ensure that the normal coordinates that we obtain preserve the position of the center of mass. They are defined by

$$\eta_1 = \sqrt{m_1}\, x_1, \eta_2 = \sqrt{m_1}\, y_1, \eta_3 = \sqrt{m_1}\, z_1, \eta_4 = \sqrt{m_2}\, x_2, \ldots, \eta_{3N} = \sqrt{m_N}\, z_N \tag{10.3}$$

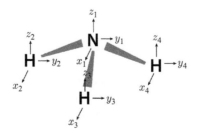

Figure 10.3 Atomic Cartesian coordinates for the ammonia molecule.

This allows the kinetic and potential energies to be expressed as

$$T = \frac{1}{2}\sum_{i=1}^{3N} \dot{\eta}_i^2 \tag{10.4}$$

$$V = \frac{1}{2}\sum_{i,j}^{3N} b_{ij}\eta_i\eta_j \tag{10.5}$$

Equation 10.5 defines the b_{ij}, which are the elements of the force constant matrix **B**. To illustrate the book-keeping associated with the **B**-matrix, here is an example element:

$$b_{18} = \frac{1}{(m_1 m_3)^{1/2}}\left(\frac{\partial^2 V}{\partial x_1 \partial y_3}\right)_0 \tag{10.6}$$

There are $3N$ classical equations of motion for this system, each of the form

$$F_{x,i} = \frac{d}{dt}\left(m\dot{x}_i\right) = \frac{-\partial V}{\partial x_i} \tag{10.7}$$

Equation 10.7 gives the force on atom i in the x direction. In terms of mass-weighted Cartesian coordinates and the force constants b_{ij}, the equations of motion are

$$\frac{d}{dt}\dot{\eta}_i + \sum_j b_{ij}\eta_j = 0 \tag{10.8}$$

The dot over x_i or η_i indicates differentiation with respect to time, so we have a set of second-order differential equations to solve. Since we expect the motion to be harmonic for a quadratic force field, let us try a solution of the form

$$\eta_i = \eta_i^0 \sin\left(\sqrt{\lambda}t + \delta\right) \tag{10.9}$$

where $\sqrt{\lambda} \equiv 2\pi v = \omega$, η_i^0 is the amplitude and δ the phase of the oscillation. We do not yet know the vibrational frequencies v, and η_i^0 and δ are constants of integration that depend on the boundary conditions. On substituting Equation 10.9 into 10.8, the result is

$$-\lambda\eta_i^0 + \sum_j b_{ij}\eta_j^0 = 0 \tag{10.10}$$

for $i = 1$ to $3N$. These $3N$ equations can all be written as one matrix equation:

$$(\mathbf{B} - \lambda\mathbf{I})\eta^0 = 0 \tag{10.11}$$

where **I** is the identity matrix: a matrix with ones along the diagonal and zeros elsewhere. η^0 is a column vector containing the amplitudes of the mass-weighted Cartesian coordinates. A nontrivial solution to the set of equations expressed by Equation 10.11 exists only if the determinant of the square matrix is zero: $\det(\mathbf{B} - \lambda\mathbf{I}) = 0$.

In matrix notation, the potential energy is succinctly expressed as $V = (1/2)\eta^T \mathbf{B}\eta$, where η^T is the transpose of η. We now seek to diagonalize the potential energy by means of the transformation to normal coordinates, Q_i, where

$$Q_i = \sum_{k=1}^{3N} l_{ki}\eta_k \tag{10.12}$$

Equation 10.12 expresses each normal coordinate as a linear combination of the mass-weighted Cartesian coordinates. The l_{ki}'s are coefficients to be determined. The $3N$ transformations of the type given in Equation 10.2 can be expressed in one matrix equation.

$$\mathbf{Q} = \mathbf{L}^T \eta \tag{10.13}$$

where $\mathbf{Q}$ is a column vector containing the normal coordinates and $\mathbf{L}^T$ is the transpose of $\mathbf{L}$. If the elements l_{ki} are considered to form the columns of a matrix $\mathbf{L}$, then they constitute the rows of $\mathbf{L}^T$. We seek the matrix $\mathbf{L}$ which diagonalizes the force constant matrix, as follows:

$$\mathbf{L}^T \mathbf{B} \mathbf{L} = \Lambda = \begin{pmatrix} \lambda_1 & 0 & 0 & \cdots & 0 \\ 0 & \lambda_2 & 0 & \cdots & 0 \\ 0 & 0 & \lambda_3 & \cdots & 0 \\ \vdots & \vdots & \vdots & \ddots & \vdots \\ 0 & 0 & 0 & \cdots & \lambda_{3N} \end{pmatrix} \tag{10.14}$$

$\mathbf{L}$ is a unitary matrix, meaning its inverse is equal to its transpose: $\mathbf{L}^T = \mathbf{L}^{-1}$. Thus the matrix equation $\mathbf{Q} = \mathbf{L}^T\eta$ can be rearranged to give $\eta = \mathbf{L}\mathbf{Q}$. The potential energy in terms of the normal coordinates is then

$$V = \frac{1}{2}\mathbf{Q}^T \mathbf{L}^T \mathbf{B} \mathbf{L} \mathbf{Q} = \frac{1}{2}\mathbf{Q}^T \Lambda \mathbf{Q} = \frac{1}{2}\sum_{i=1}^{3N}\lambda_i Q_i^2 \tag{10.15}$$

The identity $(\mathbf{L}\mathbf{Q})^T = \mathbf{Q}^T\mathbf{L}^T$ has been used. It is significant that the potential energy is now diagonal and depends on the normal mode frequencies through $\sqrt{\lambda_i} = 2\pi\nu_i$. By rewriting Equation 10.4 in terms of the normal coordinates, it is seen that the kinetic energy is still diagonal as well.:

$$T = \frac{1}{2}\dot{\mathbf{Q}}^T\dot{\mathbf{Q}} = \frac{1}{2}\sum_{i=1}^{3N}\dot{Q}_i^2 \tag{10.16}$$

The uncoupled equations of motion are now solved by $Q_i = Q_i^0 \sin\left(\lambda_i^{1/2}t + \delta\right)$. That is, each normal coordinate oscillates independently.

10.2.2 Example: Normal modes of a linear triatomic

The procedure for finding the normal coordinates and their frequencies begins by solving $\det(\mathbf{B} - \lambda\mathbf{I}) = 0$. The $3N$ roots to this equation include five or six zero-frequency ($\lambda = 0$) roots, for linear and nonlinear molecules, respectively, corresponding to rotation and translation. Each value of λ can then be substituted into Equation 10.11 to find the amplitudes of the normal coordinates expressed as linear combinations of mass-weighted Cartesian coordinates. The details of the matrix algebra and some of the chemical intuition employed to simplify the problem will be illustrated in the following example. We consider for simplicity a symmetric linear triatomic A–B–A in one dimension, as shown in Figure 10.2. The simplest harmonic function that we can write for this molecule would depend on a single force constant k for each A–B bond.

$$V = \frac{1}{2}k(x_2 - x_1)^2 + \frac{1}{2}k(x_3 - x_2)^2 \tag{10.17}$$

A more accurate and still harmonic potential would contain a cross-term of the type $k'(x_2 - x_1)(x_3 - x_2)$, but we will omit this term to make the problem more tractable. Comparing Equations 10.17 and 10.6, it is seen that the elements of the **B** matrix are

$$b_{11} = b_{33} = \frac{k}{m_A} \tag{10.18}$$

$$b_{22} = \frac{2k}{m_B} \tag{10.19}$$

$$b_{12} = b_{23} = \frac{-k}{(m_A m_B)^{1/2}} \tag{10.20}$$

$$b_{13} = 0 \tag{10.21}$$

The determinant takes the form

$$\det \begin{pmatrix} b_{11}-\lambda & b_{12} & 0 \\ b_{12} & b_{22}-\lambda & b_{12} \\ 0 & b_{12} & b_{11}-\lambda \end{pmatrix} \tag{10.22}$$

This can be expanded to give $(b_{11} - \lambda)^2(b_{22} - \lambda) - 2(b_{11} - \lambda)b_{12}^2 = 0$. The roots of this equation are

$$\lambda_1 = \frac{k}{m_A}, \quad v_1 = \frac{1}{2\pi}\sqrt{\frac{k}{m_A}} \tag{10.23}$$

$$\lambda_2 = \frac{k}{m_A} + \frac{2k}{m_B}, \quad v_2 = \frac{1}{2\pi}\sqrt{\frac{k}{m_A} + \frac{2k}{m_B}} \tag{10.24}$$

and a third root, $\lambda_3 = 0$, corresponding to the zero-frequency translational motion. As we shall see, the first and second roots correspond to the symmetric and asymmetric stretching modes, respectively. Note that the latter is higher in frequency than the former, a common occurrence in modes involving symmetric and asymmetric stretches of equivalent bonds. Now, each root can be substituted into Equation 10.11 in order to solve for the vector η corresponding to a particular normal mode. This will be illustrated for the first root.

$$\begin{pmatrix} b_{11}-\lambda & b_{12} & 0 \\ b_{12} & b_{22}-\lambda & b_{12} \\ 0 & b_{12} & b_{11}-\lambda \end{pmatrix}\begin{pmatrix} \eta_1^0 \\ \eta_2^0 \\ \eta_3^0 \end{pmatrix} = \begin{pmatrix} 0 \\ 0 \\ 0 \end{pmatrix} \tag{10.25}$$

On substituting the appropriate values of b_{11}, b_{22}, b_{12}, and λ_1, we find that $\eta_2^0 = 0$ and $\eta_1^0 = -\eta_3^0$. By imposing normalization of the l vector, $\sum_k l_{ki}^2 = 1$, we get the result

$$l_1 = \left(1/\sqrt{2} \quad 0 \quad -1/\sqrt{2}\right) \tag{10.26}$$

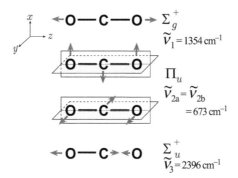

Figure 10.4 Normal modes of CO_2.

Thus the normal coordinate for the symmetric stretch is $Q_1 = (m_A/2)^{1/2}(x_1 - x_3)$. Notice that the mass and position of the central atom B do not contribute to this coordinate. It is a general feature of normal modes that the motion preserves the position of the center of mass, so in this case that means that atom B does not move during symmetric stretching.

The calculation can be repeated using the roots λ_2 and λ_3 in place of λ_1. The reader should verify that the results are

$$l_2 = \left((m_B/2M)^{1/2} \quad -2(m_A/2M)^{1/2} \quad (m_B/2M)^{1/2} \right) \tag{10.27}$$

$$l_3 = \left((m_A/M)^{1/2} \quad (m_B/M)^{1/2} \quad (m_A/M)^{1/2} \right) \tag{10.28}$$

where $M = 2M_A + M_B$ is the total mass. The negative sign on the middle element of the l_2 vector in Equation 10.27 means that the displacement of the central atom is in the direction opposite to that of the end atoms. This is the asymmetric stretch. In the third mode, all atoms move in the same direction, representing translational motion. (Not being a normal mode, the center of mass *does* move during translation.)

As an illustration, the normal coordinates and frequencies of CO_2 are shown in Figure 10.4. The two bending vibrations are degenerate, as expected from symmetry, and the symmetric and asymmetric stretches explored in the one-dimensional example are now expressed as in-phase and out-of-phase combinations of the internal coordinates for the C=O bonds. The four vibrational degrees of freedom expected for this molecule are spanned by two nondegenerate normal modes and one doubly degenerate normal mode. CO_2 belongs to the $D_{\infty h}$ point group, and the symmetry labels Σ_g^+, Π_u, and Σ_u^+ are irreducible representations of this group. Group theoretical labels for normal modes will be considered in Section 10.4.

The starting point for the above approach was the definition of the mass-weighted Cartesian coordinates for each individual atom. This is not a very appealing coordinate system from a chemical point of view, since we think of atoms as moving concertedly rather than independently. In the next section, we briefly describe a point of view which takes advantage of molecular symmetry and bonding patterns.

10.2.3 THE WILSON F AND G MATRICES

This approach begins by defining the symmetry coordinates of the molecule in terms of $3N - 6$ bond lengths and angles. Let us return to the linear triatomic of the previous example and consider its vibrational motion in three dimensions. We *could* set up the problem in terms of nine atomic coordinates, realizing that five of the nine calculated vibrational frequencies would be zero anyway. Instead, we begin with a set of four internal coordinates representing displacements in the two bond lengths and bending in two orthogonal directions. We call the vector containing these internal coordinates **R**, and for this example we write $\mathbf{R}^T = (\Delta r_1, \Delta r_2, \Delta \theta_1, \Delta \theta_2)$. The Δ's represent the displacements of the coordinates from their values at equilibrium. The internal coordinates can be expressed as linear combinations of the mass-weighted Cartesian coordinates, and the transformation is defined as $\mathbf{R} = \mathbf{D}\eta$. The matrix $\mathbf{D}$ is not square, but rather of dimension $(3N - 5)$ by $(3N)$. Note that it has elements with the dimensions of the inverse square root of mass. The procedure for writing internal coordinates is discussed in detail in [1].

The potential energy, written with the help of chemical intuition, might look like the following:

$$V = \frac{1}{2}(k_r \Delta r_1^2 + k_r \Delta r_2^2 + k_\theta \Delta \theta_1^2 + k_\theta \Delta \theta_2^2) \tag{10.29}$$

This is the simplest harmonic potential function that we could write. A more flexible one would add terms such as $k_{rr}\Delta r_1 \Delta r_2$ and $k_{r\theta}\Delta r_1 \Delta \theta_1$, and the following procedure would be the same. Together with the internal coordinates, Equation 10.29 defines the force constant matrix $\mathbf{F}$ such that $2V = \eta^T \mathbf{B}\eta = \mathbf{R}^T \mathbf{F}\mathbf{R}$. The kinetic energy expression of Equation 10.16 can also be recast to give $2T = \dot{\eta}^T \dot{\eta} = \dot{\mathbf{R}}^T (\mathbf{D}\mathbf{D}^T)^{-1}\dot{\mathbf{R}}$. The G matrix (G stands for geometry), the calculation of which is a major step in the Wilson FG approach, is defined by

$$\mathbf{G} = \mathbf{D}\mathbf{D}^T \tag{10.30}$$

which leads to $2T = \dot{\eta}^T \dot{\eta} = \dot{\mathbf{R}}^T \mathbf{G}^{-1}\dot{\mathbf{R}}$. The G matrix contains elements which depend on the inverse atomic masses and bond lengths and angles pertaining to the equilibrium geometry. After some manipulation, the previous expression, $\det(\mathbf{B} - \lambda\mathbf{I}) = 0$, gives way to the new one:

$$\det(\mathbf{F}\mathbf{G} - \lambda\mathbf{I}) = 0 \tag{10.31}$$

The roots of this equation and the corresponding normal coordinates can be found using computer calculations. The elements of the G matrix are somewhat tedious to find. Fortunately, there are convenient tables that can be used to construct the G matrix, for example, see [2].

A great advantage of this approach is obtained by starting with particular linear combinations of the internal coordinates which conform to the symmetry species of the molecular point group. When these symmetry coordinates are used as a basis, the $(3N - 6)$ by $(3N - 6)$ dimensional F and G matrices can be broken down into smaller blocks corresponding to different irreducible representations of the group. For a very readable account of how to use symmetry to set up the F and G matrices, see [2]. Symmetry coordinates will be considered in Section 10.4.

10.2.4 GROUP FREQUENCIES

In general, normal modes are indeed combinations of motions of all atoms in a molecule. Nevertheless, some types of bonds and functional groups give rise to characteristic frequencies which are remarkably constant from one molecule to the next, regardless of the nature of the rest of the molecule. This gives rise to the powerful concept of group frequencies. In the qualitative analysis of organic molecules, for example, common functional groups can be readily determined from vibrational spectra. For example, the CH_2 group of aliphatic hydrocarbons gives rise to symmetric and asymmetric stretches observed at about 2850 and 2930 cm^{-1}, respectively, in a wide range of compounds. Carbonyl groups, C=O, on the other hand, can be found anywhere from about 1540 to 1870 cm^{-1}. Why is the methylene group vibration more predictable than that of the carbonyl group? When a characteristic vibrational frequency is referred to as a group frequency, that means that most of the potential energy of that mode is concentrated in a particular bond or group. In the case of C–H stretches, the high frequency of the local vibration of the C–H bond tends to uncouple that motion from that of the rest of the molecule. If you picture a large molecule as being connected by springs at every chemical bond and imagine that each spring has a similar force constant, then it is easy to see that plucking one spring will set the entire molecule in motion. On the other hand, if one spring between two atoms is of much higher force constant than the others, then that bond can vibrate separately from the rest of the system, and the motion can be called a local mode. The C–H group, like the hydroxyl group, O–H, observed at 3580 to 3650 cm^{-1}, has a rather high local mode frequency because of the small mass of hydrogen. The group frequency referred to as the "carbonyl stretch," on the other hand, is much more likely to involve motion of atoms other than the C=O group. In addition, the bond

Table 10.1 Characteristic group frequencies of organic compounds

Group	Frequency range, cm^{-1}
C–H stretch	2700–3100
C–H$_3$ bend	1375, 1450
C–H$_2$ scissor	1465
C–H$_2$ rock	720
C–H$_2$ wag, twist	1150–1350
C=C (unconjugated)	1640–1670
C≡C	2100–2260
C–H (aromatic, in plane bend)	1000–1300
C–H (aromatic, out of plane bend)	675–900
Skeletal vibrations of aromatics	1400–1500, 1585–1600
O–H stretch (nonhydrogen bonded)	3580–3650
O–H bend	1330–1420
C–O (alcohols, phenols)	1000–1260
C=O	1540–1870

strength and hence the force constant of a nominal C=O bond can vary greatly from one molecule to the next. For example, conjugated carbonyl bonds can be viewed in terms of contributions from resonance structures in which the carbon and oxygen are connected by single bonds, and this leads to lower frequencies (1600 to 1625 cm⁻¹) for the carbonyl stretching frequency. For comparison, the carbonyl stretch in acetone is found at 1707 cm⁻¹ in the neat liquid and 1723 cm⁻¹ in the gas phase. The frequency of the carbonyl stretch is also quite solvent dependent, red-shifting by tens of wavenumbers on going from a nonpolar to polar solvent. The rather large dipole moment derivative of a typical C=O stretch (about 2 D Å⁻¹ amu⁻¹/²) makes the vibrational frequency very sensitive to the polarity of the solvent.

Table 10.1 lists the common ranges of some of the group frequencies of organic compounds. The table does not indicate how much of the total potential energy of a normal mode is actually invested in that group or bond. This information must be obtained from a normal mode calculation for a particular molecule.

10.3 QUANTUM MECHANICS OF POLYATOMIC VIBRATIONS

Up to this point, our discussion of normal modes has employed the language of classical mechanics. A fortunate outcome of normal coordinate analysis is an exactly solvable vibrational Schrödinger equation, thanks to separation of variables. The vibrational Hamiltonian in terms of normal coordinates is

$$\hat{H}_{vib} = \sum_{i=1}^{3N-6} \left(\frac{-\hbar^2}{2} \frac{\partial^2}{\partial Q_i^2} + \frac{1}{2} \lambda_i Q_i^2 \right) \tag{10.32}$$

This is the sum of one-dimensional harmonic oscillator Hamiltonians. Thus the total wavefunction is a product of harmonic oscillator terms,

$$\psi_{vib}(Q_1, Q_2, \ldots Q_{3N-6}) = \chi_{v_1}(Q_1) \chi_{v_2}(Q_2) \cdots \chi_{v_{3N-6}}(Q_{3N-6}) \tag{10.33}$$

and the total energy is a sum.:

$$E_{vib} = \left(v_1 + \frac{1}{2} \right) h\nu_1 + \left(v_2 + \frac{1}{2} \right) h\nu_2 + \cdots \left(v_{3N-6} + \frac{1}{2} \right) h\nu_{3N-6} \tag{10.34}$$

Each one-dimensional harmonic oscillator wavefunction $\chi_{v_i}(Q_i)$ is of the form

$$\chi_{v_i}(Q_i) = N_{v_i} H_{v_i}(\alpha_i^{1/2} Q_i) \exp(-\alpha_i Q_i^2/2) \tag{10.35}$$

where N_{v_i} is a normalization constant, $H_{v_i}(\alpha_i^{1/2} Q_i)$ is a Hermite polynomial, and $\alpha_i = 2\pi\nu_i/\hbar$ (see Equation 9.17). A vibrational state can be designated by specifying all $3N-6$ quantum numbers, $(v_1, v_2, \ldots, v_{3N-6})$, each of which varies independently over the range $0, 1, 2, \ldots, \infty$. The raising and lowering operators defined previously can also be employed, where the commutator relationship is amended to give $[a_i, a_j^+] = \delta_{ij}$, and the labels i and j designate normal modes.

The existence of degenerate normal modes, as in CO_2, leads to degenerate vibrational states. A vibrational state of this molecule could be designated by $(v_1, v_{2a}, v_{2b}, v_3)$ or by (v_1, v_2, v_3), where the total number of quanta in the degenerate bend is $v_2 = v_{2a} + v_{2b}$. Thus a state having, for example, $v_2 = 2$ is actually triply degenerate, since the same total energy can be achieved by the three possible combinations: $(v_{2a}, v_{2b}) = (1, 1), (2, 0)$, or $(0, 2)$. As will be seen in a later section, anharmonic couplings have the potential to remove this type of degeneracy.

In order to consider selection rules and the possibility of anharmonic mixing, the symmetries of vibrational states need to be determined. The determination of these symmetries requires the group theoretical tools discussed in the next section.

10.4 GROUP THEORETICAL TREATMENT OF VIBRATIONS

10.4.1 FINDING THE SYMMETRIES OF NORMAL MODES

The calculation of normal mode frequencies for large molecules is conveniently handled by computer code and can provide a powerful tool for assigning observed vibrational frequencies. In this section we discuss a pencil-and-paper approach, based on group theory, for predicting the symmetries, and thus the Raman and IR activity, of the normal modes. Such an approach can be extremely useful in combination with the output of a normal coordinate calculation, or even in the absence of computational results. As an example, consider the ammonia molecule NH_3, which has enough symmetry (a threefold rotation axis) to illustrate the occurrence of degenerate normal modes. If the reader is unfamiliar with the language and practice of group theory, it would be a good idea to review the material in Appendix C before proceeding.

NH_3 belongs to the C_{3v} point group, whose character table is reproduced in Table 10.2. This is a group of order $h = 6$. There are thus six symmetry operations: the identity operation $\hat{E}$, two $\hat{C}_3$ rotations, and three $\hat{\sigma}_v$ reflections. It can be checked that the sum of the squares of the dimensions l_i of the irreducible representations (ir. reps., for short), A_1, A_2, and E, is $l_1^2 + l_2^2 + l_3^2 = 1^2 + 1^2 + 2^2 = 6$, which is equal to the order h of the group as required by theory. As per the usual notation, the doubly degenerate ir. rep. is labeled E, which should not be confused with the identity operation $\hat{E}$.

The two rightmost columns of a character table (see Appendix C) list some important functions that transform according to particular ir. reps. under the operations of the group. These functions provide useful information in the consideration of selection rules. Table 10.2 lists these functions for the C_{3v} point group. We now seek to find the characters of the reducible representation, Γ_{red}, given in the last line of Table 10.2. The basis for this reducible representation is the set of 12 Cartesian coordinates for the atoms of NH_3, as shown in Figure 10.3.

Table 10.2 C_{3v} character table

C_{3v}	$\hat{E}$	$2\hat{C}_3$	$3\hat{\sigma}_v$	
A_1	1	1	1	z, $x^2 + y^2$, z^2
A_2	1	1	-1	R_z
E	2	-1	0	(x, y) (R_x, R_y) $(x^2 - y^2, xy)$ (xy, yz)
Γ_{red}	12	0	2	

Γ_{red} spans the translations, rotations, and vibrations of the molecule, and can be decomposed into ir. reps. corresponding to these degrees of freedom. The three translational degrees of freedom belong to the same ir. reps. as the Cartesian coordinates x, y, and z, while the rotational degrees of freedom belong to those for the rotations: R_x, R_y, and R_z. Inspection of the character table allows the symmetries of the translations and rotations to be discerned. These are then subtracted from the decomposed Γ_{red} to get Γ_{vib}, the symmetries of the $3N-6$ normal coordinates. We now illustrate this procedure for the ammonia molecule.

Recall that the character for each operation is the trace of the matrix which represents that operation. For example, the identity operation is represented by a diagonal unit matrix.

$$
\begin{pmatrix}
1 & 0 & 0 & \cdots & 0 \\
0 & 1 & 0 & \cdots & 0 \\
0 & 0 & 1 & \cdots & 0 \\
\vdots & \vdots & \vdots & \ddots & \vdots \\
0 & 0 & 0 & \cdots & 1
\end{pmatrix}
\begin{pmatrix}
x_1 \\ y_1 \\ z_1 \\ \vdots \\ z_4
\end{pmatrix}
=
\begin{pmatrix}
x_1' \\ y_1' \\ z_1' \\ \vdots \\ z_4'
\end{pmatrix}
\tag{10.36}
$$

The primed quantities indicate the coordinates after the operation, which in the case of the identity operation are the same as the unprimed coordinates. The trace of this matrix is 12, so the character of the identity operation is 12, $\chi_{red}(\hat{E})=12$, as shown in the last line of Table 10.2. In figuring the character for the $\hat{C}_3$ operation, we follow the general rule that only atoms which are not moved by the symmetry operation contribute to the trace. Since the hydrogen atoms are interchanged by the rotation, there are no diagonal elements corresponding to the effect of $\hat{C}_3$ on their positions. The effect on the nitrogen atom, however, is given by

$$
\begin{pmatrix}
\cos(2\pi/3) & \sin(2\pi/3) & 0 \\
-\sin(2\pi/3) & \cos(2\pi/3) & 0 \\
0 & 0 & 1
\end{pmatrix}
\begin{pmatrix}
x_1 \\ y_1 \\ z_1
\end{pmatrix}
=
\begin{pmatrix}
x_1' \\ y_1' \\ z_1'
\end{pmatrix}
\tag{10.37}
$$

Since $\cos(2\pi/3)=-1/2$, the trace of the matrix which represents $\hat{C}_3$ is zero. Note that there are two $\hat{C}_3$ rotations ($\hat{C}_3$ and $\hat{C}_3^2$), but the character is the same for all operations within a class, so we do not need to consider the second rotation.

There are three operations $\hat{\sigma}_v$, one for each of the three reflection planes containing an N–H bond. Let us consider the one that interchanges the hydrogens labeled 2 and 4, in Figure 10.3, so that we need only consider the effect of $\hat{\sigma}_v$ on the coordinates of atoms 1 and 3. Reflection will preserve the x and z coordinates of the two atoms that lie in the reflection plane, and reverse the sign of the y coordinates. Thus the operation is represented by

$$
\begin{pmatrix}
1 & 0 & 0 & 0 & 0 & 0 \\
0 & -1 & 0 & 0 & 0 & 0 \\
0 & 0 & 1 & 0 & 0 & 0 \\
0 & 0 & 0 & 1 & 0 & 0 \\
0 & 0 & 0 & 0 & -1 & 0 \\
0 & 0 & 0 & 0 & 0 & 1
\end{pmatrix}
\begin{pmatrix}
x_1 \\ y_1 \\ z_1 \\ x_3 \\ y_3 \\ z_3
\end{pmatrix}
=
\begin{pmatrix}
x_1 \\ -y_1 \\ z_1 \\ x_3 \\ -y_3 \\ z_3
\end{pmatrix}
\tag{10.38}
$$

The trace of the matrix representation of $\hat{\sigma}_v$ is 2. If we chose one of the other two reflection planes, both the x and y coordinates for the two stationary atoms would be at angles to the reflection plane. You should convince yourself that you would get the same trace for the matrix representation of $\hat{\sigma}_v$ using one of these other two reflections.

Recall that Γ_{red} spans the vibrations, rotations, and translations of the molecule. The next task is to decompose this reducible representation and subtract out the ir. reps. that correspond to rotation and translation.

A reducible representation Γ_{red} is equivalent to a linear combination $\sum a_i \Gamma_i$ of ir. reps. Γ_i. The number of times, a_i, that the i-th ir. rep. contributes to Γ_{red} is found from:

$$a_i = \frac{1}{h}\sum \chi_{red}(\hat{R})\chi_i(\hat{R}) \tag{10.39}$$

where the sum is over the operations of the group $\hat{R}$. Applying this to the problem at hand, we find:

$$a_{A_1} = \frac{1}{6}\left[(12)(1)+2(0)(1)+3(2)(1)\right] = 3 \tag{10.40}$$

$$a_{A_2} = \frac{1}{6}\left[(12)(1)+2(0)(1)+3(2)(-1)\right] = 1 \tag{10.41}$$

$$a_E = \frac{1}{6}\left[(12)(2)+2(0)(-1)+3(2)(0)\right] = 4 \tag{10.42}$$

Note that the factors of 2 and 3 in the second and third terms in each sum come from the number of $\hat{C}_3$ and $\hat{\sigma}_v$ operations, respectively. Thus $\Gamma_{red} = 3A_1 + A_2 + 4E$. From the character table, the Cartesian coordinate z transforms according to the totally symmetric (A_1) representation, and x and y transform as a pair according to the E representation. Thus $\Gamma_{trans} = A_1 + E$. Similarly, by locating the rotations R_x, R_y, and R_z we conclude that $\Gamma_{rot} = A_2 + E$. The vibrations are accounted for by $\Gamma_{vib} = \Gamma_{red} - \Gamma_{trans} - \Gamma_{rot} = 2A_1 + 2E$. The $3(4) - 6 = 6$ vibrational degrees of freedom of NH_3 comprise two totally symmetric (A_1) and two doubly degenerate (E) modes. For each doubly degenerate mode there are two sets of atomic displacements having the same vibrational frequency. The normal modes and their frequencies are pictured in Figure 10.5. For each E symmetry mode, only one set of displacements is shown. The totally symmetric stretch and bend are referred to as parallel modes, as they result in a net change in dipole moment parallel to the symmetry axis z. Conversely, the degenerate stretch and the degenerate bend are perpendicular modes. All the NH_3 fundamentals are IR active, as will be discussed in a later section.

A more straightforward way to find Γ_{vib} is to use a set of internal coordinates as a basis for the reducible representation. In the case of ammonia, we could choose the three bond lengths r_1, r_2, r_3 and the three H–N–H bond angles $\theta_1, \theta_2, \theta_3$. The reader should verify that the characters of the resulting reducible representation are

	$\hat{E}$	$2\hat{C}_3$	$3\hat{\sigma}_v$
Γ_{red}	6	0	2

(10.43)

In contrast to the previously considered reducible representation, Γ_{red} as given in 10.43 spans only the vibrations. Using Equation 10.39, we find $\Gamma_{red} = \Gamma_{vib} = 2A_1 + 2E$, the same result that was obtained previously. The use of internal coordinates as a basis is a much more direct way to find the normal mode symmetries.

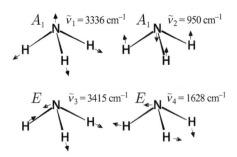

Figure 10.5 Normal modes of NH_3.

Table 10.3 Fundamental frequencies of methane-h_4 and -d_4

	CH$_4$, $\tilde{v}$ in cm^{-1}	CD$_4$, $\tilde{v}$ in cm^{-1}
$\tilde{v}_1, A_1$	2914	2085
$\tilde{v}_2, E$	1526	1054
$\tilde{v}_3, T_2$	3020	2258
$\tilde{v}_4, T_2$	1306	996

In the case of a planar molecule, it is necessary to include angles for out-of-plane motion in order to obtain the contribution of out-of-plane modes.

Some molecules present problems in the form of *redundancy* of the internal coordinates. For example, CH$_4$ has nine normal modes, one nondegenerate C–H stretch, a doubly degenerate bend, a triply degenerate bend, and a triply degenerate stretch. See Table 10.3 for a list of the fundamental frequencies in CH$_4$ and CD$_4$. Note the general trend displayed there: stretching vibrations are higher in frequency than bending modes. A natural choice for the internal coordinates is the set of four C–H bond length displacements and six HCH bond angle displacements. This adds up to ten coordinates, therefore one of the group theoretically derived symmetry species must be discarded as redundant. The choice of which one to discard can be made by recognizing that the six bond angles may not vary independently: the displacements must sum to zero. This condition on the sum of the angle displacements is totally symmetric, so we expect to get an extra coordinate of that symmetry. Proceeding as above in the ammonia example, the ten internal coordinates yield a reducible representation Γ_{red} that decomposes to $2A_1 + E + 2T_2$. The redundant coordinate which should be thrown away is the totally symmetric combination of HCH bond angle displacements. This coordinate is discarded because it is impossible for a molecule with T_d symmetry to have a totally symmetric bending mode.

The use of internal coordinates provides a natural point of reference for finding the symmetry coordinates referred to in the previous section. Projection operators can be employed, as described in [2], to find linear combinations of internal coordinates which correspond to the correct symmetries. As will be shown below, in the case of ammonia these are found to be

$$S_1 = \frac{1}{\sqrt{3}}\left(\Delta r_1 + \Delta r_2 + \Delta r_3\right) \tag{10.44}$$

$$S_2 = \frac{1}{\sqrt{3}}\left(\Delta\theta_1 + \Delta\theta_2 + \Delta\theta_3\right) \tag{10.45}$$

$$S_{3a} = \frac{1}{\sqrt{6}}\left(2\Delta r_1 - \Delta r_2 - \Delta r_3\right) \tag{10.46}$$

$$S_{3b} = \frac{1}{\sqrt{2}}\left(\Delta r_2 - \Delta r_3\right) \tag{10.47}$$

$$S_{4a} = \frac{1}{\sqrt{6}}\left(2\Delta\theta_1 - \Delta\theta_2 - \Delta\theta_3\right) \tag{10.48}$$

$$S_{4b} = \frac{1}{\sqrt{2}}\left(\Delta\theta_2 - \Delta\theta_3\right) \tag{10.49}$$

These equations illustrate how a doubly degenerate vibrational mode comprises two different sets of displacements. The projection operator approach for obtaining these symmetry coordinates is described here in brief. The operator

$$\hat{P}_j = \frac{l_j}{h}\sum_R \chi_j(\hat{R})\hat{R} \tag{10.50}$$

is applied to a basis for the representation of a group to project out the j-th ir. rep. For NH_3, let us take the vectors r_1, r_2, and r_3 as a basis. The effect of operating on r_1 with each of the six operations of the C_{3v} group is as follows:

$$\hat{E}r_1 = r_1 \quad \hat{C}_3r_1 = r_2 \quad \hat{C}_3^2r_1 = r_3$$
$$\hat{\sigma}_v r_1 = r_1 \quad \hat{\sigma}_v' r_1 = r_3 \quad \hat{\sigma}_v'' r_1 = r_2 \tag{10.51}$$

where the reflection planes $\sigma_v, \sigma_v', \sigma_v''$ pass through r_1, r_2, r_3, respectively. Applying Equation 10.50 to the basis vector r_1, and using Equations 10.51, we can project out the A_1, A_2, and E components:

$$\hat{P}_{A_1}r_1 = \frac{1}{6}\left(r_1 + r_2 + r_3 + r_1 + r_3 + r_2\right) = \frac{1}{3}\left(r_1 + r_2 + r_3\right)$$

$$\hat{P}_{A_2}r_1 = \frac{1}{6}\left(r_1 + r_2 + r_3 - r_1 - r_3 - r_2\right) = 0 \tag{10.52}$$

$$\hat{P}_{E}r_1 = \frac{2}{6}\left(2r_1 - r_2 - r_3\right) = \frac{1}{3}\left(2r_1 - r_2 - r_3\right)$$

Equations 10.52 are not quite complete: the projected coordinates are not normalized and only one of the two E-symmetry coordinates has been obtained. These problems are taken care of by requiring that the symmetry coordinates be orthonormal. For example, the totally symmetric coordinate can be defined as $S_1 = (r_1 + r_2 + r_3)/\sqrt{3} = \sum_i C_{1i}r_i$, where the coefficients C_{ji} form the elements of a transformation matrix which converts from internal coordinates to symmetry coordinates. In matrix form, this transformation (like that from mass-weighted Cartesian to normal coordinates) is written compactly: $\mathbf{S} = \mathbf{CR}$, where $\mathbf{S}$ is a column vector containing the symmetry coordinates and $\mathbf{R}$ is one containing the internal coordinates. Since the transformation is unitary, $\mathbf{C}^T\mathbf{C}$ is equal to the unit matrix. Normalization of the symmetry coordinates requires $\sum_k C_{k1}C_{1k} = 1$ or, in matrix notation,

$$\mathbf{C}_1^T\mathbf{C}_1 = \left(\frac{1}{\sqrt{3}} \quad \frac{1}{\sqrt{3}} \quad \frac{1}{\sqrt{3}}\right)\begin{pmatrix} \frac{1}{\sqrt{3}} \\ \frac{1}{\sqrt{3}} \\ \frac{1}{\sqrt{3}} \end{pmatrix} = 1 \tag{10.53}$$

Similarly, the E coordinate can be normalized to give $S_2 = (2r_1 - r_2 - r_3)/\sqrt{6}$. There is one more E coordinate, call it S_3, that is not obtained directly using this approach, but it can be determined by requiring that S_3 be normalized and orthogonal to S_2. Using the function $S_3 = (r_2 - r_3)/\sqrt{2}$, these requirements are met. For example,

$$\mathbf{C}_2^T\mathbf{C}_3 = \frac{1}{\sqrt{12}}(2 \quad -1 \quad -1)\begin{pmatrix} 0 \\ 1 \\ -1 \end{pmatrix} = 0 \tag{10.54}$$

By replacing the distances r_i by the displacements Δr_i and repeating for the angles $\Delta\theta_i$, the symmetry coordinates of Equations 10.44 through 10.49 are obtained.

10.4.2 SYMMETRIES OF VIBRATIONAL WAVEFUNCTIONS

Selection rules for vibrational spectroscopy and anharmonic coupling of normal modes are examples of problems for which the symmetries of vibrational states are required. Once the symmetries of the normal modes are known, the symmetry of any given vibrational state can be deduced. In general, for a state having v_1

quanta in the first normal mode, v_2 quanta in the second normal mode, and so forth, the overall symmetry of the vibrational state specified by the quantum numbers $\{v\} \equiv (v_1, v_2, ..., v_{3N-6})$ is the direct product of the symmetries of the one-dimensional vibrational states, $\psi_{v_1}(Q_1), \psi_{v_2}(Q_2), ..., \psi_{v_{3N-6}}(Q_{3N-6})$. This direct product can be represented by

$$\Gamma_{\{v\}} = \Gamma_{v_1} \times \Gamma_{v_2} \times \cdots \Gamma_{v_{3N-6}} \tag{10.55}$$

Let us first consider the symmetries of the one-dimensional wavefunctions. The ground state $v = 0$ of any vibration is always totally symmetric, regardless of the symmetry of the mode. This may seem to go against intuition, but recall that the ground state vibrational wavefunction has the form $N \exp(-\alpha Q^2/2)$. *The normal coordinate Q has the symmetry of the vibrational mode*; it transforms according to one of the ir. reps. of the group, as discussed above. But its square, and thus $\exp(-\alpha Q^2/2)$, transforms according to the representation which is the direct product of the ir. rep. of Q with itself, $\Gamma_Q \times \Gamma_Q$. The direct product of an ir. rep. with itself is equal to (in the case of a one-dimensional ir. rep.) or contains (in the case of a degenerate ir. rep.) the totally symmetric representation. The ground vibrational state is *always* totally symmetric.

Next consider *nondegenerate* normal modes. The form of a one-dimensional harmonic oscillator wavefunction is $H_v(\alpha^{1/2}Q) \exp(-\alpha Q^2/2)$ (see Equation 10.35), where $H_v(\alpha^{1/2}Q)$ is a polynomial that contains either even or odd powers of Q according to whether the quantum number v is even or odd. Direct products of one-dimensional representations can only result in other one-dimensional representations. We conclude that states with even numbers of quanta v are totally symmetric, regardless of the symmetry of Q; i.e., even powers of ± 1 are always $+1$. States having odd numbers of quanta have the same symmetry as the normal mode. Summarizing, the symmetries of nondegenerate normal modes alternate between totally symmetric and the symmetry of the mode, according to whether v is even or odd.

For example, the water molecule H_2O has three nondegenerate normal modes. The symmetric stretch and the bend are totally symmetric (A_1), and the asymmetric stretch corresponds to B_2, as shown in Figure 10.6. We are using a coordinate system in which the $\hat{C}_2$ rotation axis coincides with the z direction and the molecule lies in the yz plane. (If one chooses instead to have the molecule lie in the xz plane, keeping the z axis as the symmetry axis, then the asymmetric stretch vibration is designated B_1 instead of B_2. To be meaningful, symmetry labels must be referred to a particular choice of the molecular coordinate system.) The total vibrational state is specified by three quantum numbers (v_1, v_2, v_3). Because all the vibrational states of the symmetric stretch and the bend are totally symmetric, the overall symmetry is decided by Γ_{v_3}, which is either A_1 if v_3 is even or B_2 if v_3 is odd.

Degenerate vibrations are somewhat more complicated. It is still true that the ground state is totally symmetric, and the $v = 1$ state has the symmetry of the normal mode. Notice that the symmetry species of a state reflects its quantum mechanical degeneracy. Consider a triply degenerate vibration, such as the asymmetric stretch of a tetrahedral AX_4 molecule, corresponding to the T_2 symmetry species. The ground state $(v_a, v_b, v_c) = (0, 0, 0)$ is nondegenerate, as any totally symmetric state must be. The first excited level belongs to T_2 and is threefold degenerate because the states $(1, 0, 0)$, $(0, 1, 0)$, and $(0, 0, 1)$ have the same energy. The energy level having two quanta in the asymmetric stretch includes the states $(1, 1, 0)$, $(1, 0, 1)$, $(0, 1, 1)$, $(2, 0, 0)$,

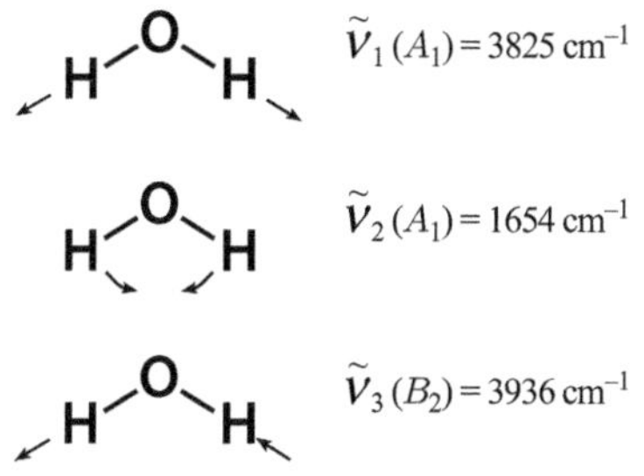

$$\tilde{V}_1 (A_1) = 3825 \text{ cm}^{-1}$$

$$\tilde{V}_2 (A_1) = 1654 \text{ cm}^{-1}$$

$$\tilde{V}_3 (B_2) = 3936 \text{ cm}^{-1}$$

Figure 10.6 Normal modes of H_2O.

(0, 2, 0), and (0, 0, 2) and is therefore sixfold degenerate. There are no six-dimensional representations in the T_d point group, so the representation to which this degenerate state belongs must be reducible.

Here we will consider doubly degenerate normal modes in states having $v = v_a + v_b \geq 2$. The characters of the corresponding reducible representation $\chi_v(R)$ are found using the formula (see [3]):

$$\chi_v(\hat{R}) = \frac{1}{2}\left[\chi(\hat{R})\chi_{v\text{-}1}(\hat{R}) + \chi(\hat{R}^v) \right] \tag{10.56}$$

where $\chi(\hat{R})$ is the character of the ir. rep. of the mode for the operation $\hat{R}$, $\chi_{v\text{-}1}(\hat{R})$ is the character of the representation to which the $v - 1$ state belongs, and $\chi(\hat{R}^v)$ is the character of the ir. rep. of the mode for the operation $\hat{R}^v$. This is illustrated by the following example for the $v_2 = 2$ state of the asymmetric stretch of NH_3. We expect this energy level to be triply degenerate; the states $(v_{2a}, v_{2b}) = (1, 1)$, $(2, 0)$ and $(0, 2)$ have the same energy. When Equation 10.56 is applied to this state, $\chi_{v-1}(\hat{R}) = \chi_1(\hat{R}) = \chi(\hat{R})$ because the symmetry of the $v = 1$ state is that of the mode. The characters of the reducible representation are found as follows:

$$\chi_2(\hat{E}) = \frac{1}{2}\left\{ \left[\chi(\hat{E}) \right]^2 + \chi(\hat{E}^2) \right\} = \frac{1}{2}(4+2) = 3 \tag{10.57}$$

$$\chi_2(\hat{C}_3) = \frac{1}{2}\left\{ \left[\chi(\hat{C}_3) \right]^2 + \chi(\hat{C}_3^2) \right\} = \frac{1}{2}(1-1) = 0 \tag{10.58}$$

$$\chi_2(\hat{\sigma}_v) = \frac{1}{2}\left\{ \left[\chi(\hat{\sigma}_v) \right]^2 + \chi(\hat{\sigma}_v^2) \right\} = \frac{1}{2}(0+2) = 1 \tag{10.59}$$

Using Equation 10.39, this is decomposed to give $\Gamma_{red} = A_1 + E$. As a check on the calculation, note that the threefold degeneracy of the state is accounted for by the sum of a singly degenerate and a doubly degenerate representation. Now that the symmetries of the $v = 2$ state are known, those of the state having $v = 3$ can be obtained from Equation 10.56, using the characters $\chi_2(R)$, as given in Equations 10.57 through 10.59, for $\chi_{v-1}(\hat{R})$ and so on to find the symmetries of higher energy states. This is illustrated in Problem 5.

10.5 SELECTION RULES FOR INFRARED ABSORPTION AND RAMAN SCATTERING: GROUP THEORETICAL PREDICTION OF ACTIVITY

IR and Raman spectra are generally dominated by the fundamental transitions, in which the number of quanta in one normal mode increases (or decreases) by one and all other normal modes remain in the same state. It is often the case that the initial state is the ground vibrational state; thus transitions such as $(0, 0, 0, \ldots) \to (1, 0, 0, \ldots)$ are most intense. When the Boltzmann population of an excited vibrational state is significant, hot bands can be observed; i.e. $(1, 0, 0, \ldots) \to (2, 0, 0, \ldots)$, as well as anti-Stokes Raman scattering $(1, 0, 0, \ldots) \to (0, 0, 0, \ldots)$. Anharmonicity of either the electrical or mechanical variety can result in the appearance of overtones, such as $(0, 0, 0, \ldots) \to (2, 0, 0, \ldots)$ and combination bands, such as $(0, 0, 0, \ldots) \to (1, 0, 1, \ldots)$. Difference bands also become allowed; for example, the transition $(1, 1, 0, \ldots) \to (2, 0, 0, \ldots)$, provided the initial state is populated. The absence of overtones and combinations in the harmonic case is the result of taking the transition operators for IR and Raman to be

$$\hat{\mu} = \sum_{i=1}^{3N-6} \left(\frac{\partial \mu}{\partial Q_i} \right)_0 Q_i \tag{10.60}$$

$$\hat{\alpha}_{\rho\sigma} = \sum_{i=1}^{3N-6} \left(\frac{\partial \alpha_{\rho\sigma}}{\partial Q_i} \right)_0 Q_i \tag{10.61}$$

Since the total wavefunction is a product of $3N - 6$ one-dimensional orthonormal wavefunctions, the operators of Equations 10.60 and 10.61 can only connect product states which differ by the quantum number for *one* normal mode. The position operator Q_i is proportional to $\left(a_i^+ + a_i\right)$. Thus operators which are linear in Q_i can only connect states which differ by one quantum of excitation in a single normal mode, and the selection rule $\Delta v = \pm 1$ is obtained.

Next we consider the group theoretical basis for Raman and IR activity of fundamentals. The tremendous advantage of group theory is the following: matrix elements of an operator $\hat{O}$ with respect to initial and final states, i.e. $\langle \psi_i | \hat{O} | \psi_f \rangle$, are guaranteed to vanish *unless* the triple direct product $\Gamma_i \times \Gamma_{\hat{o}} \times \Gamma_f$ equals or contains the totally symmetric representation. Thus it is necessary to find the symmetry species to which the wavefunctions (Γ_i, Γ_f) and the operator $\left(\Gamma_{\hat{o}}\right)$ correspond. Neglecting both mechanical and electrical anharmonicity, we need only consider fundamental transitions, and we will assume that the initial state is the ground vibrational state. Thus Γ_i is the totally symmetric representation, and Γ_f is the ir. rep. of the normal mode. A fundamental transition will thus be allowed if $\Gamma_{\hat{o}} = \Gamma_f$, since this results in $\Gamma_{\hat{o}} \times \Gamma_f$ equal to (or containing) the totally symmetric representation.

The dipole moment operator has three components corresponding to the three Cartesian coordinates. Thus the fundamental of a normal mode is IR active if it belongs to the same ir. rep. as one or more of the Cartesian coordinates. The group theory prediction is in perfect agreement with reasoning based on whether or not a normal mode results in a changing dipole moment. For example, all the normal modes of NH_3 shown in Figure 10.5 are IR active because they result in a change in the dipole moment in a direction either parallel or perpendicular to the symmetry axis. The A_1 modes share the symmetry species of the coordinate z and are parallel polarized. The E modes coincide with the symmetry of the (x, y) coordinates, and they are polarized in the perpendicular direction.

The polarizability operator has components that transform as the products and squares of Cartesian coordinates. Thus a fundamental is Raman active if the normal coordinate belongs to the same ir. rep. as one of the functions x^2, y^2, xy, etc. For example, a glance at the C_{3v} character table (Table 10.2) reveals that all the fundamentals of NH_3 are Raman active.

An important and powerful consequence of group theory is easily applied whenever a molecule has a center of symmetry, in which case one of the operations of the group is the inversion operation $\hat{i}$. In centrosymmetric molecules, the coordinates x, y, and z belong to ungerade ir. reps. Products of Cartesian coordinates thus have gerade symmetry, because $u \times u = g$. This results in the IR and Raman fundamentals being mutually exclusive. Thus vibrational spectroscopy readily reveals whether a molecule possesses inversion symmetry. An example of such a case is the CO_2 molecule, where the asymmetric stretch and the bend are IR active and Raman inactive, while the symmetric stretch is IR forbidden and Raman allowed. Benzene, SF_6, ethylene, and ethane in the staggered confirmation are examples of molecules having inversion symmetry.

Chemical intuition is readily applied to deduce IR activity, as one can imagine bond dipole moments which move during the vibration, and figure out whether the net dipole moment is changed when the bond dipole changes are added vectorially. It may be less obvious how to use similar reasoning to deduce Raman activity. Aren't all molecules polarizable, and shouldn't this polarizability change during any vibrational mode? Yes, and yes! However, the Raman activity depends on $(\partial \alpha / \partial Q)_0$, which is evaluated at the equilibrium geometry. Consider the two graphs of polarizability versus normal coordinate as shown in Figure 10.7. The slope of the graph in Figure 10.7a is zero at the equilibrium geometry, so the mode is Raman inactive. Such a situation arises, for example, in the bending mode and in the symmetric stretch of CO_2. The polarizability might increase

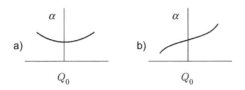

Figure 10.7 Polarizability versus normal coordinate for (a) Raman inactive and (b) Raman-active normal modes.

(or decrease) when the molecule bends (or stretches) in one direction, and it will do the same when bending (or stretching) in the other direction, resulting in a slope $(\partial\alpha/\partial Q)_0$ which is zero at the symmetric linear geometry. The polarizability during the symmetric stretch, on the other hand, resembles Figure 10.7b, where the polarizability is assumed to increase with increasing bond length. We reach the conclusion that $(\partial\alpha/\partial Q)_0 \neq 0$ for this type of vibration, and it is thus Raman active.

These considerations do not take into account the fact that the Raman transition polarizability is a second-rank tensor. The symmetry of the Raman tensor dictates the polarization properties of the scattered light. Polarization in Raman scattering is a powerful way to deduce whether a mode is totally symmetric. This will be discussed in Chapter 12.

Group theory can only tell us if a vibrational mode is active in the Raman or IR based on symmetry; it cannot address how intense a band will be once it is known to be allowed. For example, although all three fundamentals of H_2O are Raman allowed, they are not very intense because the polarizability derivatives are rather small. Polarizability correlates with electron density, and in the water molecule the electron density is concentrated on the oxygen atom. Since the position of the center of mass is nearly coincident with the oxygen atom, the magnitude of the change in polarizability during a water molecule vibration is small. On the other hand, water is a strong IR absorber, due to the large dipole moment derivatives for stretching and bending.

These symmetry considerations are also based on the assumption of harmonic normal modes and transition operators which are linear in Q_i. When these assumptions break down, transitions other than the fundamental appear: overtone and combination bands. As in the case of diatomic molecules, perturbation theory can be used to treat these "forbidden" transitions, and electrical and/or mechanical anharmonicity may be responsible for their appearance. For example, the following component of the dipole operator

$$\hat{\mu} = \left(\frac{\partial^2 \mu}{\partial Q_1 \partial Q_2} \right)_0 Q_1 Q_2 \tag{10.62}$$

permits the combination band at frequency $\nu_1 + \nu_2$. This is an example of electrical anharmonicity.

10.6 ROTATIONAL STRUCTURE

Vibrational spectroscopy of gas-phase molecules provides another approach, in addition to pure rotational spectroscopy, for observing rotational transitions. As for pure rotational spectra, the rotational structure of vibrational bands can be quite complex for asymmetric tops. The discussion here is limited to IR spectra of linear and symmetric top molecules. The factorization of the wavefunction into vibrational and rotational parts enables us to recycle some of the selection rules for rotational transitions that were obtained in the previous chapter. However, we must take into account whether the transition dipole is parallel or perpendicular to the symmetry axis. For parallel transitions, the selection rules are identical to those for pure rotational transitions: $\Delta J = 0, \pm 1$ and $\Delta K = 0$, with $\Delta J = 0$ prohibited for $K = 0$ and thus also for $J = 0$. Thus for parallel transitions of closed-shell linear molecules, where K is always zero, we observe the rotational structure typical of a diatomic molecule, i.e., P and R branches. For symmetric tops, the structure is more complex because for each branch pertaining to a change in the quantum number J, there is a series of lines for different initial values of K. If the vibrational transition is polarized perpendicular to the symmetry axis, whether linear or symmetric top, the selection rule is $\Delta J = 0, \pm 1$ and $\Delta K = \pm 1$, with $\Delta J = 0$ still prohibited when J or K is zero.

The transition dipole moment for the asymmetric stretch of CO_2, at about 2350 cm^{-1}, is parallel to the symmetry axis. Thus the IR band for this vibration (Figure 10.8) lacks a Q branch. In contrast, the bending vibration of CO_2, at about 670 cm^{-1}, is perpendicularly polarized and the Q branch is prominent, as shown in Figure 10.8.

Next consider the example IR spectrum of Figure 10.9, showing the rotational structure of the CO stretch of methanol. Methanol is a near-prolate symmetric top, and the transition dipole for the CO stretch is coincident with the near-symmetry axis. The barrier to internal rotation in CH_3OH is about 400 cm^{-1}, so at room temperature the methyl and hydroxyl groups do not rotate freely with respect to one another. However, the

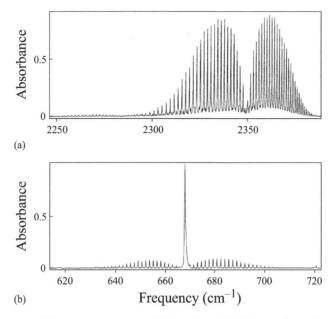

(a)

(b)

Figure 10.8 IR spectrum of CO_2, asymmetric stretch (a) and bend (b). (From *Infrared Spectra for Quantitative Analysis of Gases*, P. L. Hanst and S.T. Hanst, copyright 1992 Infrared Analysis, Inc., Anaheim, CA. With permission.)

small contribution of the off-axis hydrogen leads to $\tilde{B} \approx \tilde{C} \approx 0.8$ cm^{-1}, while the rotational constant for rotation about the "symmetry axis" is $\tilde{A} \approx 4$ cm^{-1}. Recall that the rotational energy of a prolate symmetric rotor is $E_{JK}/h = B_{[v]}J(J + 1) + (A_{[v]} - B_{[v]})K^2$, where $A_{[v]} > B_{[v]}$, and [v] represents the set of vibrational quantum numbers on which the rotational constants depend. The vibration–rotation transition frequencies are thus given by

$$v = v_0 + B'J'(J'+1) - B''J''(J''+1) + (A'-B')K'^2 - (A''-B'')K''^2 \qquad (10.63)$$

Double-primed quantities represent initial state values and single-primed ones designate final state values, and v_0 is the frequency of the vibrational transition. The Q branch transitions in the case of a parallel polarized vibration, for which $J' = J'' \equiv J$ and $K' = K'' \equiv K$, are found at

$$v_Q = v_0 + (B' - B'')J(J+1) + [(A'-B') - (A''-B'')]K^2 \qquad (10.64)$$

Since the rotational constants A' and B' in the upper vibrational state are expected to be smaller than their counterparts in the lower vibrational state, Equation 10.64 predicts that the Q branch lines move to lower frequency with increasing $|K|$ or J. Each of these Q branches, for a particular value of K, serves as an origin for a series of P and R branch lines, called a *subband*. Were it not for vibration–rotation coupling, the separation of

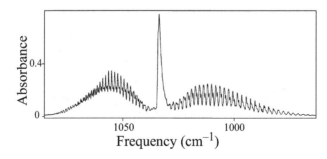

Figure 10.9 IR vibration–rotation spectrum of the CO stretch of CH_3OH. (From *Infrared Spectra for Quantitative Analysis of Gases*, P. L. Hanst and S.T. Hanst, copyright 1992 Infrared Analysis, Inc., Anaheim, CA. With permission.)

adjacent lines in each subband would be $2B$. The spacing is only approximately constant, due to the difference in B' and B''. Since the spacing within the P or R branch is typically larger than that of the subband origins, under low resolution the spectrum looks like that of a linear molecule with an allowed Q branch. There are still subtle differences, however. Besides being displaced in frequency, the subbands are not identical by virtue of the fact that J is always greater than or equal to $|K|$. Thus, as K increases, the P and R branches of the subbands start out at larger values of J.

In the methanol spectrum of Figure 10.9, there is also structure that results from nuclear spin statistics. The presence of three equivalent hydrogens results in alternation of the statistical weights of various K states. As discussed in [3], states for which K is an integral multiple of three have twice the weight of the others, leading to the intensity pattern 2, 1, 1, 2, 1, 1... for $|K| = 0, 1, 2, 3, 4, 5....$ All these factors make analysis of the low-resolution vibration/rotation spectrum of methanol very interesting.

The rotational structure of perpendicularly polarized transitions of symmetric top molecules is even more interesting, due to the selection rule $\Delta K = \pm 1$. In this case there is less resemblance of the low-resolution spectrum to that for a linear molecule, as the $\Delta K = +1$ and $\Delta K = -1$ subbands are displaced to opposite sides of the main Q branch. See [3] and [4] for more discussion of this problem.

What becomes of the rotational structure as a gas is cooled and eventually condensed to a liquid? As the temperature of the gas is decreased, the maxima in the P and R branches shift to smaller J, and thus the width of the spectrum spanned by the rotational transitions decreases. On going to the liquid phase, the rotational structure is lost, but reorientational motion continues to contribute to the spectral width of the IR band, as discussed in Chapter 5. The rotational contribution to the IR lineshape derives from the reorientational motion of the transition moment. In addition, vibrational relaxation (dephasing and population relaxation) contributes to the width of the IR band. The measurement of the IR spectrum alone does not permit the contributions of vibrational and rotational relaxation to be separated. In favorable cases, this separation can be accomplished by measuring the polarized and depolarized Raman spectra. See Chapter 5 for a discussion of this.

10.7 ANHARMONICITY

Perturbation theory can be used as described in the previous chapter to account for the expected decrease in the spacing of vibrational energy levels at higher energies. The perturbed vibrational energy are written as

$$E_{vib} = \sum_i \left(v_i + \frac{1}{2} \right) h\nu_i + \sum_{i \leq j} hx_{ij} \left(v_i + \frac{1}{2} \right) \left(v_j + \frac{1}{2} \right) + \cdots \qquad (10.65)$$

where the anharmonicity correction factors x_{ij}, unlike x_e for diatomics, are defined such that they are generally negative and have units of frequency. Experimental determination of the x_{ij}'s requires the observation of a suitable number of overtone and combination bands. The vibrational spectrum of water vapor, for example, has been measured to great accuracy, and overtones and combinations extending into the visible region of the spectrum can be observed. The harmonic frequencies and anharmonic correction factors for the water molecule are given in Table 10.4, and the normal modes are depicted in Figure 10.6.

Two special cases of mechanical anharmonicity are *Coriolis coupling* and *Fermi resonance*. As it turns out, both of these effects are displayed in the vibrational spectrum of CO_2, which will be used to illustrate the basic principles.

Table 10.4 Vibrational constants for water

$\tilde{\nu}_1 = 3825.32$ cm^{-1}	$x_{11} = -43.89$ cm^{-1}	$x_{12} = -20.02$ cm^{-1}
$\tilde{\nu}_2 = 1653.91$ cm^{-1}	$x_{22} = -19.5$ cm^{-1}	$x_{13} = -155.06$ cm^{-1}
$\tilde{\nu}_3 = 3935.59$ cm^{-1}	$x_{33} = -46.37$ cm^{-1}	$x_{23} = -19.81$ cm^{-1}

Source: B. T. Darling and D. M. Dennison, The water vapor molecule, *Phys. Rev.* 57, 128 (1940).

Figure 10.10 Vibrational angular momentum in the bending overtone of CO_2.

Coriolis coupling is a splitting observed in vibrational states having more than one quantum of excitation in a degenerate bending vibration. It is a form of vibration–rotation coupling. From a classical point of view, the Coriolis force, like the centrifugal force, is an apparent force associated with motion in a rotating coordinate system. If an object in a rotating reference frame is displaced radially, a sidewise force (the Coriolis force) must be exerted. In the molecular example at hand, the rotational motion derives from excitation of the bending overtone of CO_2, which leads to angular momentum about the bond.

Consider the triply degenerate $v = 2$ level of the bending mode of CO_2, which consists of the states $(v_{2a}, v_{2b}) = (1, 1)$, $(2, 0)$, and $(0, 2)$. Figure 10.10 illustrates the nuclear motion in the case where the two perpendicular bending motions are $90°$ out of phase. The net result is that each atom executes circular motion and there is angular momentum about the bond axis. A rigid linear molecule has no such angular momentum, but in the case where the bending vibration is excited, *vibrational angular momentum* results. To consider this from a quantum mechanical point of view, take the zero-order wavefunctions for the three states to have the form:

$$\psi_1 \propto 4\alpha_2 Q_a Q_b \exp\left[-\alpha\left(Q_a^2 + Q_b^2\right)/2\right] \tag{10.66}$$

$$\psi_2 \propto \left(4\alpha_2 Q_a^2 - 2\right)\exp\left[-\alpha\left(Q_a^2 + Q_b^2\right)/2\right] \tag{10.67}$$

$$\psi_3 \propto \left(4\alpha_2 Q_b^2 - 2\right)\exp\left[-\alpha\left(Q_a^2 + Q_b^2\right)/2\right] \tag{10.68}$$

which correspond respectively to $(1, 1)$, $(2, 0)$, and $(0, 2)$ states. It is convenient to make a change of coordinates from Q_a and Q_b to $r = \left(Q_a^2 + Q_b^2\right)^{1/2}$ and $\varphi = \tan^{-1}\left(Q_b/Q_a\right)$. Then, after expressing the wavefunctions in terms of these new coordinates, we are free to take linear combinations of them, since they are degenerate. As shown in [5], one particular set of zero-order wavefunctions which are linear combinations of the ψ_i's can be found, after some algebra, to be

$$\psi_1' \propto r^2 \exp(2i\varphi) \tag{10.69}$$

$$\psi_2' \propto r^2 \exp(-2i\varphi) \tag{10.70}$$

$$\psi_3' \propto r^2 - 1 \tag{10.71}$$

The significance of this result is that the first two wavefunctions ψ_1' and ψ_2' depend on the angle φ, and indeed all three of them are eigenfunctions of the angular momentum operator $L_z = -i\hbar\left(\partial/\partial\varphi\right)$ (where φ is the angle of rotation about the bond axis), having eigenvalues $m = 2\hbar, -2\hbar$, and 0. Anharmonic coupling in the form of vibration–rotation interaction mixes the zero-order states ψ_1, ψ_2, and ψ_3 resulting in $v_2 = v_{2a} + v_{2b}$ no longer being a good quantum number. In general, Coriolis coupling of the v_2 level results in perturbed states having quantum numbers m ranging from v_2 to $-v_2$ in steps of two. The energy depends on the absolute value of m because it is insensitive to the sense of rotation. Thus Coriolis coupling results in the triply degenerate $v_2 = 2$ level being split into two levels, one which is nondegenerate, having $|m| \equiv l = 0$, and the other doubly degenerate with $l = 2$. Using the formula given in Equation 10.56, one finds that the $v_2 = 2$ level spans the Σ_g^+ and Δ_g

representations of the $D_{\infty h}$ point group, which correspond respectively to the $l = 0$ and 2 states. To completely specify the vibrational state of CO_2, the notation (v_1, v_2^l, v_3) is used.

Since Coriolis coupling leads to angular momentum about the bond, it is reflected in the rotational fine structure of the vibrational spectrum. The rotational energies resemble those of a prolate rotor:

$$E_{rot} = hB_{[v]}\left[J(J+1) - l^2 \right] + A_{[v]}hl^2 \tag{10.72}$$

Here again, the rotational constants $B_{[v]}$ and $A_{[v]}$ are functions of all the vibrational quantum numbers $[v] = [v_1, v_2, \ldots v_{3N-6}]$.

The Σ_g^+ component of the bending overtone of CO_2 is further perturbed by an anharmonic interaction with the symmetric stretch. The overtone is expected to be observed at a frequency close to $2 \times 673\ cm^{-1} = 1346\ cm^{-1}$, and being of $u \times u = g$ symmetry it might appear as a weak band in the Raman rather than the IR. In reality, the Raman spectrum of CO_2 shows two strong bands at 1285 and 1388 cm^{-1}. These two bands are the result of Fermi resonance, an anharmonic mixing of two vibrational states that have the same symmetry and are close in energy. The symmetry restriction results from the requirement that the matrix element of the perturbation operator, $\langle \psi_1 | \hat{H}' | \psi_2 \rangle$, not vanish. Since the Hamiltonian, including any perturbation correction $\hat{H}'$, must be totally symmetric, this means that ψ_1 and ψ_2 must transform according to the same ir. rep. if they are to mix. In the example at hand, the perturbation operator is $\hat{H}' = V_{12}Q_1Q_2^2$, where

$$V_{12} = \left(\frac{\partial^3 V}{\partial Q_1 \partial Q_2^2} \right)_0 \tag{10.73}$$

This operator can connect the states $(1, 0^0, 0)$ and $(0, 2^0, 0)$, which have zero-order energies 1354 and 1346 cm^{-1}, respectively. The first of these is the totally symmetric stretch, and the second is the $l = 0$ component of the bending overtone; both of these states have Σ_g^+ symmetry.

The end result of the Fermi resonance interaction is that the states representing the symmetric stretch and the bending overtone are mixed, and the energies of the two perturbed states are split apart from one another, as shown in Figure 10.11. The higher energy state contains some of the character of the bending overtone, but can still be qualitatively regarded to be the symmetric stretch. Similarly, the lower energy transition is to a state which is mostly the bending overtone, with some symmetric stretch mixed in. In quantum mechanical terms, the coefficient a is much larger than b in Figure 10.11. The state mixing leads to intensity borrowing such that the otherwise weak overtone becomes strong by taking on some of the character of the stretching mode. There is an overall conservation of the total intensity of the two bands: the stretch losing intensity to the bending overtone, but the change in intensity of the "bending overtone" appears more dramatic because its zero-order transition strength is small.

Another example of anharmonicity at work is displayed by the vibrational spectrum of water vapor and is called Darling–Dennison coupling [6]. In this case, the near equality of the frequencies v_1 and v_3 is responsible. Since the symmetry of the asymmetric stretch alternates between A_1 and B_1 according to whether the number of quanta v_3 is even or odd, a state (v_1, v_2, v_3) has the same symmetry and nearly the same energy as the state $(v_1 - 2, v_2, v_3 + 2)$, and coupling is permitted. The result is that many of the strong overtone and combination bands in the water spectrum appear in pairs: e.g., (003) and (201) at about 11,032 and 10,613 cm^{-1}, respectively. Note that these frequencies are in the near-IR, and higher overtones extend into the visible! This is something to think about the next time you gaze at a very deep blue lake.

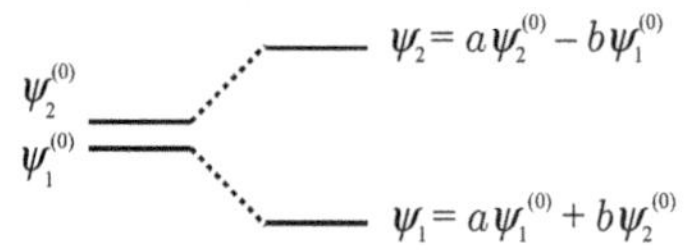

Figure 10.11 State mixing and level splitting in Fermi resonance.

10.8 SELECTION RULES AT WORK: BENZENE

As an example of a moderately large molecule with high symmetry, consider benzene, C_6H_6. The low-resolution IR spectra of gas-phase benzene and toluene are compared in Figure 10.12. The polarized and depolarized Raman spectra of liquid benzene and toluene are shown in Figures 10.13 and 10.14, respectively. Since toluene lacks a center of symmetry, there are vibrational bands common to both the Raman and IR spectra, whereas benzene adheres to the rule of mutual exclusion.

The 30 normal modes of benzene comprise ten nondegenerate and ten doubly degenerate symmetry species (ir. reps.). Due to inversion symmetry, the Raman active fundamentals are IR inactive, and vice versa.

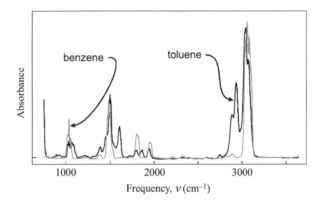

Figure 10.12 IR spectra of benzene and toluene in the gas phase.

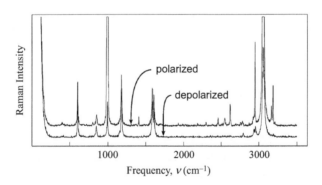

Figure 10.13 Polarized and depolarized Raman spectra of liquid benzene.

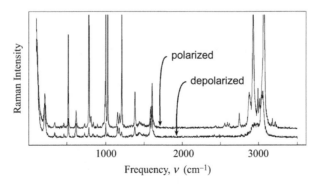

Figure 10.14 Polarized and depolarized Raman spectra of liquid toluene.

Table 10.5 Benzene fundamentals

	Raman	Infrared
Assignment	$\tilde{v}$, cm^{-1}	$\tilde{v}$, cm^{-1}
$\tilde{v}_1$ (CH) A_{1g}	3062	
$\tilde{v}_2$ (CC) A_{1g}	992	
$\tilde{v}_{11}$ (o.o.p.CH) E_{1g}	849	
$\tilde{v}_{15}$ (CH) E_{2g}	3047	
$\tilde{v}_{16}$ (CC) E_{2g}	1585	
$\tilde{v}_{17}$ (i.p.CCH) E_{2g}	1178	
$\tilde{v}_{18}$ (i.p.CCC) E_{2g}	606	
$\tilde{v}_4$ (o.o.p.CCH) A_{2u}		676
$\tilde{v}_{12}$ (CH) E_{1u}		3072
$\tilde{v}_{13}$ (CC, i.p.CCC) E_{1u}		1480
$\tilde{v}_{14}$ (i.p.CCH) E_{1u}		1036

The vibrational spectra of benzene provided early evidence of D_{6h} symmetry. Group theoretical analysis can be applied to determine the normal mode symmetries. The result is

$$\Gamma_{vib} = 2A_{1g} + A_{2g} + 4E_{2g} + 2B_{1u} + 2B_{2u} + 3E_{1u}$$
$$+ 2B_{2g} + E_{1g} + A_{2u} + 2E_{2u} \qquad (10.74)$$

The symmetry species listed in the first line of Equation 10.74 correspond to in-plane vibrations, while those on the second line designate out-of-plane vibrations. We deduce this merely by looking at the sign of the character for the σ_h operation in the D_{6h} character table (see Appendix C). In-plane vibrations are symmetric with respect to reflection through the plane containing the molecule, and out-of-plane vibrations are anti-symmetric. With the help of the D_{6h} character table, it is concluded that a number of the fundamentals are "silent"; that is, they are active in neither the Raman nor the IR. Of the Raman active fundamentals ($2A_{1g}$, E_{1g} and $4E_{2g}$), two are totally symmetric (A_{1g}) and are expected to give rise to polarized Raman bands. As will be shown in Chapter 12, these have depolarization ratio $\rho < 3/4$, while nontotally symmetric vibrations have $\rho = 3/4$. The totally symmetric modes are the ring-breathing vibration $\tilde{v}_2$ at 992 cm^{-1} and the totally symmetric C–H stretch $\tilde{v}_1$ at 3062 cm^{-1}. The IR-active modes are A_{2u} and $3E_{1u}$. A frequently used convention is to number the modes in order of decreasing symmetry (A_{1g}, A_{2g},...) and then in order of decreasing frequency within each symmetry species. The frequencies of Raman and IR fundamentals of benzene are given in Table 10.5. Note that additional weak bands are also observed and assigned to overtones or combinations, not listed in the table. The major contributions from internal coordinates are listed for each normal mode in Table 10.5, where CC and CH are stretching coordinates, and the three-letter designations represent in-plane (i.p.) or out-of-plane (o.o.p.) bending of those three-atom groups.

10.9 SOLVENT EFFECTS ON INFRARED SPECTRA

Solvent effects on vibrational spectra can appear as perturbations to the intensity, frequency, and linewidth, and may arise from specific intermolecular interactions (such as dipolar forces and hydrogen bonding) or from the bulk dielectric properties of the medium. Although the analysis of solvent effects in vibrational spectra presents a rich area of possibilities for extracting intermolecular interactions from spectra, practical quantitative approaches for interpreting such effects remain elusive. Many commonly observed solvent effects are well understood qualitatively, but are difficult to describe quantitatively.

For example, the hydrogen bonding interaction, $X - H \cdots B$, where X and B are generally quite electronegative atoms such as O, N or F, presents clear signatures. The hydrogen bonded X–H stretch is red-shifted and broadened compared to the free molecule. The red-shift is attributed to withdrawal of X–H bonding electron density through formation of the hydrogen bond. The increased linewidth has long been assumed to be inhomogeneous in nature, due to a variety of hydrogen bond environments, but recent evidence suggests homogeneous broadening may also be at work, perhaps due to an underlying progression in the low-frequency stretch of the hydrogen bond. Hydrogen bond formation also leads to intensity enhancement of the X–H stretch, in the Raman and IR. Figure 10.15 shows the influence of hydrogen bonding on the IR band due to the C–H stretch of chloroform, hydrogen bonded to dimethylsulfoxide-d6. The sharp peak apparent at low dimethylsulfoxide concentration is the C–H stretch of free chloroform, while the broad component is the hydrogen bonded C–H stretch.

Less clear-cut than hydrogen bonding are interactions due to dipolar, induction, and dispersion forces. As discussed in [7], attractive forces tend to cause red shifts and repulsive forces blue shifts in vibrational spectra, and though red shifts of vibrational frequencies are often observed on going from gas to liquid phase, blue shifts are also possible. Attractive forces are of longer range than repulsive forces, while the latter reflect the shape of molecules and determine the packing in the liquid. Vibrational modes having large dipole-moment derivatives (strong IR intensity, e.g., carbonyl stretches) are particularly sensitive to solvent polarity. Modes having large polarizability derivatives (strong Raman activity), on the other hand, are subject to frequency shifts due to dispersion forces. Intermolecular forces can also influence the peak frequencies in Raman, and IR spectra in different ways, causing small frequency shifts from the IR to the Raman and frequency differences between the isotropic and anisotropic Raman components [8–9]. Accounting for these sorts of effects requires knowledge of the liquid structure.

In neat liquids, the dispersion in the refractive index across a strong IR band is significant and may cause the peak frequency to differ from the value observed in the gas phase, even in the absence of specific intermolecular interactions. Bertie et al. [10–13] have employed a generalized Lorenz–Lorentz equation (Equation 3.75), where both the polarizability and the refractive index are complex,

$$\frac{N\hat{\alpha}(v)}{3\varepsilon_0 V} = \frac{\hat{n}^2(v)-1}{\hat{n}^2(v)+2} \tag{10.75}$$

to analyze IR bands in neat liquids and in solutions. The carets over the frequency dependent polarizability ($\hat{\alpha}$) and refractive index ($\hat{n}$) indicate that these are complex quantities. Recall from Chapter 3 that the real and imaginary parts of $\hat{n}$ are the (real) refractive index n_r and the absorption coefficient κ. The Beer's law molar absorptivity $\varepsilon_M(v)$ is proportional to $v\kappa(v)$. Bertie and coworkers exploited the Kramers–Kronig relationship to obtain the frequency-dependent refractive index n_r from the measured absorption coefficient, determined using transmission or attenuated total reflection IR spectroscopy. In the case of strong IR bands, and especially at lower frequencies, the observed absorbance spectrum $\varepsilon_M(v)$ differs in shape and peak frequency from that of the absorption coefficient $\kappa(v)$. In Figure 10.16, the spectrum of liquid methanol is represented in terms of the molar absorptivity $\varepsilon_M(\tilde{v})$, the absorption coefficient $\kappa(\tilde{v})$, and the frequency-dependent

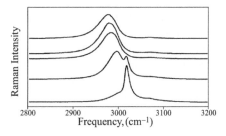

Figure 10.15 Infrared spectra of the C–H stretch of chloroform in the presence of increasing amounts of dimethylsulfoxide-d6, from bottom to top. (Reprinted with permission from D. C. Daniel and J. L. McHale, *Journal of Physical Chemistry A, 101*, 3070, (1997), Copyright 1997, American Chemical Society.)

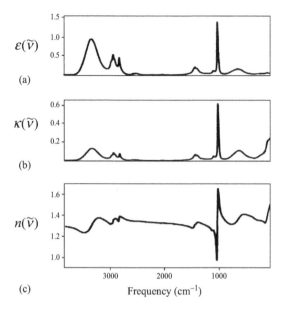

$\varepsilon(\widetilde{\nu})$

(a)

$\kappa(\widetilde{\nu})$

(b)

$n(\widetilde{\nu})$

(c)

Frequency (cm^{-1})

Figure 10.16 Infrared spectrum of liquid methanol: molar absorptivity in units of 10^5 cm^2 mol^{-1} (a), absorption coefficient (b), and refractive index (c). (Reprinted with permission from J. E. Bertie, et al., *Vibrational Spectroscopy 8*, 215, (1995), Elsevier Science - NL, Sara Burgerhartstraat 25, 1055 KV Amsterdam, The Netherlands.)

refractive index, $n(\widetilde{\nu})$. Note the dispersion in the refractive index, particularly in the vicinity of the strongest bands. Considerations such as those outlined in [10–13] are important in order to separate bulk dielectric effects in vibrational spectra from those due to specific interactions and to obtain the true peak frequency in the solution phase.

10.10 SUMMARY

The normal mode approach to the analysis of polyatomic vibrational motion has been exposed. The concept of normal modes is only as strong as the harmonic oscillator approximation, yet it provides us with a powerful vantage point from which the symmetries of molecules and vibrational modes can be discerned. Normal coordinates are linear combinations of mass-weighted Cartesian coordinates for atomic displacements. These normal coordinates can potentially involve all the atoms in the molecule, but when most of the potential energy of the vibration is invested in a small number of atoms, the concept of a group frequency applies. In symmetric molecules, normal mode symmetries and selection rules for IR and Raman activity are readily deduced using group theory. For larger, less symmetric molecules the group frequency approach can be of value. Anharmonic effects result in perturbations to vibrational frequencies and the appearance of overtone and combination bands, and in addition permit couplings between vibrational states that would otherwise be independent. These couplings can result in intensity borrowing and the appearance of forbidden transitions.

The conventional applications of IR and Raman spectroscopy observe vibrational motion within the ground electronic state. In Chapter 11, we will see that electronic spectroscopy permits the vibrational modes of excited electronic states to be investigated. Not all vibrational modes contribute to the electronic absorption and emission spectra, though, and we will have to figure out the relevant selection rules. It is useful to divide all normal modes into two groups, totally symmetric and nontotally symmetric. These two kinds of modes contribute differently to electronic spectra, as will be shown in Chapter 11.

PROBLEMS

1. Use group theoretical arguments to predict the number of fundamentals observed in the Raman and IR spectra of a triatomic molecule AB_2 for each of the possible structures: linear symmetric, linear asymmetric, bent symmetric, and bent asymmetric.
2. Find the symmetries of the normal modes of a planar AB_4 molecules (D_{4h} point group). Predict the Raman and IR activities of each fundamental. Repeat the analysis for the case where the molecule is tetrahedral (T_d point group).
3. Find the D matrix for CO_2.
4. Explain how vibrational spectra can distinguish between *cis* or *trans* dichloroethylene.
5. Find the symmetries of the states associated with the second overtone of the bending mode of CO_2.
6. Use the data in Table 10.4 to calculate the frequencies of the transitions $(000) \to (003)$ and $(000) \to (201)$ for water. Compare to the values given in Section. 10.7.
7. A general harmonic potential function for water is

$$V = \frac{1}{2}k_r \left(\Delta r_1\right)^2 + \frac{1}{2}k_r \left(\Delta r_2\right)^2 + \frac{1}{2}k_\theta \left(r\Delta\theta\right)^2$$
$$+ k_{rr}\Delta r_1 \Delta r_2 + k_{r\theta}r\Delta r_1 \Delta\theta + k_{r\theta}r\Delta r_2 \Delta\theta$$

The second line contains off-diagonal force constants, while the first three terms are diagonal. In matrix form, this can be expressed as $2V = \mathbf{R}^T \mathbf{F} \mathbf{R}$, where $\mathbf{R} = \left(\Delta r_1 \quad \Delta r_2 \quad \Delta\theta\right)$ is the vector whose elements are the internal coordinates. Find the symmetry coordinates S_1, S_2, and S_3 for water, and the block diagonal force constant matrix $\mathbf{f}$ which permits the potential energy to be written as $\mathbf{S}^T \mathbf{f} \mathbf{S}$.
8. Evaluate the upper state symmetry for each of the following combination bands in benzene: $v_{16} + v_2$, $v_{10} + v_{13}$, $v_2 + v_{16} + v_{18}$, and $2v_2 + v_{18}$. In each case, what operator could result in Raman or IR activity of the transition? Assume that the initial state is the ground vibrational state. Note that the v_{10} mode has B_{2u} symmetry.
9. Assign the bands in the benzene spectra displayed in Figures 10.12 and 10.13.

REFERENCES

1. E. B. Wilson, J. C. Decius, and P. C. Cross, *Molecular Vibrations* (Dover Publications, New York, 1955).
2. F. A. Cotton, *Chemical Applications of Group Theory* (Wiley-Interscience, New York, 1971).
3. G. Herzberg, *Molecular Structure and Molecular Spectra II. Infrared and Raman Spectra of Polyatomic Molecules* (Van Nostrand Reinhold, New York, 1945).
4. P. F. Bernath, *Spectra of Atoms and Molecules* (Oxford University Press, New York, 1995).
5. I. N. Levine, *Molecular Spectroscopy* (Wiley-Interscience, New York, 1975).
6. B. T. Darling and D. M. Dennison, The water vapor molecule, *Phys. Rev.* 57, 128 (1940).
7. K. S. Schweizer and D. Chandler, Vibrational dephasing and frequency shifts of polyatomic molecules in solution, *J. Chem. Phys.* 76, 2296 (1982).
8. J. L. McHale, The influence of angular dependent intermolecular forces on vibrational spectra of solution phase molecules, *J. Chem. Phys.* 75, 30 (1981).
9. H. Torii and M. Tasumi, Local order and transition dipole coupling in liquid methanol and acetone as the origin of the Raman noncoincidence effect, *J. Chem. Phys.* 99, 8459 (1993).
10. J. E. Bertie, L. Z. Zhang, and C. D. Keefe, Measurement and use of absolute absorption intensities of neat liquids, *Vib. Spec.* 8, 215 (1995).
11. J. E. Bertie and L. Z. Zhang, Infrared transitions of liquids. XV. Infrared refractive indices from 8000–350 cm⁻¹, Absolute integrated intensities, transition moments, and dipole moment derivatives of methanol-d at 25°, *Appl. Spec.* 48, 176 (1994).

12. J. E. Bertie, L. Z. Zhang, and C. D. Keefe, Infrared intensities of liquids. XVI. Accurate determination of molecular band intensities from infrared refractive index and dielectric constant spectra, *J. Mol. Struct.* *324*, 157 (1994).

13. J. E. Bertie and S. L. Zhang, Infrared intensities of liquids. XVII. Infrared refractive indices from 8000–350 cm^{-1}, absolute intensities, transition moments and dipole moment derivatives of methan-d_3-ol and methanol-d_4 at 25°, *J. Chem. Phys.* *101*, 8364 (1994).

14. S. A. Steiger, C. Li, C. Campana, and N. R. Natale, Lanthanide and asymmetric catalyzed syntheses of sterically hindered 4-isoxazolyl-1,4-dihydropyridines and 4-isoxazolyl-quinolines, *Tet. Lett.* *57*, 423 (2016).

Electronic spectroscopy

11.1 INTRODUCTION

Because electronic energy levels are more widely spaced than vibrational and rotational levels, the study of electronic spectroscopy encompasses many of the previously discussed concepts concerning rotational and vibrational spectra. In electronic spectra of gas-phase molecules, analysis of rovibrational transitions within an electronic absorption or emission band can provide useful structural information, particularly when the pure rotational or vibrational spectrum is forbidden. In condensed phases, discrete rotational structure is not observed, but vibrational transitions contributing to the linewidth are often resolved. For example, the electronic absorption spectrum of benzene, shown in Figure 11.1, displays vibrational structure which will be analyzed in Section 11.5.3. In larger molecules, the high density of vibrational states serves to blur the vibrational structure. In solution, the line-broadening influence of the solvent may prevent individual vibronic transitions from being resolved, but they contribute to the linewidth nonetheless. Consider the absorption spectra of I_2, shown in Figure 11.2. In the vapor phase, vibrational transitions are resolved as separate peaks. In solution, however, the vibrational transitions of I_2 continue to contribute to the breadth of the spectrum, but are not resolved as distinct features. In this chapter, we will concentrate on the analysis of vibrational contributions to electronic spectra. These hold the key to determining the difference in equilibrium geometries of the ground and excited electronic states. Electronic spectra can also reveal some of the features of the excited state potential surface, as well as dynamics such as dissociation, isomerization, and radiationless decay.

We begin by examining the theory of electronic absorption spectroscopy, in which a molecule in an initial vibronic (vibrational plus electronic) state makes transitions to a number of different vibrational levels within an excited electronic state. The commonly employed technique of UV-visible spectroscopy is based on the transitions of valence electrons, while core electron excitations may be found at higher frequencies. Quantum mechanics teaches us that electrons confined to bigger boxes have more closely spaced energy levels than those in smaller boxes. Thus when we compare, for example, the $\pi - \pi^*$ transitions of a series of conjugated molecules, we find the absorption wavelength increases with the size of the π system. While common dyes are large conjugated organic molecules, smaller molecules can also absorb visible light. Consider the purple color of I_2 or the blue color of an aqueous solution of Cu^{2+} ion. It may come as a surprise that the well-known colors of many transition metal complexes often result from electronic transitions that are forbidden to a first approximation! In Sections 11.5.3 and 16.6, we consider how electronic transitions which are forbidden to a first approximation ("symmetry forbidden") become allowed through the participation of nontotally symmetric vibrations.

Once a molecule is promoted to an excited electronic state, it may return to the ground state either radiatively or nonradiatively. The former pathway leads to emission spectroscopy. The fluorescence and phosphorescence spectra of molecules are discussed in Sections 11.3.2 and 11.7. These experimental techniques complement the information obtained from absorption spectroscopy. Nonradiative decay of excited electronic states is considered in Section 11.8.

The starting point for describing electronic transitions is the Born–Oppenheimer approximation. As discussed in Chapter 9, this approximation permits the separation of electronic and nuclear (rovibrational) energies. In the case of a diatomic molecule, each electronic state is characterized by a potential energy function $V(R)$, for example the curves shown in Figure 9.1. In polyatomics, these curves are replaced by $(3N - 6)$-dimensional surfaces, as the potential energy $V(Q_1, Q_2, \ldots, Q_{3N-6})$ depends on $3N - 6$ normal coordinates. The harmonic approximation permits these surfaces to be further decomposed: $V = \sum_{i=1}^{3N-6} V_i$; so that one can draw separate potential energy curves $V_i(Q_i)$ for each normal coordinate, in a particular electronic state.

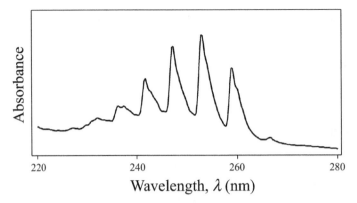

Figure 11.1 Electronic absorption spectrum of benzene in $CHCl_3$ solution.

Electronic transitions that are forbidden within the Born–Oppenheimer approximation are frequently observed, requiring that we go beyond the limits of this approximation. The corrected picture allows for coupling of the zero-order Born–Oppenheimer (BO) electronic states, and new selection rules come into play. Effects permitted by the breakdown in the BO approximation, such as vibronic activation of forbidden transitions, predissociation (defined in Section 11.3.3), internal conversion, and symmetry lowering, are of wide-ranging importance in chemistry and spectroscopy.

In this chapter, we begin with a study of diatomic molecules in order to introduce topics which apply to polyatomics as well. Starting from the electronic configuration for a molecule, we can determine the electronic states associated with that configuration. Excited state configurations are generated by promoting electrons from occupied to virtual orbitals. Group theory can then be exploited to determine which state-to-state transitions are permitted by E1, E2, and M1 selection rules. Vibrational (Franck–Condon) progressions within electronic absorption and emission bands depend on the change in equilibrium geometry. For polyatomic molecules, vibrational progressions are shown to be most important for totally symmetric vibrations. The appearance of nontotally symmetric vibrations is often an indication of Born–Oppenheimer breakdown and is of interest in both this chapter and the next.

11.2 DIATOMIC MOLECULES: ELECTRONIC STATES AND SELECTION RULES

Diatomic molecules may be homonuclear ($D_{\infty h}$ point group) or heteronuclear ($C_{\infty v}$ point group). The character tables for these two cases are given in Appendix C, and are reproduced in Tables 11.1 and 11.2 for reference. These are infinite groups, due to the existence of an infinite number of reflection planes σ_v through the bond and the infinite order of the rotation axis containing the bond. An additional designation of the irreducible representations (ir. reps. or symmetry species) in the case of the $D_{\infty h}$ point group is the label g or u for even

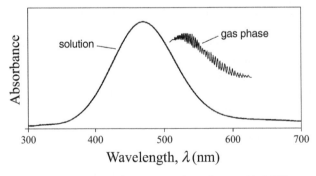

Figure 11.2 Electronic absorption spectrum of I_2 vapor, and in solution with $CHCl_3$.

Table 11.1 $C_{\infty v}$ character table

$C_{\infty v}$	E	$2C(\varphi)$	σ_v		
Σ^+	1	1	1	z	$x^2 + y^2, z^2$
Σ^-	1	1	-1	R_z	
Π	2	$2\cos\varphi$	0	$(x,y)(R_x, R_y)$	(xz, yz)
Δ	2	$2\cos2\varphi$	0		$(x^2 - y^2, xy)$
Φ	2	$2\cos3\varphi$	0		

Table 11.2 $D_{\infty h}$ character table

$D_{\infty h}$	E	$2C(\varphi)$	σ_v	i	$2S(\varphi)$	C_2		
Σ_g^+	1	1	1	1	1	1		$x^2 + y^2, z^2$
Σ_g^-	1	1	-1	1	1	-1	R_z	
Π_g	2	$2\cos\varphi$	0	2	$-2\cos\varphi$	0	(R_x, R_y)	(xz, yz)
Δ_g	2	$2\cos2\varphi$	0	2	$2\cos2\varphi$	0		$(x^2 - y^2, xy)$
Σ_u^+	1	1	1	-1	-1	-1	z	
Σ_u^-	1	1	-1	-1	-1	1		
Π_u	2	$2\cos\varphi$	0	-2	$2\cos\varphi$	0	(x,y)	
Δ_u	2	$2\cos2\varphi$	0	-2	$-2\cos2\varphi$	0		

or odd inversion symmetry. The functions listed in the two far-right columns of the $D_{\infty h}$ and $C_{\infty v}$ character tables are important in the consideration of selection rules. The molecular orbitals and electronic states can be characterized using symmetry labels corresponding to the irreducible representations of the group. Note that in both point groups the Σ ir. reps. are nondegenerate and the Π and Δ ir. reps. are doubly degenerate. The usual convention is to use lowercase letters σ, π, δ, and ϕ to denote the symmetries of molecular orbitals, reserving capital letters for state symmetries. The subscript g or u is appended to the orbital label in the case of homonuclear diatomics. The label of a molecular orbital (MO) also conveys information about the angular momentum. In molecules, as opposed to atoms, the total electronic angular momentum L is no longer a good quantum number. In diatomics and linear polyatomics, however, the component of orbital angular momentum about the bond is quantized. Calling the z direction that of the bond, the angle of rotation about the bond is φ and the projection of the vector $\vec{L}$ onto the bond is $L_z = m_l \hbar$. The labels σ, π, δ, ϕ, ... for molecular orbitals designate the quantum number $m_l = 0, \pm1, \pm2, \pm3, ...$. Knowing the occupied MOs (the electronic configuration), one can generate term symbols which designate electronic states and their degeneracies. This procedure is described in the next section.

11.2.1 MOLECULAR ORBITALS AND ELECTRONIC CONFIGURATIONS FOR DIATOMICS

Molecular orbitals are conveniently represented as linear combinations of a limited number of atomic orbitals. This LCAO-MO (linear combination of atomic orbitals to get molecular orbitals) approach is based on the variation theorem, which permits the hydrogen-like or other atomic orbitals to be used as a basis for an approximate wavefunction for the molecule. In general, one takes as a basis set a number of atomic orbitals, χ_1, χ_2, etc., and writes trial wavefunctions (molecular orbitals) as $\psi = c_1\chi_1 + c_2\chi_2 + ...$. The coefficients c_i are found using linear variation theory, as described in a number of quantum chemistry texts [1,2]. More accurate wavefunctions are generated with larger basis sets, at the cost of greater computational complexity and less intuitive physical pictures. While these sorts of calculations are of great importance in the study of electronic structure and spectra, in the present discussion we are concerned with qualitative ideas about the symmetry and energy level ordering, so we will employ the simplest possible description of MOs that is capable of capturing the physical picture.

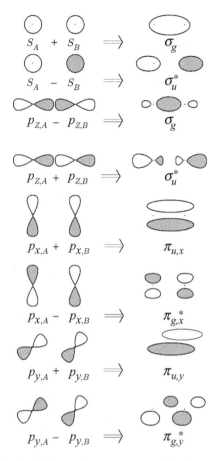

Figure 11.3 Molecular orbitals formed from overlapping atomic orbitals.

Let us begin with the example of diatomic carbon, C_2. The MOs required to describe the molecular electronic configuration can be formed from linear combinations of the $1s$, $2s$, $2p_x$, $2p_y$, and $2p_z$ orbitals on each of the two carbon atoms. Alternatively, we could use the p_1, p_0, and p_{-1} orbitals instead of the real p_x, p_y, and p_z orbitals. The complex p orbitals have the advantage of being eigenfunctions of $\hat{L}_z$, while the real p orbitals are more easily visualized; so we chose the latter for now. Figure 11.3 depicts the bonding and antibonding MOs formed from combining these basis AOs. Recall that bonding MOs have enhanced electron density in the internuclear region, while antibonding orbitals have nodal planes between the nuclei. Antibonding orbitals are distinguished by a superscript star (*), which should not be confused with the symbol for complex conjugation. There are some very general principles that are basic to the LCAO-MO approach. AOs combine to give MOs if they are similar in energy and if they overlap. The number of resulting MOs is always equal to the number of AOs in the basis set. Molecular orbitals designated by the letter σ have cylindrical symmetry about the bond. There is no angular momentum about the bond for a σ-type orbital; m_l is zero and the MO is independent of φ. Orbitals of π symmetry have nodal planes containing the nuclei. The bonding and antibonding π-type orbitals occur in degenerate pairs which differ only in the orientation of the nodal plane containing the bond. If formed from p_1 and p_{-1} atomic orbitals, π-type MOs have quantized angular momentum about the bond: $m_l = \pm 1$. Note that in this example the inversion symmetry leads to the labels g and u to denote MOs which are even or odd with respect to inversion. Bonding sigma orbitals are even with respect to inversion, while bonding pi orbitals are odd. The p_0 and p_z AOs are equivalent, so the σ MOs formed from combining p_z AOs have $m_l = 0$. The π_x and π_y MOs are linear combinations of the π_{+1} and π_{-1} MOs. The latter have $\exp(im_l\varphi) = \exp(\pm i\varphi)$ as part of the wavefunction. The energy of an MO does

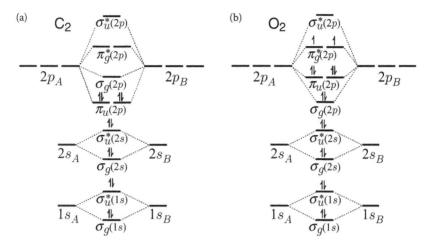

Figure 11.4 Energy levels of molecular orbitals in homonuclear diatomics and ground electronic configurations of C_2 and O_2.

not depend on the sense of rotation of electrons, so it is to be expected that orbitals that differ only in the sign of m_l are degenerate.

Arranging the MOs on an energy level diagram provides the basis for using the Aufbau principle to build up electronic configurations of homonuclear diatomics. The MO energy level order depends on the molecule, just as the order of atomic orbital energies depends on atomic number. The diagram shown in Figure 11.4a works for homonuclear diatomics H_2 through C_2, while that of Figure 11.4b can be used for homonuclear diatomics heavier than C_2. Let us use Figure 11.4a to generate the ground configuration of C_2. Loading twelve electrons into these MOs in accordance with the Pauli principle results in the electronic configuration shown in the figure. One can write this concisely as

$$(\sigma_g 1s)^2(\sigma_u^* 1s)^2(\sigma_g 2s)^2(\sigma_u^* 2s)^2(\pi_u 2p)^4$$

It is important to remember that a configuration does not necessarily represent a state, since there are as many states associated with a configuration as there are ways to achieve it, as in the case of atomic configurations. In the case at hand, however, there is only one way to generate this closed-shell configuration, so there is just a single state, the ground electronic state, arising from this configuration. It can be represented by a single Slater determinant.

The bond order b associated with a configuration is calculated from the number n of electrons in bonding orbitals and the number n^* in antibonding orbitals: $b = (n - n^*)/2$. In the case of ground state C_2, the bond order is $b = 2$, that is, a double bond. The MO result is in agreement with the Lewis dot structure, :C=C:, where only the eight valence electrons are included in this picture. This dot structure would be troubling to a beginning chemistry student, having been taught that atoms "like" being surrounded by complete octets!

In the case of a heteronuclear diatomic, the procedure for forming MOs is similar to that for a homonuclear diatomic, but the starting AO energies are not equal. Figure 11.5 illustrates the case of CO, where the greater electronegativity of oxygen is associated with the lower energy of the oxygen AOs on the right-hand side of the diagram. The ground state configuration of CO, another closed-shell molecule, represents a nondegenerate state with a bond order of three, in accordance with the triply-bonded Lewis dot structure, :C≡O:, drawn to satisfy the octet rule.

The blue color of a flame results from the emission of electronically excited C_2 molecules. To consider some of the low-lying excited states of C_2, we can start with the diagram of Figure 11.4 and promote electrons from the highest occupied MO (the HOMO) to the lowest unoccupied MO (the LUMO) to get configurations such as

$$(\sigma_g 1s)^2(\sigma_u^* 1s)^2(\sigma_g 2s)^2(\sigma_u^* 2s)^2(\pi_u 2p)^3(\sigma_g 2p)^1$$

There are $4 \times 2 = 8$ ways to arrive at this configuration: four ways to achieve a vacancy in the $\pi_u 2p$ level and two ways to put a single electron in the $\sigma_g 2p$ level. There are thus eight states arising from this configuration, but how many energy levels are there? To answer this question, we need to develop the idea of term symbols.

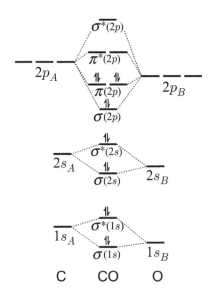

Figure 11.5 Ground electronic configuration of CO.

11.2.2 TERM SYMBOLS FOR DIATOMICS

Like term symbols for atoms, those for molecules keep track of the good quantum numbers. In diatomics and linear molecules, M_L, the quantum number for the projection of the orbital angular momentum onto the bond, is a good quantum number. The total spin S is also a good quantum number if spin–orbit coupling is weak, and it is found by summing up the spin quantum numbers of individual electrons as discussed in Chapter 7 for atoms. The uppercase letter M_L designates the sum $\Sigma_i m_{l,i}$ of the angular momentum quantum numbers of occupied MOs. In analogy to the use of the letters $S, P, D, F,\ldots$ for atomic term symbols, the Greek letters $\Sigma, \Pi, \Delta, \Phi,\ldots$ are used to designate $|M_L| = 0,1,2,3\ldots$ for linear molecules. Note that these are the *same* as the labels for the symmetry species of the $C_{\infty v}$ and $D_{\infty h}$ character tables, and these states do indeed belong to the corresponding ir. reps. of these point groups. The symbol Λ is used to denote the value of $|M_L|$, and the term symbol for a state is written $^{2S+1}\Lambda$. The superscript $2S + 1$ indicates the spin degeneracy as usual. A closed-shell configuration always leads to the $^1\Sigma_g^+$ term symbol for homonuclear diatomics, and just plain $^1\Sigma^+$ for heteronuclear diatomics. The values of m_l for the occupied orbitals cancel when the shell is full, leading to $|M_L| = 0$ and thus a Σ state. The right-superscript (+) sign gives the symmetry with respect to a reflection plane containing the molecule. States having Π or Δ symmetry have the character zero for this reflection operation, meaning that one component of the degenerate pair is (−) and the other (+) with respect to reflection. The degeneracy of a term $^{2S+1}\Lambda$ is $2S + 1$ when $\Lambda = 0$ and $2(2S + 1)$ when $\Lambda \neq 0$. Note that this degeneracy is the product of the orbital and spin degeneracies.

Consider the ground configuration of O_2, shown in Figure 11.4b, which can be written as

$$(\sigma_g 1s)^2(\sigma_u^* 1s)^2(\sigma_g 2s)^2(\sigma_u^* 2s)^2(\sigma_g 2p)^2(\pi_u 2p)^4(\pi_g^* 2p)^2$$

As in figuring atomic term symbols, we need only consider the open shells. There are six π^2 configurations, shown in Figure 11.6. We expect the ground state to be a triplet, and we shall see that this state comprises three of the diagrams of Figure 11.6. In this section, we choose π orbitals for which m_l is a good quantum number in order to compute the net M_L. All states must be gerade because the partially filled orbitals are gerade. The direct product of the open-shell orbital symmetries is $\pi_g \times \pi_g = \Sigma_g^+ + \Sigma_g^- + \Delta_g$, so these are the state symmetries. The spin multiplicity of each term must be decided in accordance with the requirement that the total wavefunction be antisymmetric with respect to exchange of two electrons. Since states having

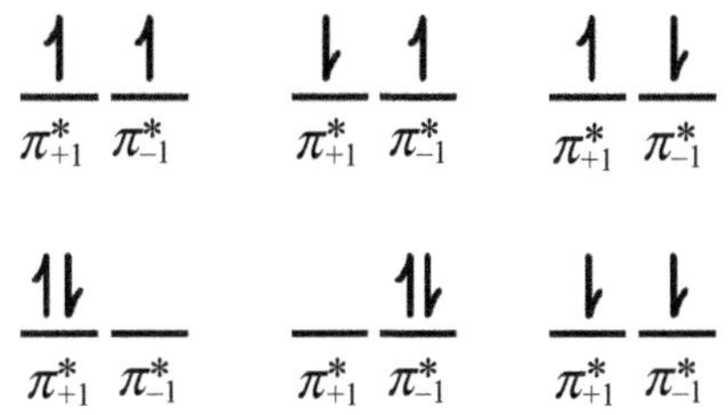

Figure 11.6 Six electronic configurations associated with π^2.

$S = 1/2 + 1/2 = 1$ must place two electrons in different orbitals, having opposite signs for m_l, the net value of Λ is zero for the triplet state. Thus there is a $^3\Sigma_g$ term, which is triply degenerate by virtue of the spin multiplicity, but it is not immediately obvious whether this state has Σ_g^+ or Σ_g^- symmetry. To pursue this question, consider the spatial part of the wavefunction for this state, which has to be written as the proper symmetry-adapted linear combination of two terms: $\pi_{+1}(1)\pi_{-1}(2)$ and $\pi_{+1}(2)\pi_{-1}(1)$, where 1 and 2 are the coordinates of the two electrons in the π_g^* orbitals, which are distinguished here by giving the quantum number m_l as a subscript. The superscript (*) is omitted for typesetting convenience, as it does not affect our symmetry conclusions. The overall wavefunction must be antisymmetric with respect to exchange of the two electrons, so the symmetric (+) and antisymmetric (−) linear combinations $\pi_{+1}(1)\pi_{-1}(2) \pm \pi_{+1}(2)\pi_{-1}(1)$ must be combined with antisymmetric and symmetric spin functions, respectively. The triplet state necessarily has three symmetric spin functions associated with it, thus the wavefunction for the triplet state is

$$\Psi\,(^3\Sigma_g^-) = \frac{1}{\sqrt{2}}\left\{\pi_{+1}(1)\pi_{-1}(2) - \pi_{+1}(2)\pi_{-1}(1)\right\}\begin{Bmatrix} \alpha(1)\alpha(2) \\ (1/\sqrt{2})[\alpha(1)\beta(2) + \alpha(2)\beta(1)] \\ \beta(1)\beta(2) \end{Bmatrix} \tag{11.1}$$

The spatial function $\pi_{+1}(1)\pi_{-1}(2) - \pi_{+1}(2)\pi_{-1}(1)$ changes sign on reflection σ_v, as you will show in one of the homework problems, so the triplet state is $^3\Sigma_g^-$. The symmetric spatial function $\pi_{+1}(1)\pi_{-1}(2) + \pi_{+1}(2)\pi_{-1}(1)$ must be combined with the antisymmetric spin function: $[\alpha(1)\beta(2) - \alpha(2)\beta(1)]/\sqrt{2}$. Since the state is symmetric with respect to σ_v, this gives rise to the $^1\Sigma_g^+$ term. There are two configurations having net $M_L = \pm 2$ and net spin $S = 0$. The spatial functions for these two states are of the form $\pi_{+1}(1)\pi_{+1}(2)$ and $\pi_{-1}(1)\pi_{-1}(2)$. Each is symmetric with respect to exchange, and must be combined with the antisymmetric spin function. There is therefore a $^1\Delta_g$ state, which is doubly degenerate by virtue of the orbital degeneracy. The lowest energy state is the $^3\Sigma_g^-$ state, which is the most stable due to its spin multiplicity (Hund's rule). The $^1\Delta_g$ is 0.98 eV higher than the ground state, and the $^1\Sigma_g^+$ is 1.6 eV above ground. The bottom line is that the ground *configuration* of O_2 comprises three energy *levels* and six *states*, and the ground level, $^3\Sigma_g^-$, is triply degenerate.

By convention, the ground electronic state of a molecule is called $\tilde{X}$, and excited states of the same spin multiplicity as the ground state are labeled $\tilde{A}, \tilde{B}, \tilde{C},\ldots$ in order of increasing energy. The tilde is not always used, but it is helpful to distinguish these symbols from symmetry labels A and B used for polyatomic molecules. Excited states of different spin multiplicity than the ground state are labeled similarly, but with lowercase letters: $\tilde{a}, \tilde{b}, \tilde{c},\ldots$, etc. For example, one refers to the $\tilde{X}(^3\Sigma_g^-)$, $\tilde{a}(^1\Delta_g)$ and $\tilde{b}(^1\Sigma_g^+)$ states of O_2.

Returning to diatomic carbon, promotion of an electron from the $\pi_u 2p$ to the $\sigma_g 2p$ MO generates the excited configuration:

$$(\sigma_g 1s)^2(\sigma_u^* 1s)^2(\sigma_g 2s)^2(\sigma_u^* 2s)^2(\pi_u 2p)^3(\sigma_g 2p)^1$$

The eight states that derive from this configuration are all *ungerade*, since the direct product of the inversion symmetries of the partially occupied orbitals is $u \times g = u$. We expect to obtain both singlets and triplets, because the spin of the "hole" in the $\pi_u 2p$ level may be either parallel or antiparallel to the spin of the electron in $\sigma_g 2p$. Adding together $m_l = \pm 1$ and $m_l = 0$ can only lead to Π terms, so we have $^1\Pi_u$ and $^3\Pi_u$, which have degeneracies of two and six, respectively, accounting for all eight states.

Electronic states of diatomics can also be characterized by the energies of the atomic states in the dissociation limit. The ground state of C_2 is expected to dissociate to two ground state carbon atoms, each having the term symbol ^3P. The $^1\Pi_u$ and $^3\Pi_u$ excited electronic states also dissociate to ground state carbon atoms. Still higher energy molecular electronic states yield to one or more excited state carbon atoms on dissociation. This is illustrated nicely by the numerous potential energy curves shown in Figure 9.1.

Spin–orbit coupling can lift the degeneracy of a term $^{2S+1}\Lambda$. The letter Ω is used to designate the quantum number for the z component of total electronic (orbital plus spin) angular momentum in diatomics, in the Russell–Saunders or weak-coupling scheme. This quantum number takes on the values $\Omega = \Lambda + S, \Lambda + S - 1, \ldots, \Lambda - S$ (see Section 8.3). The weak coupling case is illustrated in Figure 8.6a for Hund's case (a) coupling. The revised term symbol is then $^{2S+1}\Lambda_\Omega$. For example, a $^3\Delta$ state is split by spin–orbit coupling into three doubly degenerate levels: $^3\Delta_3$, $^3\Delta_2$ and $^3\Delta_1$.

In heavier diatomics, the spin–orbit interaction is quite strong and Hund's case (a) does not apply. A good example is I_2, where strong spin–orbit coupling gives rise to rather complex visible absorption spectra. Some of the low-lying potential energy curves for I_2 are shown in Figure 11.7. The ground state is $^1\Sigma_g^+$, as expected for

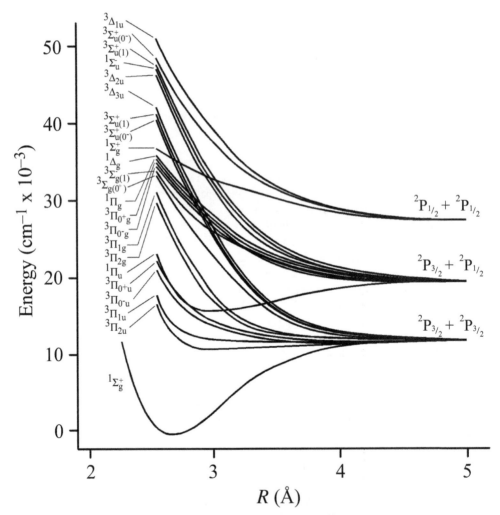

Figure 11.7 Potential energies of I₂ electronic states. (Reprinted with permission from R. S. Mulliken, Iodine Revisited, *The Journal of Chemical Physics* 55, 288 (1971), copyright 1971, American Institute of Physics.)

this closed-shell molecule, and this state dissociates into two ground state $^2P_{3/2}$ iodine atoms. Since the degeneracy of each $^2P_{3/2}$ atomic state is $2J + 1 = 4$, there are $4 \times 4 = 16$ molecular states that also dissociate to two ground state iodine atoms. There are fewer than 16 energy levels because many of them are degenerate. The valence excited states of iodine adhere approximately to Hund's case (c), discussed in Chapter 8, and a different scheme than that used above for lighter diatomics is required. The lowest excited configuration of I_2 is $\ldots(\sigma_g 5p)^2(\pi_u 5p)^4(\pi_g^* 5p)^3(\sigma_u^* 5p)^1$. In the absence of spin–orbit coupling, one expects $^1\Pi_u$ and $^3\Pi_u$ states to arise from this configuration. Spin–orbit coupling splits the latter into four new levels, labeled, $^3\Pi_{0^+u}, {}^3\Pi_{0^-u}, {}^3\Pi_{1u}$ and $^3\Pi_{2u}$ in order of decreasing energy, where the numerical subscript designates the quantum number Ω. These symbols, though convenient for quantum mechanical bookkeeping, are only approximate because the strong spin–orbit coupling means that Λ and S are not good quantum numbers. States having $\Omega \geq 1$ are doubly degenerate, while the $^3\Pi_{0u}$ level splits into two levels having (+) or (–) symmetry with respect to the σ_v reflection operation. The $^3\Pi_{0^+u}$ state dissociates to $^2P_{1/2}$ plus $^2P_{3/2}$, while the other three $^3\Pi$ states dissociate to two $^2P_{3/2}$ (ground state) iodine atoms. The $^1\Pi_u$ state, relabeled $^1\Pi_{1u}$ to denote the value of Ω, is not split by spin–orbit coupling (since $S = 0$). It also dissociates to ground state iodine atoms. Thus the lowest excited configuration gives rise to seven states, out of the total of 16 expected, that dissociate to two ground state iodine atoms. Other excited configurations, such as $\ldots(\sigma_g 5p)^2(\pi_u 5p)^3(\pi_g^* 5p)^4(\sigma_u^* 5p)^1$, also correlate with ground state iodine atoms.

Note that not all the excited state potential energy curves shown in Figure 11.7 and Figure 9.1 are bound, i.e.; have a minimum energy. For example, the states arising from the configuration $\ldots (\sigma_g 5p)^2(\pi_u 5p)^3(\pi_g^* 5p)^4(\sigma_u^* 5p)^1$ of I_2 are dissociative, as might be expected since the bond order is zero.

11.2.3 SELECTION RULES

With the help of term symbols to designate state symmetries, we can now consider selection rules for electronic transitions. As usual, the transition from state i to state f is permitted if the integral $\langle \Psi_i | \hat{O} | \Psi_f \rangle$ is nonzero, where $\hat{O}$ is the appropriate operator: the dipole moment operator $\hat{\mu}$ for E1 transitions, the quadrupole moment operator $\hat{\Theta}$ for E2 transitions, and the angular momentum operator $\hat{L}$ for M1 transitions. The wavefunctions Ψ_i and Ψ_f are state wavefunctions, which can be expressed as Slater determinants or linear combinations thereof. The symmetry of the state function is found from the direct product of the occupied orbitals. The transition $i \leftrightarrow f$ will be allowed if the triple direct product $\Gamma_i \times \Gamma_{\hat{O}} \times \Gamma_f$ equals or contains the totally symmetric representation, which is Σ_g^+ or Σ^+ for homo- or heteronuclear diatomics, respectively.

11.2.3.1 ELECTRIC DIPOLE TRANSITIONS (E1)

The electric dipole moment operator transforms according to the same ir. reps. as the Cartesian coordinates x, y, z. In heteronuclear diatomics, the z coordinate belongs to the totally symmetric representation, as it is invariant to all symmetry operations of the group. The x and y directions are equivalent; they transform as a pair according to the doubly degenerate Π representation. In homonuclear diatomics, z transforms according to Σ_u^+, while x and y belong to the Π_u ir. rep. Even without consulting the character table, clearly the Cartesian coordinates are odd under inversion symmetry, in $D_{\infty h}$. So in the case of homonuclear diatomics, we arrive at the selection rule known as the LaPorte rule: $g \leftrightarrow u$ is dipole allowed but transitions between states having the same inversion symmetry are forbidden. This selection rule also holds for polyatomics having centers of symmetry, for example benzene. In the case of Σ states, the symmetry with respect to one of the σ_v reflection planes may be either (+) or (–), and the dipole moment operator for a z polarized (parallel polarized) transition belongs to the Σ^+ or Σ_u^+ representation. Therefore, the transitions $\Sigma^+ \leftrightarrow \Sigma^+$ and $\Sigma^- \leftrightarrow \Sigma^-$ are allowed, but transitions between (+)-states and (–)-states are forbidden. There is also a selection rule restricting the change in the quantum number M_L. The φ-dependence of the state wavefunction is $\exp(-iM_L\varphi)$, where φ is the angle of rotation about the bond. In the case of parallel transitions, the dipole operator has no φ dependence. Recall the orthogonality condition:

$$\int_0^{2\pi} \exp(iM_L\varphi)\exp(-iM_L'\varphi)d\varphi = 2\pi\delta_{M_L, M_L'} \tag{11.2}$$

This condition means that transitions having $\Delta\Lambda = 0$ are permitted, e.g., $\Sigma \leftrightarrow \Sigma$, $\Pi \leftrightarrow \Pi$, etc. The components of the dipole moment along the x and y directions (perpendicularly polarized transitions) must also be considered. These transform according to Π (or Π_u) representations. The x and y components of the dipole operator contain the function $\exp(\pm i\varphi)$, which is then included in the integral shown in Equation 11.2. The result is that transitions having $\Delta\Lambda = \pm 1$ are also permitted, such as $\Sigma \leftrightarrow \Pi$, $\Pi \leftrightarrow \Delta$, etc. The net selection rule is thus $\Delta\Lambda = 0, \pm 1$. For light molecules, in which spin–orbit coupling is weak, we expect the selection rule for the spin quantum number to be $\Delta S = 0$, however in heavier molecules this will be violated due to mixing of the zero-order states. The selection rule on the quantum number Ω is $\Delta\Omega = 0, \pm 1$. The spin selection rule is independent of whether the transition is E1, E2, or M1 since none of the transition operators depends on spin.

11.2.3.2 ELECTRIC QUADRUPOLE TRANSITIONS (E2)

The electric quadrupole moment operator has components with the same symmetry as products and squares of the coordinates, e.g., z^2, $x^2 - y^2$, yz, etc. These functions belong to g representations in the $D_{\infty h}$ character table, so the transitions $g \leftrightarrow g$ and $u \leftrightarrow u$ are allowed under E2 selection rules. The representations of these functions are Σ, Π, Δ, and they thus have the φ-dependence of MOs of the same symmetry. The selection rule for changes in the orbital angular momentum quantum number is therefore $\Delta\Lambda = 0$, ± 1, ± 2. Since the functions $x^2 + y^2$ and z^2 belong to Σ^+, the transitions $\Sigma^+ \leftrightarrow \Sigma^+$ and $\Sigma^- \leftrightarrow \Sigma^-$ are permitted by E2 selection rules.

11.2.3.3 MAGNETIC DIPOLE TRANSITIONS (M1)

The operator for M1 transitions is that for orbital angular momentum, $\vec{L}$. The components of this operator transform according to the same ir. reps. as the rotations R_x, R_y and R_z listed on the right-hand side of the character tables. In the $D_{\infty h}$ point group, these functions belong to gerade representations, so we immediately obtain the selection rule $g \leftrightarrow g$ and $u \leftrightarrow u$ for M1 transitions of molecules with inversion symmetry. In the case of transitions allowed by the z component of angular momentum, the relevant function is the rotation R_z, which belongs to the Σ^- representation. In this case the transition $\Sigma^+ \leftrightarrow \Sigma^-$ is permitted. The functions R_x and R_y transform as a degenerate pair according to the Π or Π_g ir. rep.; therefore, transitions having $\Delta\Lambda = \pm 1$ are permitted, along with the transitions having $\Delta\Lambda = 0$, which are permitted by the z component of $\vec{L}$.

11.2.3.4 EXAMPLES OF SELECTION RULES AT WORK: O$_2$ AND I$_2$

As was shown previously, the ground configuration of O_2 gives rise to three terms: $^3\Sigma_g^-$, $^1\Sigma_g^+$, and $^1\Delta_g$. The spin selection rule forbids transitions between the triplet ground state and either of the excited singlets. In addition, these states derive from the same electronic configuration, so transitions among them are forbidden or weak. The $^1\Delta_g$ state is in fact metastable, with a lifetime ranging from microseconds to tenths of a second, depending on the environment. The decay of the singlet excited states takes place mainly via nonradiative transitions. These excited states of O_2, once formed, stick around for long enough to act as important photochemical intermediates. They can be formed by energy transfer from excited triplet states of other molecules to ground state O_2.

In the electronic spectrum of I_2, large spin–orbit coupling means that the $\Delta S = 0$ selection rule is easily violated. The purple color of iodine is the result of visible absorption due to several overlapping transitions, two of which involve transitions from the singlet ground to triplet excited states. The strongest absorption in I_2 is due to the transition $^1\Sigma_g^+ \to {}^3\Pi_{0^+ u}$, at about 630 nm. It is made possible by Hund's case (c) coupling, which mixes some triplet character into the singlet states and vice versa. Weaker absorption in the visible is due to transitions from the ground state to the $^3\Pi_{1u}$ and $^1\Pi_{1u}$ states. Note the selection rule $\Delta\Omega = 0, \pm 1$ is obeyed in all three visible transitions.

11.3 VIBRATIONAL STRUCTURE IN ELECTRONIC SPECTRA OF DIATOMICS

11.3.1 ABSORPTION SPECTRA

Unlike the line spectra of atoms, electronic spectra of molecules can be bands, consisting of a series of vibrational (and rotational) transitions within the electronic transition. In electronic absorption spectra, a series of vibrational transitions from one initial vibrational state (in the lower electronic state) to a number of different final vibrational states (within the final electronic state) is called a Franck–Condon (FC) progression, and the intensity distribution within this progression is determined by the difference in equilibrium bond length of the molecule in the two electronic states. Figure 11.8 illustrates a common situation: the bond length in the excited electronic state is longer than in the ground. This is to be expected in the case of a transition for which an electron is promoted from a bonding to an antibonding orbital. By convention, double-primed quantities are used to designate properties of the ground state, and single-primed symbols are used for excited state quantities. In Figure 11.8, the case $R_e'' < R_e'$ is shown, but it is also possible to have $R_e'' > R_e'$ and $R_e'' \approx R_e'$. In the spirit of the Born–Oppenheimer approximation, the most probable transition in the absorption spectrum is the one indicated by the vertical line in Figure 11.8. The aptly named "vertical transition" would be the only one observed if, during the electronic transition, the nuclei were frozen at the distance R_e''. In the semiclassical Franck–Condon picture, the most probable transition is from the bottom of the lower well to a classical turning point at $R = R_e''$ on the upper well, in keeping with the idea that the sluggish nuclei remain fixed during the electronic transition. The quantum mechanical version of this approach recognizes that the molecule begins in the zero-point level of the lower electronic state and makes a transition to a state having the same nuclear kinetic energy and a probability distribution similar to that of the initial ground vibrational state. This translates into the vertical transition being that for which the overlap of the initial and final vibrational states is maximized, as shown in Figure 11.8. In contrast, if the ground and excited state potential wells are

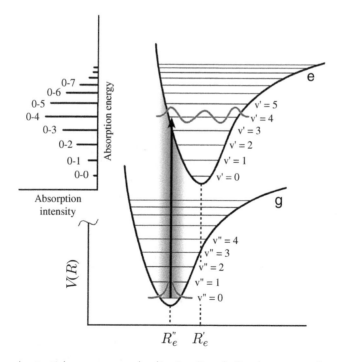

Figure 11.8 Displaced potential energy curves leading to a Franck–Condon progression.

identical except for vertical displacement in energy, then states having different vibrational quantum numbers have no overlap, and only transitions for which $\Delta v = 0$ are allowed.

Let us call the final vibrational quantum number for the vertical transition v_{vert}, e.g., $v' = 4$ in Figure 11.8. The larger the shift in bond length between the ground and excited state, the higher the value of v_{vert}, and the smaller the overlap of the $v'' = 0$ and $v' = 0$ states. As v' increases from zero, the amplitude of the vibrational wavefunction shifts toward the turning points, leading to more favorable overlap with the $v'' = 0$ state. In general, the greater the change in bond length upon excitation, the greater the number of vibrational states contributing to the progression. The width of this progression is a function of the slope of the upper potential well in the vicinity of R_e''. As illustrated in Figure 11.8, the width of the FC progression can be estimated from the range of transition energies for internuclear distances within the turning points of the initial vibrational state. This is called the "reflection principle." The steeper the slope of the upper well, the wider the range of final energy levels in absorption.

In the Born–Oppenheimer approximation, the ground and excited state wavefunctions are written as products of electronic and vibrational wavefunctions.

$$\Psi_{gv''} = \psi_g(r;R)\chi_{v''}^g(R) \tag{11.3}$$

$$\Psi_{ev'} = \psi_e(r;R)\chi_{v'}^e(R) \tag{11.4}$$

The electronic wavefunctions $\psi_g(r;R)$ and $\psi_e(r;R)$ depend parametrically on internuclear distance R, and the vibrational wavefunctions $\chi_{v''}^g(R)$ and $\chi_{v'}^e(R)$ depend on the potential energy function for the ground and excited states. The lowercase r symbolizes the coordinates of all the electrons. The transition moment for an E1-allowed transition is evaluated as follows:

$$\mu_{gv'',ev'} = \int dR \int d\vec{r}\, \Psi_g(r;R)\chi_{v''}^g(R)\mu\Psi_e(r;R)\chi_{v'}^e(R) \tag{11.5}$$

We must proceed carefully, because the dipole moment operator depends on electronic as well as nuclear coordinates. As discussed in Chapter 3, this operator is

$$\hat{\mu} = -e\sum_i \vec{r}_i + e\sum_\alpha Z_\alpha \vec{R}_\alpha$$
$$= \hat{\mu}_{el} + \hat{\mu}_{nuc} \tag{11.6}$$

Consider the second term, $\hat{\mu}_{nuc}$, which depends on positions of the stationary nuclei. When this operator is sandwiched between the ground and excited state wavefunctions in Equation 11.5, the *electronic* wavefunctions can be separated out, and on integrating over electronic coordinates, we get $\int d\vec{r}\,\psi_g(r;R)\psi_e(r;R) = 0$ due to orthogonality of the ground and excited electronic states. Thus we need only worry about the electronic part of the transition moment operator: $\hat{\mu}_{el} = -e\sum_i \vec{r}_i$, because the nuclear part cannot connect the two Born–Oppenheimer states. We imagine integrating first over the electronic coordinates and then over the internuclear distance.

$$\mu_{gv'',ev'} = \int dR \left\{ \int d\vec{r}\, \Psi_g(r;R)\mu\Psi_e(r;R) \right\} \chi_{v''}^g(R)\chi_{v'}^e(R)$$
$$= \int dR\, \mu_{ge}(R)\chi_{v''}^g(R)\chi_{v'}^e(R) \tag{11.7}$$

The electronic transition moment, $\mu_{ge}(R)$, depends on R due to the parametric dependence of the electronic wavefunctions on internuclear distance. It is common to make the *Condon approximation*, in which the electronic transition moment for absorption is evaluated at the equilibrium geometry of the ground electronic state: $\mu_{ge}^0 \equiv \mu_{ge}(R_e'')$. This amounts to expanding $\mu_{ge}(R)$ in a Taylor series about $R = R_e''$ and dropping all but the first term. Alternatively, μ_{ge}^0 can be replaced by the mean value $\bar{\mu}_{ge}$, considered to be an average over the range of internuclear distances sampled by the ground vibrational state wavefunction. In either case, the electronic transition moment is no longer dependent on R and can be factored out of the integral. The transition moment for a vibronic transition is then written as

$$\mu_{gv'',ev'} = \mu_{ge}^0 \int dR \, \chi_{v''}^g(R) \chi_{v'}^e(R)$$

$$\equiv \mu_{ge}^0 \langle v''|v' \rangle \tag{11.8}$$

where the overlap of the vibrational wavefunctions has been defined as $\langle v''|v' \rangle$. Now, *if* the two vibrational states $\chi_{v'}^e(R)$ and $\chi_{v''}^g(R)$ are eigenfunctions of the same Hamiltonian, they are orthonormal, and we get $\langle v''|v' \rangle = \delta_{v'v''}$. This is exactly what happens when there is no difference between the ground and excited state potential surfaces, except that one is higher in energy than the other. If the vibrational energy levels are widely spaced compared to thermal energy, only the ground vibrational state is populated and only the so-called $0-0$ transition is expected. So in the absence of a change in bond length or vibrational frequency, no FC progression is observed.

Consider the two potential surfaces shown in Figure 11.8, having characteristic vibrational frequencies v_e'' and v_e'. The absorption spectrum consists of a series of lines separated by the frequency difference of adjacent vibrational states in the excited electronic state. If the upper well is not harmonic, then this spacing is not constant but decreases as the vibrational quantum number of the upper state increases. (The emission spectrum, on the other hand, to be discussed in Section 11.3.2, exhibits a progression in which the vibrational spacing is that of the ground electronic state.) Each peak may be labeled by the initial and final quantum numbers as shown in Figure 11.8. The relative intensities of vibronic transitions $v'' \leftrightarrow v'$ are proportional to the Franck–Condon factor, which is the square of the overlap of the initial and final vibrational wavefunctions.

$$F_{v'',v'} \equiv \left| \langle v''|v' \rangle \right|^2 \tag{11.9}$$

The value of $F_{v'',v'}$ ranges from zero to unity, and the following sum rule is readily obtained:

$$\sum_{v'} F_{v'',v'} = \sum_{v'} \langle v''|v' \rangle \langle v'|v'' \rangle = \langle v''|v'' \rangle = 1 \tag{11.10}$$

Thus the total intensity of a transition originating in a particular initial vibrational state v'' and integrated over all final vibrational states is constant. The FC progression merely distributes the intensity among various final vibrational states.

In the special case where the upper and lower potential functions are harmonic and have the same frequency, $v_e'' = v_e'$, a simple closed form expression for the Franck–Condon factor exists:

$$F_{0v} = \left| \langle 0|v \rangle \right|^2 = \frac{1}{v!} \left(\frac{\Delta^2}{2} \right)^v \exp\left(-\Delta^2/2 \right) \tag{11.11}$$

where the dimensionless displacement Δ has been introduced:

$$\Delta \equiv \left(\frac{\mu\omega}{h} \right)^{1/2} \left(R_e' - R_e'' \right) \tag{11.12}$$

Notice that the intensity of a $0-v$ transition depends on the square of the displacement, so one cannot tell from the FC progression alone whether the bond is longer in the ground or excited state. Equation 11.11 gives $F_{0v} = \delta_{0v}$ when Δ is zero.

For initial vibrational quantum numbers other than zero, the FC factors can be generated starting from Equation 11.11 and employing recursion formulas [3]. There are also formulas in the literature that enable $F_{v'',v'}$ to be calculated in the case of displaced harmonic potentials with different frequencies, or for anharmonic potentials [4]. While it is possible to obtain an FC progression from a mere change in vibrational frequency, it turns out that in the absence of a change in R_e, a change in frequency is rather ineffective in causing a vibrational progression. For similar equilibrium positions $R_e' \approx R_e''$, it has been shown [5] that the ratio of the intensity of the $0 - 0$ transition to the sum of the intensities of all other $0 - v'$ transitions is

$$\frac{I_{00}}{\sum_{v'} I_{0v'}} = \frac{\sqrt{v_e' v_e''}}{\frac{1}{2}(v_e' + v_e'')} \tag{11.13}$$

Suppose that v_e' is as small as one-half the value of v_e''; the ratio given by Equation 11.13 is about 0.94, meaning that even for a significant frequency shift, in the absence of a change in bond length, most of the intensity is concentrated in the $0 - 0$ band.

It is common, but not necessarily correct, to neglect the change in vibrational frequency between the ground and excited state when calculating FC factors. This assumption runs the risk of being inconsistent, since weaker (longer) bonds generally have lower vibrational frequencies as well. For example, the vibrational frequency of I_2 is 214 cm^{-1} in the ground electronic state and about 120 cm^{-1} in the lowest excited electronic state. The I_2 molecule also provides an example of sequences, in which a series of FC progressions is observed. At room temperature, the populations of the first few vibrational levels of I_2 are significant, and a Franck–Condon progression in absorption is built on each thermally populated initial state.

11.3.2 EMISSION SPECTRA

Once a molecule is prepared in an excited electronic state, whether by a radiative transition or some other means, the possibility of a radiative transition to the ground electronic state is presented. As discussed in previous chapters, this emission may be either stimulated or spontaneous, and conventional emission spectroscopy exploits the latter, as it is more easily achieved than stimulated emission in the visible and ultraviolet regions of the spectrum. In the case where there is no change in spin quantum number, $\Delta S = 0$, spontaneous emission is referred to as fluorescence. For many molecules, having closed-shell ground electronic states, fluorescence is a transition from an excited singlet state to the ground singlet state. Phosphorescence, on the other hand, is emission in which there is a change in spin quantum number. The relationship between these two types of emission is discussed in more detail in Section 11.7. Our interest here is in describing vibrational structure of emission spectra in general.

Consider the Franck–Condon state achieved immediately after a vertical transition. The greater the displacement of the upper potential surface, the higher is the energy of the FC state relative to the ground vibrational level of the excited electronic state. The energy difference between the FC state and the ground vibrational state $v' = 0$ is called the *reorganization energy*. The reorganization energy is the energy lost by the molecule as it relaxes from the vertically excited to the equilibrium configuration. If emission occurs before this vibrational relaxation takes place, the emission spectrum is identical to the absorption spectrum. This is the case in the effect known as *resonance fluorescence*, and it may be observed for small molecules (such as diatomics) in the gas phase. However, consider the fact that typical radiative lifetimes in electronic spectroscopy are on the nanosecond timescale. Except in low pressure gases, this is a time within which many energy-transferring collisions can take place. *Relaxed fluorescence* is observed when emission takes place from the equilibrium geometry of the excited electronic state. The transition from the FC to the ground vibrational state, within the excited electronic state, involves loss of energy to the surroundings or to other internal degrees of freedom (such as other normal modes in the case of polyatomic molecules). This nonradiative relaxation step takes place readily in large molecules and in condensed phases, where relaxed emission is more common than emission from higher v' states.

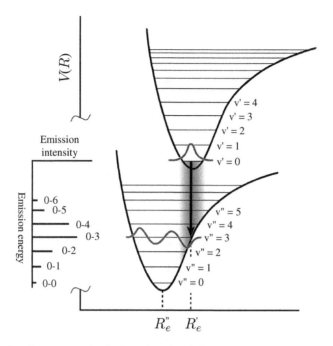

Figure 11.9 Franck–Condon progression in the relaxed emission spectrum.

Figure 11.9 illustrates the vibronic transitions associated with relaxed emission. As in the case of absorption spectra, the most intense peak in the vibrational progression in emission is that due to the downward vertical transition, from $v' = 0$ to $v'' = v_{vert}$, where $v_{vert} = 3$ in Figure 11.9. The width of the emission spectrum increases with displacement of the ground and excited state potentials or, alternatively, the slope of the ground state potential in the vicinity of $R = R'_e$. The emission spectrum is at longer wavelength than absorption. The difference in the absorption and fluorescence frequency maxima (the vertical transitions) is called the Stokes shift. It too depends on the displacement. Note that the absorption and emission intensity overlap at the frequency of the 0 − 0 transition, but for large displacements this transition may have an unfavorable FC factor. In order to extract the 0 − 0 energy from the crossing point of the absorption and emission spectra, the intensity of each must be scaled by correcting for the different frequency dependence of absorption (proportional to v) and emission (proportional to v^3).

11.3.3 DISSOCIATION AND PREDISSOCIATION

Even in high-resolution spectra of gas-phase molecules, discrete vibrational structure is not always observed. For example, a transition may take the molecule to a terminal state having more energy than required for dissociation, as shown in Figure 11.10, in which case a continuous absorption spectrum is expected. On the other hand, if one of the Franck–Condon allowed states has energy similar to that where the bound ($e1$) and dissociative ($e2$) excited state potentials cross, as in Figure 11.11, the molecule may make a transition from one potential surface to another. This is called predissociation, and it is an example of a nonadiabatic transition, to borrow vocabulary from the next section. In this case, vibrational structure may still be observed, but transitions to states in the vicinity of the curve crossing are broadened due to the shortened lifetime of the directly excited state. This crossing from one potential curve to another is the result of mixing of Born–Oppenheimer states, discussed in the next section.

In the case of direct dissociation, the dissociation energy of the excited state D'_0 may be obtained from the energy of the onset of the continuum E_{cont} in absorption and the 0 − 0 energy E_{00}.

$$E_{cont} = E_{00} + D'_0 \qquad (11.14)$$

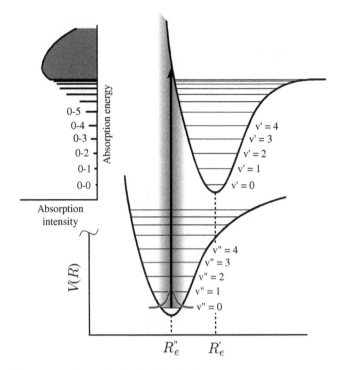

Figure 11.10 Potential energy surfaces leading to dissociation.

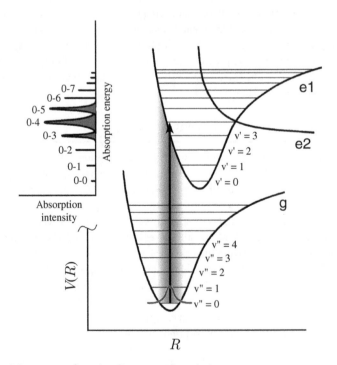

Figure 11.11 Potential energy surfaces leading to predissociation.

Conversely, the dissociation energy of the ground electronic state D_0'' can be obtained from the energy of the onset of the continuum in emission. In addition, one may know the energy difference of the atomic states in the dissociation limit, $E_{atoms}' - E_{atoms}''$, in which case the following relationship can be used to extract D_0''.

$$E_{00} + D_0' = D_0'' + E_{atoms}' - E_{atoms}'' \tag{11.15}$$

Even if continuous absorption is not observed, one may still derive dissociation energies of excited states from the pattern of vibrational spacings in absorption. As was discussed in Chapter 9, vibrational levels converge as the dissociation limit is approached, and extrapolation procedures can be employed to derive dissociation energies from vibrational progressions.

11.4 BORN–OPPENHEIMER BREAKDOWN IN DIATOMIC MOLECULES

The very idea of drawing potential energy curves (or surfaces) for electronic states is founded on the Born–Oppenheimer (BO) approximation. This approximation is based on the physically appealing premise that electrons are swifter than nuclei and thus the nuclei experience the potential energy of the averaged electronic distribution. Still, the BO approximation can break down whenever electronic energy differences are not large compared to vibrational spacings. When the BO approximation fails, it is no longer possible to separate electronic and nuclear energies, and the timescales of electronic and nuclear motion are not separable. Unexpected spectral features are often observed, such as a departure from the regular FC progression or appearance of symmetry-forbidden transitions.

Figure 11.12 illustrates a typical situation leading to BO breakdown, the crossing of two potential energy curves of the same symmetry. The perturbation operator $\hat{H}'$ can connect two states if the matrix element $\langle \Psi_1 | \hat{H}' | \Psi_2 \rangle \equiv H_{12}$ is nonzero. The Hamiltonian, including any perturbation terms, is totally symmetric, and thus the nonvanishing of H_{12} requires that Ψ_1 and Ψ_2 be of the same symmetry. This leads to the *noncrossing rule*, which states that the potential energy curves for states of the same symmetry cannot cross. Rather, in the vicinity of the crossing, the two BO states mix with one another and split apart, as shown in Figure 11.12, resulting in an "avoided crossing."

What is the nature of the perturbation? The answer to this question depends on the basis states that one uses to describe the perturbed system. In principle, the exact wavefunctions could be expanded in terms of the BO wavefunctions.

$$\Psi = \sum_i \psi_i (r; R) \chi_i (R) \tag{11.16}$$

However, this is an infinite series. In favorable cases, such as that shown in Figure 11.12, just two electronic states, call them i and j, interact strongly. One can then solve the perturbation problem in the basis of these

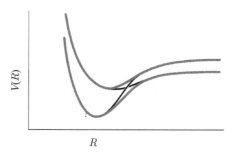

Figure 11.12 Avoided crossing resulting from Born–Oppenheimer breakdown.

two states. The BO approximation neglects the operator for the nuclear kinetic energy $\hat{T}_N$. The so-called adiabatic curves, which do not cross, have energies given by

$$E_{ad}(R) = V_i(R) + \langle \psi_i | \hat{T}_N | \psi_i \rangle \tag{11.17}$$

The second term in Equation 11.17 has been averaged over the electronic coordinates but still depends on R. The adiabatic energy E_{ad} neglects the off-diagonal matrix element $\langle \psi_i | \hat{T}_N | \psi_j \rangle = H_{ij}$, which further perturbs and mixes the adiabatic states.

It is sometimes convenient to describe the unperturbed states in terms of what is called the *diabatic* representation. These are not exact eigenfunctions, and so their potential functions *are* allowed to cross. They are chosen to have nonzero matrix elements of the electronic part of the Hamiltonian, $\hat{H}_{el}$, which contains the operators for electronic kinetic energy as well as those for the electron-electron, electron-nuclear, and nuclear-nuclear potential energies. In this case the perturbation matrix element is $\langle \psi_1^{diab} | \hat{H}_{el} | \psi_2^{diab} \rangle \equiv V_{12}$, which is a function of R. The splitting of the two diabatic curves to give the adiabatic energies is $2V_{12}$. See [6] and references therein for more details on corrections to the Born–Oppenheimer picture.

The ground and first excited-state potential curves for NaCl provide an example of the noncrossing rule. At internuclear distances near the ground state bond distance R_e'', the ground state is ionic (Na^+Cl^-) and the excited state is covalent (NaCl). In the dissociation limit, however, clearly the ionic state $Na^+ + Cl^-$ is higher in energy than the separated atoms, by about 30,000 cm^{-1}, an amount equal to the ionization potential of sodium (5.1 eV) less the electron affinity of chlorine (1.5 eV). The potential curves for the covalent and ionic states, in the Born–Oppenheimer approximation, cross near 10 Å, and the true state for each of the adiabatic surfaces is a mixture of the covalent and ionic states. There is a characteristic time associated with the adjustment of the wavefunctions to the non-BO perturbation, on the order of $h/\Delta E$, where $\Delta E = 2V_{12}$ is the splitting of the adiabatic surfaces. If the ground state Na^+Cl^- molecule could be dissociated very slowly compared to $h/\Delta E$, it would travel along the lower adiabatic state and end up going over to the neutral atoms. This is a manifestation of the Ehrenfest adiabatic principle, which states that a system will remain in a definite state if a perturbation is applied infinitely slowly. Such a process, taking place on a single potential surface, is called adiabatic, as opposed to a nonadiabatic process, in which a jump from one surface to another occurs. On the other hand, if dissociation is rapid compared to the time it takes for the wavefunctions to adjust, then a jump from the lower to the upper adiabatic surface may happen, and dissociation yields ions rather than neutral atoms.

Nonadiabatic effects can be important in spectroscopy, charge transfer processes, and photochemistry. The previously discussed case of predissociation is an example of a nonadiabatic transition, as are internal conversion and intersystem crossing, defined in Section 11.7.

11.5 POLYATOMIC MOLECULES: ELECTRONIC STATES AND SELECTION RULES

The molecular orbitals of a molecule correspond to ir. reps. of the molecular point group, and the electronic states associated with a given electronic configuration are found from the direct product of the symmetries of the occupied orbitals, taking the Pauli exclusion principle into account. The electronic state is designated by the so-called Mulliken symbol, an uppercase letter designating the state symmetry with a left-hand superscript having the value $2S + 1$. The total spin quantum number S is as good a quantum number in polyatomics as it is in diatomics and atoms; that is, it is strictly good in the absence of spin–orbit coupling. Most stable molecules have closed-shell (singlet) ground electronic states, belonging to the totally symmetric representation of the group. In the case of open shells, the state symmetry can be derived from the direct product of the singly occupied molecular orbital symmetries. This will be illustrated below by means of several examples. We continue to use the convention that lowercase letters represent the symmetry of molecular orbitals and uppercase letters that of electronic states.

11.5.1 MOLECULAR ORBITALS AND ELECTRONIC STATES OF H_2O

The water molecule provides a tractable example of symmetry considerations in molecular electronic structure and a prototype for other bent triatomic systems. A molecular orbital of water may be labeled a_1, a_2, b_1,
or b_2 to designate the representation to which it belongs. (See the C_{2v} character table of Appendix C.) The
minimal basis set of AOs required for a qualitative description of the bonding is the set $1s$, $2s$, $2p_x$, $2p_y$ and
$2p_z$ on oxygen and the $1s$ orbitals on the two hydrogens. This is a basis set of seven AOs, so seven MOs will be
obtained. The oxygen atom $1s$, $2s$, and $2p_z$ AOs transform according to the totally symmetric representation
a_1, while the $2p_x$ and $2p_y$ orbitals belong to b_1 and b_2, respectively. (The symmetry axis is the z direction and the
molecule lies in the yz plane.) To find these symmetries, consider the effect of each operation of the group on a
given AO; i.e., construct a table of plus and minus signs according to whether the operation preserves the AO
or changes its sign. The pattern should match that of one ir. rep. of the C_{2v} group. Now, it is quite general that
only atomic orbitals of the same symmetry may interact, so it is convenient to start with symmetry-adapted
linear combinations (SALCs) of the hydrogen atom $1s$ functions. In general, one can use the projection operator method described in the previous chapter to form SALCs from a given set of basis AOs. In this example,
we can derive the desired SALCs by inspection, by taking in-phase and out-of-phase combinations of the
hydrogen AOs. The symmetric and antisymmetric combinations $\psi_s = 1s_A + 1s_B$ and $\psi_a = 1s_A - 1s_B$ belong to
the a_1 and b_2 representations, respectively, where A and B label the hydrogen atoms. The symmetric SALC,
ψ_s, can combine with the three a_1 AOs on oxygen, and the SALC of symmetry b_2, ψ_a, can combine with the
sole b_2 orbital on oxygen. There should thus be four totally symmetric MOs and two of symmetry b_2. To a
good approximation, the $1s$ orbital on oxygen is little perturbed by the bonding, and we can consider it to be
the lowest energy core MO, of a_1 symmetry, $(1a_1)$. The b_1 symmetry p_x orbital, which is perpendicular to the
molecular plane, is the wrong symmetry to interact with either ψ_s or ψ_a; it is therefore a nonbonding orbital
$(1b_1)$. Figure 11.13 shows the MOs that result from combining these AOs. The ground configuration of H_2O is

$$(1a_1)^2(2a_1)^2(1b_2)^2(3a_1)^2(1b_1)^2(4a_1)^0(2b_2)^0$$

The unoccupied $4a_1$ and $2b_2$ orbitals are antibonding, as shown in Figure 11.13. We can get an approximate
picture of the water molecule's excited states by promoting electrons to these virtual orbitals. The ground
state is a totally symmetric singlet state, because all orbitals are doubly occupied, designated by $^1A_1$. Since

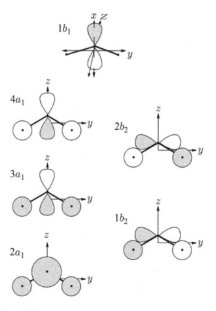

Figure 11.13 Molecular orbitals for H_2O.

the dipole operator transforms under A_1, B_1, and B_2 representations for the z, x, and y components, respectively, of the C_{2v} point group, these are also the symmetries of the excited states that can be reached from the ground state by E1 selection rules. The state symmetry of a configuration such as $(1a_1)^2(2a_1)^2(1b_2)^2$ $(3a_1)^2(1b_1)^1(4a_1)^1$ is equal to the direct product of the singly occupied orbitals: $b_1 \times a_1 = B_1$. Since the two electrons may have either parallel or antiparallel spins, both $^1B_1$ and $^3B_1$ states arise from this configuration. The transition from the ground $^1A_1$ to the excited $^1B_1$ state is allowed by E1 selection rules and is polarized in the x direction.

As for diatomics, the ground state configuration is labeled $\tilde{X}$, and higher-lying electronic states of the same spin multiplicity are designated by the letters $\tilde{A}, \tilde{B}, \tilde{C}, \ldots$, etc., in order of increasing energy. The lowest-energy electronic transition observed in H_2O is $\tilde{X}(^1A_1) \rightarrow \tilde{A}(^1B_1)$, and it turns out that it is a special type of transition known as a *Rydberg transition*. Rydberg electronic states are hydrogen-like states resulting from promotion of an electron to an orbital in which its average distance from the center of the molecule is much larger than that of the other electrons. From the point of view of the distant electron, the molecule is little different from an ion with a +1 charge, since the net nuclear charge is screened by all the other electrons. The energies of Rydberg states follow the formula

$$E_n = I - \frac{R}{(n-\delta)^2} \tag{11.18}$$

where I is the ionization energy of the orbital from which the Rydberg electron was promoted, R is the Rydberg constant, δ is a number which depends on the quantum number l for the Rydberg orbital (the Rydberg correction factor), and n is the principle quantum number. Many Rydberg states having $n \geq 3$ have been observed for water. The first member of the ns series is the $\tilde{X}(^1A_1) \rightarrow \tilde{A}(^1B_1)$ transition observed from about 186 to 145 nm. The excited state configuration is $\ldots(1b_1)^1(3sa_1)^1$; i.e., an electron from the nonbonding b_1 orbital is promoted to a hydrogen-like $3s$ orbital. At higher energies, transitions to $4s$, $5s$, $6s$, $\ldots$ Rydberg states are observed, as well as a series involving p-type Rydberg states.

The second lowest energy absorption of water, the $\tilde{X}(^1A_1) \rightarrow \tilde{B}(^1A_1)$ transition, has an upper state with the configuration $\ldots(3a_1)^1(1b_1)^2(3sa_1)^1$. This band stretches from 140 to 125 nm and has vibrational structure with a spacing of about 800 cm^{-1}, due to an excited state bending mode. The $\tilde{B}$ state of water is linear, and the vibrational progression in the $\tilde{X}(^1A_1) \rightarrow \tilde{B}(^1A_1)$ transition results from this change in symmetry. In the next section, we discuss the origin of this sort of vibrational structure.

11.5.2 FRANCK–CONDON PROGRESSIONS IN ELECTRONIC SPECTRA OF POLYATOMICS

The occurrence of Franck–Condon progressions in polyatomic molecules depends on the difference in the equilibrium geometries of the two electronic states involved in the transition. Consider a transition from the ground to the excited electronic state, with no change in molecular symmetry. The overall geometry change can be resolved into contributions along each normal coordinate Q_i, for $i = 1$ to $3N - 6$. The normal coordinates can be divided into two groups: totally symmetric and nontotally symmetric. If there is no change in point group on excitation, only totally symmetric modes are displaced in the excited state. For example, a symmetric stretch can be displaced in the excited state by virtue of a change in the equilibrium bond lengths, as shown in Figure 11.14a. The potential function for a nontotally symmetric vibration, such as the bend of a linear molecule or an asymmetric stretch of equivalent bonds, is always a symmetric function of normal coordinate, as shown in Figure 11.14b. As an example, consider the asymmetric stretch of H_2O, for which the ground state potential is of course a minimum when the two bond lengths are equal. The potential energy must be a symmetric function of the difference in the two O–H bond distances; i.e., $V(r_1 - r_2) = V(r_2 - r_1)$. We expect the two bond distances are also equal at the equilibrium geometry of the excited electronic state. Barring a change in symmetry, there can be no change in the equilibrium position for a nontotally symmetric mode. The Franck–Condon active normal modes are thus those which mimic the difference in geometry of

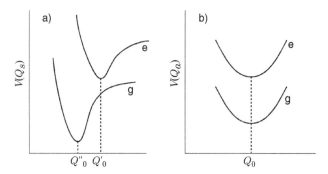

Figure 11.14 Ground- and excited- state potentials along (a) a totally symmetric and (b) a nontotally symmetric normal coordinate.

the two electronic states. In the absence of a change in symmetry in the excited electronic state, these are necessarily totally symmetric vibrational modes.

When a nontotally symmetric vibration derives FC activity through a large frequency shift (however, see Equation 11.13 and the discussion that follows it), there are selection rules that restrict the value of Δv. For example, the state symmetries of a nondegenerate, nontotally symmetric mode alternate between totally symmetric (for even v) and the symmetry of the mode (for odd v). This results in the selection rule $\Delta v = 0, \pm 2, \pm 4, \pm 6, \ldots$, etc., because the FC factor $F_{v'',v'}$ vanishes unless the states v' and v'' have the same symmetry. In the case of a degenerate vibration, even overtones *always* contain the totally symmetric representation, so the transitions having $\Delta v = 0, \pm 2, \pm 4, \pm 6, \ldots$ are again allowed. In certain point groups, such as D_{3h}, C_{3v}, and T_d, odd overtones of degenerate vibrations also contain the totally symmetric representation, so the selection rules are more permissive: $\Delta v = 0, \pm 1, \pm 2, \pm 3, \pm 4, \ldots$, except that the $0 - 1$ and $1 - 0$ transitions are not allowed.

For each FC-active normal mode, the one-dimensional FC factor $F_{v'',v'}$ can be calculated using formulas like Equation 11.11, but with a redefinition of the dimensionless displacement Δ.

$$\Delta_i \equiv \left(\frac{\omega}{\hbar} \right)^{1/2} \left(Q'_{0i} - Q''_{0i} \right) \tag{11.19}$$

In Equation 11.19, Q'_{0i} (Q''_{0i}) is the equilibrium position of the normal coordinate for mode i in the excited (ground) electronic state.

It often happens that more than one normal mode is Franck–Condon active, and vibrational structure in an electronic spectrum can be complex. A progression is a series of lines corresponding to all possible changes in quantum number for a given initial state of one normal mode. A sequence, on the other hand, is a series of vibrational transitions all having the same value of Δv for a certain mode. The observation of a sequence in absorption is contingent upon having significant Boltzmann population of states having $v'' > 0$. In addition, the frequency separation of sequences depends on the difference in the ground and excited state vibrational frequencies.

What about the case where the molecule undergoes a change in symmetry, in other words, the ground and excited states belong to different point groups? There are many examples where this is the case. The $\tilde{B}$ excited state of H_2O is linear and the absorption spectrum exhibits a long progression in the bending mode at about 800 cm⁻¹. The $\tilde{A}(^1\Delta_u)$ state of CO_2 has an equilibrium bond angle of 122 degrees, compared to 180 degrees in the ground state. These two situations are illustrated in Figure 11.15. In the emission spectrum corresponding to Figure 11.15a, and in the absorption spectrum corresponding to Figure 11.15b, the slope of the potential for the final electronic state is nonzero in the FC region, and we expect to get a progression in the vibration which correlates with the geometry change. When an electronic transition is accompanied by a change in point group, the normal coordinates of one state are linear combinations of those of the other state, and one must take into account this mixing in order to account for the intensities of vibronic transitions. This mixing of normal modes is called Duschinsky rotation. Group theory may still be useful in this case, but one must use only the symmetry elements common to both the ground and excited states.

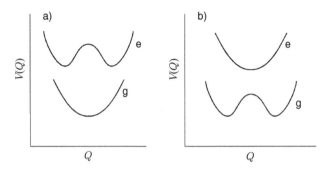

Figure 11.15 Ground- and excited-state potential energy curves for a molecule undergoing a change in symmetry. The lower curve in (a) represents a nontotally symmetric vibration of a symmetric molecule (e.g., the bending mode of a linear triatomic), while the upper curve represents a totally symmetric vibration of a molecule with two equivalent distorted geometries (e.g., a bent triatomic). In (b) the upper state is symmetric and the lower state distorted.

Even in the absence of a symmetry change, there are other ways that nontotally symmetric vibrations can show up in absorption and emission profiles. One way is for the FC activity to derive from a frequency change, but as previously mentioned this is not nearly as effective in producing FC progressions as a change in geometry. The other way is for nontotally symmetric modes to appear by means of what is called *vibronic coupling*. This is a non-Born–Oppenheimer effect which permits mixing of BO states via nontotally symmetric vibrations. It is discussed and illustrated in the next section, using benzene as an example.

11.5.3 BENZENE: ELECTRONIC SPECTRA AND VIBRONIC ACTIVITY OF NONTOTALLY SYMMETRIC MODES

The ultraviolet absorption spectrum of benzene derives from transitions of π electrons and can therefore be discussed within simple molecular orbital pictures which treat only these electrons. Benzene belongs to the D_{6h} point group. The MO energy levels and π bonding patterns of benzene are shown in Figure 11.16. These are obtained using the six carbon atom p_z orbitals as a basis, where z is considered to be the direction perpendicular to the plane of the molecule. The energies and symmetries can be found using, for example, Hückel theory [7,8]. Alternatively, as we are interested only in selection rules and not attempting to calculate energies, the projection operator techniques described in Chapter 10 can be used to find the symmetries of the MOs (a_{2u}, e_{1g}, e_{2u}, and b_{2g}), which can then be ordered on the basis of the number of nodal planes perpendicular to the plane of the molecule. As shown in Figure 11.16, the MO energy increases with the number of these nodes. The labels on the left-hand side of the diagram are the MO energies obtained from a Hückel calculation, a semiempirical method which treats only the π electrons. α is the energy of an isolated carbon atom $2p_z$ electron, and β (a negative quantity) is the resonance integral connecting $2p_z$ orbitals on neighboring carbons. The two lowest

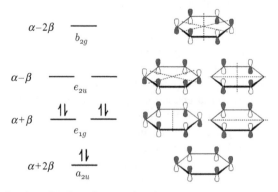

Figure 11.16 Benzene molecular orbitals and energy levels.

energy levels, the nondegenerate a_{2u} and the doubly degenerate e_{1g}, are fully occupied in the ground electronic configuration of benzene. These are all bonding π MOs, although the lowest energy MO is more bonding than the pair of e_{1g} orbitals. Also note that the two e_{1g} MOs differ only in the position of the nodal plane, and similarly the two antibonding π^* MOs of e_{2u} symmetry differ in the positions of the two nodal planes.

The ground configuration of benzene $a_{2u}^2 e_{1g}^4$ gives rise to a totally symmetric singlet state, for which the Mulliken term is $^1A_{1g}$. Excited state configurations may be generated by promoting electrons from bonding (π) orbitals to antibonding (π^*) orbitals. Consider the HOMO–LUMO transition to form the configuration $a_{2u}^2 e_{1g}^3 e_{2u}^1$, from which the lowest-lying excited states are derived. There are $4 \times 4 = 16$ states associated with this configuration. The state symmetries are found by taking the direct product of the singly occupied MO symmetry with that of the triply occupied MO. When decomposed into ir. reps., one obtains $e_{1g} \times e_{2u} = E_{1u} + B_{2u} + B_{1u}$. Each of these terms can be associated with either a singlet or a triplet spin state, thus there are $^1E_{1u}$, $^3E_{1u}$, $^1B_{2u}$, $^3B_{2u}$, $^1B_{1u}$, and $^3B_{1u}$ states. The E_{1u} states are doubly degenerate with respect to the spatial part of the wavefunction, and the B_{2u} and B_{1u} states are nondegenerate.

The electric dipole moment operator transforms according to the A_{2u} and E_{1u} representations of D_{6h}. Transitions allowed by the A_{2u} component are polarized perpendicular to the plane of the molecule, while those permitted by the E_{1u} component are polarized in the plane. Since the ground state is totally symmetric, E1-allowed transitions are to states of A_{2u} and E_{1u} symmetry, so we predict that the transition from the ground $^1A_{1g}$ to the excited $^1E_{1u}$ state should be allowed and polarized in the plane. This transition is observed in the ultraviolet, in the vicinity of 185 nm, and it is quite strong, having an oscillator strength close to one.

Now for the surprises: two lower-energy UV transitions, much weaker than the 185 nm band, are also observed. The excited states active in these two bands are of B_{1u} and B_{2u} symmetry, absorbing at about 208 and 260 nm, respectively. These would be forbidden according to the selection rules we have discussed so far, which are based on the validity of the Born–Oppenheimer separation of electronic and vibrational states. When this approximation breaks down, states which are forbidden according to BO selection rules can mix with states (such as those of E_{1u} symmetry) which are allowed. This state mixing can only take place if there is a vibration of appropriate symmetry that can couple the two pure electronic states. The coupling vibration must be a nontotally symmetric species, and one way to regard the altered selection rules is in terms of lowered symmetry of the vibrating molecule, resulting in the appearance of otherwise symmetry-forbidden transitions. This type of electronic transition is said to be allowed by *vibronic coupling*.

In the case where the BO approximation fails, we need to discuss the symmetry of a vibronic state, which can be designated as Γ_{ev}. The transition moment $\mu_{e'v',e''v''}$ refers to that connecting two vibronic states. For this transition moment to be nonzero, the direct product, $\Gamma_{e'v'} \times \Gamma_\mu \times \Gamma_{e''v''}$ must contain or equal the totally symmetric representation, in which case the transition $e'v' \leftrightarrow e''v''$ is allowed. Each vibronic symmetry species is in turn the direct product of the electronic and vibrational parts: $\Gamma_{ev} = \Gamma_e \times \Gamma_v$, where Γ_v is the symmetry of the vibrational *state* and not necessarily that of the mode. In the case where absorption originates from the vibrational ground state, $\Gamma_{e''v''} = \Gamma_{e''} = A_{1g}$ (as in the case of benzene). Thus vibronically activated excited states of benzene must become allowed by participation of a vibrational state whose symmetry satisfies $\Gamma_{e'} \times \Gamma_{v'} = \Gamma_\mu$. The 260 nm band of benzene derives its activity from the E_{2g} in-plane bending mode (ν_{18} in Table 10.5), and it can be checked that $\Gamma_{e'} \times \Gamma_{v'} = B_{2u} \times E_{2g} = E_{1u} = \Gamma_\mu$, where we have assumed that $v' = 1$; thus the symmetry of the final vibrational state is that of the mode. This is based on the fact that the observed vibrational structure of the $^1A_{1g} \rightarrow {}^1B_{2u}$ absorption band displays transitions to vibrational levels having one quantum of excitation in the bending mode and some number of quanta n in the totally symmetric breathing mode, $\nu_2(A_{1g})$. The frequency of the bending vibration in the ground electronic state is 606 cm^{-1}, but it shifts to 522 cm^{-1} in the excited $^1B_{2u}$ state. Similarly, the breathing mode shifts from 992 to 923 cm^{-1} in the $^1B_{2u}$ excited state. These shifts are consistent with the $\pi - \pi^*$ nature of the transition. The peaks observed in the progression of the 260 nm band of benzene can be labeled $2_0^n 18_0^1$, where the 2 and the 18 are the mode numbers,* and the subscripts and superscripts respectively denote initial (v'') and final (v') vibrational quantum numbers. The long FC progression in the ring-breathing mode of benzene derives from the increase in the C–C bond

---

* In some books, the notation of Wilson, Decius, and Cross is employed to number the benzene vibrations. In this notation, the breathing mode is ν_1 and the bending mode is ν_6.

distance, by about 0.04 Å, in the excited electronic state. Note that the $0 - 0$ transition of the bending mode, $2_0^n 18_0^0$, is not observed, as the vibronic activation requires that the final state have one quantum of excitation in the nontotally symmetric mode. The vapor-phase spectrum of benzene also shows transitions of the type $2_0^n 18_1^0$. At higher temperatures, the $v''_{18} = 1$ state of the bending mode is populated in the ground electronic state, and transitions to states having $v'_{18} = 0$ in the excited electronic state also satisfy the symmetry requirements. In the solution phase spectrum of Figure 11.1, the progression in the breathing vibration is apparent, but separate peaks due to v_{18} are not resolved. They are readily observed in the gas phase, however.

The 208 nm band of benzene, $^1A_{1g} \rightarrow {}^1B_{1u}$, is also made possible by the vibronic activity of the E_{2g} vibration. The direct product of B_{1u} and E_{2g} is E_{2u}. (The direct products $B_{1u} \times E_{2g}$ and $B_{2u} \times E_{2g}$ are equal because the only characters in B_{1u} which are different from those of B_{2u} are those of operations for which the E_{2g} characters are zero. See the character table in Appendix C to verify this.) Similar vibrational structure to that of the 260 nm band is observed for this band. But the following question is worth raising: If symmetry is the key to understanding vibronic activity, why are none of the other three E_{2g} vibrations of benzene, such as the in-plane CCH bend, at 1178 cm^{-1} in the ground state, observed in the structure of these two bands? To probe this question, let us consider the quantum mechanical basis for the appearance of these transitions.

The electronic Hamiltonian in the BO approximation is independent of the normal coordinates, as the nuclear positions are considered fixed. The following perturbation operator serves as a correction to the BO approximation:

$$\hat{H}' = \sum_{i=1}^{3N-6} \left(\frac{\partial \hat{H}_{el}}{\partial Q_i} \right)_0 Q_i \tag{11.20}$$

This operator is totally symmetric, so the derivative $(\partial \hat{H}_{el}/\partial Q_i)_0$ must have the symmetry of the normal mode Q_i. Two electronic states, labeled e and s, can be coupled by this perturbation if their zero-order energies are similar and the matrix element H_{es} is nonzero:

$$H_{es} \equiv \langle \psi_e | \hat{H}' | \psi_s \rangle = \sum_{i=1}^{3N-6} \left\langle \psi_e \left| \left(\frac{\partial \hat{H}_{el}}{\partial Q_i} \right)_0 \right| \psi_s \right\rangle Q_i \tag{11.21}$$

where the brackets indicate integration over the electronic coordinates only. This integral can exist only if the triple direct product $\Gamma_e \times \Gamma_{Q_i} \times \Gamma_s$ equals or contains the totally symmetric representation. In this case, at nuclear positions where the energies of state e and s are similar, then these two electronic states will be vibronically coupled. If, to a first approximation, the transition from the ground g to the excited electronic state e is forbidden, while $g \rightarrow s$ is allowed, the mixing of e and s enables the state e to borrow intensity from state s. But in addition to the symmetry and energy requirements, such coupling takes place only if the electronic Hamiltonian depends on the normal coordinate for the coupling vibration. The perturbed states are then linear combinations of the zero-order states, e.g., $\psi' = c_e \psi_e + c_s \psi_s$. In the case of benzene, the B_{2u} and B_{1u} excited states derive their intensity from mixing with the nearby allowed state of E_{1u} symmetry. It seems reasonable that the matrix element H_{es} connecting two $\pi - \pi^*$ excited states depends more strongly on vibrations which distort the pi framework than those which do not.

Another way to account for vibronic activity is to go beyond the Condon approximation and account for the normal coordinate dependence of the transition moment:

$$\mu_{ge} = \mu_{ge}^0 + \sum_{i=1}^{3N-6} \left(\frac{\partial \mu_{ge}}{\partial Q_i} \right)_0 Q_i + \cdots \tag{11.22}$$

The so-called non-Condon term $(\partial \mu_{ge}/\partial Q_i)_0$ is capable of connecting the states g and e if the direct product $\Gamma_g \times \Gamma_{Q_i} \times \Gamma_e$ equals or contains Γ_μ. This approach leads to the same conclusions as derived above, as far as symmetry requirements for vibronic activity are concerned. We will reexamine the issue of vibronic coupling in Chapter 12, where it is shown to play a role in selection rules for Raman scattering.

11.6 TRANSITION METAL COMPLEXES: FORBIDDEN TRANSITIONS AND THE JAHN–TELLER EFFECT

Transition metal complexes provide familiar examples of color in chemistry. Although metal–ligand systems may be found in a variety of symmetries (tetrahedral, square planar, linear…), we choose the classic case of octahedral symmetry, assuming that the metal is surrounded by six equivalent ligands, ML_6, as the basis for the examples of this section. The optical spectra of transition metal complexes derive from transitions of d electrons, and the conventional way to describe them often invokes a theory called crystal field theory. In this theory, electrostatic interaction of the metal ion with the surrounding ligands, assumed to be point charges, lifts the degeneracy of the five d atomic orbitals on the metal. The approach of this section, however, begins with molecular orbital theory, using a basis of the five d, three p, and one s orbitals on the metal, plus six equivalent orbitals on the ligands. These ligand orbitals can be called σ AOs, because they direct electron density toward the metal, resulting in orbitals which are axially symmetric about the metal-ligand bonds.

In the free ion, the nd AOs are occupied and the $(n + 1)p$ and $(n + 1)s$ AOs are unoccupied. Each ligand is assumed to contribute one pair of electrons to the system. The five d orbitals and the coordinate system employed are shown in Figure 11.17. With the help of the O_h character table of Appendix C, it can be shown that the d_{xy}, d_{xz}, and d_{yz} AOs transform according to the triply degenerate t_{2g} representation. Clearly these three d orbitals are equivalent under octahedral symmetry, so they must belong to a triply degenerate representation. They are also even under inversion symmetry. To distinguish whether these AOs transform as t_{1g} or t_{2g}, consider the effect of one of the six C_2 operations, which changes the phase of the d_{xy}, d_{xz}, or d_{yz} orbital; hence, they must belong to t_{2g}, for which the character of the C_2 operation is -1. Similarly, it can be concluded that the d_{z^2} and $d_{x^2-y^2}$ AOs transform according to the doubly degenerate e_g ir. rep., while the metal p_x, p_y, and p_z orbitals belong to t_{1u}, and the metal s orbital is totally symmetric, a_{1g}. The six equivalent ligand AOs can be combined into symmetry-adapted linear combinations, using projection operator techniques. The resulting symmetries are a_{1g}, e_g, and t_{1u}, as shown in Figure 11.18. The metal AOs are combined with ligand SALCs of the same symmetry, to form the bonding and antibonding MOs. The resulting energy levels are depicted in Figure 11.19. In a class by themselves and having zero overlap with ligand SALCs, the metal t_{2g} orbitals are nonbonding. The metal and ligand e_g orbitals combine to give a degenerate pair of bonding MOs, and a degenerate pair of antibonding MOs. Similarly, the metal and ligand a_{1g} AOs give bonding and nonbonding AOs as shown in Figure 11.19.

The splitting of the t_{2g} and e_g^* AOs is called Δ, and it is a function of the strength of the metal–ligand interaction. The value of Δ tends to increase in the ligand series $I^- < Br^- < Cl^- < NO_3^- < F^- < OH^- < H_2O < NH_3 < H_2NCH_2CH_2NH_2 < NO_2^- < CN^- < CO$, the so-called spectrochemical series. For example, the value of Δ is about 18,000 cm^{-1} in $[Co(H_2O)_6]^{3+}$, 23,000 cm^{-1} in $[Co(NH_3)_6]^{3+}$, and 33,500 cm^{-1} in $[Co(CN)_6]^{3+}$. Ligands that cause small (large) splitting are called weak (strong) field ligands. The splitting also depends on the metal and tends to increase with oxidation state.

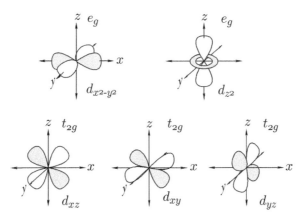

Figure 11.17 d Orbitals and their symmetries in an octahedral complex.

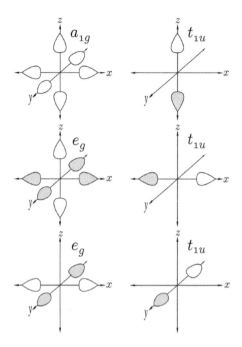

Figure 11.18 Symmetry-adapted linear combinations of ligand orbitals in O_h symmetry.

The ground electronic configuration of the complex is determined by first loading the 12 ligand electrons into the lowest energy (bonding) MOs: a_{1g}, t_{1u}, and e_g. The metal ion d electrons are then placed in the t_{2g} and/ or e_g^* MOs, but for certain numbers of d electrons one must compare the energy difference Δ to the cost of placing two electrons in the same MO with their spins paired. For example, a d^5 metal, such as Fe^{3+}, may have one unpaired electron in a strong octahedral field or five unpaired electrons if Δ is small. The former situation is referred to as a low-spin complex, and the latter is called the high-spin form.

The splitting Δ is sometimes called $10Dq$, a holdover from the old crystal field theory. In this theory, electrostatic interactions cause the d_{xy}, d_{xz}, and d_{yz} orbitals to be stabilized by an amount $4Dq$ and the d_{z^2} and $d_{x^2-y^2}$ orbitals to be destabilized by $6Dq$, with respect to the energy of the free ion d orbitals. This emphasizes the

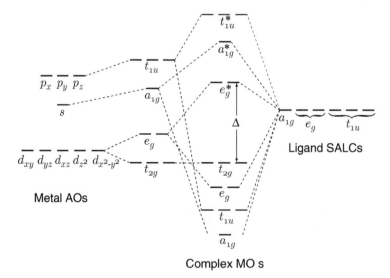

Figure 11.19 Molecular orbital energy levels in an octahedral transition metal complex.

Table 11.3 Correlation of free ion terms with symmetry species in an octahedral complex

Spherical symmetry	Octahedral symmetry
S	A_{1g}
P	T_{1g}
D	$E_g + T_{2g}$
F	$A_{2g} + T_{1g} + T_{2g}$
G	$A_{1g} + E_g + T_{1g} + T_{2g}$

net conservation of energy of the orbitals after splitting; i.e., a completely filled d^{10} or high-spin d^5 ion would experience no net stabilization or destabilization by the "crystal field." This statement obviously neglects interelectronic repulsion. Besides this, the crystal field picture is physically unsatisfying because it invokes repulsion between ligand and metal electrons as the basis for destabilization of the d_{z^2} and $d_{x^2-y^2}$ orbitals, yet experiment shows that many anions are weak ligands (the halide ions) while neutral species such as CO can be strong ligands. Nevertheless, as far as optical spectra are concerned, crystal field theory focuses on the energy levels of interest, those of the d electrons. And since it gives the correct symmetries of the molecular orbitals, it is useful for deducing selection rules.

Figuring out the state symmetries that arise from a particular d electron configuration can be rather complicated. For example, consider a free ion having a d^2 configuration, such as Ti^{2+}. The term symbols which derive from this configuration are ^3F, ^1D, ^3P, ^1G, and ^1S, a total of 45 states! On going from the spherical symmetry of the free ion to octahedral symmetry for six-fold coordination, the terms designated by S, P, D,... correlate with symmetries of the O_h point group as given in Table 11.3. In the complex, the ground configuration is t_{2g}^2, which gives rise to both triplet and singlet spin states and orbital symmetry species found from the direct product $t_{2g} \times t_{2g} = A_{1g} + E_g + T_{1g} + T_{2g}$. There are $(6 \times 5) \div 2 = 15$ states associated with the t_{2g}^2 configuration. It is not immediately obvious how to associate each of the four terms with singlet and/or triplet spin states or how to order the term energies. Fortunately, these sorts of group theoretical considerations have been worked out for a large number of d-electron configurations, and are summarized in the form of Tanabe–Sugano diagrams (see for example [9] or [10]). One such diagram, for the d^2 case, is shown in Figure 11.20. These diagrams display the energies of the terms relative to the ground state as a function of the splitting parameter Dq or Δ. Energies are expressed in units of B, a parameter dependent on the interelectronic repulsion in the complex. In the strong-field case, the right-hand side of the diagram, the terms are $^3T_{1g}$, $^1T_{2g}$, $^1E_{1g}$, and $^1A_{1g}$, in order of increasing energy. Note that these account for all 15 states deriving from t_{2g}^2. States deriving from the excited configuration $t_{2g}^1 e_g^1$ are $^3T_{2g}$, $^3T_{1g}$, $^1T_{2g}$, and $^1T_{1g}$. All of these states have g symmetry and thus cannot be connected by electric dipole selection rules.

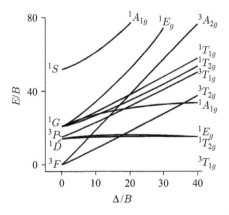

Figure 11.20 Tanabe–Sugano diagram for d^2. (Adapted from F. A. Cotton, *Chemical Applications of Group Theory*, Wiley-Interscience, New York 1971.)

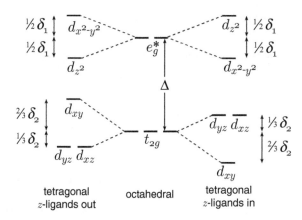

Figure 11.21 Splitting of d orbitals on distortion from octahedral to tetragonal symmetry. The drawing is not to scale; $\delta_{1,2} \ll \Delta$.

In fact, any $d - d$ transition of an octahedral complex, in which an electron is promoted from the t_{2g} level to the e_g^* level, *must* be dipole forbidden to a first approximation, as per the LaPorte rule. In spite of the well-known colors of transition metal complexes, they are often the result of forbidden transitions, as evidenced by the rather low molar absorptivities, ε_M, less than about 50 L mol^{-1} cm^{-1}. We must look to the possibility of vibronic activity to explain the absorption of visible light.

The transition dipole moment transforms according to the T_{1u} representation of the O_h point group. Forbidden transitions may borrow intensity via vibronic coupling to dipole-allowed electronic states via vibrations of the appropriate symmetry. A complex of the type ML$_6$ has $3(7) - 6 = 15$ normal modes involving metal–ligand motion. Group theoretical analysis shows that these modes have the symmetries A_{1g}, E_g, $2T_{1u}$, T_{2g}, and T_{2u}. In the case of a d^2 complex, the ground state is $^3T_{1g}$, and there is one excited state from the excited configuration having the same symmetry and spin multiplicity. Therefore, a vibration having the symmetry of the transition dipole moment, T_{1u}, is capable of vibronic activation of the $^3T_{1g}(t_{2g}^2) \rightarrow {}^3T_{1g}(t_{2g}^1 e_g^1)$ transition. As it turns out, however, there is even more to this story.

Many transition metal complexes provide classic examples of the Jahn–Teller theorem, which states that a nonlinear symmetric molecule in an orbitally degenerate state will be unstable to distortion to a lower symmetry state. The ground state of an octahedral complex such as the d^2 complex of Figure 11.20 is orbitally degenerate. According to the Jahn–Teller theorem, such a complex can lower its energy by distorting to a tetragonal geometry, having D_{4h} symmetry. The splitting of the t_{2g} and e_g^* orbitals as a result of moving two ligands in or out along the z-axis is shown in Figure 11.21. In either case, the distortion allows the d electrons to be placed in a lower-energy orbital.

The theorem as stated here does not address the size of the distortion-induced splitting. Consider a doubly degenerate electronic state. Jahn and Teller showed that there is always at least one nontotally symmetric normal mode that is capable of lifting the electronic degeneracy. This is illustrated in Figure 11.22, which shows the splitting of the electronic state as a function of a nontotally symmetric normal coordinate. The crossing

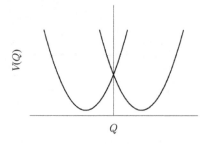

Figure 11.22 Splitting of a degenerate electronic state along a nontotally symmetric normal coordinate.

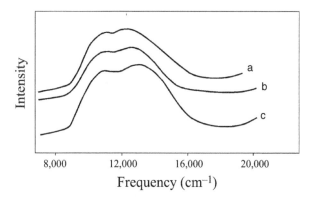

Figure 11.23 Electronic spectra of (a) $Rb_2 NaTiCl_6$ (b) Cs_2KTiCl_6, and (c) Rb_3TiCl_6 showing Jahn–Teller splitting. (Reprinted with permission from R. Ameis, S. Kremer, and D. Reinen, *Inorganic Chemistry 24*, 2751 (1985), copyright 1985, American Chemical Society.)

point corresponds to the symmetric configuration where the electronic state is doubly degenerate. If the barrier to interconversion of the two distorted forms is less than the zero-point energy, the molecule merely vibrates back and forth between the two forms. In this case, the dynamic Jahn–Teller effect is operative. The vibration in question can participate in vibronic activation of forbidden electronic transitions, but the molecule remains symmetric on the average. In the other extreme, if the barrier is large compared to thermal energy, the molecule will permanently distort to the less symmetric form. This is the static Jahn–Teller effect, and it leads to the splitting of electronic transitions which would be degenerate under O_h symmetry.

Figure 11.23 shows some example electronic spectra that reveal Jahn–Teller splitting of a d^1 transition metal complex. In this example, the expected octahedral symmetry of the $TiCl_6^{3-}$ ion is lowered to D_{4h} by means of a slight compression of the ligands along the z direction. The $^2T_{2g}$ ground state is split by about 400 cm^{-1} and the ^2E_g excited state by about 1800 cm^{-1}.

In addition to $d-d$ transitions, the electronic spectra of transition metal complexes may involve other orbitals as well. Transitions in which a d electron is moved to an orbital primarily localized on the ligand are called metal-to-ligand charge-transfer transitions. Ligand-to-metal charge-transfer transitions involve the promotion of lower-lying bonding electrons to one of the d electron levels. These transitions can adhere to the LaPorte rule, and thus they may be stronger than the $d-d$ transitions.

11.7 EMISSION SPECTROSCOPY OF POLYATOMIC MOLECULES

What is the fate of a molecule that has been promoted to an excited electronic state? Of course, what goes up, must come down, but there is more than one return path to equilibrium. Though excited electronic states may be prepared in a number of different ways, it is convenient for the present discussion to consider emission that follows absorption of light in order to see how absorption and emission spectra are related. Following vertical excitation from the ground electronic state, the molecule in its Franck–Condon state is electronically excited, but has the geometry of the ground electronic state at equilibrium. This directly excited state has excess vibrational energy invested in the FC-active modes, but in polyatomics this energy may be rapidly redistributed into other vibrational modes by a radiationless process called *intramolecular vibrational redistribution* (IVR). The rate at which the excess vibrational energy is randomized among all the normal modes depends on the molecule and is a strong function of the size. For example, in trans-stilbene, transfer of the initial nonthermal population in the FC-active modes into lower frequency vibrations takes place in about 2 to 4 ps [11]. In the fluorescent dye coumarin-153, IVR takes place on a timescale of less than 100 fs [12]. As the vibrational energy and number of modes increase, the density of vibrational states increases and so do anharmonic couplings, facilitating the shuffling of energy between modes. This IVR process leaves the molecule in a state that is still at a higher temperature than the surroundings, but with the excess vibrational energy randomly partitioned among the modes. The ensuing process of equilibration, in which excess vibrational energy is lost to the surroundings (e.g., the solvent) in the form of heat, is called *vibrational energy transfer* (VET). This form of vibrational relaxation, like IVR, takes place on a subpicosecond or picosecond timescale in typical dyes, and when it is complete, the molecule finds itself in the equilibrium geometry of the excited electronic state. The timescale for IVR and VET in large polyatomics should be contrasted with the vibrational population relaxation times of diatomic molecules in solution, which can be as long as milliseconds! Neglecting solvation, the molecule would be ready to undergo relaxed emission following IVR and VET. In solution, however, the rearrangement of solvent molecules from *their* Franck–Condon positions, equilibrated to the ground electronic state, takes place over a range of timescales, about 0.1 to 20 ps, before fully relaxed emission is observed. The timescales for IVR, VET, and solvent relaxation vary from one system to another, and the processes are not necessarily sequential.

Since the above-mentioned events are rapid compared to a typical radiative lifetime, $10^{-8} - 10^{-9}$ s, emission from the directly excited electronic state is often relaxed. In the case of relaxed emission, the molecule undergoes a nonradiative transition, from the FC state to the solvated ground vibrational state of the upper electronic state, before undergoing the radiative transition. The total energy lost by the chromophore to the surroundings during intramolecular vibrational relaxation is the internal reorganization energy ΔE_{int}, and it may be resolved into contributions from the various Franck–Condon active normal modes. The contribution of each normal mode in turn depends on its displacement Δ_i (Equation 11.19) in the excited state:

$$\Delta E_{int} = \sum_{i=1}^{3N-6} \frac{1}{2} \Delta_i^2 \hbar \omega_i \tag{11.23}$$

Equation 11.23 is correct for harmonic normal modes having the same vibrational frequency (ω_i) in the ground and excited electronic states. More generally, ΔE_{int} is the energy difference between the vertical and ground vibrational states in the upper electronic state, so it contributes to the difference between the $0-0$ frequency ν_{00} and the frequency of maximum absorption ν_{max}. In the absence of solvent effects, $h\nu_{max} = h\nu_{00} + \Delta E_{int}$, but in solution, solvent–solute interactions contribute to reorganization as well, as mentioned above. This effect is depicted in Figure 11.24, where the arrows depict the dipole moment of a spectroscopically active solute molecule and those of surrounding (polar) solvent molecules. Electronic excitation results in an instantaneous change in the dipole moment of the solute molecule, but it takes time for the surrounding solvent dipoles to equilibrate to the excited state charge distribution. Immediately following excitation to the excited electronic state, the solvent molecules are still in positions which favor the ground electronic state of the chromophore. The energy lost as solvent molecules rearrange to solvate the excited electronic state is the solvent reorganization energy ΔE_{solv}. The total reorganization energy is $\Delta E_{reorg} = \Delta E_{int} + \Delta E_{solv}$, and the energy of maximum absorption is $h\nu_{max} = h\nu_{00} + \Delta E_{reorg}$.

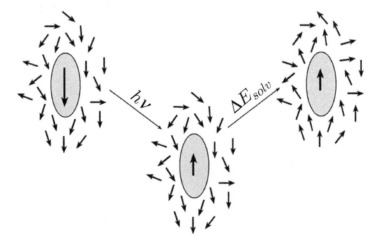

Figure 11.24 Solvent reorganization in response to an electronic transition. The arrows represent the dipole moments of the spectroscopically active solute and surrounding solvent molecules.

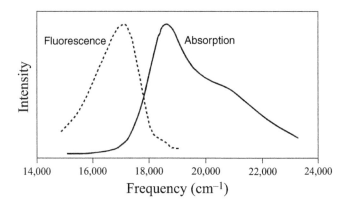

Figure 11.25 Absorption spectrum and fluorescence spectrum (excited at 21,900 cm⁻¹) of parafuschin in aqueous solution.

Let us suppose, as is often the case, that the ground electronic state is a singlet, call it S_0, and that the absorption transition which precedes emission is to the lowest energy excited singlet state S_1. In this case, the radiative transition $S_1 \rightarrow S_0$ is fluorescence. The nonradiative transition $S_1 \rightarrow S_0$ is called *internal conversion*, a term used to describe radiationless transitions from one electronic state to another of the same spin multiplicity. This transition competes with the radiative one and causes the fluorescence yield φ_{fluor} to be less than unity. The fluorescence yield, introduced in Chapter 6, can be expressed as the ratio of the number of emitted photons n_f to the number absorbed n_{abs}, $\varphi_{fluor} = n_f/n_{abs}$. It can also be viewed in terms of the relative contributions of radiative and nonradiative relaxation rates: $1/\tau_{rad}$ and $1/\tau_{nonrad}$, as expressed in Equations 6.28 and 6.29. When emission is preceded by a transition to an electronic state higher in energy than S_1, e.g., $S_0 \rightarrow S_2$, and if the transition $S_1 \rightarrow S_0$ is allowed, emission is usually observed from the lower energy excited electronic state, rather than from S_2 to S_0. This effect, sometimes called Kasha's rule, is the result of rapid nonradiative relaxation (internal conversion) from S_2 to S_1. Figure 11.25 illustrates Kasha's rule as well as the frequently observed mirror symmetry between absorption and fluorescence spectra. The absorption spectrum of parafuschin comprises two overlapping transitions to the S_1 and S_2 states, while emission is from S_1 to S_0 only. The emission spectrum mirrors the lower energy absorption band.

The difference in frequency of the absorption and fluorescence maxima is called the Stokes shift:

$$\Delta \nu_{stokes} = \nu_{max,abs} - \nu_{max,fluor} \tag{11.24}$$

In the special case where the ground and excited state vibrational potentials for each active normal mode are harmonic and the same shape, then the Stokes shift $h\Delta\nu_{stokes}$ is twice the total reorganization energy ΔE_{reorg}, as defined above. In general, the greater the difference in the equilibrium and Franck–Condon geometries is, the larger the difference in the absorption and fluorescence maxima.

A fluorescence spectrum such as that shown in Figure 11.25 is obtained by irradiating a molecule within the electronic absorption band and collecting the emitted light as a function of emission frequency or wavelength. If, instead of scanning the wavelength of emitted light, one scans the excitation wavelength λ_{exc} (or frequency) and collects the total emitted radiation as a function of λ_{exc}, then the resulting spectrum is called the excitation spectrum. If the absorption and emission transitions involve the same excited state, the excitation spectrum resembles the absorption spectrum. Excitation spectra are useful to determine if fluorescence is intrinsic to the molecule of interest or results from a fluorescent impurity.

Excited triplets are lower in energy than singlets having the same occupied orbitals, due to lower interelectronic repulsion in the triplet states. There is a good chance that these singlet and triplet excited states cross at a particular nuclear configuration, as shown in Figure 11.26. Spin–orbit coupling then allows the singlet and triplet states to mix with one another. This presents the possibility of a nonadiabatic transition called *intersystem crossing* (ISC), in which the initially excited singlet electronic state crosses over to the triplet state. In the absence of the spin–orbit perturbation the two states would have different spin symmetry and could not interact. Since spin–orbit coupling is roughly proportional to Z^4, where Z is the atomic number, the rate of

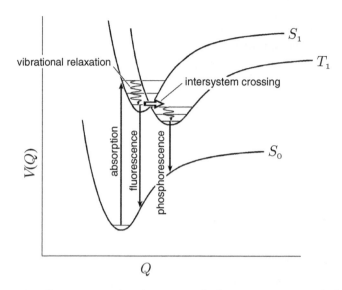

Figure 11.26 Absorption, fluorescence, phosphorescence (radiative transitions) and vibrational relaxation and intersystem crossing (nonradiative transitions).

intersystem crossing can be enhanced by the presence of heavy atoms, even if these atoms are not part of the molecule undergoing emission (the external heavy atom effect). After making the nonradiative transition to the excited triplet state, labeled T_1 in Figure 11.26, the molecule can then emit light in the form of phosphorescence. Since transitions having $\Delta S \neq 0$ are forbidden to a first approximation, the rate and thus the intensity of phosphorescence is usually lower than that for fluorescence. In rigid media, such as solids and glasses, phosphorescence lifetimes can be quite long, and emission can continue for minutes after illumination ceases.

11.8 NONRADIATIVE RELAXATION OF POLYATOMIC MOLECULES

While the radiative relaxation of molecules is of great interest to spectroscopy, radiationless decay is also an important deactivation channel for excited electronic states. Fast internal conversion from higher lying excited states (S_n) to the lowest lying singlet (S_1) is common, leading to Kasha's rule. The conversion, or "degradation," of absorbed radiation to heat is a common occurrence in everyday life. What is the perturbation that enables deactivation of an excited electronic state without emission of light? We have seen that the radiative transition rate can be understood within the Golden Rule expression of Chapter 4, where we considered the time-dependent perturbation $\hat{H}' = -\vec{\mu} \cdot \vec{E}(t)$ to be responsible for inducing transitions. We can also use the Golden Rule to treat nonradiative transitions that take place via departure from the Born–Oppenheimer approximation. While such transitions may involve surrounding molecules as energy acceptors, radiationless transitions also take place in rare gases. We explicitly treat *intramolecular* radiationless transitions in this section, in order to reveal the role of vibrations in meeting the criterion of energy conservation. The perturbation operator responsible for nonradiative transitions is the part of the Hamiltonian that was neglected in the BO approximation. This is a time-independent perturbation that mixes otherwise separate electronic states. On an energy scale, we can view a nonradiative transition as a horizontal transition from one BO potential energy surface to another, as compared to a vertical radiative transition. In order for a nonradiative transition from an excited electronic state to the ground electronic state to occur, it is clear that a number of excited vibrational quanta are necessary to soak up the energy of the relaxing excited state. We consider internal conversion, in which an excited electronic state e decays to a ground state g of the same spin multiplicity. Experimentally, radiationless transition rates are often seen to decay exponentially with increasing energy difference of initial and final states. This so-called exponential energy gap law may seem counterintuitive at first, but the treatment here exposes why larger energies are dissipated less efficiently.

To proceed, recall the expression for the transition rate given in Equation 4.38, from Fermi's Golden Rule. We make two adjustments to treat nonradiative rather than radiative transitions. We replace the perturbation operator $-\vec{\mu} \cdot \vec{E}(t)$ by $\hat{T}_N(Q)$, the nuclear kinetic energy operator, which was neglected in making the BO approximation. Further, we set the frequency ω of the perturbation to zero. The rate w_{if} of transitions between the initial and final states is then

$$w_{if} = \frac{2\pi}{\hbar^2} |V_{fi}|^2 \, \delta(\omega_{fi}) \tag{11.25}$$

The delta function ensures that the energy of the final state matches that of the initial state. It could be replaced by the density of energy conserving final states, if desired. The matrix element of the perturbation is taken to be

$$V_{fi} = \langle \Psi_f | \hat{T}_N | \Psi_i \rangle \tag{11.26}$$

The nuclear kinetic energy operator depends on the normal coordinates Q_k:

$$\hat{T}_N = \sum_{k=1}^{3N-6} \frac{-\hbar^2}{2} \frac{\partial^2}{\partial Q_k^2} \tag{11.27}$$

Recall from Chapter 10 that the effective mass of the mode k is absorbed in Q_k. In the case where there is a large energy difference in the zero point levels of the initial and final electronic states, we expect the adiabatic approximation to hold. Assuming harmonic vibrational states, we can thus write initial and final states as follows:

$$\Psi_i(r;Q) = \psi_e(r;Q)\chi_{v'}^e(Q) = \psi_e(r;Q)\chi_{v_1'}^e(Q_1)\chi_{v_2'}^e(Q_2)\ldots\chi_{v_{3N-6}'}^e(Q_{3N-6}) \tag{11.28}$$

$$\Psi_f(r;Q) = \psi_g(r;Q)\chi_v^g(Q) = \psi_g(r;Q)\chi_{v_1}^g(Q_1)\chi_{v_2}^g(Q_2)\ldots\chi_{v_{3N-6}}^g(Q_{3N-6}) \tag{11.29}$$

Here, Q (with no subscript) represents a collection of $3N - 6$ normal coordinates and v' and v designate collections of vibrational quantum numbers in the excited and ground electronic states, respectively. Subscripted v's and Q's are reserved for one-dimensional harmonic oscillator wavefunctions. We are ignoring any change in symmetry that would mix the normal coordinates. The electronic wavefunctions depend parametrically on the normal coordinates Q and explicitly on the electronic coordinates, denoted here by r. Using the shorthand version of the vibronic wavefunctions for now, we evaluate the matrix element V_{fi}.

$$V_{fi} = \frac{-\hbar^2}{2} \sum_k \left\langle \psi_g \chi_v^g \left| \frac{\partial^2}{\partial Q_k^2} \right| \psi_e \chi_{v'}^e \right\rangle \tag{11.30}$$

The angle brackets in Equation 11.30 imply integration over electronic and nuclear coordinates. To distinguish these, let us show the averages over nuclear coordinates as explicit integrals over Q and use Dirac notation for integration over the electronic coordinates. On evaluating the second derivative above, we obtain a sum of three terms:

$$V_{fi} = \frac{-\hbar^2}{2} \sum_k \int dQ \left\{ \chi_v^{g*}(Q) \left\langle \psi_g \left| \frac{\partial^2}{\partial Q_k^2} \right| \psi_e \right\rangle \chi_{v'}^e(Q) + 2\chi_v^{g*}(Q) \left\langle \psi_g \left| \frac{\partial}{\partial Q_k} \right| \psi_e \right\rangle \frac{\partial \chi_{v'}^e(Q)}{\partial Q_k} \right. $$
$$\left. + \langle \psi_g | \psi_e \rangle \chi_v^{g*}(Q) \frac{\partial^2 \chi_{v'}^e(Q)}{\partial Q_k^2} \right\} \tag{11.31}$$

The third term is clearly zero since the ground- and excited-state wavefunctions are orthogonal. The first term can be neglected for small departures from the BO approximation. Thus we consider the second term in Equation 11.31 as the main contribution to the nonradiative transition from i to f.

$$V_{fi} = -\hbar^2 \sum_k \int dQ \chi_v^{g*}(Q) \left\langle \psi_g \left| \frac{\partial}{\partial Q_k} \right| \psi_e \right\rangle \frac{\partial \chi_{v'}^e(Q)}{\partial Q_k} \tag{11.32}$$

If vibrational relaxation within the excited state is fast enough, and vibrational energy levels widely spaced compared to $k_B T$, we can take the initial state to be the zero-point level of the excited electronic state: $\chi_{v'}^e(Q) = \chi_0^e(Q) = \chi_0^e(Q_1)\chi_0^e(Q_2)\ldots\chi_0^e(Q_{3N-6})$. Then we have

$$\frac{\partial \chi_0^e(Q)}{\partial Q_k} = \frac{\partial \chi_0^e(Q_k)}{\partial Q_k} \prod_{j \neq k} \chi_0^e(Q_j) \tag{11.33}$$

Casting Equation 11.32 in terms of multimodal vibrational wavefunctions leads to

$$V_{fi} = -\hbar^2 \sum_k \int dQ_1 \int dQ_2 \cdots \int dQ_{3N-6} \left\langle \psi_g \left| \frac{\partial}{\partial Q_k} \right| \psi_e \right\rangle \left(\chi_{v_k}^{g*}(Q_k) \frac{\partial \chi_0^e(Q_k)}{\partial Q_k} \right) \prod_{j \neq k} \left(\chi_{v_j}^{g*}(Q_j) \chi_0^e(Q_j) \right) \tag{11.34}$$

The operator $\partial/\partial Q_k$ is proportional to the momentum operator. With the help of raising and lowering operators, Equation 1.38, it is a simple matter to show that

$$\frac{\partial \chi_0^e(Q_k)}{\partial Q_k} \propto \chi_1^e(Q_k) \tag{11.35}$$

The so-called promoting mode Q_k is a dominant vibrational mode that enables nonradiative decay. For the electronic matrix element $\left\langle \psi_g \left| \partial/\partial Q_k \right| \psi_e \right\rangle$ to be nonzero, the normal coordinate Q_k must belong to an irreducible representation that is the product of the irreducible representations of the ground and excited electronic states. These are the same symmetry constraints encountered for vibronic coupling, which is the same kind of coupling that makes the nonradiative transition possible. (Problem 10 explores this connection.) If we neglect the dependence of $\left\langle \psi_g \left| \partial/\partial Q_k \right| \psi_e \right\rangle$ on nuclear coordinates, V_{fi} is proportional to the product of one-dimensional vibrational overlaps:

$$V_{fi} \propto \left\langle \psi_g \left| \frac{\partial}{\partial Q_k} \right| \psi_e \right\rangle$$
$$\times \int dQ_1 \chi_{v_1}^{g*}(Q_1)\chi_0^e(Q_1) \int dQ_2 \chi_{v_2}^{g*}(Q_2)\chi_0^e(Q_2)\ldots \int dQ_k \chi_{v_k}^{g*}(Q_k)\chi_1^e(Q_k)\ldots \int dQ_{3N-6} \chi_{v_{3N-6}}^{g*}(Q_1)\chi_0^e(Q_1) \tag{11.36}$$

Since the rate w_{if} depends on the square modulus of V_{fi}, nonradiative relaxation is a function of previously discussed Franck–Condon factors. Energy conservation (and the delta function of Equation 11.25) is satisfied when the 0 – 0 energy of the excited electronic state matches the vibrational energy of the final state: $E_{00} = \Sigma_j v_j \hbar \omega_j$, where ω_j is the vibrational frequency of mode j in the ground electronic state. Modes capable of accepting the energy of the relaxing state must have favorable Franck–Condon factors. These Franck–Condon factors fall off as the number of quanta increases, therefore higher frequency modes such as C–H and O–H stretches are better accepting modes than lower frequency vibrations. The decrease in Franck–Condon factors with increasing vibrational quantum number, on its own, would make the rate of nonradiative decay decrease as the energy gap increases. Countering that trend somewhat is the increasing density of vibrational states at higher energy, which makes it easier to achieve energy matching. In condensed phases, in addition to intramolecular vibrational modes considered in the above analysis, intermolecular modes can accept the energy, as can solvent intramolecular vibrations that are displaced in the excited electronic state of the solute. The above treatment reveals why nonradiative lifetimes, and thus fluorescence yields, depend on isotopic substitution. Consider deuteration of O–H or C–H bonds that participate in acceptor modes. Higher vibrational quantum numbers are required for O–D or C–D bond vibrations to match the excited state energy, resulting in a decreased rate of radiationless decay compared to that seen in the presence of O–H or C–H bonds. For example, the fluorescence lifetime of the laser dye LDS750 is 220 ps in CH_3OH solution and 300 ps in CD_3OD solution, a result of faster nonradiative relaxation in CH_3OH compared to CD_3OD.

For more details on the theory of nonradiative transitions, including intersystem crossing induced by spin–orbit coupling, the reader is encouraged to consult the seminal work by Bixon and Jortner [13]. In addition to internal conversion and intersystem crossing, additional pathways for nonradiative relaxation include photochemical processes such as electron transfer, dissociation, and isomerization.

11.9 CHROMOPHORES

A chromophore is a molecule (or part of a molecule) responsible for absorption of light. For practical and esthetic reasons, chemists are fascinated by chromophores which absorb visible light and therefore give rise to color. Some common chromophores that exemplify various types of electronic transitions are illustrated in Figure 11.27. Electronic transitions can be classified according to the types of molecular orbitals whose

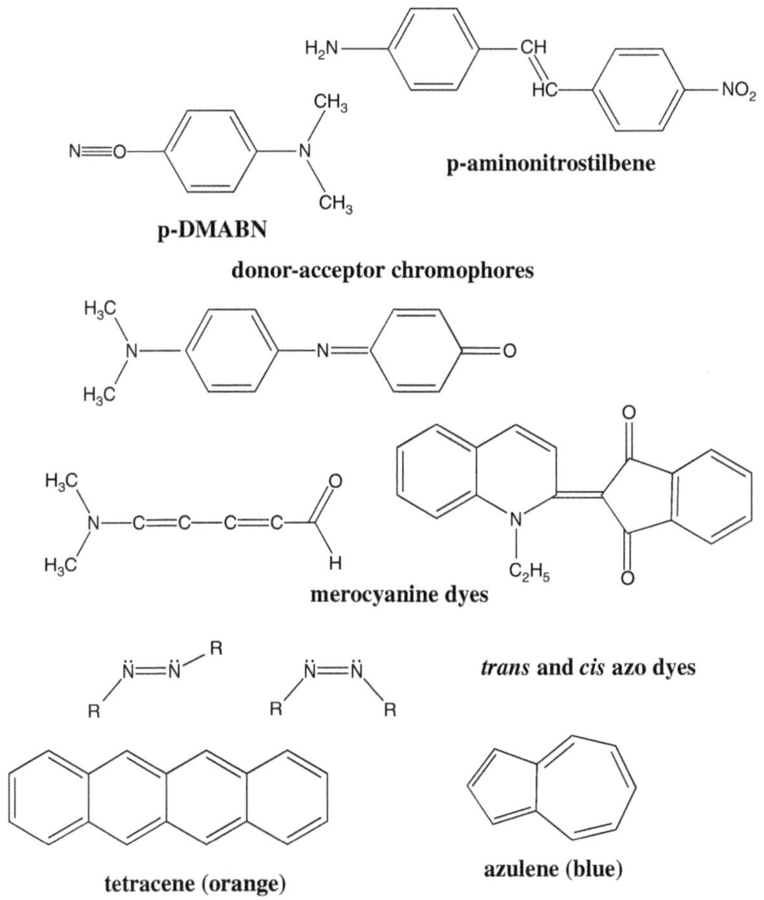

Figure 11.27 Some chromophores.

occupation numbers differ in the configurations of the ground and excited states. For example, a $\sigma - \sigma^*$ transition is one in which an electron is promoted from a bonding MO to an antibonding MO, both of sigma symmetry. Similarly, we have $\pi - \pi^*$ transitions in molecules containing double and triple bonds, and $n - \sigma^*$ and $n - \pi^*$ transitions in molecules having lone pairs, such as acetone or pyridine. (The letter n stands for nonbonding electron.) The $n - \pi^*$ transitions of conjugated molecules containing heteroatoms are found at lower energy than the $\pi - \pi^*$ transitions and are often much weaker. The low intensity is a consequence of poor overlap between the n and π^* orbitals, which leads to a small transition moment.

Molecules containing electron donor and electron acceptor groups may have intramolecular charge-transfer transitions. An example is dimethylaminonitrostilbene, in which the visible transition involves transfer of electronic charge from the amino to the nitro group. The well-studied molecule p-dimethylami-nobenzonitrile (p-DMABN) undergoes a $\pi - \pi^*$ transition followed by twisting of the dimethylamino group to form a "twisted intramolecular charge transfer" state (TICT state). Emission from the highly polar TICT state shifts to the red with increasing solvent polarity. Organic complexes called electron donor-acceptor complexes are formed from aromatic electronic donors such as alkylbenzenes and electron acceptors such as quinones, halogens, and substituted ethylenes. The signature of complex formation is the presence of a strong charge-transfer electronic transition, usually in the visible region of the spectrum, where neither the donor nor the acceptor absorbs. See [14] for more discussion of the wide variety of chromophores found in nature.

11.10 SOLVENT EFFECTS IN ELECTRONIC SPECTROSCOPY

The theory of solvent effects on the frequency and intensity of electronic transitions has long held the interest of chemists and continues to be the subject of much research. The influence of the solvent on the electronic spectrum of the solute may result from bulk solution properties such as the dielectric constant ε_s and refractive index n. More difficult to treat are the inherently quantum mechanical effects of specific intermolecular interactions. Attempts to account for the local field in solution take the nonspecific interactions into account, and are based on ideas such as the Onsager approach described in Chapter 3. Quantum and statistical mechanical approaches employ computer simulations of specific systems. A number of empirical solvent polarity scales are based on the solvent shift of the absorption maximum of a particular solvatochromic probe molecule [15]. The π^* scale is based on the $\pi - \pi^*$ transition of nitroanisole and the ET-30 scale is based on the intramolecular charge-transfer transition of 2,6-diphenyl(2,4,6-triphenyl-1-pyridinio)phenolate (or betaine-30, for short). The solvent sensitivity of the absorption maximum (solvatochromism) of betaine-30 is particularly dramatic; λ_{max} is about 810 nm in diphenyl ether and 450 nm in water! In this section we explore the basis for this behavior.

The shift in the frequency of the absorption or emission maximum on going from the gas to solution phase is the result of the difference in the solvation energies of the ground and excited electronic states: $h(\nu_{sol} - \nu_{gas}) = h\Delta\nu = \Delta E_s'' - \Delta E_s'$. The solvation energies of the ground ($\Delta E_s''$) and excited ($\Delta E_s'$) states are taken to be positive if the solvent stabilizes the solute. Thus, as shown in Figure 11.28, if the ground

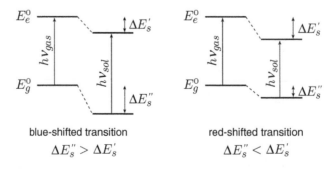

Figure 11.28 Effect of ground- and excited-state solvation energies on the energy of an electronic transition.

state is stabilized more than the excited state, the electronic transition will shift to the blue ($v_{sol} > v_{gas}$), and if the excited state is stabilized more than the ground, then the transition will shift to the red. Because electronic transitions are fast compared to the time it takes for solvent molecules to translate or rotate, the solvation energies must be defined with care. Consider absorption for the moment. The initial state is the equilibrium geometry of the ground electronic state, surrounded by solvent molecules equilibrated to the solute in its ground state. The equilibrium geometry of the excited solute favors a different distribution of solvent molecules, adapted to the excited-state charge distribution. Immediately following excitation, however, only the electronic part of the solvent polarization has adjusted to the new charge distribution. The solvent molecule positions and orientations are said to be in a state of strain [16–17]. This strain has been defined to consist of two parts, packing strain and orientation strain, corresponding to translational and orientational positions of solvent molecules. In figuring the solvent shift of absorption Δv_{abs}, the solvation energies of the ground and excited states must be evaluated at the equilibrium and Franck–Condon geometries, respectively.

$$h\Delta v_{abs} = \Delta E''_{s,eq} - \Delta E'_{s,fc} \tag{11.37}$$

When the lifetime of the excited electronic state is much longer than the time it takes for solvent molecules to translate and rotate, the initial state in emission is fully relaxed with respect to the solvent and solute configurations. The final state finds the solvent molecules in a state of strain with respect to the ground state of the solute. Thus the solvent shift in the emission spectrum (assumed here to be fluorescence) is given by

$$h\Delta v_{fluor} = \Delta E''_{s,fc} - \Delta E'_{s,eq} \tag{11.38}$$

One particularly important motif in solvent shift studies is the consideration of the permanent dipole moments, μ_g and μ_e, of the ground and excited states. The solvent shift due to dipolar forces, in the Onsager cavity approach, is given by [18,19].

$$h\Delta v_{abs}(\text{dipolar}) = \frac{\vec{\mu}_g \cdot (\vec{\mu}_g - \vec{\mu}_e)}{4\pi\varepsilon_0 a^3}\left[\frac{2(\varepsilon_s - 1)}{2\varepsilon_s + 1} - \frac{2(n^2 - 1)}{2n^2 + 1}\right]$$
$$= \frac{\vec{\mu}_g \cdot (\vec{\mu}_g - \vec{\mu}_e)}{4\pi\varepsilon_0 a^3}\left[f(\varepsilon_s) - f(n^2)\right] \tag{11.39}$$

where a is the cavity radius, ε_s is the static dielectric constant, n is the refractive index, and $f(x) \equiv 2(x-1)/(2x+1)$. In the case of fluorescence, the solvent shift due to dipolar interactions can be obtained from Equation 11.39 by interchanging the labels g and e. In a nonpolar solution, $\varepsilon_s \approx n^2$, and the shift given by Equation 11.39 is approximately zero. One still has interactions between permanent dipole moments of the solute and induced dipole moments of the solvent, as well as dispersion interactions due to induced dipoles of both solute and solvent molecules. The induction contribution to the solvent shift depends on the refractive index of the solvent.

$$h\Delta v_{abs}(\text{induction}) = \frac{(\vec{\mu}_g^2 - \vec{\mu}_e^2)}{4\pi\varepsilon_0 a^3}\left[\frac{2(n^2 - 1)}{2n^2 + 1}\right] \tag{11.40}$$

Various approximate formulas for the dispersion contribution to the solvent shift have been obtained [16–18]. The solvent shift due to dispersion is proportional to the difference $\alpha_g - \alpha_e$ between the ground and excited state polarizabilities. The excited electronic state tends to be more polarizable than the ground state, and the dispersion term generally leads to a red-shift in the absorption spectrum with increasing solvent refractive index.

The poorly defined cavity radius is a serious drawback to solvent shift models based on the Onsager theory. While it is possible to generalize the approach to nonspherical cavities, these models still suffer from treating

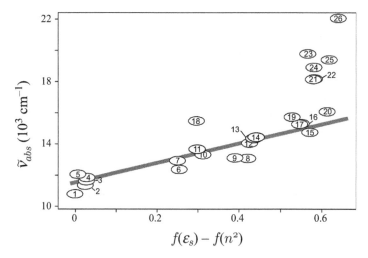

Figure 11.29 Transition frequency of betaine-30 as a function of solvent polarity. See Table 11.4 for the solvent numbering scheme.

the solvent outside the cavity as a dielectric continuum. Nevertheless, for many molecules the frequency of the absorption maximum, in a series of solvents of similar refractive index, has been found to be roughly proportional to $f(\varepsilon_s)$ as expected from Equation 11.39. This is illustrated for betaine-30 in Figure 11.29. (See also Table 11.4.) The positive slope of the graph of $\tilde{\nu}_{abs}$ versus $f(\varepsilon_s) - f(n^2)$ is a consequence of the ground state dipole moment of betaine-30 being larger than that of the excited state. This is referred to as negative solvatochromism. Note the large deviation of alcohols, water, and other hydrogen-bond donors from the main trend. The additional blue shift is a consequence of hydrogen bonding, which is stronger in the ground than the excited state. It is clear that Equation 11.39 does not account for this effect. Nevertheless, Onsager-based approaches such as this are qualitatively useful: one can immediately deduce the direction, if not the precise magnitude, of the change in dipole moment on electronic excitation. For example, the absorption maximum of the lowest energy $\pi - \pi^*$ transition of benzene is independent of solvent dielectric constant, indicating that $\mu_e = \mu_g$ and thus the excited state dipole moment, like that of the ground state, is zero.

More sophisticated approaches to solvent effects must account for the solvation structure in the vicinity of the solute as well as the sort of long-range effects that continuum theories are capable of treating. The hydrogen-bonding interaction is a well-known case of a specific interaction which affects spectra in ways that are not revealed by the bulk dielectric properties of the solution. For example, $n - \pi^*$ transitions are shifted to the blue in protic solvents such as alcohols, due to hydrogen-bonding between the solute lone pair and the solvent proton. In the more delocalized excited state, the hydrogen bond is weaker, thus $\Delta E_s'' > \Delta E_s'$ and a blue shift results.

Quantum mechanical calculations of solvent effects often invoke a hybrid approach, in which the nearest neighbor solvent molecules are included in a quantum mechanical "supermolecule" calculation, and more distant solvent molecules are considered part of the continuum. Hush, Reimer and colleagues [20–21] have employed computer simulations to calculate solution structure and electronic spectra of dilute aqueous solutions of pyrimidine. Zerner and coworkers [22–24] developed semiempirical quantum mechanical approaches which incorporate the reaction field calculated using the quantum mechanically derived charge distribution of the solute in the ground and excited states. Polarizable continuum models use a quantum mechanical treatment, for example time-dependent density functional theory, of a solute molecule enclosed in a molecule-shaped cavity [25] surrounded by a polarizable continuum. Mixed quantum/classical calculations treat the solvent molecules using molecular mechanics or molecular dynamics. Effective fragment potentials account for solvent effects using discrete solvent molecules with distributed multipoles and polarizability tensors derived from *ab initio* calculations [26]. While harder to implement than the simple continuum-based

Table 11.4 Absorption maximum of betaine-30 in various solvents

	Solvent	Dielectric constant	$\tilde{\nu}_{max}$, cm^{-1}, of betaine-30
1	n-hexane	1.9	10,800
2	carbon tetrachloride	2.2	11,400
3	m-xylene	2.4	11,600
4	toluene	2.4	11,800
5	benzene	2.3	11,900
6	diethylamine	3.6	12,400
7	m-dichlorobenzene	5.0	12,900
8	tetrahydrofuran	7.6	13,100
9	n-propylacetate	6.0	13,100
10	fluorobenzene	5.4	13,300
11	chloroform	4.8	13,700
12	pyridine	12.4	14,100
13	dichloromethane	8.9	14,400
14	acetophenone	17.4	14,400
15	acetone	20.7	14,700
16	N,N-dimethylacetamide	37.8	15,300
17	N,N-dimethylformamide	36.7	15,300
18	aniline	6.9	15,500
19	dimethylsulfoxide	46.7	15,700
20	acetonitrile	37.5	16,100
21	ethanol	24.5	18,100
22	N-methylacetamide	191.3	18,200
23	formamide	111.0	19,800
24	N-methylformamide	182.4	18,900
25	methanol	32.7	19,400
26	water	78.4	22,100

Source: C. Reichardt, *Molecular Interactions* Vol. 3, ed. H. Ratajczak and W. J. Orville–Thomas, John–Wiley and Sons, Chichester, 1982.

equations presented above, computational approaches to solvent effects are better able to account for solvatochromic shifts of electronic spectra. Solvent effects on the lineshape of electronic spectra also continue to be of interest [27]. We will examine the topic of solvent contributions to the lineshape in the next chapter, where the relationship between absorption and resonance Raman spectra is presented.

11.11 SUMMARY

In this chapter we have used group theory to derive selection rules for electronic transitions of molecules. The basis for the approach is the approximate description of electronic states in terms of occupied molecular orbitals of definite symmetry. The state symmetry is derived from the direct product of the symmetries of the occupied orbitals. We found that the totally symmetric vibrations contribute to an electronic transition, as Franck–Condon progressions, whenever the excited state geometry is displaced along that normal coordinate, relative to the ground state. The point group of a molecule is based on its equilibrium geometry, and while selection rules derived on this basis are emphasized, the participation of nontotally symmetric vibrations permits transitions that would be forbidden in the rigid molecule. Nontotally symmetric vibrations are also active in electronic transitions that are accompanied by a change in symmetry; i.e., when the ground and excited state geometries belong to different point groups. Transitions which derive from a breakdown in the Born–Oppenheimer approximation are said to be vibronically allowed. They result from coupling of electronic states via nontotally symmetric vibrations. In some cases, such coupling results in splitting of otherwise degenerate states, and the molecule may distort to a less symmetric geometry (the Jahn–Teller effect).

We have also considered the many paths taken by a molecule after it has been promoted to an excited electronic state. Nonradiative transitions back to the ground electronic state, resulting from perturbations to the Born–Oppenheimer approximation, compete with emission. Vibrational relaxation within the excited electronic state, along with solvent relaxation, often precedes emission, leading to the Stokes shift between the absorption and emission maxima. In the event that excited state potentials of different states cross, nonadiabatic transitions may take place, leading to effects such as predissociation and intersystem crossing. A myriad of dynamical effects can exert their influences on the absorption and emission spectra. In the next chapter, we will examine a formalism that enables us to consider these events in the time domain, and explore the connection between electronic spectroscopy and resonance Raman spectroscopy.

PROBLEMS

1. Find the state symmetries that derive from the ground and first excited electronic configurations of N_2, F_2, and CO. What transitions are possible between states?

2. Verify that the wavefunction $\psi = \pi_{+1}(1)\pi_{-1}(2) - \pi_{+1}(2)\pi_{-1}(1)$, discussed in Section 11.2.2, is odd with respect to reflection through the xz plane. This can be done by first showing that the $\hat{\sigma}_{xz}$ operation converts the angle φ to $-\varphi$. The $\pi_{\pm 1}$ MOs contain the function $\exp(\pm i\varphi)$.

3. The bond length of CO is 1.128 Å in the ground electronic state and 1.370 Å in the first excited state. The vibrational frequency $\bar{\nu}_e$ is 2170 cm^{-1} in the ground state and 1172 cm^{-1} in the excited state. Sketch the Franck–Condon profile of the CO absorption band, indicating the relative intensities and peak frequencies.

4. The absorption spectrum of I_2 has $E_{00}/hc = 15{,}677$ cm^{-1}, and the onset of the continuum is at 19,735 cm^{-1}. The excited state dissociates to $I(^2P_{3/2}) + I(^2P_{1/2})$, which is 7589 cm^{-1} above the energy of two ground state iodine atoms. Find the dissociation energies D_0' and D_0''.

5. The SO_2 molecule, analogous to H_2O, has a $^1A_1$ ground electronic state with the valence electronic configuration $...(1a_2)^2(4b_2)^2(6a_1)^2(2b_1)^0$ and low-lying excited singlet states of symmetry $^1B_1$, $^1A_2$, and $^1B_2$. Account for the configurations of these excited states. Which of these states are E1-allowed? Experiment reveals that weak absorption in the range 280 to 340 nm is due to the transition $^1A_1 \rightarrow {}^1A_2$. How does this transition become allowed?

6. The ground electronic state of Cr^{3+} in an octahedral field is $^4A_{2g}$, which derives from a t_{2g}^3 configuration. Is this state subject to Jahn–Teller distortion?

7. The $Ti(H_2O)_6^{3+}$ ion shows an electronic transition which is split by the Jahn–Teller effect. Derive the symmetries of the metal–ligand vibrations of an ML_6 complex of D_{4h} symmetry. Suppose that $Ti(H_2O)_6^{3+}$ is distorted by virtue of moving the two ligands along the z direction closer to the metal ("z-ligands in distortion"). What is the symmetry of the ground state configuration? What are the symmetries of the states deriving from the excited-state configurations in the distorted complex? What vibrations are responsible for vibronic activity of the $^2B_{2g} \rightarrow {}^2A_{1g}$ and $^2B_{2g} \rightarrow {}^2B_{1g}$ electronic transitions?

8. Verify that the normal mode symmetries of octahedral ML_6 are as given in Section 11.6.

9. The compound shown below has $\lambda_{max} = 736$ nm in hexane and 553 nm in water. Explain the basis for the solvent shift.

10. Using perturbation theory, derive the following expression for the matrix element pertinent to radiationless relaxation (Section 11.8).

$$\left\langle \psi_g \left| \frac{\partial}{\partial Q_k} \right| \psi_e \right\rangle = \frac{\left\langle \psi_g \left| \frac{\partial \hat{H}_{el}}{\partial Q_k} \right| \psi_e \right\rangle}{E_e(Q) - E_g(Q)}$$

How does this result compare to the vibronic coupling responsible for activation of nominally forbidden electronic transitions? Does this problem explain how the neglect of the matrix element $\left\langle \psi_g \left| \partial^2 / \partial Q_k^2 \right| \psi_e \right\rangle$ might be justified? ([13] may be helpful for this problem.)

REFERENCES

1. I. N. Levine, *Quantum Chemistry, 5th ed.* (Prentice Hall, Upper Saddle River, NJ, 1999).
2. C. Dykstra, *Quantum Chemistry and Molecular Spectroscopy* (Prentice Hall, Englewood Cliffs, NJ, 1991).
3. C. Manneback, Computation of the intensities of vibrational spectra of electronic bands of diatomic molecules, *Physica 17*, 1001 (1951) and *Physica 20*, 497 (1954).
4. E. J. Heller, R. L. Sundberg, and D. Tannor, Simple aspects of Raman scattering, *J. Phys. Chem. 86*, 1822, (1982).
5. G. Herzberg, *Electronic Spectra of Polyatomic Molecules* (Van Nostrand Reinhold, New York, 1966).
6. H. Köppel, W. Domcke, and L. S. Cederbaum, Multimode molecular dynamics beyond the Born-Oppenheimer approximation, *Adv. Chem. Phys. 57*, 59 (1984).
7. L. Salem, *Molecular Orbital Theory of Conjugated Systems* (W. A. Benjamin, Reading, MA, 1972).
8. A. Streitwieser, *Molecular Orbital Theory for Organic Chemists* (Wiley, New York, 1961).
9. J. E. Huheey, *Inorganic Chemistry* (Harper & Row, New York, 1972).
10. F. A. Cotton, *Chemical Applications of Group Theory* (Wiley-Interscience, New York 1971).
11. S. L. Schultz, J. Qian, and J. M. Jean, Separability of intra- and intermolecular vibrational relaxation processes in trans-stilbene, *J. Phys. Chem. A 101*, 1000, (1997).
12. M. L. Horng, J. A. Gardecki, A. Papazyan, and M. Maroncelli, Subpicosecond measurements of polar solvation dynamics: Coumarin 153, *J. Phys. Chem 99*, 17211, (1995).
13. M. Bixon and J. Jortner, Intramolecular radiationless transitions, *J. Chem. Phys. 48*, 715 (1968).
14. K. Nassau, *The Physics and Chemistry of Color* (Wiley, New York, 1983).
15. C. Reichardt, Solvatochromic dyes as solvent polarity probes, *Chem. Rev. 94*, 2319 (1994).
16. N. S. Bayliss and E. G. McRae, Solvent effects in organic spectra: Dipole forces and the Franck-Condon principle, *J. Phys. Chem. 58*, 1002 (1954).
17. J. E. Brady and P. W. Carr, An analysis of dielectric models of solvatochromism, *J. Phys. Chem. 89*, 5759 (1985).
18. A. T. Amos and B. L. Burrows, Solvent-shift effects on electronic spectra and excited state dipole moments, *Adv. Quantum Chem. 7*, 289 (1973).
19. P. Suppan, Solvatochromic shifts: The influence of the medium on the energy of electronic states, *J. Photochem. Photobiol. A 50*, 293 (1990).
20. J. Zeng, J. S. Craw, N. S. Hush, and J. R. Reimers, Solvent effects on molecular spectra. I. normal pressure and temperature Monte Carlo simulations of the structure of dilute pyrimidine in water, *J. Chem. Phys. 99*, 1483 (1993).
21. N. S. Hush, and J. R. Reimers, Solvent effects on the electronic spectra of transition metal complexes, *Chem. Rev. 100*, 775 (2000).
22. M. Karelson, T. Tamm, and M. C. Zerner, Multicavity reaction field method for the solvent effect description in flexible molecular systems, *J. Phys. Chem 97*, 11901 (1993).

23. N. Rösch and M. C. Zerner, Calibration of dispersion energy shifts in molecular electronic spectra, *J. Phys. Chem. 98*, 5817 (1994).

24. J. Motta de Neto, R. B. Alencastro, and M. C. Zerner, Solvent effects on the electronic spectrum of Reichardt's dye, *Int. J. Quantum. Chem. Symp 28*, 361 (1994).

25. M. Cossi, V. Barone, Time-dependent density functional theory for molecules in liquid solutions, *J. Chem. Phys. 115*, 4708 (2001).

26. A. DeFusco, N. Minezawa, L. V. Slipchenko, F. Zahariev, M. S. Gordon, Modeling solvent effects on electronic excited states, *J. Phys. Chem. Lett. 2*, 2184 (2011).

27. M. D. Stephens, J. G. Saven, and J. L. Skinner, Molecular theory of electronic spectroscopy in nonpolar fluids: Ultrafast solvation dynamics and absorption and emission line shapes, *J. Chem. Phys 106*, 2129 (1997).

Raman and resonance Raman spectroscopy

12.1 INTRODUCTION

The subject of Raman scattering has already been investigated in several chapters. Here, we want to tie together various theoretical approaches and introduce the topic of resonance Raman scattering, in which the frequency of the exciting radiation falls within an electronic absorption band. The selection rules and polarization properties of Raman scattered light are considered for both off-resonance and resonance Raman experiments. We also introduce the very interesting topic of surface-enhanced Raman spectroscopy (SERS). We are concerned here with vibrational Raman scattering in the absence of resolved rotational subbands, but we do consider the influence of rotational motion and vibrational dynamics on the Raman lineshape. In effect, Raman spectroscopy is a two-dimensional experiment, in that the intensity depends on both the incident and scattered light frequency. The Raman spectrum is represented by the intensity of scattered light as a function of the frequency shift $\Delta v = v_0 - v_s$ between the incident (0) and scattered (s) radiation. Peaks in this spectrum correspond to transitions between vibrational levels within the ground electronic state. On the other hand, the *Raman excitation profile* (REP) is the intensity of a particular normal mode as a function of the incident frequency v_0. The dependence of the Raman intensity on the incident and scattered light frequency is illustrated in Figure 12.1. As the incident frequency is tuned, the intensities of bands in the Raman spectrum vary, but the frequency shifts remain the same. The intensity profiles of the various Raman bands, as a function of v_0, reflect dynamics taking place in the excited electronic state.

In this chapter, we introduce two approaches that are useful in the interpretation of resonance Raman scattering. One of these theories, called transform theory, is based on the Kramers–Kronig relationship between the real (scattering) and imaginary (absorption) parts of the polarizability. The other is a time-dependent approach, sometimes called wave packet theory, useful for describing spectra which involve more than one potential energy surface: electronic absorption and emission spectra and Raman excitation profiles. Both of these approaches take advantage of the relationship between Raman excitation profiles and absorption and emission profiles involving the resonant electronic state.

We begin with the Kramers–Heisenberg–Dirac (KHD) expression for the polarizability as expressed in Chapter 4 and follow the presentation of Albrecht [1] to derive the selection rules for Raman scattering both on- and off-resonance. We then analyze the depolarization ratio, which is the relative intensity of depolarized and polarized light scattering. (See Chapter 6 for a discussion of the experimental arrangement for measuring these.) This will show how off-resonance Raman scattering can be used to determine symmetries of modes and molecules, as well as dynamics in the ground electronic state. In resonance Raman spectroscopy, depolarization ratios reveal details of the resonant excited electronic states, as well as mode symmetries. The lineshape of a band in the Raman spectrum depends on dynamics that take place in the ground electronic state, while the shape of the Raman excitation profile reveals dynamics in the excited electronic state. In Section 12.6, we examine how the electromagnetic fields associated with electronic resonances of metal nanoparticles lead to enhanced Raman scattering and even single molecule detection. In Chapter 13, we will show that Raman scattering can be viewed as a nonlinear (third-order) spectroscopic technique and we will examine a unified formalism that encompasses Raman and fluorescence emission.

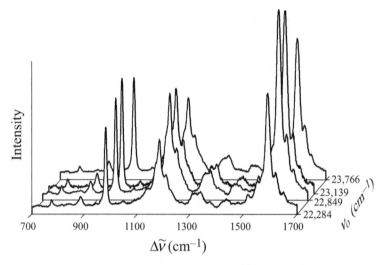

Figure 12.1 Raman spectra of 4-[2-(4-dimethylaminophenyl)ethenyl]-1-methylpyridinium iodide in water, at several excitation frequencies. (Courtesy of Dr. Xuan Cao, PhD thesis, University of Idaho, 1998.)

12.2 SELECTION RULES IN RAMAN SCATTERING

In Chapter 10, it was revealed that fundamental transitions in vibrational Raman spectroscopy are allowed through the polarizability derivative with respect to normal coordinate, $\alpha' \equiv (\partial\alpha/\partial Q)_0$, also called the derived polarizability. This operator is actually a second-rank tensor, the form of which dictates the polarization of the scattered light relative to that of the incident light. As discussed in Chapter 6, a Raman scattering experiment entails the measurement of both the polarized (I_{pol}) and depolarized (I_{dep}) components of the intensity, where the scattered light is detected having polarization vector polarized parallel or perpendicular to the incident polarization vector. (Alternatively, these are sometimes designated as $I_{\parallel}$ and $I_{\perp}$.) I_{pol} and I_{dep} are in turn proportional to the differential cross-sections $(d\sigma/d\Omega)_{\parallel}$ and $(d\sigma/d\Omega)_{\perp}$. We focus on the transition polarizability tensor α_{if} for which the square modulus gives the intensity. The components of this tensor are matrix elements of the polarizability operator connecting initial and final states. The differential Raman cross-section for the $i \rightarrow f$ transition is related to the polarizability tensor in the laboratory frame of reference.

$$\left(\frac{d\sigma_{if}}{d\Omega}\right)_{s,0} = \frac{\omega_s^3\omega_0}{c^4}\left|\hat{e}_s\cdot\alpha_{if}\cdot\hat{e}_0\right|^2 \tag{12.1}$$

The unit vectors $\hat{e}_0$ and $\hat{e}_s$ point along the polarization directions of the incident and scattered radiation; their role is to project out the part of the tensor α_{if} relevant to the experimental situation. ω_0 and ω_s are the angular frequencies of the incident and scattered radiation. Equation 12.1 contains the Cartesian tensor α_{if} (in units of volume) having elements such as $(\alpha_{XX})_{if}$, $(\alpha_{XY})_{if}$, etc., where X, Y, and Z are laboratory directions. If the polarizability is in MKS units, the right-hand side of Equation 12.1 should be divided by $(4\pi\varepsilon_0)^2$.

The KHD expression, on the other hand, is conveniently related to the polarizability in the molecule frame of reference.

$$\left(\alpha_{\rho\sigma}\right)_{if} = \frac{1}{\hbar}\sum_n\left[\frac{\langle i|\mu_\rho|n\rangle\langle n|\mu_\sigma|f\rangle}{\omega_0+\omega_{nf}+i\Gamma_n} - \frac{\langle i|\mu_\sigma|n\rangle\langle n|\mu_\rho|f\rangle}{\omega_0-\omega_{ni}-i\Gamma_n}\right]$$
$$\equiv\langle i|\hat{\alpha}_{\rho\sigma}|f\rangle \tag{12.2}$$

Equation 12.2 expresses the transition polarizability, a function of the incident frequency ω_0, as a matrix element of the polarizability operator connecting initial and final states. The transition polarizability is written here as a Cartesian tensor where ρ and σ are directions x, y, z in the molecule frame. Equation 12.2 is taken from Equation 4.57 with the addition of the level dependent damping term $i\Gamma_n$, which accounts for the finite lifetime $\tau \propto 1/\Gamma_n$ of the intermediate state (see Section 4.4.2). The factor Γ_n is the homogeneous linewidth of the electronic transition, that is, the population decay rate due to radiative and nonradiative mechanisms. We have several tasks at hand to connect the above expressions to experiment: we need to obtain the tensor α_{if} in the molecular frame from Equation 12.2 and then project out the desired components to find the differential cross-section $d\sigma_{if}/d\Omega$ appropriate to the experimental scattering geometry. This treatment exposes how the polarizability derivative and the transition polarizability are related.

Let us look more closely at the KHD equation. The ground, intermediate and final vibronic states of Equation 12.2 can be written according to the Born–Oppenhiemer (BO) approximation as

$$|i\rangle = |\psi_g(r;Q)\rangle |\chi_0^g(Q)\rangle \equiv |g0\rangle \tag{12.3}$$

$$|n\rangle = |\psi_e(r;Q)\rangle |\chi_v^e(Q)\rangle \equiv |ev\rangle \tag{12.4}$$

$$|f\rangle = |\psi_g(r;Q)\rangle |\chi_{v''}^g(Q)\rangle \equiv |gv''\rangle \tag{12.5}$$

Equations 12.3–12.5 consider a Raman transition from the ground vibronic state to a given final vibrational level v'' within the ground electronic state. As per the usual notation, r and Q denote the electronic and nuclear coordinates, respectively. The vibrational wavefunctions are assumed to be separable into $3N - 6$ one-dimensional harmonic oscillator wavefunctions. Thus the pertinent states are more completely expressed as

$$|g0\rangle = |\psi_g(r;Q)\rangle |\chi_0^g(Q_1)\chi_0^g(Q_2)...\chi_0^g(Q_{3N-6})\rangle \tag{12.6}$$

$$|ev\rangle = |\psi_e(r;Q)\rangle |\chi_{v_1}^e(Q_1)\chi_{v_2}^e(Q_2)...\chi_{v_{3N-6}}^e(Q_{3N-6})\rangle \tag{12.7}$$

$$|gv''\rangle = |\psi_g(r;Q)\rangle |\chi_{v''}^g(Q_1)\chi_0^g(Q_2)...\chi_0^g(Q_{3N-6})\rangle \tag{12.8}$$

The electronic wavefunctions depend parametrically on the set of normal coordinates, and the Raman active mode is arbitrarily chosen to be mode 1. The index "v" in Equation 12.7 represents a set of vibrational quantum numbers $\{v_1, v_2,...\}$ within the intermediate excited electronic state e. We are assuming that only one normal mode undergoes a transition, but all normal modes can potentially contribute to the intermediate state. Combination bands and hot bands can be treated similarly, but here we will just examine the possibility of fundamentals and overtones originating in the ground vibrational state. Using Equations 12.6 through 12.8 in Equation 12.2, the matrix elements of $\hat{\mu}$ can be evaluated first with respect to the electronic wavefunctions to get the normal-coordinate dependent transition moment $\mu_{ge}(Q)$. This can be expanded in terms of the normal coordinates.

$$\mu_{ge}(Q) = \mu_{ge}^0 + \sum_{i=1}^{3N-6}\left(\frac{\partial\mu_{ge}}{\partial Q_i}\right)_0 Q_i + \cdots \tag{12.9}$$

where the first term μ_{ge}^0 is the Condon approximation to the transition moment. If Equation 12.9 is truncated after the linear term, then Equation 12.2 contains terms such as

$$\langle g0|\mu|ev\rangle = \mu_{ge}^0 \langle \chi_0^g(Q_1)\chi_0^g(Q_2)...|\chi_{v_1}^e(Q_1)\chi_{v_2}^e(Q_2)...\rangle$$
$$+ \sum_{i=1}^{3N-6}\left(\frac{\partial\mu_{ge}}{\partial Q_i}\right)_0 \langle \chi_0^g(Q_1)\chi_0^g(Q_2)...|Q_i|\chi_{v_1}^e(Q_1)\chi_{v_2}^e(Q_2)...\rangle \tag{12.10}$$

Alternatively, the Herzberg–Teller (HT) formulation of the electronic transition moment can be employed. In this approach the perturbed wavefunction for state e is

$$|e\rangle' = |e\rangle + \sum_i \sum_{r \neq e} \frac{H_{er}^i Q_i}{E_e - E_r} |r\rangle \tag{12.11}$$

where the sum is over all the normal modes i and all excited states except e, and $|e\rangle$ and $|r\rangle$ are the unperturbed electronic states having energies E_e and E_r. (The prime in Equation 12.11 indicates a first-order perturbed quantity, not a derivative.) The vibronic coupling matrix element is

$$H_{er}^i \equiv \left\langle \psi_e \left| \left(\frac{\partial \hat{H}}{\partial Q_i} \right)_0 \right| \psi_r \right\rangle \tag{12.12}$$

The angle brackets in Equation 12.12 indicate an average over electronic coordinates. The transition moment connecting states g and e is thus

$$\mu_{ge} = \mu_{ge}^0 + \sum_i \sum_{r \neq e} \frac{H_{er}^i \mu_{gr}^0}{E_e - E_r} Q_i \tag{12.13}$$

In the Herzberg–Teller approach, the coordinate dependence of the electronic transition moment μ_{ge} arises from intensity borrowing from other electronic states, through the dependence of the Hamiltonian on normal coordinate. The operator $(\partial H / \partial Q_i)_0$ is a perturbation to the BO approximation. In either approach, Equation 12.9 truncated at the second term or Equation 12.13, the transition moment is linear in the normal coordinates, leading to the same selection rules. We continue with Equation 12.9 for now, but at times it is convenient to make the following correspondence:

$$\left(\frac{\partial \mu_{ge}}{\partial Q_i} \right)_0 = \sum_r \frac{H_{er}^i \mu_{gr}^0}{E_e - E_r} \tag{12.14}$$

The BO approximation also allows the electronic and vibrational contributions in the energy denominator to be separated:

$$\omega_{ni} = \omega_{eg} + \sum_{j=1}^{3N-6} \mathrm{v}_j \omega_j^e \tag{12.15}$$

$$\omega_{nf} = \omega_{eg} + \sum_{j=1}^{3N-6} \mathrm{v}_j \omega_j^e - \mathrm{v}'' \omega_1 \tag{12.16}$$

where ω_{eg} is the origin (0 – 0 transition) of the electronic transition from g to e, ω_1 is the frequency of the Raman mode in the ground electronic state, and ω_j^e is the frequency of normal mode j in electronic state e. The ground and excited states are assumed here to be harmonic.

12.2.1 Off-resonance Raman scattering

Let us suppose that the frequency ω_0 of the exciting radiation is far removed from the resonance frequency, e.g., $\omega_0 \ll \omega_{ge}$, so that the vibrational energy terms of Equations 12.15 and 12.16 can be neglected compared to $\omega_{eg} - \omega_0$ and $\omega_{eg} + \omega_0$ in the energy denominators of Equation 12.2. The damping term can also be neglected in this limit, as it is on the order of a vibrational frequency. Keeping only the Condon approximation transition moments then leads to

$$(\alpha_{\rho\sigma})_{0v''} = \frac{1}{\hbar} \sum_{ev} (\mu_{ge}^0)_\rho (\mu_{ge}^0)_\sigma \left[\frac{\langle 0|v\rangle\langle v|v''\rangle}{\omega_0 + \omega_{eg}} - \frac{\langle 0|v\rangle\langle v|v''\rangle}{\omega_0 - \omega_{eg}} \right] \tag{12.17}$$

The quantities $\langle 0|v\rangle$ and $\langle v|v''\rangle$ are vibrational overlaps, the squares of which are the Franck–Condon (FC) factors discussed in Chapter 11. Again, the summation index v is shorthand notation for the set of intermediate state vibrational quantum numbers $\{v\} = \{v_1 v_2 ... v_{3N-6}\}$. In the harmonic approximation, the multimodal overlaps are factored into $3N - 6$ one-dimensional overlap integrals.

$$\langle 0|v\rangle \equiv \left\langle \chi_0^g(Q_1) \middle| \chi_{v_1}^e(Q_1) \right\rangle \left\langle \chi_0^g(Q_2) \middle| \chi_{v_2}^e(Q_2) \right\rangle ... \left\langle \chi_0^g(Q_{3N-6}) \middle| \chi_{v_{3N-6}}^e(Q_{3N-6}) \right\rangle \tag{12.18}$$

Similarly,

$$\langle v|v''\rangle \equiv \left\langle \chi_{v_1}^e(Q_1) \middle| \chi_{v''}^g(Q_1) \right\rangle \left\langle \chi_{v_2}^e(Q_2) \middle| \chi_0^g(Q_2) \right\rangle ... \left\langle \chi_{v_{3N-6}}^e(Q_{3N-6}) \middle| \chi_0^g(Q_{3N-6}) \right\rangle \tag{12.19}$$

Having removed the denominator's dependence on the summation over intermediate vibrational states, we can get rid of the identity term, $1 = \sum_v |v\rangle\langle v|$. This gives

$$(\alpha_{\rho\sigma})_{0v''} = \frac{1}{\hbar} \sum_e (\mu_{ge}^0)_\rho (\mu_{ge}^0)_\sigma \left[\frac{2\omega_{eg}\delta_{0v''}}{\omega_{eg}^2 - \omega_0^2} \right] \tag{12.20}$$

$$(\alpha_{\rho\sigma})_{00} = \frac{1}{\hbar} \sum_e (\mu_{ge}^0)_\rho (\mu_{ge}^0)_\sigma \left[\frac{2\omega_{eg}}{\omega_{eg}^2 - \omega_0^2} \right] \tag{12.21}$$

We have used $\langle 0|v''\rangle = \delta_{0v''}$, which follows from that fact that the states 0 and v'' both belong to the ground state potential surface. Equation 12.21 describes Rayleigh scattering, for which the initial and final vibrational states are the same. Note that this expression poses no restrictions in terms of selection rules: all molecules are active in Rayleigh scattering.

To obtain Raman scattering, we need the part of the transition moment that depends on normal coordinate. The contribution of the polarizability which is linear in Q_i has terms in the numerator such as

$$\langle g0|\mu_\rho|ev\rangle\langle ev|\mu_\sigma|gv''\rangle \Rightarrow (\mu_{ge}^0)_\rho (\mu_{ge}^{(i)})_0 \langle 0|v\rangle\langle v|Q_i|v''\rangle \tag{12.22}$$

where the derivative of the transition moment $(\partial\mu_{ge}/\partial Q_i)_0$ is defined as $\mu_{ge}^{(i)}$. Again, the summation over intermediate vibrational states can be removed when the exciting frequency is far from resonance, yielding

$$(\alpha_{\rho\sigma})_{0v''} = \frac{1}{\hbar} \sum_e \left[\frac{(\mu_{ge}^0)_\rho (\mu_{ge}^{(i)})_\sigma \langle 0|Q_i|v''\rangle}{\omega_0 + \omega_{eg}} - \frac{(\mu_{ge}^0)_\sigma (\mu_{ge}^{(i)})_\rho \langle 0|Q_i|v''\rangle}{\omega_0 - \omega_{eg}} \right] \tag{12.23}$$

$$\equiv \alpha_{\rho\sigma}^{(i)} \langle 0|Q_i|v''\rangle$$

For harmonic oscillators, the matrix element $\langle 0|Q_i|v''\rangle$ vanishes unless $v'' = 1$, in keeping with the previously introduced selection rules for Raman scattering. The term $\alpha_{\rho\sigma}^{(i)}$ can now be associated with the polarizability derivative, $(\partial\alpha_{\rho\sigma}/\partial Q_i)_0$.

More generally, allowing for a vibrationally excited initial state $v' \neq 0$, the matrix element of the normal coordinate contributing to $(\alpha_{\rho\sigma})_{v'v''}$ is

$$\langle v' | Q_i | v'' \rangle = \sqrt{\frac{\hbar(v'+1)}{2\omega_i}} \delta_{v'',v'+1}, \quad \text{for Stokes scattering}$$

$$\langle v' | Q_i | v'' \rangle = \sqrt{\frac{\hbar v'}{2\omega_i}} \delta_{v'',v'-1}, \quad \text{for anti-Stokes scattering}$$

(12.24)

The transition polarizability $(\alpha_{\rho\sigma})_{v'v''}$ is related to the polarizability derivative through

$$(\alpha_{\rho\sigma})_{v'v''} = \alpha_{\rho\sigma}^{(i)} \langle v' | Q_i | v'' \rangle$$

(12.25)

where

$$\alpha_{\rho\sigma}^{(i)} = \frac{1}{\hbar} \sum_e \left[\frac{(\mu_{ge}^0)_\rho (\mu_{ge}^{(i)})_\sigma}{\omega_0 + \omega_{eg}} - \frac{(\mu_{ge}^0)_\sigma (\mu_{ge}^{(i)})_\rho}{\omega_0 - \omega_{eg}} \right]$$

(12.26)

The selection rule $\Delta v = \pm 1$, which applies when both mechanical and electrical anharmonicity are absent, derives from the matrix element of the normal coordinate in Equation 12.25. The condition that the polarizability derivative $\alpha_{\rho\sigma}^{(i)}$ be nonzero leads to the gross selection rule and is a function of the symmetry of the mode and of the molecule. Raman activity of a normal mode requires that $\alpha_{\rho\sigma}^{(i)}$ not vanish for at least one combination of molecule-frame Cartesian directions ρ and σ. Now there are always allowed electonic states for which $(\mu_{ge})_\rho$ is nonzero, but what are the conditions for which $(\partial\mu_{ge}/\partial Q_i)_0$ is nonzero? To examine this more closely, let us return to the HT formalism of Equation 12.14. We require the matrix element H_{er}^i (Equation 12.12) connecting two E1-allowed excited states to be nonzero. To be allowed, the states e and r must transform according to irreducible representations to which the coordinates x, y, and z belong. The operator $(\partial H/\partial Q_i)_0$ belongs to the same ir. rep. as the normal coordinate Q_i. Therefore, Raman active fundamentals are those for which the normal coordinates transform as quadratic functions of the Cartesian coordinates, such as $x^2 - y^2$, xy, and other products and squares of x, y, z. These functions are found on the right-hand side of the character table. These are the same selection rules that were introduced in Chapter 10 on the basis of the symmetry of the polarizability tensor. In Section 12.3 we examine how the molecular symmetry dictates the form of the molecule-frame tensor and thus the polarization properties of the scattered light.

12.2.2 RESONANCE RAMAN SCATTERING

Figure 12.2 shows a Raman spectrum of I_2 in cyclohexane, excited at a wavelength within the electronic absorption band of I_2. The concentration of I_2 in this sample is on the order of 10^{-3} M, yet the Raman intensities of I_2 are comparable to those of the solvent. Several overtones are observed; thus it is apparent that the selection rule $\Delta v = \pm 1$ does not hold for resonance Raman. Particularly for small molecules, such as the spectrum of liquid Br_2 in Figure 9.8, resonance excitation of the Raman spectrum leads to activity of high-order overtones. In this section, we examine the selection rules operative in resonance Raman spectra like those of Figures 9.8 and 12.2.

Let us back up to the point just before Equation 12.17, and *not* make the assumption that the incident frequency ω_0 is far from resonance. We assume that ω_0 is close to the origin ω_{eg} of an allowed electronic transition. Then we can neglect the anti-resonance term (the first term in Equation 12.2), because it is small compared to the term with the small energy denominator. We shall also limit the sum over intermediate states to vibrational states v within the resonant excited electronic state e. We are thus ignoring the possibility

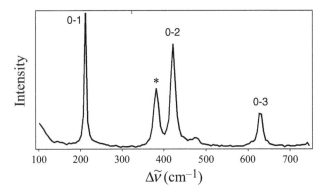

Figure 12.2 Raman spectrum of I_2 in cyclohexane, excited at 514.5 nm. The fundamental and first and second overtones of the I–I stretch are observed. The (*) marks a solvent vibration.

of simultaneous resonance with more than one electronic transition. Keeping the terms up to linear in Q_i, we may write the transition polarizability $(\alpha_{\rho\sigma})_{v'v''}$ as a sum of two terms:

$$(\alpha_{\rho\sigma})_{v'v''} = A_{v'v''} + B_{v'v''} \tag{12.27}$$

called the Albrecht A and B terms. The first of these comes from the Condon approximation. It is

$$A_{v'v''} = \frac{1}{\hbar} \sum_{v} (\mu_{ge}^0)_\rho (\mu_{ge}^0)_\sigma \left[\frac{\langle v' | v \rangle \langle v | v'' \rangle}{\omega_{ev,gv'} - \omega_0 - i\Gamma_e} \right] \tag{12.28}$$

Here, $\hbar\omega_{ev,gv'}$ is the energy difference between the intermediate $|ev\rangle$ and initial $|gv'\rangle$ vibronic states. Equation 12.28 already indicates that incident frequencies close to an allowed transition frequency can lead to intensity enhancement, through an increase in the transition polarizability. Typical enhancement factors, that is, intensity increases compared to off-resonance Raman, are on the order of 10^5 to 10^6, but the enhancement does not apply to all normal modes, as we shall see.

Again, the label v represents a collection of vibrational quantum numbers in the intermediate (excited electronic) state: $\{v\} = v_1, v_2,\ldots v_{3N-6}$, and v' and v'' are similar collections of vibrational quantum numbers for the initial (v') and final (v'') vibrational states, both of which are within the ground electronic state. Assuming harmonic vibrational modes that are the same frequency in the ground and excited electronic states, we can write

$$\omega_{ev,gv'} = \omega_{eg} + \sum_{j=1}^{3N-6} (v_j - v'_j)\omega_j \tag{12.29}$$

Equation 12.29 can be readily adapted to the case where the excited state vibrational frequency differs from that of the ground. In any event, the dependence of the energy denominator on intermediate vibrational states prevents the sum over v from being removed as was done in the off-resonance case. The multimodal vibrational overlaps in the numerator of Equation 12.28 (see Equations 12.18 and 12.19) are the basis for selection rules in resonance Raman scattering. Recall the Franck–Condon factors introduced in Chapter 11, which are most important for totally symmetric normal modes that are displaced in the excited electronic state. If a particular mode j is not Franck–Condon active, then the (one-dimensional) vibrational overlaps reduce to $\langle v'_j | v_j \rangle = \delta_{v_j,v_j}$ and $\langle v_j | v''_j \rangle = \delta_{v_j,v''_j}$. Consider for example a transition from $\{v'\} = (0,0,0\ldots)$ to $\{v''\} = (v''_1,0,0\ldots)$. The transition polarizability vanishes unless the excited state potential is displaced with respect to that of the ground state along normal coordinate Q_1. Thus vibrational modes which are not Franck–Condon active are not allowed by the A term, since Equation 12.28 in this case only permits $v' = v''$, which corresponds to Rayleigh scattering. Conversely, totally symmetric modes that contribute to the vibrational progression of

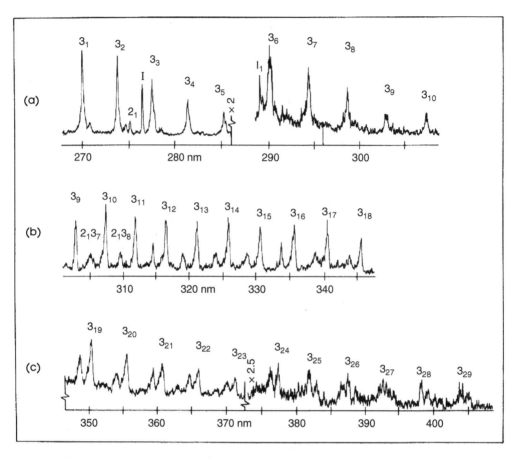

Figure 12.3 Resonance Raman spectrum of CH_3I excited at 266 nm, showing the fundamental and 28 overtones of the C–I stretch, $\tilde{\nu}_3 \approx 500$ cm^{-1}. (Reprinted with permission from Imre, D. et al., *J. Phys. Chem.* 88, 3956, 1984. Copyright 1984 American Chemical Society.)

the $g \to e$ transition in electronic absorption are active in resonance Raman, as the overlap integrals in the numerator of Equation 12.28 do not vanish. Thus the totally symmetric modes that are enhanced in the resonance Raman spectrum are those which correspond to the geometry change of the molecule. These modes are said to be "*A*-term enhanced" or "Franck–Condon enhanced." Note that the selection rule $\Delta v = \pm 1$ no longer applies in this situation: long FC progressions in absorption are associated with overtones $\Delta v = \pm 2, \pm 3, \ldots$ in the resonance Raman spectrum. As in the resonance Raman spectrum of CH_3I shown in Figure 12.3, excitation wavelengths resonant with an electronic transition can result in the appearance of many overtones of the Franck–Condon active modes. In CH_3I, the excited state is dissociative, $CH_3I \to CH_3 + I$, leading to a long progression in the C–I stretch. In larger molecules, overtones are less frequently observed in the resonance Raman spectrum. This is something we return to in Section 12.5 as it is easier to understand in the time-dependent view.

The Albrecht *B* term is obtained by employing Equation 12.9 for the transition moment and keeping terms which are linear in the normal coordinate. This results in quite different selection rules than those from the *A* term, and we show next that the *B* term is responsible for resonance Raman activity of nontotally symmetric vibrations. It is convenient to employ the Herzberg–Teller formalism in writing the *B* term.

$$B_{v'v''} = \frac{1}{\hbar} \sum_{v,r} \frac{H_{er}^1}{E_r - E_e} \left[\frac{(\mu_{ge}^0)_\rho (\mu_{gr}^0)_\sigma \langle v'|v\rangle\langle v|Q_1|v''\rangle + (\mu_{ge}^0)_\sigma (\mu_{gr}^0)_\rho \langle v'|Q_1|v\rangle\langle v|v''\rangle}{\omega_{ev,gv'} - \omega_0 - i\Gamma_e} \right] \qquad (12.30)$$

Mode 1 is the one under consideration, and we continue to use shorthand notation for the multimodal vibrational states. Equation 12.30 emphasizes the importance of allowed electronic states r close in energy to the resonant state e. Let us consider the B term as it pertains to a nontotally symmetric vibration in which the initial state is the ground vibrational state. As we have seen, nontotally symmetric modes do not contribute to vibrational progressions in electronic spectra unless there is a change in symmetry in the excited electronic state. In the absence of Franck–Condon activity, the overlap $\langle v'|v \rangle = \langle 0|v_1 \rangle \langle 0|v_2 \rangle \ldots \langle 0|v_{3N-6} \rangle$ vanishes unless $v_1 = 0$. (As for the remaining modes 2 through $3N-6$, we need only consider overlaps for those modes that are displaced in the excited state. Intermediate states having an arbitrary number of vibrational quanta invested in these modes contribute to the sum over states.) Thus the first overlap in the first term in the numerator of Equation 12.30 selects the $v_1 = 0$ intermediate state. The quantity $\langle v|Q_1|v'' \rangle$ contains the term $\langle 0_1|Q_1|v''_1 \rangle$, which vanishes unless $v''_1 = 1$. Thus this term allows for enhancement of the fundamental transition of mode 1 when the radiation is resonant with the $g0 \rightarrow e0$ transition. Similarly, the second term results in activity of the fundamental when the radiation is resonant with the $g0 \rightarrow e1$ transition. Note that the fundamental transition of a nontotally symmetric vibration is active in the resonance Raman spectrum if that vibration is responsible for vibronic coupling of two nearby electronic states. The intensities of nontotally symmetric vibrations are generally weaker than totally symmetric ones in resonance Raman, as the latter are symmetry allowed and the former depend on breakdown of the Born–Oppenheimer approximation. The Albrecht C term, not given here, employs an HT expansion of the ground electronic state. It becomes important for molecules with low-lying excited electronic states.

In the next section, we examine how the depolarization ratio can be exploited to distinguish between totally and nontotally symmetric vibrations in ordinary and resonance Raman.

12.3 POLARIZATION IN RAMAN SCATTERING

The problem at hand is to convert molecule-frame tensor elements α_{xx}, α_{xy}, etc. to the necessary lab-frame components. In the conventional 90° scattering arrangement depicted in Figure 6.5, we require α_{ZZ} for the polarized spectrum and α_{ZX} for the depolarized spectrum. We are interested here in a sample of randomly oriented molecules, as in the gas or liquid phase. One approach is to employ direction cosines to project each molecule-frame inertial axis onto the desired lab-frame direction. This approach has been employed in the older literature (see [2]), but it is somewhat clumsy because the averages (integrals) over molecular orientation have to be done on a case by case basis. In Chapter 8, the Wigner rotation matrices were introduced in order to connect the lab-frame and molecule-frame polarizabilities (see Equation 8.62). The use of Wigner rotation functions is convenient because the symmetry properties of these functions can be exploited in performing the orientational averages. The polarizability is expressed as a spherical, rather than Cartesian, tensor. The formulas for relating the spherical tensor components α_M^J to the Cartesian components are given in Chapter 8. Each spherical tensor component in the molecule frame can be rotated into the lab frame as described in Section 8.6. The equations of that section were expressed in terms of the polarizability for rotational Raman scattering, but they apply as well to the case of vibrational Raman scattering. When we write α_M^J in this chapter, it denotes a spherical tensor component of the Raman polarizability.

It is possible, as shown in [3,4], to write down three rotational invariants, Σ^J, which are linear combinations of the α_M^J that are independent of reference frame. These are

$$\Sigma^J = \sum_{M=-J}^{J} \left| \alpha_M^J \right|^2 \tag{12.31}$$

where $J = 0$, 1, and 2 and $M = 0, \pm 1, \ldots, \pm J$. Each Σ^J is called an invariant because it is independent of orientation. By way of analogy, consider how a vector, such as the dipole $\vec{\mu}$, a first-rank tensor, has a length which is independent of its orientation. The combination $\mu_x^2 + \mu_y^2 + \mu_z^2$ is a rotational invariant. As a second-rank tensor, the polarizability has three invariants, Σ^0, Σ^1, and Σ^2:

$$\Sigma^0 = \frac{1}{3}\left|\alpha_{xx} + \alpha_{yy} + \alpha_{zz}\right|^2$$

$$\Sigma^1 = \frac{1}{2}\left[\left|\alpha_{xy} - \alpha_{yx}\right|^2 + \left|\alpha_{xz} - \alpha_{zx}\right|^2 + \left|\alpha_{yz} - \alpha_{zy}\right|^2\right]$$

$$\Sigma^2 = \frac{1}{2}\left[\left|\alpha_{xy} + \alpha_{yx}\right|^2 + \left|\alpha_{xz} + \alpha_{zx}\right|^2 + \left|\alpha_{yz} + \alpha_{zy}\right|^2\right]$$
$$+ \frac{1}{3}\left[\left|\alpha_{xx} - \alpha_{yy}\right|^2 + \left|\alpha_{xx} - \alpha_{zz}\right|^2 + \left|\alpha_{yy} - \alpha_{zz}\right|^2\right]$$

(12.32)

The invariants in Equations 12.32 are written in terms of the molecule-frame (xyz) components of the polarizability. Since they are invariants, we could also have used the lab frame (XYZ) components, which relate directly to the measured intensities, in Equations 12.32. Σ^0 is called the isotropic part of the polarizability. Note that it is proportional to the square modulus of the trace of the polarizability tensor, $\Sigma^0 = |\text{Tr}\alpha|^2/3$. Σ^1 and Σ^2 are the antisymmetric and symmetric anisotropies, respectively. Σ^1 is zero in ordinary Raman scattering because α is a symmetric tensor: $\alpha_{\rho\sigma} = \alpha_{\sigma\rho}$. It can be nonzero in resonance Raman experiments, as discussed below. Σ^2 depends on the existence of off-diagonal elements of the polarizability tensor as well as differences in the diagonal components, so it represents the deviation of the polarizability from spherical symmetry.

The lab-frame components $|\alpha_{ZZ}|^2$ and $|\alpha_{ZX}|^2$ can be written as linear combinations of the invariants. As shown for example in [3],

$$|\alpha_{ZZ}|^2 = \frac{1}{3}\Sigma^0 + \frac{2}{15}\Sigma^2$$

$$|\alpha_{ZX}|^2 = \frac{1}{6}\Sigma^1 + \frac{1}{10}\Sigma^2$$

(12.33)

If the molecule-frame tensor α is known, one can write the invariants Σ^J according to Equations 12.32, and then use Equations 12.33 to predict the polarized and depolarized intensity. Next, we will examine how symmetry principles are applied to the analysis of the depolarization ratio ρ in off-resonance Raman scattering.

12.3.1 Polarization in off-resonance Raman scattering

The form of α in the molecule frame reflects the symmetry of the molecule and the symmetry of the vibration. The polarizability tensor of a molecule in its equilibrium geometry can always be diagonalized, and thus the derived polarizability tensor for a totally symmetric mode preserves this symmetry. Therefore, the Cartesian Raman tensor for any totally symmetric mode is of the form

$$\alpha = \begin{pmatrix} a & 0 & 0 \\ 0 & b & 0 \\ 0 & 0 & c \end{pmatrix} \quad \text{totally symmetric vibration}$$

(12.34)

Asymmetric top molecules have $a \neq b \neq c$, linear molecules and symmetric top molecules (those having threefold symmetry or higher) have two equal components, say $a = b \neq c$, and spherically symmetric molecules have three equivalent diagonal elements $a = b = c$. For example, the Raman polarizability tensor for any totally symmetric mode of a molecule with spherical symmetry, such as SF_6 or CCl_4, has three equivalent diagonal components and Σ^2 is zero, as is Σ^1.

In contrast, the polarizability tensor for a nontotally symmetric mode has zero trace and is not diagonal. First we define the average polarizability $\bar{\alpha} \equiv (\alpha_{xx} + \alpha_{yy} + \alpha_{zz})/3 = \text{Tr}\alpha/3$. This is necessarily a totally symmetric quantity because it is invariant to all symmetry operations. As an operator, $\bar{\alpha}$ is responsible for the activity of totally symmetric modes, since the selection rules require that $\Gamma_i \times \Gamma_\alpha \times \Gamma_f$ be totally symmetric, where Γ_i and Γ_f are the ir. reps. of the initial and final vibrational states. In the case of nondegenerate, nontotally symmetric modes for which $\Delta v = \pm 1$, Γ_i and Γ_f are never the same symmetry, so the totally symmetric part of the Raman tensor cannot connect two such states.

It is often convenient to separate out the totally symmetric and nontotally symmetric parts of the polarizability. We can always take an arbitrary molecule-frame Raman tensor α and divide it up into two parts:

$$\alpha = \bar{\alpha} \begin{pmatrix} 1 & 0 & 0 \\ 0 & 1 & 0 \\ 0 & 0 & 1 \end{pmatrix} + \beta \tag{12.35}$$

where the tensor β, called the anisotropy of the polarizability, is given by

$$\beta = \begin{pmatrix} \alpha_{xx} - \bar{\alpha} & \alpha_{xy} & \alpha_{xz} \\ \alpha_{xy} & \alpha_{yy} - \bar{\alpha} & \alpha_{yz} \\ \alpha_{xz} & \alpha_{yz} & \alpha_{zz} - \bar{\alpha} \end{pmatrix} \tag{12.36}$$

It is easy to show that $\mathrm{Tr}\,\beta = 0$. While the tensor β need not vanish for totally symmetric modes, except for molecules of spherical symmetry, for nontotally symmetric modes it is the only surviving contribution to α, since the totally symmetric part ($\mathrm{Tr}\,\alpha$) vanishes. The reason that β can be nonzero even for a totally symmetric vibration is that it reflects deviations from spherical symmetry, and these may be inherent to the molecule. Equation 12.36 assumes that the Raman tensor is symmetric, since we are treating off-resonance Raman scattering.

The depolarization ratio $\rho = I_{dep}/I_{pol}$ is related to the lab-frame components of the polarizability as follows:

$$\rho = \frac{|\alpha_{ZX}|^2}{|\alpha_{ZZ}|^2} \tag{12.37}$$

This formula assumes the scattering geometry of Figure 6.5. Using Equation 12.33, the depolarization ratio in terms of invariants is

$$\rho = \frac{5\Sigma^1 + 3\Sigma^2}{10\Sigma^0 + 4\Sigma^2} \tag{12.38}$$

Since the trace of the polarizability tensor vanishes for any nontotally symmetric mode, such modes have $\Sigma^0 = 0$. Coupled with the fact that Σ^1 vanishes in off-resonance Raman, we conclude that the depolarization ratio of any nontotally symmetric vibration must be $\rho = 3/4$. Totally symmetric vibrations have $\Sigma^0 \neq 0$, so the only general statement one can make is that $\rho < 3/4$ in this case. In off-resonance Raman scattering, ρ is never greater than 3/4. The use of polarization in Raman scattering enables totally and nontotally symmetric modes to be distinguished. For molecules symmetric enough for the x, y, and z directions to be equivalent (the tetrahedral and octahedral point groups), then clearly $\rho = 0$ for totally symmetric vibrations. Measurement of the depolarization ratio is a powerful way to assign normal modes and deduce symmetry.

The spectrum of CCl_4 shown in Figure 12.4 provides an illustration. CCl_4 is tetrahedral, and the totally symmetric C–Cl stretch at 459 cm^{-1} has a depolarization ratio near zero, about 0.01. It is not exactly zero in the liquid phase, due to interaction-induced contributions to the polarizability tensor. The remaining modes are nontotally symmetric vibrations and all have $\rho = 3/4$. Note the Fermi resonance doublet in the vicinity of 760 and 790 cm^{-1}. This pair of peaks is due to anharmonic mixing of the v_3 fundamental and the $v_1 + v_4$ combination.

For any normal mode of known symmetry, the character table can be consulted to predict the form of the polarizability tensor. We have already seen how this works for totally symmetric vibrations. Now let us look at the general case. Consider for example the normal modes of a molecule of C_{2v} symmetry. The quadratic functions of Cartesian coordinates and corresponding ir. reps. are given in Table 12.1. Modes of A_2 symmetry

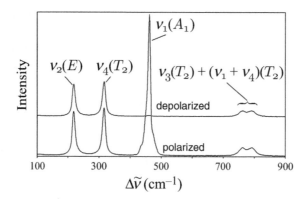

Figure 12.4 Raman spectrum of neat CCl_4, excited at 514.5 nm. The depolarized spectrum is offset for clarity.

Table 12.1 Quadratic functions in C_{2v}

C_{2v}	Function
A_1	x^2, y^2, z^2
A_2	
B_1	xz
B_2	yz

are Raman inactive, while those of A_1, B_1, and B_2 symmetry are allowed. The Raman tensors for the B_1 and B_2 modes are

$$\alpha(B_1) = \begin{pmatrix} 0 & 0 & a \\ 0 & 0 & 0 \\ a & 0 & 0 \end{pmatrix} \tag{12.39}$$

$$\alpha(B_2) = \begin{pmatrix} 0 & 0 & 0 \\ 0 & 0 & b \\ 0 & b & 0 \end{pmatrix} \tag{12.40}$$

Equations 12.39 and 12.40 merely account for the correspondence between the symmetry of a function like xz and the tensor element α_{xz}. Point groups with degenerate vibrations must be handled with care. For example, the E mode of a C_{3v} molecule encompasses the functions $(x^2 - y^2, xy)$ and (xz, yz). The parentheses enclose functions that transform as a pair under the operations of the group. There are necessarily two Raman tensors for a doubly degenerate vibration. In this case they are

$$\alpha(E) = \begin{pmatrix} a & b & 0 \\ b & -a & 0 \\ 0 & 0 & 0 \end{pmatrix}, \begin{pmatrix} 0 & 0 & a \\ 0 & 0 & b \\ a & b & 0 \end{pmatrix} \tag{12.41}$$

Nafie and Peticolas [2] have tabulated the form of the Raman tensor for a large number of point groups and mode symmetries.

12.3.2 POLARIZATION IN RESONANCE RAMAN SCATTERING

The preceding considerations assume nonresonance Raman scattering. In the resonance Raman experiment, the Raman tensor is not necessarily symmetric and the depolarization ratio can exceed the value 3/4. It is often possible to predict the depolarization ratio ρ from the form of the resonant electronic state, as shown in [3,4].

Consider the common scenario in which a totally symmetric vibration is A-term enhanced (see Equation 12.28) through resonance with a single nondegenerate electronic state. The transition moment for the $g \rightarrow e$ transition has a unique direction in the molecule frame, let us say it is the x direction. This means that the only nonzero component of the Raman tensor is α_{xx}. Using Equations 12.32 and 12.38, the value $\rho = 1/3$ is readily obtained. On the other hand, a doubly degenerate resonant electronic state results in two equal diagonal Raman tensor components, such as $\alpha_{xx} = \alpha_{yy}$, which leads to $\rho = 1/8$. When the Raman intensity derives from more than one excited electronic state with transition moments in different directions, the depolarization ratio depends on excitation wavelength. This is referred to as dispersion in the depolarization ratio.

B-term enhancement can lead to interesting effects such as anomalous polarization, in which $\rho > 3/4$. Anomalously polarized Raman bands have been observed for some heme ring vibrations of hemoglobin and ferrocytochrome c [5], and for A_{2g} modes of five-coordinate metalloporphyrins [6]. Vibronic coupling of two states leads to off-diagonal components of the Raman tensor such that $\alpha_{\rho\sigma} \neq \alpha_{\sigma\rho}$. As shown by Equation 12.38, nonzero Σ^1 can lead to a depolarization ratio in excess of 3/4.

The explanation of anomalous polarization begins with the same Herzberg–Teller coupling scheme that led to Equation 12.30. The perturbed wavefunctions are written as in Equation 12.11, and the vibrational states are included in the picture. Let us suppose that the electronic transition $g \rightarrow e$ is x polarized while $g \rightarrow r$ is y polarized, leading to Raman activity of the fundamental transition of a nontotally symmetric vibration of symmetry $\Gamma_v = \Gamma_x \times \Gamma_y$. As shown above, this transition is resonant with the $|e0\rangle$ and $|e1\rangle$ intermediate states. When electronic state e is perturbed by vibronic coupling to state r, symmetry requires that $|e0\rangle$ mix with state $|r1\rangle$ and $|e1\rangle$ with state $|r0\rangle$. The perturbed intermediate states are thus

$$|e0\rangle' = |e0\rangle + \frac{\langle r1 | \left(\frac{\partial \hat{H}}{\partial Q} \right)_0 Q |e0\rangle}{\hbar\omega_{r1} - \hbar\omega_{e0}} |r1\rangle$$

$$|e1\rangle' = |e1\rangle + \frac{\langle r0 | \left(\frac{\partial \hat{H}}{\partial Q} \right)_0 Q |e1\rangle}{\hbar\omega_{r0} - \hbar\omega_{e1}} |r0\rangle \tag{12.42}$$

where $\hbar\omega_{r1}$ is the energy of the unperturbed state $|r1\rangle$, $\hbar\omega_{e0}$ is that of $|e0\rangle$, etc. Using the above states to write the Albrecht B term leads to the following xy and yx Raman tensor components:

$$(\alpha_{xy})_{01} = \frac{1}{\hbar^2} \frac{\langle r0 | \left(\frac{\partial \hat{H}}{\partial Q} \right)_0 |e1\rangle}{\omega_{r0} - \omega_{e1}} \left[\frac{(\mu_{ge}^0)_x (\mu_{gr}^0)_y \langle 0 | Q | 1\rangle}{\omega_{e1,g0} - \omega_0 - i\Gamma_e} \right] \tag{12.43}$$

$$(\alpha_{yx})_{01} = \frac{1}{\hbar^2} \frac{\langle r1 | \left(\frac{\partial \hat{H}}{\partial Q} \right)_0 |e0\rangle}{\omega_{r1} - \omega_{e0}} \left[\frac{(\mu_{ge}^0)_x (\mu_{gr}^0)_y \langle 1 | Q | 0\rangle}{\omega_{e0,g0} - \omega_0 - i\Gamma_e} \right] \tag{12.44}$$

The α_{xy} component dominates when the incident frequency is resonant with the $g0 \rightarrow e1$ transition, while the α_{yx} term is resonant with the $g0 \rightarrow e0$ transition. If the origins of state r and e are well separated compared to a typical vibrational energy, then $\omega_{r1} - \omega_{e0} \approx \omega_{r0} - \omega_{e1} \approx \omega_r - \omega_e$. If, in addition, ω_0 is far from resonance, then $\alpha_{xy} \approx \alpha_{yx}$. Thus in nonresonance Raman scattering, the Raman tensor is symmetric, Σ^1 vanishes, and $\rho = 3/4$ for a nontotally symmetric vibration. Close to resonance with the 0 – 1 transition, on the other hand, we have $\alpha_{xy} \gg \alpha_{yx}$, while for ω_0 close to the frequency of the 0 – 0 transition, $\alpha_{yx} \gg \alpha_{xy}$. When the energy of exciting radiation is midway between the 0 – 0 and 0 – 1 resonances, the relationship $\alpha_{xy} \approx -\alpha_{yx}$ results. This leads to $\Sigma^1 \neq 0$, while $\Sigma^2 = \Sigma^0 = 0$. Consequently, ρ goes to infinity. The result is that at particular excitation frequencies Raman bands for vibrations active in vibronic coupling show up in the depolarized but not the polarized

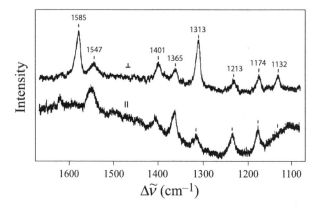

Figure 12.5 Anomalous polarization in the resonance Raman spectrum of ferrocytochrome c. (From Hamaguchi, H.: *Advances in Infrared and Raman Spectroscopy*. *1985*. Copyright Wiley-VCH Verlag GmbH & Co. KGaA. Reproduced with permission.)

spectrum. This is illustrated in the Raman spectrum of ferrocytochrome c in Figure 12.5. Note the 1585 cm^{-1} mode which is quite strong in the depolarized spectrum, but absent in the polarized. It is also apparent that the $\rho > 1$ for the 1313 cm^{-1} mode.

12.4 ROTATIONAL AND VIBRATIONAL DYNAMICS IN RAMAN SCATTERING

In this section, we examine the lineshape of a Raman transition using the time-correlation function (TCF) approach introduced in Chapter 5. The dynamics which take place in the ground electronic state determine the intensity as a function of frequency shift $\Delta v = v_0 - v_s$. For ease of illustration, we assume a molecule with axial symmetry (a linear or symmetric top molecule), for which the Raman tensor of a totally symmetric mode takes the form

$$\alpha = \begin{pmatrix} \alpha_\perp & 0 & 0 \\ 0 & \alpha_\perp & 0 \\ 0 & 0 & \alpha_\parallel \end{pmatrix} \tag{12.45}$$

The average polarizability is thus $\bar{\alpha} = (2\alpha_\perp + \alpha_\parallel)/3$. Let us define the anisotropy γ (a scalar quantity) as $\alpha_\perp - \alpha_\parallel$. We express these in terms of polarizability derivatives $\bar{\alpha}'$ and γ', where $\bar{\alpha} = \bar{\alpha}'Q$ and $\gamma = \gamma'Q$.

In Chapter 8, expressions were presented for the time-correlation functions $C_{VV}(t)$ and $C_{VH}(t)$, the Fourier transforms of which give polarized and depolarized rotational Raman scattering, respectively. The tensor components α_M^J of Section 8.6 are those for the molecule in its equilibrium geometry. Here, we reinterpret these α_M^J as components of the Raman tensor, so we can recycle Equations 8.67 and 8.71 to apply to vibrational Raman scattering. We also make use of the fact that the Wigner function $D_{00}^J(\theta\varphi\chi)$ is equivalent to the Legendre polynomial $P_J(\cos\theta)$. The required spherical tensor components (see Table 8.2) are

$$\alpha_0^0 = \sqrt{3}\bar{\alpha}$$

$$\alpha_0^2 = \frac{2}{\sqrt{6}}\gamma \tag{12.46}$$

The polarized and depolarized spectra and the corresponding time-correlation functions can be expressed in terms of isotropic and anisotropic components (see Section 8.6):

$$C_{VV}(t) = C_{iso}(t) + \frac{4}{3}C_{anis}(t)$$

$$C_{VH}(t) = C_{anis}(t)$$

(12.47)

Since the total intensity $I = \int I(\omega)d\omega$ integrated over the band is proportional to the TCF at zero time, we also have

$$I_{VV} = I_{iso} + \frac{4}{3}I_{anis}$$

(12.48)

$$I_{VH} = I_{anis}$$

(12.49)

Equations 12.48 and 12.49 again lead to the relationship $\rho = I_{VH}/I_{VV} \leq 3/4$. The isotropic and anisotropic TCFs are given by

$$C_{iso}(t) = (\bar{\alpha}')^2 \langle QQ(t) \rangle$$

(12.50)

$$C_{anis}(t) = (\gamma')^2 \langle QQ(t)P_2[\cos\theta(t)] \rangle$$

(12.51)

The angle $\theta(t)$ is that of the symmetry axis with respect to its direction at $t = 0$. More generally, the anisotropic TCF is given by

$$C_{anis}(t) = \langle \beta' \cdot \beta'(t)QQ(t) \rangle$$

(12.52)

where β' is the derived polarizability tensor from Equation 12.36.

Note that $C_{iso}(t)$ is independent of molecular orientation; the lineshape of the isotropic spectrum reflects only vibrational dynamics. The anisotropic spectrum, on the other hand, depends on both vibrational and reorientational dynamics. The angle brackets in Equations 12.50 and 12.51 indicate an equilibrium average over vibrational, rotational and translational degrees of freedom. In arriving at this result, it has been implicitly assumed that the motions of different molecules in the system are uncorrelated; for example, there are no pair terms $Q_1Q_2(t)$. These pair terms are important in liquids with strong orientation-dependent (e.g., dipolar) interactions, but are neglected here. If the vibrational and rotational motions are uncorrelated, the rotational and vibrational degrees of freedom can be separated, and the average in Equation 12.51 can be written

$$C_{anis}(t) = (\gamma')^2 \langle QQ(t) \rangle \langle P_2[\cos\theta(t)] \rangle$$

(12.53)

Equation 12.53 suggests a way to obtain the vibrational and reorientational TCFs from experiment. First, we use correlation functions $\tilde{C}(t) = C(t)/C(0)$ normalized to unity at $t = 0$. Then we can define the vibrational and rotational TCFs as follows:

$$\tilde{C}_{iso}(t) = \frac{\langle QQ(t) \rangle}{\langle |Q|^2 \rangle} \equiv C_{vib}(t)$$

(12.54)

and

$$\tilde{C}_{anis}(t) = C_{vib}(t)C_{rot}^{(2)}(t)$$

(12.55)

where the rotational correlation function is given by

$$C_{rot}^{(J)}(t) = \langle P_J[\cos\theta(t)] \rangle \tag{12.56}$$

(Note that the rotational TCF is automatically normalized to unity at time zero, since $P_2(\cos\theta) = 1$ when $\theta = 0$.) $C_{rot}^{(2)}(t)$ can be obtained from Raman scattering by Fourier transforming the isotropic (I_{iso}) and anisotropic (I_{anis}) spectra to get the corresponding correlation functions and then taking the ratio:

$$C_{rot}^{(2)}(t) = \frac{\tilde{C}_{anis}(t)}{\tilde{C}_{iso}(t)} \tag{12.57}$$

The infrared spectrum can be treated similarly. Still assuming a molecule with axial symmetry, the direction of the dipole moment derivative μ' coincides with the symmetry axis. The infrared TCF is

$$C_{IR}(t) = (\mu')^2 \langle QQ(t) \rangle \langle \hat{u} \cdot \hat{u}(t) \rangle = (\mu')^2 \langle QQ(t) \rangle \langle \cos\theta(t) \rangle \tag{12.58}$$

where $\hat{u}$ is a unit vector pointing along the symmetry axis. The normalized correlation function factors into vibrational and reorientational parts:

$$\tilde{C}_{IR}(t) = C_{vib}(t)C_{rot}^{(1)}(t) \tag{12.59}$$

There are numerous pitfalls to the determination of vibrational and rotational TCFs from Raman and infrared spectra. The neglect of vibration–rotation coupling is not valid if the vibrational motion greatly perturbs the moment of inertia, or if the intermolecular interactions are strongly dependent on orientation. In addition, theory suggests that vibrational contributions to infrared, isotropic Raman and anisotropic Raman linewidths differ [7]. Experiments have found different rotational relaxation times using different Raman lines of the same molecule [8]. This suggests that the separation of vibration and rotation in Equations 12.55 and 12.59 may not be valid even in the absence of strong vibration–rotation coupling. Also, the results of this section are based on the neglect of interaction-induced effects, which influence infrared, isotropic Raman, and anisotropic Raman spectra differently. Even when the separation of rotational and vibrational correlation functions is valid, practical considerations can limit experimental determination of TCFs. For example, the vibrational transition of interest must be well separated from other bands and should not be affected by isotope splittings or overlapping hot bands. More complicated rotational correlation functions result for molecules with less than axial symmetry (Equation 12.52, see [2]). We have also neglected inhomogeneous broadening, to be discussed below. In spite of these limitations, we can glean considerable insight about liquid-state dynamics from the analysis of Raman and infrared band shapes.

Let us now focus on the vibrational correlation function $\langle QQ(t) \rangle$ obtained from the isotropic Raman band. The topic of vibrational relaxation is still the subject of much current research. Nonlinear spectroscopic techniques, to be discussed in Chapter 14, are more well-suited to the determination of vibrational relaxation than is analysis of the band shapes of vibrational spectra. Here, we wish to consider how vibrational dynamics contribute to Raman and infrared spectra in general and reveal intermolecular interactions. The literature of vibrational relaxation distinguishes two different mechanisms, population relaxation and dephasing, introduced in Chapter 4, and their associated timescales T_1 and T_2. The rate at which the population of the excited vibrational state relaxes is $1/T_1$. Although this process includes radiative as well as nonradiative relaxation, the former is typically slow in vibrational spectra, and the major contribution to population decay is from nonradiative relaxation. In the gas phase, inelastic collisions lead to nonradiative decay of excited vibrational states. Elastic collisions, on the other hand, can perturb the phase of a vibrating molecule without depopulating the excited vibrational state. Loss of phase coherence leads to destructive interference and thereby contributes to decay of the vibrational TCF. This dephasing process perturbs the amplitude of the vibration and thus contributes to $C_{vib}(t)$. In the liquid phase, the idea of a collision is not so well defined, but we may still discuss vibrational relaxation in terms of T_1 and T_2 processes. Experimental observations have generally found T_1 to be larger than T_2, with T_2 in the range 10^{-10} to 10^{-12} s. Equation 4.100 leads to $T_2 \leq 2T_1$. The ratio of the two relaxation times varies a great deal. Perhaps the most extreme case is liquid N_2, where T_2 is 150 ps and T_1 is as long as 60 sec! (see [9,10]). For the methyl group symmetric stretch of CH_3CCl_3, on the other hand, T_1 is only about twice as large as T_2, the latter being 2.4 ps [11]. The qualitative explanation for this comparison is the following: In order for a vibrationally excited molecule to decay nonradiatively, it must lose energy to other degrees of freedom. In a diatomic, the vibrational energy is not so easily deposited into rotational and translational degrees of freedom because of the energy mismatch. Larger molecules provide more channels for the degradation of vibrational energy because other vibrational states, including overtones and combinations, can soak up the energy lost by the relaxing mode. Relaxation rates also depend on the environment due to the contribution of intermolecular energy transfer. Recall the exponential energy-gap law for nonradiative relaxation that was exemplified by the analysis of vibrational overtones of liquid Br_2, presented in Section 9.5.2.

If a vibrating molecule had a constant, undamped frequency ω_{vib}, then $Q(t)$ and $C_{vib}(t)$ would just be proportional to $\exp(i\omega_{vib}t)$, as shown in Problem 3. This would lead to a delta function spectrum: a spike at $\omega = \omega_{vib}$. In the condensed phase, however, the vibrational frequency fluctuates in time through interactions with other molecules, causing band broadening. We need to perform an average of this fluctuating frequency, using the Kubo model introduced in Chapter 6. Making an adiabatic separation of the vibrational and rotational–translational degrees of freedom, the vibrational frequency is an instantaneous function of molecular positions and orientations.* We can then consider the frequency to be a function of time, $\omega(t)$, and perform a time average inside the expression for the vibrational correlation function.

$$\langle QQ(t) \rangle \propto \left\langle \exp\left[i\int_0^t \omega(t')dt' \right] \right\rangle \tag{12.60}$$

(Equation 12.60 omits the scaling factor since we are interested in normalized correlation functions.) Now, Equation 12.60 is not a convenient expression to work with: it involves the equilibrium average of an exponential function. There is a way to replace such an average by a sum of exponential functions of what are called *cumulant averages,* but unfortunately it is an infinite series. (See [12] for a discussion of the cumulant expansion.) The good news is that it is sufficient to truncate the cumulant expansion. Considering $\omega(t)$ to be a stochastic variable, that is, a random function of time, it is valid to perform the cumulant average to second order. This gives rise to

$$\langle QQ(t) \rangle \propto \exp(i\omega_{vib}t)\exp\left[-i\int_0^t dt'(t-t')\langle \delta\omega\delta\omega(t')\rangle \right] \tag{12.61}$$

The quantity $\delta\omega(t) \equiv \omega(t) - \omega_{vib}$, and $\langle \delta\omega\delta\omega(t) \rangle \equiv M(t)$ (see Equation 6.50) is a time-correlation function for solvent-induced fluctuations in the vibrational frequency. These frequency fluctuations give rise to pure dephasing, with timescale T_2^*. Recall that pure dephasing is a loss of phase coherence without population relaxation, while the total dephasing rate $1/T_2$ includes population relaxation and pure dephasing. The quantity ω_{vib} is interpreted to be the average vibrational frequency; i. e., the first moment of the band. The model on which Equation 12.61 is based is physically and mathematically similar to the problem of Brownian motion. The vibrational frequency, like the position of a diffusing particle, is considered to be perturbed by random forces. The central limit theorem states that when a physical property is subject to a very large number of small perturbing influences, it behaves as a Gaussian random variable. This is the limit which leads to the second-order cumulant expression of Equation 12.61.

---

* The word adiabatic here is used in the same sense as in previous discussions of the separation of electronic and nuclear timescales. Here, we consider the *vibrational* motion to be much faster than *rotation and translation*, so that the vibrational frequency is a function of position and orientation. Since position and orientation depend on time, so does the vibrational frequency.

Calculation of $C_{vib}(t)$ using Equation 12.61 requires a functional form for $M(t)$. The frequently used Kubo lineshape function, introduced in Chapter 6, derives from assuming that the TCF for frequency fluctuations relaxes exponentially.

$$M(t) = \Delta^2 \exp(-t/\tau_c) \qquad (12.62)$$

where Δ is the amplitude and $\Lambda \equiv 1/\tau_c$ the rate of the frequency fluctuations.

Now we can write the vibrational TCF as follows:

$$C_{vib}(t) = \exp(i\omega_{vib}t)\varphi(t) \qquad (12.63)$$

where $\varphi(t)$ is the lineshape function of Equation 6.53. Note that, for times much less than τ_c, $\varphi(t)$ approaches a Gaussian function, while at long times the form is exponential.

The lineshape of isotropic Raman scattering is obtained by Fourier transforming Equation 12.63:

$$I_{iso}(\omega) = \frac{1}{2\pi} \int_{-\infty}^{\infty} dt \, \exp[-i(\omega - \omega_{vib})t]\varphi(t) \qquad (12.64)$$

It is apparent that the frequency variable in the Fourier transform is referenced to the average value ω_{vib}, which is the first moment of the intensity spectrum. For a symmetric band, ω_{vib} is the band center. Note that the time argument in $\varphi(t)$ should be interpreted as the absolute value $|t|$. Thus the intensity I_{iso} is an even function of $\omega - \omega_{vib}$. There are two limits in which Equation 12.64 results in a well-known functional form for the lineshape. If the fluctuations which modulate the frequency vary slowly compared to the amplitude; i.e., $\Lambda/\Delta \ll 1$, then $\varphi(t)$ goes over to a Gaussian function.

$$\varphi(t) \Rightarrow \exp(-\Delta^2 t^2/2), \text{ for } \Lambda/\Delta \ll 1 \qquad (12.65)$$

The isotropic lineshape in this limit is a Gaussian function centered at ω_{vib} and having a full width at half maximum (FWHM) of $2\sqrt{2\ln 2}\Delta$ (see Equation 6.46). Note that in this limit the rate Λ drops out, and the line broadening can be considered to result from a static distribution of frequencies with width Δ. This is the inhomogeneous broadening or *slow modulation* limit.

The opposite limit of *fast modulation* results when $\Lambda/\Delta \gg 1$, in which case an exponential TCF results.

$$\varphi(t) \Rightarrow \exp(-\Delta^2 t/\Lambda), \text{ for } \Lambda/\Delta \gg 1 \qquad (12.66)$$

On Fourier transforming Equation 12.66, a Lorentzian lineshape is obtained having a FWHM of Δ^2/Λ. In this limit, the spectral width is much less than the range of perturbed frequencies (see Figure 6.7). The fast fluctuations average out the spread in frequencies and the line is said to be motionally narrowed. This limit is exemplified by liquid-phase NMR spectra, where a similar analysis of spin relaxation can be employed. It is also often observed in vibrational spectra. For example, analysis of the isotropic Raman band of liquid N_2 has found Λ/Δ approximately equal to 30 [9,10]. In the intermediate range, the resulting lineshape is neither Gaussian nor Lorentzian, and the broadening is intermediate between homogeneous and inhomogeneous. The Kubo lineshape model permits the parameters Δ and Λ to be determined from the observed spectrum.

Figure 12.6 shows $C_{vib}(t)$ determined from the 2125 cm^{-1} isotropic Raman band of deuterated dimethyl-sulfoxide (DMSO-d$_6$) in the neat liquid and in solution with chloroform. The parameters τ_c and Δ obtained from fitting $C_{vib}(t)$ to the Kubo function $\varphi(t)$ are displayed in Table 12.2. The Raman band from which the data in Figure 12.6 were obtained is a symmetric methyl stretch, and lineshape analysis reveals that the slow modulation limit is approached, as Λ/Δ is on the order of 0.1. It is apparent that the amplitude Δ is rather insensitive to dilution, whereas the correlation time τ_c in the solution decreases compared to the value in the neat liquid. It is reasonable to speculate that dipolar interactions and hydrogen bonding (in the mixture) contribute to a lineshape that is close to the inhomogeneous limit.

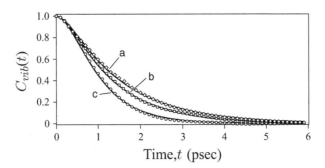

Figure 12.6 $C_{vib}(t)$ for the 2125 cm^{-1} line of DMSO-d$_6$ in CHCl$_3$, for mole fractions of DMSO: (a) 1.0, (b) 0.73, and (c) 0.53. The points are experimental data and the lines are calculated using the Kubo lineshape formula. (Courtesy of Dr. Douglas C. Daniel, PhD thesis, University of Idaho, 1996.)

Table 12.2 Kubo lineshape parameters for 2125 cm^{-1} line of DMSO-d$_6$

Mole fraction of DMSO-d6 in CHCl$_3$	τ_c, ps	Δ, ps^{-1}
1.00	0.50	26
0.73	0.33	26
0.53	0.30	25

There are situations where the source of spectral broadening is truly inhomogeneous, for example, in the glass phase where environmental perturbations persist for very long times. In liquid solutions, true inhomogeneous broadening can arise from naturally occurring isotopes, or from long-lived interactions such as hydrogen bonding. It is then necessary to consider the observed lineshape I_{inh} to result from a convolution of the homogeneous lineshape I_{homo} and the distribution function $P(\omega)$ of frequencies:

$$I_{inh}(\omega) = \int d\omega' P(\omega') I_{homo}(\omega - \omega') \qquad (12.67)$$

The normalized probability distribution $P(\omega)$ is often taken to be a Gaussian function, but other functional forms can also be associated with inhomogeneous broadening. From the theory of Fourier transforms, it is known that the FT of a convolution is the product of the FTs of the two convolved functions. Thus we can write $C_{vib}(t) = C_{inh}(t)C_{homo}(t)$, where $C_{inh}(t) = \int \exp(i\omega t)P(\omega)d\omega$. $C_{inh}(t)$ is not a true time-correlation function, just a convenient representation of inhomogeneous broadening. Time-domain nonlinear Raman echo experiments enable the homogeneous and inhomogeneous contributions to the lineshape to be separated [13].

12.5 ANALYSIS OF RAMAN EXCITATION PROFILES

We have seen that the KHD Equation 12.2 is capable of explaining resonance Raman selection rules through the Albrecht A and B terms. The KHD equation, also called the sum-over-states expression for the polarizability, has been applied to the analysis of Raman excitation profiles, in which the intensity (relative or absolute) of a Raman active vibration is determined as a function of excitation frequency. Electronic absorption and emission spectra can also be interpreted in terms of sums over Franck–Condon active vibrational states. The problem with the sum-over-states approach is that, in principle, it involves an infinite number of intermediate states. For small molecules excited with radiation resonant with a single electronic state, this is not an insurmountable problem. The vibrational overlaps eventually decrease for higher vibrational states and the sum can be truncated. But for large molecules, such as the visible chromophores which are often of interest in Raman spectroscopy, the sum-over-states approach of Equation 12.2 is unwieldy, as there are just too many terms in the sum. Especially in the condensed phase where individual eigenstates are not spectrally resolved, it is convenient to avoid assumptions about the intermediate states that go into the KHD equation. Here, we discuss two ways to avoid summing over eigenstates, both of which take advantage of the common features of Raman excitation and absorption profiles. The first of these is the transform method [14–16], and the second is a time-dependent approach due to Heller [17–20]. In either case, the goal is to account for the frequency-dependent Raman cross-section $\sigma_R(\omega_0)$ of a particular normal mode. To exploit the full power of either technique, absolute Raman cross-sections are required.

12.5.1 Transform theory of Raman intensity

The idea of transform theory is to use the information contained within the absorption spectrum to predict the form of the REP. The theory derives from the relationship of the absorption and resonance Raman intensity to the imaginary and real parts of the polarizability, respectively, which are in turn connected via the Kramers–Kronig transformation. The straightforward application of this theory is valid when a set of standard assumptions hold. These are (1) there is one resonant electronic state, (2) the Raman active vibrational modes are harmonic, (3) the vibrational frequencies of the active modes are the same in the ground and excited electronic states, (4) the symmetry of the molecule is the same in the ground and excited electronic states, and (5) there is no inhomogeneous broadening. These conditions restrict the approach to the treatment of totally symmetric (A-term enhanced) modes. The Raman cross-section is then given by

$$\sigma_R(\omega_0) = \frac{\omega_0 \omega_s^3 n^2}{8c^2\pi^3} \Delta_a^2 \left\| A(\omega_0) - A(\omega_0 - \omega_a) \right\|^2 \tag{12.68}$$

where ω_0 (ω_s) is the angular frequency of the incident (scattered) radiation, n is the refractive index, Δ_a is the dimensionless displacement (see Equation 11.19) of the active mode and ω_a its frequency. The function $A(\omega_0)$ is the transform of the absorption spectrum given by

$$A(\omega_0) = P \int_0^\infty \frac{\sigma_A(\omega)}{\omega(\omega - \omega_0)} d\omega + i\pi \frac{\sigma_A(\omega_0)}{\omega_0} \tag{12.69}$$

where P indicates the principal part of the integral. The absorption cross-section σ_A is related to the molar absorptivity ε_M through

$$\sigma_A = \frac{1000 \ln(10)\varepsilon_M}{N_A} \tag{12.70}$$

where ε_M is in L mol^{-1} cm^{-1}, N_A is Avogadro's number, and the quantity 1000 is the conversion from liters to cm^3, to give σ_A in units of cm^2 per molecule.

There are some important physical conclusions to be drawn at this point. The first is that the intensity of resonance Raman scattering scales as the square of the displacement of the mode in the excited electronic state. As our previous consideration of the Albrecht A term suggests, resonance Raman intensity is dependent on the Franck–Condon activity of the normal mode. Analysis of the resonance Raman spectrum readily provides information concerning the geometry change of the molecule in the excited electronic state, even when the absorption spectrum is too diffuse to reveal vibrational structure directly. Transform theory also reveals that diffuse (i.e., structureless) absorption profiles are less likely to be associated with strong resonance Raman scattering than those which reveal vibrational structure. The reason is that the Raman cross-section at excitation frequency ω_0 is dependent on the difference between the absorption cross-section at ω_0 and that at $\omega_0 - \omega_a$. A broad featureless absorption spectrum with a half-width large compared to ω_a is not very different when shifted by ω_a, so the transforms $A(\omega_0)$ and $A(\omega_0 - \omega_a)$ are similar. It is also apparent that the resonance Raman spectrum discriminates against low-frequency modes, again due to the small difference between $A(\omega_0)$ and $A(\omega_0 - \omega_a)$.

In practice, the transform theory is applied by calculating a predicted Raman profile $\sigma_R(\omega)$ from the observed absorption spectrum $\sigma_A(\omega)$. The displacement Δ_a is employed as a scaling factor to match the predicted to the observed absolute intensity. This requires determination of Raman cross-sections. Alternatively, if absolute Raman intensities are unavailable, the relative displacements of all Raman active modes can still be determined. One then fixes the values of the dimensionless displacements by adjusting their values to reproduce the width of the absorption spectrum. The use of Equation 12.68 is dependent on the validity of the Condon approximation, which assumes that the transition moment is independent of normal coordinate. If this assumption is not valid, the positions of the experimental and predicted REPs along the frequency axis differ. In the absence of a strong non-Condon contribution, the relative intensities of various normal modes are independent of excitation frequency, and the REP covers the same frequency range as the absorption spectrum. Champion and Albrecht [21,22] have shown how to modify the transform approach to allow for a transition moment which depends linearly on the normal coordinate, by making the following modification to Equation 12.68:

$$\left[A(\omega_0) - A(\omega_0 - \omega_a)\right] \Rightarrow \left[(1 + C_a)A(\omega_0) - (1 - C_a)A(\omega_0 - \omega_a)\right] \tag{12.71}$$

The non-Condon coefficient is defined by

$$C_a = \frac{\sqrt{2}\left(\partial\mu_{ge}/\partial Q_a\right)_0 \bar{Q}_a}{\mu_{ge}^0 \Delta_a} \tag{12.72}$$

where $\bar{Q}_a = \langle 0|Q_a^2|0\rangle^{1/2}$ is the root-mean-square displacement of the normal mode a. The effect of the non-Condon term is to shift the calculated REP to the red or the blue, according to whether the sign of C_a is positive or negative. If the signs of the displacements for all active modes can be determined, and if the normal coordinates are known, the change in internal coordinates (bond lengths and angles) on going from the ground to the excited state can be determined. This is difficult to realize in practice, due to the indeterminate sign of the displacement. For n Franck–Condon active modes, there are 2^n different excited state geometries consistent with the observed Raman data. In this case it is useful to deduce the sign of each displacement from quantum mechanical calculations on the excited electronic state.

An advantage of transform theory is that by taking into account the information within the absorption spectrum, the number of adjustable parameters is kept to a minimum. The calculated REP of a given Raman active mode includes the effects of all the other modes, which need not adhere to all the standard assumptions. The other modes may be anharmonically coupled, undergo frequency shifts, or contribute to the coordinate dependence of the transition moment. On the other hand, the existence of thermally populated initial states and inhomogeneous broadening can invalidate the direct application of transform theory. Champion and Albrecht [21,22] have discussed how to generalize the approach when these effects exert their influence.

Shreve et al. [15] have applied transform theory to the analysis of Raman excitation profiles of single-walled carbon nanotubes (CNTs) as illustrated in Figures 12.7 and 12.8. Carbon nanotubes are visualized as resulting from rolling a single-atom thick layer of graphite, called graphene, into a cylinder with a diameter on the order of a nanometer. The symmetry of the resulting tube is indexed by two integers (n,m) that specify the direction and length of the circumference of the nanotube as a combination of two lattice vectors that form a basis for the two-dimensional hexagonal graphene lattice. Raman spectroscopy and group theory have played important roles in accounting for the electronic (semiconducting versus metallic) and optical properties of carbon nanotubes. Synthesis of carbon nanotubes generally leads to a mixture of many different chiralities (values of n and m), with different diameters and different electronic properties. Until recently, separation of such mixtures has been notoriously difficult. In [15], the authors measured the Raman excitation profiles of the radial breathing modes and their first overtones for five different CNTs in their mixture. Observation of both the fundamental and overtone constrained the values of the dimensionless displacements. The radial breathing mode is a collective expansion and contraction of the nanotube in the radial direction. The frequency of this vibration decreases for increasing CNT diameter, leading to well-resolved Raman modes for different CNTs, ranging from about 150 to 300 cm⁻¹ as seen in Figure 12.7. The nanotubes in this mixture had diameters ranging, respectively, from about 1.2 to 0.88 nm. The fluorescence excitation

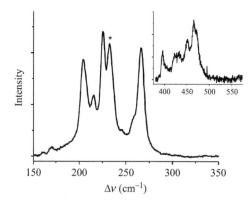

Figure 12.7 Resonance Raman spectrum of the radial breathing modes and their first overtones (inset) of a mixture of single-walled CNTs, excited at 766 nm. The asterisk marks the fundamental of the (12,1) nanotube at 236 cm⁻¹. (Reprinted with permission from Shreve, A. P. et al., *Phys. Rev. Lett. 98*, 037405, 2007. Copyright 2007 by the American Physical Society.)

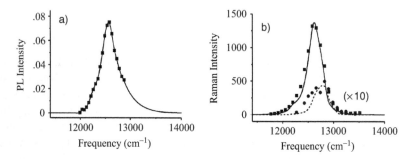

Figure 12.8 Application of transform theory to resonance Raman excitation profile of the radial breathing mode of the (12,1) CNT (see Figure 12.7). (a) PLE spectrum used as input to the transform theory calculation and (b) calculated Raman excitation profiles of the radial breathing mode fundamental (solid curve) and its first overtone (dashed curve), for the (12,1) CNT. The points are experimental Raman intensities. (Reprinted with permission from Shreve, A. P. et al., *Phys. Rev. Lett., 98*, 037405, 2007. Copyright 2007 by the American Physical Society.)

and emission spectra of CNTs with different chirality are well-separated. By monitoring the intensity of emission for a particular (n,m) nanotube as a function of excitation frequency, the authors obtained photoluminescence excitation spectra (PLE) of the distinct chiralities of CNTs in their mixture. The authors used the PLE spectra in lieu of the absorption profiles as input to the transform theory calculation. Figure 12.8 shows the result of their analysis for the (12,1) CNT, for which the radial breathing mode and its overtone are found at 236 and 473 cm⁻¹ respectively. It was found necessary to include non-Condon terms as discussed earlier. The authors determined dimensionless displacements for the radial breathing modes of the five major chiralities present in their sample.

12.5.2 TIME-DEPENDENT THEORY OF RESONANCE RAMAN AND ELECTRONIC SPECTRA

The time-dependent theory developed by Heller and coworkers [17–20] is an approach to the direct modeling of resonance Raman excitation and electronic absorption/emission spectral profiles. This technique presented here goes beyond the time-correlation function method described in Chapter 5 in that it accounts for spectra involving more than one potential surface. Thus, while the TCF approach described in Chapter 5 is applicable to the analysis of lineshapes of Raman spectra, the theory described in this section is employed to interpret Raman excitation profiles as well as absorption and emission spectra. Applications of time-dependent theory have often employed semiclassical approaches to describe the motion of "wave packets" on excited state potential energy surfaces. A wave packet is a superposition of plane waves, the center of which follows a classical trajectory in position and momentum space. The resulting pictures of excited state dynamics have a great deal of physical appeal, particularly in systems that undergo charge transfer, photodissociation, or photoisomerization. The theory readily lends itself to computational techniques developed for semiclassical systems. Let us derive the working equations and see how the theory leads to informative physical pictures.

We start with the Albrecht A term contribution to the transition polarizability for the $i \rightarrow f$ transition:

$$\alpha_{if}(\omega_0) = \frac{(\mu_{ge}^0)^2}{\hbar} \sum_v \left[\frac{\langle f|v\rangle\langle v|i\rangle}{\omega_{eg} + \omega_v - \omega_i - \omega_0 - i\Gamma} \right] \tag{12.73}$$

It is assumed that there is a single nondegenerate resonant electronic state e. The indices i and f denote the initial and final vibrational quantum numbers within the ground electronic state and the energy of the electronic transition is $\hbar\omega_{eg}$. The energy of the intermediate vibrational state is $\hbar\omega_v$, and $\hbar\omega_i$ is the energy of the initial vibrational state. The damping term Γ is taken to be independent of the vibrational state. The transition polarizability α_{if} is sometimes called the Raman amplitude; the square of its absolute value (square modulus) is proportional to the Raman cross-section.

The first trick is to recognize that the energy denominator is related to a half-Fourier transform.

$$\frac{1}{i(\omega - i\Gamma)} = \int_0^\infty e^{-i(\omega - i\Gamma)t} dt \tag{12.74}$$

Using this in Equation 12.73 results in

$$\alpha_{if}(\omega_0) = \frac{i(\mu_{ge}^0)^2}{\hbar} \int_0^\infty dt \sum_v \langle f|v\rangle\langle v|i\rangle \exp[-i(\omega_{eg} + \omega_v - \omega_i - \omega_0 - i\Gamma)t] \tag{12.75}$$

Since the state $|v\rangle$ is an eigenket of the excited electronic state, we can write

$$\langle v|\exp[-i(\omega_{eg} + \omega_v)t] = \langle v|\exp\left[-\frac{i}{\hbar}(E_{00} + E_v)t\right] = \langle v|\exp\left[-\frac{i\hat{H}_e t}{\hbar}\right] \tag{12.76}$$

Taking advantage of the resolution of the identity and using the time-evolution operators introduced in Chapter 4, we obtain

$$\sum_v \langle f|v\rangle\langle v|\exp\left[-\frac{i\hat{H}_e t}{\hbar}\right]|i\rangle = \langle f|i(t)\rangle \tag{12.77}$$

Note that the wavefunction for the initial state i evolves in time according to the Hamiltonian for the excited electronic state. We say that $|i(t)\rangle$ evolves on the excited state potential surface. Putting Equation 12.77 into 12.75 results in

$$\alpha_{if}(\omega_0) = \frac{i\left(\mu_{ge}^0\right)^2}{\hbar} \int_0^\infty \langle f|i(t)\rangle \exp[i(\omega_i + \omega_0)t] \exp[-\Gamma t] dt \tag{12.78}$$

Equation 12.78 gives the single nonzero diagonal component of the transition polarizability in the molecule frame, let us say that it is α_{yy}. With the help of Equations 12.32 and 12.33, we obtain the lab-frame component $|\alpha_{ZZ}|^2 = \alpha_{yy}^2/5$ related to the polarized intensity. Since the depolarization ratio is 1/3, the total differential cross-section is

$$\frac{d\sigma_{if}}{d\Omega} = \left(\frac{d\sigma_{if}}{d\Omega}\right)_\parallel + \left(\frac{d\sigma_{if}}{d\Omega}\right)_\perp = \frac{4}{3}\left(\frac{d\sigma_{if}}{d\Omega}\right)_\parallel \tag{12.79}$$

Using Equation 6.44 for the total cross-section gives

$$\sigma_R = \frac{8\pi}{3}\left(\frac{5}{3}\right)\left(\frac{d\sigma_{if}}{d\Omega}\right)_{\parallel} \tag{12.80}$$

Combining Equation 12.80 with Equation 12.1 and using Equation 12.78 for α_{yy}, we obtain the Raman cross-section in the time-dependent theory:

$$\sigma_{R,i\to f}(\omega_0) = \frac{8\pi\omega_s^3\omega_0\left(\mu_{ge}^0\right)^4}{9\hbar^2c^4}\left|\int_0^\infty \langle f|i(t)\rangle \exp[i(\omega_i+\omega_0)t]\exp[-\Gamma t]dt\right|^2 \tag{12.81}$$

Equation 12.81 is based on the Condon approximation, but is readily generalized to allow μ_{ge} to depend on normal coordinates. It can also be extended to allow for a thermal distribution of initial states and inhomogeneous broadening, as shown below. The Raman cross-section is represented as the square modulus of the half-Fourier transform of the damped time-dependent overlap of the final state and the initial state as the latter evolves on the excited state surface. This is easier to grasp with the help of a picture such as that of Figure 12.9.

In Figure 12.9, the incident photon is envisioned to promote the ground state wavefunction to the excited state surface. The v = 0 wavefunction behaves as a Gaussian wave packet, which evolves on a harmonic surface with no change in width. At $t = 0$, the wave packet finds itself in the Franck–Condon region of the displaced upper potential surface, and the overlap $\langle f|i(0)\rangle$ with the final state starts out at zero. The initial state is not an eigenfunction of the excited state potential, so it must evolve in time. Classically, the wave packet experiences a force and begins to move to the right as depicted in Figure 12.9. In the absence of damping, the overlap $\langle f|i(t)\rangle$ would oscillate in time at the frequency of the vibration. The damping term $\exp(-\Gamma t)$, however, causes the envelope of the time-dependent overlap to die off. In addition, the overlaps are in fact multimodal, and dynamics along other normal coordinates also contribute to moving the wave packet under consideration away from the Franck–Condon region, causing the overlap to decay. Note that modes which are not displaced in the excited state do not undergo enhancement, in agreement with the sum-over-states approach. The absence of a slope in the FC region of a nondisplaced mode means that the wavepacket will not evolve in time, and the overlap $\langle f|i(t)\rangle$ for $f \neq i$ remains zero at all times. In the case where the transition $i \to f$ is an overtone, the higher the overtone is, the longer it takes for the initial state to evolve to a point where the overlap is favorable. The larger the molecule is, the greater the chance that dynamics along other normal coordinates will move the wave packet out of the FC region before favorable overlap with highly excited final states can occur. This leads to the previously noted observation that overtones are weak or absent in resonance Raman spectra of large molecules.

Equation 12.81 is the working equation for applying the time-dependent theory, and more will be said about it shortly. First, we see how the same theoretical approach may be applied to the electronic absorption

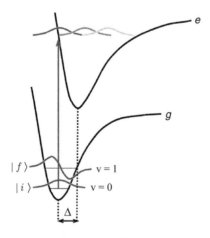

Figure 12.9 Wave packet dynamics on a displaced excited state potential surface.

spectrum. At the same level of approximation on which the Albrecht A term is based, the absorption cross-section can be written as

$$\sigma_A(\omega_0) = \frac{4\pi^2 \left(\mu_{ge}^0\right)^2 \omega_0}{3cn\hbar} \sum_v \frac{\Gamma}{\pi} \frac{\left|\langle i|v\rangle\right|^2}{(\omega_{eg} + \omega_v - \omega_i - \omega_0)^2 + \Gamma^2} \tag{12.82}$$

The absorption spectrum is expressed as a series of Lorentzian lines centered at the transition frequencies of the peaks in the Franck–Condon progression. The quantity $|\langle i|v\rangle|^2$ is the FC factor introduced in Chapter 11. Now, using the same tricks as were employed to arrive at Equation 12.78, the time-dependent expression for the absorption cross-section is

$$\sigma_A(\omega_0) = \frac{2\pi \left(\mu_{ge}^0\right)^2 \omega_0}{3\hbar cn} \int_{-\infty}^{\infty} \langle i|i(t)\rangle \exp\left[i(\omega_i + \omega_0)t\right] \exp\left[-\Gamma|t|\right] dt \tag{12.83}$$

Note that σ_A is proportional to a full Fourier transform, and the relevant overlap integral is between the time-evolving initial state and the initial state at time zero. Equation 12.83 is readily generalized to allow for an equilibrium distribution of initial states:

$$\sigma_A(\omega_0) = \frac{2\pi \left(\mu_{ge}^0\right)^2 \omega_0}{3\hbar cn} \sum_i P_i \int_{-\infty}^{\infty} \langle i|i(t)\rangle \exp\left[i(\omega_i + \omega_0)t\right] \exp\left[-\Gamma|t|\right] dt \tag{12.84}$$

where P_i is the Boltzmann probability that initial state i is occupied. Of course, Equation 12.81 can be modified in the same way:

$$\sigma_{R,i\rightarrow f}(\omega_0) = \frac{8\pi\omega_s^3\omega_0 \left(\mu_{ge}^0\right)^4}{9\hbar^2 c^4} \sum_i P_i \left|\int_0^{\infty} \langle f|i(t)\rangle \exp\left[i(\omega_i + \omega_0)t\right] \exp\left[-\Gamma t\right] dt\right|^2 \tag{12.85}$$

Next, we want to make a connection between the absorption cross-section as given in Equation 12.84 and the time-correlation function formula described in Chapter 5. We do this by recasting the expression as follows (essentially we are reversing some of the steps that led to 12.84):

$$\sigma_A(\omega_0) = \frac{2\pi\omega_0}{3\hbar cn} \sum_i P_i \int_{-\infty}^{\infty} \langle gi|\exp(i\hat{H}_g t/\hbar)\mu\exp(-i\hat{H}_g t/\hbar)\mu|gi\rangle \exp(i\omega_0 t)\exp(-\Gamma|t|) dt \tag{12.86}$$

In the case where the ground and excited state Hamiltonians are the same (as in rovibrational spectra where transitions take place within the ground electronic surface), we can write:

$$\sum_i P_i \langle gi|\exp(i\hat{H}_g t/\hbar)\mu\exp(-i\hat{H}_g t/\hbar)\mu|gi\rangle = \langle \mu(t)\mu\rangle_{eq} = \langle \mu\mu(-t)\rangle_{eq} \tag{12.87}$$

The subscript eq represents an equilibrium-averaged quantity, as in the expression for the time-correlation function $C(t)$ discussed in Chapter 5:

$$C(-t) \equiv \langle \mu\mu(-t)\rangle_{eq} = \sum_i P_i \langle gi|\mu\mu(-t)|gi\rangle \tag{12.88}$$

The absorption cross-section in this limit is given by

$$\sigma_A(\omega_0) \propto \int_{-\infty}^{\infty} C(t)\exp(-i\omega_0 t)\exp(-\Gamma|t|) dt \tag{12.89}$$

To obtain this expression, the integration variable t was replaced by $-t$, in order to compare the result to that of Equation 5.15. Except for the presence of the damping factor, the present approach leads to an identical result to that presented in Chapter 5, in the case where the transition takes place within a single electronic surface.

There are important similarities and distinctions between the time-dependent expressions for absorption and resonance Raman cross-section. Obviously, the absorption strength scales as the square of the transition moment, and the Raman intensity depends on the transition moment to the fourth power. While the Raman amplitude depends on the time-dependent overlap of the initial and final states, the absorption strength depends on the overlap of the time-evolving initial state with the same state at time zero. The fact that the absorption is given by a full Fourier transform compared to the half-Fourier transform needed for the Raman amplitude also has nontrivial consequences. A full Fourier transform can be directly inverted; e.g., the damped time-dependent overlap can be determined from Fourier inversion of the absorption cross-section. This is not possible for the half-Fourier transform. Increased damping (larger Γ) has the effect of broadening the absorption spectrum while conserving its total intensity, since the integrated absorption cross-section depends only on the time-zero value of the correlation function. Resonance Raman intensities, on the other hand, are diminished in the presence of more rapid damping.

The time evolution of the initial state in either expression is determined from the characteristics of the excited state potential. General procedures are available for evolving wave packets on arbitrary potential energy surfaces. In the case of separable harmonic ground and excited state surfaces, closed-form expressions exist for $\langle i|i(t)\rangle$ and $\langle f|i(t)\rangle$, as shown below. The calculation of these overlaps requires the displacement of each Franck–Condon active mode and the vibrational frequencies in the ground and excited electronic state. Thus a valid set of these parameters must simultaneously reproduce the absorption spectrum and all the Raman profiles. Similarly, the damping factor Γ and the transition moment μ_{ge}^0 are common to both σ_A and σ_R.

We are now in a position to examine the various contributions to $\langle i|i(t)\rangle$ and $\langle f|i(t)\rangle$ more closely. The qualitative features of these time-dependent overlaps are illustrated in Figure 12.10. The theory of Fourier transforms teaches us that fast relaxation in the time domain leads to broad features in the frequency spectrum, while the slowest time response influences the finer details in the frequency domain. Three characteristic timescales are depicted in Figure 12.10: the times $t_1 < t_2 < t_3$. The fastest event in $\langle i|i(t)\rangle$, on the timescale t_1, is the initial decrease in the overlap as the wave packet moves out of the FC region. This initial decay determines the overall width of the absorption spectrum, and it depends on the slope of the upper state potential in the Franck–Condon region. The greater the displacement, the steeper is the slope and the greater is the breadth of the absorption spectrum. We have already seen this correspondence in our previous analysis of Chapter 11. The intermediate time t_2 is the time between recurrences in the overlap. It is therefore equal to the reciprocal of the excited state vibrational frequency. The periodic structure in the overlap gives rise to vibrational structure in the absorption spectrum and Raman profile. The slowest timescale is that for the overall decay of the overlap. It is dictated by the damping factor Γ; the larger the value of Γ, the faster is the decrease in the envelope of $\langle i|i(t)\rangle$ and the broader the width of individual FC peaks in the absorption spectrum. Note that as the damping rate increases, the area under the Raman overlap $\langle f|i(t)\rangle$ decreases, and the Raman intensity decreases. For large enough damping, whatever the cause, the vibrational structure is completely blurred and an unstructured absorption profile results. Again, as predicted by transform theory, unstructured absorption profiles correlate with weaker Raman scattering. Note that the total absorption intensity is independent of Γ; as the damping rate increases, the absorption spectrum broadens but the total intensity is conserved. The different dependence of Raman and absorption intensity on the homogeneous linewidth allows this property to be distinguished from inhomogeneous line-broadening factors, which have a similar effect on the Raman and absorption profiles.

Although it is not apparent from the simplified notation that we are using, the overlaps are multidimensional functions of the normal coordinates. Let us make the convenient assumption that the vibrational modes are harmonic, so that they can be separated. For example, let the overlap $\langle f|i(t)\rangle$ be that required to calculate the $0 \to 1$ transition of mode a, keeping all other modes in the ground vibrational state. The overlap factors as follows:

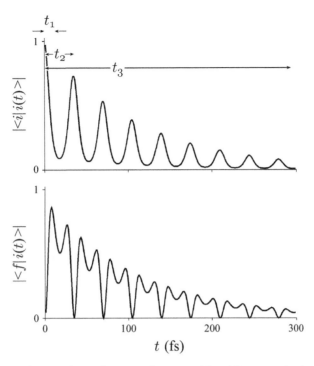

Figure 12.10 Time-dependent overlaps relevant to absorption (a) and Raman excitation profile (b).

$$\langle f\,|\,i(t)\rangle = \langle 0_1\,|\,0_1(t)\rangle\langle 0_2\,|\,0_2(t)\rangle\ldots\langle 1_a\,|\,0_a(t)\rangle\ldots\langle 0_{3N-6}\,|\,0_{3N-6}(t)\rangle$$
$$= \langle 1_a\,|\,0_a(t)\rangle\prod_{k\neq a}\langle 0_k\,|\,0_k(t)\rangle \tag{12.90}$$

The bra vectors here are $t = 0$ vibrational states within the ground electronic surface and the ket vectors propagate on the excited electronic surface. The subscript indexes the normal mode. It would appear that all $3N - 6$ modes need to be included in the calculation, but it turns out that only the displaced modes contribute. To see this, consider the overlaps in the case where the normal modes undergo no frequency shift upon excitation [23]:

$$\langle 0_k\,|\,0_k(t)\rangle = \exp\left[-\frac{\Delta_k^2}{2}(1-e^{-i\omega_k t})-\frac{i\omega_k t}{2}-i\omega_{eg}t\right]$$
$$\langle 1_k\,|\,0_k(t)\rangle = \frac{\Delta_k}{\sqrt{2}}\left(e^{-i\omega_k t}-1\right)\langle 0_k\,|\,0_k(t)\rangle \tag{12.91}$$

The displacement and frequency of mode k are denoted by Δ_k and ω_k. The 0 – 0 frequency ω_{ge} appears only once in the product over $\langle 0_i|0_i(t)\rangle$. Some authors prefer to factor $\exp(i\omega_{eg}t)$ out of the overlap and include it with the other frequency factors in the Fourier transform. Also, the $\exp(-i\omega_k t/2)$ parts conspire to cancel the $\exp(i\omega_i t)$ term in the Fourier transform, since $\hbar\omega_i = \sum_k \hbar\omega_k/2$ is the initial vibrational energy. That leaves us with an exponential function of the square of the displacement in Equation 12.91, which is replaced by unity for modes which are not Franck–Condon active. More general expressions for the time-dependent overlaps can be found in [23] and references therein.

Recall that Γ represents the homogeneous linewidth of the intermediate state. As such, it ought to allow for the exponential population relaxation of the resonant electronic state. In condensed phases, the dephasing of the electronic transition also contributes to the linewidth, and though the dephasing rate should contribute to the damping, it does not necessarily lead to a Lorentzian lineshape. In addition, inhomogeneous broadening, perhaps resulting in a Gaussian distribution of electronic transition energies, ought to be included in the calculation. The homogeneous and inhomogeneous line broadening mechanisms affect Raman and absorption profiles differently. Thus the time-dependent analysis of these spectra enables the two contributions to be disentangled. The inhomogeneous contribution to σ_A and σ_R can been accounted for by replacing ω_{eg} in Equations 12.91 by $\omega_{eg} + \delta\omega$, and averaging (integrating) the resulting expressions over a Gaussian distribution of solvent-induced frequency shifts $\delta\omega$. This incorporates the effects of solvent perturbations to the linewidth which are long-lived compared to the relaxation time of the initial vibrational state. On the other hand, if electronic dephasing is fast compared to vibrational relaxation, then frequency fluctuations need to be incorporated into the Raman amplitude rather than the cross-section. See [24,25] for further discussion. A modified form of the Kubo lineshape function has been proposed. Mukamel [26] and coworkers have employed a complex lineshape function $g(t) = g'(t) + ig''(t)$ based on the Brownian oscillator model of solvent dephasing. The real part of this function derives from the previously discussed Kubo lineshape function, where Δ and Λ are interpreted here as the amplitude and pure dephasing rate of *electronic* frequency (ω_{eg}) fluctuations. This function $\varphi(t)$ in Equation 6.53 is the same as $\exp[-g'(t)]$. The imaginary part of $g(t)$ is found using the fluctuation-dissipation theorem, and it accounts for the solvent-induced shift in the electronic absorption spectrum (see [26]):

$$g''(t) = \frac{\Delta E_{solv}}{\Lambda}\left[1 - \exp(-\Lambda t)\right] \tag{12.92}$$

The solvent reorganization energy ΔE_{solv} is equal to $\Delta^2/2k_BT$ in this model, where Δ is the displacement of the excited state potential, relative to the ground, along the solvent coordinate. The revised expressions for the absorption and Raman cross-sections are

$$\sigma_{R,i\to f}(\omega_0) \propto \sum_i P_i \int d(\delta\omega)G(\delta\omega)$$

$$\times \left| \int_0^\infty \langle f|i(t)\rangle \exp[i(\omega_i + \omega_0 + \delta\omega)t]\exp[-g(t)]\exp[-\Gamma t]dt \right|^2 \tag{12.93}$$

and

$$\sigma_A(\omega_0) \propto \sum_i P_i \int d(\delta\omega)G(\delta\omega)$$

$$\times \int_{-\infty}^\infty \langle i|i(t)\rangle \exp[i(\omega_i + \omega_0 + \delta\omega)t]\exp[-g(|t|)]\exp[-\Gamma|t|]dt \tag{12.94}$$

where $G(\delta\omega)$ is a normalized Gaussian distribution function whose width can be taken as an adjustable parameter. The electronic dephasing rate Λ and amplitude Δ can be obtained by using the above expressions to model the experimental absorption and Raman profiles. Implicit in these equations is the assumption that the solvent dynamics which perturb the ground state electronic energy are the same as those which modulate the energy of the excited electronic state. In systems with strong solvent–solute interactions and large change in solute electronic structure, this assumption is not valid [27–29].

The wave packet approach can also be applied to fluorescence emission spectra. The fluorescence intensity I_f in photons per second is

$$I_f \propto |\mu_{ge}|^2 \omega_0^3 \sum_i P_i \int d(\delta\omega)G(\delta\omega)$$

$$\times \int_{-\infty}^\infty \langle i|i(t)\rangle \exp[i(\omega_i - \omega_0 + \delta\omega)t]\exp[-g(t)]\exp[-\Gamma|t|]dt \tag{12.95}$$

The initial states are now taken to be vibrational levels within the upper electronic state, which evolve on the ground state potential energy surface under the influence of $\exp(-i\hat{H}_g t/\hbar)$. The populations of the initial states adhere to the Boltzmann distribution only for relaxed emission. As will be considered in Chapter 13, it turns out that the Raman and fluorescence emission both spring from a common theoretical foundation, and in some experiments they are not separable.

Both transform theory and time-dependent theory of resonance Raman excitation profiles spring from the dependence of Raman and absorption spectra on a set of shared physical parameters. However, for chromophores that are strongly coupled to the solvent, linear solvation response may not hold. The observation of strong solvent effects on Raman intensities, and the absence of similar solvent effects on the absorption spectrum, belie the assumptions on which common implementations of these theories are based. Such effects have been seen in absorption spectra and resonance Raman spectra of molecules with charge-transfer transitions [27–29]. When the resonant state has a very different charge distribution than the ground, the shape of the excited state potential as a function of "solvent coordinate" can differ from that of the ground state. Further, the theories we have discussed consider that solvent and internal reorganization are uncoupled. The inertial solvent response on the subpicosecond timescale runs concurrently with intramolecular reorganization, facilitating the possibility of coupling of inter- and intra-molecular reorganization.

12.6 SURFACE-ENHANCED RAMAN SCATTERING

Surface-enhanced Raman spectroscopy (SERS), first reported in the 1970s, continues to fascinate researchers eager to apply it to trace analysis and even single molecule detection. The SERS effect can be defined as a dramatic enhancement—by orders of magnitude—of Raman cross-sections of molecules adsorbed at conductive substrates. Early observations of SERS effects were seen in the intense enhancement of the Raman spectra of molecules adsorbed at roughened silver electrodes. Raman intensity enhancement has now been observed for molecules adsorbed on a variety of nanoscale metal substrates, such as colloidal suspensions, metal island films, and lithographically produced nanoparticle arrays. The technique of tip-enhanced Raman spectroscopy exploits the effect in a scanning probe configuration, without the requirement that the analyte be chemically or physically adsorbed on the metal. The emphasis on SERS enhancement by the coinage metals, Cu, Ag, and Au, hints at the role of electrical conductivity in this phenomenon. Ni, Pd, Pt and the alkali metals also give rise to Raman enhancement. The bulk of experimental observations concerning surface-enhanced Raman, and other related linear and nonlinear surface-enhanced spectroscopies, can be explained handily by electromagnetic theory. Nevertheless, vigorous discussion continues on the importance of the so-called chemical enhancement mechanism, a sort of resonance enhancement deriving from a new coupled electronic state of the adsorbate and substrate. The observation of modest Raman intensity enhancements of molecules adsorbed on semiconductor surfaces argues that there is some contribution from this mechanism. For Ag and Au substrates, on the other hand, SERS enhancement factors as large as 10^{10} can be achieved, though $\sim 10^6$ is more typical. Along with other observations discussed below, these enhancement factors and theoretical treatments confirm that SERS effects derive from enhanced electric fields associated with the surface plasmon resonance (SPR), a collective oscillation of free electrons in the metal. This transition dipole associated with this resonance radiates an electromagnetic field which is strong in the vicinity of the surface of a metal nanostructure. The electromagnetic enhancement theory of SERS is considered further in this section.

The surface plasmon resonance can be described in a classical picture, called the Drude model, in which free electrons of the metal oscillate against the fixed lattice of positively charge nuclei. This simple model is a periodic extension of the Lorentz model of the atom, treated in Chapter 3. Therefore, we use the expression for the polarizability $\alpha(\omega)$ given in Equation 3.43. Taking the polarization as $\vec{P} = N\alpha\vec{E} = \varepsilon_0(\varepsilon_r - 1)\vec{E}$, where N is the number density of electrons, the Lorentz model leads to the relative permittivity as a function of frequency:

$$\varepsilon_r(\omega) = 1 + \frac{Ne^2}{m\varepsilon_0}\sum_j \frac{f_j}{\omega_j^2 - \omega^2 - i\Gamma\omega} \tag{12.96}$$

We are interested in frequencies in the vicinity of the SPR. We take the oscillator strength $f_j = 1$ and define $\omega_p = (Ne^2/m\varepsilon_0)^{1/2}$ as the bulk plasma frequency to obtain

$$\varepsilon_r(\omega) = 1 + \frac{\omega_p^2}{\omega_0^2 - \omega^2 - i\Gamma\omega} \approx 1 - \frac{\omega_p^2}{\omega(\omega + i\Gamma)} \tag{12.97}$$

The approximation above results from recognizing that the weak binding of conduction electrons to nuclei translates to a small harmonic frequency, ω_0. The damping constant Γ in this case is the electron scattering rate and is inversely proportional to the conductivity of the metal. The Drude model can be made more general by adding to the right-hand side of Equation 12.97 a background permittivity ε_b that depends less sharply on frequency than the SPR, taking into account interband transitions in the metal:

$$\varepsilon_r(\omega) = \varepsilon_b + 1 - \frac{\omega_p^2}{\omega(\omega + i\Gamma)} \tag{12.98}$$

In bulk metal crystals the SPR is forbidden, owing to a selection rule on the conservation of momentum. Note that the k-dependence of the electromagnetic field, neglected for molecules small compared to the wavelength, must be considered when treating the collective excitations of crystalline materials. The selection rules are relaxed for nanoparticles, and for the roughened electrodes on which SERS was first observed. The SPR then becomes allowed and leads to an enhanced field near the particle when it is illuminated at incident frequencies determined by ω_p, as illustrated below. The frequency and width of the resonance depends on the shape of the nanoparticle. For the case of a spherical metal nanoparticle of radius R, the polarizability is given by

$$\alpha(\omega) = 4\pi\varepsilon_0 R^3 \frac{\varepsilon_r(\omega) - 1}{\varepsilon_r(\omega) + 2} \tag{12.99}$$

Using Equation 12.98 for the relative permittivity leads to:

$$\alpha(\omega) = 4\pi\varepsilon_0 R^3 \frac{(\varepsilon_b\omega^2 - \omega_p^2) + i\omega\Gamma\varepsilon_b}{(\varepsilon_b + 3)\omega^2 - \omega_p^2 + i\omega\Gamma(\varepsilon_b + 3)} \tag{12.100}$$

The polarizability is enhanced for frequencies near $\omega_R = \omega_p/\sqrt{\varepsilon_b + 3}$, the frequency of the dipolar surface plasmon. For Cu, Ag and Au, $\hbar\omega_p$ is on the order of 9 eV, but ω_R is in the visible. The SPR of silver is typically in the vicinity of 3.5 eV (350 nm), while that of gold is lower energy, about 2.4 eV (520 nm). Cu, though less often used as a SERS substrate owing to poor stability and weaker enhancement, has its SPR at a visible wavelength. The width of the resonance is $\Gamma(\varepsilon_b + 3)$. Electromagnetic theory predicts enhancement that is stronger when the width of the SPR is small compared to the peak frequency. Thus, better electrical conductivity fosters a more narrow resonance and stronger SERS effects. It is also desirable for the background permittivity ε_b to be small. Poor conductivity and the contribution of interband transitions to ε_b result in a broader resonance and less enhancement. This explains why transition metals other than the coinage metals, which may be good conductors but have significant d–d interband transitions, are relatively poor SERS substrates.

Continuing with the idea of a single spherical metal particle, we can explore how the SPR leads to enhancement of the field. For an incident field $\vec{E}_0$ in the z direction, the resulting field outside the metal sphere is

$$\vec{E}_{out} = E_0\hat{k} - 4\pi\alpha\varepsilon_0 E_0 \left[\frac{\hat{k}}{r^3} - \frac{3z}{r^5}(x\hat{i} + y\hat{j} + z\hat{k}) \right] \tag{12.101}$$

We rewrite the polarizability α by modifying 12.99 to account for a surrounding medium with relative permittivity given by ε_s:

$$\alpha(\omega) = 4\pi\varepsilon_0 R^3 \frac{\varepsilon_r - \varepsilon_s}{\varepsilon_r + 2\varepsilon_s} \equiv (4\pi\varepsilon_0 R^3)g \tag{12.102}$$

It must be remembered that the permittivities, and thus g, where $g = (\varepsilon_r - \varepsilon_s)/(\varepsilon_r + 2\varepsilon_s)$, are complex and frequency-dependent quantities. Equation 12.102 predicts enhanced polarizability (large g) at frequencies such that ε_r is close to $-2\varepsilon_s$. The resulting induced dipole radiates a field which is greatly enhanced when this condition is met. Since $\varepsilon_s \approx n^2$, where n is the (real) refractive index of the solvent, enhanced electric fields require that the relative permittivity of the metal particle be negative, a condition adhered to by typical SERS substrates at optical frequencies. The relative permittivity ε_r for a metal tends to negative infinity at zero frequency and becomes less negative as frequency increases, as per Equation 12.97.

Putting r equal to R and using the Drude model for the permittivity, the field on the surface of the sphere is

$$E_{out}^2 = E_0^2 \left[|1-g|^2 + 3\cos^2\theta \left(2\,\mathrm{Re}(g) + |g|^2 \right) \right] \tag{12.103}$$

Here, $\cos\theta = \hat{k} \cdot \hat{r}$ is the cosine of the angle between $\vec{E}_0$ and the vector $\vec{r}$ locating a position on the surface of the sphere. In the vicinity of the SPR where g is large, Equation 12.103 simplifies:

$$E_{out}^2 = E_0^2 |g|^2 [1 + 3\cos^2\theta] \tag{12.104}$$

The enhanced field is largest in the directions $\theta = 0°$ and $180°$, where we can take $E_{out}^2 = 4E_0^2|g|^2$. Now, in a Raman scattering experiment, both the incident and scattered fields are enhanced if their frequencies are near the surface plasmon resonance. The observed Raman intensity is proportional to the product of the incident and scattered light intensities, and the intensities are proportional to the square of the field. Thus the maximum enhancement factor EF for a Raman active molecule at the surface of a metal sphere is

$$\mathrm{EF} = \frac{\left(E_{out} \right)^2 \left(E'_{out} \right)^2}{E_0^4} = 16|g|^2 |g'|^2 \tag{12.105}$$

The quantities E'_{out} and g' pertain to the field and the value of g at the frequency of the scattered light, while unprimed quantities refer to the incident field. For a typical narrow resonance, where Γ is on the order of $\omega_p/10$, this difference becomes significant for modes with larger Stokes shifts, such as C–H stretches at about 3000 cm^{-1}. Equation 12.105 assumes the Raman tensor of the molecule is unperturbed by adsorption at the metal. It is reasonable to envision that the interaction of a molecule with a metal surface would perturb the transition polarizability for Raman scattering by introducing new coupled metal-molecule states, akin to the charge-transfer transitions of discrete metal–ligand complexes. Such effects, if present, contribute to the so-called chemical enhancement effect, but since they are not readily generalized and contribute more weakly to Raman enhancement, these are ignored here.

Now we must admit that the simple model above can only qualitatively account for SERS phenomena; it leads to predicted Raman enhancements much smaller than the factors of 10^6 to 10^{10} that are typically seen. The problem is not just that the Drude model is admittedly crude. SERS is not generally observed from isolated spherical nanoparticles but from assemblies of particles. Improving on the treatment above, one can allow for nonspherical particle shapes, which lead to larger enhancements and red-shifted plasmon resonances. For spheroidal shapes, one finds then that the enhancement of the field is largest near the regions of highest curvature, and that it increases with increasing aspect ratio. In addition, the largest SERS effects are seen for molecules located at so-called "hot spots," regions of enhanced local field in the interstices between particles or between particles and plane surfaces. Calculations on simple dimers of spherical metal nanoparticles reveal enhancements for fields directed along the interparticle axis. A practical consequence of hot spots is that aggregated noble metal colloids, achieved for example by the addition of salt to the colloidal suspension, can be used to obtain stronger SERS signals. This aggregation also leads to red-shifting and broadening of the plasmon resonance which may be beneficial for achieving a convenient wavelength range within which to excite the Raman spectrum.

Electromagnetic theory explains why selection rules for SERS differ from those in ordinary Raman spectroscopy. Surface enhancement does not boost all Raman modes by the same factor, and their relative intensities depend on the frequencies of the incident and scattered light. Near a metal surface, a molecule sees a superposition of the incident and reflected fields. The amplitude of the reflected field is governed by the Fresnel coefficients for s and p polarization, which depend on the angle of incidence and the dielectric function of the metal. At frequencies below the SPR, the net field for s-polarized light is zero, and at higher frequencies, both s- and p-polarized net fields coexist. Thus the selection rules vary with excitation wavelength. The enhancement factors for Raman modes of molecules adsorbed at a metal nanoparticle depend on the orientation of the molecule, the values of the components of the Raman tensor, and the strength of the fields in the directions normal, E_n, and tangential, E_t, to the surface. The net field strengths depend on $g = (\varepsilon_r - \varepsilon_s)/(\varepsilon_r + 2\varepsilon_s)$ as follows:

$$\overline{E_t^2} \propto 2|1-g|^2$$

$$\overline{E_n^2} \propto |1+2g|^2 \tag{12.106}$$

We define Z as the direction normal to the surface. There are three categories of SERS transitions. Those that derive from the α_{ZZ} component of the polarizability depend on $E_n^2 E_n'^2$, while off-diagonal elements mixing the normal and tangential directions, such as α_{ZX} and α_{ZY}, depend on $E_n^2 E_t'^2$, and α_{XX}, α_{YY} and α_{XY} depend on $E_t^2 E_t'^2$. To the red of the SPR, the normal component of the field dominates over the tangential component. In this case, vibrations for which there is a polarizability change in the direction perpendicular to the surface are more strongly enhanced, enabling the determination of the orientation of the molecule provided some information on the Raman tensor is available. To the blue of the SPR, on the other hand, the selection rules are more liberal owing to the coexistence of the normal and tangential components of the field. The selection rules for Raman scattering by a molecule adsorbed at a planar metal surface and on a spherical metallic particle were derived by Moskovits [30,31]. In addition to the above considerations, the existence of a non-negligible field gradient in the vicinity of a metal nanostructure can lead to the activity of normal modes not permitted by electric dipole selection rules.

Very strong Raman intensities are seen when the exciting radiation is resonant with both an electronic transition of the molecule and the surface plasmon, leading to surface-enhanced resonance Raman scattering, or SERRS. Under these conditions, enhancement factors are large enough to detect single molecules using the technique of single-molecule surface-enhanced Raman spectroscopy (SMSERS). For example, Van Duyne and colleagues have reported SMSERS of the laser dye rhodamine 6G on arrays of Ag nanotriangles. Electrodynamic calculations predict enhancements on the order of 10^8 in the vicinity of the tips of these nanotriangles. Using a mixture of two isotopomers of rhodamine 6G, Zrimsek et al. [32] proved that spectra of single molecules were obtained. In that study, the SPR of the nanotriangle array was resonant at 546 nm, while the dye absorption maximum was 527 nm. The excitation wavelength of 532 nm provided a good overlap with both these resonances.

Now, the fact that rhodamine 6G is a strongly fluorescent laser dye poses the following question: isn't the fluorescence of the molecule also enhanced at the same excitation frequency that leads to SERS? One expects that would be true, were it not for the fact that a molecule adsorbed on a metal surface undergoes enhanced nonradiative decay owing to energy transfer, and possibly charge transfer, to the metal. This serves to quench the fluorescence that would otherwise overwhelm the resonance Raman spectrum of a dye such as rhodamine 6G in the absence of SERS enhancement. We have seen above that damping of the resonant excited state also has the potential to reduce Raman intensities. However, this nonradiative relaxation is slower than the subpicosecond Raman timescale but fast enough to compete with the typically nanosecond timescale of fluorescence emission. The result is that this damping has little effect on the Raman intensity, but considerably quenches the fluorescence, making SERS attractive for Raman studies of fluorescent molecules.

Figure 12.11 illustrates the absorption spectrum of an aqueous colloidal suspension of silver nanoparticles, and the SERS spectrum of rhodamine 6G adsorbed on the colloid, excited at 457.9 nm. The colloidal suspension shows the Ag SPR at 440 nm superimposed on background absorption and scattering. (The net absorption plus scattering in this case is referred to as extinction.) The SERS spectrum was obtained using a rhodamine 6G concentration of only 10^{-8} M. For comparison, the background spectrum of 10^{-7} M aqueous rhodamine 6G is shown, revealing only a weak water Raman band at ~ 1600 cm^{-1} and slight background from fluorescence. The latter is weak owing to low absorption of the exciting light and the low concentration of rhodamine 6G.

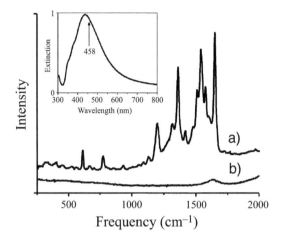

Figure 12.11 (a) SERS spectrum of 10^{-8} M rhodamine 6G adsorbed on aqueous colloidal silver and (b) Raman spectrum of 10^{-7} M rhodamine 6G in aqueous solution, both excited at 458 nm. The feature at about 1600 cm^{-1} in (b) is a Raman band of water. The inset shows the extinction spectrum of the Ag colloid, indicating the proximity of the laser wavelength to the peak in the SPR.

The same field enhancements that lead to SERS can also enhance the photochemistry of molecules adsorbed on nanoscale metals, an effect that can be a curse or a blessing depending on the goal of the experiment. In addition, enhancements of infrared absorption and of nonlinear spectroscopies such as second harmonic generation and hyper-Rayleigh scattering (a two-photon excited version of Rayleigh scattering) are also seen.

12.7 SUMMARY

This chapter has examined resonance and nonresonance Raman scattering from first principles. We have seen how the selection rules and expected depolarization ratios differ in the two experiments. Dynamical contributions to the Raman spectrum and the resonance Raman profiles have been explored. Lineshapes in the Raman spectrum depend on vibrational and rotational dynamics within the ground electronic state. The intensity profile of a given Raman band as a function of incident frequency (the Raman excitation profile) reveals dynamics of the excited electronic state. The REP is intimately linked to the electronic absorption spectrum, and the two depend on the same factors: normal mode displacements, vibronic couplings, and the lifetime and dephasing rate of the resonant electronic state. Two theoretical approaches for the analysis of Raman excitation profile were presented: transform theory and time-dependent theory. In the first of these, the absorption spectrum is used as input to a calculation of the REP. In the second, the absorption spectrum and the Raman profiles are modeled simultaneously with a common set of parameters. Though it is not always the case that the parameters for modeling absorption and Raman excitation profile should be the same, the time-dependent theory is of great conceptual value in the interpretation of Raman intensities in terms of dynamics on excited state potentials.

We have also introduced the topic of surface-enhanced Raman spectroscopy. While not without controversy, this field continues to advance in parallel with advances in nanotechnology and development of biological and material science applications. Electromagnetic theory has been shown to account for the large enhancement of Raman signals of molecules in the vicinity of metal nanostructures. The ability to detect even single molecules using SERS ensures continued interest in the technique.

PROBLEMS

1. Show that Equation 12.2 leads to a symmetric tensor, $\alpha_{\rho\sigma} = \alpha_{\sigma\rho}$, in the limit $\omega_0 \ll \omega_{eg}$.
2. How can combination bands be accounted for within the Kramers–Heisenberg–Dirac formula?

3. The correlation function for vibrational relaxation $\langle QQ(t)\rangle$ can be written in second quantized form, using dimensionless normal coordinates, $Q = (a^+ + a)/\sqrt{2}$, where $[a, a^+] = 1$. (a) Show that the Heisenberg representations of the operators a and a^+ are

$$a(t) = \exp(i\hat{H}_0 t/\hbar) a \exp(-i\hat{H}_0 t/\hbar) = \exp(-i\omega_v t) a$$

$$a^+(t) = \exp(i\hat{H}_0 t/\hbar) a^+ \exp(-i\hat{H}_0 t/\hbar) = \exp(i\omega_v t) a^+$$

where $\hat{H}_0 = \hbar\omega_v(a^+ a + 1/2)$ is the harmonic oscillator Hamiltonian. (Hint: First show that for two operators A and B, $[\exp B]\, A[\exp(-B)] = (\exp B^X)A$, where the superoperator B^X is defined by $B^X A \equiv [B, A]$. (b) Show that the correlation function $\langle QQ(t)\rangle$ is the sum of four terms $\langle a^+a(t)\rangle$, $\langle aa^+(t)\rangle$, $\langle aa(t)\rangle$ and $\langle a^+a^+(t)\rangle$. The last two terms average to zero at $t = 0$, so they do not contribute to the intensity, while the first two, in light of part (a), give rise to anti-Stokes and Stokes Raman scattering.

4. Show that the second moment of the Kubo lineshape is Δ^2. How can the second moment remain constant as the motional narrowing limit is approached?

5. Explain each of the following observations using the sum-over-states formalism (KHD equation), transform theory, and wave packet theory: (a) low intensity of low frequency modes; (b) effect of homogeneous absorption linewidth on Raman intensity; (c) weak Raman activity of high-order overtones in large molecules.

6. (a) Using a bulk plasmon frequency $\hbar\omega_p$ of 9 eV, account for the background permittivity ε_b (assumed to be real) that leads to a surface plasmon resonance $\hbar\omega_R$ of 2.4 eV. (b) Assuming ω_p/Γ is on the order of 10, use the result from part (a) to find the value of the relative permittivity at the frequency of the SPR. Is this value consistent with large enhancement (large value of g) for aqueous solution?

REFERENCES

1. A. C. Albrecht, On the theory of Raman intensities, *J. Chem. Phys.* 34, 1476 (1961).
2. L. A. Nafie and W. Peticolas, Reorientation and vibrational relaxation as line broadening factors in vibrational spectroscopy, *J. Chem. Phys.* 57, 3145 (1972).
3. O. Sonnich Mortensen and S. Hassing, Polarization and interference phenomena in resonance Raman scattering, *Advances in Infrared and Raman Spectroscopy*, 6, Chapter 1 (Wiley, New York, 1980).
4. H. Hamaguchi, The resonance Raman effect and depolarization in vibrational Raman scattering, *Advances in Infrared and Raman Spectroscopy*, 12, Chapter 6 (Wiley, New York, 1985).
5. T. G. Spiro and T. C. Strekas, Resonance Raman spectra of hemoglobin and cytochrome c: Inverse polarization and vibronic scattering, *Proc. Nat. Acad. Sci. USA* 69, 2622 (1972).
6. F. Paulat, V. K. K. Praneeth, C. Näther, and N. Lehnert, Quantum chemistry-based analysis of the vibrational spectra of five-coordinate metalloporphyrins [M(TPP)Cl], *Inorg. Chem.* 45, 2837 (2006).
7. R. Lynden-Bell, Vibrational relaxation and line widths in liquids: Dephasing by intermolecular forces, *Mol. Phys.* 33, 907 (1977).
8. A. M. Amorim da Costa, M. A. Norman, and J. H. R. Clark, Reorientational motion in liquids: A comparison of Raman and Rayleigh scattering, *Mol. Phys.* 29, 191 (1975).
9. D. W. Oxtoby, Vibrational relaxation in liquids, *Ann. Rev. Phys. Chem.* 32, 77 (1981).
10. D. W. Oxtoby, Dephasing of molecular vibrations in liquids, *Adv. Chem. Phys.* 41, 1 (1978).
11. A. Laubereau and W. Kaiser, Vibrational dynamics of liquids and solids investigated by picosecond light pulses, *Rev. Mod. Phys.* 50, 607 (1978).
12. C. H. Wang, *Spectroscopy of Condensed Media* (Academic Press, Orlando, FL, 1985).
13. J. J. Muller, D. Vanden Bout, and M. Berg, Broadening of vibrational lines by attractive forces: Ultrafast Raman echo experiments in a $CH_3I:CDCl_3$ mixture, *J. Chem. Phys.* 99, 810 (1993).
14. V. V. Hizhnyakov and I. J. Tehver, Resonance Raman profile with consideration for quadratic vibronic coupling, *Opt. Commun.* 32, 419 (1980).

15. A. P. Shreve, E. H. Haroz, S. M. Bachilo, R. B. Weisman, S. Tretiak, S. Kilina, and S. K. Doorn, Determination of electron-phonon coupling in single-walled carbon nanotubes by Raman overtone analysis, *Phys. Rev. Lett. 98*, 037405 (2007).

16. D. C. Blazej and W. Peticolas, Ultraviolet resonance Raman excitation profiles of pyrimidine nucleotides, *J. Chem. Phys. 72*, 3134 (1980).

17. S.-Y. Lee and E. J. Heller, Time-dependent theory of Raman scattering, *J. Chem. Phys 71*, 4777 (1979).

18. E. J. Heller, R. L. Sundberg, and D. Tannor, Simple aspects of Raman scattering, *J. Phys. Chem. 86*, 1822 (1982).

19. E. J. Heller, The semiclassical way to molecular spectroscopy, *Acc. Chem. Res. 14*, 368 (1981).

20. D. J. Tannor and E. J. Heller, Polyatomic Raman scattering for general harmonic potentials, *J. Chem. Phys. 77*, 202 (1982).

21. P. M. Champion and A. C. Albrecht, On the modeling of absorption band shapes and resonance Raman excitation profiles, *Chem. Phys. Lett. 82*, 410 (1981).

22. B. R. Stallard, P. M. Champion, P. R. Callis, and A. C. Albrecht, Advances in calculating Raman excitation profiles by means of the transform theory, *J. Chem. Phys. 78*, 712 (1983).

23. A. B. Myers and R. A. Mathies, Resonance Raman intensities: A probe of excited state structure and dynamics, *Biological Applications of Raman Spectroscopy*, Vol. 1, ed. T. G. Spiro (Wiley, New York, 1988).

24. A. B. Myers, Resonance Raman intensities and charge-transfer reorganization energies, *Chem. Rev. 96*, 911 (1996).

25. A. B. Myers, Relating absorption, emission, and resonance Raman spectra with electron transfer rates in photoinduced charge transfer systems, *Chem. Phys. 180*, 215 (1994).

26. S. Mukamel, *Principles of Nonlinear Optical Spectroscopy* (Oxford University Press, New York, 1995).

27. Y. Zong and J. L. McHale, Resonance Raman study of solvent dynamics in electron transfer. I. Betaine-30 in CH_3CN and CD_3CN, *J. Chem. Phys. 106*, 4963 (1997).

28. Y. Zong and J. L. McHale, Resonance Raman study of solvent dynamics in electron transfer. II. Betaine-30 in CH_3OH and CD_3OD, *J. Chem. Phys. 107*, 2920 (1997).

29. G. D. Scholes, T. Fournier, A. W. Parker, and D. Phillips, Solvent and intramolecular reorganization in 9,9′-bianthryl: Analysis of resonance Raman excitation profiles and *ab initio* molecular orbital calculation, *J. Chem. Phys. 111*, 5999 (1999).

30. M. Moskovits, Surface-enhanced spectroscopy, *Rev. Mod. Phys. 67*, 783 (1985).

31. M. Moskovits and J. S. Suh, Surface selection rules for surface-enhanced Raman spectroscopy: Calculations and applications to the surface-enhanced Raman spectrum of phthalazine on silver, *J. Phys. Chem. 88*, 5526 (1984).

32. A. B. Zrimsek, A.-I. Henry, R. P. Van Duyne, Single molecule surface-enhanced Raman spectroscopy without nanogaps, *J. Phys. Chem. Lett. 4*, 3206 (2013).

Nonlinear optical spectroscopy

13.1 INTRODUCTION

So far in this book, we have remained within the realm of *linear spectroscopy*. This means that the induced polarization, the dipole moment per unit volume, is a linear function of the electric field. This is a good approximation for electromagnetic fields which are very small compared to the internal fields of molecules, and it gives us a wealth of conventional spectroscopy techniques bound by the selection rules introduced in Chapter 4. Linear response of matter to electromagnetic radiation also gives us familiar phenomena such as reflection and refraction. In the presence of more intense fields such as those provided by pulsed lasers, non-linear effects result in many new spectroscopic techniques to probe matter. At the same time, nonlinear media, such as doubling crystals used to generate the second harmonic frequency of incident light, provide a powerful means to manipulate the properties of light beams. To enter into this new territory, we have to surrender the notion that the light–matter interaction is weak enough to leave each relatively unperturbed by the other. For example, as you will show in Problem 1 at the end of this chapter, the electric field amplitude associated with a typical pulsed laser source is not negligible compared to the internal fields of atoms and molecules.

In Chapter 4, we used time-dependent perturbation theory to describe optical spectroscopy in the weak perturbation limit that leads to linear spectroscopy. The perturbed wavefunction was expanded in the basis of the zeroth-order (dark) eigenfunctions; i.e. the stationary states. The expansion coefficients $c_m(t)$ were taken to be time-dependent, where m indexes one of these stationary states. We reached a point (Equation 4.17, reproduced below) where we had to make an approximation to find these coefficients.

$$\frac{dc_m(t)}{dt} = \frac{-i}{\hbar} \sum_n c_n(t) e^{-i\omega_{nm}t} V_{mn}(t) \tag{13.1}$$

Linear spectroscopy results from a first-order solution to Equation 13.1, where the time-dependent coefficients on the right hand side are replaced by their values at $t = 0$, leading to Fermi's Golden Rule, Equation 4.39. The first-order perturbation of the electromagnetic field was visualized to result in changes in the populations $c_i^* c_i$ and $c_f^* c_f$ of the initial and final states, respectively. Since the matrix element $V_{if}(t)$ of the perturbation operator, $\hat{H}' = -\vec{\mu} \cdot \vec{E}(t)$, is proportional to the amplitude of the electric field E_0, we obtained transition rates and thus intensities that were proportional to the square of this amplitude.

Remaining still in the weak-perturbation limit, we can improve on this approximation by adopting an iterative solution to Equation 13.1 to describe nonlinear optical spectroscopy. A second-order solution can be found by using the first-order solutions to the coefficients (Equation 4.33) on the right hand side of Equation 13.1 to solve for the coefficients $c_f^{(2)}(t)$, which could then be used to obtain $c_f^{(3)}(t)$ after the third iteration, etc. We will continue to associate terms such as $c_i^* c_i$ with populations, but we also want to discuss coherences $c_i^* c_j$ of pairs of states. It is thus much more fruitful to use a density matrix approach as introduced in Section 4.6. We will still have to iterate and for simplicity we will limit our discussion to second- and third-order nonlinear optical experiments. This leads to an array of powerful experimental techniques, discussed in this chapter and the next.

Nonlinear response is as much about optics as about spectroscopy, since matter can mediate the interaction between light beams leading to new frequencies, amplitudes and phases, and light can induce transitions and uncover dynamics that are recalcitrant to linear spectroscopy experiments. In addition to providing the basis for probing matter, nonlinear optical effects are exploited to generate light sources for spectroscopy experiments. Nonlinear spectroscopic techniques have advantages over linear methods such as improved spatial and temporal resolution. The high incident powers required to manifest nonlinearity makes pulsed laser sources necessary, hence the discussion of nonlinear effects will carry over to the next chapter where time-resolved spectroscopy is considered. For more information on nonlinear phenomena in optics and spectroscopy, the reader is referred to [1–5].

Nonlinear effects can be classified as nth order, with $n > 1$, according to the proportionality between the induced polarization and the nth power of the incident electric field, which is determined by the nth-order susceptibility $\chi^{(n)}$. In addition, we classify nonlinear processes as parametric or nonparametric on the basis of whether the matter undergoes a change in quantum state. A parametric process leaves the material system in the same energy state. A nonparametric process is associated with a change in the energy level of the material as a result of absorption or emission of photons. This nomenclature is used historically but is not very descriptive of the two types of spectroscopy; thus, some authors prefer the categorization of *active* (nonparametric) and *passive* (parametric) spectroscopies. In the case of a parametric process, $\chi^{(n)}$ is real, and for a nonparametric process it is complex. By analogy, in the linear regime we have the parametric processes of refraction and reflection, arising from the real part of $\chi^{(1)}$, versus absorption and emission, which are nonparametric processes permitted by the imaginary part of $\chi^{(1)}$.

Advances in the field of nonlinear spectroscopy were stimulated by the invention of the laser in 1960. The first reported laser-induced nonlinear effect was second harmonic generation (SHG), a second-order parametric process, observed for a ruby laser incident on a quartz crystal [6]. Two-photon absorption (TPA), a third-order nonparametric process, was observed in 1961 [7]. There are now a large number of nonlinear experiments capable of providing information that would be difficult or impossible to obtain within the confines of linear spectroscopy. Nonlinearities in the light–matter coupling can reveal dynamics, anharmonicities, and couplings of quantum states. Enhanced nonlinear effects are observed when an incident frequency is resonant with a dipole-allowed transition. We will continue to emphasize the role of the transition dipole moment, but we have to view the system as interacting at multiple times with the incident fields, leading to a visual approach to describing the theory of nonlinear polarization. These interactions lead to interference of quantum states and are the reason that coherence effects are more important in nonlinear than in linear spectroscopy experiments. Multiple field–matter interactions are responsible for the increased information content of nonlinear as compared to linear spectroscopy. As examples in this and in the following chapter will show, experimental control of the pulse sequence, propagation directions, and frequencies of the input fields enable a wide variety of nonlinear processes to be probed. In addition, the nonlinear process of interest is isolated by choice of the timing, propagation direction, and frequency of the detected signal.

This chapter presents classical and quantum mechanical formalisms that can be applied to a large number of nonlinear optical experiments. The list of second-order effects is small; consisting of second harmonic generation (SHG), sum-frequency generation (SFG), difference frequency generation (DFG), and optical rectification. A larger number of third-order experiments exists, such as transient grating experiments, photon echo measurements, two-photon absorption, and coherent anti-Stokes Raman scattering (CARS). The previously discussed spontaneous Raman spectroscopy experiment (Chapter 12) is actually a third-order process, and we revisit this topic from this vantage point in Section 13.5.2.

Historically, coherence effects have long been known in nuclear magnetic resonance (NMR) spectroscopy, owing to much longer dephasing times (milliseconds) compared to those for vibrational (~1 ps) and electronic (~10 fs) transitions. This is a result of much weaker interactions of nuclear spins with their environment compared to vibrational and electronic states. Coherence effects in NMR decay on a timescale that is longer than the Lamor frequency, permitting the observation of Rabi oscillations such as those depicted in Figure 4.3. Dephasing times in optical spectroscopy are not only shorter than in magnetic resonance, they are shorter in comparison to the frequency of coherence oscillations; hence, the observation of Rabi oscillations in optical spectroscopy is rare. Some of the language for discussing nonlinear optical spectroscopy, especially in a two-state picture, derives from pulsed NMR experiments, which were discussed briefly in Chapters 3 and 4. The reader is referred to Macomber [8] for more discussion of the analogies between pulsed optical and pulsed magnetic resonance experiments.

Relevant to our discussion of nonlinear spectroscopy effects is the distinction between homodyne and heterodyne detection of the resulting signal. In the homodyne case, the intensity of the signal is proportional to the square of the amplitude of the generated electric field: $I_s \propto |E|^2$. The heterodyne technique mixes this signal with that of a weak second light beam, referred to as a "local oscillator" for historical reasons, and the signal is proportional to the amplitude-squared of their sum: $I_s \propto |E + E_{LO}|^2$. If the nonlinear process generates a field with the same frequency and propagation direction as one of the input fields, the technique is "self-heterodyned." The importance of the heterodyne technique is through the phase information contained in the cross-term $\mathrm{Re}(E_{LO}E^*)$. Depending on the detection scheme, a nonlinear experiment can be made to reveal the real or imaginary part of the nonlinear response.

13.2 CLASSICAL APPROACHES TO NONLINEAR OPTICAL PROCESSES

13.2.1 POLARIZATION AS AN EXPANSION IN POWERS OF THE INCIDENT FIELD

In Chapter 3, we considered the polarization P, the induced electric dipole moment per unit volume, to be a linear function of the macroscopic electric field E (Equation 3.54), through the linear susceptibility χ. (The subscript e is omitted as we are limiting our discussion to electric rather than magnetic susceptibility.) To go beyond this weak field limit, the polarization is expanded in a Taylor series:

$$\vec{P} = \varepsilon_0 \left(\chi^{(1)}\vec{E} + \chi^{(2)}\vec{E}^2 + \chi^{(3)}\vec{E}^3 + ... \right)$$
$$= \vec{P}^{(1)} + \vec{P}^{(2)} + \vec{P}^{(3)} + ...$$

$$(13.2)$$

where $\chi^{(1)}$ is the linear (first-order) susceptibility of Equation 3.54, $\chi^{(2)}$ is the second-order susceptibility, $\chi^{(3)}$ the third-order susceptibility, and so on, which lead to the first-, second-, and third-order polarizations, etc. The expansion of Equation 13.2 is only valid if successive nonzero terms are smaller than the previous ones. There exist focused laser powers for which the expansion does not converge, but these will not concern us here. Also excluded are experiments for which the photon energy is sufficient to cause ionization. Though such an event does not require high incident power, the abrupt change in the wavefunction on ionization precludes treatment by perturbation theory.

Equation 13.2 is incomplete in that it does not recognize that, due to anisotropy of matter, the induced polarization does not have to be in the same direction as the incident field. To account for this, $\chi^{(1)}$ is a second-rank tensor (like the polarizability, to which it is related), $\chi^{(2)}$ is a third-rank tensor, etc. (See Appendix A for a review of tensor algebra.) Taking this into consideration, we get

$$P_i = \varepsilon_0 \left(\sum_j \chi_{ij}^{(1)} E_j + \sum_{j,k} \chi_{ijk}^{(2)} E_j E_k + \sum_{j,k,l} \chi_{ijkl}^{(3)} E_j E_k E_l + ... \right)$$

$$(13.3)$$

where $i, j, k, l...$ denote Cartesian coordinates in the laboratory frame. For an isotropic sample, changing the sign (reversing the direction) of the field must change the sign of the polarization. For example, in second order, replacing E by $-E$ preserves the square of the field. This implies that for an isotropic sample, $\chi^{(2)} = -\chi^{(2)}$, which is only satisfied if $\chi^{(2)} = 0$. We conclude that all even-order susceptibilities vanish for isotropic media. A consequence of this is that $\chi^{(2)}$ experiments such as SFG are powerful approaches to the study of interfaces and surfaces. Noncentrosymmetric crystals also possess nonzero $\chi^{(2)}$ (and other even-order) susceptibilities and are useful for optical frequency conversion. Note that the susceptibilities $\chi^{(1)}, \chi^{(2)}, \chi^{(3)}...$ above are bulk properties that correspond to molecular properties polarizability α, hyperpolarizability β, second hyperpolarizability γ, etc. The latter occur in the expansion of the molecular dipole moment μ in powers of the electric field: $\mu = \mu_0 + \alpha E + \beta^2 + \gamma E^3 + \cdots$.

In linear spectroscopy, the intensity of the spectroscopic signal (absorption or emission) is proportional to the number of molecules N that are interrogated. Nonlinear techniques, on the other hand, can result in a material response which is coherent, in that the phases of the signals emitted by different molecules are correlated. This can lead to spectral intensities which vary as N^2. To see this, consider a collection of emitting molecules in which the field responsible for the signal can be expressed as the sum of the fields: $\vec{E}_1 + \vec{E}_2 + ... \vec{E}_N$. The intensity of emitted light is proportional to the square of the total field:

$$I \propto \sum_{i=1}^{N} \vec{E}_i \cdot \sum_{j=1}^{N} \vec{E}_j = \sum_{i=1}^{N} \sum_{j=1}^{N} |\vec{E}_i| |\vec{E}_j| \cos \theta_{ij} \tag{13.4}$$

The angle θ_{ij} represents the relative phase of the oscillating field from molecules i and j. Consider the case where the field amplitudes are all the same, and separate the summation into N terms where $i = j$ and $N(N-1)$ terms where $i \neq j$:

$$I \propto |E|^2 \left(N + \sum_{i \neq j}^{N} \cos \theta_{ij} \right) \tag{13.5}$$

In the incoherent case, the cosine averages to zero, and the intensity is proportional to the number of molecules. In the opposite limit, the emitting molecules are in step with one another and we can put $\cos \theta_{ij} = 1$:

$$I \propto |E|^2 \left(N + N(N-1) \right) = N^2 |E|^2 \tag{13.6}$$

The two extremes, intensity which is proportional to N or N^2, result from random phases or correlated phases of the oscillators, respectively. Though we have taken a classical point of view here, picturing emitting molecules as dipole antennae, the picture is accounted for quantum mechanically using the density matrix. The coherence of quantum states is represented by off-diagonal elements of the density matrix. Coherent signals are highly directional. In linear spectroscopy, the off-diagonal density matrix elements average to zero, resulting in incoherent signals which are emitted isotropically.

13.2.2 Three-wave mixing (TWM)

The second-order polarization depends on $\chi^{(2)}$ and can be viewed as an output field that results from two input fields, leading to three-wave mixing (TWM) processes. Consider the situation where the field incident on the sample is the sum of two fields: $E = E_1 + E_2$, where:

$$E_1 = \frac{E_{10}}{2} \left(e^{i(k_1 r - \omega_1 t)} + e^{-i(k_1 r - \omega_1 t)} \right)$$

$$E_2 = \frac{E_{20}}{2} \left(e^{i(k_2 r - \omega_2 t)} + e^{-i(k_2 r - \omega_2 t)} \right) \tag{13.7}$$

where k_i and ω_i are the propagation vector and frequency of the ith field, and E_{i0} is the corresponding amplitude. We consider the second-order polarization, which is proportional to the square of the total field. On expanding, we find that $P^{(2)}$ is the sum of three terms, one proportional to the square of E_1, one proportional to the square of E_2, and a cross-term:

$$P^{(2)} \propto \frac{E_{10}^2}{4} \left[e^{2i(k_1 r - \omega_1 t)} + e^{-2i(k_1 r - \omega_1 t)} + 2 \right] + \frac{E_{20}^2}{4} \left[e^{2i(k_2 r - \omega_2 t)} + e^{-2i(k_2 r - \omega_2 t)} + 2 \right]$$

$$+ \frac{E_{10} E_{20}}{2} \left[e^{i[(k_1 + k_2) r - (\omega_1 + \omega_2) t]} + e^{-i[(k_1 + k_2) r - (\omega_1 + \omega_2) t]} + e^{i[(k_1 - k_2) r - (\omega_1 - \omega_2) t]} + e^{-i[(k_1 - k_2) r - (\omega_1 - \omega_2) t]} \right] \tag{13.8}$$

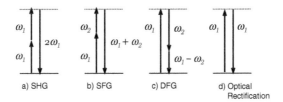

Figure 13.1 Energy level diagrams for (a) second-harmonic generation, (b) sum-frequency generation, (c) difference-frequency generation, and (d) optical rectification.

This expression encompasses the four possible $\chi^{(2)}$ experiments, shown schematically in the energy level diagrams of Figure 13.1, that result when two intense laser pulses at frequencies ω_1 and ω_2 are incident on a noncentrosymmetric sample. The following terms account for second-harmonic generation (SHG) at each of the two incident frequencies:

$$P_{SHG}^{(2)} \propto \frac{E_{10}^2}{4}\left[e^{2i(k_1 r - \omega_1 t)} + e^{-2i(k_1 r - \omega_1 t)}\right] + \frac{E_{20}^2}{4}\left[e^{2i(k_2 r - \omega_2 t)} + e^{-2i(k_2 r - \omega_2 t)}\right] \tag{13.9}$$

Equation 13.9 accounts for frequency doubling, in which the input beam results in an output beam at twice the frequency. We also see terms that account for sum-frequency generation, i.e., an output field at the frequency $\omega_1 + \omega_2$:

$$P_{SFG}^{(2)} \propto \frac{E_{10} E_{20}}{2}\left[e^{i[(k_1 + k_2)r - (\omega_1 + \omega_2)t]} + e^{-i[(k_1 + k_2)r - (\omega_1 + \omega_2)t]}\right] \tag{13.10}$$

Second-harmonic generation is a special case of SFG where the two input frequencies are the same. The difference frequency signal is

$$P_{DFG}^{(2)} \propto \frac{E_{10} E_{20}}{2}\left[e^{i[(k_1 - k_2)r - (\omega_1 - \omega_2)t]} + e^{-i[(k_1 - k_2)r - (\omega_1 - \omega_2)t]}\right] \tag{13.11}$$

Difference frequency generation is the basis for the operation of an optical parametric amplifier (OPA), an often-used source in time-resolved spectroscopy. Finally, there is a constant term representing optical rectification:

$$P_{opt\ rect}^{(2)} \propto \frac{1}{2}\left(E_{10}^2 + E_{20}^2\right) \tag{13.12}$$

In analogy to the rectification of an alternating current, this effect results from the cancellation of the two light beams to create a DC field. Optical rectification is a special case of DFG where the two input frequencies are equal. Though we have considered a superposition of two input beams of different frequency, SHG and optical rectification are achievable with a single incident beam of sufficient intensity. The above analysis does not reveal how the amplitudes of the various output beams compare to those of the input light beams. It is clear that there are limits to the efficiency of frequency conversion and light energy must be (at best) conserved. For example, in the first SHG experiment reported in 1961, red light at 694 nm from a ruby laser was doubled to 347 nm, and only about two photons in 10^8 were converted. Modern Nd:YAG lasers (see Chapter 7), on the other hand, produce output at 1064 nm that can be frequency doubled to a wavelength of 532 nm with efficiency greater than 50%, defined as the power in the doubled beam relative to that of the input beam.

Phase matching is a powerful tool in nonlinear optics and allows for the experimental conditions to be optimized in order to collect a particular type of signal. Consider sum-frequency generation with output signal at $\omega_3 = \omega_1 + \omega_2$ and propagation vector $\vec{k}_3 = \vec{k}_1 + \vec{k}_2$. Each k-vector has a magnitude depending on frequency and refractive index, $|\vec{k}_i| = n_i \omega_i / c$, where n_i is the real part of the refractive index at frequency ω_i. Constructive interference of the output beam and the source polarization leads to strong emission of

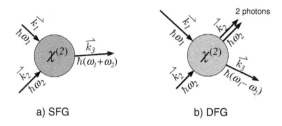

a) SFG   b) DFG

Figure 13.2 Phase-matching conditions for (a) sum-frequency generation and (b) difference-frequency generation.

light at frequency ω_3 only in the direction of the propagation vector $\vec{k}_3$. Phase-matching conditions for SFG and DFG are illustrated in Figure 13.2. The output energy as depicted in Figure 13.1 and the output momentum of Figure 13.2 both spring from Equation 13.8. Note that if dispersion in the refractive index can be neglected, the lengths of the propagation vectors shown in the diagrams are proportional to their frequencies. The output k-vectors are found from the vector addition of the input k-vectors, taking into account dispersion in the refractive index. To achieve high power densities, tightly focused Gaussian beams (see Chapter 2) are often used rather than the plane waves envisioned in drawing Figure 13.2. Nevertheless, the focal volume dimensions are larger than the wavelength, and each input Gaussian beam can be considered to have one effective propagation direction. The directional properties of the emission resulting from nonlinear polarization enable efficient collection of the desired signal and spatial rejection of signals from other simultaneously occurring nonlinear processes.

The phase-matching conditions for difference- and sum-frequency generation are advantageously applied in optical parametric amplification. An OPA uses a pump frequency ω_1 incident on a nonlinear crystal along with a "signal" frequency ω_2 at lower power than the pump. As shown in Figure 13.2, the outputs are the amplified signal at ω_2, for which two photons are generated for each incident photon, and the difference frequency at $\omega_1 - \omega_2$ (the "idler"); hence, $\omega_1 = \omega_2 + \omega_1 - \omega_2$. In other words, for each photon that is created at the idler frequency, a pump photon is destroyed and a signal photon is created. The two outputs are spatially separated by the phase-matching conditions $\vec{k}_{signal} = \vec{k}_2$ and $\vec{k}_{idler} = \vec{k}_1 - \vec{k}_2$ as illustrated in Figure 13.2b. From a quantum mechanical point of view, phase matching is a consequence of conservation of momentum of the incident and emitted photons. The parametric nature of the process is revealed in the equality of the summed frequencies of the input and output beams. Though we discuss the destruction and creation of photons, difference- and sum-frequency generation were obtained above using a purely classical treatment.

Nonlinear optical frequency conversion often employs birefringent crystals, for which the refractive index depends on the propagation direction and polarization of light. These take advantage of angle-tuning to achieve phase matching. There are also nonbirefringent crystals that are noncentrosymmetric, such as GaAs, and thus capable of second-order response. In other cases, for example an SFG signal from a surface, consideration of the polarization directions of the input and output fields leads to information about molecular orientations at the interface, and the tensor properties of $\chi^{(2)}$ are important.

The susceptibility $\chi^{(2)}$ is frequency dependent and may have both real and imaginary parts. We can write a general expression for the second-order polarization that accounts for this as follows. First, we introduce convenient notation for writing the electric field:

$$\vec{E}(r,t) = \frac{1}{2} \sum_m \vec{A}_m e^{-i\omega_m t} + \text{c.c.} \tag{13.13}$$

where the wave vector dependence has been absorbed into the complex amplitude:

$$\vec{A}_m \equiv \vec{E}_{m0} e^{-i\vec{k}_m \cdot \vec{r}} \tag{13.14}$$

A phase factor could also be absorbed into this definition as well. In Equation 13.13 and elsewhere, c.c. denotes the complex conjugate of the preceding term. The amplitude $\vec{A}_m = \vec{A}(\omega_m)$ is that of the component of the field at frequency ω_m. It is not the vector potential introduced in Chapter 2, though it is proportional to it. (Note that some authors omit the factor of ½ in Equation 13.13.) The property $A_m(-\omega_m) = A_m(\omega_m)^*$ ensures that the electric field is real and will help us keep track of positive and negative frequency contributions. We follow the convention of writing the second-order susceptibility with three frequency arguments, the first of which is the sum of the other two. Thus the component of the second-order polarization in direction i is

$$P_i^{(2)}(\omega_m + \omega_n) = \frac{\varepsilon_0}{4}\sum_{j,k}\sum_{m,n}\chi_{ijk}^{(2)}(\omega_m + \omega_n;\omega_m,\omega_n)A_j(\omega_m)A_k(\omega_n)e^{-i(\omega_m+\omega_n)t} \tag{13.15}$$

The first summation is over directions x, y and z, while the sum over m and n runs over positive and negative frequencies. $A_j(\omega_m)$ is the component of $\vec{A}(\omega_m)$ in direction j. Since i, j and k run independently over the three Cartesian coordinates, there could be as many as 3^3 or 27 elements of $\chi^{(2)}$. Fortunately, symmetry considerations limit the number of unique tensor elements. In the absence of resonance, $\chi^{(2)}$ is real. Since i, j, k and m, n are all dummy indices, we can permute j and k if we also permute m and n:

$$\chi_{ijk}^{(2)}(\omega_m + \omega_n;\omega_n,\omega_m) = \chi_{ikj}^{(2)}(\omega_n + \omega_m;\omega_m,\omega_n) \tag{13.16}$$

Care must be taken with the signs of the frequencies; for example, if we permute ijk on the left hand side of Equation 13.16 to kij, we have to make a sign change in order for the first frequency (the output signal) to be the sum of the second two frequencies:

$$\chi_{ijk}^{(2)}(\omega_m + \omega_n;\omega_n,\omega_m) = \chi_{kij}^{(2)}(\omega_m;\omega_m + \omega_n,-\omega_n) \tag{13.17}$$

The above equations apply to lossless materials, for which there is no absorption of light. They show that the second-order susceptibilities for SFG and DFG are the same. Far from resonance we can neglect the frequency dependence (dispersion) of $\chi^{(2)}$. As shown below, this leads to the susceptibility being real: $\chi_{ijk}^{(2)} = \chi_{ijk}^{(2)*}$. Under Kleinman's symmetry, valid in the frequency range where the material is transparent, the three indices can be permuted freely and the frequency dependence neglected. We invoke this symmetry in the following treatment of the second-order susceptibility for SFG:

$$P_{i,SFG}^{(2)}(\omega_1 + \omega_2) = \frac{\varepsilon_0}{4}\sum_{j,k}\chi_{ijk}^{(2)}\Big(A_j(\omega_1)A_k(\omega_2)e^{-i(\omega_1+\omega_2)t} + A_j^*(\omega_1)A_k^*(\omega_2)e^{i(\omega_1+\omega_2)t}$$

$$+ A_j(\omega_2)A_k(\omega_1)e^{-i(\omega_1+\omega_2)t} + A_j^*(\omega_2)A_k^*(\omega_1)e^{i(\omega_1+\omega_2)t}\Big) \tag{13.18}$$

By recognizing that the second and fourth terms are the complex conjugates of the first and second ones, this can be written more simply:

$$P_{i,SFG}^{(2)}(\omega_1 + \omega_2) = \frac{\varepsilon_0}{4}\sum_{j,k}\chi_{ijk}^{(2)}\Big(A_j(\omega_1)A_k(\omega_2)e^{-i(\omega_1+\omega_2)t} + A_j(\omega_2)A_k(\omega_1)e^{-i(\omega_1+\omega_2)t} + \text{c.c.}\Big) \tag{13.19}$$

Recall that we can permute j and k without changing $\chi_{ijk}^{(2)}$. Consider the terms in Equation 13.19 that occur when the incident beams are polarized in the x and y directions. We can write

$$P_{i,SFG}^{(2)}(\omega_1 + \omega_2) = \frac{\varepsilon_0}{4}\Big(\chi_{ixy}^{(2)} + \chi_{iyx}^{(2)}\Big)\Big(A_x(\omega_1)A_y(\omega_2)e^{-i(\omega_1+\omega_2)t} + A_x(\omega_2)A_y(\omega_1)e^{-i(\omega_1+\omega_2)t} + \text{c.c.}\Big) \tag{13.20}$$

Table 13.1 Contracted indices for $d_{i(jk)}$

Contracted Index	j,k
1	xx
2	yy
3	zz
4	yz, zy
5	xz, zx
6	xy, yx

It is convenient to contract the i, j, k indices on $\chi^{(2)}$ according to the scheme shown in Table 13.1, where the first index $i = 1, 2$ or 3 corresponds to x, y or z and the second pair of indices are denoted by 1 through 6. The resulting doubly subscripted susceptibility is referred to as d, where $d_{i(jk)} \equiv \chi_{ijk}^{(2)}/2$. For example, $\chi_{zxy}^{(2)} = \chi_{zyx}^{(2)} = 2d_{36}$. Using this notation, we can rewrite Equation 13.20 to express, for example, the z component of the polarization as follows:

$$P_{z,SFG}^{(2)}(\omega_1 + \omega_2) = \varepsilon_0 d_{36} \left(A_x(\omega_1)A_y(\omega_2)e^{-i(\omega_1+\omega_2)t} + A_y(\omega_1)A_x(\omega_2)e^{-i(\omega_1+\omega_2)t} + \text{c.c.} \right) \tag{13.21}$$

The full expression for the SFG signal is then:

$$\vec{P}_{SFG}^{(2)}(\omega_1 + \omega_2) = \varepsilon_0 \begin{pmatrix} d_{11} & d_{12} & d_{13} & d_{14} & d_{15} & d_{16} \\ d_{21} & d_{22} & d_{23} & d_{24} & d_{25} & d_{26} \\ d_{31} & d_{32} & d_{33} & d_{34} & d_{35} & d_{36} \end{pmatrix} \begin{pmatrix} A_x(\omega_1)A_x(\omega_2) + \text{c.c.} \\ A_y(\omega_1)A_y(\omega_2) + \text{c.c.} \\ A_z(\omega_1)A_z(\omega_2) + \text{c.c.} \\ A_y(\omega_1)A_z(\omega_2) + A_z(\omega_1)A_y(\omega_2) + \text{c.c.} \\ A_x(\omega_1)A_z(\omega_2) + A_z(\omega_1)A_x(\omega_2) + \text{c.c.} \\ A_x(\omega_1)A_y(\omega_2) + A_y(\omega_1)A_x(\omega_2) + \text{c.c.} \end{pmatrix} \tag{13.22}$$

Obviously, the use of polarized input beams limits the number of terms that need to be considered, and further symmetry considerations of the medium limit the number of nonzero components of the d-matrix. For example, they are all zero for centrosymmetric crystals, as will now be shown. The operator for inversion of the Cartesian coordinates is:

$$\hat{R}_{inv} = \begin{pmatrix} -1 & 0 & 0 \\ 0 & -1 & 0 \\ 0 & 0 & -1 \end{pmatrix} \tag{13.23}$$

For a generic symmetry operation $\hat{R}$ having matrix elements R_{ij}, the effect on the susceptibility is found as follows:

$$\chi_{ijk} = \sum_{\alpha\beta\gamma} R_{i\alpha} R_{j\beta} R_{k\gamma} \chi_{\alpha\beta\gamma} \tag{13.24}$$

where the superscript (2) on χ has been suppressed for convenience. In the presence of inversion symmetry, Equation 13.24 leads to

$$\chi_{ijk} = \sum_{\alpha\beta\gamma} (-\delta_{i\alpha})(-\delta_{j\beta})(-\delta_{k\gamma}) \chi_{\alpha\beta\gamma} = -\chi_{ijk} \tag{13.25}$$

The δ's in this equation are Kronecker deltas. The above equation of course implies $\chi_{ijk} = 0$ as previously stated.

We can use a similar approach to determine the unique, nonzero elements of the susceptibility for a crystal of a particular symmetry. The widely used nonlinear crystal β-barium borate (β-BaB$_2$O$_4$ or BBO) has a unit cell with rhombohedral symmetry belonging to the $3m$ point group, meaning that it has three-fold rotational symmetry and three corresponding mirror planes. We take the 3-fold rotation axis to be z and one of the three mirror planes to be yz. The matrix representation of the reflection operation is thus

$$\hat{R}_{ref} = \begin{pmatrix} -1 & 0 & 0 \\ 0 & 1 & 0 \\ 0 & 0 & 1 \end{pmatrix} \tag{13.26}$$

and that for 3-fold rotation about z is

$$\hat{R}_{rot} = \begin{pmatrix} -1/2 & \sqrt{3}/2 & 0 \\ -\sqrt{3}/2 & -1/2 & 0 \\ 0 & 0 & 1 \end{pmatrix} \tag{13.27}$$

(See Appendix C for more discussion of the matrix representations of symmetry operators.) With the first of these, it is simple to show that any element of χ with an odd number of x subscripts must vanish. The d-matrix of BBO has only three unique nonzero elements, d_{22} (equal to $-d_{16}$ and $-d_{21}$), d_{31} (equal to d_{24}, d_{15} and d_{32}), and d_{33}. For example, to show that $d_{15} = d_{24}$ for this symmetry, use the rotation matrix above to transforma $d_{15} = \chi_{xxz}$ as follows:

$$\chi_{xxz} = \sum_{ijk} R_{xi} R_{xj} R_{zk} \chi_{ijk} = \left(\frac{-1}{2}\right)^2 (1)\chi_{xxz} + \left(\frac{\sqrt{3}}{2}\right)^2 (1)\chi_{yyz} \tag{13.28}$$

which leads to $\chi_{xxz} = \chi_{yyz}$. Note that the terms χ_{xyz} and χ_{yxz} are absent from Equation 13.28 because they contain an odd number of x subscripts. See Powers [2] for a table listing the nonzero values of the d-matrix for all crystal symmetries.

13.2.3 FOUR-WAVE MIXING (FWM)

The third-order susceptibility $\chi^{(3)}$ does not vanish for isotropic media and is the lowest-order nonlinear susceptibility in this case. The processes that are determined by $\chi^{(3)}$ can be broadly described as four-wave mixing (FWM), in that the output field is the result of three input fields, $E = E_1 + E_2 + E_3$, where

$$E_i = \frac{E_{i0}}{2}\left(e^{i(k_i r - \omega_i t)} + e^{-i(k_i r - \omega_i t)}\right) \tag{13.29}$$

Once again, we use the definition of Equation 13.14 to define a field amplitude $\bar{A}$ which has the propagation vector and phase factor absorbed into it. The third-order polarization is expressed as

$$P_i^{(3)}(\omega_m + \omega_n + \omega_p)$$
$$= \frac{\varepsilon_0}{8} \sum_{m,n,p} \sum_{j,k,l} \chi_{ijkl}^{(3)}(\omega_m + \omega_n + \omega_p; \omega_m, \omega_n, \omega_p) \, A_j(\omega_m) A_k(\omega_n) A_l(\omega_p) e^{-i(\omega_m + \omega_n + \omega_p)t} \tag{13.30}$$

A sum over positive and negative frequencies is implied. Since each field is the sum of two terms (exponentials with plus and minus signs), the cube of the field $E = E_j + E_k + E_l$ is the sum of $6^3 = 216$ terms accounting for 108 output frequencies. Each of these occurs with a certain "degeneracy factor" equal to the number of times that frequency appears in the expansion of E^3. For example, each of the frequency tripling terms, which give output polarizations at $3\omega_1$, $3\omega_2$ or $3\omega_3$, occurs only once, while terms with various combinations of three distinct frequencies, such as $\omega = \pm\omega_1 \pm \omega_2 \pm \omega_3$ have degeneracy factors of six, since the three different fields can be ordered 3! ways. Problem 4 explores the degeneracy factors for other combinations of the input frequencies. As a fourth-rank tensor, $\chi^{(3)}$ could have as many as 81 elements, but thankfully symmetry considerations limit this to a much smaller number of independent tensor elements. Examples of $\chi^{(3)}$ processes will be considered in Section 13.5.

13.2.4 CLASSICAL CALCULATION OF $\chi^{(2)}$ AND $\chi^{(3)}$

In Chapter 3, we used a phenomenological, classical approach, the Lorentz model of the atom, to describe the polarizability on which the first-order susceptibility depends. It is mathematically straightforward to extend this approach to the calculation of $\chi^{(2)}$ and $\chi^{(3)}$. This exercise will reveal how the anharmonicity of the interaction of the electron cloud with the nucleus leads to nonlinear susceptibilities. The model is admittedly rather limiting as we consider a single resonance frequency ω_0 of the electrons of the medium. We start with an equation of motion that is slightly different from that of Equation 3.40:

$$\frac{d^2x(t)}{dt^2} + \Gamma \frac{dx(t)}{dt} + \omega_0^2 x(t) + ax^2(t) = \frac{-e}{m} E(t) \tag{13.31}$$

Recall that x is the displacement of the electron cloud, ω_0 is a harmonic frequency, Γ is the damping rate divided by the mass m, and $E(t)$ is the driving field. The new term ax^2 is the anharmonicity. Since Equation 13.31 is the force, this quadratic term represents a cubic term in the potential energy and is thus the leading perturbation to the previously considered harmonic case. In the case of a molecule with a center of symmetry, this cubic term in the potential and thus the quadratic term in the equation of motion would vanish. We see below that if $a \to 0$ the second-order susceptibility vanishes. We assume that the displacement can be expanded as

$$x = \lambda x^{(1)} + \lambda^2 x^{(2)} + \lambda^3 x^{(3)} + \dots \tag{13.32}$$

where λ is a tag to keep track of the order. Accordingly, we consider the driving field $E(t)$ to be first-order and multiply the right-hand side of Equation 13.31 by λ. We then equate the coefficients of like powers of λ to get the corresponding equation of motion for $x^{(n)}$, the nth-order correction to the displacement. We assume the electric field is a superposition of fields at the frequencies ω_1 and ω_2, and for simplicity we ignore the k-dependence of the fields and vector notation:

$$E(t) = \frac{1}{2}\left(A_1 e^{-i\omega_1 t} + A_2 e^{-i\omega_2 t} + A_1^* e^{i\omega_1 t} + A_2^* e^{i\omega_2 t}\right) = \frac{1}{2}\left(A_1 e^{-i\omega_1 t} + A_2 e^{-i\omega_2 t} + c.c.\right) \tag{13.33}$$

First-order susceptibility in the Lorentz model. To find the first-order solution $x^{(1)}$, we drop the anharmonicity ax^2, and assume a solution of the form:

$$x^{(1)} = \frac{1}{2}\left(x_1 e^{-i\omega_1 t} + x_2 e^{-i\omega_2 t} + c.c.\right) \tag{13.34}$$

Equating the coefficients of the first power of λ on the left- and right-hand sides of Equation 13.31 leads to:

$$\frac{d^2x^{(1)}}{dt^2} + \Gamma \frac{dx^{(1)}}{dt} + \omega_0^2 x^{(1)} = \frac{-eE(t)}{m} \tag{13.35}$$

On evaluating the derivatives and equating the coefficients of, for example, $\exp(i\omega_1 t)$ on the left- and right-hand sides, the following solution is obtained:

$$x_1(\omega_1) = \frac{(-e/m)A_1}{(\omega_0^2 - \omega_1^2 - i\omega_1\Gamma)} \equiv \frac{(-e/m)A_1}{D(\omega_1)} \tag{13.36}$$

The resonance denominator is defined as $D(\omega_i) = \omega_0^2 - \omega_i^2 - i\omega_i\Gamma$. A similar solution for $x_2(\omega_2)$ is obtained with ω_1 replaced by ω_2 and A_1 by A_2. If we had equated the coefficients of $\exp(-i\omega_2 t)$ instead of $\exp(i\omega_2 t)$, we would have obtained the complex conjugates of x_1 and x_2. Note that this solution is the same as Equation 3.42 of Chapter 3. The first-order polarization is expressed as follows:

$$P^{(1)}(\omega_i) = -Nex^{(1)}(\omega_i) = \varepsilon_0 \chi^{(1)}(\omega_i) E(\omega_i) \tag{13.37}$$

The first-order susceptibility is obtained:

$$\chi^{(1)}(\omega_i) = \frac{Ne^2}{m\varepsilon_0(\omega_0^2 - \omega_i^2 - i\omega_i\Gamma)} \tag{13.38}$$

Equation 13.38 can be compared to the molecular property of polarizability in Equation 3.43.

Second-order susceptibility in the Lorentz model. To find the second-order displacement $x^{(2)}$, we solve the equation of motion using the square of $x^{(1)}$ in the anharmonic term:

$$ax^2 \cong a\left(x^{(1)}\right)^2 = \frac{a}{4}\left(x_1^2 e^{-2i\omega_1 t} + x_1^{*2} e^{2i\omega_1 t} + x_2^2 e^{-2i\omega_2 t} + x_2^{*2} e^{2i\omega_2 t} + 2x_1 x_2 e^{-i(\omega_1 + \omega_2)t}\right.$$
$$\left. + 2x_1^* x_2^* e^{i(\omega_1 + \omega_2)t} + 2x_1 x_2^* e^{i(\omega_2 - \omega_1)t} + 2x_1^* x_2 e^{-i(\omega_2 - \omega_1)t} + 2x_1^* x_1 + 2x_2^* x_2\right) \tag{13.39}$$

There are five pairs of terms in Equation 13.39 that represent displacements oscillating at frequencies $2\omega_1$, $2\omega_2$, $\omega_1 + \omega_2$, $\omega_1 - \omega_2$, and 0, respectively. As above, we can focus on one of the exponential terms in finding $x^{(2)}$, for example, the SHG term that varies as $\exp(-2i\omega_1 t)$. We define

$$x^{(2)}(2\omega_1) = \frac{1}{2}\left(x_2 e^{-2i\omega_1 t} + x_2^* e^{2i\omega_1 t}\right) \tag{13.40}$$

The second-order solution is obtained by collecting the terms proportional to λ^2 in the equation of motion:

$$\frac{d^2 x^{(2)}}{dt^2} + \Gamma \frac{dx^{(2)}}{dt} + \omega_0^2 x^{(2)} + \frac{a}{4} x_1^2 e^{-2i\omega_1 t} = 0 \tag{13.41}$$

Since the input field $E(t)$ has no component at the frequency $2\omega_1$, the collection of terms on the left-hand side of 13.41 that multiply $\exp(-2i\omega_1 t)$ must sum to zero. The result is

$$x_2 = \frac{-a}{2}\left(\frac{eA_1}{m}\right)^2 \frac{1}{(\omega_0^2 - \omega_1^2 - i\omega_1\Gamma)^2} \frac{1}{(\omega_0^2 - 4\omega_1^2 - 2i\omega_1\Gamma)} \tag{13.42}$$

We can use the above expression to find $\chi^{(2)}$ for second harmonic generation with the help of the following:

$$P_{SHG}^{(2)} = -Nex_2 = \varepsilon_0 \chi^{(2)}(2\omega_1; \omega_1, \omega_1) A_1^2 \tag{13.43}$$

The second-order susceptibility for SHG is thus

$$\chi^{(2)}(2\omega_1; \omega_1, \omega_1) = \frac{Ne^3 a}{2\varepsilon_0 m^2} \frac{1}{\left(\omega_0^2 - \omega_1^2 - i\omega_1\Gamma\right)^2} \frac{1}{\left(\omega_0^2 - 4\omega_1^2 - 2i\omega_1\Gamma\right)} \tag{13.44}$$

Keeping in mind that this result was obtained with a simple phenomenological model, it reveals certain general features that will be upheld in a subsequent quantum mechanical treatment. First, we see that $\chi^{(2)}$ is a result of anharmonicity in the electronic response. We also see resonance denominators that result in enhancement of SHG when either the fundamental ω_1 or the second harmonic $2\omega_1$ is resonant with the natural frequency ω_0 of the molecule. Finally, we see that $\chi^{(2)}$ has real and imaginary parts, the latter being dependent on the damping Γ. The damping term permits the driven field resulting from the polarization to be out of phase with the driving field $E(t)$, corresponding to absorption or emission of light. When the driving frequency is far from an absorption band, the damping can be neglected and $\chi^{(2)}$ is real as previously stated.

The procedure can be repeated for other terms in the expansion of Equation 13.39. For example, we can isolate the terms for difference frequency generation (DFG) at $\omega_3 = \omega_1 - \omega_2$. The desired polarization is then

$$P_{DFG}^{(2)} = \varepsilon_0 \chi^{(2)}(\omega_3; \omega_1, -\omega_2) A_1 A_2^* \tag{13.45}$$

Note the dependence on the complex conjugate of the field A_2 is tied to the negative sign in front of ω_2. The result is

$$\chi^{(2)}(\omega_3; \omega_1, -\omega_2) = \frac{Ne^3 a}{2\varepsilon_0 m^2} \frac{1}{(\omega_0^2 - \omega_3^2 - i\omega_3\Gamma)(\omega_0^2 - \omega_2^2 + i\omega_2\Gamma)(\omega_0^2 - \omega_1^2 - i\omega_1\Gamma)}$$

$$= \frac{Ne^3 a}{2\varepsilon_0 m^2} \frac{1}{D(\omega_3)D(\omega_2)^* D(\omega_1)} \tag{13.46}$$

It is seen that resonance of ω_1, ω_2 or ω_3 with the natural frequency ω_0 will give enhanced DFG. We also obtain the result that $\chi^{(2)}(\omega_3; \omega_1, -\omega_2)$ is proportional to the product of the three relevant first-order susceptibilities:

$$\chi^{(2)}(\omega_3; \omega_1, -\omega_2) = \frac{a\varepsilon_0^2 m}{2N^2 e^3} \chi^{(1)}(\omega_1)\chi^{(1)*}(\omega_2)\chi^{(1)}(\omega_3) \tag{13.47}$$

Equation 13.47 illustrates Miller's rule:

$$\chi_{ijk}^{(2)}(\omega_3; \omega_1, \omega_2) = \Delta_{ijk} \chi_{ii}^{(1)}(\omega_3)\chi_{jj}^{(1)}(\omega_1)\chi_{kk}^{(1)}(\omega_2) \tag{13.48}$$

where Δ_{ijk} is called Miller's delta and is of similar magnitude for different materials.

Third-order susceptibility in the Lorentz model. For simplicity we consider a centrosymmetric molecule or medium for which the anharmonic term ax^2 vanishes, such that the third-order susceptibility depends on the force term bx^3 that derives from the quartic anharmonicity. To third order, the equation of motion is thus:

$$\frac{d^2 x^{(3)}}{dt^2} + \Gamma \frac{dx^{(3)}}{dt} + \omega_0^2 x^{(3)} + b\left(x^{(1)}\right)^3 = \frac{-eE(t)}{m} \tag{13.49}$$

To avoid computational complexity, the field is taken to be that of Equation 13.33. We cannot capture all of the $\chi^{(3)}$ processes this way, but we treat third harmonic generation (THG) at $3\omega_1$ as an example:

$$\left(x^{(1)}\right)^3 = \frac{1}{8} x_1^3 e^{-3i\omega_1 t} + \text{other terms} \tag{13.50}$$

We assume that the third-order displacement has the form:

$$x^{(3)}(3\omega_1) = \frac{1}{2}\left(x_3 e^{-3i\omega_1 t} + x_3^* e^{3i\omega_1 t}\right) \tag{13.51}$$

Inserting this in Equation 13.49 and setting the coefficient of $\exp(-3i\omega_1 t)$ equal to zero gives

$$x_3 = \frac{-b}{4}\left(\frac{eA_1}{m}\right)^3 \frac{1}{(\omega_0^2 - 9\omega_1^2 - 3i\omega_1\Gamma)(\omega_0^2 - \omega_1^2 - i\omega_1\Gamma)^3} \tag{13.52}$$

The third-order polarization for third harmonic generation is

$$P_{THG}^{(3)} = -Nex^{(3)} = \varepsilon_0 \chi^{(3)}(3\omega_1; \omega_1, \omega_1, \omega_1)A_1^3 \tag{13.53}$$

and thus

$$\begin{aligned}
\chi^{(3)}(3\omega_1; \omega_1, \omega_1, \omega_1) &= \frac{b}{4}\left(\frac{e}{m}\right)^3 \frac{Ne}{\varepsilon_0} \frac{1}{(\omega_0^2 - 9\omega_1^2 - 3i\omega_1\Gamma)(\omega_0^2 - \omega_1^2 - i\omega_1\Gamma)^3} \\
&= \frac{b}{4}\left(\frac{e}{m}\right)^3 \frac{Ne}{\varepsilon_0} \frac{1}{D(3\omega_1)D(\omega_1)^3}
\end{aligned} \tag{13.54}$$

The denominators reflect resonance effects when the natural frequency matches the fundamental or the tripled frequency.

13.2.5 SECOND-ORDER FREQUENCY CONVERSION IN THE SMALL SIGNAL LIMIT

Noncentrosymmetric crystals such as potassium niobate, potassium dihydrogen phosphate, and BBO are used for SHG, SFG and DFG in order to generate tunable-frequency light for spectroscopy experiments. Our goal in this section is to make some simplifying approximations that lead to tractable solutions to the wave equation for frequency conversion, solving for the driven fields as they relate to the driving field amplitudes. Assuming a simple geometry for the experiment and taking the output field to be weak compared to the input fields, we will arrive at understanding of how phase-matching and interference of the driving field with the induced polarization influences the strength of the nonlinear intensity.

In Chapter 2, we considered Maxwell's equations in free space and obtained the wave Equation 2.13, which is reproduced as follows:

$$\nabla^2 \vec{E} = \varepsilon_0 \mu_0 \frac{\partial^2 \vec{E}}{\partial t^2} \tag{13.55}$$

where the product of the permittivity ε_0 and permeability μ_0 of free space is related to the speed of light in vacuum:

$$c = \left(\frac{1}{\varepsilon_0 \mu_0}\right)^{1/2} \tag{13.56}$$

As shown for example in [2], when light propagates through a nonlinear medium, Equation 13.55 is altered as follows:

$$\nabla^2 \vec{E} - \mu_0 \varepsilon_0 \varepsilon_r \frac{\partial^2 \vec{E}}{\partial t^2} = \mu_0 \frac{\partial^2 \vec{P}^{NL}}{\partial t^2} \tag{13.57}$$

where ε_r is the (linear) relative permittivity defined in Chapter 3, equal to the square of the linear refractive index n, and $\vec{P}^{NL}$ is the nonlinear part of the polarization. The relative permittivity is taken to be real as we assume a lossless medium (no absorption or emission of light). If the right-hand side of the above equation were zero, we would obtain the result for freely propagating light waves traveling with phase velocity c/n. The nonlinear polarization acts as a source of radiation. We solve this wave equation with some simplifying approximations. Consider the field to be a superposition of three fields with frequencies ω_1, ω_2 and ω_3:

$$\vec{E} = \frac{1}{2} \sum_{m=1,2,3} \left(E_m e^{i(k_m z - \omega_m t)} + c.c \right) \tag{13.58}$$

We are considering plane waves propagating in the z direction. This permits us to write an equation just like Equation 13.57 that is valid for each frequency:

$$\nabla^2 \vec{E}_m - \frac{\varepsilon_r}{c^2} \frac{\partial^2 \vec{E}_m}{\partial t^2} = \frac{1}{\varepsilon_0 c^2} \frac{\partial^2 \vec{P}_m^{NL}}{\partial t^2} \tag{13.59}$$

where the relative permittivity is to be evaluated at the frequency ω_m. Again, we define $E_m \equiv A_m e^{ik_m z}$, and for our purposes here we keep the explicit k-dependence in the formulas. Let us consider the polarization at the sum frequency $\omega_3 = \omega_1 + \omega_2$.

$$\vec{P}_{SFG} = \frac{1}{2} P^{(2)} e^{-i\omega_3 t} + c.c \tag{13.60}$$

As shown previously, $P^{(2)}$ is related to the field amplitudes according to

$$P^{(2)} = \varepsilon_0 \chi^{(2)} A_1 A_2 e^{i(k_1 + k_2)z} \tag{13.61}$$

Consider propagation of light in the z direction. To evaluate Equation 13.59 for the field at the sum frequency, first we find:

$$\nabla^2 E_3 = \frac{d^2}{dz^2}\left(A_3 e^{ik_3 z} \right) = \left(\frac{d^2 A_3}{dz^2} + 2ik_3 \frac{dA_3}{dz} - k_3^2 A_3 \right) e^{ik_3 z} \cong \left(2ik_3 \frac{dA_3}{dz} - k_3^2 A_3 \right) e^{ik_3 z} \tag{13.62}$$

The last step in Equation 13.62 assumes that the change in amplitude of the wave occurs on a length scale which is large compared to the wavelength, consistent with the small signal limit considered here. This is equivalent to neglecting the second derivative of the field compared to k times the first derivative:

$$\left| \frac{d^2 A_3}{dz^2} \right| << \left| k_3 \frac{dA_3}{dz} \right| \tag{13.63}$$

Thus,

$$2ik_3 \frac{dA_3}{dz} e^{ik_3 z} = -\omega_3^2 \mu_0 P^{(2)} = -\omega_3^2 \mu_0 \varepsilon_0 \chi^{(2)} A_1 A_2 e^{i(k_1 + k_2)z} \tag{13.64}$$

The relationships $k = n\omega/c$ and $\varepsilon_r = n^2$ have been used. Equation 13.64 can be rearranged,

$$\frac{dA_3}{dz} = \frac{i\omega_3}{2cn_3} \chi^{(2)} A_1 A_2 e^{i\Delta k z} \tag{13.65}$$

where n_3 is the refractive index at frequency ω_3 and we have defined the wave vector mismatch as

$$\Delta k = k_1 + k_2 - k_3 \tag{13.66}$$

Since we can permute the indices 1, 2 and 3, three coupled wave equations are obtained:

$$\frac{dA_3}{dz} = \frac{i\omega_3}{cn_3} \chi^{(2)} A_1 A_2 e^{i\Delta kz}$$

$$\frac{dA_2}{dz} = \frac{i\omega_2}{cn_2} \chi^{(2)} A_1^* A_3 e^{-i\Delta kz} \qquad (13.67)$$

$$\frac{dA_1}{dz} = \frac{i\omega_1}{cn_1} \chi^{(2)} A_2^* A_3 e^{-i\Delta kz}$$

Equations 13.67 apply to SFG at $\omega_1 + \omega_2$, DFG at $\omega_3 - \omega_1$, and DFG at $\omega_3 - \omega_2$, respectively. The symmetry properties of $\chi^{(2)}$, which is the same for all three experiments, are taken into account. These equations can be solved to find how the amplitude of each field varies as it traverses the medium. In an optical parametric amplifier, for example, it is desirable for the field at the difference frequency to be amplified by the nonlinear medium. However, amplitudes may increase or decrease depending on the distance traveled and the phase-matching conditions. As shown for example in [2], the net flow of energy into or out of the medium depends on the quantity:

$$\left\langle \vec{E} \cdot \frac{d\vec{P}}{dt} \right\rangle = \text{rate of work done on the medium by the field} \qquad (13.68)$$

where the angle brackets represent an average over the rapidly varying part of the field and the polarization. Thus the phase relationship between the driven field and the time derivative of the polarization determines the sign of this expression and whether energy flows into or out of the system as the wave traverses the sample.

Let us now focus on the SFG term and make the assumption that the efficiency is low enough to neglect the change in the amplitudes A_1 and A_2. We can then integrate the first-coupled wave equation over the range 0 to L, where L is the distance traveled through the nonlinear medium. The result is

$$A_3(L) = \frac{i\omega_3}{cn_3} \chi^{(2)} A_1 A_2 \int_0^L e^{i\Delta kz} dz = \frac{i\omega_3}{cn_3} \chi^{(2)} A_1 A_2 \frac{e^{i\Delta kL} - 1}{i\Delta k} \qquad (13.69)$$

The intensity at the sum frequency is given by

$$I_{SFG} = \frac{1}{2} n_3 \varepsilon_0 c A_3^* A_3 = \frac{1}{2} \frac{\varepsilon_0 \omega_3^2}{cn_3} \left(\chi^{(2)}\right)^2 |A_1|^2 |A_2|^2 L^2 \text{sinc}^2\left(\frac{\Delta kL}{2}\right) \qquad (13.70)$$

The above equation shows that the intensity of light at the sum frequency is proportional to the product of the intensities at ω_1 and ω_2, as expected. More important is the dependence of I_{SFG} on the phase mismatch, which varies as the sinc-squared function that was previously encountered in Chapter 4. Figure 13.3 illustrates that

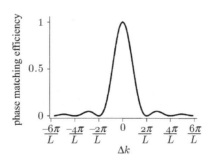

Figure 13.3 Phase-matching efficiency as a function phase mismatch.

the efficiency of SFG is sharply peaked at $\Delta k = 0$. The coherence length L_{coh} is defined as $2/\Delta k$ and represents the distance over which the output light and the polarization are in phase leading to amplification of the signal.

Perfect phase-matching, $\Delta k = 0$, is difficult to achieve. For the collinear geometry considered here, and for SFG, this implies

$$k_3 = k_1 + k_2$$

$$\frac{n_3 \omega_3}{c} = \frac{n_1 \omega_1}{c} + \frac{n_2 \omega_2}{c}$$

(13.71)

For normal dispersion, the refractive index increases with frequency, thus n_3 is greater than n_1 or n_2. This makes it impossible to satisfy Equation 13.71 at the frequency $\omega_3 = \omega_1 + \omega_2$ without additional constraints. One approach is to use birefringent crystals for which the refractive index depends on polarization and propagation direction. A negative uniaxial crystal, for example, has a larger refractive index n_e for the extraordinary ray than for the ordinary ray, for which the refractive index is n_o. The ordinary ray is polarized in the direction perpendicular to the plane defined by the propagation vector and the axis of symmetry of the crystal (the optic axis). The extraordinary ray is polarized in this plane and experiences a refractive index which depends on the angle θ between the optic axis and the propagation direction. For example, SHG can be achieved in a birefringent crystal by tuning the angle θ to achieve $n_e(2\omega) = n_o(\omega)$. More discussion of this and other phase-matching techniques can be found in [1,2].

Nonlinear experiments frequently use tightly focused Gaussian beams rather than plane waves. (See Chapter 2 for a brief discussion of Gaussian beams.) The former are preferred in order to achieve high power densities, but result in a finite range of directions for the wave vector. Simple geometric considerations, illustrated in Figure 13.4 for the case of SFG, reveal that somewhat positive Δk is favorable for efficient frequency conversion, while a negative Δk would make phase matching impossible.

The above solution to the coupled wave equations made the approximation that the amplitudes of the pump waves could be considered constant. Consider the intensity at any one of the three frequencies,

$$I_m = \frac{1}{2} n_m \varepsilon_0 c A_m^* A_m$$

(13.72)

Differentiating with respect to z and defining the total intensity as $I = I_1 + I_2 + I_3$, one can show that $dI/dz = 0$. In other words, the total intensity is conserved as expected for a parametric process. Further, with SFG in mind and $\omega_3 = \omega_1 + \omega_2$, one can derive the Manley–Rowe equations:

$$\frac{1}{\omega_3} \frac{dI_3}{dz} = -\frac{1}{\omega_2} \frac{dI_2}{dz} = -\frac{1}{\omega_1} \frac{dI_1}{dz}$$

(13.73)

Since I_i/ω_i is proportional to the flux of photons at frequency ω_i, Equation 13.73 reveals that the rate at which photons at the sum frequency are created is equal to the rate at which photons at either of the two incident frequencies are destroyed. One can also show that with imperfect phase matching the amplitudes of the field components oscillate sinusoidally on traversing the nonlinear medium. This is discussed further in [1–4].

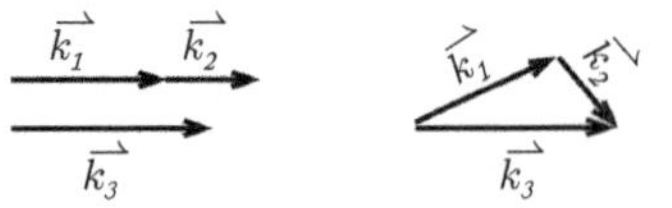

Figure 13.4 Phase matching of noncollinear beams can still be achieved when the phase mismatch Δk is positive.

13.3 QUANTUM MECHANICAL APPROACH TO NONLINEAR OPTICAL PROCESSES

13.3.1 TIME-DEPENDENT PERTURBATION THEORY APPROACH

We begin our quantum mechanical discussion of the second- and third-order susceptibilities by extending the time-dependent approach that was used to find the first-order correction to the wavefunction in Chapter 4. The result is useful in the limiting case of isolated molecules. To describe nonlinear effects of molecules interacting with their surroundings, we need a density matrix approach to introduce relaxation effects in a phenomenological way. Before doing that, the notation and approach is established by extending Equation 4.17 in an iterative fashion. We begin by calculating $\chi^{(1)}$. The nth-order correction to the perturbed wavefunction is given by

$$\left| \psi^{(n)}(t) \right\rangle = \sum_m c_m^{(n)}(t) e^{-iE_m t/\hbar} \left| m \right\rangle \tag{13.74}$$

where the sum is over the stationary states $\left| m \right\rangle$. The iterative approach to finding the coefficients is obtained as follows:

$$\frac{dc_m^{(n)}}{dt} = \frac{-i}{\hbar} \sum_l c_l^{(n-1)}(t) e^{i\omega_{ml} t} V_{ml}(t) \tag{13.75}$$

Recall that $V_{ml}(t)$ is the matrix element of the perturbation operator $\hat{H}'(t) = -\hat{\mu} \cdot \vec{E}(t)$ connecting zero-order states m and l. Previously, we considered the time-dependent field to be monochromatic. Here we consider a field with an unspecified number of frequency components ω_p:

$$\vec{E}(t) = \sum_p \vec{E}_p e^{-i\omega_p t} \tag{13.76}$$

Note that the sum runs over both positive and negative frequencies and that $E_p \equiv E(\omega_p)$ is twice the amplitude of the field that oscillates with frequency ω_p. The k-dependence is neglected for now. The matrix element of the perturbation is

$$V_{ml} = -\left\langle m \left| \hat{\mu} \right| l \right\rangle \cdot \vec{E}(t) = -\vec{\mu}_{ml} \cdot \sum_p \vec{E}_p e^{-i\omega_p t} \tag{13.77}$$

Inserting this in Equation 13.75, taking $c_l^{(0)}(t) = \delta_{lg}$, and integrating:

$$c_m^{(1)}(t) = \int_0^t \frac{dc_m^{(1)}}{dt'} dt' = \frac{i}{\hbar} \int_0^t e^{i\omega_{mg} t'} \vec{\mu}_{mg} \cdot \sum_p \vec{E}_p e^{-i\omega_p t'} dt'$$

$$= \frac{1}{\hbar} \sum_p \frac{\vec{\mu}_{mg} \cdot \vec{E}_p}{\omega_{mg} - \omega_p} \left[e^{i(\omega_{mg} - \omega_p)t} - 1 \right] \tag{13.78}$$

The first-order contribution to the induced dipole moment of the ground electronic state is:

$$\left\langle \mu_{ind}^{(1)} \right\rangle = \left\langle \psi^{(0)} \left| \hat{\mu} \right| \psi^{(1)} \right\rangle + \left\langle \psi^{(1)} \left| \hat{\mu} \right| \psi^{(0)} \right\rangle \tag{13.79}$$

The derivation is identical to that of Chapter 4, where we computed the transition polarizability α_{if}. Here, we are interested in the induced dipole moment of the ground state, and want to set up an iterative process for obtaining the higher order components of this induced dipole moment (parametric nonlinear processes). As in Chapter 4, the rotating wave approximation (RWA) allows us to drop contributions to the polarization that do not follow the field. This approximation is equivalent to dropping the lower limit of integration in Equation 13.78 and using a simpler form of the first-order coefficients in subsequent iterations:

$$c_m^{(1)}(t) = \frac{1}{\hbar} \sum_p \frac{\vec{\mu}_{mg} \cdot \vec{E}_p}{\omega_{mg} - \omega_p} e^{i(\omega_{mg} - \omega_p)t} \quad \text{(RWA)} \tag{13.80}$$

Next, we anticipate a total polarization of the following form:

$$\mathbf{P}^{(1)} = \sum_p \vec{P}^{(1)}(\omega_p) e^{-i\omega_p t} = N \left\langle \mu_{ind}^{(1)} \right\rangle \tag{13.81}$$

where the Cartesian components (*ijk*) of the polarization are $P_i^{(1)}(\omega_p) = \varepsilon_0 \Sigma_j \chi_{ij}^{(1)} E_j(\omega_p)$. Combining Equations 13.79 through 13.81, we obtain an expression for the first-order susceptibility:

$$\chi_{ij}^{(1)} = \frac{N}{\varepsilon_0 \hbar} \sum_m \left[\frac{\mu_{gm}^i \mu_{mg}^j}{\omega_{mg} - \omega_p} + \frac{\mu_{mg}^i \mu_{gm}^j}{\omega_{mg} + \omega_p} \right] \tag{13.82}$$

This expression is equivalent to Equation 4.73 of Chapter 4, with $i = f = g$. It reveals that the tensor nature of $\chi^{(1)}$ derives from the directions of the transition dipoles in the laboratory frame. Equation 13.82 and the following expressions for $\chi^{(2)}$ and $\chi^{(3)}$ take the total polarization to be the sum of the induced dipole moments of individual molecules.

Using Equation 13.80 in Equation 13.75, we obtain an expression for the second-order coefficient:

$$c_n^{(2)} = \frac{1}{\hbar^2} \sum_m \sum_{pq} \frac{\left(\vec{\mu}_{mg} \cdot \vec{E}_p \right)\left(\vec{\mu}_{nm} \cdot \vec{E}_q \right)}{(\omega_{mg} - \omega_p)(\omega_{ng} - \omega_p - \omega_q)} e^{i(\omega_{ng} - \omega_p - \omega_q)t} \tag{13.83}$$

In keeping with the RWA, the lower limit of the integration over time has been dropped in obtaining the above equation. The next iteration of Equation 13.75 gives us the third-order coefficient:

$$c_k^{(3)} = \frac{1}{\hbar^3} \sum_{ml} \sum_{pqr} \frac{\left(\vec{\mu}_{mg} \cdot \vec{E}_p \right)\left(\vec{\mu}_{lm} \cdot \vec{E}_q \right)\left(\vec{\mu}_{kl} \cdot \vec{E}_r \right)}{(\omega_{mg} - \omega_p)(\omega_{lg} - \omega_p - \omega_q)(\omega_{kg} - \omega_p - \omega_q - \omega_r)} e^{i(\omega_{kg} - \omega_p - \omega_q - \omega_r)t} \tag{13.84}$$

Next, we use Equation 13.83 to find the second-order correction to the induced dipole moment:

$$\left\langle \mu_{ind}^{(2)} \right\rangle = \left\langle \psi^{(0)} \left| \hat{\mu} \right| \psi^{(2)} \right\rangle + \left\langle \psi^{(2)} \left| \hat{\mu} \right| \psi^{(0)} \right\rangle + \left\langle \psi^{(1)} \left| \hat{\mu} \right| \psi^{(1)} \right\rangle \tag{13.85}$$

Using care with dummy indices in the summations, the following contributions are obtained:

$$\left\langle \psi^{(0)} \left| \hat{\mu} \right| \psi^{(2)} \right\rangle = \frac{1}{\hbar^2} \sum_{nm} \sum_{pq} \frac{\left(\vec{\mu}_{mg} \cdot \vec{E}_p \right)\left(\vec{\mu}_{nm} \cdot \vec{E}_q \right) \vec{\mu}_{gn}}{(\omega_{mg} - \omega_p)(\omega_{ng} - \omega_p - \omega_q)} e^{-i(\omega_p + \omega_q)t} \tag{13.86}$$

$$\left\langle \psi^{(2)} \left| \hat{\mu} \right| \psi^{(0)} \right\rangle = \frac{1}{\hbar^2} \sum_{km} \sum_{pq} \frac{\left(\vec{\mu}_{gm} \cdot \vec{E}_p \right)\left(\vec{\mu}_{mk} \cdot \vec{E}_q \right) \vec{\mu}_{kg}}{(\omega_{mg} - \omega_p)(\omega_{kg} - \omega_p - \omega_q)} e^{i(\omega_p + \omega_q)t} \tag{13.87}$$

$$\left\langle \psi^{(1)} \middle| \hat{\mu} \middle| \psi^{(1)} \right\rangle = \frac{1}{\hbar^2} \sum_{nm} \sum_{pq} \frac{\left(\vec{\mu}_{gm} \cdot \vec{E}_p \right)\left(\vec{\mu}_{ng} \cdot \vec{E}_q \right)\vec{\mu}_{mn}}{(\omega_{mg} - \omega_p)(\omega_{ng} - \omega_q)} e^{i(\omega_p - \omega_q)t} \tag{13.88}$$

Since the summations over p and q run over both positive and negative frequencies, we are free to replace ω_p and ω_q in 13.87 by $-\omega_p$ and $-\omega_q$, and similarly we replace ω_q by $-\omega_q$ in Equation 13.88. After collecting terms, the result is

$$\left\langle \mu_{ind}^{(2)} \right\rangle = \frac{1}{\hbar^2} \sum_{nm} \sum_{pq} \Bigg[\frac{\left(\vec{\mu}_{gm} \cdot \vec{E}_p \right)\left(\vec{\mu}_{mn} \cdot \vec{E}_q \right)\vec{\mu}_{ng}}{(\omega_{mg} + \omega_p)(\omega_{ng} + \omega_p + \omega_q)} + \frac{\left(\vec{\mu}_{mg} \cdot \vec{E}_p \right)\left(\vec{\mu}_{nm} \cdot \vec{E}_q \right)\vec{\mu}_{gn}}{(\omega_{mg} - \omega_p)(\omega_{ng} - \omega_p - \omega_q)}$$

$$+ \frac{\left(\vec{\mu}_{gm} \cdot \vec{E}_p \right)\left(\vec{\mu}_{ng} \cdot \vec{E}_q \right)\vec{\mu}_{mn}}{(\omega_{mg} + \omega_p)(\omega_{ng} - \omega_q)} \Bigg] e^{-i(\omega_p + \omega_q)t} \tag{13.89}$$

We can now use the expression for the second-order nonlinear polarization:

$$\mathbf{P}^{(2)} = \sum_{pq} \vec{P}^{(2)}(\omega_p + \omega_q) e^{-i(\omega_p + \omega_q)t} = N < \mu_{ind}^{(2)} > \tag{13.90}$$

and the relation

$$P_i^{(2)}(\omega_p + \omega_q) = \varepsilon_0 \sum_{jk} \chi_{ijk}^{(2)}(\omega_p + \omega_q; \omega_p, \omega_q) E_j(\omega_q) E_k(\omega_p) \tag{13.91}$$

to find the expression for the second-order susceptibility:

$$\chi_{ijk}^{(2)} = \frac{N}{\varepsilon_0 \hbar^2} \hat{P} \sum_m \Bigg[\frac{\mu_{ng}^i \mu_{mn}^j \mu_{gm}^k}{(\omega_{mg} + \omega_p)(\omega_{ng} + \omega_p + \omega_q)} + \frac{\mu_{gn}^i \mu_{nm}^j \mu_{mg}^k}{(\omega_{mg} - \omega_p)(\omega_{ng} - \omega_p - \omega_q)}$$

$$+ \frac{\mu_{nm}^i \mu_{ng}^j \mu_{gm}^k}{(\omega_{mg} + \omega_p)(\omega_{ng} - \omega_q)} \Bigg] \tag{13.92}$$

Note that the index i is associated with the direction of the resulting polarization and is derived from the direction of the transition dipoles $\vec{\mu}_{ng}$, $\vec{\mu}_{gn}$ and $\vec{\mu}_{nm}$ in the first, second and third terms inside the brackets of Equation 13.92, respectively. The permutation operator $\hat{P}$ simultaneously swaps the indices j and k, and p and q, since each of the three terms in Equation 13.92 can lead to two contributions; one taking the jth direction of the field E_p and the kth direction of E_q and vice versa.

A similar calculation can be done to obtain the third-order susceptibility, which we outline below. We start with the third-order induced dipole moment:

$$\left\langle \mu_{ind}^{(3)} \right\rangle = \left\langle \psi^{(0)} \middle| \hat{\mu} \middle| \psi^{(3)} \right\rangle + \left\langle \psi^{(3)} \middle| \hat{\mu} \middle| \psi^{(0)} \right\rangle + \left\langle \psi^{(1)} \middle| \hat{\mu} \middle| \psi^{(2)} \right\rangle + \left\langle \psi^{(2)} \middle| \hat{\mu} \middle| \psi^{(1)} \right\rangle \tag{13.93}$$

Next we write the third-order polarization as follows:

$$\mathbf{P}^{(3)} = \sum_{pqr} \vec{P}^{(3)}(\omega_p, \omega_q, \omega_r) e^{-i(\omega_p + \omega_q + \omega_r)t} = N < \mu_{ind}^{(3)} > \tag{13.94}$$

with the kth component of the polarization amplitude expressed as

$$P_k^{(3)}(\omega_p,\omega_q,\omega_r)=\varepsilon_0\sum_{hij}\sum_{pqr}\chi_{kjih}^{(3)}(\omega_s;\omega_r,\omega_q,\omega_p)E_j(\omega_r)E_i(\omega_q)E_h(\omega_p) \qquad (13.95)$$

where $\omega_s = \omega_p + \omega_q + \omega_r$. The result is

$$
\chi_{kjih}^{(3)}=\frac{N}{\varepsilon_0\hbar^3}\hat{P}\sum_{\sigma\nu\lambda}\Biggl[\frac{\mu_{vg}^h\mu_{\lambda v}^i\mu_{\sigma\lambda}^j\mu_{g\sigma}^k}{(\omega_{vg}-\omega_p)(\omega_{\lambda g}-\omega_p-\omega_q)(\omega_{\sigma g}-\omega_p-\omega_q-\omega_r)}
$$

$$
+\frac{\mu_{gv}^h\mu_{v\lambda}^i\mu_{\lambda\sigma}^j\mu_{\sigma g}^k}{(\omega_{vg}+\omega_p)(\omega_{\lambda g}+\omega_p+\omega_q)(\omega_{\sigma g}+\omega_p+\omega_q+\omega_r)}
$$

$$
+\frac{\mu_{gv}^j\mu_{\lambda g}^h\mu_{\sigma\lambda}^i\mu_{v\sigma}^k}{(\omega_{\lambda g}-\omega_p)(\omega_{\sigma g}-\omega_p-\omega_q)(\omega_{vg}+\omega_r)}
$$

$$
+\frac{\mu_{vg}^j\mu_{g\lambda}^h\mu_{\lambda\sigma}^i\mu_{\sigma\lambda}^k}{(\omega_{\lambda g}+\omega_p)(\omega_{\sigma g}+\omega_p+\omega_q)(\omega_{vg}-\omega_r)}\Biggr] \qquad (13.96)
$$

The permutation operator simultaneously permutes the fields p, q, and r and their attached indices h, i, and j, spawning six permutations from each of the four terms in Equation 13.96.

13.3.2 DENSITY MATRIX CALCULATION OF $\chi^{(2)}$ AND $\chi^{(3)}$

As introduced in Chapter 4, the density matrix approach is a phenomenological way to account for the influence of a medium in condensed phase spectroscopy experiments. It can be pursued within the Heisenberg or the Schrödinger picture (see Chapter 4), and we adopt the latter in the present discussion. As in Chapter 4, we take the time-dependent wavefunction to be a superposition of stationary states: $|\Psi(t)\rangle=\Sigma_n a_n(t)|n\rangle$. The density operator is defined as $\hat{\rho}(t)=|\Psi(t)\rangle\langle\Psi(t)|$, and an equilibrium averaged element of this matrix is $\rho_{nm}=\overline{a_n(t)a_m^*(t)}$. As discussed previously, diagonal elements of the density matrix represent populations, and off-diagonal elements are associated with coherences of pairs of states. We use phenomenological relaxation rates to describe the time dependence of ρ_{nm}.

$$\frac{\partial\rho_{nm}}{\partial t}=\frac{-i}{\hbar}\Bigl[\hat{H},\hat{\rho}\Bigr]_{nm}-\Gamma_{nm}(\rho_{nm}-\rho_{nm}^{eq}) \qquad (13.97)$$

The damping rate is given by $\Gamma_{nm}=1/2(\Gamma_n+\Gamma_m)+\Gamma_{nm}^*$. Comparing this to Equation 4.100, we see that Γ_{nm} is the total dephasing rate, Γ_n is the inverse lifetime of state n and Γ_{nm}^* is the pure dephasing rate. Recall that Γ_n is a phenomenological population relaxation rate as presented in Equation 4.76 and accounts for both radiative and nonradiative decay.

The equilibrium value of the density matrix ρ_{nm}^{eq} is zero for $n \neq m$, while ρ_{nn}^{eq} is given by Equation 1.86 (Boltzmann's law). The off-diagonal elements of the density matrix arise from coherent coupling of the basis states under the influence of the time-dependent perturbation of the radiation field, therefore they vanish at equilibrium. Since dephasing pertains to the decay of coherences, we have $\Gamma_{nn}^* = 0$ and the time-dependence of the diagonal density matrix elements is given by:

$$\frac{\partial\rho_{nn}}{\partial t}=\frac{-i}{\hbar}\Bigl[\hat{H},\hat{\rho}\Bigr]_{nn}-\Gamma_n(\rho_{nn}-\rho_{nn}^{eq}) \qquad (13.98)$$

Next we separate the Hamiltonian into a zero-order part plus the time-dependent perturbation: $\hat{H} = \hat{H}_0 + \hat{H}'(t)$, where $\hat{H}'(t) = -\hat{\mu} \cdot \vec{E}(t)$, which results in

$$\frac{\partial \rho_{nm}}{\partial t} = -i\omega_{nm}\rho_{nm} - \frac{i}{\hbar}\left[\hat{H}',\hat{\rho}\right]_{nm} - \Gamma_{nm}(\rho_{nm} - \rho_{nm}^{eq}) \tag{13.99}$$

To get a perturbative approach to finding the density matrix to Nth order, we take the following expansion:

$$\rho_{nm} = \rho_{nm}^{(0)} + \lambda\rho_{nm}^{(1)} + \lambda^2\rho_{nm}^{(2)} + \dots \tag{13.100}$$

As usual, we also multiply the perturbation Hamiltonian by λ. Inserting Equation 13.100 into Equation 13.99 leads to an iterative equation for finding the Nth order correction to the density matrix:

$$\frac{\partial \rho_{nm}^{(N)}}{\partial t} = -(i\omega_{nm} + \Gamma_{nm})\rho_{nm}^{(N)} - \frac{i}{\hbar}\left[\hat{H}',\hat{\rho}^{(N-1)}\right]_{nm} \tag{13.101}$$

Linear Susceptibility. Let us find the first-order density matrix by using $N = 1$ in the above equation and integrating over time:

$$\rho_{nm}^{(1)}(t) = \frac{-i}{\hbar}e^{-(i\omega_{nm}+\Gamma_{nm})t}\int_{-\infty}^{t}e^{(i\omega_{nm}+\Gamma_{nm})t'}[\hat{H}'(t'),\,\hat{\rho}^{(0)}]_{nm}\,dt' \tag{13.102}$$

With the substitution $\hat{H}'(t') = -\hat{\mu} \cdot \vec{E}(t') = -\hat{\mu} \cdot \Sigma_p \vec{E}(\omega_p)e^{-i\omega_p t'}$, the solution to Equation 13.102 is

$$\rho_{nm}^{(1)}(t) = \frac{1}{\hbar}(\rho_{mm}^{(0)} - \rho_{nn}^{(0)})\sum_p \frac{\vec{\mu}_{nm} \cdot \vec{E}(\omega_p)e^{-i\omega_p t}}{\omega_{nm} - \omega_p - i\Gamma_{nm}} \tag{13.103}$$

This expression is then used to get the first-order correction to the induced dipole moment:

$$\left\langle \mu_{ind}^{(1)}\right\rangle = Tr(\hat{\rho}^{(1)}\mu) = \sum_{nm}\rho_{nm}^{(1)}\mu_{mn} = \sum_p \left\langle \mu^{(1)}(\omega_p)\right\rangle e^{-i\omega_p t}$$

$$= \frac{1}{\hbar}(\rho_{mm}^{(0)} - \rho_{nn}^{(0)})\sum_p \frac{\vec{\mu}_{mn}\left[\vec{\mu}_{nm} \cdot \vec{E}(\omega_p)\right]e^{-i\omega_p t}}{\omega_{nm} - \omega_p - i\Gamma_{nm}} \tag{13.104}$$

With the first-order polarization given by $P^{(1)}(\omega_p) = N\left\langle \mu^{(1)}(\omega_p)\right\rangle = \varepsilon_0\chi^{(1)}(\omega_p) \cdot E(\omega_p)$, where N is now the number density of molecules, the linear susceptibility is found to be

$$\chi^{(1)}(\omega_p) = \frac{N}{\varepsilon_0\hbar}\sum_{nm}(\rho_{mm}^{(0)} - \rho_{nn}^{(0)})\frac{\vec{\mu}_{mn}\vec{\mu}_{nm}}{\omega_{nm} - \omega_p - i\Gamma_{nm}} \tag{13.105}$$

This can be compared to Equation 13.82, which explicitly shows the positive and negative frequency components, but lacks the population difference and damping. Equation 13.105 expresses the susceptibility as a second-rank tensor for which the ij Cartesian component is decided by the i and j components of the two transition dipole moments. Again, it is the damping term Γ_{nm} which imparts an imaginary component to the susceptibility. When the field frequency ω_p approaches that of the $m \rightarrow n$ transition, the susceptibility is dominated by a single term and becomes pure imaginary. The sign is then decided by $\rho_{mm}^{(0)} - \rho_{nn}^{(0)}$, allowing for the susceptibility to take on a positive (absorption) or negative (stimulated emission) value according to whether the population is greater in state n or m respectively.

In the absence of local fields, the susceptibility is related to the molecular polarizability through $\varepsilon_0 \chi^{(1)}(\omega) = N\alpha(\omega)$, and we previously expressed $\alpha(\omega)$ using the Kramers–Heisenberg–Dirac expression of Chapter 4. Equation 13.105 can be rearranged by splitting the population difference into two summations, swapping the indices n and m in the second sum, and using $\omega_{nm} = -\omega_{mn}$. The result is

$$\alpha_{ij}(\omega) = \frac{1}{\hbar} \sum_{nm} \rho_{mm}^{(0)} \left[\frac{\mu_{mn}^i \mu_{nm}^j}{\omega_{nm} - \omega - i\Gamma_{nm}} + \frac{\mu_{nm}^i \mu_{mn}^j}{\omega_{nm} + \omega + i\Gamma_{nm}} \right] \tag{13.106}$$

The subscript p on the frequency has been dropped. Equation 13.106 contains the resonance and antiresonance terms and can be compared to Equation 4.74. The present form of the polarizability is a state property rather than a transition polarizability, and the subscripts (and superscripts) i and j denote directions in the laboratory frame.

Second-order susceptibility. We use the first-order solution to the density matrix on the right hand side of Equation 13.101 to find the second-order correction to the density matrix:

$$\rho_{nm}^{(2)}(t) = \frac{-i}{\hbar} e^{-(i\omega_{nm} + \Gamma_{nm})t} \int_{-\infty}^{t} e^{(i\omega_{nm} + \Gamma_{nm})t'} [\hat{H}'(t'), \hat{\rho}^{(1)}]_{nm} dt' \tag{13.107}$$

The derivation is lengthy and a few intermediate steps will be shown. The commutator needed above is found to be

$$[\hat{H}'(t'), \hat{\rho}^{(1)}]_{nm} = -\sum_k (\vec{\mu}_{nk} \rho_{km}^{(1)} - \rho_{nk}^{(1)} \vec{\mu}_{km}) \cdot \vec{E}(t) \tag{13.108}$$

The field is expressed as $\vec{E}(t) = \Sigma_p \vec{E}(\omega_p) e^{-i\omega_p t}$, with the condensed notation $\vec{E}(\omega_p) = \vec{E}_p$. The result is

$$\rho_{nm}^{(2)} = \frac{1}{\hbar^2} \sum_{p,q} \sum_k e^{-i(\omega_p + \omega_q)t} \left\{ \left(\rho_{mm}^{(0)} - \rho_{kk}^{(0)} \right) \frac{(\vec{\mu}_{km} \cdot \vec{E}_p)(\vec{\mu}_{nk} \cdot \vec{E}_q)}{(\omega_{nm} - \omega_p - \omega_q - i\Gamma_{nm})(\omega_{km} - \omega_p - i\Gamma_{km})} \right.$$

$$\left. -(\rho_{kk}^{(0)} - \rho_{nn}^{(0)}) \frac{(\vec{\mu}_{nk} \cdot \vec{E}_p)(\vec{\mu}_{km} \cdot \vec{E}_q)}{(\omega_{nm} - \omega_p - \omega_q - i\Gamma_{nm})(\omega_{nk} - \omega_p - i\Gamma_{nk})} \right\} \tag{13.109}$$

$$\equiv \sum_{p,q} \sum_k e^{-i(\omega_p + \omega_q)t} R_{nmk}$$

(The definition above of R_{nmk} will save writing later when we consider the third-order expressions.) We then find the second-order polarization and susceptibility as follows (see Equations 13.90 and 13.91):

$$\mathbf{P}^{(2)}(\omega_p + \omega_q) = N \langle \mu_{ind}^{(2)} \rangle = N \sum_{n,m} \rho_{nm}^{(2)} \mu_{mn} \tag{13.110}$$

Using i, j, k as indices for the directions x, y, z and changing the indices for n, m, k in 13.109 to a, b, c respectively, we obtain

$$\chi_{ijk}^{(2)} = \frac{N}{\varepsilon_0 \hbar^2} \sum_{a,b,c} \left\{ \left(\rho_{bb}^{(0)} - \rho_{cc}^{(0)} \right) \frac{\mu_{ba}^i \mu_{ac}^j \mu_{cb}^k}{(\omega_{ab} - \omega_p - \omega_q - i\Gamma_{ab})(\omega_{cb} - \omega_p - i\Gamma_{cb})} \right.$$

$$\left. -\left(\rho_{cc}^{(0)} - \rho_{aa}^{(0)} \right) \frac{\mu_{ba}^i \mu_{cb}^j \mu_{ac}^k}{(\omega_{ab} - \omega_p - \omega_q - i\Gamma_{ab})(\omega_{ac} - \omega_p - i\Gamma_{ac})} \right\} \tag{13.111}$$

Equation 13.111 needs further manipulation in order to reflect the intrinsic permutation symmetry discussed in Section 13.2.2. We do this by adding to it an identical expression in which the indices i and j have been exchanged and similarly p and q have been swapped, dividing the sum by two. With further changes to dummy indices to get all the populations in terms of states a and b, $\chi^{(2)}$ is determined to be the sum of four terms:

$$\chi_{ijk}^{(2)}(\omega_p + \omega_q; \omega_p, \omega_q) = \frac{N}{2\varepsilon_0 \hbar^2} \sum_{abc} \left(\rho_{aa}^{(0)} - \rho_{bb}^{(0)} \right) \left(K_{1,abc} + K_{2,abc} + K_{3,abc} + K_{4,abc} \right) \qquad (13.112)$$

The four terms are

$$K_{1,abc} = \frac{\mu_{ac}^i \mu_{cb}^j \mu_{ba}^k}{(\omega_{ca} - \omega_p - \omega_q - i\Gamma_{ca})(\omega_{ba} - \omega_p - i\Gamma_{ba})}$$

$$K_{2,abc} = \frac{\mu_{ac}^i \mu_{cb}^k \mu_{ba}^j}{(\omega_{ca} - \omega_p - \omega_q - i\Gamma_{ca})(\omega_{ba} - \omega_q - i\Gamma_{ba})}$$

$$\qquad (13.113)$$

$$K_{3,abc} = \frac{\mu_{ac}^j \mu_{cb}^i \mu_{ba}^k}{(\omega_{cb} + \omega_p + \omega_q + i\Gamma_{cb})(\omega_{ba} - \omega_p - i\Gamma_{ba})}$$

$$K_{4,abc} = \frac{\mu_{ac}^k \mu_{cb}^i \mu_{ba}^j}{(\omega_{cb} + \omega_p + \omega_q + i\Gamma_{cb})(\omega_{ba} - \omega_q - i\Gamma_{ba})}$$

Note that if the a and b state populations are equal, $\chi^{(2)}$ vanishes. The first quantity in parentheses in the denominator of each term in Equation 13.113 represents resonance with the output frequency while the second term permits resonance with one of the input frequencies.

It is also possible to split Equation 13.112 into two terms, one with the population $\rho_{aa}^{(0)}$ and the other with $\rho_{bb}^{(0)}$. After changing dummy indices, the result is the sum of eight terms:

$$\chi_{ijk}^{(2)} = \frac{N}{2\varepsilon_0 \hbar^2} \sum_{a,b,c} \rho_{aa}^{(0)} \left(K_{1,abc} + K_{2,abc} + K_{3,abc} + K_{4,abc} + K_{5,abc} + K_{6,abc} + K_{7,abc} + K_{8,abc} \right) \qquad (13.114)$$

This introduces four new terms:

$$K_{5,abc} = \frac{\mu_{ac}^k \mu_{cb}^i \mu_{ba}^j}{(\omega_{bc} - \omega_p - \omega_q - i\Gamma_{bc})(\omega_{ca} + \omega_p + i\Gamma_{ca})}$$

$$K_{6,abc} = \frac{\mu_{ac}^j \mu_{cb}^i \mu_{ba}^k}{(\omega_{bc} - \omega_p - \omega_q - i\Gamma_{bc})(\omega_{ca} + \omega_q + i\Gamma_{ca})}$$

$$\qquad (13.115)$$

$$K_{7,abc} = \frac{\mu_{ac}^k \mu_{cb}^j \mu_{ba}^i}{(\omega_{ba} + \omega_p + \omega_q + i\Gamma_{ba})(\omega_{ca} + \omega_p + i\Gamma_{ca})}$$

$$K_{8,abc} = \frac{\mu_{ac}^j \mu_{cb}^k \mu_{ba}^i}{(\omega_{ba} + \omega_p + \omega_q + i\Gamma_{ba})(\omega_{ca} + \omega_q + i\Gamma_{ca})}$$

Figure 13.5 illustrates these eight contributions to $\chi^{(2)}$ on an energy level diagram under resonance conditions, for which input and output frequencies match those for dipole-allowed transitions. Note that the diagrams for K_3 is the same as that for K_6, and similarly the diagrams for K_4 and K_5 are equivalent. The arrows drawn in Figure 13.5 are in the order given by the chain of transition dipoles, reading them from right to left, reflecting transitions from a to b, from b to c and then from c back to a. Note that the transition dipole in the direction corresponding to the index i is always associated with the signal at sums and differences of the input frequencies, which can be positive or negative. For convenience, positive frequencies are assumed in drawing Figure 13.5.

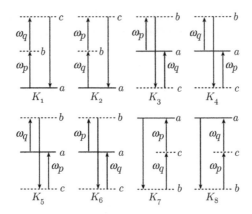

Figure 13.5 Energy level diagrams illustrating the resonances in each of the terms that contribute to $\chi^{(2)}$ in Equation 13.114.

Third-order susceptibility. The derivation of the third-order components of the density matrix and susceptibility and the resulting expressions are lengthy. Here, we present the results and refer the reader to [1,2] for more details. With the definition

$$R_{nmk} = \frac{1}{\hbar^2}\left\{(\rho_{mm}^{(0)} - \rho_{kk}^{(0)})\frac{(\vec{\mu}_{km}\cdot\vec{E}_p)(\vec{\mu}_{nk}\cdot\vec{E}_q)}{(\omega_{nm} - \omega_p - \omega_q - i\Gamma_{nm})(\omega_{km} - \omega_p - i\Gamma_{km})}\right.$$
$$\left. -(\rho_{kk}^{(0)} - \rho_{nn}^{(0)})\frac{(\vec{\mu}_{nk}\cdot\vec{E}_p)(\vec{\mu}_{km}\cdot\vec{E}_q)}{(\omega_{nm} - \omega_p - \omega_q - i\Gamma_{nm})(\omega_{nk} - \omega_p - i\Gamma_{nk})}\right\} \tag{13.116}$$

the third-order density matrix is

$$\rho_{ab}^{(3)} = \frac{1}{\hbar}\sum_{c,d}\sum_{p,q,r}e^{-i(\omega_p+\omega_q+\omega_r)t}\left\{\frac{(\vec{\mu}_{ad}\cdot\vec{E}_r)R_{dbc}}{(\omega_{ab} - \omega_p - \omega_q - \omega_r - i\Gamma_{ab})} - \frac{(\vec{\mu}_{db}\cdot\vec{E}_r)R_{adc}}{(\omega_{ab} - \omega_p - \omega_q - \omega_r - i\Gamma_{ab})}\right\} \tag{13.117}$$

Again, we relate the third-order polarization to $\chi^{(3)}$.

$$P_i^{(3)}(\omega_p + \omega_q + \omega_r) = \varepsilon_0\sum_{j,k,l}\sum_{p,q,r}\chi_{ijkl}^{(3)}(\omega_p + \omega_q + \omega_r; \omega_p, \omega_q, \omega_r)E_j(\omega_q)E_k(\omega_p)E_l(\omega_r) \tag{13.118}$$

We find $\chi^{(3)}$ to be the sum of eight terms:

$$\chi_{ijkl}^{(3)}(\omega_p + \omega_q + \omega_r; \omega_p, \omega_q, \omega_r) = \frac{N}{\varepsilon_0\hbar^3}\hat{P}\sum_{a,b,c,d}(A_1 + A_2 + B_1 + B_2 + C_1 + C_2 + D_1 + D_2)\rho_{dd}^{(0)} \tag{13.119}$$

where $\hat{P}$ is the previously used permutation operator. The terms are defined as

$$A_1 = \frac{\mu_{da}^i\mu_{ab}^j\mu_{bc}^k\mu_{cd}^l}{(\omega_{ad} - \omega_p - \omega_q - \omega_r - i\Gamma_{ad})(\omega_{bd} - \omega_p - \omega_q - i\gamma_{bd})(\omega_{cd} - \omega_p - i\Gamma_{cd})} \tag{13.120}$$

$$A_2 = \frac{\mu_{da}^l\mu_{ab}^i\mu_{bc}^j\mu_{cd}^k}{(\omega_{ba} - \omega_p - \omega_q - \omega_r - i\Gamma_{ba})(\omega_{ca} - \omega_p - \omega_q - i\Gamma_{ca})(\omega_{ad} + \omega_p + i\Gamma_{ad})} \tag{13.121}$$

$$B_1 = \frac{\mu_{da}^k \mu_{ab}^i \mu_{bc}^j \mu_{cd}^l}{(\omega_{ba} - \omega_p - \omega_q - \omega_r - i\Gamma_{ba})(\omega_{ac} + \omega_p + \omega_q + i\Gamma_{ac})(\omega_{cd} - \omega_p - i\Gamma_{cd})} \tag{13.122}$$

$$B_2 = \frac{\mu_{da}^l \mu_{ab}^k \mu_{bc}^i \mu_{cd}^j}{(\omega_{cb} - \omega_p - \omega_q - \omega_r - i\Gamma_{cb})(\omega_{bd} + \omega_p + \omega_q + i\Gamma_{bd})(\omega_{ad} + \omega_p - i\Gamma_{ad})} \tag{13.123}$$

$$C_1 = \frac{\mu_{da}^j \mu_{ab}^i \mu_{bc}^k \mu_{cd}^l}{(\omega_{ab} + \omega_p + \omega_q + \omega_r - i\Gamma_{ab})(\omega_{bd} - \omega_p - \omega_q - i\Gamma_{bd})(\omega_{cd} - \omega_p - i\Gamma_{cd})} \tag{13.124}$$

$$C_2 = \frac{\mu_{da}^l \mu_{ab}^j \mu_{bc}^i \mu_{cd}^k}{(\omega_{bc} + \omega_p + \omega_q + \omega_r + i\Gamma_{bc})(\omega_{ca} - \omega_p - \omega_q - i\Gamma_{ca})(\omega_{ad} + \omega_p + i\Gamma_{ad})} \tag{13.125}$$

$$D_1 = \frac{\mu_{da}^k \mu_{ab}^j \mu_{bc}^i \mu_{cd}^l}{(\omega_{bc} + \omega_p + \omega_q + \omega_r + i\Gamma_{bc})(\omega_{ac} + \omega_p + \omega_q - i\Gamma_{ac})(\omega_{cd} - \omega_p - i\Gamma_{cd})} \tag{13.126}$$

$$D_2 = \frac{\mu_{da}^l \mu_{ab}^k \mu_{bc}^j \mu_{cd}^i}{(\omega_{cd} + \omega_p + \omega_q + \omega_r + i\Gamma_{cd})(\omega_{bd} + \omega_p + \omega_q + i\Gamma_{bd})(\omega_{ad} + \omega_p + i\Gamma_{ad})} \tag{13.127}$$

The permutation operator gives rise to six permutations for each of the above eight terms (six orderings of the labels p, q, and r), so that an element of $\chi^{(3)}$ is a sum of 48 terms. Figure 13.6 illustrates the resonances associated with each of the eight terms.

Though the density matrix approach accounts for time-dependent interactions of the system with its environment, the screening of the fields by the medium (local field effect) is also a difficult problem in nonlinear as in linear spectroscopy. It is frequently neglected, such that bulk properties (susceptibilities) are obtained from a sum of the molecular properties, taking into account orientation of the molecular frame relative to the lab frame. For example, the molecular hyperpolarizability (Equation 3.29), β_{ijk}, is a third-rank tensor which gives rise to $\chi_{IJK}^{(2)}$ in the laboratory frame. The connection between the two requires summations over appropriate rotation matrices which depend on the Euler angles that orient the molecular frame (ijk) relative to the lab frame (IJK). Neglecting the tensor nature of both, they would be connected by $\chi^{(2)} = N\beta/\varepsilon_0$.

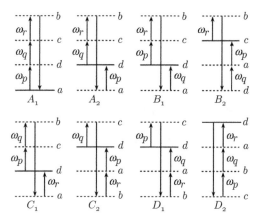

Figure 13.6 Energy level diagrams illustrating the resonances in each of the terms that contribute to $\chi^{(3)}$ in Equation 13.119.

13.4 FEYNMAN DIAGRAMS AND CALCULATION OF TIME-DEPENDENT RESPONSE FUNCTIONS

An alternative approach to finding the nonlinear corrections to the density matrix and susceptibility is to use the interaction representation, which is intermediate between the Schrödinger and Heisenberg pictures presented in Chapter 4. Recall that in the former, the wavefunctions are considered to evolve in time while the latter considers the operators to be time-dependent. In the interaction picture, the time-dependent perturbation operator $\hat{H}'(t)$ evolves in time under the influence of the time-independent zero-order Hamiltonian $\hat{H}_0$ rather than the complete Hamiltonian. The approach leads to perturbation expansions for $\rho_{ab}^{(n)}$ and $\chi^{(n)}$ which are found from time-ordered integrations over matrix elements of commutators of the dipole moment operators with the perturbation operator $\hat{H}'(t) = -\vec{\mu} \cdot \vec{E}$. The derivation of this expansion is beyond the scope of this book (see [3]), so we shall merely outline the formalism here. The theory leads to a convenient visual representation of the expansion terms in the expressions that we have obtained for $\chi^{(2)}$ and $\chi^{(3)}$, known as double-sided Feynman diagrams. In some cases, a specific Feynman diagram can be associated with a particular experiment, but more generally a sum of diagrams, each representing a particular term contributing to $\chi^{(n)}$, must be considered. Caution must be exercised when associating a single diagram with an experimental technique as cancellation can occur when diagrams are summed to get the susceptibility. Nevertheless, the diagrammatic approach is useful for introducing the concepts of coherences and populations in the discussion of nonlinear experiments. Most important, a contributing term to the time-dependent nth-order response function can be written down by inspection of the corresponding Feynman diagram, as shown here.

In the Feynman diagram picture, each interaction with the field is considered to act on the bra or ket side of the density matrix element ρ_{nm} such that the pulse sequence results in a series of populations and coherences that defines the type of experiment. What does it mean to act on the "bra" or "ket" side of the density matrix? Equation 13.101 reveals an iterative approach to calculating the nth-order perturbation to $\hat{\rho}$ from n nested commutators of $\hat{H}'$ and $\hat{\rho}^{(0)}$. On expansion, the perturbation Hamiltonians can appear on the right- or left-hand side of $\hat{\rho}^{(0)}$. These are associated with bra- and ket-side perturbations respectively. The light–matter perturbation results in a superposition of two states when the input frequency matches that of a dipole-allowed transition between them. The resulting coherence leads to polarization that oscillates at the beat frequency of these coupled states. In free induction decay, FID, the emission of light resulting from an induced polarization is observed after the field is turned off.

In Chapter 4, we introduced the Rabi frequency $\Omega = \vec{\mu}_{fi} \cdot \vec{E}_0/\hbar$, which, unlike the Bohr frequency ω_{fi}, is a function of the electric field amplitude and the transition dipole. The Rabi frequency is the frequency of *optical nutation*, which is the cycling of absorption and emission that occurs at high field strength, while the field is on. In a two-state picture such as that used in Chapter 4, a Rabi cycle occurs when a pulse sequence results in the following evolution of the density matrix: $|g\rangle\langle g| \to |e\rangle\langle g| \to |e\rangle\langle e| \to |g\rangle\langle e| \to |g\rangle\langle g|$, where the first two events (arrows) result in an absorption and the third and fourth represent stimulated emission. Though common in magnetic resonance experiments (see Chapter 3), this kind of cycling in the case of electronic states is rare, though not impossible, because electronic dephasing rates are much faster than achievable Rabi frequencies. Note that in a two-state model, the density matrix formalism leading to the cycling of ground and excited state populations was achieved without recourse to perturbation theory. This is distinct from the perturbative (partial) transfer of population from the ground to the excited state, taking place through intermediate coherences. As discussed here and in the next chapter, ultrafast time-resolved optical experiments can reveal the dynamics of population and coherence for sufficiently short pulse durations and time delays.

The interaction picture approach leads to the following response functions [3]:

$$S^{(1)}(t_1) = \frac{i}{\hbar}\theta(t_1)\overline{[\mu(t_1), \mu(0)]} \tag{13.128}$$

$$S^{(2)}(t_1, t_2) = \left(\frac{i}{\hbar}\right)^2 \theta(t_1)\theta(t_2)\overline{[[\mu(t_1 + t_2), \mu(t_1)], \mu(0)]} \tag{13.129}$$

$$S^{(3)}(t_1,t_2,t_3)=\left(\frac{i}{\hbar}\right)^3 \theta(t_1)\theta(t_2)\theta(t_3)\overline{[[[\mu(t_1+t_2+t_3),\mu(t_1+t_2)],\mu(t_1)],\mu(0)]} \tag{13.130}$$

The Heaviside step function $\theta(t)$ is equal to unity for $t > 0$, and zero for $t < 0$. The nth-order response function is pictured as resulting from a series of n field–matter interactions that occur at times 0, t_1, $t_1 + t_2$, $t_1 + t_2 + t_3$, ..., $t_1 + t_2 + t_3 + \ldots t_{n-1}$, where t_j is the time interval between the jth interaction and the one following it. In other words, t_1 is the time interval between the first and second interactions, t_2 is the time interval between the second and third interactions, as so on. The signal is detected at time $t_1 + t_2 + t_3 + \ldots t_n$. The overbar again represents an equilibrium average, using $\hat{\rho}_{eq} = \hat{\rho}^{(0)}$ as the density operator in evaluating the trace (Equation 1.91). In the interaction representation, the time-dependent dipole moment operator evolves as

$$\mu(t)=\exp\left(\frac{i\hat{H}_0 t}{\hbar}\right)\mu(0)\exp\left(\frac{-i\hat{H}_0 t}{\hbar}\right) \tag{13.131}$$

The above response functions are extensions of the time-correlation function approach taken in Chapter 5 for the linear response, e.g. Equation 5.15. Thus $S^{(1)}$ is the dipole correlation function, the Fourier transform of which yields the spectral intensity $I(\omega)$ in linear response.

The nonlinear polarization in first, second and third order is

$$P^{(1)}(t)=\int_0^\infty dt_1 S^{(1)}(t_1)E(t-t_1) \tag{13.132}$$

$$P^{(2)}(\vec{r},t)=\int_0^\infty dt_2 \int_0^\infty dt_1 S^{(2)}(t_2,t_1)E(\vec{r},t-t_2)E(\vec{r},t-t_2-t_1) \tag{13.133}$$

$$P^{(3)}(\vec{r},t)=\int_0^\infty dt_3 \int_0^\infty dt_2 \int_0^\infty dt_1 S^{(3)}(t_3,t_2,t_1)E(\vec{r},t-t_3)E(\vec{r},t-t_3-t_2)E(\vec{r},t-t_3-t_2-t_1) \tag{13.134}$$

It should be clear how the above expressions can be extended to nonlinear response functions and polarizations of any order n. The field $E(\vec{r},t)$ is the Fourier transform of the electric field which has previously been expressed as a function of k and ω. The variables conjugate to $\vec{r}$ and t are respectively $\vec{k}$ and ω, thus the Fourier transform takes the form:

$$E(\vec{r},t)=\left(\frac{1}{2\pi}\right)^4 \int_{-\infty}^\infty d\omega \int d\vec{k} \exp\left(-i\omega t + i\vec{k}\cdot\vec{r}\right)\tilde{E}(\vec{k},\omega) \tag{13.135}$$

This equation uses the common convention of distinguishing members of a Fourier transform pair by placing a tilde over one of the functions. The vector nature of the field, which would lead to different elements of the tensor $\chi^{(n)}$, is neglected here for simplicity. Similarly, the time-dependent polarization can be Fourier-transformed to give the polarization as a function of frequency and wave vector, and the above approach leads to the same expressions for $\chi^{(n)}$ as obtained earlier.

Feynman diagrams (Figures 13.7 through 13.9) provide a way to visualize the evolution of the density matrix in the time domain in terms of perturbations to the bra or ket side and give us a vocabulary for discussing experiments. The parallel vertical lines on the left and right indicate the time evolution of the ket and bra side, respectively, and time increases from bottom to top. Solid wavy arrows represent field–matter interactions at the indicated times, while the topmost dashed wavy arrow represents the signal. Arrows pointing toward the double lines represent an increase in energy level (absorption), while those pointing outward correlate to decrease in energy (stimulated emission). Each arrow is associated with a transition dipole and an exponential function of the input frequency, as summarized in Table 13.2. The frequency and wave vector of the signal are obtained by summing the input frequencies and wave vectors with appropriate signs, positive ω and k for arrows pointing to the right, and negative ω and k for arrows pointing to the left. The

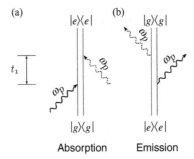

Figure 13.7 Feynman diagram for linear response: (a) absorption and (b) emission.

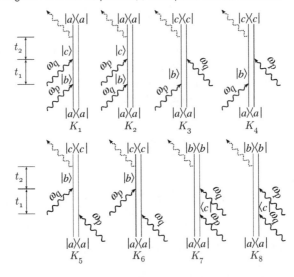

Figure 13.8 Feynman diagrams representing the eight resonances that contribute to $\chi^{(2)}$.

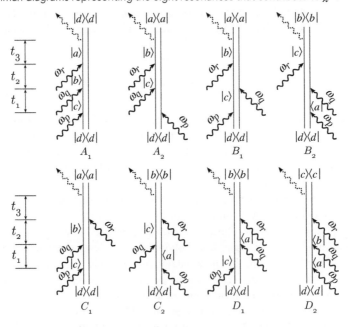

Figure 13.9 Feynman diagrams representing the eight unique resonances that contribute to $\chi^{(3)}$. The remaining 40 diagrams are obtained by permuting the labels p, q, and r.

Table 13.2 Rules for Feynman diagrams [22]

	Interaction	Diagram	Contribution to $\vec{k}_{sig}$ and ω_{sig}
Ket (left side)	Absorption $\left(\vec{\mu}_{ba}\cdot\vec{E}_i\right)e^{(i\vec{k}_i\cdot\vec{r}-i\omega_i t)}$	$|b\rangle$ _____ b $\vec{E}_i$ $|a\rangle$ _____ a	$+\vec{k}_i,+\omega_i$
	Emission $\left(\vec{\mu}_{ba}\cdot\vec{E}_i^*\right)e^{(-i\vec{k}_i\cdot\vec{r}+i\omega_i t)}$	$|b\rangle$ _____ a $\vec{E}_i^*$ $|a\rangle$ _____ b	$-\vec{k}_i,-\omega_i$
Bra (right side)	Absorption $\left(\vec{\mu}_{ba}^*\cdot\vec{E}_i^*\right)e^{(-i\vec{k}_i\cdot\vec{r}+i\omega_i t)}$	$\langle b|$ _____ b $\vec{E}_i^*$ $\langle a|$ _____ a	$-\vec{k}_i,-\omega_i$
	Emission $\left(\vec{\mu}_{ba}^*\cdot\vec{E}_i\right)e^{(i\vec{k}_i\cdot\vec{r}-i\omega_i t)}$	$\langle b|$ _____ a $\vec{E}_i$ $\langle a|$ _____ b	$+\vec{k}_i,+\omega_i$

topmost arrow represents the signal and corresponds to the trace operation to get the equilibrium average. By cyclic invariance of the trace, ρ_{eq} could appear on the left- or right-hand side. By convention, the signal is often, but not always, written as an emission on the left (ket) side of the diagram. The sign of a Feynman diagram is negative (positive) when the number of bra side interactions is odd (even). Table 13.2 also depicts the contribution to a wave-mixing energy level (WMEL) diagram. These are similar to conventional energy level diagrams with up and down arrows representing absorption and emission, respectively. Solid arrows represent ket side transitions and dashed arrows are used for bra side transitions. Not shown in Table 13.2 is the signal, which is depicted as a dashed wavy arrow in Figures 13.7 through 13.9. Each interaction, and the signal as well, introduces a transition dipole connecting two states.

The effects of dephasing are accounted for phenomenologically. Starting from Equation 4.104 and using only the zero-order Hamiltonian leads to:

$$\rho_{ab}(t) = e^{-i\hat{H}_0 t/\hbar}\rho_{ab}(0)e^{i\hat{H}_0 t/\hbar} = e^{-i\omega_{ab}t}\rho_{ab}(0) \tag{13.136}$$

Adding the phenomenological damping rate, we have:

$$\rho_{ab}(t) = e^{-i\omega_{ab}t}e^{-\Gamma_{ab}t}\rho_{ab}(0) \tag{13.137}$$

During the time-interval when the system exists in a coherence $|a\rangle\langle b|$, where $a \neq b$, the density matrix evolves according to Equation 13.137. To illustrate the use of Feynman diagrams as bookkeeping devices for calculating the response, we begin by finding $\chi^{(1)}$.

Application to Linear Response. Consider the Feynman diagrams for linear absorption and emission shown in Figure 13.7. For absorption, let us take the initial state of the system as the ground state $|g\rangle\langle g|$ and the final excited state as $|e\rangle\langle e|$. The Feynman diagram envisions a pulse at time zero acting on the ket side: $|g\rangle\langle g| \rightarrow |e\rangle\langle g|$ followed by a signal at t_1 acting on the bra side and resulting in $|e\rangle\langle g| \rightarrow |e\rangle\langle e|$. The two arrows in Figure 13.7 should not be interpreted as two photons absorbed; rather they result from the two dipole moment operators in the response function $S^{(1)}$, at times 0 and t_1. Two perturbations are required to go from the initial to the final population. Using the rules summarized in Table 13.2 for this "single pulse" example, the frequency and wave vector of the signal are the same as those of the input field, and the signal intensity is proportional to the square of the transition dipole moment. We should interpret the Feynman diagram for absorption as the creation of an output signal which interferes destructively with the input signal at the same frequency, leading to attenuation of the incident light. The absorption event could also be depicted by having the first and second interactions occur on the bra and ket sides, respectively, depicting the evolution of the density matrix from $|g\rangle\langle g|$ to $|g\rangle\langle e|$ to $|e\rangle\langle e|$. Such a diagram would be the complex conjugate of the one shown in Figure 13.7. The emission diagram is similarly explained with the outward pointing arrows representing descending energy. In this case, the diagram represents constructive interference of the input and output fields. Since the topmost arrow occurs at a frequency which is the sum of all the frequencies at preceding times, in the present case, this arrow denotes the overall transition at ω or $-\omega$ for absorption and emission, respectively. In the latter case, the frequency is provided by an external source if the emission is stimulated and by the vacuum radiation field if spontaneous. The picture is equivalent to the time-correlation function approach for describing linear spectroscopy as discussed in Chapter 5, where the spectral response depended on the dipole correlation function $\overline{\langle \mu(0)\mu(t)\rangle}$. (We used angle brackets in Chapter 5 to denote equilibrium averages. We are using an overbar here in order to reserve angle brackets for bras and kets.)

Let us use the response function of Equation 13.128 to calculate the linear susceptibility, illustrating aspects of the formalism that apply to the calculation of the nonlinear response as well. Superscripts denoting linear response and vector notation are omitted for clarity in the following expressions. We assume a cosine dependence for the incident field and write $E = E_0[\exp(i\omega_0 t) + \exp(-i\omega_0 t)]$. To ensure that the polarization is a real function, we can take

$$P(\omega) = \varepsilon_0 E_0 [\chi(\omega)e^{i\omega_0 t} + \chi^*(\omega)e^{-i\omega_0 t}] \tag{13.138}$$

This can be compared to the assumption used in the derivation of the polarizability in Section 4.4.1, where we used a wavefunction approach. It follows from the above expression that

$$P(\omega) = \varepsilon_0 E_0 [\chi'(\omega)\cos(\omega t) - \chi''(\omega)\sin(\omega t)] \tag{13.139}$$

Thus, the real part of the susceptibility χ' leads to a response which is in phase with the driving field, while the imaginary part χ'' gives rise to a $\pi/2$ phase shift. The time-dependent polarization $P(t)$ and the frequency domain polarization $P(\omega)$ are Fourier transforms of one another. The former is found using Equation 13.132. Expanding the commutator in Equation 13.128 leads to

$$S^{(1)}(t) = \frac{i}{\hbar}\theta(t)Tr\{[\rho_{eq}\mu(t)\mu(0)] - [\rho_{eq}\mu(0)\mu(t)]\} = \frac{i}{\hbar}\theta(t)[C(t) - C^*(t)] \tag{13.140}$$

We have used the fact that $C(-t) = C^*(t)$ (Chapter 5). To find this correlation function, refer to the Feynman diagram for absorption shown in Figure 13.7. For now, we assume a single initial state g with a Boltzmann population $\rho_{gg}^{(0)}$. The correlation function $C(t)$ applies to this picture while its complex conjugate (not shown) would indicate an interaction on the bra side instead of the ket, accounting for the minus sign in front of $C^*(t)$. By inspection and using the rules of Table 13.2, we write

$$C(t) = \rho_{gg}^{(0)}\mu_{ge}e^{-i\omega_{eg}t}e^{-\Gamma_{eg}|t|}\mu_{eg} \tag{13.141}$$

Reading from right to left, the interaction on the ket side gives the transition dipole μ_{eg}. The resulting coherence $|e\rangle\langle g|$ evolves as $e^{-i\omega_{eg}t}e^{-\Gamma_{eg}|t|}$, which is the product of phase evolution and damping due to the total dephasing rate Γ_{eg}. (Since the integration is over positive times only, we could omit the absolute value signs around t.) The signal then introduces the transition dipole μ_{ge} associated with the absorption of light at $\omega = \omega_{eg}$, and the probability $\rho_{gg}^{(0)}$ is associated with taking the trace to get the equilibrium average. The reader should verify that in the case where there is excited-state population at equilibrium, $\rho_{gg}^{(0)}$ should be replaced by $\rho_{gg}^{(0)} - \rho_{ee}^{(0)}$ in Equation 13.141. Using $E = E_0[\exp(i\omega_0 t) + \exp(-i\omega_0 t)]$ for the field, the time-dependent polarization is then

$$P(t) = \frac{i}{\hbar} E_0 \rho_{gg}^{(0)} |\mu_{ge}|^2 \left\{ -e^{-i\omega_0 t} \int_0^\infty dt_1 e^{i(\omega_{eg} - \omega_0)t_1} e^{-\Gamma_{ge}|t_1|} + e^{i\omega_0 t} \int_0^\infty dt_1 e^{-i(\omega_{eg} - \omega_0)t_1} e^{-\Gamma_{ge}|t_1|} \right\} \tag{13.142}$$

The two integrals above are half-Fourier transforms of an exponential and have both real and imaginary parts [9]:

$$\int_0^\infty e^{-|x|} e^{-2\pi i s x} dx = \frac{1 - 2\pi i s}{1 + (2\pi s)^2} \tag{13.143}$$

With the help of this integral, we arrive at

$$P(t) = \frac{E_0 \rho_{gg}^{(0)} |\mu_{ge}|^2}{\hbar \Gamma_{ge}} \left\{ e^{-i\omega_0 t} \left(\frac{i + (\omega_0 - \omega_{eg})/\Gamma_{ge}}{1 + [(\omega_0 - \omega_{eg})/\Gamma_{ge}]^2} \right) + e^{i\omega_0 t} \left(\frac{-i + (\omega_0 - \omega_{eg})/\Gamma_{ge}}{1 + [(\omega_0 - \omega_{eg})/\Gamma_{ge}]^2} \right) \right\} \tag{13.144}$$

Finally, we find $P(\omega) = (1/2\pi) \int_{-\infty}^\infty e^{-i\omega t} P(t) dt$ and use the Fourier representation of the delta function (Equation 5.9). The first exponential term in Equation 13.144 gives the delta function $\delta(\omega + \omega_0)$, and the second one gives $\delta(\omega - \omega_0)$, in other words, antiresonance and resonance terms. We keep the latter, consistent with its much larger contribution for positive frequency, and on comparing to Equation 13.138 arrive at the following expression for the linear susceptibility:

$$\chi(\omega) = \frac{\rho_{gg}^{(0)} |\mu_{ge}|^2}{\varepsilon_0 \hbar \Gamma_{ge}} \left(\frac{-i + (\omega_0 - \omega_{eg})/\Gamma_{ge}}{1 + [(\omega_0 - \omega_{eg})/\Gamma_{ge}]^2} \right) \tag{13.145}$$

We see that the imaginary part gives the expected Lorentzian lineshape, and the real part gives the expected dispersion. The above treatment took the point of view that the incident light would be resonant with the $g \to e$ transition. More generally, there is a sum over all initial (populated) and final states.

Second-order response. Figure 13.8 shows the eight Feynman diagrams associated with the terms $K_{1,abc}$ through $K_{8,abc}$ of Equation 13.114. In this case, the first pulse occurs at time 0 and the second pulse arrives a time t_1 later. The emitted frequency (topmost arrow) at $\omega_s = \pm \omega_p \pm \omega_q$ is detected at $t_1 + t_2$. For example, the diagram of Figure 13.8 that correlates to $K_{1,abc}$ (see also Figure 13.5), is interpreted as follows. The system starts out in state a, with density matrix $\rho_{aa}^{(0)} = |a\rangle\langle a|$. The first pulse of frequency ω_p results in the coherence $|b\rangle\langle a|$. Note the second term in the denominator of $K_{1,abc}$ permits resonance enhancement of this term when the (absorptive) transition from a to b is allowed. The second pulse at time t_1 later with frequency ω_q similarly acts on the ket side to result in the coherence $|c\rangle\langle a|$ and is enhanced when the allowed transition from b to c is resonant with input frequency ω_q. The output signal at $\omega_s = \omega_p + \omega_q$, again indicated by the dashed wavy arrow, occurs at time $t_1 + t_2$. Note that the changes to the ket along the time line represent the chain of three transition moments in the numerator of the expression for $K_{1,abc}$, read from right to left. In the example at hand, the system returns to the original state at the end (parametric process) and the diagram describes sum-frequency generation.

Contrast this diagram with the third one in Figure 13.8, which goes with $K_{3,abc}$. Here the first pulse again acts on the ket side with ω_p at time 0, taking $|a\rangle\langle a|$ to $|b\rangle\langle a|$ by absorption. The pulse at time t_1 then carries $|b\rangle\langle a|$ to $|b\rangle\langle c|$ by absorption of light at ω_q. We now have a signal at $\omega_s = \omega_p - \omega_q$, where the minus sign arises from the bra side interaction with the field. Note that eight Feynman diagrams for $\chi^{(2)}$ processes occur in pairs in which the indices p and q are swapped. Though we started from an expression for the electric field which was summed over positive and negative frequencies, we have adopted the convention that input frequencies are considered positive for the purpose of drawing both the energy level diagrams and the Feynman diagrams. Feynman diagrams can be drawn to represent many different experimental situations as will be seen in the next section and in Chapter 14. Note that whereas the energy level diagrams for K_3 and K_6, and for K_4 and K_5, are equivalent, the corresponding Feynman diagrams are not.

To see how the response function of Equation 13.129 leads to various diagrams for $\chi^{(2)}$, let us consider one of the eight terms that results when the commutators are expanded, for example:

$$\overline{\mu(0)\mu(t_1+t_2)\mu(t_1)} = \sum_m \langle m|\hat{\rho}^{(0)}\mu(0)\mu(t_1+t_2)\mu(t_1)|m\rangle \tag{13.146}$$

We wish to time order the transition moments by taking into account the invariance of the trace to a cyclic permutation.

$$\overline{\mu(0)\mu(t_1+t_2)\mu(t_1)} = \sum_m \langle m|\mu(t_1+t_2)\mu(t_1)\hat{\rho}^{(0)}\mu(0)|m\rangle \tag{13.147}$$

This shows that the perturbation at time zero acts on the bra side of the density operator while that at t_1, and the signal, occurs on the ket side. With the help of the resolution of the identity and Equation 13.131, and after changing some dummy indices,

$$\overline{\mu(0)\mu(t_1+t_2)\mu(t_1)} = \sum_m \sum_{a,b,c} \langle m|\hat{\rho}^{(0)}|a\rangle\langle a|\mu(0)|c\rangle\langle c|\mu(t_1+t_2)|b\rangle\langle b|\mu(t_1)|m\rangle$$

$$= \sum_{a,b,c} \rho_{aa}^{(0)} \mu_{ac} e^{i\omega_{cb}(t_1+t_2)} \mu_{cb} e^{i\omega_{ba}t_1} \mu_{ba} \tag{13.148}$$

$$= \sum_{a,b,c} \rho_{aa}^{(0)} \mu_{ca}^* \mu_{cb} \mu_{ba} e^{-i\omega_{bc}t_2} e^{i\omega_{ca}t_1}$$

Equation 13.148 neglects the dephasing terms, which we now add to the expression. The term we have selected corresponds to the K_5 diagram of Figure 13.8, a "bra–ket" term. Starting from population $|a\rangle\langle a|$, the first interaction at time 0 (μ_{ca}) causes the transition $|a\rangle\langle a| \rightarrow |a\rangle\langle c|$; it operates on the bra side. This coherence evolves during the interval t_1, so we need a dephasing term Γ_{ca} during this interval. The $|b\rangle\langle c|$ coherence that results from the second interaction, on the ket side, evolves during t_2, so we also need to include Γ_{bc} dephasing. The result is

$$\overline{\mu(0)\mu(t_1+t_2)\mu(t_1)} = \sum_{a,b,c} \rho_{aa}^{(0)} \mu_{ca}^* \mu_{cb} \mu_{ba} e^{-i\omega_{bc}t_2 - \Gamma_{bc}t_2} e^{i\omega_{ca}t_1 - \Gamma_{ca}t_1} \tag{13.149}$$

This is one term in the sum of eight. Using the rules of Table 13.2, each term can be written by inspection, and the labor is shortened by recognizing that the expressions (and diagrams) occur in pairs which are complex conjugates of one another, e.g. K_5 and K_3, representing DFG at frequencies $\omega_q - \omega_p$ and $\omega_p - \omega_q$ respectively.

It is also possible to associate a term in Equation 13.114 (or Equation 13.119 for $\chi^{(3)}$) with a particular Feynman diagram as follows. First, find the energy level diagram such as one of those shown in Figure 13.5

by looking at the transitions from state a to b to c in the numerator of a term in Equation 13.114. The transition with superscript i is associated with the output signal, and a resonance denominator with a single input frequency is associated with one of the other two transition moments. The remaining transition dipole is associated with the second input and gives rise to resonance of the output signal with the sum of the two input frequencies. Inputs that go with transition dipoles to the right (left) of that associated with the output signal, in the numerators of the expressions for $K_{i,abc}$, are drawn as arrows to the left (right) of the output transition in energy level diagrams such as those of Figures 13.5 and 13.6. The energy level diagram is translated into a Feynman diagram by drawing the input fields as acting on the ket or bra side according to whether the corresponding arrows in the energy level diagram occur to the left or right of the output arrow, respectively. Since a, b, c are dummy indices, the order of the energy levels is arbitrary, but the transitions must originate in state a owing to the dependence on the population of that state. For the example above (term K_5), the associated diagram corresponds to DFG and thus the Feynman diagram shows a signal which results in a final state different from the initial state, even though the chain of transition moments in the sum-over-states formula begins and ends in state a.

Similarly, when the nested commutators needed for the third-order response (Equation 13.130) are expanded, there are eight terms, each of which is a product of four dipole moment operators. Three of these correspond to light pulses acting at 0, t_1, and $t_1 + t_2$, while the fourth goes with the signal detected at $t_1 + t_2 + t_3$. The eight unique Feynman diagrams for $\chi^{(3)}$ processes are shown in Figure 13.9. Since there are 3! permutations of the three input frequencies, there are actually 48 terms in the expansion of $\chi^{(3)}$. Specific experiments invoking some of these terms are considered in Section 13.5. Feynman diagrams provide insight that is not apparent from energy level diagrams like those in Figures 13.5 and 13.6 by showing how the density matrix evolves through perturbations to populations and coherences.

13.5 EXPERIMENTAL APPLICATIONS OF NONLINEAR PROCESSES

We next survey a few of the many spectroscopic experiments that spring from $\chi^{(2)}$ and $\chi^{(3)}$. Since high laser powers are required to observe nonlinear effects, laser sources for these experiments are typically pulsed and many experiments are performed in the time domain to acquire dynamical information. Further discussion of time-resolved experiments will be taken up in the next chapter, and examples in this chapter emphasize experiments where the signal is often detected in the frequency domain.

13.5.1 EXAMPLES OF $\chi^{(2)}$ EXPERIMENTS

Vibrational Sum-Frequency Generation Spectroscopy. Owing to the vanishing even-order nonlinear terms in isotropic media, $\chi^{(2)}$ experiments are particularly useful as interface-specific spectroscopic tools and in microscopy [10–13]. Vibrational sum-frequency (VSF) spectroscopy is a surface-specific technique that, unlike other surface techniques such as photoelectron spectroscopy, does not require high vacuum. It is therefore used to probe gas/liquid, solid/liquid and liquid/liquid interfaces. In this experiment, shown schematically in Figure 13.10, a visible beam and a tunable infrared (IR) beam are incident on the interface and an output signal at the sum frequency $\omega_{vis} + \omega_{IR}$ is observed when the IR beam is resonant with a molecular vibration. For the geometry shown in Figure 13.10, where signals are detected in reflection, the propagation vector of the output signal is close to, but not exactly collinear with, the reflected visible beam. As ω_{IR} is tuned to achieve resonance with a vibrational transition that is both Raman and IR active, an enhanced signal at $\omega_{SF} = \omega_{vis} + \omega_{IR}$ is detected. Were it not for the interface, this selection rule would preclude VSF activity of centrosymmetric molecules. In fact, intermolecular interactions and symmetry breaking at the interface result in nonzero VSF intensities of molecules which possess inversion symmetry when isolated.

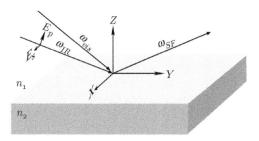

Figure 13.10 Experimental configuration for vibrational sum-frequency spectroscopy using S or P polarized incident beams.

The relevant susceptibility for vibrational SFG is obtained starting from Equation 13.112 and letting all equilibrium population reside in the ground state, $\rho_{gg}^{(0)} = 1$. The terms which are resonant when the IR frequency is tuned to a vibrational transition $g \to v$ can be collected:

$$\chi_{IJK}^{(2)}(\omega_{SF};\omega_{vis},\omega_{IR}) = \frac{N}{\varepsilon_0 \hbar^2} \sum_e \left\{ \frac{\mu_{ge}^I \mu_{ev}^J}{(\omega_{eg} - \omega_{SF} - i\Gamma_{ge})} + \frac{\mu_{ge}^J \mu_{ev}^I}{(\omega_{eg} + \omega_{SF} + i\Gamma_{ge})} \right\} \frac{\mu_{vg}^K}{(\omega_{vg} - \omega_{IR} - i\Gamma_{vg})} \quad (13.150)$$

Equation 13.150 is the resonant signal. There is in addition a broad background resulting from the nonresonant terms. The term in the curly brackets is the IJ component of the Raman tensor (transition polarizability) for the vibrational transition and the last term contains the transition dipole moment μ_{vg}^K that determines the IR activity of the vibration with transition frequency ω_{vg}. Alternatively, we can write the molecular hyperpolarizability as

$$\beta_{ijk} = \frac{\langle g|\alpha_{ij}|v\rangle \langle v|\mu_k|g\rangle}{\omega_{IR} - \omega_{vg} + i\Gamma_{vg}} \quad (13.151)$$

where the numerator is the product of the Raman and IR transition amplitudes. The molecular hyperpolarizability is written in the molecule frame $ijk = xyz$ while the bulk measurement is dependent on lab-frame directions $IJK = XYZ$.

In either representation, it is apparent that the SFG signal is resonantly enhanced when the IR input is tuned to the frequency of a vibrational transition (at ω_{vg}) that is both Raman and IR active. Detection of the output signal is facilitated by its occurrence at a visible rather than IR wavelength. Owing to the availability of tunable IR sources with wavelengths in the vicinity of 3 μm, VSF studies of higher frequency vibrational modes such as OH and CH stretches are prevalent in the literature. The pulsed IR and visible input beams are typically spatially and temporally overlapped at the surface. A signal can still be observed using a visible pulse that is delayed in time with respect to the IR pulse, provided the delay is shorter than the decay of the vibrational coherence. Such a scheme can reject the more rapidly decaying nonresonance signal. Experiments can be done by scanning the IR frequency to achieve resonance or by using a broadband IR source.

Figure 13.10 defines S and P polarization for the input and signal beams, which refer to light which is respectively polarized perpendicular or parallel to the plane of incidence. Phase matching for detection of the signal at $\omega_{SF} = \omega_{IR} + \omega_{vis}$ is met by the condition:

$$n_{SF}\omega_{SF} \sin\theta_{SF} = n_{vis}\omega_{vis} \sin\theta_{vis} \pm n_{IR}\omega_{IR} \sin\theta_{IR} \quad (13.152)$$

The plus or minus sign is used according to whether the two input beams propagate from the same or from opposite directions (the former is depicted in Figure 13.10). The angles θ_{SF}, θ_{vis}, and θ_{IR} are the angles between the designated propagation vectors and the surface normal, and n_{SF}, n_{vis} and n_{IR} are refractive indices for the medium through which the beams propagate, evaluated at the relevant frequencies. The intensities of the reflected or transmitted beams depend on the linear and nonlinear Fresnel coefficients, as discussed in [10–12].

Molecules adsorbed at isotropic surfaces such as liquids are axially symmetric with respect to the surface normal (Z), since the X and Y directions are equivalent. The average orientation of molecules at the interface can be determined by exploiting the tensor nature of $\chi^{(2)}$. Symmetry considerations such as those in Section 13.2.2 enable us to determine that many of the 27 possible values of $\chi^{(2)}_{IJK}$ are zero. Since X is equivalent to $-X$, and Y is equivalent to $-Y$, tensor elements of $\chi^{(2)}$ with an odd number of X and/or Y subscripts must vanish, e.g. $\chi^{(2)}_{XZZ} = -\chi^{(2)}_{-XZZ} = -\chi^{(2)}_{XZZ} = 0$. This leaves only four unique tensor elements: $\chi^{(2)}_{ZZZ}$, $\chi^{(2)}_{XXZ} = \chi^{(2)}_{YYZ}$, $\chi^{(2)}_{XZX} = \chi^{(2)}_{YZY}$, and $\chi^{(2)}_{ZXX} = \chi^{(2)}_{ZYY}$. Four unique SFG experiments are done using SSP, SPS, PSS and PPP polarizations. These polarizations are listed in the order sum frequency, visible, IR, such that the last index correlates to the direction of the vibrational transition moment. Recall that S-polarized light has an electric field perpendicular to the plane of incidence, defined by the incident and reflected beams, whereas P-polarized light is polarized in the plane of incidence. Thus the intensity of the signal in an SSP experiment depends on the component of the IR transition moment which is parallel to the plane of incidence, while the visible and sum-frequency beams are polarized perpendicular. The only surviving component for the SSP configuration is thus $\chi^{(2)}_{XXZ} = \chi^{(2)}_{YYZ}$. The SSP experiment sees vibrations for which the transition moment has a component perpendicular to the interface. The SPS ($\chi^{(2)}_{XZX} = \chi^{(2)}_{YZY}$) and PSS ($\chi^{(2)}_{ZXX} = \chi^{(2)}_{ZYY}$) experiments reveal the extent to which this transition moment is perpendicular to the plane of incidence, or alternatively, *in* the plane of the interface. Since P-polarized IR has components in both the Y and Z directions, the PPP experiment probes all four nonzero tensor elements and thus sees both parallel and perpendicular components of the vibrational transition moment.

A number of VSF studies have looked at the orientation of hydrocarbons such as surfactants at the air/water or water/organic liquid interface. In [13], polarized VSF spectra of the isomers 1-, 2-, 3- and 4-hexadecanol were obtained at the air/water interface as shown in Figure 13.11, revealing the order of the molecular alignment

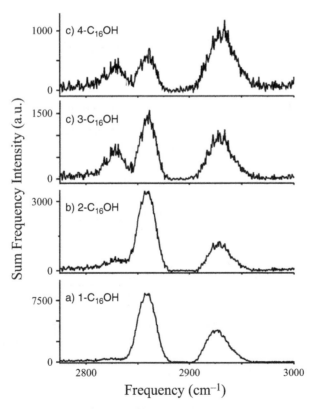

Figure 13.11 Vibrational sum-frequency spectra with *SSP* polarization of monolayers of 1-, 2-, 3-, and 4-hexadecanol on water. (Reprinted with permission from Can, S.Z. et al., Structure and organization of hexadecanol isomers adsorbed to the air/water interface, *Langmuir*, 22, 8043, 2006. Copyright 2006 American Chemical Society.)

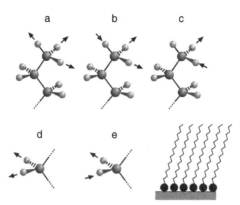

Figure 13.12 Vibrations of an all-trans hydrocarbon chain. (a) symmetric methyl stretch, (b) antisymmetric methyl stretch, in-plane component, (c) antisymmetric methyl stretch, out-of-plane component, (d) symmetric methylene stretch, and (e) antisymmetric methylene stretch. The cartoon at bottom right shows a monolayer of a film of hydrocarbon molecules, in the all-trans configuration, adsorbed at a surface through a polar head-group. (Adapted from Shen, Y.R. and Ostroverkhov, V., *Chem. Rev.*, 106, 1140–1154, 2006.)

at the interface. The alignment is the result of a preference for hydrogen bonding between the –OH group of the alcohol and surface water, and van der Waals interactions which favor close packing of the hydrophobic hydrocarbon chains. For 1-hexadecanol, these interactions work together to favor the all-trans alignment pictured in the cartoon of Figure 13.12. The isomers for which the hydroxyl group is in the 2, 3 and 4 position, on the other hand, have to form kinks, known as *gauche* deformations, to optimize hydrophilic and hydrophobic interactions, reducing the packing density at the surface and increasing alignment disorder. Can et al. [13] measured VSF spectra using *SSP* and *SPS* polarization to reveal molecular vibrations with IR transition moments respectively perpendicular and parallel to the interface. Figure 13.12 shows the displacement vectors for (a) the symmetric methyl stretch r$^+$, (b and c) the degenerate methyl stretch r$^-$, (d) the symmetric methylene stretch d$^+$, and (e) the asymmetric methylene stretch d$^-$. For the perfectly ordered all-trans chain depicted for 1-hexadecanol, the d$^+$ stretch at 2841 cm^{-1} would not appear in *SSP* polarization, since its transition dipole moment would be parallel to the interface. As seen in *SSP* spectra of Figure 13.11, this feature is quite weak for 1-hexadecanol and increases in relative intensity on moving the –OH group to the 2, 3, and 4 positions. All spectra show contributions from r$^+$ at 2872 cm^{-1}, reflecting the component of the symmetric stretch transition dipole moment, which is coincident with the local C_3 axis, in the direction perpendicular to the interface. The intensity ratio of r$^+$ to d$^+$ is taken as an order parameter, and indeed it decreases on moving the hydroxyl group away from the chain terminus, as kinks are introduced. In addition to the r$^+$ and d$^+$ features, the spectra in Figure 13.11 all show a broad band at about 2930 cm^{-1} that contains contributions from r$^-$ and from a bending overtone of the methyl group which borrows intensity from the symmetric stretch r$^+$ via Fermi resonance.

13.5.2 Examples of $\chi^{(3)}$ experiments

The number of four-wave mixing (FWM) experiments is quite large, and some of them will be discussed in Chapter 14. We consider first some experiments which are often done in the frequency domain, though pulsed variations are also possible. We also revisit the topic of spontaneous Raman scattering (Chapter 12), which can be viewed as a FWM experiment where one of the three input fields is that of the vacuum field of radiation.

13.5.2.1 TWO-PHOTON AND MULTIPHOTON ABSORPTION

Absorption of *n* photons, for which the sum of photon energies matches the energy difference E_{eg} between the ground and excited electronic state, is a nonparametric process. While one-photon transitions depend on $\chi^{(1)}$, those for two-, three-, and four-photon transitions arise from $\chi^{(3)}$, $\chi^{(5)}$, and $\chi^{(7)}$, etc., respectively. It is possible to achieve multiphoton absorption of photons of the same or different wavelengths. Two-photon absorption (TPA) is referred to as degenerate or nondegenerate according to whether the two photons are the same or different

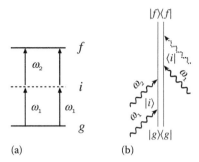

Figure 13.13 (a) Energy level diagram and (b) Feynman diagram for two-photon absorption, a $\chi^{(3)}$ process.

color, respectively. In either case, the process is determined by the imaginary part of $\chi^{(3)}$. TPA was predicted theoretically by Maria Goeppert-Mayer in the 1930s but not observed experimentally until 1961. The dependence of TPA on the square of the incident light intensity, or product of light intensities in the nondegenerate case, presents experimental advantages for applications such as microscopy, where the fluorescence resulting from TPA is detected with higher spatial resolution in comparison to fluorescence excited by linear absorption. The spatial resolution of ordinary fluorescence-based microscopy cannot exceed the diffraction limit, which is typically on the order of half the wavelength. Fluorescence microscopy based on two-photon excitation exceeds this resolution limit because the strongest signals are obtained from the most intense part of the diffraction limited focal volume. Fluorescence detection of TPA is also advantageous because the incident light is weakly attenuated, and in highly scattering media lower frequency incident light contributes less background to the desired signal.

It may seem puzzling that absorption of two photons should be a third-order rather than a second-order nonlinear process. As shown in the energy level diagram of Figure 13.13a, the transition from ground state g to excited electronic state f takes place via intermediate states i. Though this diagram emphasizes the resonance condition, where $\omega_1 + \omega_2 = \omega_{fg}$, a sum-over-states formula for TPA, similar to that for the Raman effect, includes *all* intermediate states, not just those with energies bracketed by those of the ground and excited states. Referring to the Feynman diagram of Figure 13.13b, TPA is understood as resulting from two interactions with real photons and one with a virtual photon. In the nondegenerate case, the virtual photon can have frequency ω_1 or ω_2 resulting in attenuation of the input beam at either ω_2 or ω_1, respectively. TPA spectra are often graphed as a function of the sum of the input frequencies for ease of comparison to the one-photon spectra, in the case where the latter is also allowed.

When the intermediate state i is real rather than virtual, a resonance term dominates the sum-over-states formula for TPA. It should be emphasized that TPA is not the same phenomenon as excited state absorption, to be discussed in the next chapter. The latter is a two-step process which is sensitive to relaxation of the intermediate state, whereas TPA is a simultaneous two-photon event. The reader should verify that in the diagram of Figure 13.13b, three field–matter interactions cause the density matrix to evolve as follows: $|g\rangle\langle g| \rightarrow |i\rangle\langle g| \rightarrow |f\rangle\langle g| \rightarrow |f\rangle\langle i|$. The system never passes through a population term $|i\rangle\langle i|$, and the second interaction results in a coherence that oscillates at the frequency ω_{fg}, which is larger than either of the two driving frequencies. In the case of excited-state absorption, on the other hand, the density matrix passes through such a population term, connected by coherences that oscillate at frequencies of the driving fields (we look at this more closely in the next chapter). This distinction between TPA and excited state absorption is analogous to that between the simultaneous two-photon process of resonance Raman scattering as compared to sequential absorption and emission of photons in a fluorescence experiment.

Considering the case of degenerate TPA with excitation at frequency ω, the attenuation of incident light intensity I as it traverses the sample is given by

$$-\frac{dI}{dx} = N\delta I^2/\hbar\,\omega \qquad (13.153)$$

where N is the number of molecules per unit volume, I is the intensity of light, and δ is the TPA cross-section. δ is given in Göppert-Mayer units, GM, where 1 GM $= 10^{-50}$ cm^4 s/photon. The TPA signal has a Lorentzian lineshape with maximum cross-section given by [14,15]:

$$\delta_{max} = \frac{\omega^2 L^4}{\varepsilon_0^2 n^2 c^2 \Gamma} \left[\sum_i \frac{\langle \mu_{gi} \mu_{if} \rangle}{\omega_{ig} - \omega} \right]^2 \equiv \frac{\omega^2 L^4}{\varepsilon_0^2 n^2 c^2 \Gamma} S_{fg} \tag{13.154}$$

where $L = (n^2 + 2)/3$ is the local field factor, n the refractive index, and Γ is the halfwidth of the Lorentzian. The angle brackets represent an orientational average. When this average is performed with the approximation that all transition moments are parallel, the sum S_{fg} can be separated into two terms:

$$S_{fg} = \frac{1}{5} \left\{ \left[\frac{(\mu_{ff} - \mu_{gg})\mu_{gf}}{\omega} \right]^2 + \sum_{i \neq f, g} \frac{\mu_{gi}^2 \mu_{if}^2}{(\omega_{ig} - \omega)^2} \right\} \tag{13.155}$$

The first term above, called the D-term (for "dipolar") depends on the difference in the state dipole moments of the ground and final states. This sort of contribution is not operative in linear absorption, but it can dominate in TPA when the $g \to f$ transition results in a large change in dipole moment. The D-term also requires that the $g \to f$ transition be allowed in linear absorption. Thus there is much emphasis on the strong charge-transfer transitions of dyes with donor–acceptor character. Note there is no resonance enhancement for the D-term. The second term, called the T-term or "two-photon" term, does permit resonance enhancement when the incident photon matches the ground-to-intermediate state transition frequency. For this term to contribute, both the $g \to i$ and $i \to f$ transitions have to be allowed by electric dipole selection rules. For molecules with inversion symmetry, the D-term must vanish as neither the ground nor excited state can possess a permanent dipole moment. In this case, the T-term results in selection rules opposite to that of linear absorption in that TPA permits transitions between states of the same inversion symmetry: a $u \to u$ transition takes place via an intermediate state of *gerade* symmetry and a $g \to g$ transition goes through an *ungerade* intermediate state.

Two-photon absorption is often detected by observing the fluorescence rather than the weak attenuation of the incident light. The latter measurement is required when one wants to determine the TPA cross-section, typically using the "z-scan" measurement. The sample is translated along the path of a focused laser beam to vary the incident intensity and the transmitted light is detected after passing through a large aperture. The technique requires care to eliminate artifacts from nonlinear refraction, the dependence of the real part of the refractive index on light intensity, known as the optical Kerr effect.

Chromophores which exhibit large TPA cross-sections are often of the D-π-A type, dipolar dyes consisting of donor and acceptor moieties bridged by a conjugated chain. Centrosymmetric TPA dyes are quadrupolar analogues with D-π-A-π-D or A-π-D-π-A structure. Clearly, these dipolar and quadrupolar dyes invoke the D- and T-terms respectively. Examples of these two kinds of donor-acceptor dyes are shown in Figure 13.14.

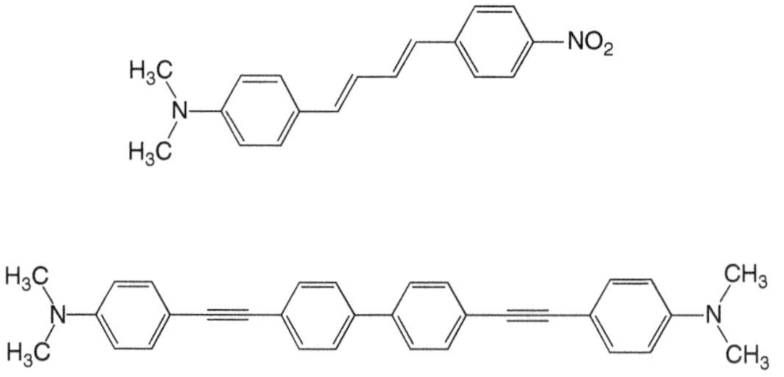

Figure 13.14 (a) Example of a D-π-A molecule and (b) a D-π-A-π-D molecule.

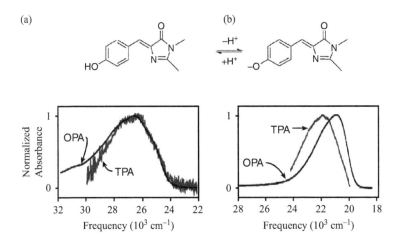

(a)   (b)

Figure 13.15 Neutral (a) and ionized (b) forms of HDBI and their one-photon (OPA) and two-photon (TPA) absorption spectra in DMSO. (Reprinted from Hosoi, H. et al., Solvent dependence of two-photon absorption spectra of the enhanced green fluorescence protein (eGFP) chromophore, *Chem. Phys. Lett.*, 2015, *630*, 32. Copyright 2015, with permission from Elsevier. .)

Figure 13.15 shows one- and two-photon absorption spectra of the molecule 4'-hydroxybenzylidene-3-dimethylimidazolinone (HBDI) in its neutral and anionic forms, taken from [16]. HBDI serves as a model for the chromophore in green fluorescent protein, which is of interest for biological fluorescence microscopy applications. The data were gathered with two beams, a fixed frequency near-IR beam and a white light continuum ranging from 450 to 750 nm. In the case of the neutral form, the one-photon absorption (OPA) coincides with the TPA spectrum graphed as a function of the sum of the two input frequencies, as expected. In contrast, the ionized form of HBDI shows a blue shift of the TPA spectrum relative to OPA. Hosoi et al. [16] interpreted this to be the result of a second excited state of the ionized form that is allowed in the TPA spectrum but not the OPA spectrum. The putative $S_0 \rightarrow S_2$ transition of the ionized form was concluded to have a large solvatochromism. We have seen in Chapter 11 that solvatochromism can result from a change in the dipole moment on going from the ground to the excite state, as in the D-term of Equation 13.155.

13.5.2.2 COHERENT RAMAN SPECTROSCOPY

A number of different nonlinear spectroscopy methods invoke the Raman effect and are powerful approaches to the determination of vibrational mode frequencies and lifetimes. A few are briefly mentioned in passing before going into more detail about the popular CARS experiment. One such nonlinear effect is *hyperRaman scattering*. With a sufficiently intense incident laser pulse of frequency ω_0, hyperRaman scattering is observed at frequencies shifted from twice the incident frequency by integral multiples of a vibrational frequency ω_{vib}, i.e., $2\omega_0 \pm n\omega_{vib}$. Note that unlike vibrational overtones, the peaks are shifted by exact multiples of the fundamental frequency, and both Stokes and anti-Stokes peaks are observed. For the Stokes sides, for example, the incident laser creates a beam at $2\omega_0 - \omega_{vib}$, which acts as a source for further Stokes Raman scattering at $2\omega_0 - 2\omega_{vib}$, which acts as a source for scattering at $2\omega_0 - 3\omega_{vib}$, etc. This experiment can be considered the two-photon counterpart to ordinary Raman scattering, and the intensity of scattered light is very weak without resonance enhancement. Since the theoretical foundations of hyperRaman spring from the first hyperpolarizability β, it is actually a second-order nonlinear phenomenon and as such is closely related to second harmonic generation, though the scattered light is not coherent.

Raman experiments can be classified as spontaneous (incoherent) or coherent, according to whether the molecular vibrations are driven by a single input field, as is the case for spontaneous Raman, or by two input fields, as in stimulated Raman spectroscopy. Coherence in this context represents the phase relationship of different molecules viewed as dipole emitters in the sample. In spontaneous Raman, the vibrations of different molecules are uncorrelated and the signal is proportional to the number of molecules. In stimulated

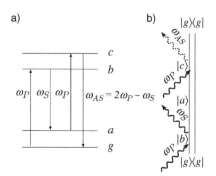

Figure 13.16 (a) CARS energy levels and (b) Feynman diagram.

Raman spectroscopy experiments, the coherence of the input fields results in molecular vibrations being driven with definite phase relationships, and depending on detection scheme the signal may be proportional to the square of the number of molecules (see [17]). A single representative Feynman diagram can be drawn for both spontaneous or coherent Raman scattering (Figure 13.19b); it is the origin of the second beam, a real or virtual photon, that differs. The state diagrams often drawn to depict Raman scattering, such as Figure 4.2, obscure the fact that the effect results from three field–matter interactions. A more informative energy level scheme is identical to that depicted for CARS in Figure 13.16.

Stimulated Raman Spectroscopy requires two input beams at frequencies ω_1 and ω_2. With $\omega_1 > \omega_2$, the ω_1 and ω_2 beams are referred to as the pump and the Stokes beams, respectively. When the beat frequency of the two input beams is resonant with a vibrational frequency, $\omega_1 - \omega_2 = \omega_{vib}$, four nonlinear effects occur simultaneously: stimulated Raman gain (SRG), stimulated Raman loss (SRL), coherent Stokes Raman (CSRS), and coherent anti-Stokes Raman (CARS). In the SRG experiment, the induced field interferes constructively with the input field E_2 and, as the name implies, an increase in intensity at the Stokes frequency is observed. In contrast, for SRL there is destructive interference between the induced field and E_1 and the experiment measures the attenuation of the pump beam at ω_1. CARS and CSRS result from scattering at the frequencies $\omega_1 + \omega_{vib}$ and $\omega_2 - \omega_{vib}$, respectively. SRG, SRL, CSRS and CARS are all third-order nonlinear phenomena.

The CARS experiment is an advantageous way to obtain Raman spectra with high sensitivity and enhanced spatial resolution compared to ordinary (spontaneous) Raman scattering. The directionality of the response (phase matching) and the observation of a signal at shorter wavelength than that of the incident beams result in rejection of the fluorescence signal that often swamps that from ordinary Raman scattering. We shall refer to two input beams at the frequencies ω_P (the "pump beam") and ω_S (the "Stokes beam"). Interaction of the sample with these two beams drives coherent oscillations at the beat frequency $\omega_P - \omega_S$, while a second interaction with the Stokes beam probes this coherence. The resulting polarization is resonantly enhanced when this beat frequency matches a vibrational frequency ω_{vib}. A second interaction with the pump beam results in anti-Stokes Raman scattering. The energy level diagram and Feynman diagram for CARS resonances are shown in Figure 13.16. Both show that the frequency of the signal is the anti-Stokes frequency $\omega_{AS} = 2\omega_P - \omega_S = \omega_P + \omega_{vib}$. Phase matching permits the desired CARS spectrum to be isolated from the signals due to other simultaneously occurring nonlinear responses. Figure 13.17 shows the typical "folded boxCARS" arrangement of the propagation vectors used to achieve this phase matching.

A typical CARS experiment uses a monochromatic pump beam (such as the 532 nm pulsed output of a Nd:YAG laser) and a broadband dye laser which provides the Stokes beam. The latter can be scanned to bring the difference frequency into resonance with a vibrational frequency. Alternatively, the broadband dye laser is pulsed and the Raman spectrum of the sample is obtained in a single shot. The relevant susceptibility is [18]:

$$\chi_{CARS}^{(3)} = \frac{4\pi^2 \varepsilon_0 c^4 (N_g - N_a)\left(\dfrac{d\sigma}{d\Omega}\right)\omega_{vib}}{\hbar \omega_S^4 [\omega_{vib}^2 - (\omega_P - \omega_S)^2 + i\Gamma(\omega_P - \omega_S)]} \tag{13.156}$$

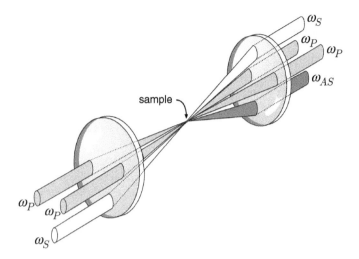

Figure 13.17 Experimental phase-matching configuration for folded box CARS.

The susceptibility is seen to depend on the differential cross-section for ordinary Raman scattering, $(d\sigma/d\Omega)$, thus the usual selection rules apply. It is also proportional to the population difference $N_g - N_a$ between the lower and upper vibrational levels. The signal depends on the square of the susceptibility and on the intensities of the pump and Stokes beams:

$$I_{CARS} \propto \left[\chi^{(3)}_{CARS} \right]^2 I_P^2 I_S \tag{13.157}$$

As in the VSF experiment, a nonresonant background is present in addition to the desired signal. This background decays more rapidly than that of the vibrational coherences, which have lifetimes in the picosecond range. Thus the nonresonant contribution can be partly rejected by selecting a sufficiently long delay time between the pump and Stokes pulses.

The dependence of the CARS signal on vibrational populations renders it a useful probe of vibrational (or rotational) populations. Figure 13.18 shows an example of a CARS spectrum of N_2 taken at a series of times following an electric discharge. The discharge creates a highly nonequilibrium population of vibrational levels, and numerous hot bands $v \to v + 1$ are observed to increase in time following the discharge, permitting an estimate of the vibrational populations as a function of time. The CARS signal can be detected using

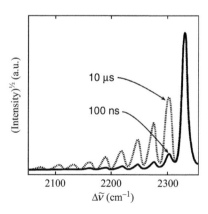

Figure 13.18 CARS spectrum of N_2 at 100 ns and 10 μs following electric discharge. (Reprinted from Lempert, W.R. and Adamovich, I.V., Coherent anti-Stokes Raman scattering and spontaneous Raman scattering diagnostics of nonequilibrium plasmas and flows, *J. Phys. D. Appl. Phys*, 47, 433001, 2014. Copyright IOP Publishing. Reproduced with permission. All rights reserved.)

either homodyne or heterodyne techniques. The heterodyne case permits the real and imaginary parts of the susceptibility to be determined.

13.5.2.3 SPONTANEOUS RAMAN SCATTERING AS A THIRD-ORDER NONLINEAR PROCESS

In this section, the origins of Raman spectroscopy as a third-order nonlinear experiment are revealed. When one views the state diagram for the Raman event (Figure 13.19a), the question arises as to how to distinguish resonance Raman scattering from absorption followed by resonance fluorescence. While there are clear experimental distinctions between resonance Raman and relaxed fluorescence, to be discussed below, one can embrace both effects with the concept of spontaneous light emission (SLE). In ordinary cases where the fluorescence is emitted following vibrational relaxation of the excited electronic state, SLE is the sum of easily distinguished Raman and fluorescence components. The Raman peaks are spectrally sharp and appear at constant frequency shift with respect to the exciting radiation, while the broader fluorescence is emitted at a fixed range of absolute frequencies. In the gas phase, on the other hand, vibrational SLE can take place before excited state thermalization, leading to unrelaxed fluorescence, also called "hot luminescence."

Raman emission is coherent, while spontaneous fluorescence emission is not. In the present context, the word coherent refers to the phase relationship between the driving field and the induced polarization, rather than the coherence of the response from different molecules in the sample. The relative yield of fluorescence and Raman emission depends strongly on the dephasing rate of the coupled electronic states. In general, the two effects are not separable in that the total SLE is not the sum of Raman and fluorescence contributions. Our previously employed Kramers–Heisenberg–Dirac expression for the polarizability (Chapter 4) did not include the effects of dephasing, and as such is applicable to isolated molecules. Here, we present without derivation a more general treatment of SLE [3] which highlights the role of dephasing.

Consider the energy levels involved in both types of SLE (Figure 13.19a), where a and c are the ground and excited vibrational states within the ground electronic state, and b is an intermediate excited vibronic state. The Feynman diagram in Figure 13.19b can be considered typical of the Raman experiment, while Figure 13.19c is representative of fluorescence. The distinction is that in the latter case the system passes through a bb (i.e., $|b\rangle\langle b|$) population: $aa \rightarrow ab \rightarrow bb \rightarrow bc \rightarrow cc$, as it must for the two-step process of absorption/fluorescence. The path of Figure 13.19b never goes through a population, and while it is "typical" of Raman scattering, this diagram also contributes to fluorescence. In both cases, the third interaction leads to the bc coherence associated with the signal. The diagram of Figure 13.19c only contributes to fluorescence. Figure 13.19b can be considered to depict either coherent or spontaneous Raman scattering according to whether the ω_2 photon is real or virtual, respectively.

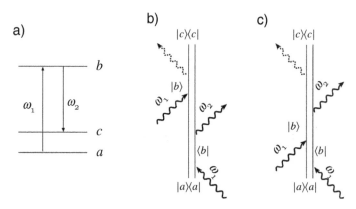

Figure 13.19 Energy level diagram (a) and Feynman diagrams typical of (b) spontaneous or coherent Raman scattering and (c) fluorescence. See text for discussion.

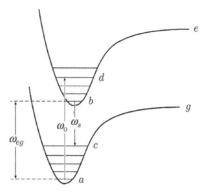

Figure 13.20 State diagram for discussion of spontaneous light emission.

The following treatment of SLE refers to the state diagram shown in Figure 13.20, which includes an additional intermediate vibronic state d. As discussed in [3], the cross section for SLE is defined as follows:

$$\sigma_{SLE}(\omega_0,\omega_s)=\frac{4\omega_0\omega_s^3}{9\hbar^2c^4}S(\omega_0,\omega_s) \tag{13.158}$$

where

$$S(\omega_0,\omega_s)=\sum_{a,c}P_a\left[K_I+K_{II}+K_{III}\right] \tag{13.159}$$

P_a is the probability that state a is occupied. The K terms are defined as follows:

$$K_I=-i\sum_{b,d}\mu_{ab}\mu_{bc}\mu_{cd}\mu_{da}\left(\frac{1}{\omega_{ba}-\omega_0+i\gamma}\right)\left(\frac{1}{\omega_{bd}+i\Gamma}\right)\left(\frac{1}{\omega_{bc}-\omega_s+i\gamma}\right)+c.c. \tag{13.160}$$

$$K_{II}=-i\sum_{b,d}\mu_{ab}\mu_{bc}\mu_{cd}\mu_{da}\left(\frac{1}{\omega_{bd}+i\Gamma}\right)\left(\frac{1}{\omega_{ba}-\omega_0+i\gamma}\right)\left(\frac{1}{\omega_{cd}-\omega_s+i\gamma}\right)+c.c. \tag{13.161}$$

$$K_{III}=-i\sum_{b,d}\mu_{ab}\mu_{bc}\mu_{cd}\mu_{da}\left(\frac{1}{\omega_{ba}-\omega_0+i\gamma}\right)\left(\frac{1}{\omega_{ca}+\omega_s-\omega_0+i\varepsilon}\right)\left(\frac{1}{\omega_{cd}+\omega_s+i\gamma}\right)+c.c. \tag{13.162}$$

We are using Γ here for the inverse lifetime of the excited electronic state, and $\gamma=\gamma_{eg}$ for the electronic dephasing rate, considered independent of vibrational level. The above three equations assume that the inverse lifetimes of the a and c states are zero: $\Gamma_a=\Gamma_c=0$. This is based on the longer lifetime of vibronic states within the ground electronic state, compared to those of the excited electronic state. The inverse lifetimes of the b and d states are assumed equal: $\Gamma_b=\Gamma_d=\Gamma$, and $\gamma=\Gamma/2+\gamma^*$ where γ^* is the electronic pure dephasing rate. Lastly, the term $i\varepsilon$ in the second energy denominator of K_{III} prevents the cross-section from diverging when $\omega_0-\omega_s$ is resonant with the vibrational transition at ω_{ca}.

How do Equations 13.160 through 13.162 account for common experimental observations concerning Raman and fluorescence emission? The former consists of sharp lines centered at frequencies such that $\omega_0-\omega_s=\omega_{ca}$. Clearly, these must originate from the K_{III} term, but this term also includes fluorescence. The two processes can be distinguished experimentally by tuning the incident frequency ω_0. Raman lines appear at constant frequency shifts $\omega_0-\omega_s$ while fluorescence appears at constant emission frequency ω_s. In addition, the fluorescence is broad compared to the Raman lines, consistent with the neglect of the a and c state inverse lifetimes compared to those of b and d. This is illustrated in Figure 13.21, which shows the variation in the total emission of methyl iodide, excited within the UV absorption band, as the incident radiation is tuned [23]. Note that the wavelength of the sharp Raman-like peak shifts with the excitation wavelength, while the fluorescence-like component remains fixed.

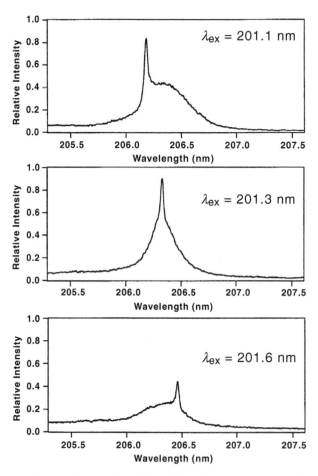

Figure 13.21 Spontaneous light emission of methyl iodide at three excitation wavelengths. (Reprinted with permission from Ziegler, L., On the difference between resonance Raman scattering and resonance fluorescence: An experimental view, *Acc. Chem. Res.*, 27, 1, 1994. Copyright 1994, American Chemical Society.)

Within the limit of fast solvent-induced dephasing, the expression for spontaneous light emission, $S(\omega_0,\omega_s)$, can be written as the sum of two terms: $S_{Raman} + S_{fluor}$. The result is

$$S_{Raman}(\omega_0,\omega_S) = 2\pi \sum_{a,c} P_a \left| \frac{\mu_{cb}\mu_{ba}}{\omega_0 - \omega_{ba} + i\gamma} \right|^2 \delta(\omega_0 - \omega_S - \omega_{ca}) \tag{13.163}$$

and

$$S_{fluor}(\omega_0,\omega_S) = \sum_{a,b,c,d} P_a \mu_{ab}\mu_{bc}\mu_{cd}\mu_{da} \left(\frac{2\gamma^*}{\omega_{bd} + i\Gamma} \right) \left(\frac{1}{\omega_{ba} - \omega_0 + i\gamma} \right) \left(\frac{1}{\omega_{da} - \omega_0 - i\gamma} \right)$$

$$\times \left\{ \left(\frac{1}{\omega_{dc} - \omega_S - i\gamma} \right) - \left(\frac{1}{\omega_{bc} - \omega_S + i\gamma} \right) \right\} \tag{13.164}$$

The delta function in Equation 13.163 produces sharp lines at the Raman frequencies, and the resonance denominator causes enhancement when the laser frequency is tuned to the $a \rightarrow b$ electronic transition. Note that the fluorescence emission S_{fluor} is proportional to the electronic pure dephasing rate, γ^*. When the dephasing rate goes to zero, the fluorescence vanishes and all the radiation is emitted as Raman scattering. This is illustrated by the data in Figure 13.22, which shows the spontaneous light emission of methyl iodide

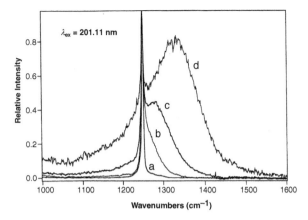

Figure 13.22 Spontaneous light emission of methyl iodide in the presence of increasing buffer gas pressure: (a) ~ 0 atm, (b) 14 atm, (c) 27 atm, and (d) 55 atm. (Reprinted with permission from Ziegler, L., On the difference between resonance Raman scattering and resonance fluorescence: An experimental view, *Acc. Chem. Res.*, 27, 1, 1994. Copyright 1994, American Chemical Society.)

in the presence of increasing amounts of methane [33]. As the pressure of the buffer gas increases, so does the rate of pure dephasing and the contribution of the broad fluorescence emission compared to the Raman line.

The partitioning $S_{SLE} = S_{Raman} + S_{fluor}$ is obtained when the vibrational width is small compared to the electronic width, as is generally the case for larger molecules. If this relationship between the widths does not hold, then the total emission is *not* the sum of Raman and fluorescence components, and in general there is a mixing term which need not even be positive. Thus resonance Raman and resonance fluorescence processes may interfere in smaller molecules, and the categorization of the emitted light is not so clear-cut. The question of Raman versus fluorescence should not create any confusion when the fluorescence is in the form of relaxed emission. Relaxed fluorescence is completely dephased and clearly distinguished from Raman emission.

13.5.2.4 FEMTOSECOND STIMULATED RAMAN SPECTROSCOPY

As a final example of a third-order nonlinear spectroscopy technique, the femtosecond stimulated Raman experiment (FSRS or "fissors") is an elegant time-resolved technique capable of following vibrationally resolved excited state dynamics [19–21]. FSRS can provide dynamical information at fast timescales and with high spectral resolution. Typical time-resolved vibrational spectroscopic techniques, discussed further in the next chapter, employ picosecond pulses in order to achieve sufficient spectral resolution. FSRS uses a series of pulses as follows (see Figure 13.23). An actinic visible pulse of duration ~30 fs prepares the molecule in an excited electronic state at time zero. Alternatively, this pulse is omitted and the FSRS spectrum of the ground electronic state is observed. At a delay time Δt later, two pulses arrive, both centered at a near-IR wavelength: an intense 3 ps Raman pulse at 800 nm, and a weaker 20 fs broadband probe pulse spanning 830 to 960 nm. The much larger time duration of the Raman pulse compared to the broadband probe pulse means that it is the latter which serves as the clock for following dynamics of the state prepared by the actinic pulse. The probe pulse excites vibrational coherences at beat frequencies of the Raman and probe pulses, which appear as photons transferred from the intense Raman pulse to the weak broadband pulse (Figure 13.23). These decay with typical vibrational dephasing times on the order of picoseconds or less, and it is this lifetime which limits the width of the sharp features riding atop the transmitted broadband pulse. The detected probe pulse is not time-resolved but rather is dispersed in the frequency domain into a multichannel detector. On subtracting the broadband continuum, the Raman spectrum as a function of the time interval Δt is obtained.

We can understand the presence of the sharp Raman-shifted peaks riding atop the broad background with the help of four-wave mixing energy level (WMEL) diagrams such as those shown in Figure 13.24. These diagrams convey the same information as the Feynman diagrams (see Table 13.2), with the additional advantage of using shading to indicate that a range of energies is accessed by the spectrally broad fs probe pulse. Time increases from left to right in these WMEL diagrams, and interactions on the ket and bra side are indicated

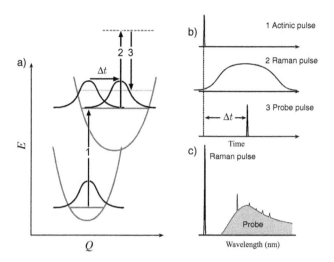

Figure 13.23 Pulse sequence (a and b) and resulting spectrum (c) for femtosecond stimulated Raman spectroscopy.

by solid and dashed arrows, respectively, followed by a wavy line indicating the signal. Neglecting the actinic pulses which prepare the excited electronic state, the FSRS experiment is a $\chi^{(3)}$ experiment in which the three interactions derive from two interactions with the ps pump pulse and one with the fs probe. We have seen that there are eight unique diagrams for such an experiment, of which only two are shown for illustration. The WMEL diagram for Stokes-shifted Raman lines, Figure 13.24a, contributes to the broad baseline, while Figure 13.24b accounts for the sharp Stokes-shifted Raman lines. In the former, the last interaction before the signal accesses a range of intermediate states, and the coherence associated with this signal derives from a range of upper states, leading to a broad background. Figure 13.24b, in contrast, depicts a third interaction with the spectrally narrow ps pump pulse, which lifts the ket side to a narrow range of excited levels which then "emit" to the $v = 1$ level of the ground electronic state, resulting in Stokes-shifted Raman lines. The ability to determine sharp Raman lines at early time delays does not conflict with the time–energy uncertainty principle, since the frequency resolution and time delay are not connected by a Fourier transform relationship. (The properties of transform limited pulses will be considered in the next chapter.)

Mathies and co-workers [19,20] have used the technique to study the excited electronic state manifold of accessory photosynthetic pigments such as β-carotene. As summarized in Figure 13.25, the actinic pulse at 495 nm carries the molecule to the optically allowed S_2 excited state. The rapid relaxation of this state to the forbidden S_1 state broadens the Raman bands at early times as seen in the figure. The study ruled out the existence of states intermediate in energy between S_1 and S_2, in contrast to previous proposals.

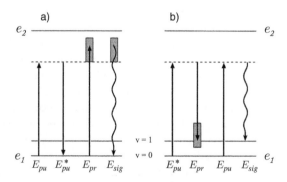

Figure 13.24 Representative wave-mixing energy level (WMEL) diagrams leading to (a) broad background and (b) sharp Stokes-shifted Raman lines, in femtosecond stimulated Raman spectroscopy. The shaded bar indicates a range of frequencies spanned by the fs probe pulse.

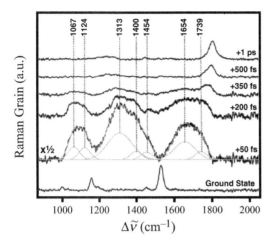
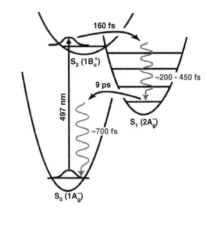

Figure 13.25 Time-dependent FSRS spectra of β-carotene, and deduced dynamics in the S_2 and S_1 excited states. (Reprinted with permission from Kukura, P. et al., Femtosecond time-resolved stimulated Raman spectroscopy of the S_2 (1B$_u^+$) excited state of β-carotene, *J. Phys. Chem. A*, 108, 5921. Copyright 2004, American Chemical Society.)

13.6 SUMMARY

The field of nonlinear optics, like the spectral breadth of an ultrafast laser pulse, is very broad. In this chapter, we have attempted to highlight the theory and some of the many examples that spring from a perturbative expansion of the polarization in powers of the time-varying electric field. Though it is possible to approach this problem within a wavefunction picture that builds on the time-dependent perturbation theory of Chapter 4, the inclusion of interactions of the molecule with its surroundings is more readily handled within the density matrix formalism. The drawback of the perturbation approach is the increasingly larger number of terms in the expansion of $\chi^{(n)}$ as the order n increases. We have introduced the prevalent practice of drawing Feynman diagrams to represent the resonances that emphasize certain terms when an input frequency matches a dipole-allowed transition. We have cautioned the reader that while representative diagrams reveal resonances typical of a particular experiment, it is not in general possible to avoid summation of all terms. Nevertheless, the diagrams serve to visualize experimental pulse sequences and are useful in the discussion of time-resolved nonlinear experiments. In the next chapter, we will build on the present formalism to discuss time-resolved spectroscopy both in the linear and nonlinear regimes.

PROBLEMS

1. Calculate the amplitude of the electric field associated with a 100 nJ pulse from a Ti-sapphire laser with a duration of 10 fs, focused to circular spot with a radius of 1 μm. Compare this to the electric field at a distance of 0.5 nm from an electron.
2. Use the anharmonic Lorentz model at second order to find $\chi^{(2)}(\omega_1 + \omega_2; \omega_1, \omega_2)$ for sum frequency generation. Does the expression you obtain agree with $\chi^{(2)}(2\omega_1; \omega_1, \omega_1)$ for SHG in the limit that $\omega_1 = \omega_2$?
3. Deduce the MKS units of $\chi^{(2)}$ and of $\chi^{(3)}$. The Lorentz model for $\chi^{(2)}$ can be used to estimate an order of magnitude value of

$$\chi^{(2)} \approx \frac{e^3}{\varepsilon_0 m^2 \omega_0^4 d^4}$$

where d is the interatomic spacing. Estimate $\chi^{(2)}$ using $\omega_0 = 10^{16}$ s^{-1} and $d = 0.5$ nm.

4. Derive a matrix equation analogous to Equation 13.22 but for DFG rather than SFG.
5. Determine the degeneracy factors for FWM experiments resulting in the following output frequencies: (a) $2\omega_1 + \omega_3$, and (b) $\omega_2 + \omega_1 - \omega_1$. The latter is an example of a "self-induced process," where the output frequency matches one of the input frequencies.
6. Consider a crystal belonging to the *3m* point group, possessing the symmetry operations given in Section 13.3. (a) Prove that all elements of $\chi^{(2)}$ with an odd number of x subscripts vanish. State the elements of the d tensor that are zero as a result. (b) Prove that d_{23} and d_{24} also vanish. (c) Prove that $d_{31} = d_{32}$.
7. Derive the expression for $c_k^{(3)}$, Equation 13.84.
8. (a) Verify that the response function $S^{(1)}(t_1)$ given in Equation 13.128 is real. (See Chapter 5.) (b) Write down the expression for the fourth-order response function $S^{(4)}(t_1, t_2, t_3, t_4)$. How many terms derive from expanding the commutator in this case?
9. Expand the commutator in the expression for the second-order nonlinear response (Equation 13.129) and choose one of the eight terms. Assume that the zero-order density matrix is $\rho_{mm}^{(0)} = \delta_{gm}$ (only the ground state is populated at equilibrium). With help from the resolution of the identity, obtain a contribution to $S^{(2)}(t_1, t_2)$ in terms of transition dipole moments and frequencies and associate that term with the appropriate Feynman diagram.
10. (a) Assign the hot bands observed in the CARS spectrum of Figure 13.18. (b) Use the approximate relative intensities of the four strongest bands to try to estimate the temperature of the sample at 10 µs. Are the vibrational populations equilibrated? Use data from Chapter 9 for the vibrational frequency and anharmonicity of N_2.
11. Deduce the energy level diagram and Feynman diagram for CSRS, taking into account the response is a signal at $2\omega_S - \omega_P$.
12. Deduce the phase-matching criterion for two-photon absorption.
13. Derive the Manley–Rowe equations (Equation 13.73) for DFG. Convert them from intensities to photon flux and verify that the process is parametric.
14. For continuous-wave incident radiation, the second-order susceptibility is found from the Fourier transform of the second order response function:

$$\chi^{(2)}(\omega_1 + \omega_2) = \int_{-\infty}^{\infty} dt_2 \int_{-\infty}^{\infty} dt_1 e^{i(\omega_2 + \omega_1)t_2} e^{i\omega_1 t_1} S^{(2)}(t_2, t_1)$$

Starting from Equation 13.129, derive the K_7 contribution to $\chi^{(2)}$, assuming two continuous-wave electric fields with frequencies ω_1 and ω_2 are incident on the sample. Consider all the population to be in the ground state.
15. Verify that Equation 13.152 follows from phase matching and Snell's law.
16. Explain why the absolute intensity of the VSF spectra of Figure 13.11 decreases in the series 1-, 2-, 3-, 4-hexadecanol.

REFERENCES

1. R. W. Boyd, *Nonlinear Optics, 3rd ed.* (Academic Press, Amsterdam, 2008.)
2. P. E. Powers, *Fundamentals of Nonlinear Optics* (Taylor and Francis, Boca Raton, 2011).
3. S. Mukamel, *Principles of Nonlinear Optical Spectroscopy* (Oxford University Press, New York, 1995).
4. Y. R. Shen, *The Principles of Nonlinear Optics* (John Wiley & Sons, New York, 1984).
5. J. C. Wright, Nonlinear spectroscopy and coherent multidimensional spectroscopy, in *Lasers in Chemistry* 1, Ed. M. Lackner (Wiley-VCH, Weinheim, 2008).
6. P. A. Franken, A. E. Hill, C. W. Peters, G. Weinreich, Generation of optical harmonics, *Phys. Rev. Lett.* 7, 118 (1961).
7. W. Kaiser, C. G. B. Garret, Two-photon excitation in CaF_2:Eu^{2+}, *Phys. Rev. Lett.* 7, 229 (1961).
8. J. D. Macomber, *The Dynamics of Spectroscopic Transitions* (Wiley, New York, 1976).

9. R. N. Bracewell, *The Fourier Transform and Its Applications* (McGraw Hill, New York, 1978).

10. G. L. Richmond, Molecular bonding and interactions at aqueous surfaces as probed by vibrational sum frequency generation, *Chem. Rev. 102*, 2693 (2002).

11. Y. R. Shen, V. Ostroverkhov, Sum-frequency vibrational spectroscopy on water interfaces: Polar orientation of water molecules at interfaces, *Chem. Rev. 106*, 1140 (2006).

12. H.-F. Wang, L. Velarde, W. Gan, L. Fu, Quantitative sum-frequency generation vibrational spectroscopy of molecular surfaces and interfaces: Lineshape, polarization and orientation, *Ann. Rev. Phys. Chem. 66*, 189 (2015).

13. S. Z. Can, D. D. Mago, R. A. Walker, Structure and organization of hexadecanol isomers adsorbed to the air–water interface, *Langmuir 22*, 8043 (2006).

14. F. Terenziani, C. Katan, E. Badaeva, S. Tretiak, M. Blanchard-Desce, Enhanced two-photon absorption of organic chromophores: Theoretical and experimental assessments, *Adv. Mater. 20*, 4641 (2008).

15. M. Pawlicki, H. A. Collins, R. G. Denning, H. L. Anderson, Two-photon absorption and the design of two-photon dyes, *Angew. Chem. Ed. 48*, 3244 (2009).

16. H. Hosoi, R. Tayama, S. Takeuchi, T. Tahara, Solvent dependence of the two-photon absorption spectra of the enhanced green fluorescence protein (eGFP) chromophore, *Chem. Phys. Lett. 630*, 32 (2015).

17. E. O. Potma, S. Mukamel, Theory of coherent Raman scattering, in *Coherent Raman Scattering Microscopy*, Ed. J.-X. Cheng, X. S. Xie (CRC Press/Taylor & Francis Group, Boca Raton 2013).

18. W. R. Lempert, I. V. Adamovich, Coherent anti-Stokes Raman scattering and spontaneous Raman scattering diagnostics of nonequilibrium plasmas and flows, *J. Phys. D. Appl. Phys. 47*, 433001 (2014).

19. P. Kukura, D. W. McCamant, R. A. Mathies, Femtosecond stimulated Raman spectroscopy, *Ann. Rev. Phys. Chem. 58*, 461 (2007).

20. P. Kukura, D. W. McCamant, R. A. Mathies, Femtosecond time-resolved stimulated Raman spectroscopy of the S_2 ($1B_u^+$) excited state of β-carotene, *J. Phys. Chem. A 108*, 5921 (2004).

21. Z. Bin, N. Kai, X. T. Li, S.-Y. Lee, Simple aspects of femtosecond stimulated Raman spectroscopy, *Science China, 54*, 1989 (2011).

22. A. Tokmakoff, Time-dependent quantum mechanics and spectroscopy, lecture notes, *http://tdqms.uchicago.edu/page/tdqms-notes.* (accessed March 15, 2016)

23. L. Ziegler, On the difference between resonance Raman scattering and resonance fluorescence: An experimental view, *Acc. Chem. Res., 27*, 1 (1994).

Time-resolved spectroscopy

14.1 INTRODUCTION

We have seen in previous chapters that spectral lineshapes $I(\omega)$ are related via Fourier transformation to time-correlation functions that depend on the dynamics of molecular motion coupled to the spectrum under consideration. For example, as shown in Chapter 5, the lineshape of an infrared absorption band is determined by the vibrational and reorientational dynamics of the dipole moment derivative with respect to normal coordinate. In the time-dependent theory of electronic spectroscopy, discussed in Chapter 12, the frequency distribution of the absorption spectrum and Raman excitation profile derives from the motion of vibrational wavepackets propagating on displaced excited electronic state potential surfaces. Solvent-induced dephasing, population relaxation, vibrational and reorientational motion, and static inhomogeneous broadening can all contribute to the frequency distribution $I(\omega)$. Separating these effects without models or assumptions is simply not tenable in the linear regime. Time-dependent spectroscopy, on the other hand, provides insight into the same dynamics that contribute to the lineshape, but are not readily uncovered in the frequency domain. As an extreme example, consider that a typical excited electronic state lifetime of 10 ns translates into a breadth of only 0.003 cm⁻¹, according to the time-energy uncertainty principle, $\Delta \nu \Delta t \geq 1$. This contribution to the spectral width is insignificant compared to other line broadening effects in the condensed phase, but a 10 ns (or much shorter) lifetime is readily measured in the time domain. Coherent time-domain experiments with a variety of pulse sequences, wavelengths and detection schemes enable the separation of homogeneous and inhomogeneous contributions to the spectral width. Using coherent excitation of vibrational states which beat against one another, the wavepacket motion that underlies electronic and resonance Raman spectroscopy is revealed in real time. Following pulsed excitation, the effect of the solvent environment on the time-evolving transition frequency, known as spectral diffusion, is seen. Spectral diffusion derives from fluctuations in the transition frequency of a molecule, fluctuations like that of the general dynamic variable depicted in the cartoon of Figure 5.1, as a result of interactions with its environment. These dynamics give rise to the frequency fluctuation correlation function (FFCF), $\langle \delta\omega(0)\delta\omega(t) \rangle$, that influences linear and nonlinear spectra. Time-dependent experiments enable the determination of the effects of solvent and internal dynamics on spectroscopic transitions. Nonlinear and time-resolved experiments offer more information content than steady-state linear spectroscopy, as a result of multiple field–matter interactions. Though we limit our discussion in this chapter to the realm of third-order response, the number of experimental configurations embraced by $\chi^{(3)}$ is quite large. We focus here on some of the more widely used techniques.

The availability of pulsed laser sources at infrared, visible and ultraviolet wavelengths enables dynamic studies of vibrational and electronic transitions. Time-resolved spectroscopy measurements are used to determine excited-state lifetimes, study photochemical processes such as vision and photosynthesis, uncover structural changes that accompany electronic excitation, and reveal dynamics such as reorientational motion, energy transfer, and solvent-induced dephasing. Historically, the measurement of fluorescence lifetimes, discussed in the next section, probably provides the first example of a time-dependent spectroscopy measurement. Early millisecond time-resolved absorption experiments, referred to as "flash photolysis," have evolved to ultrafast timescales, providing the powerful technique of pump-probe spectroscopy, introduced in Section 14.3.2. Later sections consider other nonlinear optical techniques in the time domain, such as photon echo and two-dimensional spectroscopy. Examples are presented to show the utility of time-resolved spectroscopy in biological systems, materials science, and fundamental chemical physics.

14.2 TIME-RESOLVED FLUORESCENCE SPECTROSCOPY

Time-resolved fluorescence spectroscopy uses a short light pulse (the pump pulse) to excite an electronic state and measures the intensity of emitted light as a function of time delay following the pump. As long ago as 1960, gated photomultipliers were coupled with pulsed flashlamp sources to determine lifetimes in the nanosecond regime [1]. Later, Strickler and Berg [2] used a chopped light source and phase-sensitive detection to verify lifetimes calculated from absorption and fluorescence spectra as described in Chapter 6. Recall that integrated absorption and fluorescence spectra can be combined to reveal the radiative lifetime τ_{rad} through the Strickler–Berg relation, Equation 6.35. Unless the quantum yield of fluorescence is unity, however, this radiative lifetime is not the same as the fluorescence lifetime τ_{fluor}. (See Equations 6.28 and 6.29.) The fluorescence lifetime is the inverse of the rate of population decay, which is the sum of the rates of radiative and nonradiative relaxation: $1/\tau_{fluor} = 1/\tau_{rad} + 1/\tau_{nonrad}$. The radiative lifetime τ_{rad} is a function of the strength of the spectroscopic transition (transition dipole moment squared). The nonradiative lifetime τ_{nonrad} depends on the dissipation of the excited state energy by energy transfer to other degrees of freedom of the molecule and its environment. In this section we consider three variations on the theme of time-resolved fluorescence: determination of the lifetime of the excited state through measurement of the intensity as a function of time delay, investigation of reorientational relaxation as revealed by the time-resolved polarization of the emitted light, and determination of excited state relaxation along solvent and internal coordinates through measurement of the time-evolving frequency of emitted light.

As previously stated, the time-resolved fluorescence experiment uses an initial pulse to prepare the excited state. The intensity of fluorescence is determined as a function of time t following this pulse and is proportional to the instantaneous population N of excited state molecules. In the ideal case, the decay of the population N is exponential:

$$\frac{-dN(t)}{dt} = \left(\frac{1}{\tau_{rad}} + \frac{1}{\tau_{nonrad}} \right) N(t) = \left(\frac{1}{\tau_{fluor}} \right) N(t) \tag{14.1}$$

Since the intensity of fluorescence is proportional to the excited state population, the intensity also decays exponentially:

$$I = I_0 \exp(-t/\tau_{fluor}) \tag{14.2}$$

Nonexponential dynamics may be observed when the lifetime is influenced by the effects of intermolecular interactions, collisions, intramolecular motion such as internal rotation, and photochemical processes. An average lifetime τ_{ave} can be found from

$$\tau_{ave} = \frac{\displaystyle\int_0^\infty tN(t)dt}{\displaystyle\int_0^\infty N(t)dt} \tag{14.3}$$

It is straightforward to show that $\tau_{ave} = \tau_{fluor}$ for decay via a single exponential.

Several experimental approaches exist for the determination of τ_{fluor}. Time-correlated single photon counting determines the arrival time of emitted photons following a pulse of exciting light. Fluorescence upconversion experiments use sum-frequency mixing of a gate pulse with the emitted light within a nonlinear crystal. The gate pulse is delayed with respect to the excitation pulse to measure the emission intensity as a function of time. By measuring the time-dependent emission at different wavelengths, for example, by tuning the angle of the nonlinear crystal, the spectrum can be constructed as a function of time. Another approach is to modulate the intensity of a continuous-wave source and detect the phase delay and demodulation of the emitted signal, as discussed in detail in [3]. Fluorescence lifetimes vary in accordance with fluorescence yields. Highly fluorescent molecules such as laser dyes have little contribution from nonradiative decay, hence $\tau_{fluor} \approx \tau_{rad}$ is typically a few nanoseconds. Molecules which undergo more significant nonradiative decay, through internal dynamics or intermolecular energy or electron transfer, are more likely to have shorter (picosecond or subpicosecond) fluorescence lifetimes.

14.2.1 POLARIZATION IN TIME-RESOLVED FLUORESCENCE SPECTROSCOPY

The measurement of the polarization of emitted light relative to that of the excitation beam holds useful information on reorientational motion of solution phase molecules. Similar to polarized Raman scattering, we define $I_\perp$ and $I_\parallel$ as the intensities of emitted (rather than scattered) light which is polarized respectively perpendicular or parallel to the polarization of the incident light. The fluorescence anisotropy r is defined as

$$r = \frac{I_\parallel - I_\perp}{I_\parallel + 2I_\perp} \tag{14.4}$$

The anisotropy r is zero when the emitted light is completely depolarized ($I_\parallel = I_\perp$), and $r = 1$ for completely polarized emission ($I_\perp = 0$). As shown below, it is also possible for the perpendicular component of the emitted light to exceed the parallel component, resulting in a negative value of r. The anisotropy is time-dependent as a result of rotation of the transition dipole moment during the lifetime of the excited state. $r(t)$ decays from an initial maximum value, r_0, which is decided by the angle between the absorption and emission transition dipoles in the molecular frame. The anisotropy decay may be exponential but is generally more complex for molecules of lower than spherical symmetry. Let us consider how determination of $r(t)$ might reveal internal or external rotation of the emission dipole of a fluorescent molecule. We are considering here an ensemble of randomly oriented molecules such as dye molecules in liquid or glass solution.

Consider a sample of molecules in a laboratory coordinate system XYZ where the incoming light is polarized along Z and propagates in the Y direction. We consider absorption with a transition dipole moment direction specified by the unit vector $\hat{u}_{abs}$, and let $\hat{u}_{emis}$ specify the direction of the transition dipole for emission. For randomly oriented chromophores, we can average over the orientation of the molecules with respect to the laboratory frame, but the absorption and emission transition dipoles are fixed in the frame of the molecule, so we cannot average over their orientations separately. We specify α as the angle between the transition dipoles for absorption and emission: $\hat{u}_{abs} \cdot \hat{u}_{emis} = \cos\alpha$. Figure 14.1 illustrates the geometry of the transition dipoles in the laboratory frame and shows a sketch of the time-dependent anisotropy $r(t)$ in fluid solution. The two transition dipoles are not always parallel as a result of reorganization in the excited state or emission from a lower lying state than that reached in absorption. The intensities for the parallel and perpendicular components of the fluorescence are given by

$$I_\parallel = I_{ZZ} \propto \left\langle (\hat{u}_{abs} \cdot \hat{Z})^2 (\hat{u}_{emis} \cdot \hat{Z})^2 \right\rangle \tag{14.5}$$

$$I_\perp = I_{ZX} \propto \left\langle (\hat{u}_{abs} \cdot \hat{Z})^2 (\hat{u}_{emis} \cdot \hat{X})^2 \right\rangle \tag{14.6}$$

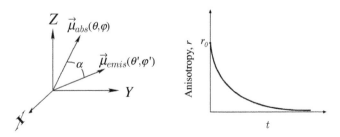

Figure 14.1 Orientation of absorption and emission transition dipoles in the laboratory frame and typical anisotropy decay curve.

The omitted proportionality constants depend on the inherent strength of the absorption and emission (Einstein B and A coefficients, respectively), and experimental details such as incident light power and collection angle. We take these to be the same for both parallel and perpendicular components, so they will not contribute to the ratio that is the anisotropy. Let us take θ and φ as the polar and azimuthal angles that orient μ_{abs} in the lab frame, while θ' and φ' specify the orientation of μ_{emis}. The angle brackets represent orientational averaging. With this notation, we have

$$I_{ZZ} \propto \left\langle (\cos^2\theta)(\cos^2\theta') \right\rangle \tag{14.7}$$

$$I_{ZX} \propto \left\langle (\cos^2\theta)(\sin^2\theta')(\cos^2\varphi') \right\rangle \tag{14.8}$$

The sample is axially symmetric about the Z direction, so $I_{ZX} = I_{ZY}$. This permits us to write:

$$
\begin{aligned}
I_{ZX} &= \frac{1}{2}(I_{ZX} + I_{ZY}) \\
&= \frac{1}{2}\left[\left\langle (\cos^2\theta)(\sin^2\theta')(\cos^2\varphi') \right\rangle + \left\langle (\cos^2\theta)(\sin^2\theta')(\sin^2\varphi') \right\rangle \right] \\
&= \frac{1}{2}\left[\left\langle \cos^2\theta \right\rangle - \left\langle (\cos^2\theta)(\cos^2\theta') \right\rangle \right]
\end{aligned}
\tag{14.9}
$$

In Equation 14.9, we can replace $\left\langle \cos^2\theta \right\rangle$ by 1/3, as we are free to average over the orientation of the absorption dipole moment. A quick way to compute this average is to recognize that for this isotropic sample, the averages $\left\langle \cos^2\theta \right\rangle$, $\left\langle \sin^2\theta \cos^2\varphi \right\rangle$, and $\left\langle \sin^2\theta \sin^2\varphi \right\rangle$ are equal to one another and sum to one. Next, we express $\left\langle (\cos^2\theta)(\cos^2\theta') \right\rangle$ in terms of the relative orientation of the two transition dipoles. Wigner D-functions, $D_{mn}^{j}(\Omega)$, used in Section 8.6 and discussed further in Appendix A, are convenient for problems where relative rotations are of interest. Recall that Ω is shorthand for the set of Euler angles (φ, θ, χ), and in this case we only need the first two to specify the dipole direction. We use the relationship $D_{00}^{j}(\Omega) = P_j(\cos\theta)$ for $j = 2$, where the second Legendre polynomial is $P_2(\cos\theta) = (3\cos^2\theta - 1)/2$. Using this to solve for each of the cosine-squared functions, the angular function in Equation 14.9 is manipulated as follows:

$$
\begin{aligned}
(\cos^2\theta)(\cos^2\theta') &= \frac{1}{9}\left[2P_2(\cos\theta) + 1 \right]\left[2P_2(\cos\theta') + 1) \right] \\
&= \frac{1}{9}\left[4D_{00}^{2}(\Omega)D_{00}^{2}(\Omega') + 2D_{00}^{2}(\Omega) + 2D_{00}^{2}(\Omega') + 1 \right]
\end{aligned}
\tag{14.10}
$$

Next, we consider the orientation Ω' of the emission dipole to be the result of successive rotations by Ω, then $\delta\Omega = \alpha$, and use the addition theorem Equation A.56 to write

$$D_{00}^{2}(\Omega') = \sum_{m} D_{0m}^{2}(\delta\Omega)D_{m0}^{2}(\Omega) \tag{14.11}$$

With this substitution, we can average over the orientation Ω by integrating over $d\Omega$ and dividing by $\int d\Omega = 8\pi^2$. We use the fact that the integral $\int D_{mn}^{j}(\Omega)d\Omega$ is zero unless $j = m = n = 0$. Thus

$$
\begin{aligned}
\left\langle (\cos^2\theta)(\cos^2\theta') \right\rangle &= \frac{1}{9}\left[\frac{1}{8\pi^2} \sum_{m} 4D_{0m}^{2}(\delta\Omega)\int D_{00}^{2}(\Omega)D_{m0}^{2}(\Omega)d\Omega + 1 \right] \\
&= \frac{1}{9}\left[\frac{1}{8\pi^2} 4D_{00}^{2}(\delta\Omega)\frac{8\pi^2}{5} \right] + \frac{1}{9} = \frac{1}{9}\left[\frac{4}{5}P_2(\cos\alpha) + 1 \right]
\end{aligned}
\tag{14.12}
$$

We have used the orthonormality of the Wigner functions as defined in A.49. Now we are ready to evaluate the initial anisotropy r_0.

$$r_0 = \frac{I_{ZZ} - I_{ZX}}{I_{ZZ} + 2I_{ZX}} = \frac{\frac{3}{2}\langle(\cos^2\theta)(\cos^2\theta')\rangle - \frac{1}{2}\langle\cos^2\theta\rangle}{\langle\cos^2\theta\rangle}$$

(14.13)

$$r_0 = \frac{2}{5}P_2(\cos\alpha) = \frac{1}{5}(3\cos^2\alpha - 1)$$

Equation 14.13 predicts the maximum anisotropy r_0 for motionless, randomly oriented molecules, for example in a glass sample. When the absorption and emission transition dipoles are parallel, $r_0 = 0.4$, and when they are perpendicular, $r_0 = -0.2$. At the "magic angle" of 54.7°, $\cos^2\alpha = 1/3$ leading to $r_0 = 0$. In time-resolved fluorescence anisotropy measurements, the parallel and perpendicular components of the emitted light are determined to find the time-dependent anisotropy $r(t)$. Reorientational motion or internal rotation of the emitting molecule within the lifetime of the excited state further reduces the anisotropy from the initial value r_0 as a result of the time-dependent change in direction of the emission dipole. In the simplest case that the reorientational motion is diffusional and the molecule is isotropic, the anisotropy decays via a single exponential:

$$r(t) = r_0\exp(-t/\phi)$$

(14.14)

In the Equation 14.14, ϕ is called the reorientational correlation time. Clearly, if ϕ is long compared to the lifetime, the anisotropy will not be much smaller than the maximum value r_0. Defining $\bar{r}$ as the time averaged value of the anisotropy, one gets:

$$\bar{r} = \frac{\int I(t)r(t)dt}{\int I(t)dt} = \frac{r_0}{1 + (\tau_{fluor}/\phi)}$$

(14.15)

In the case that internal rotation changes the dipole direction, Equation 14.14 is altered:

$$r(t) = (r_0 - r_\infty)\exp(-t/\phi) + r_\infty$$

(14.16)

The above expressions assume the anisotropy decays by a single exponential. Rotational diffusion by anisotropic molecules can be more complex than single exponential, because different rotational relaxation times apply to different inertial axes. Nonetheless, the single exponential relaxation is convenient to apply and gives an estimate of the angle through which the transition dipole rotates during the excited state lifetime. If the excitation and emission polarizers in the experiment are set so that the polarization of the emitted light is at the magic angle with respect to that of the exciting light, the measurement is insensitive to rotational motion.

Fluorescence lifetime and anisotropy measurements have been advantageously adapted for imaging cellular microenvironments. One example is the use of a fluorescent probe to determine viscosity in live cancer cells [4]. In this study, the rotational correlation time was taken to be $\phi = \eta V/k_B T$, where η is the viscosity and V is the hydrodynamic volume. The molecular probe used in [4] undergoes internal rotation in its excited electronic state, and this rotation is hindered in more viscous environment. As a result, both the fluorescence yield and lifetime increase with increase in viscosity.

14.2.2 TIME-RESOLVED FLUORESCENCE STOKES SHIFT

Consider the events that take place following electronic excitation of a molecule in solution. Immediately after photon absorption, at $t = 0$, the molecule finds itself in the vertical (Franck–Condon) state, a nonequilibrium geometry of the excited state. At the earliest timescales, as discussed in Chapter 11, intramolecular vibrational redistribution (IVR) randomizes the vibrational energy, initially invested in the Franck–Condon active modes, through anharmonic coupling with other vibrational modes. The timescale for IVR varies greatly but for typical dye molecules it is a subpicosecond process. IVR preserves the total vibrational energy but alters the partitioning of this energy among vibrational states. Internal (or vibrational) reorganization, on the other hand, is the loss of excess vibrational energy as heat when the molecule relaxes from the Franck–Condon position to its equilibrium geometry. Solvent reorganization represents a similar thermalization as solvent molecules adjust to their new equilibrium positions that minimize the energy of the excited electronic state. In steady-state absorption and emission spectra, internal and solvent reorganization determine the magnitude of the Stokes shift, the energy difference between absorption and emission maxima. Time-dependent fluorescence Stokes shift spectroscopy (TDFSS) permits the observation of internal and solvent reorganization in real time by measuring the dynamic red-shift of the emission spectrum. Below are some examples of the use of TDFSS to study solvent and internal reorganization. Additional applications of TDFSS include the study of protein and DNA dynamics.

We first consider the influence of solvent dynamics in TDFSS. We assume for now that the internal reorganization is fast compared to the measurement timescale, such that the TDFSS experiment sees only the effects of solvent reorganization. Recall that solvatochromic molecules exhibit absorption and emission spectra that depend on solvent polarity as a result of different ground- and excited-state charge distributions. Various types of intermolecular interactions can perturb the electronic transition energy: dispersion forces, hydrogen-bonding, dipole-induced dipole, dipole–dipole, etc. Dispersion forces result from distortions of the molecular electronic distribution and are too fast to be perceived by TDFSS, while the permanent dipole moments of the solute and solvent exert a strong influence. Solvatochromic molecules used as probes of solvent relaxation undergo a significant change in dipole moment on electronic excitation. Thus in polar solvents the reorientational motion of the solvent dipoles is highlighted, and the electronically excited molecule acts as a probe of this dynamics. This leads to some connections with the determination of solvent dynamics via lineshape effects on infrared spectroscopy and depolarized Rayleigh and Raman scattering, as discussed in Chapter 5. Figure 14.2 depicts the assumptions of linear solvent response, in which the ground and excited state potential surfaces of the molecule are taken as displaced harmonic functions of a collective solvent coordinate. Linear solvent response assumes that the shapes of these potential surfaces do not depend on

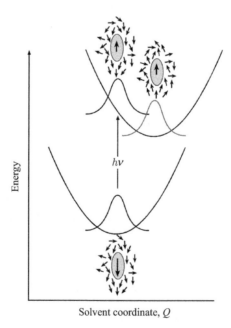

Solvent coordinate, Q

Figure 14.2 Assumptions of linear solvent response, highlighting reorientation of solvent dipole moments to adjust to the excited-state dipole moment of the solute.

whether the molecule is in the ground or excited state. When light is absorbed, there is an instantaneous change in the state dipole moment from μ_g to μ_e, and the surrounding solvent molecules find themselves in a nonequilibrium geometry. The ensuing motion of the solvent molecules, especially the rotation of their permanent dipole moments, results in an emission spectrum which red-shifts with time. As illustrated for example in Figure 14.3, the time-resolved fluorescence spectrum of the solvatochromic molecule coumarin 153 (C153) exhibits the influence of solvent reorganization on a picosecond timescale [5]. In addition to the increasing red-shift with time, the overall decay of the excited state population results in a decrease in the emission intensity. The ability to observe the solvent dynamics in TDFSS hinges on the relative timescales for solvent motion and excited state population relaxation. In the case of C153, for example, the fluorescence spectrum extrapolated to infinite time is more red-shifted than the experimental steady-state fluorescence, revealing that the ~6 ns fluorescence lifetime is not long enough for the surrounding solvent to relax to the equilibrium geometry of the excited electronic state. This means that the steady-state fluorescence spectrum and Stokes shift result from an excited state that is not fully relaxed with respect to the surrounding solvent.

With the assumption of linear solvent response, the solvent reorganization energy ΔE_{solv} (often called λ_{solv} in the literature) is given by

$$\Delta E_{solv} = \frac{1}{2}(h\nu_0 - h\nu_\infty)$$

(14.17)

where ν_0 is the peak frequency at time zero, and ν_∞ is that extrapolated to infinite time. A spectral response function $S_\nu(t)$ can be defined as follows:

$$S_\nu(t) = \frac{\nu(t) - \nu(\infty)}{\nu(0) - \nu(\infty)} = \frac{\langle \delta\nu(0)\delta\nu(t)\rangle}{\langle \delta\nu^2\rangle}$$

(14.18)

The first equality above is a definition and the second follows from the assumption of linear solvent response. The angle brackets represent an equilibrium average. The correlation function $\langle \delta\nu(0)\delta\nu(t)\rangle$ is the same FCFF discussed previously in Chapter 6. (Note that the use of angular frequencies ω in Equation 14.18 would not affect the value of $S_\nu(t)$). Calculations show a relation between $S_\nu(t)$ and the dipole correlation function $\langle \hat{\mu}(0)\cdot \hat{\mu}(t)\rangle$ that we encountered in Chapter 5:

$$S_\nu(t) \cong \langle \hat{\mu}(0)\cdot\hat{\mu}(t)\rangle^\alpha$$

(14.19)

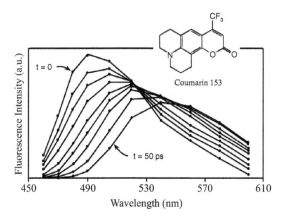

Figure 14.3 Time-resolved fluorescence spectrum of coumarin 153 in formamide from 0 to 50 ps. (Reprinted with permission from Horng, M. L. et al. Sub-picosecond measurements of polar solvation dynamics: Coumarin 153 Revisited, *J. Phys. Chem.* 99, 17311 (1995). Copyright 1995 American Chemical Society.)

Here, $\hat{\mu}$ is a unit vector in the direction of the permanent dipole moment of a solvent molecule. The parameter α can be found from the dipole density and the dielectric constant of the solvent, ranging from 0 for nonpolar solvents to a value of about 20 for very polar solvents such as water and acetonitrile. Recall from Chapter 5 that we expect a Gaussian response on a subpicosecond timescale from the inertial motion of solvent molecules, while at later times an exponential response results from rotational diffusion. This is the basis for the following functional form used to fit $S_v(t)$:

$$S_v(t) = f_G e^{-\frac{1}{2}\omega_G^2 t^2} + (1 - f_G)e^{-(t/\tau)^\beta} \tag{14.20}$$

where f_G is the fraction of the response contributed by inertial motion and $1 - f_G$ is the fraction contributed by diffusion. The latter is represented as a stretched exponential and accounts for a range of relaxation times. The initial Gaussian response at early times leads to an initial slope of zero for the function $S_v(t)$, corresponding to the sluggish inertial response of the solvent immediately following excitation. The exponential character of the response at longer times is seen only after a sufficient number of reorientational steps have taken place to treat the motion as diffusional. In the absence of sufficient time resolution, the early Gaussian decay may not be observed. $S_v(t)$ is sometimes fit to a sum of exponentials with various amplitudes a_i and relaxation times τ_i.

$$S_v(t) = \sum_i a_i e^{-t/\tau_i} \tag{14.21}$$

Some of values of the amplitudes and relaxation times for several solvents, taken from [5], are listed in Table 14.1 to illustrate the range of timescales for solvent relaxation. For example, the fast-relaxing solvent acetonitrile, CH_3CN, has a large contribution ($a_1 = 0.686$) from a 0.089 ps component, with the remainder of the relaxation taking place on a timescale of 0.63 ps. Note the complex multiexponential nature of the solvent response of alcohols, for which there are additional dynamics of hydrogen bonds. The dynamics reported in Table 14.1 were derived using C153 as a probe. C153 undergoes an increase in dipole moment from about 7 to 14 Debye on electronic excitation, with a small change in dipole direction. If linear solvent response is operative, solvent dynamics should be independent of the probe molecule; however, effects of internal and external rotation of the probe may need to be considered. Recent applications of TDFSS have investigated the dynamics of ionic liquids [6], finding a subpicosecond component that reflects the inertial translational motion of the ions.

While most TDFSS measurements have emphasized dynamical solvent reorganization, they also have the potential to reveal internal dynamics such as IVR, vibrational relaxation, and excited state conformational changes. Sufficiently fast time resolution opens up a window on these picosecond or subpicosecond processes; however, the timescales for internal and solvent reorganization are not in general separable, and models and theory are needed to interpret the kinetics observed in time-resolved fluorescence. An example of excited state dynamics addressable by time-resolved fluorescence is that of twisted intramolecular charge-transfer, or "TICT-state" formation. In this case, internal rotation of the molecule is accompanied by charge separation, leading to dual fluorescence from both the unrelaxed locally excited (LE) and relaxed charge-transfer (CT) conformations. The large dipole moment of the latter state highlights the coupling of solvent relaxation

Table 14.1 Examples of solvent relaxation times and amplitudes obtained using TDFSS with coumarin 153 as the probe

Solvent	a_1	τ_1 (ps)	a_2	τ_2 (ps)	a_3	τ_3 (ps)	a_4	τ_4 (ps)
Acetonitrile	0.686	0.089	0.314	0.63				
Dimethylsulfoxide	0.500	0.214	0.408	2.29	0.092	10.7		
Chloroform	0.356	0.285	0.644	4.15				
Ethanol	0.085	0.030	0.230	0.39	0.182	5.03	0.502	29.6

Source: Horng, M. L. et al. *J. Phys. Chem.* **99**, 17311, 1995.

to internal rotation. A molecule that has provided a prototype of this motif is 4-(dimethylamino)benzonitrile (DMABN). The molecule has a characteristic donor–acceptor structure with the dimethylamino and cyanine groups serving as electron donor (D) and acceptor (A), respectively. DMABN has an absorption maximum at 290 nm. The locally excited state emits at 350 nm, while the emission of the TICT state is observed at a longer wavelength that shifts to the red in more polar solvents. For example, Park et al. [7] followed the conversion of the LE to the TICT state using time-resolved fluorescence. By using the fast-relaxing solvent acetonitrile, the authors attempted to distinguish the intramolecular dynamics of the LE → TICT conversion from that due to solvent reorganization. They concluded that internal reorganization within the LE state takes place in less than 30 fs, after which the molecules may convert to a partially twisted state on a timescale of 160 fs or to a totally twisted state on a timescale of 3.3 ps.

14.3 TIME-RESOLVED FOUR-WAVE MIXING EXPERIMENTS

The time-resolved fluorescence experiments discussed above did not require us to consider the nonlinear response of the system to the incident light. We now consider time-resolved measurements that derive from the third-order nonlinear response. Since the second-order nonlinear response vanishes for centrosymmetric samples, experiments that spring from $\chi^{(3)}$ constitute the bulk of nonlinear spectroscopy measurements. These are broadly defined as time-resolved four-wave mixing (TD-FWM), where the signal results from three field–matter interactions. Variations in the timing, propagation vectors, and frequencies of these fields lead to a wide array of experiments.

First, we wish to make some general statements about the theory and practice of TD-FWM. On the theory side, we review the diagrammatic approach to calculation of the third-order polarization $P^{(3)}(t)$ in the next section. With care, we can associate the experimental pulse sequence with a small number of Feynman diagrams: where E_1, E_2, and E_3 are the fields acting at the times 0, t_1, and $t_1 + t_2$, respectively, while the signal is detected at $t_1 + t_2 + t_3$. In the impulsive limit, the incident fields are considered to be delta function pulses; a valid approximation when their duration is short compared to the dynamics of the system. In this limit, the polarization $P^{(3)}(t)$ is directly proportional to the response function $S^{(3)}(t_3, t_2, t_1)$. (See Equation 13.132 through 13.134 and Problem 8 of this Chapter.) More generally, the pulse characteristics are taken into consideration, and it may be necessary to calculate the polarization by numerical integration.

The duration τ_p of the pulse is inversely related to the width of its frequency distribution via Fourier transformation. We account for finite duration of the field using a time-dependent envelope $\vec{E}_0(\vec{r}, t)$ for the amplitude of the electric field, multiplied by the spatial ($\vec{k}$) and time-dependent (ω_0) parts of the carrier wave:

$$\vec{E}(\vec{r}, t) = \frac{\vec{E}_0(\vec{r}, t)}{2} \exp[i(\omega_0 t - \vec{k} \cdot \vec{r} + \varphi)] + c.c \tag{14.22}$$

For example, $\vec{E}_0(\vec{r}, t)$ could be a Gaussian function of both space and time. Beams for which the intensity is a Gaussian function of radial distance from the beam axis were considered in Chapter 2. A pulse with a Gaussian temporal profile contributes a time-dependence $\exp(-t^2/2\tau_p^2)$ to the pulse envelope $\vec{E}_0(\vec{r}, t)$.

$$\vec{E}_0(\vec{r}, t) \propto \frac{1}{(2\pi\tau_p^2)^{1/2}} e^{-t^2/2\tau_p^2} \tag{14.23}$$

The factor in front of the exponent normalizes the time integral of the Gaussian function. The intensity of the pulse is proportional to the square of the electric field. Thus $I \propto \exp(-t^2/\tau_p^2)$, and the full width at half-maximum (FWHM) of the pulse is $\Delta\tau = 2(\ln 2)^{1/2}\tau_p = 1.665\tau_p$. The limited time duration results in a spread of frequencies about the carrier frequency ω_0. The Fourier transform of the temporal pulse envelope is a Gaussian function of frequency:

$$\vec{E}_0(\vec{r}, \omega) \propto \frac{\tau_p}{(2\pi)^{1/2}} e^{-(\omega-\omega_0)^2\tau_p^2/2} \tag{14.24}$$

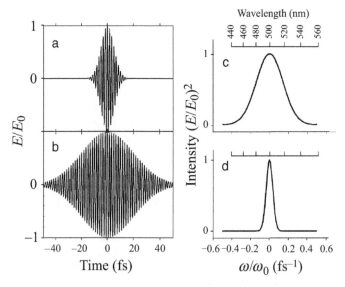

Figure 14.4 Transform-limited Gaussian pulses in the time (a, b) and frequency (c, d) domain, for pulses with a carrier frequency ω_0 corresponding to a wavelength of 500 nm, for 5 fs (a and c) and 20 fs (b and d) pulse durations. The x axis in c and d is the frequency shift ω relative to ω_0.

Again, we square this function to get the intensity and find the FWHM to be $\Delta\omega = 2(\ln 2)^{1/2}/\tau_p$. Dividing by 2π gives $\Delta v = (\ln 2)^{1/2}/\pi\tau_p = 0.265/\tau_p$ for the frequency distribution. Thus the frequency distribution and temporal width of this Gaussian pulse satisfy the following expression [8]:

$$\Delta v \Delta \tau = 0.441 \tag{14.25}$$

This represents the minimum product of $\Delta\tau$ and Δv for a so-called transform-limited pulse. Other functional forms give different values of the product, but in any case we see that as expected, the shorter the pulse the bigger the frequency spread. Figure 14.4 illustrates 5 fs and 20 fs pulses, both with a carrier wavelength of 500 nm, and their frequency distributions. Note that we arrived at Equation 14.25 using a purely classical treatment of light, rather than a quantum mechanical uncertainty principle. Recall that the classical treatment of light is sufficient to describe most spectroscopy experiments, particularly laser experiments for which the incident photon flux is large.

The spectral breadth of ultrashort (femtosecond) pulses creates additional experimental concerns compared to spectroscopy on longer timescales. Dispersion of the refractive index causes temporal broadening of the pulse owing to the frequency dependence of the transit times, as the beam passes through various optical elements. This dispersion can be compensated, for example using pairs of prisms, or avoided, using reflective optics.

The timescale for interaction of the molecule with the field is of course also determined by the dynamics of the molecule itself. In Chapter 4, we introduced phenomenological relaxation rates of diagonal ($1/T_1$) and off-diagonal ($1/T_2$) elements of the density matrix, related by $1/T_2 = 1/2T_1 + 1/T_2^*$. In what follows, we use the notation $\Gamma_{aa} = 1/T_1$ and $\Gamma_{ab} = \Gamma_{ba} = 1/T_2$ for the relaxation rates of the aa and ab elements of the density matrix. $1/T_2$ comprises the homogeneous linewidth of the spectrum. The population relaxation rate $1/T_1$ is the sum of irreversible radiative and nonradiative relaxation rates, while pure dephasing, with timescale T_2^*, results from environmentally-induced frequency fluctuations. In addition, inhomogeneous broadening, such as that envisioned for chromophores immobilized in a rigid glass, leads to a static distribution of spectral frequencies. It is frequently the case, particularly in electronic spectroscopy, that the homogeneous linewidth is buried under the inhomogeneous distribution as illustrated in Figure 14.5. In the Kubo line shape model introduced in Chapter 6, solvent-induced dephasing is accounted for using $\langle \delta\omega(0)\delta\omega(t) \rangle = \Delta^2 \exp(-t/\tau_c)$ for the FFCF.

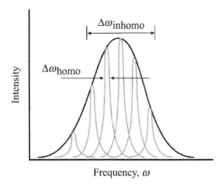

Figure 14.5 Homogeneous and inhomogeneous broadening.

(Compare to Equation 14.18.) An inhomogeneous (Gaussian) limit for the lineshape is obtained when the rate of frequency fluctuations $1/\tau_c$ is slow compared to the spread of frequencies, Δ. A tremendous advantage of TD-FWM experiments is their ability to separate homogeneous and inhomogeneous contributions to the line shape. In time-domain spectroscopy, the distinction between homogeneous and inhomogeneous broadening depends on the duration and temporal separation of light pulses. We refer to static inhomogeneity when the timescale for the frequency fluctuations is long compared to that of the experiment.

Hole-burning spectroscopy is a nonlinear approach to unveiling the homogeneous line width $1/T_2$ buried under the inhomogeneous distribution. The sample is irradiated with a high power pump laser having a peak frequency and bandwidth within the absorption band. The absorption band is observed with a weaker light source, the probe, with frequencies that span the inhomogeneous spectrum. Sufficiently intense pump light tends to cause saturation of the transition; that is, more nearly equal populations in the ground and excited states, reducing the net spectral intensity at the pump frequency. If the band were purely homogeneously broadened, the pump laser would diminish the overall intensity of the whole spectrum. In the presence of inhomogeneous broadening, on the other hand, the pump laser pokes a hole in the spectrum with a width on the order of $1/T_2$. In the transient hole burning experiment, the hole evolves in time after the pump pulse owing to spectral diffusion, and eventually "heals" on the timescale T_1. Hole-burning experiments are often done at low temperature to slow the decay of the hole. The ground state bleach contribution to pump-probe spectroscopy, discussed below, is an example of dynamic hole-burning. The photon echo and two-dimensional experiments described below will be shown to provide another approach to separating homogeneous and inhomogeneous contributions.

14.3.1 THIRD-ORDER NONLINEAR RESPONSE FUNCTION

We saw in Chapter 13 that the third-order response function $S^{(3)}$ depends on three nested commutators of the transition dipole, which give rise to eight terms. Upon inspection these are seen to occur in pairs which are complex conjugates of one another. The third-order response function (Equation 13.130) can then be expressed somewhat succinctly as follows [9]:

$$S^{(3)}(t_1,t_2,t_3) = \left(\frac{i}{\hbar}\right)^3 \theta(t_1)\theta(t_2)\theta(t_3)Tr\left\{[[[\,\mu(t_1+t_2+t_3),\mu(t_1+t_2)],\mu(t_1)],\mu(0)]\rho_{eq}\right\}$$

$$= \left(\frac{i}{\hbar}\right)^3 \theta(t_1)\theta(t_2)\theta(t_3)\sum_{\alpha=1}^{4}\left[R_\alpha(t_3,t_2,t_1) - R_\alpha^*(t_3,t_2,t_1)\right]$$

(14.26)

Here, we are using the Heaviside step function $\theta(t)$, which is equal to unity for $t > 0$ and zero for $t < 0$, leading to integration over positive times in the calculation of $P^{(3)}$. The four functions R_1, R_2, R_3 and R_4 are terms in the expansion in which the perturbations to the density matrix occur in the order ket–ket–ket, bra–ket–bra, bra–bra–ket, and ket–bra–bra, respectively. In the eigenstate picture, these are expressed as a sum over intermediate states:

$$R_1 = \sum_{a,b,c,d} p_a \mu_{ad}(t_1+t_2+t_3)\mu_{dc}(t_1+t_2)\mu_{cb}(t_1)\mu_{ba}(0)$$

$$R_2 = \sum_{a,b,c,d} p_a \mu_{ad}(0)\mu_{dc}(t_1+t_2)\mu_{cb}(t_1+t_2+t_3)\mu_{ba}(t_1)$$

$$R_3 = \sum_{a,b,c,d} p_a \mu_{ad}(0)\mu_{dc}(t_1)\mu_{cb}(t_1+t_2+t_3)\mu_{ba}(t_1+t_2)$$

$$R_4 = \sum_{a,b,c,d} p_a \mu_{ad}(t_1)\mu_{dc}(t_1+t_2)\mu_{cb}(t_1+t_2+t_3)\mu_{ba}(0)$$

(14.27)

The terms R_1, R_2, R_3 and R_4 are represented by the Feynman diagrams A_1, C_2, B_2 and D_1 of Figure 13.9, respectively, with a change in the summation indices. The reader should verify that complex conjugates R_α^* also appear in Figure 13.9. For example, R_1^* corresponds to the bra–bra–bra sequence of interactions. The R_α^* terms enter into Equation 14.26 with a minus sign as a consequence of the rule that the sign of a Feynman diagram is $(-1)^n$, where n is the number of bra-side interactions. These terms will be used to deduce the response for particular experiments discussed below, where the sum-over-states in Equation 14.27 will be narrowed down to specific terms for resonance with the input fields. Four is the maximum number of states that contribute to a term in the expansion of the third-order polarization, and in some cases a smaller number of states may participate.

14.3.2 Pump-probe spectroscopy

Pump-probe spectroscopy, also called transient absorption spectroscopy, is a powerful tool for preparing excited states of molecules and materials and following the ensuing dynamics. This experiment employs an ultrafast light pulse (the "pump" pulse) to prepare an excited electronic or vibrational state, followed by a probe pulse that is delayed by time τ. A schematic of the experiment in shown in Figure 14.6. The time delay is made possible using a variable path length for the probe pulse compared to the pump. The time delay for a given path length difference derives from the speed of light, about 300 nm/fs. The extra length of the probe path relative to the pump path may be as long as 30 cm to achieve a delay of 1 ns, or as short as 3 μm for a delay of 10 fs. Computer-controlled translation stages can vary the path length to within about 1 μm, providing a time resolution as short as 3.3 fs.

The first pump-probe experiments were done by Sir George Porter and colleagues in the millisecond time regime using a flashlamp with a rotating sector to provide the time delay between the pump and probe flashes [10]. The development of what was then called flash photolysis earned Porter the Nobel Prize in 1967, shared

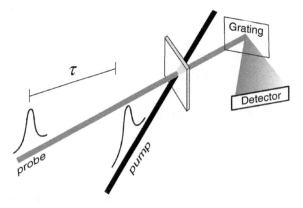

Figure 14.6 Pump-probe experimental geometry.

with M. Eigen and R. G. W. Norrish. A myriad of experimental discoveries permitted by this technique led to concepts that are now standard content in textbooks. These include the detection of free radicals, observation of triplet–triplet absorption spectra, charge transfer, radiationless decay, vibrational relaxation, photodissociation, and initial forays into the fundamental mechanism of photosynthesis. Since its inception, the time resolution of pump-probe spectroscopy has evolved from the millisecond to the femtosecond regime and beyond, and applications span a wide range of fields including biological systems, semiconductors and nanoparticles. The available time resolution provides a window on photon-triggered dynamics such as bond rearrangements, electron transfer, proton transfer, and isomerization. While kinetic studies often employ time-resolved detection at a single wavelength, acquisition of complete time-dependent spectra at a range of probed wavelengths is often used. With only two light pulses, it may not be obvious that pump-probe is a third-order nonlinear experiment. However, as shown below, the sample is considered to interact twice with the pump pulse and once with the (weaker) probe pulse.

In femtosecond broad-band pump-probe experiments, the probe is a spectrally broadened, white light pulse. As shown in Figure 14.6, the pump and probe pulses impinge on the sample in a near-collinear geometry, and the transmission of the probe pulse is recorded in the direction of the probe beam after being spectrally resolved by a grating or prism. The polarization of the probe beam is sometimes set at the magic angle of 54.7° with respect to that of the pump beam to negate the effect of the rotation of the molecular transition dipole moment. Alternatively, reorientational motion can be perceived by using a probe pulse which is polarized perpendicular to the pump. The pump-probe experiment can follow electronic or vibrational state dynamics depending on the wavelength of the pulses and their time delay. A pump pulse at a visible or UV wavelength prepares the molecule in an excited electronic state. A probe beam at a visible or UV wavelength is used to follow the dynamics of excited electronic (or vibronic) states, while a probe at an infrared wavelength reveals vibrations within the excited electronic state. Alternatively, time-resolved vibrational spectroscopy of excited electronic states employs resonance Raman scattering of a visible or UV probe pulse which is resonant with an allowed transition from the pumped electronic state to a higher energy electronic state. In one of the examples below, both the pump and probe pulses are in the IR region in order to follow vibrational dynamics within the ground electronic state. Before discussing examples of the above experiments, let us look at some general aspects of pump-probe spectroscopy.

The transmission of the probe beam at time τ following the pump pulse at time zero is compared to the transmission of the probe in the absence of the pump pulse. This difference can be converted to an absorbance change ΔA, which may be positive or negative. The pump-induced depletion of the ground state population leads to negative ΔA, referred to as a ground state bleach (GSB). GSB is a "hole" in the spectrum resulting from transfer of ground state population to the excited state. The hole is as broad as the steady-state absorption spectrum when the spectrum is homogeneously broadened. In the limit of inhomogeneous broadening, on the other hand, the breadth of the hole depends on the frequency distribution of the pump. Like GSB, stimulated emission (SE) also leads to negative ΔA. In this case, the pump creates a population inversion which results in emission of light in the direction of the probe beam. This emission adds to the intensity and appears as a negative absorption change. Positive features, $\Delta A > 0$, represent excited state absorption (ESA) resulting from transitions from the pumped excited state to higher lying states. Pump-probe spectra are sometimes represented as the relative change in transmittance $\Delta T/T_0$, or intensity $\Delta I/I_0$. In this representation, ESA appears as a negative feature while GSB and SE result in positive peaks.

The pump-probe experiment is said to be self-heterodyned. To explain this, consider for the moment linear spectroscopy. The incident electric field generates a first-order polarization that results in a signal field E_{sig} with the same frequency and propagation direction as the incident field E_0, but not necessarily the same phase. The measured intensity is proportional to $|E_0 + E_{sig}|^2$. In the case of light absorption, destructive interference of the incident and signal fields results in attenuation of the transmitted beam, while stimulated emission corresponds to constructive interference and amplification of the transmitted beam. In either case, the experiment qualifies as heterodyned by virtue of the mixing of the signal with the incident field. The pump-probe experiment is similar except that the probe field E_{pr} interacts with a sample that has already interacted twice with the pump field at time τ earlier. The intensity $I(\tau)$ of the transmitted probe in the presence of the pump and for delay time τ is proportional to the square of the sum of the fields for the signal and the transmitted probe:

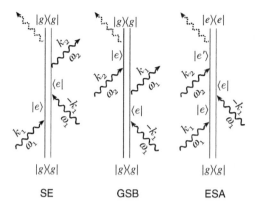

Figure 14.7 Three of the six Feynman diagrams for pump-probe, accounting for stimulated emission (SE), ground state bleach (GSB) and excited state absorption (ESA). The first two of these are rephasing diagrams. The construction of three more nonrephasing diagrams is left as an exercise for the reader.

$$I(\tau) \propto \left| E_{pr} + E_{sig} \right|^2 \tag{14.28}$$

where E_{sig} is proportional to the imaginary part of $P^{(3)}$. The intensity above is compared to that of the transmitted probe in the absence of the pump to get the change in intensity ΔI:

$$\Delta I(\tau) \propto \left| E_{pr} + E_{sig} \right|^2 - \left| E_{pr} \right|^2 \propto \mathrm{Re}(E_{pr}^* E_{sig}) \tag{14.29}$$

Equation 14.29 uses the fact that the amplitude of the probe pulse is much larger than that of the signal. The intensity change ΔI can be converted to a difference in absorbance ΔA.

The phase-matching condition for pump-probe requires that the two interactions with the pump field contribute with wave vectors $\vec{k}_{pu}$ and $-\vec{k}_{pu}$. This leads to six Feynman diagrams, three of which are depicted in Figure 14.7 and three more left as an exercise for the reader. The diagrams shown in Figure 14.7 depict the pump-pump-probe sequence of interactions leading to SE, GSB, and ESA. Those drawn in the figure are "rephasing" diagrams in that, for SE and GSB, the coherence resulting from the first pulse is reversed by the second pulse. (More will be said about rephasing and nonrephasing diagrams in Section 14.4.) The excitation density of the pump pulse is much larger than that of the probe, and the third-order polarization in this experiment is proportional to $E_{pu}^2 E_{pr}$, where E_{pu} and E_{pr} are the electric field amplitudes of the pump and probe, respectively. We are interested here in diagrams in which the two pump interactions precede the interactions with the probe, and other orderings will lead to different nonlinear experiments to be discussed later. Recall that interactions represented by arrows that point to the right are associated with $\exp(ikr - i\omega t)$ phase factors while arrows pointing to the left imply $\exp(-ikr + i\omega t)$. Thus we have a signal at the frequency $\omega_{sig} = \omega_{pu} - \omega_{pu} + \omega_{pr} = \omega_{pr}$ and the phase-matching condition is $\vec{k}_{pu} - \vec{k}_{pu} + \vec{k}_{pr} = \vec{k}_{pr}$. The signal is detected in the same propagation direction as that of the probe pulse, leading to the "self-heterodyned" nature of the detection. The GSB and SE diagrams differ in that following the two pump interactions the molecule finds itself in an excited state population $|e\rangle\langle e|$ for SE and in the ground state population $|g\rangle\langle g|$ for GSB. In both cases the signal represents evolution of the density matrix from the coherence $|e\rangle\langle g|$ to the population $|g\rangle\langle g|$. In contrast, for the ESA diagram the system is in the population $|e\rangle\langle e|$ following the two interactions with the pump, and the final arrow (the signal) represents the evolution from $|e'\rangle\langle e|$ to $|e\rangle\langle e|$, where e' is a higher lying excited state. Note that the sign of the diagram for ESA is opposite to that for GSB and SE.

The distinction between the conventional pump-probe experiment and other third-order phenomena depends on the pulse sequence as well as the pump-probe time delay and the dephasing time of the pumped state. When the time delay is sufficiently long compared to dephasing and the states e and e' are well-separated in energy compared to the spectral width of the pump pulse, the two pump interactions leave the molecule

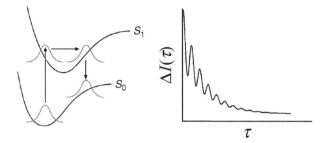

Figure 14.8 Wavepacket motion and oscillations from quantum beats.

in a population rather than a coherence, as seen for the Feynman diagrams discussed above. In this limit, the pump pulse prepares the system in a particular state, and the transmittance of the probe pulse is determined by the linear susceptibility of the optically prepared state. We can then consider pump-probe as a conventional spectroscopy experiment in which the initial state just happens to be an excited state. In the limit that the pump and probe pulses are not temporally overlapped and are separated by a time interval longer than that for dephasing, the time dependence of the spectrum then reflects population dynamics and reveals the kinetics of photophysical and photochemical processes.

On the other hand, if the pump pulse is broad enough to excite more than one state, coherence effects are seen which show up as oscillations in the transmittance of the probe at the beat frequency of the two states. These effects are observable within the pump-pump-probe scheme as seen in one of the examples presented below. For example, pump pulses with durations on the order of a typical vibrational period are capable of exciting vibrational coherences $|v\rangle\langle v'|$ within the ground or excited electronic state. These appear as oscillations, also called quantum beats, in pump-probe spectra and persist for delay times which are less than the dephasing time. Figure 14.8 depicts wavepacket motion leading to quantum beats. These beats occur for displaced potential surfaces, exactly the condition for Franck–Condon activity of normal modes. The time-dependent overlap of the initial and propagating wavepackets modulates the intensity of the transmitted probe beam with a frequency given by that of the vibration.

When the pump and probe pulses are separated by a time that is long compared to dephasing, the time dependence of the transmitted intensity reflects population dynamics only. For a two-level system where the pulses are resonant with the $g \to e$ transition, the transmitted signal is

$$\Delta I(\tau) \propto |\mu_{ge}|^4 e^{-\Gamma_{ee}\tau} \tag{14.30}$$

$\Gamma_{ee} = 1/T_1$ is the population relaxation of the upper state. In a two-level system the rate of decay of the upper state is equal to the rate of filling the hole left in level g: $\Gamma_{ee} = \Gamma_{gg}$. More generally, population relaxation of additional states connected to g and e contributes to the kinetics of bleach recovery.

To understand the occurrence of quantum beats, we use a four-level system as shown in Figure 14.9 with vibrational levels a and c within the ground electronic state and b and d within the excited electronic state. Figure 14.9 shows that the diagrams for R_1 and R_2 correspond to vibrational coherences within the ground and excited electronic states, respectively. For example, the R_1 diagram shown in Figure 14.9 reveals that the system is in a $|c\rangle\langle a|$ coherence after the two time-coincident pump pulses. Thus the density matrix evolves as $\exp(-i\omega_{ca}\tau - \Gamma_{ca}\tau)$ following the pump, revealing an oscillatory component at the vibrational frequency and an exponential damping $\Gamma_{ca} = 1/T_2$ like that shown in Figure 14.8. The diagram for R_2 on the other hand, shows that the pump pulses produce the coherence $|d\rangle\langle b|$ which then evolves as $\exp(-i\omega_{db}\tau - \Gamma_{db}\tau)$, and the oscillations correspond to vibrational motion within the excited electronic state. Note that it is no coincidence that the diagram for the R_1 term in Figure 14.9 is identical to that shown for CARS in Figure 13.16. The terms R_3 and R_4 correspond to vibrational coherence within the ground and excited electronic state, respectively, as explored in one of the homework problems.

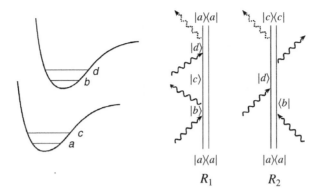

Figure 14.9 Ground and excited electronic state potential surfaces and Feynman diagrams giving rise to vibrational coherences in the ground (R_1) and excited (R_2) electronic state.

When the envelopes of the pump and probe pulses are not well-separated in time, so-called coherence artifacts result. These are not really artifacts, but rather signals that derive from "pump-probe-pump" and "probe-pump-pump" orderings that are not of interest in the pump-probe (or "pump-pump-probe") experiment. We next consider some examples of pump-probe experiments that highlight how populations and coherences are revealed.

14.3.2.1 TRANSIENT ABSORPTION SPECTRA OF EXCITED ELECTRONIC STATES

Pump-probe spectroscopy is widely used to follow photophysical and photochemical dynamics taking place in excited electronic states. These experiments take advantage of the ability to monitor time-dependent populations of excited states. A large number of studies have been aimed at determining the kinetics of fundamental processes such as electron transfer, a recurring motif in fields ranging from biology to materials science. One such example is the study of electron transfer from electronically excited dyes to the conduction band of nanocrystalline metal oxides, primarily TiO_2. TiO_2 is a wide band gap semiconductor with a separation $E_g = 3.2$ eV between the valence and conduction bands. As such, it absorbs UV but not visible light. In a dye-sensitized solar cell, nanoparticulate TiO_2 is sensitized to visible light via adsorption of a monolayer of dye (the sensitizer) capable of excited-state electron transfer to the conduction band of the semiconductor. This electron transfer is the primary process in the conversion of sunlight to electricity in a dye-sensitized solar cell. Electron injection is thermodynamically favored when the excited state reduction potential of the dye, found by subtracting E_{00}/e from the ground state redox potential, is more negative than the conduction band redox potential. Early transient absorption spectroscopy measurements of dyes adsorbed on nano-TiO_2 revealed electron injection to take place on a subpicosecond timescale, faster than the thermalization of the vibrational levels within the excited electronic state. In order for dye-sensitization to lead to an efficient photovoltaic device, the collection of injected electrons in the external circuit has to be faster than recombination of injected electrons with oxidized dye, a feature which is facilitated by efficient regeneration of the original oxidation state by a redox mediator in the electrolyte. In addition, successful sensitizer dyes have donor–acceptor moieties that are spatially separated. Proximity of the LUMO to the semiconductor surface favors faster forward electron transfer while a greater separation of the HOMO from the surface favors slower recombination of the injected electron with the sensitizer dye.

Transient absorption measurements have permitted the detailed kinetics of electron injection, recombination, and dye regeneration to be determined for many sensitizers on TiO_2. Using the Ru-based metalorganic dye popularly known as N3, Heimer et al. [11] used transient absorption at visible and IR wavelengths to uncover the dynamics of forward (injection) and reverse (recombination) electron transfer. Their analysis was aided by comparison of the transient spectra of N3 adsorbed on nanocrystalline TiO_2 and ZrO_2. The latter has a conduction band for which the reduction potential is more negative than that of excited state N3, precluding excited state electron injection. Figure 14.10 shows transient absorption spectra for N3 on ZrO_2 and on TiO_2 generated by a pump at 532 nm, which excites the $S_0 \rightarrow S_1$ transition of N3. In both samples there is an obvious GSB resulting from depletion of the ground electronic state of the dye. On ZrO_2 there is,

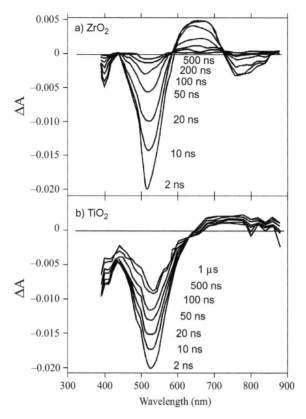

Figure 14.10 Transient absorption spectra of Ru(4,4'-dicarboxylic acid–2,2'-bipyridine)$_2$(NCS)$_2$ adsorbed (a) on ZrO$_2$ and (b) on TiO$_2$. (Reprinted with permission from Heimer, T. A. et al. Electron Injection, Recombination, and Halide Oxidation Dynamics at Dye-Sensitized Metal Oxide Interfaces, *J. Phys. Chem. A 104*, 4256 (2000). Copyright 2000 American Chemical Society.)

in addition, ESA of the dye centered at 650 nm, and a negative going feature near 760 nm assigned to stimulated emission of N3. These latter two features are not apparent in the transient absorption spectra of N3 on TiO$_2$ because electron transfer to the semiconductor effectively quenches the dye excited state. Instead, for N3 on TiO$_2$, the ESA in the near-IR region is assigned to the metal-to-ligand charge-transfer transition of oxidized N3 and absorption by conduction band electrons in TiO$_2$. The authors interpreted the kinetics of the bleach recovery to determine the rate at which the dye is regenerated by reaction of the oxidized form with a redox mediator.

Consider next an example where a sufficiently short pump pulse excites superpositions of vibrational states, leading to quantum beats in pump-probe spectroscopy. This process, known as impulsive stimulated Raman scattering, reveals wavepacket motion on either the ground or excited electronic state potential surface, depending on whether the beats are measured in the vicinity of the GSB, ESA or SE. Dean et al. [12] used transient absorption with 16 fs pulses and a broadband probe to investigate the excited state dynamics of methylene blue in aqueous solution, as shown in Figure 14.11. With an absorption band at 664 nm and fluorescence at 690 nm, the dye is of interest for photodynamic therapies which require red emission wavelength to minimize scattering by biological tissues. Oscillations in the transmitted intensity were observed to persist for up to 2 ps. Note that the traces presented there were obtained by subtracting the background due to population dynamics, fitted to a biexponential function. The time-dependent intensity changes were monitored at the wavelengths of the fluorescence, the ground state bleach, its vibronic side band, and the excited state absorption, and in each case were Fourier transformed to give a vibrational spectrum in good agreement with results from a quantum mechanical calculation. Oscillations in the time-dependent intensity change $\Delta I/I$ were dominated by the beating of the two most intense (highest Franck–Condon factor) modes at 450 and 500 cm^{-1}, where the period of the beat is $1/c\Delta\tilde{v} \approx 650$ fs, in good agreement with the dominant peak separations of the time-dependent intensity.

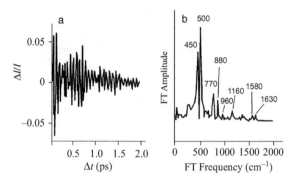

Figure 14.11 (a) Quantum beats resulting from vibrational coherences in broad-band transient absorption of aqueous methylene blue, observed at a probe wavelength near the fluorescence maximum. (b) Vibrational frequencies obtained by Fourier transformation of the data in (a). (Reprinted with permission from Dean, J. C. et al. Broadband Transient Absorption and Two-Dimensional Electronic Spectroscopy of Methylene Blue, *J. Phys. Chem. A 119*, 9098 (2015). Copyright 2015 American Chemical Society.)

An important photochemical process that was the subject of some of the earliest femtosecond pump-probe experiments is the cis–trans isomerization of bacteriorhodopsin (BR). Molecules in the rhodopsin family undergo fast torsional motion in their excited S_1 electronic states. Isomerization about the C_{13}–C_{14} bond of the retinal chromophore of the rhodopsin protein is the first step in a series of events that lead to stimulation of the optic nerve. In the purple membrane of the photosynthetic bacterium *Halobacterium salinarum*, the bacteriorhodopsin molecule acts a light-driven proton pump. Due to its stability and ease of preparation, this pigment has served as a convenient model for the rhodopsin family of molecules. The Mathies lab at UC Berkeley [13–15] has studied the light-driven isomerization of light-adapted bacteriorhodopsin, BR_{568}, in which the retinal chromophore adopts the all-trans form and has an absorption maximum of 568 nm. Using 6 fs probe and 60 fs pump pulses, Mathies et al. looked at the pathway for excited state isomerization to the 13-cis form. In [13,14], the third-order polarization was calculated using a wavepacket approach like that introduced in Chapter 12, where $\langle i|i(t)\rangle$ is the time-dependent overlap of the initial state with that evolving on the excited state surface. As in the treatment of resonance Raman spectroscopy, the wavepacket description of $P^{(3)}(t)$ shown below is an alternative to the sum-over states expression used in Equations 14.27. In the time-dependent view, the polarization was found to be the sum of eight terms as follows:

$$P_{pr}^{(3)}(t) = \mu^4 \left(\frac{i}{\hbar}\right)^3 \int_{-\infty}^{t} dt_3 \int_{-\infty}^{t_3} dt_2 \int_{-\infty}^{t_2} dt_1 e^{(i\omega_1 - 1/T_2)(t-t_3)} e^{-(t_3-t_2)/T_1} e^{(i\omega_1 - 1/T_2)(t_2-t_1)}$$

$$\times \Big\{ E_{pr}(t_3) E_{pu}^*(t_2) E_{pu}(t_1) \langle i|i(t-t_3)\rangle \langle i|i(t_2-t_1)\rangle$$

$$+ E_{pr}(t_3) E_{pu}(t_2) E_{pu}^*(t_1) \langle i|i(t-t_3)\rangle \langle i|i(t_2-t_1)\rangle^*$$

$$+ E_{pr}(t_3) E_{pu}^*(t_2) E_{pu}(t_1) \langle i|i(t-t_1)\rangle \langle i|i(t_3-t_2)\rangle^*$$

$$+ E_{pr}(t_3) E_{pu}(t_2) E_{pu}^*(t_1) \langle i|i(t-t_2)\rangle \langle i|i(t_3-t_1)\rangle^*$$

$$+ E_{pu}(t_3) E_{pr}(t_2) E_{pu}^*(t_1) \langle i|i(t-t_3)\rangle \langle i|i(t_2-t_1)\rangle^*$$

$$+ E_{pu}(t_3) E_{pr}(t_2) E_{pu}^*(t_1) \langle i|i(t-t_2)\rangle \langle i|i(t_3-t_1)\rangle^*$$

$$+ E_{pu}(t_3) E_{pu}^*(t_2) E_{pr}(t_1) \langle i|i(t-t_3)\rangle \langle i|i(t_2-t_1)\rangle$$

$$+ E_{pu}(t_3) E_{pu}^*(t_2) E_{pr}(t_1) \langle i|i(t-t_1)\rangle \langle i|i(t_3-t_2)\rangle \Big\}$$

(14.31)

In this expression, $E_{pu}(t)$ and $E_{pr}(t)$ are the time-dependent pump and probe fields. Each of the eight terms above can be associated with one of the eight terms in the sum-over-states description of $P^{(3)}(t)$. The time t_i in the above expression is the absolute time of the ith pulse rather than the interval between two pulses. Equation 14.31 was obtained assuming both fields oscillate at a frequency ω_1, and the vibrational dephasing rate for intermediate vibronic levels of the S_1 excited state was ignored in comparison to electronic dephasing. Since these terms focus on two states resonant with the frequency $\omega_1 = \omega_{eg}$, only GSB and SE diagrams (Figure 14.7) are relevant. With the help of these diagrams, we can rationalize the exponential terms in the integrand as follows. In the time interval $t_2 - t_1$ between the first and second interactions the system is in a ge coherence so the density matrix evolves as $\exp[(i\omega_1 - 1/T_2)(t_2 - t_1)]$. After the second interaction, the Feynman diagram tells us we have a population (either ee or gg) and the density matrix evolves as $\exp[(-1/T_1)(t_3 - t_2)]$. Finally, the third pulse at t_3 creates the eg coherence and between t_3 and the detection time t, the density matrix evolves as $\exp[(i\omega_1 - 1/T_2)(t - t_3)]$. Note that though the peaks of the two pump pulses are time coincident, the integration is over their corresponding temporal profiles, as well as over that of the probe pulse.

The first four terms in Equation 14.31 are of the pump-pump-probe type and are associated with the conventional pump-probe experiment. The fifth and sixth terms derive from the sequence pump-probe-pump and are important at early delay times during the temporal overlap of the pump and probe, giving rise to the previously mentioned coherence artifact. The last two terms are probe-pump-pump and lead to perturbed free induction decay. In the limit that the first two pulses are time-coincident delta function pulses, the integrals over t_1 and t_2 collapse, and the argument of the rightmost exponential term vanishes, taking its dependence on the dephasing rate with it. At the same time, when the time interval $t - t_3$ is long comparing to the dephasing rate, the contribution of dephasing to the left-most exponential becomes insignificant. These are the conditions for which population dynamics rather than dephasing controls the dynamics (Equation 14.30). More generally, however, the dephasing rate does influence the pump-probe spectrum as seen above.

Figure 14.12 shows the differential transmittance $\Delta T/T_0$ of BR_{568} at a range of times from −54 to 998 fs. Figure 14.13 is a sketch of the S_0, S_1 and S_n potential surfaces along the torsional coordinate for cis-trans isomerization. The 6 fs probe pulse spans the wavelength range 560 to 700 nm, while the 60 fs pump pulse with peak at 618 nm encompasses a more narrow spectral range as seen in the top trace of Figure 14.12. Positive features, $\Delta T > 0$, at 618 nm and 593 nm show up at negative time delays. (Negative time delays may seem contradictory for the pump-probe experiment, but the delay time τ corresponds to the difference in the peaks of the pump and probe temporal profiles. When the pump and probe are not well-separated in time, the temporal widths

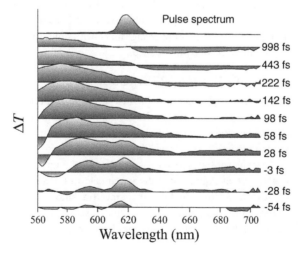

Figure 14.12 Pump-probe spectrum of bacteriorhodopsin, BR_{568}. The top trace is the spectrum of the pump pulse, and the others are the transient spectra at the indicated times. (Reprinted from Pollard, W. T. et al. Direct Observation of the Excited State cis-trans Photoisomerization of Bacteriorhodopsin: Multilevel Line Shape Theory for Femtosecond Hole Burning and Its Application, *J. Chem. Phys. 90*, 199 (1989) with the permission of AIP Publishing.)

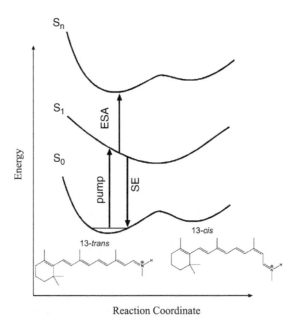

Figure 14.13 Potential energy surfaces for photoisomerization of BR_{568}. (From Pollard, W. T. et al. *J. Chem. Phys.* 90, 199, 1989.)

of the pulses permit some of the pump pulse to precede some of the probe pulse when $\tau < 0$.) These positive features represent GSB, but there is more structure displayed at early times than is observed in the steady-state absorption spectrum. The latter is further broadened by dephasing of the electronic state, which is evidently slow compared to the time interval where the structured "holes" in the absorption spectrum are seen. As the pump-probe delay increases, one sees the emergence of ESA that blue-shifts from about 580 nm to ~560 nm, eventually moving out of the window of the probe spectrum. At the same time, SE is evident at wavelengths to the red of about 640 nm. This SE decays in concert with the ESA as the molecule moves along the excited state potential, which is dissociative with respect to the torsional coordinate. Beyond 222 fs, the negative features at a broad range of red wavelengths result from the absorption of the 13-cis photoproduct. The potential curves in Figure 14.13 account for wavepacket motion along the S_1 excited state and qualitatively explain the blue shift of the $S_1 \rightarrow S_n$ absorption with time, while the SE $S_1 \rightarrow S_0$ shifts to the red. Note that from 58 to 142 fs, the GSB is much broader than the pump spectrum and does not change much with time. This is taken as evidence that the absorption spectrum of BR_{568} is homogeneously broadened.

14.3.2.2 TIME-RESOLVED VIBRATIONAL SPECTROSCOPY

For determination of molecular structural dynamics, transient vibrational spectroscopy offers several advantages over time-resolved electronic spectroscopy. The diffuse bands frequently observed in the latter conceal underlying vibrational progressions, and overlapping electronic transitions may be unresolvable. Vibrational transitions on the other hand, often lead to bands which are only several cm⁻¹ in breadth and provide snapshots of molecular structure in transient states. (Exceptions arise; for example, the broader vibrational bands of hydrogen-bonded molecules such as liquid water, discussed below.) Frequency shifts in transient IR and Raman spectra provide windows into vibrational cooling, since transitions $v \rightarrow v \pm 1$ are increasingly redshifted for higher vibrational quantum numbers v. Raman spectroscopy has additional advantages through the observation of transient anti-Stokes transitions, the time-dependence for which reveals vibrational population decay times.

Ultrafast transient IR spectroscopy has been of great utility in the study of liquid water, where absorption bands of stretching and bending modes span several hundred cm⁻¹. Hydrogen bonding of the O–H group of liquid water leads to red-shifting and broadening of the O–H stretch, relative to the spectrum of water vapor. The red-shift arises from decreased electron density (bond weakening) as a result of partial bond formation

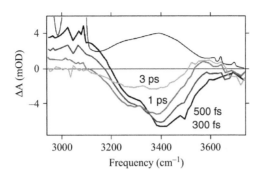

Figure 14.14 Transient IR spectrum of water in reverse micelles with $w_0 = 3$. The top trace is the steady-state absorption spectrum and shows the broad water absorption and sharp peaks from benzene. (Reprinted with permission from Costard, R. et al. Ultrafast Vibrational Dynamics of Water Confined in Phospholipid Reverse Micelles, *J. Phys. Chem. B 116*, 5752 (2012). Copyright 2012 American Chemical Society.)

with the acceptor: O–H---O. The breadth of the O–H stretch is generally considered to result from inhomogeneous broadening owing to a distribution of H-bond strengths. However, the lifetime and dephasing rate of the H-bond can also contribute to the spectral width. Femtosecond pump-probe spectroscopy using tunable mid-IR pulses has been applied to the study of water in confined media such as reverse micelles [16]. Reverse micelles are nanometer-sized spherical assemblies of surfactant molecules in organic solvent, with water molecules encapsulated within. The surfactant molecules possess polar head groups which orient inward toward the water pool and nonpolar hydrocarbon tails which present to the nonpolar solvent. The number of water molecules in the pool is controlled by the diameter of the micelles which can be varied by changing the ratio w_0 of the water concentration to that of the surfactant. The micellar environment provides a model for interfacial water in biological systems. Water confined in biological cavities may play a role in protein dynamics; it is therefore of interest to understand how confinement perturbs the dynamics of water molecules.

Costard et al. [16] used reverse micelles of the phospholipid dioleoylphosphatidylcholine (DOPC) suspended in benzene with femtosecond IR pump and probe pulses to investigate the effects of confinement on the dynamics of water. Phospholipids are components of cell membranes; thus this system serves as a simple model for water in cells. The experiments were performed as a function of variable water content in the range of $w_0 = 1$ to 16, corresponding to micelle diameters from about 4 to 8 nm. In the authors' core-shell model, the shell consists of water which strongly interacts with the polar interface while water molecules in the core behave more like bulk water for sufficiently large micelle size. Figure 14.14 shows the absorbance change ΔA for time delays ranging from 300 fs to 3 ps, compared to the steady-state absorption spectrum. Note that the sharp lines in the steady-state spectrum are C–H stretches of benzene, and the broad peak at ~3400 cm⁻¹ is the O–H stretch of water. In the transient spectra, a negative peak in the vicinity of 3400 cm⁻¹ is the result of GSB of the $v = 0 \rightarrow v = 1$ transition superimposed on the SE for $v = 1 \rightarrow v = 0$. Hot bands arising from $v = 1 \rightarrow v = 2$ give rise to positive features (ESA) on the red edge of the spectral window. The breadth of the main negative feature reflects the distribution of O–H stretching frequencies resulting from water molecules in different environments; i.e., inhomogeneous broadening. The authors attributed the dynamic red-shift of this feature to spectral diffusion on a timescale of about 1.4 ps or more, slower than the spectral diffusion time in bulk water. The decay time of the GSB/SE depends on the micelle size owing to several mechanisms for the population relaxation of the $v = 1$ state. One is the coupling of the $v = 1$ state of the stretch to the overtone of the bend at a similar frequency. The authors found the timescale for decay of the $v = 1$ state of the stretch to match the rise time of the $v = 2$ state of the bend. Fluctuations in the frequencies of these modes bring them into resonance and enhance the energy transfer. These fluctuations and the decay rate increase as the size of the water pool increases. These and other time-resolved studies of water in confined media reveal significant differences in the dynamics compared to those found in the bulk.

An example of the application of time-resolved IR spectroscopy to the study of a photochemical process is provided by [17], which used a UV pump and mid-IR probe to study the ring opening reaction of the spiropyran 1′, 3′, 3′-trimethylspiro-[-2H-1-benzopyran-2,2′-indoline], or BIPS. As illustrated in Figure 14.15,

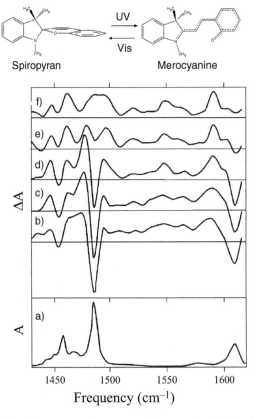

Figure 14.15 UV-pump mid-IR probe spectra of spiropyran ring opening. (a) Steady-state spectrum, and transient absorption spectra at (b) 2 ps, (c) 10 ps, (d) 20 ps, (e) 51 ps, and (f) 100 ps, after UV excitation. (Reprinted with permission from Rini, M. et al. Ultrafast UV-mid-IR Investigation of the Ring Opening Reaction of a Photochromic Spiropyran, *J. Amer. Chem. Soc. 125*, 3028 (2003). Copyright 2003 American Chemical Society.)

BIPS undergoes a ring-opening reaction in its excited S_1 electronic state, which results in the formation of a merocyanine molecule which absorbs at 540 nm, while BIPS absorbs at about 300 nm. Photochromic molecules of this type are of interest for information storage and optical switches, motivating spectroscopic studies of the pathway for interconverting the two forms. Reference [17] employed 70 fs UV pump pulses at 316 nm to create the S_1 excited state of BIPS followed by broadband 100 fs mid-IR pulses spanning about 2000 cm^{-1} (generated by SFG) with a center frequency of 1530 cm^{-1}. Characteristic vibrational fundamentals of BIPS, observed at 1458, 1486 and 1610 cm^{-1} in the ground state IR spectrum, are bleached at early times, and positive absorption changes at 1461, 1489 and 1591 cm^{-1} are assigned to the product merocyanine. Stepwise vibrational cooling within both the ground and excited states of the closed form affects the dynamics in the vicinity of the bleach. The timescale for conversion from the closed to the open form was reported as 28 ps.

Time-resolved resonance Raman (TR^3) spectroscopy is a form of pump-probe, in which the Raman spectrum of the pumped excited electronic state is measured by observing the scattering excited by a probe pulse which is resonant with a transition to a higher-lying excited state. The time and spectral resolutions offered are a compromise because the Raman bands are at least as spectrally broad as the probe pulse, which is typically in the ns or ps regime. (This compromise is avoided in the FSRS experiment described in the previous chapter, where the temporal and spectral resolutions are not decided by a Fourier transform relation.) An example application is provided by the picosecond TR^3 measurements of Wang and Tauber [18], who used the technique to investigate singlet fission in aggregates of the carotenoid molecule 3R,3′R-zeaxanthin (Figure 14.16). Singlet fission is the decay of a singlet excited state to form two excited triplet states having energies no greater than half that of the singlet state. It has been observed in organic crystals, conjugated

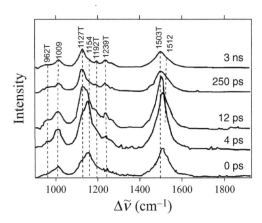

Figure 14.16 Transient Raman spectra of zeaxanthin aggregates in 90:10 tetrahydrofuran: water, obtained at a pump wavelength of 415 nm and probe wavelength of 551 nm. Raman bands of the triplet state are labeled T and remaining bands are from the singlet excited state. (Reprinted with permission from Wang, C. and Tauber, M. J. High Yield Singlet Fission in a Zeaxanthin Aggregate Observed by Picosecond Resonance Raman Spectroscopy, *J. Amer. Chem. Soc. 132*, 13988 (2010). Copyright 2010 American Chemical Society.)

polymers, and in light-harvesting complexes of photosynthetic organisms. In the latter, carotenoids are polyenes that assist with energy transfer to the photosynthetic reaction center, and singlet fission is a possible mechanism along the pathway for this energy transfer. Recent interest in singlet fission is motivated by the desire to exploit the process in solar energy conversion, since a single photon could create two excited electronic states that would hopefully be capable of electron transfer. Though the process is considered the formation of two triplet states from one singlet, the total spin is conserved because the resulting state after fission consists of pairs of triplets with a net spin of zero:

$$\psi_{^1(TT)} = \frac{1}{\sqrt{3}}\left[\psi_{T_1}(1)\psi_{T_{-1}}(2) + \psi_{T_{-1}}(1)\psi_{T_1}(2) + \psi_{T_0}(1)\psi_{T_0}(2)\right] \tag{14.32}$$

This state is represented by the symbol $^1(TT)$ to designate a singlet state formed by mixing two triplets. The subscripts on T indicate the spin quantum number $M_S = 1, 0,$ or -1 for the triplet state T and the numbers in parenthesis designate different molecules in the case of intermolecular fission, or different parts of the same molecule in intramolecular fission.

Wang and Tauber measured the ps TR³ spectrum of self-assembled rod-shaped aggregates of 3R,3′R-zeaxanthin in a mixed solvent of water and tetrahydrofuran, as seen in Figure 14.16. The $S_0 \rightarrow S_1$ transition of symmetric polyenes is symmetry forbidden, while the $S_0 \rightarrow S_2$ transition is allowed. The S_2 excited state lies 14,500 cm⁻¹ above the ground state and the spin-forbidden T_1 state is estimated at 7000 cm⁻¹; thus, singlet fission is energetically possible. The transient resonance Raman spectrum was determined at wavelengths resonant with the $S_0 \rightarrow S_2$ transition of the monomer as well as the red-shifted absorption band of the aggregate. The C=C stretch of the monomer and aggregate ground state, at 1519 cm⁻¹, was found to disappear on a ~4 ps timescale, and the triplet state C=C stretch at 1502 cm⁻¹ was observed. The authors examined the dependence of the ground state depletion as a function of pump power for both the monomer and the aggregate, and used the result to determine the efficiency of singlet fission.

14.4 TRANSIENT GRATING AND PHOTON ECHO EXPERIMENTS

In this section, we consider the closely related transient grating (TG) and photon echo (PE) spectroscopy experiments [9,19,20]. Both of these techniques can be visualized with the schematic shown in Figure 14.17. Three pulses with propagation vectors $\vec{k}_1, \vec{k}_2$ and $\vec{k}_3$ are overlapped in the sample. Using standard notation, the interval

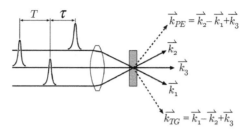

Figure 14.17 Pulse sequence and phase-matching directions for photon echo (PE) and transient grating (TG) experiments.

between the first and second pulses is designated as τ, the coherence time, while the interval between the second and third pulses is called T, the waiting time or population time, and t is the time following the third pulse when the signal is detected. These are respectively the time intervals we have referred to as t_1, t_2 and t_3. Figure 14.17 shows a configuration in which all the propagation vectors are coplanar, but boxcar geometries are also used. The Feynman diagrams for this pulse sequence are shown in Figure 14.18. Figures 14.18a–c show the stimulated emission, ground state bleach, and excited state absorption contributions to photon echo, respectively, while Figures 14.18d–f show SE, GSB and ESA for the transient grating phase-matching condition. In Figures 14.18a and b, the diagram reveals that the first pulse creates the coherence $|g\rangle\langle e|$, while the third pulse creates the coherence $|e\rangle\langle g|$. These are referred to as rephasing diagrams. As shown below, this phase reversal leads to

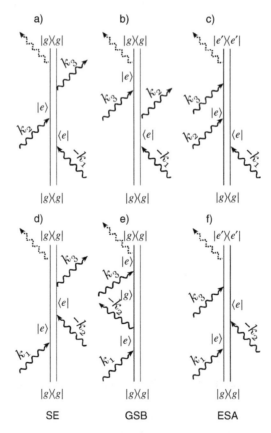

Figure 14.18 Feynman diagrams relevant to photon echo (a–c) and transient grating (d–f) experiments, and for stimulated emission (a and d), ground state bleach (b and e) and excited state absorption (c and f).

the occurrence of a photo echo. No such phase reversal is possible for the ESA diagram Figure 14.18c. There is no rephasing in any of the diagrams associated with the TG experiment, Figures 14.18d–f. In Figures 14.18d and e, the coherence initiated by the first pulse is re-established by the third pulse. In Figure 14.18f, the first pulse creates the coherence $|e\rangle\langle g|$ and the third pulse creates $|e'\rangle\langle g|$. In all cases, the second pulse leads to a population, either $|e\rangle\langle e|$ for SE and ESA diagrams or $|g\rangle\langle g|$ for GSB diagrams. The diagrams relevant to TG and PE can be compared to those for pump-probe (Figure 14.7). The difference is that in the pump-probe case the first two interactions come from the same pulse ($\tau = 0$) and have the same propagation direction. In the PE and TG experiments discussed below, on the other hand, the first two pulses are noncollinear.

14.4.1 TRANSIENT GRATING SPECTROSCOPY

Whenever two optical pulses with stable phases are overlapped on a sample, they create an interference pattern that can act as a grating and scatter a third pulse. We begin with a discussion of the state of the sample following two time-coincident plane wave pulses ($\tau = 0$) of the same frequency, $\omega_1 = \omega_2 = \omega$. The propagation vectors of the two pulses are at an angle 2θ and may be resonant with a spectroscopic transition. The interference of the two beams causes the intensity of the light to be modulated along the length of the sample, which is specified to be the Z direction, as illustrated in Figure 14.19. We can understand the grating formation by considering the wave fronts of the two incident plane-polarized beams, which propagate along opposite diagonal directions inside the sample. A snapshot of the resulting nodal pattern displays a lattice of constructive and destructive interference. As time progresses, the nodes in the vertical direction move from left to right as the beams propagate, but the horizontal nodes are stationary. The time-averaged net electric field displays an interference pattern with horizontal peaks and valleys, forming a diffraction grating in the Z direction. This grating is depicted in Figure 14.19 as a series of constant intensity planes aligned in a direction perpendicular to Z. The intensity modulation $I(Z)$ and grating spacing Λ are given by

$$I(Z) = 2I\left[1 + \cos\left(\frac{2\pi Z}{\Lambda}\right)\right] \tag{14.33}$$

$$\Lambda = \frac{\lambda}{2n\sin\theta} = \frac{\pi c}{\omega n\sin\theta} \tag{14.34}$$

The interference pattern is a volume grating, like that used to create a hologram. The modulation of the light intensity leads to modulation of the complex refractive index, $n = n_r + i\kappa$, the real and imaginary parts of which govern dispersion and absorption/emission, respectively. The spatial modulation of n_r results in a phase grating while that of κ leads to an amplitude grating. In the absence of resonance, the real part n_r determines the grating properties. Resonant pulses lead to a spatial modulation of the ground and excited state populations, known as a population grating. The population grating causes a change in the absorbance

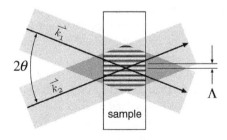

Figure 14.19 Formation of a transient grating from two overlapped pulses.

A that is modulated with the same spacing Λ. The real refractive index is modulated along with this change in absorbance, thus the population grating also results in a phase grating. In addition, the refractive index is modulated by local heating as the populations relax, leading to density changes, or because of volume changes caused by photoproduct formation. Density changes with intensity can also result from electrorestriction, increased local density in regions of high electric field strength. This effect does not require resonance. Another nonresonant effect comes from the nonlinear contribution to the refractive index, proportional to the square of the electric field. We refer to Δn_r and ΔA as the depth of the modulation in the refractive index and absorption, respectively, and express the phase and amplitude gratings as:

$$n_r = n_{r,0} + \Delta n_r \cos\left(\frac{2\pi Z}{\Lambda}\right)$$

$$A = A_0 + \Delta A \cos\left(\frac{2\pi Z}{\Lambda}\right)$$

(14.35)

The subscript zero indicates the average quantity. The diffraction efficiency of the grating depends on the modulation depths. A third type of grating results when the two pump beams are polarized in orthogonal directions, resulting in modulation of the polarization though circular, elliptical and linear polarizations along the length of the grating. Regardless of the physical mechanism for grating formation, a third pulse that arrives before the grating relaxes is partially diffracted at the Bragg angle. The Bragg angle of the transient grating is consistent with the phase-matching condition shown in Figure 14.17 and given below in Equation 14.36. The formation of a grating can still occur if the two pulses are temporally separated as long as the polarization from the first pulse has not decayed before the arrival of the second. Two pulses which are coincident in time but differ in frequency, $\omega_1 \neq \omega_2$, result in a grating that oscillates at the beat frequency $\omega_1 - \omega_2$. This is the case for CARS, where the pump beam ω_1 and the Stokes beam ω_2 differ by a vibrational frequency. When all three input beams are the same frequency, the experiment is referred to as degenerate four-wave mixing.

The general phase-matching constraint for the transient grating (TG) experiment is

$$\vec{k}_{TG} = \vec{k}_1 - \vec{k}_2 + \vec{k}_3$$

(14.36)

Here, we take the first two pulses to be coincident in time ($\tau = 0$) and frequency ($\omega_1 = \omega_2$), impinging on the sample as in Figure 14.17. These two pulses are referred to together as the pump pulse. In the standard experiment we have $\vec{k}_{TG} = \vec{k}_1 - \vec{k}_{1'} + \vec{k}_2$ where the pump pulses 1 and 1' have different angles but the same frequency and arrival time at the sample. The probe pulse $\vec{k}_2$ is then scattered at an angle which depends on its frequency. Resonant pulses induce a population grating, such that the decay of the signal with time delay of the probe reveals population relaxation. In the multiplexed TG experiment, a small angle of incidence is used for the pump pulses and the probe is a white light pulse. The smaller angle θ translates to a larger value of Λ, diminishing the spectral resolution of the grating and permitting the detection of a larger range of diffracted wavelengths. In creating an amplitude grating, relaxation of excited state populations releases heat and causes a phase grating from modulations in the density and thus the refractive index. The density modulations result in acoustic waves launched in counter propagating directions with wave vector magnitude $k_{ac} = 2\pi/\Lambda$, which show up as oscillations of the time-dependent signal. The frequency of these waves, v_{ac}, satisfies $2\pi v_{ac} = k_{ac} v_{sound}$, where v_{sound} is the speed of sound in the medium. Thus the frequency of acoustic waves depends on the angle of incidence of the pump pulses.

An excellent illustration of broadband TG spectroscopy is seen in the study of [21] and shown in Figure 14.20. Vauthey et al. sought to determine the rate of back electron transfer in an exciplex formed between the electron acceptor 9,10-dicyanoanthracene (DCA) and a series of aromatic electron donors. An exciplex is an excited state complex of two molecules that is not bound in the ground state. Exciplexes generally have some charge-transfer character D^+A^- such that back electron transfer is a significant nonradiative relaxation pathway. For multiplex TG experiments, the authors created a population grating using 355 nm

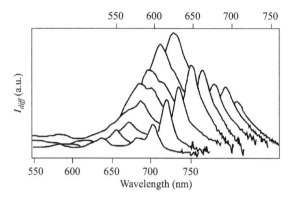

Figure 14.20 Transient grating spectra of the exciplex of 9,10-dicyanoanthracene and 1,2,4,5-tetramethyl-benzene in acetonitrile solution, measured at time delays (from back to front) of 60, 300, 470, 600, 750, 1600, 2300, and 3300 ps. The wavelength axis on the top (bottom) refers to the spectrum in the back (front). I_{diff} is the intensity of diffracted light. (Reprinted with permission from Vauthey, E. et al. Direct Investigation of the Dynamics of Charge Recombination Following the Fluorescence Quenching of 9,10-Dicyanoanthracene by Various Electron Donors in Acetonitrile, *J. Phys. Chem. A 102*, 7362 (1998). Copyright 1998 American Chemical Society.)

pulses overlapped at an angle of 0.3° and probed the sample with a white light pulse. Figure 14.20 shows the resulting TG spectra of a solution of DCA with the electron donor 1,2,4,5-tetramethylbenzene (durene) in acetonitrile at a range of delay times from 60 to 3300 ps. The early spectra show the 620 nm excited state absorption of the S_1 state of DCA. At increasing times this band diminishes and transitions of the radical anion DCA·⁻ are seen at 580 640, 685 and 705 nm. Using a probe wavelength of 681 nm, the authors measured the rate of decay of the TG signal and accounted for it in terms of relaxation of the population of the radical ion, the quenching of excited state DCA by electron transfer from the donor, and decay of the dissociated D⁺ and A⁻ ions. It is apparent from this example that TG spectroscopy shares similarities with pump probe experiments. Both are "pump-pump-probe" orderings of the interactions. The typical pump-probe experiment can be considered as a variation of TG with $\theta = 0$. In the TG experiment, the diffracted signal intensity is measured directly and not compared to the probe beam intensity; thus, the signal is always positive. This makes it difficult to assign features to GSB, SE and ESA contributions in a transient grating experiment. Though not as susceptible to errors from fluctuations in the probe beam intensity, the TG experiment is susceptible to scattering artifacts. The TG experiment can also be done with a nonzero time separation of the first two pulses, as will be considered further below after examination of the photon echo experiment.

14.4.2 PHOTON ECHO SPECTROSCOPY

Photon echo experiments are described by the Feynman diagrams in Figures 14.18a–b. As for the TG case, there are diagrams corresponding to ground state bleach, stimulated emission, and excited state absorption. During the dephasing and rephasing time periods, the density matrix evolves as $\exp(-i\omega_{eg}t_1)$ and $\exp(i\omega_{eg}t_3)$, respectively, leading to a net cancellation of the transition frequency in the overall phase when $t_1 = t_3$. Note that the TG and PE experiments can be interconverted by reversing the order of the first two pulses, which reverses the signs for their wave vectors. The phase-matching condition for the photon echo signal is

$$\vec{k}_{PE} = -\vec{k}_1 + \vec{k}_2 + \vec{k}_3 \tag{14.37}$$

The signal is generally integrated over the detection time t and may be gathered as a function of coherence time τ or waiting time T. We consider three implementations of the method, denoted by the acronyms two-pulse photon echo (2PE), three-pulse stimulated photon echo (3PSE), and three-pulse photon-echo peak shift (3PEPS).

In the *two-pulse photon echo* (2PE) experiment, the second and third pulses derive from the same beam. Hence the phase-matching condition is $\vec{k}_{2PE} = -\vec{k}_1 + 2\vec{k}_2$ and the waiting time T is zero. This implies a response that is proportional to $E_1^0 (E_2^0)^2$, where $E_{1,2}^0$ is the amplitude of the first or second pulse; thus, the experiment can be categorized as probe-pump-pump. In the usual notation where t_i is the time interval between pulse i and pulse $i + 1$, we define $t_1 = \tau$, $t_2 = T = 0$, and t_3 is the time following the third pulse when the signal is detected. We consider pulses resonant with the $g \to e$ transition. For a two-level system, we consult the rephasing diagrams for GSB and SE, which are equivalent when $t_2 = 0$. We write the response function using rules from the previous chapter as follows:

$$S^{(3)}(\omega_{eg}) = |\mu_{eg}|^4 \rho_{gg}^{(0)} e^{-i\omega_{eg}(t_1 - t_3)} e^{-\Gamma_{eg}(t_1 + t_3)} \tag{14.38}$$

Equation 14.38 represents the homogeneous response function. We are interested in the situation where there is a Gaussian distribution of transition frequencies with average $\langle \omega_{eg} \rangle$ and width Δ:

$$G(\omega_{eg}) = \frac{1}{N} \exp\left[\frac{-\left(\omega_{eg} - \langle \omega_{eg} \rangle\right)^2}{2\Delta^2} \right] \tag{14.39}$$

We account for inhomogeneity by multiplying the response function $S^{(3)}(\omega_{eg})$ by this distribution function and integrating over ω_{eg}, to get

$$S^{(3)}(t_1, t_3) = |\mu_{eg}|^4 \rho_{gg}^{(0)} e^{-i\langle \omega_{eg} \rangle (t_1 - t_3)} e^{-\Gamma_{eg}(t_1 + t_3)} e^{-(t_1 - t_3)^2 \Delta^2 / 2} \tag{14.40}$$

In the case where the inhomogeneous linewidth Δ is much larger than the homogeneous width $\Gamma_{eg} = 1/T_2$, the Gaussian function of time is sharply peaked compared to the dephasing term and can be replaced by a delta function: $e^{-(t_1 - t_3)^2 \Delta^2 / 2} \approx \delta(t_1 - t_3)$. The response function is then sharply peaked at $t_1 = t_3 = \tau$, giving rise to an echo that follows the second pulse (which interacts twice with the sample) by the coherence time τ. The first exponential in Equation 14.40 represents dephasing during the time period t_1 that is exactly reversed during t_3. In the limit of large inhomogeneous linewidth, then, the echo signal is observed when $t_3 = t_1$, and the contributions of $\langle \omega_{eg} \rangle$ and Δ in Equation 14.40 disappear. The integrated intensity of the echo decays exponentially:

$$I_{sig}(\tau) \propto |\mu_{eg}|^8 e^{-4\Gamma_{eg}\tau} \tag{14.41}$$

Equation 14.41 represents the echo decay curve that results when one measures the intensity as a function of the coherence time τ in the inhomogeneous limit. The factor of 4 in the exponential of Equation 14.41 results from putting $t_1 + t_3 = 2\tau$ in the dephasing term, then squaring the field E_{sig} to get the intensity I_{sig}. It is remarkable that though the distribution of slow frequency fluctuations dominates the breadth of the linear spectrum, only the rapid fluctuations affect the echo decay. The impulsive limit has been assumed here, meaning that the pulse durations are shorter than the timescales of the system, such that the polarization is proportional to the response function. Hence, Equation 14.41 does not capture the initial increase in the photon echo signal near $\tau = 0$. For small values of τ, less than a full echo is realized because only those parts of the first pulse that precede parts of the second pulse can give rise to an echo. Note that although the echo signal decays with a timescale characteristic of the inverse homogeneous width, it is the *presence* of inhomogeneous broadening that leads to the echo. In the absence of inhomogeneous broadening, $\Delta = 0$, and Equation 14.38 leads to $I_{sig}(\tau) \propto \exp(-2\Gamma_{eg}\tau)$. These limits and the intermediate case are explored further below.

We next examine how the 2PE signal varies as a function of detection time $t_3 \equiv t$. In the impulsive limit, we have $P^{(3)}(\tau, t) \propto S^{(3)}(\tau, t)$. This gives

$$|P^{(3)}(\tau, t)| \propto e^{-\Gamma_{eg}(\tau + t)} e^{-(\tau - t)^2 \Delta^2 / 2} \tag{14.42}$$

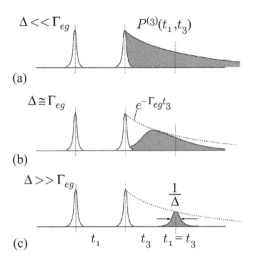

Figure 14.21 Polarization in the two-pulse photon echo (2PE) experiment in the homogeneous (a), intermediate (b), and inhomogeneous (c) limits. (From Tokmakoff, A. *Time-Dependent Quantum Mechanics and Spectroscopy* (University of Chicago), *http://tdqms.uchicago.edu/page/nonlinear-and-two-dimensional-spectroscopy-notes* (Accessed March 19, 2016). License: Creative commons BY-NC-SA. With permission.)

Figure 14.21 depicts the time-dependent polarization following the second pulse in the homogeneous ($\Delta \ll \Gamma_{eg}$) and inhomogeneous ($\Delta \gg \Gamma_{eg}$) limits as well the intermediate ($\Delta \approx \Gamma_{eg}$) case. In the homogeneous limit, the polarization falls of exponentially with detection time t, $P^{(3)} \propto \exp(-\Gamma_{ge}t)$, with no reversal of coherence. In the opposite (inhomogeneous) limit, the echo signal appears as a Gaussian-shaped pulse with a peak at $t = \tau$ and a temporal width $1/\Delta$. The intermediate case leads to a peak at a time $t < \tau$ with a shape reflecting both homogeneous and inhomogeneous timescales. The ability of the system to rephase and give rise to an echo depends on the relative strengths of the homogeneous damping, Γ_{eg}, and the inhomogeneous broadening, Δ.

The signal in the 2PE experiment is homodyne detected by integrating over the detection time t:

$$I_{sig}(\tau) \propto \int_0^\infty \left|P^{(3)}(\tau,t)\right|^2 dt = e^{-4\Gamma_{eg}\tau} e^{-\Gamma_{eg}^2/\Delta^2} erfc\left(-\Delta\tau + \frac{\Gamma_{eg}}{\Delta}\right) \tag{14.43}$$

Here, $erfc(x) = 1 - erf(x)$ is the complementary error function.*

The limits of this function for $\Delta \gg \Gamma_{eg}$ and $\Delta \ll \Gamma_{eg}$ give rise to signals which decay exponentially as $\exp(-4\Gamma_{eg}/\tau)$ and $\exp(-2\Gamma_{eg}/\tau)$, respectively, as previously noted. In the intermediate case, on the other hand, the function $erfc(-\Delta\tau + \Gamma_{eg}/\Delta)$ increases with τ, offsetting the decay from the first exponential term. The result is that the signal maximizes at a coherence time $\tau^* > 0$, called the peak shift. Note that this initial increase in $I_{sig}(\tau)$ for $\tau > 0$ is obtained in the impulsive limit. It is not caused by the finite widths of overlapping first and second pulses; rather, it results from integrating over the detection time. The value of the peak shift as an indication of the ability of the system to rephase is further explored below in the discussion of three-pulse photo echo experiments.

Before discussing the three-pulse photo echo experiments, we consider how the 2PE experiment for a two-level system can be compared to the spin-echo experiment of magnetic resonance presented in Chapter 3. How is a $\pi/2$ pulse in NMR analogous to creating a coherence in a photo echo experiment? Recall that for a spin-1/2 particle, the Z-component of the spin operator is diagonal in the basis of the two spin states $|1/2\rangle$ and $|-1/2\rangle$, which are split by the magnetic field B_0 in the Z-direction. In the optical case, the ground and excited

---

* The error function is $erf(x) = (2/\sqrt{\pi}) \int_0^x \exp(-t^2)dt$.

states are already "split" by the zero-order Hamiltonian. The application of the oscillating magnetic field B_1 in a direction perpendicular to B_0 is responsible for M1-allowed transitions between spin states. This is so because the transitions in magnetic resonance are induced by magnetic dipoles which are proportional to the spin angular momentum. The X and Y components of the spin operators are nondiagonal, thus they connect the two different spin states just as the electronic transition dipole moment permits optical transitions. Thus the magnetization in the X and Y directions, as suggested by the optical Bloch equations of Chapter 4, is a consequence of off-diagonal elements of the density matrix. In the two-state case, we can define an optical $\pi/2$ pulse as one for which the product of the Rabi frequency Ω and the pulse duration τ_p is equal to $\pi/2$. Recall that under the influence of a resonant pulse, the populations of the upper and lower states oscillate back and forth (as seen in Figure 4.3) at the Rabi frequency. A $\pi/2$ pulse results in a superposition state which is an equal mixture of the two coupled states, $\left(1/\sqrt{2}\right)(|g\rangle+|e\rangle)$, i.e. $\left|c_g\right|^2=\left|c_e\right|^2$. This is far from a weak perturbation, which should only slightly change the equilibrium populations, $\left|c_g\right|^2\approx 1$, $\left|c_e\right|^2\approx 0$. But we did not use any perturbation theory to arrive at the optical Bloch equations, so let us continue with this analogy with the caveat that we cannot rigorously extend it to the case of a weakly perturbed, multistate system. The nonzero value of $c_g^* c_e \propto \exp(-i\omega_{eg}t)$ represents the coherence. Following a $\pi/2$ pulse, the precessing spins spread out in the rotating coordinate system as a result of an inhomogeneous magnetic field B_0 (static inhomogeneity). There are also frequency fluctuations resulting from spin–spin interactions (T_2 processes). A subsequent π pulse converts the system to another 50:50 mixture of the two states but of opposite phase: $c_g^* c_e \propto \exp(i\omega_{eg}t)$. The π pulse reverses the directions of precessional motion, and the spins rephase after a time equal to that of the first coherence period. The greater the static inhomogeneity, the greater the spread in precessional frequencies and the larger is the temporal width of the echo pulse. As for the case of photo echos, the appearance of a spin echo itself requires inhomogeneity. In its absence, the spin vectors would remain aligned with one another, undergoing only free-induction decay with rate $1/T_2$. The formation of an echo requires frequency fluctuations that affect different molecules differently; i.e., inhomogeneity.

Three-pulse stimulated echo experiment. In a three-pulse, or stimulated, photo echo experiment, 3PSE, the waiting time T is nonzero and the second and third pulses may be of different frequency. The phase-matching condition is now

$$\vec{k}_{3PSE} = -\vec{k}_1 + \vec{k}_2 + \vec{k}_3 \qquad (14.44)$$

Comparison of the phase-matching conditions and Feynman diagrams for both the three-pulse PE and TG reveals that, at least for SE and ESA, the two experiments differ by exchanging the time ordering of the first two pulses. (This is not possible for the GSB diagram because we cannot start out with a de-excitation of the ground state.) Experimentally, we can go from 3PSE to TG by changing the coherence time from τ to $-\tau$ and observing the signal at the appropriate phase-matched direction. The two experiments are equivalent when the coherence time τ is zero. The integrated intensity using homodyne detection is $I_{3PSE}(\tau,T)=\int\left|P^{(3)}(\tau,T,t)\right|^2 dt$. This signal is measured as a function of the coherence time τ and waiting time T. As seen above, for $T=0$, the interplay of inhomogeneous and homogeneous broadening results in a peak shift $\tau^* > 0$ for which the integrated intensity is a maximum. On introducing a waiting time T between the second and third pulses, spectral diffusion and homogeneous dephasing degrade the memory of the coherence established by the first pulse, hindering echo formation and causing the peak in the integrated intensity versus τ, $I_{3PSE}(\tau,T)$, to tend toward $\tau=0$. In order for rephasing to occur, the response of the system during the time $t_3 = t$ has to be correlated to that during the interval $t_1 = \tau$. This is the basis for the 3PEPS experiment discussed next.

The *three-pulse echo peak shift* (3PEPS) experiment incorporates the phase-matching conditions for PE and TG and measures the signal as a function of positive and negative τ for fixed time T. As seen in the example of Figure 14.22, taken from [22], the 3PSE signal as a function of coherence time τ results in two Gaussian functions symmetrically displaced with respect to the origin, one for each of the phase-matched directions $\vec{k}_{sig}=-\vec{k}_1+\vec{k}_2+\vec{k}_3$ and $\vec{k}_{sig}=\vec{k}_1-\vec{k}_2+\vec{k}_3$. The difference in the maxima of these two Gaussians is twice the peak shift τ^*. By recording the signal for both the rephasing and nonrephasing directions, the peak shift can be determined with subfemtosecond precision. The peak shift is maximum when $T=0$ (2PE), and shifts toward the origin as T increases. At sufficiently long population times, the system has lost memory of the coherence

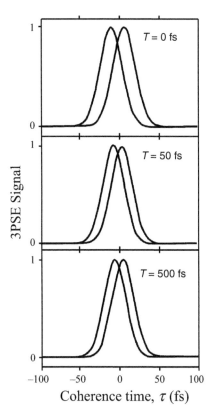

Figure 14.22 Three-pulse photon echo signal of 1.8 nm radius CdSe nanocrystals as a function of coherence time τ for three different population times T. Curves on the right-hand side correspond to the phase-matching condition $k_2 - k_1 + k_3$ and those on the left to $k_1 - k_2 + k_3$. (Reprinted from Salvador, M. R. et al. Exciton-Bath Coupling and Inhomogeneous Broadening in the Optical Spectroscopy of Semiconductor Quantum Dots, *J. Chem. Phys. 118*, 9380 (2003), with permission of AIP Publishing.)

that existed before the arrival of the second pulse. There is then no rephasing, only free-induction decay, when T is longer than the timescale for frequency fluctuations. The shift in the peak τ^* toward the origin reflects the transition from inhomogeneous to homogeneous broadening as defined by a timescale compared to T. A graph of τ^* versus T reveals dephasing dynamics directly. Figure 14.23 displays such a graph for two samples of CdSe nanocrystals and for a solution of rhodamine 6G. Superimposed on the overall decay in τ^* are oscillations that result from vibrations that are coupled to the resonant electronic transition, as seen clearly in the 3PEPS data for CdSe nanocrystals. Static inhomogeneous broadening on the timescale of the experiment results in a nonzero asymptotic value of τ^* as $T \to \infty$. The data for CdSe quantum dots reveals such an effect, resulting from a static distribution of quantum dot size that in turn influences the electronic transition energy. Note that the asymptotic value of τ^* is larger for the CdSe sample with a larger size distribution. On the other hand, the peak shift τ^* for a solution of rhodamine 6G does decay to zero at large T, indicating the absence of true static inhomogeneity.

Fleming and Cho [23] have shown that the dependence of τ^* on T is directly related to the FFCF, defined here as $C(t) = \langle \delta\omega(0)\delta\omega(t) \rangle$. 3PEPs experiments provide another window, often with higher time-resolution, into the same solvent dynamics that are manifested in time-dependent fluorescence Stokes shift measurements. In a two-state picture, the peak shift as a function of population time is expressed as

$$\tau^*(T) = \frac{\Delta \operatorname{Re} C(t)}{\sqrt{\pi}\left\{ \Delta^4 - \left[\operatorname{Re} C(t) \right]^2 \right\}} \tag{14.45}$$

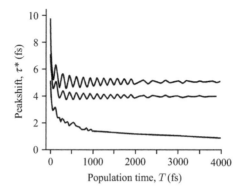

Figure 14.23 Peak shift τ^* versus population time T for CdSe nanocrystals with broad size distribution (top), with a narrow size distribution (middle), and for rhodamine-6G (bottom). (Reprinted from Salvador, M. R. et al. Exciton-Bath Coupling and Inhomogeneous Broadening in the Optical Spectroscopy of Semiconductor Quantum Dots, *J. Chem. Phys. 118*, 9380 (2003), with permission of AIP Publishing.)

Here, $\Delta = \langle (\delta \omega)^2 \rangle^{1/2}$ is the amplitude of the frequency fluctuations. In addition to revealing solvent dynamics, 3PEPS has been used to follow the dynamics of protein motion coupled to a chromophore, for example in light-harvesting photosynthetic complexes.

14.5 TWO-DIMENSIONAL SPECTROSCOPY

Two-dimensional (2D) electronic and vibrational spectra reveal the correlations between pairs of spectroscopic transitions [24,25]. If two spectroscopic transitions are coupled to one another, pumping at the frequency of one transition influences the response at the probe frequency of the second transition. A 2D spectrum represents the intensity of the signal as a function of pump (or excitation) and probe (or emission) frequencies as a contour map as seen in the cartoon of Figure 14.24. A linear spectrum with two transitions ω_a and ω_b is sketched along the top and right side of the figure. These correlate to the diagonal peaks at (ω_a, ω_a) and (ω_b, ω_b). If these two transitions were uncoupled, for example, if they were merely from two noninteracting components of a mixture, there would be no cross-peaks (ω_a, ω_b) and (ω_b, ω_a). Cross-peaks arise when the transitions at ω_a and ω_b are correlated.

Two-dimensional NMR spectroscopy, which reveals the coupling of nuclear spins in a molecule, is the historical and conceptual precedent for 2D optical spectroscopy. The first demonstration of two-dimensional

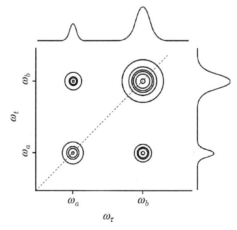

Figure 14.24 Two-dimensional spectrum for two coupled transitions with frequencies ω_a and ω_b. The curves at the top and right depict the linear spectrum.

electronic spectroscopy was reported by Jonas and co-workers in 1998 [24]. In 2D infrared spectroscopy, couplings of fundamental transitions derive from anharmonicity, and the technique can be used to deduce molecular structure. A prominent example of coupling in 2D electronic spectroscopy is excitonic coupling, which derives from the interaction of electronic transition dipoles in molecular assemblies. Evolution of the 2D spectrum with population time T is used to reveal the nature of the couplings and their timescale. 2D spectra as a function of time T, called relaxation experiments, can reveal population dynamics, spectral diffusion, coherence times, and energy flow.

A 2D optical spectrum could be generated by measuring the pump-probe spectrum as a function of pump frequency and for various pump-probe delays. This is the so-called dynamic hole-burning experiment, a frequency domain analogue to the time-domain measurement of 2D spectra. For example, dynamic hole-burning IR spectra were measured by Hamm et al. [26] on a series of peptides, using pulses resonant with the amide-I vibration at about 1650 cm^{-1}. The coupling of local modes on adjacent amide groups makes the frequency of this vibration dependent on the tertiary structure of the peptide. In addition, the strong transition dipole $d\mu/dQ$ of this mode leads to partial delocalization over a number of peptide units and contributes to the 30–40 cm^{-1} width of the linear IR spectrum. Hamm et al. used a pump pulse sufficiently broad to excite the entire manifold of amide-I vibrations. Thus it was not possible to discern which frequency in the pump pulse was responsible for modifying a feature in the probe spectrum. If one attempts to circumvent this limitation using a more narrow range of pump frequencies, the time-resolution is accordingly diminished. 2D spectroscopy uses Fourier transformation of the signal obtained as a function of pulse delay times to obtain frequency resolved information. The technical and mathematical details of the approach are explained in detail in [25]. Below, we consider some of the general features of 2D spectroscopy to highlight the type of information that can be obtained.

Two-dimensional spectra are obtained from time-resolved 3PSE data that is Fourier transformed with respect to coherence time τ and detection time t at a fixed value of population time T. Heterodyne detection enables phase information to be obtained; i.e., the real and imaginary parts of the response. The 2D spectrum at a given T is

$$I(\omega_t, \omega_\tau, T) \propto \int_{-\infty}^{\infty} d\tau \int_{0}^{\infty} dt\, e^{i\omega_t t} e^{-i\omega_\tau \tau} \hat{E}_{LO} \cdot iP^{(3)}(t, T, \tau) \tag{14.46}$$

where ω_τ and ω_t are the excitation and emission frequencies, respectively, and $\hat{E}_{LO}$ is a unit vector in the direction of the field of the local oscillator. Alternatively, the Fourier transform involving ω_t is performed optically by spectrally resolving the emitted light with a prism or grating. The spectrum above is the sum of rephasing and nonrephasing contributions. As described in [27], addition of the rephasing ($\tau > 0$) and nonrephasing ($\tau < 0$) contributions leads to absorptive signals, which enable excited state absorption to be distinguished from oppositely signed ground state bleach and stimulated emission.

The resulting 2D spectra are gathered as a function of waiting time T and displayed as contour maps such as those shown in the cartoon of Figure 14.24. Diagonal peaks, where $\omega_\tau = \omega_t$, represent populations, and off-diagonal features (cross-peaks), where $\omega_\tau \neq \omega_t$, arise from couplings. In some literature, ω_τ is called the pump frequency ω_{pu}, and ω_t is referred to as the probe frequency ω_{pr}. Alternatively, these are called ω_1 and ω_3 to indicate they are respectively conjugate to the time delays t_1 and t_3. When ω_τ and ω_t are presented on the x and y axes, respectively, vertical slices of the graph represent the pump-probe spectrum at a given pump frequency. Diagonal and antidiagonal features represent positive and negative correlations, respectively, of the pump and probe frequencies.

Like the 3PSE experiment, 2D spectra permit the separation of inhomogeneous and homogeneous broadening. Static broadening on the timescale of the experiment results in a distribution of diagonal frequencies (Δ), while homogeneous broadening contributes to the width in the antidiagonal direction (Γ), as shown in Figure 14.25a. When the population time T is much larger than the dephasing timescale, information on homogeneous dephasing is lost and the diagonal peak evolves into a more symmetric shape, as sketched in Figure 14.25b. In this limit, the extent of the signal in both the horizontal and vertical directions is determined by the inhomogeneous width Δ. In the case of an electronic 2D spectrum, horizontal and vertical slices

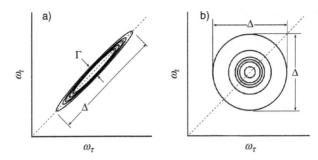

Figure 14.25 Diagonal peak in a two-dimensional spectrum at (a) short and (b) long waiting time, T.

of the diagonal peaks at sufficiently long waiting time T resemble the steady-state excitation and emission spectra, respectively, and a Stokes shift ($\omega_t < \omega_\tau$) is sometimes seen.

2D electronic spectroscopy has been applied to the study of light-harvesting molecular aggregates of photosynthetic organisms. The excited electronic state of chlorophyll or bacteriochlorophyll molecules within a supramolecular assembly is delocalized to some extent by transition–dipole moment coupling. This coupling leads to a manifold of collective excited electronic states, known as excitons, which differ in the phase relationship of the component monomer wavefunctions. The excited electronic state of an excitonically coupled aggregate of n molecules is split into n different excitonic states, with a bandwidth proportional to the coupling strength J of adjacent molecules. In principle, these excited states are delocalized over the entire assembly, but static and dynamic disorder tend to localize the excitation. Excitonic coupling within light-harvesting aggregates of photosynthetic organisms enhances energy transfer, funneling energy to the reaction center where electron transfer takes place. In addition to excitonic states, charge-transfer excited states exist in which the electron and hole reside on different molecules.

2D spectroscopy is well-suited to uncover the electronic couplings and energy flow within a molecular aggregate [28,29]. Cross-peaks observed at early times T reveal the coupling strength J. Indeed, in the absence of such coupling the cross-peaks vanish as a result of cancellation of different Feynman pathways. The coupling strength depends on the relative orientation of transition dipole moments. Cross-peaks resulting from coherences oscillate at the beat frequency of the two states, while those from energy transfer are nonoscillatory and arise at longer waiting T. Recently, the observation of vibrational coherences in photosynthetic complexes has attracted a great deal of excitement, and the role of coherent vibrational motion in efficient energy transfer is being debated [29].

For the rest of this section, we turn our attention to 2D IR spectroscopy and use an example from the literature to illustrate the information that can be obtained. Cross-peaks in 2D IR spectra reveal vibrations that are coupled by mechanical or electrical anharmonicity. We have seen that mechanical anharmonicity leads, for example, to a red-shift of the excited state absorption $v = 1 \rightarrow v = 2$ compared to the $v = 0 \rightarrow v = 1$ transition. In the absence of anharmonicity, cross-peaks connecting these two transitions cancel exactly, as shown in one of the homework problems. As we have seen in the previous chapter, nonlinear signals in general depend on some kind of anharmonicity.

Khalil et al. [27] used 2D IR spectroscopy to study the strongly coupled symmetric (ω_s) and asymmetric (ω_a) carbonyl stretches of $Rh(CO)_2C_5H_7O_2$, or RDC, dissolved in hexane and chloroform. The linear IR spectrum of RDC in either solvent displays fundamental transitions of the asymmetric and symmetric stretch at 2015 and 2084 cm^{-1}, respectively. (Strong coupling of the two CO groups attached to Rh leads to a significantly larger frequency for the symmetric stretch, in contrast to the common observation of a larger frequency for the asymmetric stretch.) Interestingly, the spectral widths (FWHM) are much smaller in hexane (< 3 cm^{-1} for both ω_s and ω_a) than in chloroform (about 9 and 15 cm^{-1} for ω_s and ω_a, respectively). This suggests stronger coupling of the molecule to the solvent in chloroform. Using pulses with a center frequency of 2050 cm^{-1} and a FWHM of 160 cm^{-1}, the 2D experiment sees the effects of transitions connecting the ground state $(v_a, v_s) = (0,0)$ to the excited states $(1,0)$ and $(0,1)$, as well as those connecting the single quantum excited states $(1,0)$ and $(0,1)$ to the double quantum excited states $(1,1)$, $(2,0)$, and $(0,2)$.

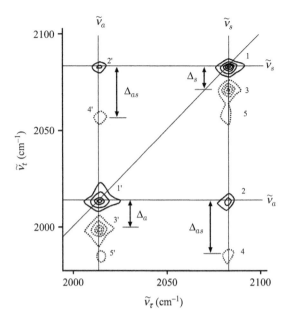

Figure 14.26 2D IR spectrum of Rh(CO)$_2$C$_5$H$_7$O$_2$ in hexane at $T = 0$. (Reprinted with permission from Khalil, M. et al. Coherent 2D IR spectroscopy: Molecular Structure and Dynamics in Solution, *J. Phys. Chem. A 107*, 5258 (2003). Copyright 2003 American Chemical Society.)

Figure 14.26 illustrates the 2D IR spectrum of RDC in hexane at $T = 0$. The diagonal peaks 1 and 1′ represent the fundamental transitions of the symmetric and asymmetric stretch, respectively, while the cross-peaks 2 and 2′ derive from coupling of these fundamentals. Alternatively, one can say that the cross-peaks result from the transfer of coherence from one fundamental transition to the other. Positive features, for which the contours are drawn with full lines, derive from coherences of the ground and single quantum vibrational states, while negative features (dashed contours) come from coherences connecting one- and two-quantum states. As indicated in Figure 14.26, the peaks 1, 1′, 2 and 2′ are positive in sign, while those labeled 3, 3′, 4, 4′, 5, and 5′ are negative. The separation of each member of the latter set of peaks from a diagonal or cross-peak reveals diagonal and off-diagonal anharmonicities as defined in Equation 10.65. (Here, we follow the notation of [27] and use the symbol Δ_{ij} instead of x_{ij} for the anharmonicity. Recall that the anharmonicity for polyatomic molecules is defined to have units of cm^{-1}.) Peak 3 arises from the transition between the overtone of the symmetric stretch, $(v_a, v_s) = (0,2)$ and its fundamental (0,1). It is red-shifted along the ω_t axis by the (diagonal) anharmonicity Δ_s. Similarly, peak 3′ reveals the diagonal anharmonicity of the asymmetric stretch, Δ_a. Peaks 4 and 4′ result from transitions between the (1,1) state and the symmetric (0,1) and asymmetric (1,0) fundamentals, respectively. Both features are shifted from the cross-peak by the off-diagonal anharmonicity Δ_{as}. The weak features labeled 5 and 5′ derive from the nominally forbidden transitions $(0,1) \rightarrow (2,0)$ and $(1,0) \rightarrow (0,2)$, respectively.

The symmetric diamond shapes of the peaks in Figure 14.26 reveal that the vibrational transitions of RDC in hexane are homogeneously broadened; i.e., the linear IR spectrum is in the motional narrowing limit. In contrast, the 2D IR spectrum of RDC in chloroform displays peaks which are elongated and tilted in the diagonal direction at $T = 0$. This is consistent with inhomogeneous broadening, which evidently leads to the larger linewidth in the linear spectrum of RDC in chloroform. At longer waiting times (2D spectra were measured with $T = 2.9$ and 6.2 ps), the peaks are not as tilted and elongated. The authors of [27] simulated this effect using a Kubo lineshape model: $\langle \delta\omega(0)\delta\omega(t) \rangle = \Delta^2 \exp(-t / \tau_c)$. They found the product $\Delta\tau_c$ to be 2.5, consistent with the slow modulation limit. The frequency fluctuations become uncorrelated on a picosecond timescale. Simulated 2D spectra were obtained using a model that included the six states (0,0), (1,0), (0,1), (1,1), (2,0) and (0,2). This led to a response function which was the sum of 66 resonant Feynman diagrams.

14.6 SUMMARY

In this chapter, we have explored time-resolved measurements that derive from both the linear and nonlinear response of the material. Compared to steady-state, linear experiments, time-domain spectroscopy provides additional information content and direct observations of molecular dynamics. Enhanced structural information, such as that revealed by vibrational couplings and anharmonicity, is also provided in pulsed experiments. The examples of nonlinear and time-resolved techniques presented in this chapter and the previous one are a mere sampling of this rapidly evolving branch of spectroscopy. We have emphasized those techniques that dominate when the incident fields are resonant with dipole-allowed transitions. There are in addition nonresonant, nonlinear responses that influence the real part of the refractive index. One such effect is the optical Kerr effect (OKE), which derives from the nonlinear part of refractive index and has been used to study dynamics in liquids. We have also not discussed important applications of nonlinear spectroscopy to imaging. For example, CARS is exploited in microscopy to zero in on particular functional groups with vibrational modes resonant with the frequency difference of two input beams. The formalism that we have used to understand second- and third-order nonlinear spectroscopy is readily extended to higher orders (particularly fifth and higher odd orders). Fifth-order multidimensional coherence spectroscopies have been developed, for example three-dimensional infrared spectroscopy [30]. The Raman echo experiment is a seventh-order technique [31]. Though cross-sections tend to smaller values as the order increases, higher-order nonlinear techniques offer increasing information content. The technique of high harmonic generation is pushing the limits of spatial and temporal resolution to subfemtosecond and subÅngstrom regimes. Applications of spectroscopy also continue to grow, in fields such as materials science and biology. It is the author's hope that the background provided by this book will serve to encourage future work in this vibrant field.

PROBLEMS

1. Derive an expression relating the anisotropy r to the depolarization ratio ρ used in Raman spectroscopy. What is the value of ρ for completely depolarized scattered light? Is such a situation possible in Raman scattering? Is it possible to achieve $r = 0$ in a fluorescence experiment? Is it possible to achieve $r = 1$ in a fluorescence experiment? Comment on the difference in the anisotropy of scattered versus emitted light.

2. In [4], the long-time limit of the anisotropy r_∞ was found to be approximately 0.12. Find the angle between the absorption and emission transition dipole moments.

3. Would a non-Condon contribution to the transition dipole moment, $(d\mu_{ge}/dQ)_0 Q$, influence the measurement of fluorescence lifetime and time-resolved anisotropy? Explain.

4. Using standard bond distances and angles for acetonitrile, CH_3CN, estimate the moment of inertia for rotation of the dipole moment. Use this value to estimate the time constant for inertial (Gaussian) relaxation at room temperature. (See Chapter 5.) Compare this value to the fast response of CH_3CN from TDFSS as shown in Table 14.1.

5. Convert Equation 14.25 to units of cm^{-1} fs. Find the frequency distribution $\Delta\tilde{\nu}$ for 1 fs, 1 ps, and 1 ns Gaussian pulses.

6. Construct the Feynman diagrams for stimulated emission and ground state bleach in pump-probe spectroscopy, analogous to those of Figure 14.7, but of the "nonrephasing" variety.

7. Construct the Feynman diagrams that correspond to the response functions R_3 and R_4 for a four-level system leading to quantum beats. Write down the mathematical expressions for these contributions.

8. Show that for delta function pulses, i.e. $\vec{E}(t - \tau) = E_0 \delta(t - \tau)$ the third-order polarization is proportional to the response function.

9. Verify Equations 14.33 and 14.34 by computing the intensity I which is proportional to $|\vec{E}_1 + \vec{E}_2|^2$.

10. Consider the transient grating experiment described in [21] and depicted in Figure 14.20, where 355 nm pulses were used at an angle of $0.3°$ to create the grating. Use phase-matching considerations to determine the scattering angles for probe light at 550 and 750 nm, incident normal to the grating.

11. Hamm et al. [26] used dynamic hole-burning spectroscopy to show that the homogeneous bandwidth of the amide I vibration is 10 cm^{-1} and that the population relaxation time is $T_1 = 1.7$ ps. Find the pure dephasing time T_2^*.

12. Write the contributions to the response function for the three diagrams Figures 14.18a–c, using the vibrational levels v = 0, 1 and 2. Show that the sum of the three is zero for a harmonic oscillator. Hint: $\mu_{12} = \sqrt{2}\,\mu_{01}$ for a harmonic oscillator.

13. In the presence of true static inhomogeneity with width Σ, the expression for the peak shift in Equation 14.45 can be modified by replacing Δ^2 by $\Delta^2 + \Sigma^2$. Show that this leads to a nonzero value of the peak shift τ^* as $T \to \infty$.

14. Derive the form of the 2PE signal, Equation 14.43, in the slow ($\Delta \gg \Gamma_{eg}$) and fast ($\Delta \ll \Gamma_{eg}$) modulation limits.

REFERENCES

1. R. G. Bennett, Instrument to measure fluorescence lifetimes in the micromillisecond region, *Rev. Scient. Instrument.* 31, 1275 (1960).

2. S. J. Strickler, and R. A. Berg, Relationship between absorption intensity and fluorescence lifetimes of molecules, *J. Chem. Phys.* 37, 818 (1962).

3. J. R. Lakowicz, Principles of fluorescence spectroscopy, 3rd ed. (Springer, New York, 2006).

4. J. A. Levitt, M. K. Kuimova, G. Yahioglu, P.-H. Chung, K. Suhling, and D. Phillips, Membrane-bound molecular rotors measure viscosity in live cells via fluorescence lifetime imaging, *J. Phys. Chem. C* 113, 11634 (2009).

5. M. L. Horng, J. A. Gardecki, A. Papazyan, and M. Maroncelli, Sub-picosecond measurements of polar solvation dynamics: Coumarin 153 revisited, *J. Phys. Chem.* 99, 17311 (1995).

6. X.-X. Zhang, M. Liang, N. P. Ernsting, and M. Maroncelli, Complete solvation response of coumarin 153 in ionic liquids, *J. Phys. Chem. B* 117, 4291 (2013).

7. M. Park, C. H. Kim, and T. Joo, Multifaceted ultrafast intramolecular charge transfer dynamics of 4-(dimethylamino)benzonitrile (DMABN). *J. Phys. Chem. A* 117, 370 (2013).

8. W. W. Parson, *Modern Optical Spectroscopy* (Springer, New York, 2007).

9. A. Tokmakoff, *Time-Dependent Quantum Mechanics and Spectroscopy* (University of Chicago), *http://tdqms.uchicago.edu/page/nonlinear-and-two-dimensional-spectroscopy-notes* (Accessed March 19, 2016). License: Creative commons BY-NC-SA.

10. G. Porter, Flash photolysis and some of its applications, *Science 160*, 1299 (1968).

11. T. A. Heimer, E. J. Heilweil, C. A. Bignozzi, and G. J. Meyer, Electron injection, recombination, and halide oxidation dynamics at dye-sensitized metal oxide interfaces, *J. Phys. Chem. A 104*, 4256 (2000).

12. J. C. Dean, S. Rafiq, D. G. Oblinsky, E. Casette, C. C. Jumper, and G. D. Scholes, Broadband transient absorption and two-dimensional electronic spectroscopy of methylene blue, *J. Phys. Chem. A 119*, 9098 (2015).

13. W. T. Pollard, S. L. Dexheimer, Q. Wang, L. A. Peteanu, C. V. Shank, and R. A. Mathies, Theory of dynamic absorption spectroscopy of nonstationary states. 4. Application to 12-fs resonant impulsive Raman spectroscopy, *J. Phys. Chem. 96*, 6147 (1992).

14. S. L. Dexheimer, Q. Wang, L. A. Peteanu, W. T. Pollard, R. A. Mathies, and C. V. Shank, Femtosecond impulsive excitation of nonstationary vibrational states in bacteriorhodopsin, *Chem. Phys. Lett. 188*, 61 (1992).

15. W. T. Pollard, C. H. B. Cruz, C. V. Shank, and R. A. Mathies, Direct observation of the excited state cis-trans photoisomerization of bacteriorhodopsin: Multilevel line shape theory for femtosecond hole burning and its application, *J. Chem. Phys. 90*, 199 (1989).

16. R. Costard, N. E. Levinger, E. T. J. Nibbering, and T. Elsaesser, Ultrafast vibrational dynamics of water confined in phospholipid reverse micelles, *J. Phys. Chem. B 116*, 5752 (2012).

17. M. Rini, A.-K. Holm, E. T. J. Nibbering, and H. Fidder, Ultrafast UV-mid-IR investigation of the ring opening reaction of a photochromic spiropyran, *J. Amer. Chem. Soc. 125*, 3028 (2003).

18. C. Wang and M. J. Tauber, High yield singlet fission in a zeaxanthin aggregate observed by picosecond resonance Raman spectroscopy, *J. Amer. Chem. Soc. 132*, 13988 (2010).

19. K. Duppen and D. A. Wiersma, Picosecond multiple-pulse experiments involving spatial and frequency gratings: A unifying approach. *J. Opt. Soc. Am. B 3*, 614 (1986).

20. A. M. Weiner, S. De Silvestri, and E. P. Ippen, Three-pulse scattering for femtosecond dephasing studies: Theory and experiment. *J. Opt. Soc. Am. B 2*, 654 (1984).

21. E. Vauthey, C. Hogemann, and X. Allonas, Direct investigation of the dynamics of charge recombination following the fluorescence quenching of 9,10-dicyanoanthracene by various electron donors in acetonitrile, *J. Phys. Chem. A 102*, 7362 (1998).

22. M. R. Salvador, M. A. Hines, and G. D. Scholes, Exciton-bath coupling and inhomogeneous broadening in the optical spectroscopy of semiconductor quantum dots, *J. Chem. Phys. 118*, 9380 (2003).

23. G. R. Fleming and M. Cho, Chromophore-solvent dynamics, *Ann. Rev. Phys. Chem. 47*, 109 (1996).

24. J. D. Hybl, A. W. Albrecht, S. M. Gallagher Faeder, and D. M. Jonas, Two-dimensional electronic spectroscopy, *Chem. Phys. Lett. 297, 307* (1998).

25. M. Cho, Two-dimensional optical spectroscopy, Taylor & Francis group, Boca Raton, 2009.

26. P. Hamm, M. Lim, and R. M. Hochstrasser, Structure of amide I band of peptides measured by femtosecond nonlinear-infrared spectroscopy, *J. Phys. Chem. B 102*, 6123 (1998).

27. M. Khalil, N. Demirdoven, and A. Tokmakoff, Coherent 2D IR spectroscopy: Molecular structure and dynamics in solution, *J. Phys. Chem. A 107*, 5258 (2003).

28. N. S. Ginsberg, Y.-C. Cheng, and G. R. Fleming, Two-dimensional electronic spectroscopy of molecular aggregates, *Acc. Chem. Res. 42*, 1352 (2009).

29. E. Romero, R. Augulis, V. I. Novoderezhkin, M. Ferretti, J. Thieme, D. Zigmantas, and R. van Grondelle, Quantum coherence in photosynthesis for efficient solar-energy conversion, *Nature Physics 10*, 676 (2014).

30. S. Garret-Roe and P. Hamm, What can we learn from three-dimensional infrared spectroscopy? *Acc. Chem. Res. 42*, 1412 (2009).

31. M. Berg and D. A. Vanden Bout, Ultrafast Raman echo measurements of vibrational dephasing and the nature of solvent-solute interactions, *Acc. Chem. Res. 65*, 30 (1997).

Appendix A: Math review

A.1 VECTORS AND TENSORS IN THREE DIMENSIONS

We are interested here in functions in three-dimensional space. A scalar function $F(x, y, z)$ may depend on three spatial coordinates, but has only a single value at a given point x, y, z. A vector function $\vec{F}(x, y, z)$, on the other hand, has both a value and a direction at any given point in space, and thus we require three numbers (components) to describe the function. We will discriminate between these two types of functions merely by placing an arrow over the vector function. Each of the three components of the vector function depends on the position; thus

$$\vec{F}(x,y,z) = \hat{i}F_x(x,y,z) + \hat{j}F_y(x,y,z) + \hat{k}F_z(x,y,z) \tag{A.1}$$

where $\hat{i}$, $\hat{j}$, and $\hat{k}$ are unit vectors in the x, y and z directions, respectively. A vector may also be considered to be a tensor of rank one.

The dot product of two vectors, say $\vec{F}$ and $\vec{G}$, where $\vec{G} = \hat{i}G_x(x,y,z) + \hat{j}G_y(x,y,z) + \hat{k}G_z(x,y,z)$, is given by

$$\vec{F} \cdot \vec{G} = |\vec{F}||\vec{G}|\cos\theta = F_xG_x + F_yG_y + F_zG_z \tag{A.2}$$

where θ is the angle between the two vectors. The result is a scalar. The cross-product, on the other hand, is a vector quantity:

$$\vec{F} \times \vec{G} = \hat{i}(F_yG_z - F_zG_y) + \hat{j}(F_zG_x - F_xG_z) + \hat{k}(F_xG_y - F_yG_x) \tag{A.3}$$

The magnitude of the cross-product of $\vec{F}$ and $\vec{G}$ is $|\vec{F}||\vec{G}|\sin\theta$, and the direction is perpendicular to the plane of $\vec{F}$ and $\vec{G}$ as given by the right-hand rule. If you imagine using the fingers of your right hand to push $\vec{F}$ into $\vec{G}$, through the angle between them which is less than 180°, your thumb will point in the direction of $\vec{F} \times \vec{G}$. Another way to express the cross-product uses the following determinant:

$$\vec{F} \times \vec{G} = \begin{vmatrix} \hat{i} & \hat{j} & \hat{k} \\ F_x & F_y & F_z \\ G_x & G_y & G_z \end{vmatrix} \tag{A.4}$$

Determinants are discussed in Section A.2.

The "del" or "grad" operator, specified by the symbol ∇, can operate on a scalar function and return a vector quantity. It is defined by

$$\nabla = \hat{i}\frac{\partial}{\partial x} + \hat{j}\frac{\partial}{\partial y} + \hat{k}\frac{\partial}{\partial z} \tag{A.5}$$

For example, operating on the scalar function F, we get

$$\nabla F(x,y,z) = \hat{i}\frac{\partial F}{\partial x} + \hat{j}\frac{\partial F}{\partial y} + \hat{k}\frac{\partial F}{\partial z} \equiv \vec{G}(x,y,z) \tag{A.6}$$

The resulting vector, arbitrarily defined as $\vec{G}$ in the above expression, points in the direction of the greatest rate of change of F, and has a magnitude equal to the slope of the function.

When the ∇ operator is applied to a vector function, the result is a second-rank tensor, $\nabla\vec{F} = \mathbf{G}$, having components $G_{xx} = \partial F_x/\partial x$, $G_{xy} = \partial F_y/\partial x$, etc., as follows:

$$
\begin{aligned}
\nabla\vec{F} &= \left(\hat{i}\frac{\partial}{\partial x} + \hat{j}\frac{\partial}{\partial y} + \hat{k}\frac{\partial}{\partial z} \right)\left(\hat{i}F_x + \hat{j}F_y + \hat{k}F_z \right) \\[2mm]
&= \hat{i}\hat{i}\frac{\partial F_x}{\partial x} + \hat{i}\hat{j}\frac{\partial F_y}{\partial x} + \hat{i}\hat{k}\frac{\partial F_z}{\partial x} \\[2mm]
&\quad + \hat{j}\hat{i}\frac{\partial F_x}{\partial y} + \hat{j}\hat{j}\frac{\partial F_y}{\partial y} + \hat{j}\hat{k}\frac{\partial F_z}{\partial y} \\[2mm]
&\quad + \hat{k}\hat{i}\frac{\partial F_x}{\partial z} + \hat{k}\hat{j}\frac{\partial F_y}{\partial z} + \hat{k}\hat{k}\frac{\partial F_z}{\partial z}
\end{aligned}
\tag{A.7}
$$

$\mathbf{G}$ is a second-rank Cartesian tensor having $3 \times 3 = 9$ components. A good example of such a tensor is the electric field gradient $\nabla\vec{E}$ discussed in Chapter 3. The trend may be continued: $\nabla\nabla\vec{F}$ is a third-rank tensor, with 27 components of the type G_{xxx}, G_{xxy}, etc. Operations involving both Cartesian and spherical tensors are considered in Section A.3.

The divergence of a vector function, $\text{div}\vec{F} \equiv \nabla\cdot\vec{F}$, is given by

$$
\nabla\cdot\vec{F}(x, y, x) = \frac{\partial F_x}{\partial x} + \frac{\partial F_y}{\partial y} + \frac{\partial F_z}{\partial z}
\tag{A.8}
$$

As usual for a dot product, the result is a scalar. The curl of a function, $\text{curl}\vec{F} \equiv \nabla\times\vec{F}$, can be written using Equations A.2 or A.3 with ∇ given by A.5. For example,

$$
\nabla\times\vec{F} = \begin{vmatrix} \hat{i} & \hat{j} & \hat{k} \\ \dfrac{\partial}{\partial x} & \dfrac{\partial}{\partial y} & \dfrac{\partial}{\partial z} \\ F_x & F_y & F_z \end{vmatrix} \equiv \vec{H}
\tag{A.9}
$$

where the components of the vector are defined as $\vec{H} = \hat{i}H_x + \hat{j}H_y + \hat{k}H_z$ and

$$
\begin{aligned}
H_x &= \left(\frac{\partial F_z}{\partial y} - \frac{\partial F_y}{\partial z} \right) \\[2mm]
H_y &= \left(\frac{\partial F_x}{\partial z} - \frac{\partial F_z}{\partial x} \right) \\[2mm]
H_z &= \left(\frac{\partial F_y}{\partial x} - \frac{\partial F_x}{\partial y} \right)
\end{aligned}
\tag{A.10}
$$

Another important operator is the Laplacian $\nabla^2 = \nabla\cdot\nabla$. By operating on the scalar function F twice with the del operator, we get

$$
\nabla^2 F = \frac{\partial^2 F}{\partial x^2} + \frac{\partial^2 F}{\partial y^2} + \frac{\partial^2 F}{\partial z^2}
\tag{A.11}
$$

A.2 MATRICES

A matrix is just an array of numbers. Let us say that the matrix $\mathbf{A}$ has n rows and m columns. Then $\mathbf{A}$ is referred to as an $n \times m$ matrix, and the element a_{ij} is the number in the i-th row and j-th column of the matrix.

$$\mathbf{A} = \begin{pmatrix} a_{11} & a_{12} & \cdots & a_{1m} \\ a_{21} & a_{22} & \cdots & a_{2m} \\ \vdots & \vdots & \ddots & \vdots \\ a_{n1} & a_{n2} & \cdots & a_{nm} \end{pmatrix} \tag{A.12}$$

Two matrices can be added (or subtracted) only if they have the same dimensions, that is, the same number of rows and the same number of columns. The result of adding two such matrices, $\mathbf{A} + \mathbf{B} = \mathbf{C}$, is obtained merely by adding corresponding elements: $a_{ij} + b_{ij} = c_{ij}$.

A square matrix has $n = m$. A diagonal matrix is a square matrix with nonzero elements only along the diagonal: $a_{ij} = 0$ if $i \neq j$. The unit matrix $\mathbf{I}$ is a diagonal matrix with ones along the diagonal; i.e., the elements of $\mathbf{I}$ are δ_{ij}. The delta function δ_{ij} is equal to one when $i = j$ and zero when $i \neq j$. A column vector is a matrix of dimension $n \times 1$, while a row vector is one of dimension $1 \times n$.

In order to multiply two matrices: $\mathbf{AB} = \mathbf{C}$, the number of columns in the matrix on the left must be the same as the number of rows in the matrix on the right. If $\mathbf{A}$ is a $k \times l$ matrix and $\mathbf{B}$ is $l \times m$, then $\mathbf{C}$ has dimension $k \times m$. Each element of $\mathbf{C}$ is obtained as

$$c_{ij} = \sum_{r=1}^{l} a_{ir} b_{rj} \tag{A.13}$$

The product of a row matrix and column matrix, where the row matrix is on the left, results in a scalar.

$$\begin{pmatrix} a_1 & a_2 & \cdots & a_n \end{pmatrix} \begin{pmatrix} b_1 \\ b_2 \\ \vdots \\ b_n \end{pmatrix} = a_1 b_1 + a_2 b_2 + \cdots a_n b_n \tag{A.14}$$

On the other hand, if an $n \times 1$ matrix multiplies a $1 \times n$ matrix from the left, the result is an $n \times n$ matrix:

$$\begin{pmatrix} b_1 \\ b_2 \\ \vdots \\ b_n \end{pmatrix} \begin{pmatrix} a_1 & a_2 & \cdots & a_n \end{pmatrix} = \begin{pmatrix} b_1 a_1 & b_1 a_2 & \cdots & b_1 a_n \\ b_2 a_1 & b_2 a_2 & \cdots & b_2 a_n \\ \vdots & \vdots & \ddots & \vdots \\ b_n a_1 & b_n a_2 & \cdots & b_n a_n \end{pmatrix} \tag{A.15}$$

The transpose of a matrix, $\mathbf{C}^T$, is obtained by interchanging rows and columns. That is, if the matrix $\mathbf{B} = \mathbf{C}^T$, then $b_{ij} = c_{ji}$. Note that the transpose of a column vector is a row vector and vice versa. The inverse of a square matrix $\mathbf{C}^{-1}$ is defined such that $\mathbf{CC}^{-1} = \mathbf{C}^{-1}\mathbf{C} = \mathbf{I}$, where $\mathbf{I}$ is the unit matrix.

The determinant of a square matrix (let us say it is $n \times n$) is symbolized by enclosing the matrix in vertical lines: $\det\mathbf{A} = |\mathbf{A}|$. It is a number that results from summing $n!$ products of the n elements of the matrix. Each product contains elements taken one from each row, or one from each column. The sign of each term in this sum depends on the permutation of these elements. The determinant of a 2×2 matrix is the following:

$$\begin{vmatrix} a_{11} & a_{12} \\ a_{21} & a_{22} \end{vmatrix} = a_{11}a_{22} - a_{21}a_{12} \tag{A.16}$$

For larger dimensions, the determinant can be reduced to a sum involving lower-order determinants. This procedure employs the *minor* of a matrix element, which is the determinant of the matrix that is obtained when the row and column containing that element are deleted. Let us define A^{ij} to be the minor of element a_{ij}. The determinant of **A** is obtained by summing across any row or column of **A** as follows:

$$|\mathbf{A}| = \sum_{i=1}^{n} (-1)^{i+j} a_{ij} A^{ij}, \text{ for any } j$$

$$|\mathbf{A}| = \sum_{j=1}^{n} (-1)^{i+j} a_{ij} A^{ij}, \text{ for any } i$$

(A.17)

For example, the determinant of a 3×3 matrix is expanded as follows:

$$\begin{pmatrix} a & b & c \\ d & e & f \\ g & h & i \end{pmatrix} = a(ei - fh) - b(di - fg) + c(dh - eg)$$

(A.18)

The form of Equation A.18 results from summing across the first row, but in the end it does not matter which row or column is chosen. Determinants have the property that if any two rows or two columns of the matrix are interchanged, the determinant changes sign. Also, if two rows or two columns are identical, the determinant vanishes. If one row or one column contains only zeros, then the determinant is zero.

Determinants are important in the solution of linear simultaneous equations. Suppose that we have n linear equations for n unknowns. In matrix form, we write these n equations as

$$\begin{pmatrix} c_{11} & c_{12} & \cdots & c_{1n} \\ c_{21} & c_{22} & \cdots & c_{2n} \\ \vdots & \vdots & \ddots & \vdots \\ c_{n1} & c_{n2} & \cdots & c_{nn} \end{pmatrix} \begin{pmatrix} x_1 \\ x_2 \\ \vdots \\ x_n \end{pmatrix} = \begin{pmatrix} b_1 \\ b_2 \\ \vdots \\ b_n \end{pmatrix}$$

(A.19)

The coefficients c_{ij} are numbers and the x_i's are unknowns. In matrix notation, Equation A.19 is written more concisely: $\mathbf{CX} = \mathbf{B}$. To proceed to find the x_i's, we define the matrix $\mathbf{C}_i$ as that which is obtained by replacing the i-th column of **C** by the column vector **B**. For example

$$\mathbf{C}_2 = \begin{pmatrix} c_{11} & b_1 & \cdots & c_{1n} \\ c_{21} & b_2 & \cdots & c_{2n} \\ \vdots & \vdots & \ddots & \vdots \\ c_{n1} & b_n & \cdots & c_{nn} \end{pmatrix}$$

(A.20)

The unknown x_i is found from the ratio of two determinants.

$$x_i = \frac{|\mathbf{C}_i|}{|\mathbf{C}|}$$

(A.21)

The above approach fails when all the b_i's are zero, in which case $|\mathbf{C}_i| = 0$ and the only solution obtained is the trivial one: $x_1 = x_2 = \ldots = x_n = 0$. A nontrivial solution to this set of equations,

$$\begin{pmatrix} c_{11} & c_{12} & \cdots & c_{1n} \\ c_{21} & c_{22} & \cdots & c_{2n} \\ \vdots & \vdots & \ddots & \vdots \\ c_{n1} & c_{n2} & \cdots & c_{nn} \end{pmatrix} \begin{pmatrix} x_1 \\ x_2 \\ \vdots \\ x_n \end{pmatrix} = \begin{pmatrix} 0 \\ 0 \\ \vdots \\ 0 \end{pmatrix}$$

(A.22)

exists only if the determinant of the coefficients, $|\mathbf{C}|$, is zero. This is the case in matrix eigenvalue problems. Suppose that we have $\mathbf{CX} = \lambda\mathbf{X}$, where λ is the eigenvalue (constant) and $\mathbf{X}$ is an eigenvector. This matrix expression can be rearranged to get

$$\begin{pmatrix} c_{11} - \lambda & c_{12} & \cdots & c_{1n} \\ c_{21} & c_{22} - \lambda & \cdots & c_{2n} \\ \vdots & \vdots & \ddots & \vdots \\ c_{n1} & c_{n2} & \cdots & c_{nn} - \lambda \end{pmatrix} \begin{pmatrix} x_1 \\ x_2 \\ \vdots \\ x_n \end{pmatrix} = \begin{pmatrix} 0 \\ 0 \\ \vdots \\ 0 \end{pmatrix} \tag{A.23}$$

A nontrivial solution exists only if

$$\begin{vmatrix} c_{11} - \lambda & c_{12} & \cdots & c_{1n} \\ c_{21} & c_{22} - \lambda & \cdots & c_{2n} \\ \vdots & \vdots & \ddots & \vdots \\ c_{n1} & c_{n2} & \cdots & c_{nn} - \lambda \end{vmatrix} = 0 \tag{A.24}$$

Expanding the determinant in Equation A.24 results in a polynomial of order n in λ. Thus one obtains n values of λ by setting this determinant equal to zero. For each root λ, Equation A.23 can be solved for the corresponding eigenvector $\mathbf{X}$. The x_i's of each eigenvector are determined only to within a multiplicative constant, so one may impose normalization:

$$1 = \sum_i x_i^2 \tag{A.25}$$

to fix the values of the x_i's.

A.3 OPERATIONS WITH CARTESIAN AND SPHERICAL TENSORS

A tensor property depends on the orientation of the system. In a Cartesian coordinate system, an ordinary vector is such a tensor. A vector such as $\vec{F} = \hat{i}F_x + \hat{j}F_y + \hat{k}F_z$ is a first-rank tensor; each of its three components is indexed by one of the three directions in space. A second-rank tensor, on the other hand, requires two indices for each component, and since each of these runs over the directions x, y, z there are $3 \times 3 = 9$ components of a second-rank tensor. We could think of a second-rank tensor property, such as polarizability, as a square matrix, and the math would be the same as employed in matrix problems. The tensor notation is more general and can be applied when vectors and matrices do not provide enough dimensions for the physical property.

For example, consider expanding the dipole moment of a molecule in a power series in the electric field.

$$\vec{\mu} = \vec{\mu}^{(0)} + \vec{\mu}^{(1)} + \vec{\mu}^{(2)} + \vec{\mu}^{(3)} + \cdots$$

$$\vec{\mu} = \vec{\mu}^{(0)} + \alpha \cdot \vec{E} + \frac{1}{2}\beta : \vec{E}\vec{E} + \frac{1}{6}\gamma \vdots \vec{E}\vec{E}\vec{E} + \cdots \tag{A.26}$$

where $\mu^{(0)}$ is the permanent dipole moment, a first-rank tensor. The second-rank tensor α, the polarizability, leads to the part of the induced dipole moment $\vec{\mu}^{(1)}$ which is linear in the field. The third-rank tensor β, the hyperpolarizability, results in an induced moment $\vec{\mu}^{(2)}$ which depends quadratically on the field, and so on. The quantity $\vec{E}\vec{E}$ is itself a second-rank tensor having components E_xE_x, E_xE_y, etc. Similarly, γ is a fourth-rank tensor, and $\vec{E}\vec{E}\vec{E}$ is of third rank. The single and double dots represent tensor contractions, to be defined below. The relation $\vec{\mu}^{(1)} = \alpha \cdot \vec{E}$ can be written in matrix form as follows:

$$
\begin{pmatrix} \mu_x \\ \mu_y \\ \mu_z \end{pmatrix} = \begin{pmatrix} \alpha_{xx} & \alpha_{xy} & \alpha_{xz} \\ \alpha_{yx} & \alpha_{yy} & \alpha_{yz} \\ \alpha_{zx} & \alpha_{zy} & \alpha_{zz} \end{pmatrix} \begin{pmatrix} E_x \\ E_y \\ E_z \end{pmatrix} \tag{A.27}
$$

We can express any particular component of $\vec{\mu}^{(1)}$ as follows:

$$
\mu_i^{(1)} = \sum_j \alpha_{ij} E_j \tag{A.28}
$$

where i and j range over x, y, and z. Equations such as A.28 are readily generalized when higher-rank tensors are involved. The part of the induced moment that derives from the hyperpolarizability is written

$$
\mu_i^{(2)} = \beta : \vec{E}\vec{E} = \sum_{j,k} \beta_{ijk} E_j E_k \tag{A.29}
$$

Similarly, the third-order component of the dipole moment is

$$
\mu_i^{(3)} = \gamma : \vec{E}\vec{E}\vec{E} = \sum_{j,k,l} \gamma_{ijkl} E_j E_k E_l \tag{A.30}
$$

Note that the summation always runs over the repeated indices.

It is often preferable to work with spherical rather than Cartesian tensors. The spherical forms are convenient when one is interested in how a tensor transforms under a rotation, to be discussed in Section A.5. A spherical tensor of rank l is denoted by T_m^l, where the index m takes on the $2l + 1$ values: $-l, -l+1, ..., l-1, l$. The spherical tensor components are just combinations of the Cartesian components, as will be illustrated below for the case of a first-rank tensor (a vector). Let us call the elements of the Cartesian tensor T_x, T_y, and T_z. The spherical tensor components are

$$
T_1^1 = \frac{-1}{\sqrt{2}}(T_x + iT_y)
$$

$$
T_0^1 = T_z \tag{A.31}
$$

$$
T_{-1}^1 = \frac{1}{\sqrt{2}}(T_x - iT_y)
$$

A second-rank Cartesian tensor such as α has irreducible spherical tensor components having $l = 0$, 1, and 2. Each of these has $2l + 1$ components, so the $1 + 3 + 5 = 9$ components of α are accounted for. Each spherical tensor component α_m^l is a linear combination of the Cartesian components. (See Table 8.2.) An arbitrary second-rank Cartesian tensor $\mathbf{T}$ is decomposed into the following irreducible spherical tensor components [1]:

$$
T_0^0 = \frac{-1}{\sqrt{3}}(T_{xx} + T_{yy} + T_{zz}) \tag{A.32}
$$

$$
T_0^1 = \frac{i}{\sqrt{2}}(T_{xy} - T_{yx}) \tag{A.33}
$$

$$
T_{\pm 1}^1 = \frac{1}{2}[(T_{zx} - T_{xz}) \pm i(T_{zy} - T_{yz})] \tag{A.34}
$$

$$T_0^2 = \frac{1}{\sqrt{6}}[2T_{zz} - T_{xx} - T_{yy}] \tag{A.35}$$

$$T_{\pm1}^2 = \mp\frac{1}{2}[(T_{xz} + T_{zx}) \pm i(T_{yz} + T_{zy})] \tag{A.36}$$

$$T_{\pm2}^2 = \frac{1}{2}[(T_{xx} - T_{yy}) \pm i(T_{xy} + T_{yx})] \tag{A.37}$$

A.4 SPHERICAL HARMONICS

The spherical harmonics $Y_{lm}(\theta\varphi)$ appear in problems having spherically symmetric potential functions, such as the hydrogen atom or rigid rotor. The orientation of a linear object in three dimensions can be specified by the polar and azimuthal angles, θ and φ, described in Figure 1.3. The spherical harmonics $Y_{lm}(\theta\varphi)$ form a complete set in this space. Thus an arbitrary function $f(\theta,\varphi)$ in this coordinate system can be expanded in terms of the spherical harmonics. The $Y_{lm}(\theta\varphi)$ are products of functions of θ and φ, as follows:

$$Y_{lm}(\theta\varphi) = (-1)^m \left(\frac{(2l+1)(l-m)!}{4\pi(l+m)!}\right)^{1/2} P_{lm}(\cos\theta)e^{im\varphi} \tag{A.38}$$

Equation A.38 applies for $m \geq 0$. For negative values of m, the following relationship is useful:

$$Y_{l\bar{m}}(\theta\varphi) = (-1)^m Y_{lm}(\theta\varphi)^* $$

The notation $\bar{m} \equiv -m$ is often used. The number l can be any nonnegative integer, and m ranges from $-l$ to l in steps of one. $P_{lm}(\cos\theta)$ is an associated Legendre function, which is defined in terms of the Legendre polynomial $P_l(\cos\theta)$ as follows:

$$P_{lm}(x) = (1-x^2)^{m/2}\left(\frac{d}{dx}\right)^m P_l(x) \tag{A.39}$$

The first few Legendre polynomials are tabulated in Table A.1. The recursion relation

$$(l+1)P_{l+1} = (2l+1)xP_l - lP_{l-1} \tag{A.40}$$

may be used to generate higher order P_l's. The recursion relation for the associated Legendre functions is

$$(l-m+1)P_{l+1,m}(x) = (2l+1)xP_{lm}(x) - (l+m)P_{l-1,m}(x) \tag{A.41}$$

Table A.1 Some Legendre polynomials

$P_0(x) = 1$
$P_1(x) = x$
$P_2(x) = \frac{1}{2}(3x^2 - 1)$
$P_3(x) = \frac{1}{2}(5x^3 - 3x)$

Table A.2 Some spherical harmonics

l	m	$Y_{l,m}(\theta,\varphi)$
0	0	$1/\sqrt{4\pi}$
1	0	$\dfrac{1}{2}\sqrt{\dfrac{3}{\pi}}\cos\theta$
	±1	$\mp\dfrac{1}{2}\sqrt{\dfrac{3}{2\pi}}\sin\theta\, e^{\pm i\varphi}$
2	0	$\dfrac{1}{4}\sqrt{\dfrac{5}{\pi}}\left(3\cos^2\theta-1\right)$
	±1	$\mp\dfrac{1}{2}\sqrt{\dfrac{15}{2\pi}}\cos\theta\,\sin\theta\; e^{\pm i\varphi}$
	±2	$\dfrac{1}{4}\sqrt{\dfrac{15}{2\pi}}\sin^2\theta\, e^{\pm 2i\varphi}$

Equation A.41 is useful in the derivation of selection rules. The spherical harmonics and the Legendre polynomials obey the following orthonormality relations:

$$\int_0^{2\pi} d\varphi \int_0^{\pi} \sin\theta\, d\theta Y_{lm}^*(\theta\varphi) Y_{l'm'}(\theta\varphi) = \delta_{ll'}\delta_{mm'}$$

$$\int_0^{\pi} \sin\theta\, d\theta P_l(\cos\theta) P_{l'}(\cos\theta) = \frac{2\delta_{ll'}}{2l+1}$$

(A.42)

The spherical harmonics (Table A.2) are eigenfunctions of the angular momentum operators $\hat{L}^2$ and $\hat{L}_z$. Using Dirac notation, $Y_{lm} = |lm\rangle$, the eigenvalue relations are

$$\hat{L}^2|lm\rangle = l(l+1)\hbar^2|lm\rangle$$

$$\hat{L}_z|lm\rangle = m\hbar|lm\rangle$$

(A.43)

A.5 WIGNER ROTATION FUNCTIONS AND SPHERICAL TENSORS

When we need to define the orientation of a three-dimensional object, three angles are needed. The Euler angles defined in Figure 8.2 are convenient, and the Wigner rotation functions $D_{mn}^l(\varphi\theta\chi) \equiv D_{mn}^l(\Omega)$ form a complete set in this space. The Wigner functions are useful when we need to rotate a space-dependent quantity or operator from one reference frame to another. Often, we are interested in the orientation of a molecule relative to the laboratory. We imagine a coordinate system (xyz) embedded in the molecule, the so-called body-fixed frame, which rotates with the molecule. The laboratory coordinate system (XYZ) is referred to as the space-fixed frame. (Throughout this book, lower case letters are employed for the body-fixed frame and upper case letters for the space-fixed frame.) If we consider a specific direction z in the body-fixed frame, then the angles φ and θ orient this axis in the lab, and these angles are equivalent to those used in the linear molecule case. The additional angle χ is the rotation of the object about its own z axis. If the molecule has enough symmetry, it is convenient to take the z direction as one of the symmetry axes, for example the n-fold rotation axis of a molecule belonging to a C_{nv} point group. The Euler angles pictured in Figure 8.2 describe the three-step process which rotates the molecule frame from XYZ to xyz. The first step ($XYZ \rightarrow X'Y'Z'$) is a rotation by the angle φ about the space-fixed Z axis. The second step ($X'Y'Z' \rightarrow X''Y''Z''$) is rotation by θ about the Y' axis, and finally rotation about the Z'' axis by χ results in the final orientation $X'''Y'''Z''' = xyz$.

There is an operator which accomplishes this three-step process that we have envisioned. The rotation operator $\hat{D}(\Omega) = \hat{D}(\varphi\theta\chi)$ may be applied to a function in the original coordinate system XYZ in order to convert it to the new reference frame $X'''Y'''Z'''$. This operator must be equivalent to the product of three successive rotation operators.

$$\hat{D}(\varphi\theta\chi) = \hat{D}_{Z''}(\chi)\hat{D}_{Y'}(\theta)\hat{D}_{Z}(\varphi) \tag{A.44}$$

It can be shown (see for example, [2] or [3]) that the operation of Equation A.44 can be cast in terms of rotations about axes in the original coordinate system, and that each of the three required rotation operators is an exponential function of an angular momentum operator.

$$\hat{D}(\varphi\theta\chi) = \hat{D}_{Z}(\varphi)\hat{D}_{Y}(\theta)\hat{D}_{Z}(\chi)$$
$$= \exp\left(\frac{-i\varphi\hat{L}_{Z}}{\hbar}\right)\exp\left(\frac{-i\theta\hat{L}_{Y}}{\hbar}\right)\exp\left(\frac{-i\chi\hat{L}_{Z}}{\hbar}\right) \tag{A.45}$$

The Wigner functions $D_{mn}^{l}(\varphi\theta\chi)$ are defined as matrix elements of the above operator.

$$D_{mn}^{l}(\varphi,\theta,\chi) = \langle lm|\hat{D}(\varphi\theta\chi)|ln\rangle \tag{A.46}$$

The matrix elements in Equation A.46 are taken with respect to the angular momentum operators in the exponents.

$$\langle lm|\hat{L}_{Z}|ln\rangle = m\hbar\delta_{nm} \tag{A.47}$$

$$\langle lm|\exp\left(\frac{-i\alpha L_{Z}}{\hbar}\right)|ln\rangle = \exp(-i\alpha m)\delta_{nm} \tag{A.48}$$

and

$$d_{mn}^{l}(\theta) \equiv \langle lm|\exp\left(\frac{-i\theta\hat{L}_{Y}}{\hbar}\right)|ln\rangle \tag{A.49}$$

The form of the Wigner functions is then

$$D_{mn}^{l}(\varphi\theta\chi) = e^{-im\varphi}d_{mn}^{l}(\theta)e^{-in\chi} \tag{A.50}$$

where $d_{mn}^{l}(\theta) = D_{mn}^{l}(0\theta0)$ is given by

$$d_{mn}^{l}(\theta) = \left[(l+m)!(l-m)!(l+n)!(l-n)!\right]^{1/2}$$
$$\times \sum_{k}\frac{(-1)^{k}(\cos\theta/2)^{2l+m-n-2k}(-\sin\theta/2)^{2k-m+n}}{(l+m-k)!(l-n-k)!k!(k-m+n)!} \tag{A.51}$$

The sum in Equation A.51 runs over values of k for which the argument of the factorial is nonnegative. Note that the values of m and n each range from $-l$ to l in integral steps. Thus for every value of l there are $(2l+1)^{2}$

Wigner functions. This is consistent with the physical interpretation of l, m, and n as angular momentum quantum numbers. The D_{mn}^l's are in fact eigenfunctions of the operators $\hat{L}^2$, $\hat{L}_Z$, and $\hat{L}_z$:

$$\hat{L}^2 D_{mn}^l(\Omega) = l(l+1)\hbar^2 D_{mn}^l(\Omega)$$

$$\hat{L}_Z D_{mn}^l(\Omega)^* = m\hbar D_{mn}^l(\Omega)^*$$ (A.52)

$$\hat{L}_z D_{mn}^l(\Omega)^* = n\hbar D_{mn}^l(\Omega)^*$$

The complex conjugate $D_{mn}^l(\Omega)^*$ satisfies the following equation:

$$D_{mn}^l(\Omega)^* = (-1)^{m-n} D_{\bar{m}\bar{n}}^l(\Omega)$$ (A.53)

The Wigner functions obey the orthogonality relation

$$\int d\Omega D_{mn}^l(\Omega) D_{m'n'}^{l'}(\Omega)^* = \left(\frac{8\pi^2}{2l+1}\right)\delta_{ll'}\delta_{mm'}\delta_{nn'}$$ (A.54)

where $\int d\Omega$ is a shorthand notation for the three-dimensional integral

$$\int d\Omega = \int_0^{2\pi} d\varphi \int_0^{\pi} \sin\theta\, d\theta \int_0^{2\pi} d\chi$$ (A.55)

and integration over all Euler angle space gives $\int d\Omega = 8\pi^2$.

Some other important integral relations are the following:

$$\int d\Omega D_{mn}^l(\Omega) = \left(\frac{8\pi^2}{2l+1}\right)\delta_{l0}\delta_{m0}\delta_{n0}$$ (A.56)

and

$$\int d\Omega D_{m_3 n_3}^{l_3}(\Omega)^* D_{m_2 n_2}^{l_2}(\Omega) D_{m_1 n_1}^{l_1}(\Omega)$$

$$= \left(\frac{8\pi^2}{2l_3+1}\right) C\left(l_1 l_2 l_3; m_1 m_2 m_3\right) C\left(l_1 l_2 l_3; n_1 n_2 n_3\right)$$ (A.57)

The first of these follows from $D_{00}^0(\Omega) = 1$ and Equation A.54. The Clebsch–Gordan coefficients which appear in Equation A.57 arise in angular momentum coupling problems. For now, consider them to be numbers which can be found in tables. They are discussed further in Section A.6 and also in Chapter 7. They are proportional to the $3j$ symbols used in Equation A.58.

The product of two Wigner functions adheres to

$$D_{m_1 n_1}^{l_1}(\Omega) D_{m_2 n_2}^{l_2}(\Omega) = \sum_{lmn}(2l+1)\begin{pmatrix} l_1 & l_2 & l \\ m_1 & m_2 & m \end{pmatrix}\begin{pmatrix} l_1 & l_2 & l \\ n_1 & n_2 & n \end{pmatrix} D_{mn}^l(\Omega)^*$$ (A.58)

In the case where either m or n is zero, the Wigner function reduces to a spherical harmonic.

$$D_{0m}^l(\varphi\theta\chi) = (-1)^m \left(\frac{4\pi}{2l+1}\right)^{1/2} Y_{lm}^*(\theta\chi)$$

$$D_{m0}^l(\varphi\theta\chi) = \left(\frac{4\pi}{2l+1}\right)^{1/2} Y_{lm}^*(\theta\varphi)$$ (A.59)

When both subscripts are zero, a Legendre polynomial is obtained.

$$D_{00}^l(\varphi\theta\chi) = P_l(\cos\theta) \tag{A.60}$$

The addition theorem applies when the orientation Ω is considered to result from the rotation Ω_1 followed by Ω_2, in which case,

$$D_{mn}^l(\Omega) = \sum_k D_{mk}^l(\Omega_2) D_{kn}^l(\Omega_1) \tag{A.61}$$

The components of an irreducible spherical tensor transform under a rotation in the same fashion. Letting T_m^l denote the components of the tensor in the new reference frame and T_k^l those in the old reference frame, the transformation relation is

$$T_m^l = \sum_k T_k^l D_{km}^l(\Omega) \tag{A.62}$$

where Ω is the set of Euler angles that rotates the old coordinate system into the new one. This relationship is used in Chapter 8 (Equation 8.62) to relate the lab-frame and molecule-frame components of the polarizability tensor.

A.6 THE CLEBSCH–GORDAN SERIES AND 3j SYMBOLS

The quantum mechanical addition of angular momentum vectors is a ubiquitous problem in spectroscopy. Consider the coupling of two generalized angular momentum eigenfunctions $|j_1m_1\rangle$ and $|j_2m_2\rangle$. The index j represents any type of angular momentum quantum number: orbital, electron or nuclear spin, rotational angular momentum, or any coupled angular momentum such as $J = L + S$. The magnitude of the angular momentum vector is $j(j + 1)\hbar$ and the z component is $m\hbar$, where m ranges from $-j$ to j in steps of one. If a perturbation causes the states $|j_1m_1\rangle$ and $|j_2m_2\rangle$ to mix to give new states having quantum numbers J and M, the perturbed states can be expressed as linear combinations of products of these states.

$$|JM;j_1j_2\rangle = \sum_{m_1m_2} C(j_1j_2J;m_1m_2M)|j_1m_1\rangle|j_2m_2\rangle \tag{A.63}$$

The permitted values of J range from $j_1 + j_2$ to $|j_1 - j_2|$ in steps of one, and the permitted values of M satisfy $m_1 + m_2 = M$. The Clebsch–Gordan coefficient $C(j_1j_2J;m_1m_2M)$ vanishes unless the triangle rule is satisfied, and the z component of the resulting angular momentum state is the sum of those for the two added states. The 3j symbols are often preferred over the Clebsch–Gordan coefficients because they have useful symmetry properties. They are defined by

$$\begin{pmatrix} j_1 & j_2 & J \\ m_1 & m_2 & -M \end{pmatrix} \equiv \frac{(-1)^{j_1-j_2-M}}{\sqrt{2J+1}} C(j_1j_2J;m_1m_2M) \tag{A.64}$$

The sum of the elements in the second row of the 3j symbol must equal zero or the quantity vanishes. Also, any even permutation of the columns leaves the value unchanged.

$$\begin{pmatrix} j_1 & j_2 & j_3 \\ m_1 & m_2 & m_3 \end{pmatrix} = \begin{pmatrix} j_3 & j_1 & j_2 \\ m_3 & m_1 & m_2 \end{pmatrix} = \begin{pmatrix} j_2 & j_3 & j_1 \\ m_2 & m_3 & m_1 \end{pmatrix} \tag{A.65}$$

An odd permutation of the columns multiples the value by $(-1)^{j_1+j_2+j_3}$.

$$\begin{pmatrix} j_1 & j_2 & j_3 \\ m_1 & m_2 & m_3 \end{pmatrix} = (-1)^{j_1+j_2+j_3} \begin{pmatrix} j_2 & j_1 & j_3 \\ m_2 & m_1 & m_3 \end{pmatrix}$$

$$= (-1)^{j_1+j_2+j_3} \begin{pmatrix} j_1 & j_3 & j_2 \\ m_1 & m_3 & m_2 \end{pmatrix} \tag{A.66}$$

$$= (-1)^{j_1+j_2+j_3} \begin{pmatrix} j_3 & j_2 & j_1 \\ m_3 & m_2 & m_1 \end{pmatrix}$$

Also, changing the signs of all the m_i results in multiplication by $(-1)^{j_1+j_2+j_3}$.

$$\begin{pmatrix} j_1 & j_2 & j_3 \\ m_1 & m_2 & m_3 \end{pmatrix} = (-1)^{j_1+j_2+j_3} \begin{pmatrix} j_1 & j_2 & j_3 \\ -m_1 & -m_2 & -m_3 \end{pmatrix} \tag{A.67}$$

A consequence of Equation A.67 is that the $3j$ symbol vanishes if $m_1 = m_2 = m_3 = 0$ and the sum of the j values is odd:

$$\begin{pmatrix} j_1 & j_2 & j_3 \\ 0 & 0 & 0 \end{pmatrix} = 0, \text{ if } j_1 + j_2 + j_3 \text{ is odd} \tag{A.68}$$

The properties of the $3j$ symbols are used to derive selection rules for atomic and rotational spectra in Chapters 7 and 8, respectively.

REFERENCES

1. C. G. Gray and K. E. Gubbins, *Theory of Molecular Fluids, Volume I, Fundamentals* (Oxford University Press, New York, 1984).
2. R. N. Zare, *Angular Momentum, Understanding Spatial Aspects in Chemistry and Physics* (Wiley, New York, 1988).
3. A. R. Edmonds, *Angular Momentum in Quantum Mechanics* (Princeton University Press, Princeton, NJ, 1974).

Appendix B: Principles of electrostatics

B.1 UNITS

With a few exceptions, the MKS (meters–kilograms–seconds) system of units is used in formulas of electrostatics throughout this book. The use of MKS units is apparent in expressions which contain the permittivity of free space: $\varepsilon_0 = 8.85419 \times 10^{-12}$ F m^{-1}. The farad (F) is a unit of capacitance, equivalent to C V^{-1}, or C^2 J^{-1}. So the fundamental units of ε_0 are C^2 N^{-1} m^{-2}. By definition, the value of ε_0 is found from

$$\frac{1}{4\pi\varepsilon_0} = 10^{-7} c^2 \tag{B.1}$$

where c is the speed of light in m s^{-1} and the implied units of the number 10^{-7} are N/A^2. The permeability of free space is $\mu_0 = 4\pi \times 10^{-7}$ N/A^2. Therefore Equation B.1 is equivalent to the expression $c^2 = 1/(\varepsilon_0\mu_0)$. In the MKS system of units, Coulomb's law is written

$$F = \frac{q_1 q_2}{4\pi\varepsilon_0 r^2} \tag{B.2}$$

where the force F is in Newtons, the charges q_1 and q_2 are in Coulombs, and the distance r is in meters. In working problems, it is convenient to use the fact that $1/4\pi\varepsilon_0$ is approximately equal to 9×10^9 N m^2 C^{-2}.

In the centimeter–gram–second (cgs) system of units, the permittivity of free space is $\varepsilon_0 = 1/4\pi$, and the unit of charge is the statcoulomb (statC) or esu (for electrostatic unit). For example, the charge on an electron is $e = 1.60218 \times 10^{-19}$ C in the MKS system and $e = 4.80320 \times 10^{-10}$ esu in the cgs system. In the cgs system, we write Coulomb's law as follows:

$$F = \frac{q_1 q_2}{r^2} \tag{B.3}$$

where the force is in dynes, the charge in esu, and the distance in cm. Note that 1 dyne is equal to 1 esu^2 cm^{-2}, a convenient conversion factor. Some authors prefer the cgs system in expressions like B.3 because it is cleaner. The cgs-system Debye unit for the dipole moment, equivalent to 10^{-18} esu cm, is also used prevalently. In what follows, the expressions are given in the MKS system, but can be converted to cgs by replacing ε_0 by $1/4\pi$.

B.2 SOME APPLICATIONS OF GAUSS' LAW

The basic formulas of electrostatics relate the electric field $\vec{E}$ to the charge density ρ. Two key expressions are

$$\nabla \cdot \vec{E} = \frac{\rho}{\varepsilon_0} \tag{B.4}$$

$$\nabla \times \vec{E} = 0 \tag{B.5}$$

These may be compared to two of Maxwell's equations, presented in Chapter 2. In the absence of a time-dependent magnetic field, the curl of the electric field, $\nabla \times \vec{E}$, is zero. Equation B.4 is called the differential form of Gauss' law. The integral form is

$$\int \vec{E} \cdot \hat{n} dA = \frac{Q}{\varepsilon_0} \tag{B.6}$$

where the integral is over a closed surface, and $\hat{n}$ is a unit vector perpendicular to the surface. $Q = \int \rho d\vec{r} = \sum q_i$ is the total charge enclosed by the surface.

The electric field is the negative gradient of the electrical potential φ.

$$\vec{E} = -\nabla \varphi \tag{B.7}$$

The potential is related to the work W done to move a unit charge q between two points,

$$W = q(\varphi_2 - \varphi_1) = -\int_1^2 \vec{E} \cdot d\vec{s} \tag{B.8}$$

where the integral in Equation B.8 is a line integral. Equation B.5 follows from Equation B.7, because for any scalar function φ, $\nabla \times \nabla \varphi = 0$. The electric field is the force per unit charge, and the potential is the work per unit charge.

Another way to relate the charge distribution to the potential is through the Poisson equation.

$$\nabla^2 \varphi = \frac{-\rho}{\varepsilon_0} \tag{B.9}$$

which follows from combining Equation B.7 with Equation B.4.

The integral form of Gauss' law leads directly to some important principles of electrostatics. For example, we know that the field due to a uniformly charged sphere behaves as if all the charge were concentrated at the center of the sphere. We can show this by drawing a Gaussian surface in the form of a concentric sphere of radius r surrounding a charged sphere of radius a. The field is radial, so it is normal to this surface. The charge enclosed by the surface is $Q = (4/3)\pi a^3 \rho$. Applying Equation B.6, we have

$$E(4\pi r^2) = \frac{\frac{4}{3}\pi a^3 \rho}{\varepsilon_0} = \frac{Q}{\varepsilon_0} \tag{B.10}$$

$$E = \frac{Q}{4\pi r^2 \varepsilon_0} \tag{B.11}$$

This is the same field that would be obtained at a distance r from a point charge of magnitude Q.

B.2.1 The Lorentz model of the atom

In Chapter 3, a classical approach to the frequency-dependent polarizability is presented, based on the Lorentz model of the atom. This model considers the atom to consist of a positive point charge $+Q$ surrounded by a uniform sphere of total negative charge $-Q$. Let us say that the radius of this sphere is a, and ask what is the force when the nucleus is displaced by the distance d from the central position, under the action of an external electric field $\vec{E}$. (We could just as easily consider the positive charge to be stationary, and the electron cloud to be displaced by the distance d.) The force due to the external field, QE, is balanced by the attractive Coulombic force. According to Gauss' law, we can calculate the field due to the negative charge within the *smaller sphere* of radius d as follows:

$$E(4\pi d^2) = \frac{d^3}{a^3} \frac{Q}{\varepsilon_0} \tag{B.12}$$

As far as the force on the nucleus is concerned, we only need to consider the charge within the smaller sphere, because only the enclosed charge contributes to the field at distance d. Equation B.12 follows from the fact that the charge within the sphere of radius d is equal to the total charge Q times the ratio of the volumes d^3/a^3. Solving for the electric field, we get

$$E = \frac{Qd}{4\pi\varepsilon_0 a^3} = \frac{\rho d}{3\varepsilon_0} \tag{B.13}$$

Note that the displaced nucleus experiences a restoring force proportional to the displacement:

$$F = \frac{Q\left(\dfrac{d^3 Q}{a^3}\right)}{4\pi\varepsilon_0 d^2} = \frac{Q^2 d}{4\pi\varepsilon_0 a^3} \tag{B.14}$$

At equilibrium, the force due to the *external field* (Equation B.13 gives the field due the internal charges) is equal to the Coulombic force of Equation B.14:

$$QE = \frac{Q^2 d}{4\pi\varepsilon_0 a^3} \tag{B.15}$$

The polarizability α is the induced dipole Qd divided by the field E. Thus we have

$$\alpha = 4\pi\varepsilon_0 a^3 \tag{B.16}$$

Equation B.16 confirms the intuitive notion that polarizability scales as the volume of the electron cloud.

B.2.2 ELECTRIC FIELD WITHIN A CAPACITOR

In the study of dielectrics, the parallel plate capacitor is an important concept. Let us first imagine that we have a single infinite sheet of charge with surface charge density $\sigma = Q/A$. We surround a portion of this sheet with a rectangular box with surface area A in the direction normal to the field. The electric field is directed outward from the sheet of charge, and the box is oriented with two faces parallel to the sheet. The electric field penetrates these two faces, and there is no component of E through the remaining four faces. Thus, by Gauss' law, we have

$$2EA = \frac{\sigma A}{\varepsilon_0} \tag{B.17}$$

$$E = \frac{\sigma}{2\varepsilon_0} \tag{B.18}$$

Now, if there are two oppositely charged sheets, the fields within the capacitor (between the sheets) add, while the fields outside the capacitor cancel. Well within the capacitor (so we can neglect edge effects), the field is

$$E = \frac{\sigma}{\varepsilon_0} \tag{B.19}$$

This equation is employed in Chapter 3 in the discussion of dielectrics.

B.3 SOME MATHEMATICAL DETAILS

The potential due to a collection of charges is discussed in Chapter 3. It is often of interest to know the potential φ at a distance R which is large compared to the extent of the charge distribution which gives rise to the potential. This leads naturally to expansions of the type given in Equation 3.12, involving $\nabla(1/R)$, $\nabla\nabla(1/R)$, etc. The function $1/R$ is a scalar, and the result of operating with the grad operator ∇ is a vector, or first-rank tensor. Similarly, $\nabla\nabla(1/R)$ is a second-rank tensor. The form of each can be obtained as follows. Consider the first-rank tensor, defined by

$$T^1 = \nabla\left(\frac{1}{R}\right) \tag{B.20}$$

Using $R = (x^2 + y^2 + z^2)^{1/2}$ and the grad operator as defined in Appendix A, we obtain

$$
\begin{aligned}
T^1 &= (\hat{i}\frac{d}{dx} + \hat{j}\frac{d}{dy} + \hat{k}\frac{d}{dz})\left(\frac{1}{(x^2+y^2+z^2)^{1/2}}\right) \\
&= \hat{i}\frac{-x}{(x^2+y^2+z^2)^{3/2}} + \hat{j}\frac{-y}{(x^2+y^2+z^2)^{3/2}} + \hat{k}\frac{-z}{(x^2+y^2+z^2)^{3/2}} \\
&= \frac{-\vec{R}}{R^3} = \frac{-\hat{R}}{R^2}
\end{aligned} \tag{B.21}
$$

where $\hat{R}$ is a unit vector in the direction of $\vec{R} = (\hat{i}x + \hat{j}y + \hat{k}z)$. In general, we have

$$\nabla\left(\frac{1}{R^n}\right) = \frac{-n\hat{R}}{R^{n+1}} \tag{B.22}$$

From the Taylor series expansion of Equation 3.12, the potential due to a dipole $\vec{\mu}$ is

$$\varphi(\vec{R}) = \frac{-1}{4\pi\varepsilon_0}\vec{\mu}\cdot\nabla\left(\frac{1}{R}\right) = \frac{1}{4\pi\varepsilon_0}\frac{\vec{\mu}\cdot\vec{R}}{R^3} \tag{B.23}$$

in agreement with Equation 3.7.

The potential due to the quadrupole moment Θ depends on $T^2 = \nabla\nabla(1/R)$. To find the second-rank tensor T^2, we first recognize that $\nabla\vec{R}$ is found as follows:

$$
\begin{aligned}
\nabla\vec{R} &= (\hat{i}\frac{d}{dx} + \hat{j}\frac{d}{dy} + \hat{k}\frac{d}{dz})(\hat{i}x + \hat{j}y + \hat{k}z) \\
&= \mathbf{I}
\end{aligned} \tag{B.24}
$$

where $\mathbf{I}$ is a unit tensor:

$$\mathbf{I} = \begin{pmatrix} 1 & 0 & 0 \\ 0 & 1 & 0 \\ 0 & 0 & 1 \end{pmatrix} \tag{B.25}$$

Then, using the results from Equations B.22 and B.24, we get

$$T^2 = \nabla\left(\frac{-\vec{R}}{R^3}\right) = \frac{-\mathbf{I}}{R^3} - \vec{R}\left(\frac{-3\hat{R}}{R^4}\right)$$

$$= \frac{1}{R^3}\left(3\hat{R}\hat{R} - \mathbf{I}\right) \tag{B.26}$$

Writing Equation B.26 as an array, we have

$$T^2 = \frac{1}{R^5}\begin{pmatrix} 3x^2 - R^2 & 3xy & 3xz \\ 3yx & 3y^2 - R^2 & 3yz \\ 3zx & 3zy & 3z^2 - R^2 \end{pmatrix} \tag{B.27}$$

Note that the trace of this tensor is zero.

The tensor T^2 is often called the dipolar tensor. The interaction energy of two dipoles can be found from

$$W = -\vec{\mu}_1 \cdot T^2 \cdot \vec{\mu}_2 \tag{B.28}$$

Equation B.28 can be compared to Equation 3.9.

Returning to the potential due to a quadrupole, it was stated in Chapter 3 that this potential is unchanged if a term is added to each diagonal element of Θ. We show here why this is true. Suppose that the quadrupole moment is altered as follows:

$$\Theta' = \Theta + c\mathbf{I} \tag{B.29}$$

where c is a scalar. The potential due to the quadrupole moment is then

$$\varphi' = \frac{1}{8\pi\varepsilon_0}(\Theta + c\mathbf{I}):\nabla\nabla\left(\frac{1}{R}\right)$$

$$= \frac{1}{8\pi\varepsilon_0}\left[\Theta:\nabla\nabla\left(\frac{1}{R}\right) + c\mathbf{I}:\nabla\nabla\left(\frac{1}{R}\right)\right] \tag{B.30}$$

$$= \frac{1}{8\pi\varepsilon_0}\left[\Theta:\nabla\nabla\left(\frac{1}{R}\right)\right]$$

The last line of Equation B.30 follows from

$$c\mathbf{I}:\nabla\nabla\left(\frac{1}{R}\right) = c\mathrm{Tr}\left(T^2\right) = 0 \tag{B.31}$$

The addition of a constant term to each diagonal element of the quadrupole tensor has no effect on the potential due to the quadrupole moment. Thus, by convention, the quadrupole tensor is chosen to have zero trace, as defined in Equation 3.16.

Appendix C: Group theory

The classification of molecules into point groups on the basis of symmetry provides a platform on which much of the discussion of spectroscopy is based. While group theory in itself can be the subject of purely mathematical discourse, the chemical applications are straightforward and concrete. After introducing some basic theorems and language, group theory can be employed to deduce selection rules and to derive symmetries of molecular orbitals and normal modes. All that is needed is a pencil, some scratch paper, and a set of character tables. The starting point is the classification of molecules into point groups. It is hoped that the reader has had some previous introduction to basic concepts of molecular symmetry. If not, the book by Cotton [1] or the chapter on symmetry in Levine [2] should be consulted. The principles of symmetry are reviewed below in order to relate them to group theoretical tools needed for the study of spectroscopy. It should be kept in mind that the discussions here apply to molecules in their equilibrium geometries.

C.1 POINT GROUPS AND SYMMETRY OPERATIONS

The symmetry of a molecule is specified by a set of operations that leave the molecule unchanged. Consider a water molecule. Though the hydrogens may be labeled, they are of course indistinguishable. If you turned your back and someone rotated the water molecule about the axis that bisects the bond angle, you could not tell. We say that the rotation operator $\hat{C}_2$ is a symmetry operator for the water molecule. Symmetry operators commute with the Hamiltonian and thus share a common set of eigenfunctions. In the example at hand, the eigenvalue relationship is $\hat{C}_2\psi = \pm\psi$. In other words, the rotation either preserves the wavefunction or changes its sign. Either way, the probability $\psi^*\psi$ is unaffected by the symmetry operation.

There are five symmetry operations which a molecule can possibly possess. Each one, except for the identity operation, is performed with respect to a particular symmetry element. The operations and elements are as follows:

1. *Identity operation $\hat{E}$:* This operation does nothing! It is required by the theory, and every molecule is of course symmetric with respect to this operation.
2. *Rotation operation $\hat{C}_n$:* If a molecule possesses an n-fold rotation axis (the symmetry element), then rotation by $2\pi/n$ (the operation) about this axis leaves the molecule unchanged. The water molecule has one C_2 rotation axis.
3. *Reflection operation $\hat{\sigma}$:* The reflection operation is performed with respect to a plane. The water molecule has two reflection planes: one containing the molecule (the yz plane) and a second one perpendicular to the first and containing the C_2 axis (the xz plane).
4. *Inversion $\hat{i}$:* This operation is performed with respect to a point: the center of symmetry. If one considers this point to be the origin of a Cartesian coordinate system, the inversion operation consists of replacing every point (x, y, z) by $(-x, -y, -z)$. Since symmetry operations leave the center of mass unmoved, the center of symmetry, if it exists, must be at the center of mass of the molecule. Water molecule lacks inversion symmetry.
5. *Improper rotation $\hat{S}_n$:* There is not really anything improper about this rotation, so some books refer to it as the rotation–reflection operation, which is more descriptive. There are two symmetry elements and two steps to this operation. First the molecule is rotated by $2\pi/n$ about an axis, followed by a reflection through a plane perpendicular to the rotation axis. Water molecule lacks this symmetry operation.

A molecular orbital or normal mode can be characterized by the set of eigenvalues for each of the group operations. This defines the symmetry of the MO (or other function) in a precise way. We say that the MO belongs to a particular *irreducible representation* of the group. These "ir. reps." are listed in the left-hand column of the character table. Before we can use them, we have to know which character table to use for the molecule of interest.

To assign a molecule to a molecular point group, one needs to know the set of symmetry operations which apply. Consider a tetrahedral molecule such as CH_4. A three-fold rotation axis can be aligned with any one of the four C–H bonds. For each of these C_3 axes, there are two such operations that can be performed: $\hat{C}_3$ and $\hat{C}_3^2$. The $\hat{C}_3^2$ operation results from applying $\hat{C}_3$ twice; that is, it is rotation by $2 \times 2\pi/3$. While it is true that the operation $\hat{C}_3^3$ also leaves the molecule unchanged, this operation is equivalent to the identity operation, which is already included. So there are eight $\hat{C}_3$ operations in this *class*, two for each of the four three-fold rotation axes. The idea of a class is discussed further below. There are also three mutually perpendicular C_2 axes, and associated $\hat{C}_2$ operations, for CH_4. There are six planes of symmetry, called dihedral planes σ_d. (This notation denotes a reflection plane which bisects the angle between two rotation axes. A vertical reflection plane σ_v, on the other hand, contains the axis of highest symmetry, called the principal axis, while a horizontal reflection plane σ_h is perpendicular to the principal axis.) Each σ_d reflection plane bisects one H–C–H bond and contains the other. Finally, there are six $\hat{S}_4$ operations: three axes of rotation and an $\hat{S}_4$ and $\hat{S}_4^2$ operation for each. The rotation is performed with respect to an axis which bisects one of the H–C–H bonds. This set of operations relegates the CH_4 molecule to the T_d point group. The top row of the T_d character table (see Section C.4) lists the operations that we have just enumerated.

For simple molecules, it is straightforward to count the symmetry operations and assign the point group on that basis. See [1] or [2] for a flow chart that can be used to find the point group in the general case.

C.2 INFORMATION CONVEYED BY CHARACTER TABLES

Consider the character table for the C_{2v} point group, shown below in Table C.1. There are four symmetry operations: the identity, the twofold rotation, and two mutually perpendicular reflection planes. The letters A_1, A_2, B_1 and B_2 denote the irreducible representations. In this section we want to understand the meaning of the entries in the table, called characters.

The operations of the group $\hat{R}_1, \hat{R}_2, \hat{R}_3, \ldots$ obey a set of multiplication rules which define a group. In particular, each operation must have an inverse, such that $\hat{R}_i \hat{R}_i^{-1} = \hat{E}$. The product of two operations of the group must be equivalent to another operation of the group. Multiplication is not necessarily commutative: i.e., $\hat{R}_1 \hat{R}_2$ may not be the same as $\hat{R}_2 \hat{R}_1$. In the C_{2v} point group, it turns out that each operation is its own inverse.

The operations of the group can be applied to any function, not just the molecular framework. These operations may be represented by matrices which obey the same multiplication rules as the operations of the group. A set of functions on which the symmetry operations are considered to act is called a basis for a representation of the group. We will write the basis as a column vector. Consider a simple example. Let the point x, y, z form a basis for the representation of the C_{2v} point group. This is the point group to which the water molecule belongs, and we imagine setting up a coordinate system with its origin at the center of mass (close to the oxygen atom). The z axis is coincident with the C_2 rotation axis, and the molecule lies in the zy plane. The identity operation is represented by the identity matrix:

$$\begin{pmatrix} 1 & 0 & 0 \\ 0 & 1 & 0 \\ 0 & 0 & 1 \end{pmatrix} \begin{pmatrix} x \\ y \\ z \end{pmatrix} = \begin{pmatrix} x \\ y \\ z \end{pmatrix} \tag{C.1}$$

The $\hat{C}_2$ operation leaves the z coordinate alone and reverses the sign of x and y:

$$\begin{pmatrix} -1 & 0 & 0 \\ 0 & -1 & 0 \\ 0 & 0 & 1 \end{pmatrix} \begin{pmatrix} x \\ y \\ z \end{pmatrix} = \begin{pmatrix} -x \\ -y \\ z \end{pmatrix} \tag{C.2}$$

Reflection through the xz plane, $\hat{\sigma}(xz)$, is represented as follows:

$$\begin{pmatrix} 1 & 0 & 0 \\ 0 & -1 & 0 \\ 0 & 0 & 1 \end{pmatrix} \begin{pmatrix} x \\ y \\ z \end{pmatrix} = \begin{pmatrix} x \\ -y \\ z \end{pmatrix} \qquad \text{(C.3)}$$

And the reflection through the yz plane, $\hat{\sigma}(yz)$, is written:

$$\begin{pmatrix} -1 & 0 & 0 \\ 0 & 1 & 0 \\ 0 & 0 & 1 \end{pmatrix} \begin{pmatrix} x \\ y \\ z \end{pmatrix} = \begin{pmatrix} -x \\ y \\ z \end{pmatrix} \qquad \text{(C.4)}$$

You can verify that these four square matrices obey the same multiplication rules as the operators that they represent. For example, the square of any operation belonging to the C_{2v} point group gives the identity operation. The product of the operations $\hat{C}_2 \times \hat{\sigma}_{xz}$ is equal to $\hat{\sigma}_{yz}$, and so forth.

The choice of a representation is not unique. For the water molecule example, we could just as easily take the two O–H bonds, call them r_1 and r_2, and the bond angle θ, as a basis. The operations are still represented by 3×3 matrices, but not the same as those given above. For example, the effect of the $\hat{C}_2$ rotation is represented as follows:

$$\begin{pmatrix} 0 & 1 & 0 \\ 1 & 0 & 0 \\ 0 & 0 & -1 \end{pmatrix} \begin{pmatrix} r_1 \\ r_2 \\ \theta \end{pmatrix} = \begin{pmatrix} r_1 \\ r_2 \\ -\theta \end{pmatrix} \qquad \text{(C.5)}$$

Notice that the trace of the 3×3 matrix in Equation C.5, -1, is the same as that of the matrix representation of the $\hat{C}_2$ operation given in Equation C.2. It is quite general that the trace of a matrix is independent of representation. This is proven as follows. Consider a set of operations which obey the group multiplication rules, say $AB = C$, $CD = A$, etc. It is always possible to apply a similarity transform to the matrices (operations) as follows:

$$A' = S^{-1}AS, \quad B' = S^{-1}BS, \quad C' = S^{-1}CS \qquad \text{(C.6)}$$

where $S^{-1}S = SS^{-1} = I$, the identity matrix. These transformed matrices obey the same multiplication rules as the original ones:

$$A'B' = (S^{-1}AS)(S^{-1}BS) = S^{-1}ABS = S^{-1}CS = C' \qquad \text{(C.7)}$$

In addition, since the trace of a product of matrices is invariant to a cyclic permutation,

$$\text{Tr}C' = \text{Tr}(S^{-1}CS) = \text{Tr}(CSS^{-1}) = \text{Tr}C \qquad \text{(C.8)}$$

the trace of the matrix is unaffected by a similarity transform.

What if we had used a larger basis for the C_{2v} point group? For example, we could take the set of Cartesian coordinates locating the three atoms in H_2O. The matrix representations of the operations would be of dimension 9×9, and the traces would be different from those obtained previously. For example, we would get a trace of nine for the identity operation rather than three as obtained above. So obviously the trace is only independent of representation if we compare representations of the same dimension. This leads us back to the concept of *irreducible representations*, which is based on the set of *smallest* matrices capable of representing the group operations. In the case of water molecule, the smallest matrices that multiply like the operations of the group are mere numbers, or 1×1 matrices if you prefer. We refer to A_1, A_2, B_1, and B_2 as one-dimensional ir. reps. The trace of each matrix is the number itself.

The character $\chi_i(R)$ of the ith irreducible representation for operation $\hat{R}$ is the trace of the matrix which represents that operation. We write this in general as

$$\chi_i(R) = \sum_n [\Gamma_i(R)]_{nn} \tag{C.9}$$

where $\Gamma_i(R)$ is the matrix representation of the operation R in the ith irreducible representation.

In the C_{2v} point group, there are four irreducible representations, and we cannot find all of them using the three-dimensional representation given previously. How do we know there are four ir. reps.? There is a theorem which states that the number of ir. reps. is equal to the number of classes. In the C_{2v} point group, each of the four operations is in a class by itself; we conclude there are four ir. reps. Another theorem is that the order of the group, h, which is equal to the number of operations of the group, is equal to the sum of the squares of the dimensions of the ir. reps. This relation is symbolized by

$$h = \sum_i l_i^2 \tag{C.10}$$

where the sum is over the ir. reps. and l_i is the dimension of the ith representation. In any molecule having less than threefold rotation symmetry, all ir. reps. are one-dimensional. (The reason for this will be exposed below.) Again, we conclude that there are four ir. reps. in C_{2v}. We can fill in a table (Table C.2) based on what we obtained from the matrices of Equations C.1 through C.4 to generate three ir. reps. which we call Γ_1, Γ_2, and Γ_3. A fourth one, Γ_4, can be figured out as described below.

The entries in the first three rows of Table C.2 are the first, second, and third diagonal elements, respectively, of the square matrices in Equations C.1 through C.4, for the corresponding operation in the heading of the table. The irreducible representations of a group, Γ_1 through Γ_4 in this example, form an orthonormal set. This means that the following relation is obeyed:

$$\sum_R [\chi_i(R)]^* \chi_j(R) = h\delta_{ij} \tag{C.11}$$

In other words, take any two rows from Table C.2 and sum the products of the characters. The result is zero, as can be checked. So, based on the form of the first three ir. reps., the characters for the fourth one are deduced.

Table C.2 does not display the ir. reps. in the conventional manner, as is apparent from comparison to Table C.1. We have presented it to give the reader some appreciation for the origin of character tables. It is not necessary to know how to produce character tables, but it is certainly good to know what meaning they convey.

Table C.1 C_{2v} character table

C_{2v}	E	C_2	$\sigma_v(xz)$	$\sigma'_v(yz)$		
A_1	1	1	1	1	z	x^2, y^2, z^2
A_2	1	1	−1	−1	R_z	xy
B_1	1	−1	1	−1	x, R_y	xz
B_2	1	−1	−1	1	y, Rx	yz

Table C.2 Irreducible representations of C_{2v}

C_{2v}	E	$C_2(z)$	σ_{xz}	σ_{yz}
Γ_1	−1	1	1	−1
Γ_2	1	−1	−1	1
Γ_3	1	1	1	1
Γ_4	1	1	−1	−1

Table C.3 C_{3v} character table

C_{3v}	E	$2C_3$	$3\sigma_v$		
A_1	1	1	1	z	$x^2 + y^2, z^2$
A_2	1	1	-1	R_z	
E	2	-1	0	$(x, y)\ (R_x, R_y)$	$(x^2 - y^2, xy)\ (xz, yz)$

As an example, consider the character table for the C_{3v} point group, Table C.3, to which NH_3 belongs. There are three classes of operations in this point group, and thus three ir. reps. In addition to the identity operation, there are two C_3 rotations that belong to the same class, and three vertical reflection planes σ_v that belong to the same class. Operations in the same class all have the same character. Think of ammonia molecule as an example. There are three reflection planes: each contains one N–H bond and bisects the angle between the other two bonds. The threefold rotation axis permits rotation by 120° and 240°, hence two rotations in the class. The order of the group is six, so three ir. reps. must have dimensions such that $l_1^2 + l_2^2 + l_3^2 = 6$. Thus there are two one-dimensional representations and one two-dimensional representation. Ir. reps. represented by the letter A or B are always one dimensional, while the E representation is doubly degenerate.

To see how doubly degenerate representations come about, consider the effect of a $\hat{C}_n$ operation on a point (xyz).

$$\begin{pmatrix} \cos(2\pi/n) & -\sin(2\pi/n) & 0 \\ \sin(2\pi/n) & \cos(2\pi/n) & 0 \\ 0 & 0 & 1 \end{pmatrix} \begin{pmatrix} x \\ y \\ z \end{pmatrix} = \begin{pmatrix} x' \\ y' \\ z' \end{pmatrix} \tag{C.12}$$

In the case where $n = 3$, the matrix representation of the rotation operation is

$$\begin{pmatrix} \cos(2\pi/3) & -\sin(2\pi/3) & 0 \\ \sin(2\pi/3) & \cos(2\pi/3) & 0 \\ 0 & 0 & 1 \end{pmatrix} = \begin{pmatrix} -1/2 & -\sqrt{3}/2 & 0 \\ \sqrt{3}/2 & 1/2 & 0 \\ 0 & 0 & 1 \end{pmatrix} \tag{C.13}$$

This matrix cannot be diagonalized by a similarity transform. It is, however, in block diagonal form. The two-dimensional matrix within the 3×3 matrix represents the fact that the x and y coordinates transform as a pair under a C_3 rotation. This means that the rotated values x' and y' are linear combinations of the initial values x and y. In molecules having only twofold rotation symmetry, the matrix of Equation C.12 reduces to being diagonal ($\sin(\pi)$ is zero), so no two-dimensional representation arises. The dimension of an ir. rep. is revealed by the character of the identity operation.

The two right-hand columns of the character table denote functions which transform according to the corresponding ir. rep. For example, the z coordinate in C_{3v}, which is coincident with the rotation axis, transforms according to the *totally symmetric representation*. The first ir. rep. listed in any character table is always the totally symmetric representation, and each character of this representation is equal to one. The z coordinate is unaffected by any of the operations of the group, so it is totally symmetric. The rotation operation R_z, on the other hand, transforms according to the A_2 representation. To understand why, imagine an arrow that curls around the C_3 axis, say, in a clockwise direction. This arrow is unaffected by the identity and rotation operations, but any one of the σ_v reflections changes the sense of the rotation. This corresponds to the signs of the characters for each operation in the A_2 representation. Note that a number of pairs of functions transform according to the E representation.

The functions listed in these right-hand columns are extremely useful in deducing selection rules. We can also use the character tables to find the symmetries (that is, the ir. reps.) of a given vibrational mode or molecular orbital. Consider the asymmetric stretch of H_2O, depicted in Figure 10.6. To confirm that this vibration transforms as B_2, imagine performing each of the four operations of C_{2v} and keep track of whether the arrows representing the vibration stay the same or are reversed. It is not hard to see that the $\hat{\sigma}(xz)$ and $\hat{C}_2$ operations reverse the direction of the bond displacements, meaning the character is -1. The other two operations leave the bond vectors pointing in the same direction, meaning that the character is 1. This combination of signs reveals that the mode belongs to the B_2 ir. rep. Note that our conclusions depend on how we set up the coordinate system. By convention, the z direction coincides with the principal axis, and in most of the literature the water molecule is considered to lie in the yz plane.

C.3 DIRECT PRODUCTS AND REDUCIBLE REPRESENTATIONS

Perhaps the most powerful tool provided by group theory is the ability to decide whether an integral vanishes. Consider a one-dimensional function defined from $-\infty \le x \le \infty$. An odd function $f(x)$ is one for which $f(-x) = -f(x)$, while an even function obeys $g(x) = g(-x)$. Any integral of an odd function over all x (i.e., from $-\infty$ to $+\infty$) must vanish, since the negative and positive contributions to the area cancel. And if we take a product of two functions, say $f(x)g(x)$, the integral over all x vanishes unless they are both even or both odd, since in either case the product (the integrand) is even.

These simple ideas can be extended to integrals involving functions that transform according to the various ir. reps. of a group. Using $d\tau$ as a generic integration variable, we say that the integral $\int f d\tau$ vanishes unless the function f is totally symmetric. If the integrand consists of a product of two functions, $\int f_A g_B d\tau$, then we need to know whether the product of the functions f_A and g_B is totally symmetric, based on the knowledge that the two functions transform according to the Γ_A and Γ_B representations, respectively. Let us call $\Gamma_{A \times B}$ the representation to which the function $f_A \times g_B$ belongs. The characters of the $\Gamma_{A \times B}$ representation are

$$\chi_{A \times B}(R) = \chi_A(R)\chi_B(R) \tag{C.14}$$

We say that $\Gamma_{A \times B}$ is the direct product of the representations Γ_A and Γ_B. If Γ_A and Γ_B are both one-dimensional representations, then so is $\Gamma_{A \times B}$. In general, the dimension of $\Gamma_{A \times B}$ is equal to the product of the dimensions of Γ_A and Γ_B. It is often the case that $\Gamma_{A \times B}$ is a reducible representation, in which case it can be decomposed into contributions from irreducible representations as described below.

Let us take a simple example from the C_{3v} point group, the direct product of the E representation with itself. Since this has to be a four-dimensional representation, it is reducible. The characters of $\Gamma_{E \times E} = \Gamma_E \times \Gamma_E$ can be entered into the character table as follows:

C_{3v}	E	$2C_3$	$3\sigma_v$
A_1	1	1	1
A_2	1	1	-1
E	2	-1	0
$\Gamma_{E \times E}$	4	1	0

$\Gamma_{E \times E}$ is a linear combination of the A_1, A_2 and E representations, and the sum of the dimensions of the contributing ir. reps. must be four. In general, a reducible representation Γ_{red} may be decomposed into ir. reps. Γ_i:

$$\Gamma_{red} = \sum_i a_i \Gamma_i \tag{C.15}$$

where each character of the reducible representation is given by

$$\chi_{red} = \sum_i a_i \chi_i \tag{C.16}$$

and the coefficients a_i are integers. While the coefficients a_i can often be found by inspection, more generally we can use the formula

$$a_i = \frac{1}{h} \sum_R \chi_{red}(R)\chi_i(R) \tag{C.17}$$

We now use Equation C.17 to decompose $\Gamma_{E\times E}$:

$$a_{A_1} = \frac{1}{6}\left[(1)(1)(4)+(2)(1)(1)+(3)(1)(0)\right]=1 \tag{C.18}$$

$$a_{A_2} = \frac{1}{6}\left[(1)(1)(4)+(2)(1)(1)+(3)(-1)(0)\right]=1 \tag{C.19}$$

$$a_E = \frac{1}{6}\left[(1)(2)(4)+(2)(-1)(1)+(3)(0)(0)\right]=1 \tag{C.20}$$

The result is $\Gamma_{E\times E} = A_1 + A_2 + E$. Note that Equations C.18 through C.20 take into consideration the number of operations in each class (the first number in parentheses in each term). Also, clearly the dimensions sum to four as they must. One final point is important and universal: the direct product of any representation with itself always equals or contains the totally symmetric representation. The word "equals" applies for one-dimensional representations, while the direct product of a degenerate representation with itself always contains a contribution from the totally symmetric representation.

The idea of the direct product can be extended to any number of terms in the product. In spectroscopy, we are often interested in the triple direct product, in order to see if the matrix element of an operator is zero or not. Consider two states whose wavefunctions transform according to the Γ_i and Γ_f representations, and an operator $\hat{O}$ which transforms according to the Γ_O representation. We want to know if the integral $\int \psi_i \hat{O} \psi_f d\tau$ exists. Group theory enables us to conclude that this integral vanishes unless the triple direct product $\Gamma_i \times \Gamma_O \times \Gamma_f$ contains (or equals) the totally symmetric representation. Suppose that we first take the direct product of Γ_i and Γ_f to get $\Gamma_{i\times f}$. In order for the triple direct product to contain the totally symmetric representation, we require that $\Gamma_{i\times f}$ and Γ_O belong to the same representation. It does not matter in what order we take the direct product. We could just as easily require that $\Gamma_i \times \Gamma_O = \Gamma_f$. If the triple direct product does *not* contain the totally symmetric representation, then we are assured that the integral in question is zero. Many examples of this sort of analysis occur in the study of spectroscopy.

C.4 CHARACTER TABLES

The following is a partial list of character tables taken from [1]:

C_s	E	σ_h		
A'	1	1	x, y, R_z	x^2, y^2, z^2, xy
A''	1	−1	z, R_x, R_y	yz, xz

C_i	E	i		
A_g	1	1	R_z, R_x, R_y	$x^2, y^2, z^2, xy, yz, xz$
A_u	1	−1	x, y, z	

C_2	E	C_2		
A	1	1	z, R_z	x^2, y^2, z^2, xy
B	1	−1	x, y, R_x, R_y	yz, xz

C_3	E	C_3	C_3^2		$\varepsilon = \exp(2\pi i/3)$
A	1	1	1	z, R_z	$x^2 + y^2, z^2$
E	$\left\{\begin{array}{l}1\\1\end{array}\right.$	$\begin{array}{l}\varepsilon\\\varepsilon^*\end{array}$	$\left.\begin{array}{l}\varepsilon^*\\\varepsilon\end{array}\right\}$	$(x, y)(R_x, R_y)$	$(x^2 - y^2, xy)(yz, xz)$

C_4	E	C_4	C_2	C_4^3		
A	1	1	1	1	z, R_z	$x^2 + y^2, z^2$
B	1	-1	1	-1		$x^2 - y^2, xy$
E	$\left\{\begin{array}{l}1\\1\end{array}\right.$	$\begin{array}{l}i\\-i\end{array}$	$\begin{array}{l}-1\\-1\end{array}$	$\left.\begin{array}{l}-i\\i\end{array}\right\}$	$(x, y)(R_x, R_y)$	(yz, xz)

D_2	E	$C_2(z)$	$C_2(y)$	$C_2(x)$		
A_1	1	1	1	1		x^2, y^2, z^2
B_1	1	1	-1	-1	z, R_z	xy
B_2	1	-1	1	-1	y, R_y	xz
B_3	1	-1	-1	1	x, R_x	yz

D_3	E	$2C_3$	$3C_2$		
A_1	1	1	1		$x^2 + y^2, z^2$
A_2	1	1	-1	z, R_z	
E	2	-1	0	$(x, y)(R_x, R_y)$	$(x^2 - y^2, xy)(xz, yz)$

D_4	E	$2C_4$	C_2	$2C_2'$	$2C_2''$		
A_1	1	1	1	1	1		$x^2 + y^2, z^2$
A_1	1	1	1	-1	-1	z, R_z	
B_1	1	-1	1	1	-1		$x^2 - y^2$
B_2	1	-1	1	-1	1		xy
E	2	0	-2	0	0	$(x, y)(R_x, R_y)$	(xz, yz)

D_5	E	$2C_5$	$2C_5^2$	$5C_2$		
A_1	1	1	1	1		$x^2 + y^2, z^2$
A_2	1	1	1	-1	z, R_z	
E_1	2	$2\cos 72°$	$2\cos 144°$	0	$(x, y)(R_x, R_y)$	(xz, yz)
E_2	2	$2\cos 144°$	$2\cos 72°$	0		$(x^2 - y^2, xy)$

D_6	E	$2C_6$	$2C_3$	C_2	$3C_2'$	$3C_2''$		
A_1	1	1	1	1	1	1		$x^2 + y^2, z^2$
A_2	1	1	1	1	-1	-1	z, R_z	
B_1	1	-1	1	-1	1	-1		
B_2	1	-1	1	-1	-1	1		
E_1	2	1	-1	-2	0	0	$(x, y)(R_x, R_y)$	(xz, yz)
E_2	2	-1	-1	2	0	0		$(x^2 - y^2, xy)$

C_{2v}	E	C_2	$\sigma_v(xz)$	$\sigma'_v(yz)$		
A_1	1	1	1	1	z	x^2, y^2, z^2
A_2	1	1	-1	-1	R_z	xy
B_1	1	-1	1	-1	x, R_y	xz
B_2	1	-1	-1	1	y, R_x	yz

C_{3v}	E	$2C_3$	$3\sigma_v$		
A_1	1	1	1	z	$x^2 + y^2, z^2$
A_2	1	1	-1	R_z	
E	2	-1	0	$(x, y) (R_x, R_y)$	$(x^2 - y^2, xy) (xz, yz)$

C_{4v}	E	$2C_4$	C_2	$2\sigma_v$	$2\sigma_d$		
A_1	1	1	1	1	1	z	$x^2 + y^2, z^2$
A_2	1	1	1	-1	-1	R_z	
B_1	1	-1	1	1	-1		$x^2 - y^2$
B_2	1	-1	1	-1	1		xy
E	2	0	-2	0	0	$(x, y)(R_x, R_y)$	(xz, yz)

C_{5v}	E	$2C_5$	$2C_5^2$	$5\sigma_v$		
A_1	1	1	1	1	z	$x^2 + y^2, z^2$
A_2	1	1	1	-1	R_z	
E_1	2	$2\cos 72°$	$2\cos 144°$	0	$(x, y)(R_x, R_y)$	(xz, yz)
E_2	2	$2\cos 144°$	$2\cos 72°$	0		$(x^2 - y^2, xy)$

C_{6v}	E	$2C_6$	$2C_3$	C_2	$3\sigma_v$	$3\sigma_d$		
A_1	1	1	1	1	1	1	z	$x^2 + y^2, z^2$
A_2	1	1	1	1	-1	-1	R_z	
B_1	1	-1	1	-1	1	-1		
B_2	1	-1	1	-1	-1	1		
E_1	2	1	-1	-2	0	0	$(x, y)(R_x, R_y)$	(xz, yz)
E_2	2	-1	-1	2	0	0		$(x^2 - y^2, xy)$

C_{2h}	E	C_2	i	σ_h		
A_g	1	1	1	1	R_z	x^2, y^2, z^2, xy
B_g	1	-1	1	-1	R_x, R_y	xz, yz
A_u	1	1	-1	-1	z	
B_u	1	-1	-1	1	x, y	

C_{3h}	E	C_3	C_3^2	σ_h	S_3	S_3^5		$\varepsilon = \exp(2\pi i/3)$
A'	1	1	1	1	1	1	R_z	$x^2 + y^2, z^2$
E'	$\begin{cases} 1 \\ 1 \end{cases}$	$\begin{matrix} \varepsilon \\ \varepsilon^* \end{matrix}$	$\begin{matrix} \varepsilon^* \\ \varepsilon \end{matrix}$	$\begin{matrix} 1 \\ 1 \end{matrix}$	$\begin{matrix} \varepsilon \\ \varepsilon^* \end{matrix}$	$\left. \begin{matrix} \varepsilon^* \\ \varepsilon \end{matrix} \right\}$	(x, y)	$(x^2 - y^2, xy)$
A''	1	1	1	-1	-1	-1	z	
E''	$\begin{cases} 1 \\ 1 \end{cases}$	$\begin{matrix} \varepsilon \\ \varepsilon^* \end{matrix}$	$\begin{matrix} \varepsilon^* \\ \varepsilon \end{matrix}$	$\begin{matrix} -1 \\ -1 \end{matrix}$	$\begin{matrix} -\varepsilon \\ -\varepsilon^* \end{matrix}$	$\left. \begin{matrix} -\varepsilon^* \\ -\varepsilon \end{matrix} \right\}$	(R_x, R_y)	(xz, yz)

C_{4h}	E	C_4	C_2	C_4^3	i	S_4^3	σ_h	S_4		
A_g	1	1	1	1	1	1	1	1	R_z	x^2+y^2, z^2
B_g	1	−1	1	−1	1	−1	1	−1		x^2-y^2, xy
E_g	$\begin{cases}1\\1\end{cases}$	$\begin{matrix}i\\-i\end{matrix}$	$\begin{matrix}-1\\-1\end{matrix}$	$\begin{matrix}-i\\i\end{matrix}$	$\begin{matrix}1\\1\end{matrix}$	$\begin{matrix}i\\-i\end{matrix}$	$\begin{matrix}-1\\-1\end{matrix}$	$\begin{matrix}-i\\i\end{matrix}$	(R_x, R_y)	(xy, yz)
A_u	1	1	1	1	−1	−1	−1	−1	z	
B_u	1	−1	1	−1	−1	1	−1	1		
E_u	$\begin{cases}1\\1\end{cases}$	$\begin{matrix}i\\-i\end{matrix}$	$\begin{matrix}-1\\-1\end{matrix}$	$\begin{matrix}-i\\i\end{matrix}$	$\begin{matrix}-1\\-1\end{matrix}$	$\begin{matrix}-i\\i\end{matrix}$	$\begin{matrix}1\\1\end{matrix}$	$\begin{matrix}i\\-i\end{matrix}$	(x, y)	

D_{2h}	E	$C_2(z)$	$C_2(y)$	$C_2(x)$	i	$\sigma(xy)$	$\sigma(xz)$	$\sigma(yz)$		
A_g	1	1	1	1	1	1	1	1		x^2, y^2, z^2
B_{1g}	1	1	−1	−1	1	1	−1	−1	R_z	xy
B_{2g}	1	−1	1	−1	1	−1	1	−1	R_y	xz
B_{3g}	1	−1	−1	1	1	−1	−1	1	R_x	yz
A_u	1	1	1	1	−1	−1	−1	−1		
B_{1u}	1	1	−1	−1	−1	−1	1	1	z	
B_{2u}	1	−1	1	−1	−1	1	−1	1	y	
B_{3u}	1	−1	−1	1	−1	1	1	−1	x	

D_{3h}	E	$2C_3$	$3C_2$	σ_h	$2S_3$	$3\sigma_v$		
A_1'	1	1	1	1	1	1		x^2+y^2, z^2
A_2'	1	1	−1	1	1	−1	R_z	
E'	2	−1	0	2	−1	0	(x, y)	(x^2-y^2, xy)
A_1''	1	1	1	−1	−1	−1		
A_2''	1	1	−1	−1	−1	1	z	
E''	2	−1	0	−2	1	0	(R_x, R_y)	(xz, yz)

D_{4h}	E	$2C_4$	C_2	$2C_2'$	$2C_2''$	i	$2S_4$	σ_h	$2\sigma_v$	$2\sigma_d$		
A_{1g}	1	1	1	1	1	1	1	1	1	1		x^2+y^2, z^2
A_{2g}	1	1	1	−1	−1	1	1	1	−1	−1	R_z	
B_{1g}	1	−1	1	1	−1	1	−1	1	1	−1		x^2-y^2
B_{2g}	1	−1	1	−1	1	1	−1	1	−1	1		xy
E_g	2	0	−2	0	0	2	0	−2	0	0	(R_x, R_y)	(xz, yz)
A_{1u}	1	1	1	1	1	−1	−1	−1	−1	−1		
A_{2u}	1	1	1	−1	−1	−1	−1	−1	1	1	z	
B_{1u}	1	−1	1	1	−1	−1	1	−1	−1	1		
B_{2u}	1	−1	1	−1	1	−1	1	−1	1	−1		
E_u	2	0	−2	0	0	−2	0	2	0	0	(x, y)	

D_{5h}	E	$2C_5$	$2C_5^2$	$5C_2$	σ_h	$2S_5$	$2S_5^3$	$5\sigma_v$		
A_1'	1	1	1	1	1	1	1	1		x^2+y^2, z^2
A_2'	1	1	1	−1	1	1	1	−1	R_z	
E_1'	2	$2\cos 72°$	$2\cos 144°$	0	2	$2\cos 72°$	$2\cos 144°$	0	(x, y)	
E_2'	2	$2\cos 144°$	$2\cos 72°$	0	2	$2\cos 144°$	$2\cos 72°$	0		(x^2-y^2, xy)
A_1''	1	1	1	1	−1	−1	−1	−1		
A_2''	1	1	1	−1	−1	−1	−1	1	z	
E_1''	2	$2\cos 72°$	$2\cos 144°$	0	−2	$-2\cos 72°$	$-2\cos 144°$	0	(R_x, R_y)	(xz, yz)
E_2''	2	$2\cos 144°$	$2\cos 72°$	0	−2	$-2\cos 144°$	$-2\cos 72°$	0		

D_{6h}	E	$2C_6$	$2C_3$	C_2	$3C_2'$	$3C_2''$	i	$2S_3$	$2S_6$	σ_h	$3\sigma_d$	$3\sigma_v$		
A_{1g}	1	1	1	1	1	1	1	1	1	1	1	1		x^2+y^2, z^2
A_{2g}	1	1	1	1	−1	−1	1	1	1	1	−1	−1	R_z	
B_{1g}	1	−1	1	−1	1	−1	1	−1	1	−1	1	−1		
B_{2g}	1	−1	1	−1	−1	1	1	−1	1	−1	−1	1		
E_{1g}	2	1	−1	−2	0	0	2	1	−1	−2	0	0	(R_x, R_y)	(xz, yz)
E_{2g}	2	−1	−1	2	0	0	2	−1	−1	2	0	0		(x^2-y^2, xy)
A_{1u}	1	1	1	1	1	1	−1	−1	−1	−1	−1	−1		
A_{2u}	1	1	1	1	−1	−1	−1	−1	−1	−1	1	1	z	
B_{1u}	1	−1	1	−1	1	−1	−1	1	−1	1	−1	1		
B_{2u}	1	−1	1	−1	−1	1	−1	1	−1	1	1	−1		
E_{1u}	2	1	−1	−2	0	0	−2	−1	1	2	0	0	(x, y)	
E_{2u}	2	−1	−1	2	0	0	−2	1	1	−2	0	0		

D_{2d}	E	$2S_4$	C_2	$2C_2'$	$2\sigma_d$		
A_1	1	1	1	1	1		x^2+y^2, z^2
A_2	1	1	1	−1	−1	R_z	
B_1	1	−1	1	1	−1		x^2-y^2
B_2	1	−1	1	−1	1	z	xy
E	2	0	−2	0	0	$(x, y)(R_x, R_y)$	(xz, yz)

D_{3d}	E	$2C_3$	$3C_2$	i	$2S_6$	$3\sigma_d$		
A_{1g}	1	1	1	1	1	1		x^2+y^2, z^2
A_{2g}	1	1	−1	1	1	−1	R_z	
E_g	2	−1	0	2	−1	0	(R_x, R_y)	$(x^2-y^2, xy)(xz, yz)$
A_{1u}	1	1	1	−1	−1	−1		
A_{2u}	1	1	−1	−1	−1	1	z	
E_u	2	−1	0	−2	1	0	(x, y)	

S_4	E	S_4	C_2	S_4^3		
A	1	1	1	1	R_z	x^2+y^2, z^2
B	1	−1	1	−1	z	x^2-y^2, xy
E	$\begin{cases}1 \\ 1\end{cases}$	$\begin{matrix}i \\ -i\end{matrix}$	$\begin{matrix}-1 \\ -1\end{matrix}$	$\begin{matrix}-i \\ i\end{matrix}$	$(x, y)(R_x, R_y)$	(yz, xz)

T_d	E	$8C_3$	$3C_2$	$6S_4$	$6\sigma_d$		
A_1	1	1	1	1	1		$x^2+y^2+z^2$
A_2	1	1	1	-1	-1		
E	2	-1	2	0	0		$(2z^2-x^2-y^2, x^2-y^2)$
T_1	3	0	-1	1	-1	(R_x, R_y, R_z)	
T_2	3	0	-1	-1	1	(x, y, z)	(xy, xz, yz)

O_h	E	$8C_3$	$6C_2$	$6C_4$	$3C_2$	i	$6S_4$	$8S_6$	$3\sigma_h$	$6\sigma_d$		
A_{1g}	1	1	1	1	1	1	1	1	1	1		$x^2+y^2+z^2$
A_{2g}	1	1	-1	-1	1	1	-1	1	1	-1		
E_g	2	-1	0	0	2	2	0	-1	2	0		$(2z^2-x^2-y^2, x^2-y^2)$
T_{1g}	3	0	-1	1	-1	3	1	0	-1	-1	(R_x, R_y, R_z)	
T_{2g}	3	0	1	-1	-1	3	-1	0	-1	1		(xz, yz, xy)
A_{1u}	1	1	1	1	1	-1	-1	-1	-1	-1		
A_{2u}	1	1	-1	-1	1	-1	1	-1	-1	1		
E_u	2	-1	0	0	2	-2	0	1	-2	0		
T_{1u}	3	0	-1	1	-1	-3	-1	0	1	1	(x, y, z)	
T_{2u}	3	0	1	-1	-1	-3	1	0	1	-1		

$C_{\infty v}$	E	$2C(\varphi)$	σ_v		
Σ^+	1	1	1	z	x^2+y^2, z^2
Σ^-	1	1	-1	R_z	
Π	2	$2\cos\varphi$	0	$(x, y)(R_x, R_y)$	(xz, yz)
Δ	2	$2\cos 2\varphi$	0		(x^2-y^2, xy)
Φ	2	$2\cos 3\varphi$	0		

$D_{\infty h}$	E	$2C(\varphi)$	σ_v	i	$2S(\varphi)$	C_2		
Σ_g^+	1	1	1	1	1	1		x^2+y^2, z^2
Σ_g^-	1	1	-1	1	1	-1	R_z	
Π_g	2	$2\cos\varphi$	0	2	$-2\cos\varphi$	0	(R_x, R_y)	(xz, yz)
Δ_g	2	$2\cos 2\varphi$	0	2	$2\cos 2\varphi$	0		(x^2-y^2, xy)
Σ_u^+	1	1	1	-1	-1	-1	z	
Σ_u^-	1	1	-1	-1	-1	1		
Π_u	2	$2\cos\varphi$	0	-2	$2\cos\varphi$	0	(x, y)	
Δ_u	2	$2\cos 2\varphi$	0	-2	$-2\cos 2\varphi$	0		

REFERENCES

1. F. A. Cotton, *Chemical Applications of Group Theory* (Wiley-Interscience, New York, 1971).
2. I. N. Levine, *Quantum Chemistry, 5th ed.* (Prentice Hall, Upper Saddle River, NJ, 1999).

Index